Keith Gough

ELECTRICAL ENGINEERING

CONCEPTS AND APPLICATIONS

ELECTRICAL ENGINEERING

CONCEPTS AND APPLICATIONS

A. Bruce Carlson

David G. Gisser

RENSSELAER POLYTECHNIC INSTITUTE

With contributions from Fred K. Manasse, UNIVERSITY OF NEW HAMPSHIRE

ADDISON-WESLEY PUBLISHING COMPANY

Reading, Massachusetts • Amsterdam • London
Manila • Singapore • Sydney • Tokyo

WORLD STUDENT SERIES EDITION

This book is in
ADDISON-WESLEY SERIES IN ELECTRICAL ENGINEERING

SPONSORING EDITOR: *Tom Robbins*
DESIGNER: *Marshall Henrichs*
ILLUSTRATOR: *Oxford Illustrators Ltd*
COVER DESIGN: *Richard Hannus*

ISBN 0-201-10299-4
CDEFGHIJKL-DA-898765

PREFACE

The purpose of this book is to introduce basic concepts of electrical engineering in four general areas: linear circuits, electronic devices and circuits, analog and digital systems, and energy systems. To stimulate interest and to acquaint students with useful design ideas, all concepts are related to practical applications. We give primary emphasis to elementary analysis methods using simplified models of electrical, electronic, and electromechanical devices. However, we also discuss the limitations of those methods and models to alert students to situations where more sophisticated techniques would be needed.

The material is arranged in a modular structure that permits subject selection to fit various curriculum settings and courses of different lengths and scopes. For beginning students in electrical engineering and electrical technology, we hope to present a motivating and unifying overview of the profession. For students in other engineering disciplines or in the sciences, we hope to provide a working vocabulary of circuits, electronics, instrumentation, and machinery. For all students, we hope to illustrate the significance of integrated circuits and the systems approach in modern technology.

PREREQUISITES

The text is written at a level suitable for students who have completed at least one term of college physics and mathematics. We have class-tested major portions with freshmen, sophomores, and juniors at Rensselaer Polytechnic Institute and with technology students at the University of New Hampshire.

The prerequisite physics consists of elementary mechanics, dimensional analysis, and some familiarity with electric charge, atomic structure, and sinusoidal waves. Electrical energy, conduction processes, and magnetic phenomena are discussed when introduced.

The prerequisite mathematics consists of trigonometric and exponential functions, quadratic equations, simultaneous linear equations, and elementary calculus. For the most part, the use of calculus is restricted to the slope and rate interpretation of differentiation and the area interpretation of integration. A few stated results draw upon differentiation to find a maximum value.

CONTENTS AND ORGANIZATION

We have organized the subject matter into four major parts. Part I, Linear Circuits, includes controlled sources, ideal transformers, amplifier models, op-amps, transfer functions, and frequency response, along with steady-state analysis and transients. Part II, Electronic Devices and Circuits, emphasizes the external characteristics of diodes, transistors, and selected IC units; FETs and BJTs receive equal attention, as do linear and nonlinear electronic circuits. Part III, Analog and Digital Systems, develops the block-diagram concept and describes applications ranging from automatic control and communication to signal processing and digital computation. Part IV, Energy Systems, introduces magnetics and electromechanics in relation to electric-power generation and conversion.

All four parts can be covered in a year-long course of about 90 one-hour class sessions. We have, however, attempted to provide flexible options for shorter courses. In particular, Part I establishes the essential background for each of the other three parts, which are otherwise self-contained. Furthermore, several sections and chapters in Parts II–IV may be taken out of context if desired.

Thus, for instance, a one-term survey course might consist of Part I followed by selected sections from later chapters. Or Parts I and IV could be used for a course on circuits and machines. An introduction to circuits and electronics might consist of Parts I and II, perhaps augmented with topics from Part III. Parts I and III could be used for a systems-oriented instrumentation course.

To facilitate course planning, we have indicated in the table of contents the minimum prerequisite material for each section. In addition, the symbol † marks optional sections that may be omitted. A few of the optional sections are descriptive in nature, intended for enrichment. Other optional sections involve quantitative material that goes somewhat beyond the level of a first-year course.

TEACHING AND LEARNING AIDS

Each chapter begins with a list of learning objectives identified by section. Each section includes worked examples and exercises designed to help students master the objectives. Selected answers to the exercises are at the back of the book. There are more than 130 examples and 230 exercises.

Many examples and exercises involve numerical computations with realistic parameter values. We assume that students will use calculators for these exercises, especially when performing vector calculations. (Our numerical values in the examples are usually rounded off to three significant figures; consequently, some third-digit discrepancies appear between intermediate and final results.)

Besides the exercises, we have provided about forty problems at the end of each chapter. The problems include simple drills, analysis and design-oriented calculations, derivations, and occasional extensions of ideas presented in the text. Since the problems are organized by section, and not necessarily in order of increasing difficulty, we use the symbol ‡ to mark the more challenging problems. The same symbol identifies problems based on optional topics, and problems that require calculus. We expect students to refer to the tables at the back of the book for the mathematical relations needed in certain problems.

NOTATION AND UNITS

Pedagogical considerations have led us to adopt certain notations that, although not unprecedented, may be new to some instructors. Except for voltage sources, reference voltage polarities are indicated by arrows rather than by plus and minus signs. We find that this notation makes it easier for students to apply Kirchhoff's voltage law; the arrow notation also goes well with electronic circuit topology and node equations. Different arrowheads distinguish between voltages and currents. We emphasize the special nature of controlled sources by using a diamond-shaped symbol.

We have purposely avoided boldface type and special symbols that would be difficult for instructors to reproduce by hand. Phasors and other two-dimensional quantities are indicated by an underbar, such as $\underline{V}$; and absolute-value signs denote phasor magnitudes, such as $|V|$. Script letters are limited to $\mathcal{E}$ (electric field), $\mathcal{R}$ (reluctance), and $\mathcal{F}$ (magnetomotive force). The "loop ell" (ℓ) is used to prevent possible confusion with the numeral 1.

SI units are, of course, employed throughout. However, we refer to conductance in mhos (℧) rather than siemens (S), since the former still dominates in the literature. For the same reason, rotational speed of machines is often expressed in revolutions per minute (rpm).

ACKNOWLEDGE-MENTS

We gratefully acknowledge the support and encouragement of Lester A. Gerhardt, Chairman of the Electrical and Systems Engineering Department, Rensselaer Polytechnic Institute. Discussions with Charles M. Close, Allan N. Greenwood, Thornton S. Lauber, Kenneth Rose, and other Rensselaer colleagues were most helpful, as were comments from our students and teaching assistants. Photographs were graciously provided by M. Harry Hesse and Andrew J. Steckl. Special appreciation goes to the several reviewers for their thoughtful suggestions and criticisms.

Finally, we express our thanks to Nancy Case, Rose Rafun, and Donna Taurinskas for secretarial services; to the staff at Addison-Wesley for expertise and prodding; and to our families for patience and understanding.

Troy, New York

A. Bruce Carlson

David G. Gisser

CONTENTS

The numbers in parentheses after each section name identify the previous sections that constitute the minimum prerequisite material. The symbol † identifies optional material that may be omitted without loss of continuity.

8 TRANSIENT RESPONSE 235

PART II ELECTRONIC DEVICES AND CIRCUITS

9 DIODES AND SEMICONDUCTORS 271

10 TRANSISTORS AND INTEGRATED CIRCUITS 305

PART III ANALOG AND DIGITAL SYSTEMS

1 INTRODUCTION

Stop for a moment and look around. Close at hand you'll probably find several products of electrical engineering: a lamp that illuminates this page, the alarm clock that awakened you, a hi-fi system, TV set and transistor radio for entertainment and news, the telephone for communication, and of course your calculator. Surely ours is a "wired world" in which the quality of daily life depends to a significant degree upon the use of electrical phenomena.

This textbook addresses the underlying concepts and methods of electrical engineering behind applications ranging from consumer products and biomedical electronics to giant computers and industrial control systems. Obviously, we cannot deal with all aspects of these areas. Instead, we emphasize general ideas and techniques that, we hope, will provide the necessary background for you to pursue specific topics in more detail. Consistent with that philosophy, this introductory chapter presents an overview of electrical engineering and a prospectus of the chapters that follow.

1.1 THE SCOPE OF ELECTRICAL ENGINEERING

It seems appropriate to begin with a definition of electrical engineering. In the words of Professor H. H. Woodson*:

> Electrical engineering is a profession whose practitioners exploit electromagnetic phenomena and electrical and magnetic properties (and sometimes mechanical, thermal, chemical, and other properties) of matter to do useful things, i.e., someone will pay the engineers to do them. The "useful things" usually involve the processing of information or energy and sometimes both, as in the processing of information about energy in the relay protection of a power system; and the engineer's activities generally, but not always, involve equipment.

This definition quite properly focuses on *information* and *energy*, two crucial commodities of modern society. They are central to electrical engineering by virtue of the fact that electrical embodiments of information and energy generally result in more effective processing, control, transmission, and distribution. Looking to the future, those advantages loom with even more importance as we seek greater efficiency in energy utilization and the substitution of low-energy activities (such as long-distance communication) for high-energy activities (such as long-distance travel).

Our definition also emphasizes doing "useful things" with electromagnetic phenomena and, possibly, other properties of matter. Certainly, all engineers should do useful things and, as technology advances, we find increasing interaction between the traditional branches of engineering and science. On the one hand, the fabrication of integrated circuits, for instance, involves chemical, mechanical, metallurgical, and thermal considerations along with electronic expertise. On the other hand, electronic instrumentation and microprocessors have become vital parts of the measurement and control systems employed by aeronautical, biomedical, chemical, civil, mechanical, and nuclear engineers and scientists.

Without belaboring the point, modern technology demands a team approach in which "EEs" and "non-EEs" work together and share a common vocabulary. Accordingly, the goals of this book include sensitizing electrical engineering students to the relevance of nonelectrical topics, and introducing other students to the language of electrical engineers. Both goals will be furthered by considering the scope of electrical engineering.

Perhaps the best indication of that scope is the variety of periodicals published by The Institute of Electrical and Electronics Engineers. The IEEE (I-triple-E) has a membership of about 200,000, making it the larg-

---

* From "What is Electrical Engineering?," *IEEE Transactions on Education*, vol. E-22, May 1979, p. 35. (Incidentally, this is the only page cluttered by a footnote. All other references will be found under the heading Supplementary Reading.)

est engineering society in the world. Its publications include two general magazines (*IEEE Spectrum* and *Proceedings of the IEEE*) and 38 special-interest transactions and journals listed in Table 1.1–1. The listing has, somewhat arbitrarily, been divided into four categories.

The category headed "Devices" consists of specializations that deal primarily with electromagnetic fields and charged particles and their interaction with matter, often at the microscopic level. The "Circuits and Electronics" category includes several traditional EE activities and some newer ones, essentially characterized by the task of connecting individual devices to achieve a desired circuit behavior. The "Systems" category reflects the fact that many applications are now so complex that the designer must concentrate on the overall functions rather than the specific details of the constituent circuits that make up a large system.

Roughly speaking, the headings at the top of Table 1.1–1 cover a spectrum from the applied physics of device development to the applied mathematics of systems engineering. But all areas of electrical engineering require a working knowledge of physics and mathematics along with engineering methodology and supporting skills in communication and human relations. Thus, the fourth category at the bottom of the table includes specializations that cut across the other categories in one way or another, and several extend into territory not ordinarily associated with electrical engineering.

TABLE 1.1–1 Transactions and Journals of the IEEE

Devices	*Circuits and electronics*	*Systems*
Antennas and Propagation	Broadcasting	Aerospace and Electronic Systems
Electrical Insulation	Circuits and Systems	Automatic Control
Electron Devices	Consumer Electronics	Communications
Magnetics	Geoscience Electronics	Computers
Microwave Theory and Techniques	Industrial Electronics and Control Instrumentation	Information Theory
Nuclear Science	Instrumentation and Measurement	Pattern Analysis and Machine Intelligence
Plasma Science	Solid-State Circuits	Software Engineering
Quantum Electronics and Applications		Systems, Man, and Cybernetics
Cross-disciplinary		
Acoustics, Speech, and Signal Processing		Industry Applications
Biomedical Engineering		Oceanic Engineering
Cable TV		Power Apparatus and Systems
Components, Hybrids, and Manufacturing Technology		Professional Communication
Education		Reliability
Electromagnetic Compatibility		Sonics and Ultrasonics
Engineering Management		Technology and Society
		Vehicular Technology

1.2 HISTORICAL PERSPECTIVE

Having described the present scope of electrical engineering, we should give at least some attention to its historical evolution. For that purpose, the following chronology lists selected discoveries, inventions, and theoretical developments over a span of more than 200 years. Many of the names and events are well known, while others will be unfamiliar to you. Their importance should emerge from the pertinent sections of the text, so you may find it useful to return here on occasion.

Year	Event
1750–1831	*The Beginnings:* Franklin and Coulomb study electric charge; Volta discovers the battery; Fourier and Laplace develop mathematical theories; Ampere, Weber, and Henry conduct experiments on current and magnetism; Ohm's law is stated; Faraday publishes his theory of induction.
1838–1866	*Telegraphy:* Morse perfects his telegraph, and commercial service begins; multiplexing is invented; the first transatlantic cables are laid.
1845	Kirchhoff's circuit laws are stated.
1864	Maxwell's equations predict electromagnetic radiation.
1876–1887	*Telephony:* Bell devises an acoustic transducer; the first telephone exchange, which has eight lines, is constructed in New Haven, CT; Strowger invents the automatic switch.
1879–1882	*DC Power Systems:* Edison finds a suitable lamp filament; he establishes in New York City the first electric utility with 59 customers.
1885–1895	*AC Power Systems:* Stanley develops a practical transformer; Steinmetz conceives of phasors for AC circuit analysis; Westinghouse promotes AC systems; Tesla builds an induction motor; in Germany, the first long-distance, three-phase power line is constructed; generators are installed at Niagara Falls.
1887–1897	*Radio:* Hertz verifies Maxwell's prediction; Marconi patents a complete wireless telegraph system.
1904–1920	*Vacuum-Tube Electronics:* Fleming and DeForest build vacuum tubes; electronic amplifiers make possible a transcontinental telephone line; the superheterodyne radio receiver is perfected by Armstrong; KDKA in Pittsburgh is the first AM broadcasting station.
1923–1938	*Television:* Farnsworth and Zworykin devise electronic image formation; DuMont markets cathode-ray tubes; experimental broadcasting begins.
1934	Black invents the negative-feedback amplifier.

Year	Event
1938	Shannon applies Boolean algebra to switching circuits.
1938–1945	*World War II:* Major advances take place in electronics, instrumentation, and theory; radar and microwave systems are developed; operational amplifiers are incorporated in analog computers; FM communication systems are perfected for military applications.
1945–1948	*Systems Theory:* Papers by Bode, Shannon, Wiener, and others establish the basis for systems engineering.
1946	The ENIAC vacuum-tube digital computer is constructed at the University of Pennsylvania.
1948	Long-playing microgroove records are introduced.
1948–1955	*Transistor Electronics:* Schockley, Bardeen, and Brattain invent the point-contact and junction transistor (Bell Laboratories); the development of the surface-barrier transistor improves manufacturing techniques; transistor radios go into mass production.
1951–1958	*Digital Computers:* UNIVAC I is installed at the US Census Bureau; IBM markets the popular Model 650; the programming language FORTRAN is developed to facilitate scientific use of computers; the transistorized Philco 2000 marks the "second generation" of computing equipment.
1957	The first commercial nuclear power plant becomes operational at Shippingport, PA.
1958–1961	*Microelectronics:* Hoerni invents the planar transistor (Fairchild Semiconductor); integrated circuits are developed by Kilby (Texas Instruments) and others; commercial production begins.
1960	Laser demonstrations by Maiman.
1962	Telstar I is launched as the first communications satellite.
1962–1966	*Digital Communication:* Commercial data transmission service begins; experiments prove the feasibility of pulse-code modulation for voice and TV; electronic switching systems become operational; practical implementations of error-control coding are devised.
1969	765,000-volt AC power lines are constructed.
1971–1975	*Microcomputers:* MOS technology permits large-scale integration; Hewlett-Packard markets the HP-35 calculator; Intel introduces the 8080 microprocessor chip; charge-coupled devices are used for memory.

Year	Event
1975–?	*State of the Art:* Interactive computer-graphics systems become commercially available; fusion plasmas show potential for power generation; microprocessors are used in consumer products and adaptive control and communication systems; fiber optics and optoelectronics are developed; high-voltage DC power transmission systems are installed; software engineering emerges as a new discipline; power electronics are applied to AC motor control.

1.3 PROSPECTUS

The text of this book is divided into four major parts. Part I, Linear Circuits, develops the basic circuit concepts and analysis methods needed for the remaining parts, starting in Chapter 2 with a review of electrical quantities and the properties of elementary circuits. Chapter 3 presents important modeling techniques in the context of resistive circuits, while Chapter 4 extends those models to include ideal transformers, amplifiers, and the versatile "op-amp." Energy-storage elements (capacitance and inductance) are introduced in Chapter 5 and described in terms of impedance. Chapters 6 through 8 use the impedance concept to investigate AC circuits, frequency response, and transients.

Part II, Electronic Devices and Circuits, begins in Chapters 9 and 10 with a discussion of the external characteristics of diodes and transistors and a brief look at semiconductor physics and fabrication. Chapters 11 and 12 use the device and circuit models of previous chapters to explain the operating principles of practical electronic switching and amplification circuits.

Part III, Analog and Digital Systems, shifts the emphasis from circuit diagrams to the block diagrams of systems engineering. The analog transfer function developed in Chapter 7 is applied in Chapter 13 to the study of feedback and automatic control systems, and in Chapter 14 to signal processing and communication systems. Chapter 15 introduces digital logic concepts and building blocks that are used for the instrumentation and computer systems of Chapter 16.

Part IV, Energy Systems, returns to devices and circuits with the introduction of magnetics and electromechanical transducers in Chapter 17. Chapter 18 expands the coverage of AC circuits to include power analysis, practical transformers, and three-phase systems, while Chapter 19 surveys the properties and applications of rotating machines.

Parts II through IV are largely self-contained, and some chapters and sections may be taken out of sequence if desired. In this regard, the table of contents indicates the minimum prerequisite material for each section. The symbol † identifies optional material that can be omitted without loss of continuity.

Every chapter has two features intended to guide your study: a list of objectives stated at the beginning, and practice exercises that should help you master the objectives. We recommend that you read the objectives and work the exercises as you come to them, then reread the objectives by way of review. Answers to the exercises are given at the back of the book, along with tables of useful mathematical relationships. Additional problems at the end of each chapter are arranged by section. Problems marked with the symbol ‡ are more challenging than the others.

If you need further help with a particular topic, you may wish to consult one of the other textbooks listed as Supplementary Reading. Under that heading you will also find more advanced references on the subjects introduced here.

PART I

Transient waveforms produced by a pulse train (top) applied to an *RC* circuit.

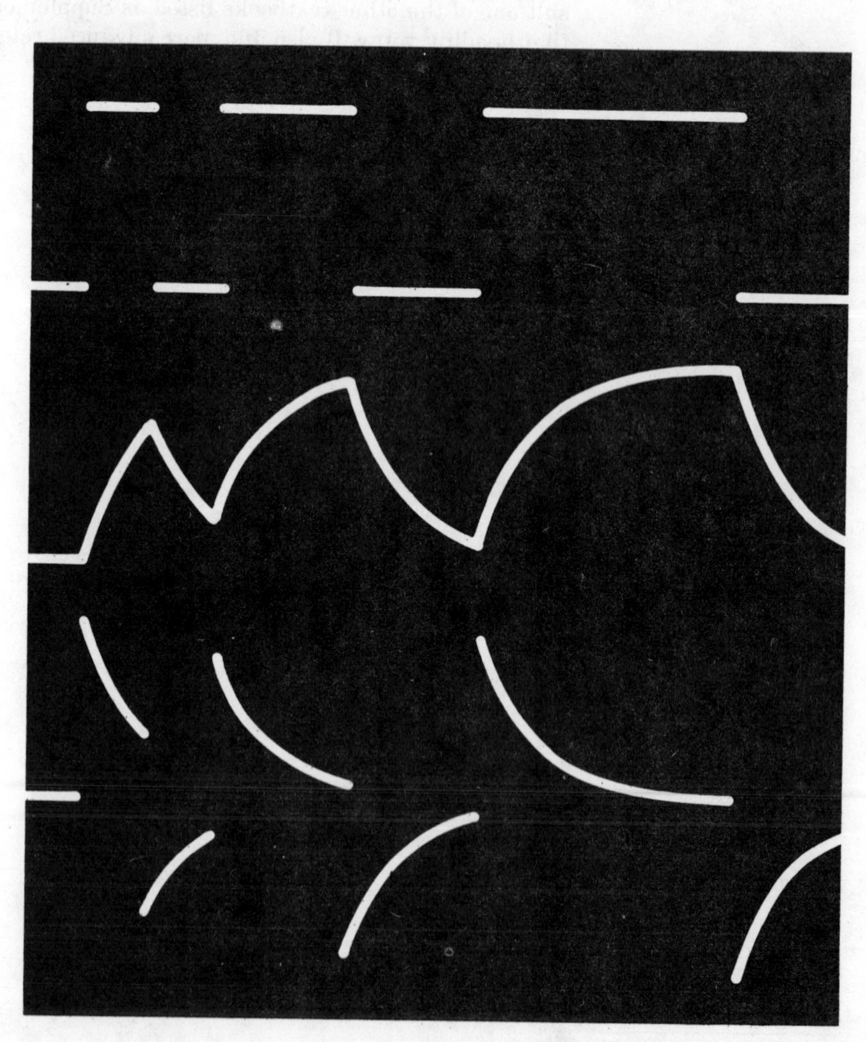

LINEAR CIRCUITS

2

CIRCUIT CONCEPTS

Whether a simple flashlight or part of a large digital computer, all electrical circuits involve at least four things: energy sources, current, voltage, and resistance. Energy sources provide the driving force that produces currents and voltages in the circuit. Current and voltage, in turn, are the dependent variables that carry out whatever task the circuit has been designed to do. The cause-and-effect relationship between sources and variables is dictated by the devices or elements that make up the circuit and the way they are connected. Of the many different kinds of circuit elements, resistance is the most common.

So we begin by considering these simple but basic circuit concepts. We'll define current and voltage and ideal sources, and develop their relationship to energy and power. We then introduce three important "laws": Ohm's law, which describes resistive circuit elements, and Kirchhoff's current and voltage laws, which govern the interconnection of circuit elements. These three laws are sufficient for analyzing and designing simple but illustrative practical circuits.

OBJECTIVES

After studying this chapter and working the exercises, you should be able to do each of the following:

- Define and give the units for current, voltage, and power (Section 2.1).

- Draw the symbols and current-voltage curves for ideal sources and resistance (Section 2.2 and 2.3).
- Use Ohm's law to calculate current, voltage, resistance, and power, expressing the units with appropriate magnitude prefixes (Sections 2.1 and 2.3).
- Identify loops and nodes in a given circuit, and apply Kirchhoff's laws to them (Section 2.4).
- Solve for the voltages and currents in a simple circuit, using the branch-current or node-voltage method (Sections 2.3 and 2.4).

2.1 CURRENT, VOLTAGE, AND POWER

Figure 2.1–1 represents the essential parts of a flashlight—a battery, bulb, switch, and connecting wires. Each element has two electrical contact points or terminals, and turning the switch to the ON position forms a complete loop or closed circuit. The battery's voltage then causes current to go from one terminal, through the bulb and wires, and back to the other battery terminal. As current passes through the bulb, electrical power is converted to heat and light. The current thus serves as a vehicle of power transfer from battery to bulb. Actually, we know that the current consists of moving charges. These charges acquire energy from the battery and deliver energy to the bulb.

This section reviews the fundamental relationships between charge, current, energy, and voltage, and develops the expression for electrical power. Thereafter, we will use current, voltage, and power as our primary electrical quantities.

CHARGE AND CURRENT

Electrical charge has two characteristics: The amount or magnitude measured in *coulombs* (C), and the sign or polarity. The charge carried by an electron, for instance, is

$$q_e = -1.60 \times 10^{-19} \text{ C} \tag{1}$$

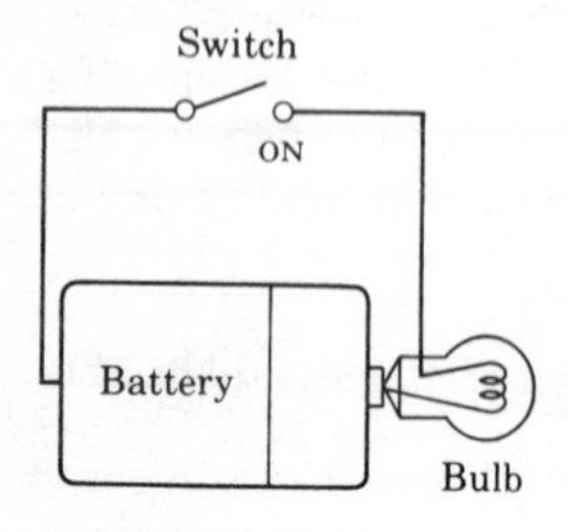

FIGURE 2.1–1
A flashlight circuit.

which has *negative* polarity. A proton has the same amount of charge but with *positive* polarity. The polarity names were assigned by early experimenters and bear no particular significance. What is significant is the fact that charges of opposite sign tend to neutralize each other. Thus the net charge on a larger particle, such as an ion, is the difference between the amounts of positive and negative charge.

Current exists whenever there is net transfer of charge through a given area in a given time. To illustrate, suppose a positive charge q_1 and a negative charge q_2 both pass from left to right through the shaded area in Fig. 2.1–2, while charges q_3 and q_4 travel in the opposite direction.

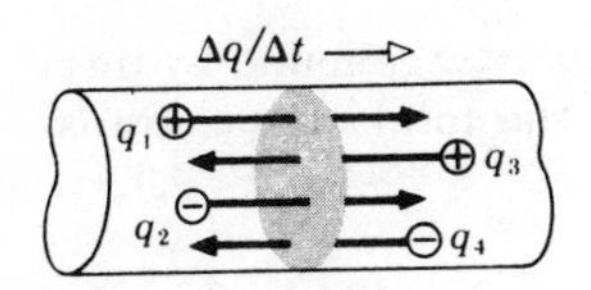

FIGURE 2.1–2
Charge transfer through an area.

(Situations like this actually occur in a semiconductor diode.) The net charge transfer from left to right will be

$$\begin{aligned}\Delta q &= (q_1 + q_2) - (q_3 + q_4)\\ &= (q_1 + |q_4|) - (q_3 + |q_2|)\end{aligned}$$

where the absolute-value notation emphasizes that the negative quantity q_4 going from right to left is equivalent to positive quantity $|q_4|$ going the other way—and similarly for q_2. If this charge transfer takes place in a time interval Δt, the average current is

$$I = \frac{\Delta q}{\Delta t}$$

with direction shown by the triangular-headed arrow assuming Δq is positive. Generalizing using differential quantities, we define *instantaneous current* as

$$i \triangleq \frac{dq}{dt} \tag{2}$$

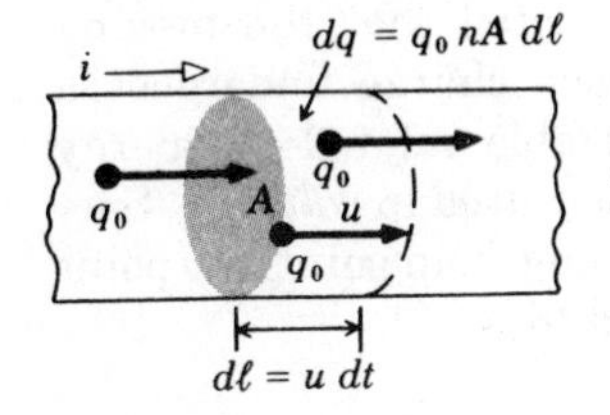

FIGURE 2.1–3
Current composed of like charge carriers q_0 with density n and average velocity u.

in which the symbol $\triangleq$ identifies a definition. We measure current in *amperes* (A), called *"amps"* for short, and one ampere equals the transfer of one coulomb per second (C/s). The unit equation is then 1 A = 1 C/s.

Frequently, current consists of a large number of identically charged particles, each carrying charge q_0 and moving in the same direction at *average velocity u*, Fig. 2.1–3. The charge flow through the cross-sectional area A may then be written as

$$i = \frac{dq}{dt} = \frac{dq}{d\ell}\frac{d\ell}{dt} = \frac{dq}{d\ell}\,u \tag{3a}$$

where $d\ell = u\,dt$ is the distance traveled by a carrier in time dt and dq equals the total amount of charge in length $d\ell$. If the *carrier density* is n particles per unit volume, the volume $A\,d\ell$ contains $nA\,d\ell$ carriers at any instant of time so $dq = q_0 nA\,d\ell$. Hence, $dq/d\ell = q_0 nA$ and

$$i = q_0 nuA \tag{3b}$$

which expresses the current in terms of the carriers' charge, volume density, and velocity.

Note that current i has a negative value when q_0 is negative. In such cases we can view it as positive current going in the opposite direction. This happens to be an important observation, because electrons (with negative charge) constitute most electrical currents.

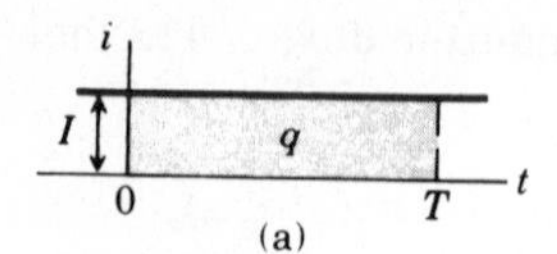

(b)

FIGURE 2.1–4
Total charge transferred in $0 < t < T$ when (a) $i = I$ (constant), (b) $i_{av} = I$.

Occasionally we will be interested in the *total charge* found by integrating current. In particular, since $dq = i\,dt$, the total charge transported in some time interval, say 0 to T, is

$$q = \int_0^T i\,dt \tag{4a}$$

meaning that q equals the *area* under the curve of i versus time from $t = 0$ to $t = T$. If i has the constant value I, then

$$q = IT \tag{4b}$$

as shown in Fig. 2.1–4a. This expression also holds when I stands for the *average value* of i over the time interval in question, defined by

$$I = i_{av} = \frac{1}{T}\int_0^T i\,dt$$

and illustrated in Fig. 2.1–4b.

ENERGY AND VOLTAGE

Returning to the flashlight in Fig. 2.1–1, we noted that the moving charges gave up energy to the bulb; hence, each charge undergoes a change in potential energy. The electrical variable related to energy change is called *potential difference* or *voltage*, measured in *volts* (V). Specifically, if charge dq gives up energy dw when going from point a to point b, then the voltage across those points is defined to be

$$v \triangleq \frac{dw}{dq} \tag{5}$$

with point a at the higher potential if dw/dq is positive. Expressing energy in joules (J), we have the voltage unit equation 1 V = 1 J/C.

Polarity again enters the picture in view of the fact that either dq or dw can be negative quantities. If dq and dw have the same sign, then energy is *delivered* by a positive charge going from a to b or a negative charge going the other way. Conversely, charged particles *gain* energy inside a source where dq and dw have opposite polarities.

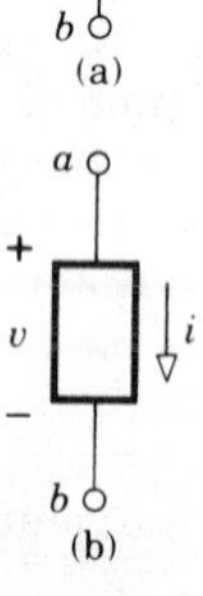

FIGURE 2.1–5
Symbols for voltage polarity across an element assumed to be receiving energy.

Figure 2.1–5 shows an arbitrary circuit element with two symbolic conventions used to indicate voltage polarity. The open-headed arrow in Fig. 2.1–5a points to the end of the element assumed to be at the higher potential, while Fig. 2.1–5b has a plus sign at the higher-potential end and a minus sign at the other end. If both i and v are positive quantities, this element is *receiving* energy because a positive charge goes in the direction of the current arrow from higher to lower potential. If either i or v has a negative value—implying its actual polarity is the reverse of the symbol—then the element must be *supplying* energy.

Generally, we use the arrow notation for voltage polarity (Fig. 2.1–5a) with the type of arrowhead distinguishing voltage and current. Thus, an energy-absorbing element has the arrows pointed in *opposite* directions. The other polarity notation will be reserved for voltage sources, in which case we can omit the redundant minus sign at the lower-potential end. A source voltage is sometimes called an *electromotive force* or *emf* to convey the notion that it is the "force" that "drives" the current through the circuit.

ELECTRIC POWER

Instantaneous power p is defined as the rate of doing work or the rate of change of energy, dw/dt. The electric power in *watts* (W) consumed or produced by a circuit element simply equals its voltage-current product, that is

$$p = vi \tag{6}$$

so 1 W = 1 V·A. This equation follows from the fact that voltage is work per unit charge while current is charge transfer per unit time; thus $vi = (dw/dq)(dq/dt) = dw/dt = p$.

One way of measuring the power consumed by a circuit element is diagrammed in Fig. 2.1–6. We connect a current-sensing instrument called an *ammeter* (AM) so that the current i flows through both the element and the meter. We also connect a *voltmeter* (VM) to read the voltage v across the element. The product of the two meter readings then equals the power—providing the voltmeter is ideal so that all of the current i measured by the ammeter passes through the element.

Electric power is the stock-and-trade of electric utility companies, and many electrical devices are characterized by their power rating. But your service bill is for *total energy* consumed, not power. Total energy over a time interval is found by integrating power

$$w = \int_0^T p\,dt \tag{7a}$$

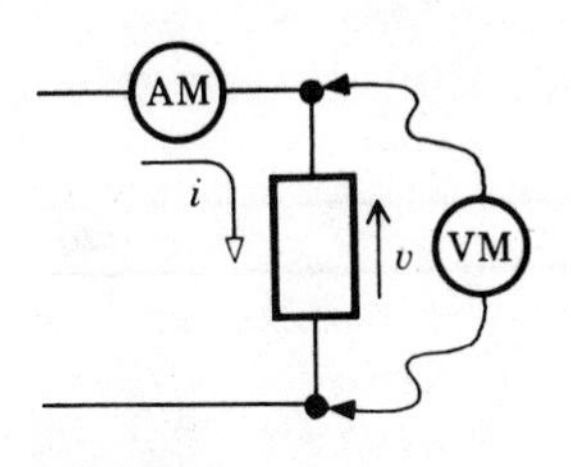

FIGURE 2.1–6
Measuring power $p = vi$ with an ammeter and voltmeter.

or, if p has the constant or average value P,

$$w = PT \tag{7b}$$

whose units are watt-seconds or joules. Utility bills are commonly expressed in terms of the *kilowatthour* (kWh),

$$1\ \text{kWh} = 3.6 \times 10^6\ \text{J} \tag{8}$$

which equals the total energy delivered in one hour when $P = 1000$ W.

Example 2.1–1
Capacity of a battery

A typical 12-V automobile battery stores about 5×10^6 J of energy, or slightly over one kilowatthour. If the battery is connected to a 4-A headlight, the power delivered to the bulb will be

$$p = 12 \text{ V} \times 4 \text{ A} = 48 \text{ W}$$

Assuming v and i remain constant, the energy consumed in one minute of operation is

$$w = 48 \text{ W} \times 60 \text{ s} = 2880 \text{ J}$$

The total charge that has passed through the bulb during this period is

$$q = 4 \text{ A} \times 60 \text{ s} = 240 \text{ C}$$

equivalent to $240/(1.6 \times 10^{-19}) = 1.5 \times 10^{21}$ electrons.

Incidentally, the maximum charge storage or "capacity" of auto batteries is often rated in ampere-hours (Ah), one ampere-hour being $1 \text{ C/s} \times 3600 \text{ s} = 3600$ C. For the battery in question, the capacity equals $5 \times 10^6 \text{ J}/12 \text{ V} = 4.17 \times 10^5 \text{ C} = 116$ Ah.

Exercise 2.1–1

Find the current when the above battery is connected to a 50-W headlight. Assuming v and i remain constant and there is lossless energy transfer from battery to bulb, how long can the headlight be operated before the battery is completely discharged?

MAGNITUDE PREFIXES

Current, voltage, power, and other electrical quantities come in very small to very large values. An electronic device, for instance, might have a current of 10^{-6} A and dissipate 10^{-4} W. At the other extreme, a high-voltage transmission system might handle 10^7 W at 10^5 V. Instead of writing powers of 10 all the time, certain *magnitude prefixes* have been adopted to represent them. Table 2.1–1 lists the common prefixes we will be using. We then write 2×10^3 W as 2 kW (kilowatts) while 80 μA (micro-amps) stands for 80×10^{-6} A.

TABLE 2.1–1 Magnitude Prefixes

Prefix	*Abbreviation*	*Magnitude*
giga-	G	10^{+9}
mega-	M	10^{+6}
kilo-	k	10^{+3}
milli-	m	10^{-3}
micro-	μ	10^{-6}
nano-	n	10^{-9}
pico-	p	10^{-12}

Example 2.1–2

Suppose a 50-kV source is rated at a maximum of 20 W. The corresponding current will be

$$i = 20 \text{ W}/50 \text{ kV} = 0.4 \times 10^{-3} \text{ A} = 0.4 \text{ mA} = 400 \ \mu\text{A}$$

an answer we could have obtained with fewer intermediate steps by noting that, as far as magnitudes are concerned, $1/1 \text{ k} = 10^{-3} = 1 \text{ m} = 1000 \ \mu$.

Exercise 2.1–2

Calculate p when the above source produces $i = 0.3 \ \mu\text{A}$.

2.2 SOURCES AND SINKS

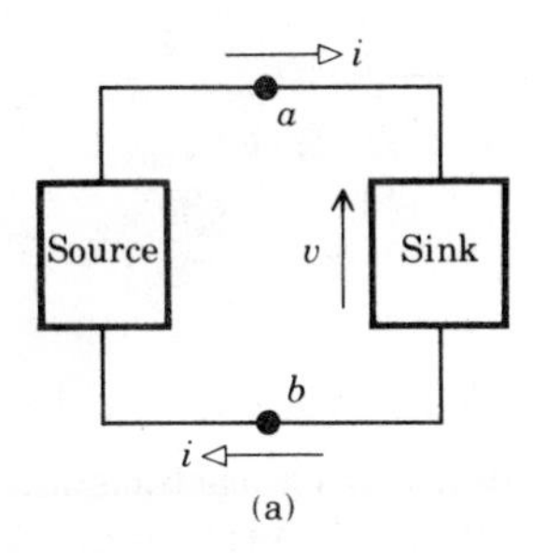

(a)

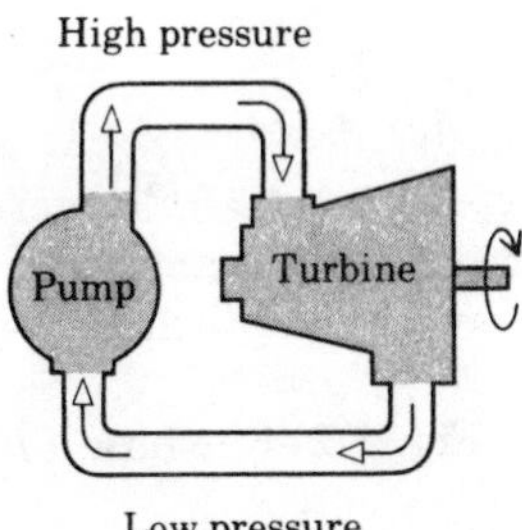

(b)

FIGURE 2.2–1
(a) A source-sink circuit. (b) Analogous hydraulic system.

Figure 2.2–1a shows another way of drawing the flashlight circuit, with the battery represented by a box marked "source" and the bulb by a box marked "sink," meaning energy sink or consumer of power. Positive current i passes through the sink from a to b, and point a is at a higher potential than b. We say that a voltage "drop" v exists across the sink from a to b because the potential decreases in the direction of current flow. The power consumed by the sink is $p = vi$. On the other side of the circuit, the source provides power, and a voltage "rise" exists across the source from b to a because the potential increases in the direction of current flow.

Two important observations should be drawn from this discussion. First, a voltage *rise* indicates an electrical *source,* with the charge being raised to a higher potential, while a voltage *drop* indicates a *sink,* with charge going to a lower potential. Second, voltage can be thought of as an "across" variable, since we speak of the voltage *across* a circuit element; and current, by the same token, can be thought of as a "through" variable, since current flows *through* a circuit element.

A little confusing? Perhaps a rough analogy will help: We'll briefly describe the operation of the hydraulic system of Fig. 2.2–1b, indicating the analogous circuit concepts in parentheses. A pump (the source) forces water flow (the current) through pipes (connecting wires) to drive a turbine (the sink). The water pressure (potential) is higher at the inlet port of the turbine than at the output, so a pressure drop (voltage drop) exists across the turbine in the direction the water flows. The pump, on the other hand, raises the pressure (voltage rise). Water flow (current) is a "through" variable, while pressure difference (voltage) is an "across" variable.

In this section we further develop the distinction between sources and sinks through consideration of their current-voltage relationships. We also introduce the concept of ideal sources and discuss the significance of idealized device models.

i-v CURVES

We have seen that the values and relative polarities of current and voltage characterize the power produced or consumed by a circuit element. Many types of two-terminal elements have a direct relationship between current and voltage that can be expressed as an equation or plotted as a graph called the *i-v* curve. These curves provide useful information about the nature and behavior of the device. They can be determined experimentally with the help of an adjustable source and the meter arrangement previously shown in Fig. 2.1–6.

By way of illustration, consider a typical flashlight bulb and its *i-v* curve, Fig. 2.2–2. At every point on the curve, i and v have the same sign—both positive or both negative. Hence, the bulb is an electrical sink that consumes but never produces power. (We already knew this, of course.) Such devices are said to be *passive,* meaning they have no energy supply of their own.

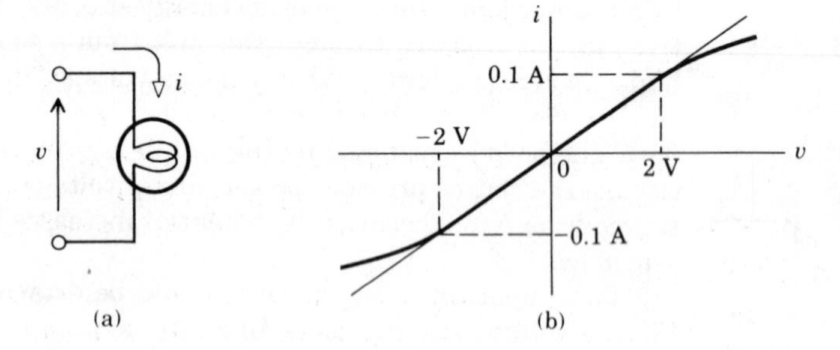

FIGURE 2.2–2
A flashlight bulb and its *i-v* curve.

The curve also reveals an almost straight-line or *linear* characteristic near the origin, with slope $\Delta i/\Delta v = 0.1 \text{ A}/2 \text{ V} = 0.05 \text{ A/V}$. Accordingly, for small values of voltage and current we have the simple approximation

$$i \approx 0.05\ v$$

which holds when $-2 \text{ V} \leq v \leq 2 \text{ V}$ and $-0.1 \text{ A} \leq i \leq 0.1 \text{ A}$. At larger current or voltage values, we would have to take account of the curvature or *nonlinearity* of the *i-v* plot. The result would be a more complicated expression.

A more intriguing example is the *photodiode,* Fig. 2.2–3, whose *i-v* equation has the form

$$i = I_0(e^{40v} - 1) - I_p$$

where I_0 and I_p are constants, the latter depending upon the light intensity falling on the device. We are not concerned here with the physics underlying this somewhat awesome expression, but rather with its inter-

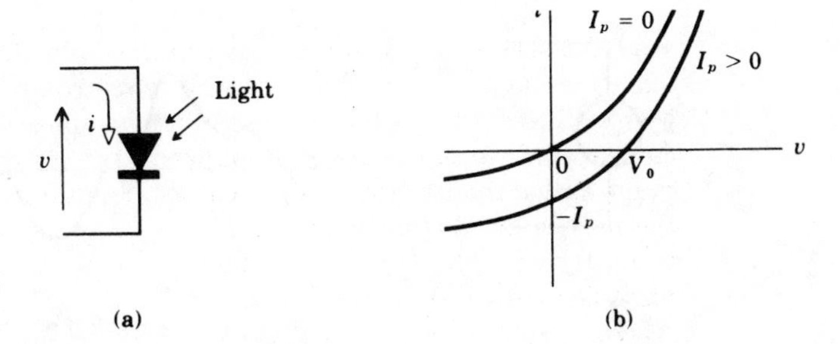

FIGURE 2.2-3 A photodiode and its i-v curves.

pretation from the i-v curve. Actually, we must examine two curves, corresponding to the presence or absence of light.

With no light on the diode, I_p equals zero and the curve has an asymmetrical shape passing through the origin. Under this condition, the device is passive but draws more current in one direction than the other for the same voltage magnitude. With $I_p > 0$, the curve crosses the vertical axis at $i = -I_p$ and the horizontal axis at the point $v = V_0$. Between these points the device is *active* and *produces power,* since positive current goes from lower to higher potential if $i < 0$. Where does the power come from? From the incident light! The photodiode has become a *solar cell.*

By plotting i versus v in these figures we convey the notion that i is a function of v or that v is the *cause* and i is the *effect.* Sometimes the reverse will be true and it is better to plot v versus i, a v-i curve. Since cause and effect often depend on the specific circuit arrangement, you should be prepared to deal with both i-v and v-i curves.

IDEAL SOURCES

Batteries and AC outlets are familiar electrical sources. Both can be classified as *voltage sources* in the sense that the voltage is essentially independent of the current—though it may vary with time, as is true for an AC source. Formalizing this concept, we say that

> An *ideal voltage source* is one whose terminal voltage v_s is a specified function of time, regardless of the current i through the source.

Figure 2.2-4a shows the symbol we will use for an ideal voltage source with arbitrary time variation. As indicated, v_s may have negative values at various times. In this case the terminal marked + is actually at a lower potential than the other terminal, and the source tends to force current in the opposite direction.

An ideal *battery* has a constant voltage with respect to time, represented by Fig. 2.2-4b where we use the capitol symbol V_s to emphasize its

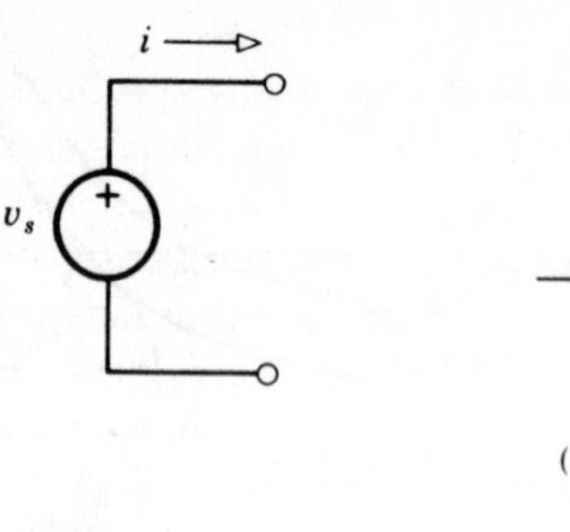

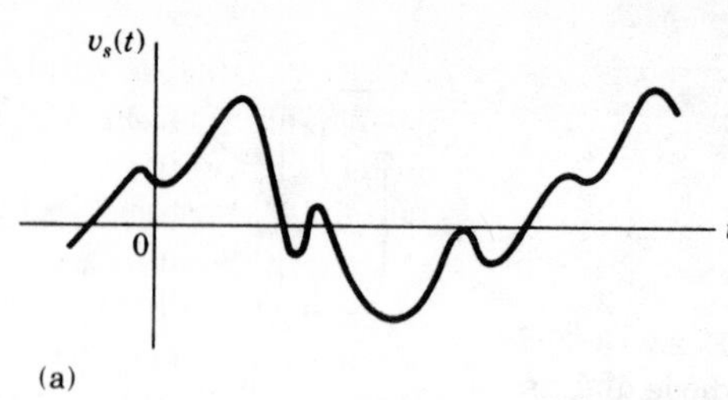

(a)

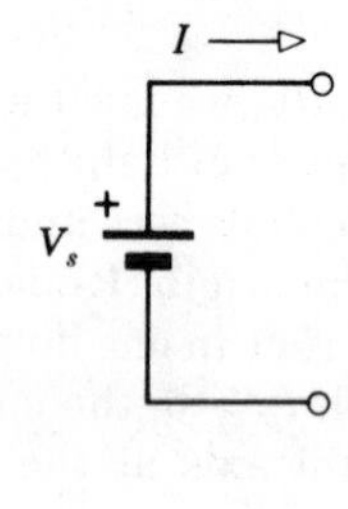

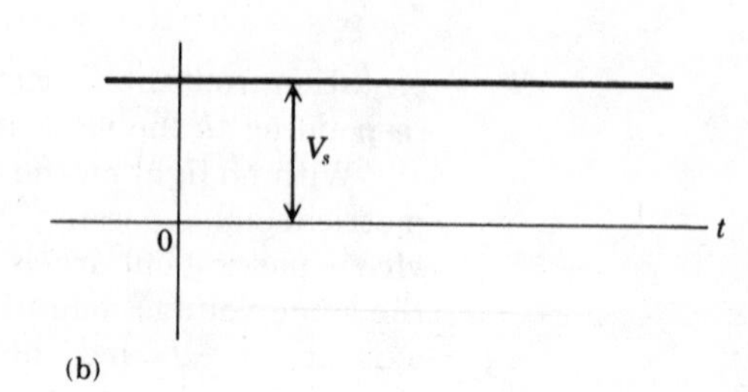

(b)

FIGURE 2.2–4
(a) Ideal voltage source.
(b) Ideal battery.

constancy. In normal operation, this source produces a constant or direct current $i = I$ flowing out of the + terminal. The value of I depends on both V_s and the external circuitry connected to the source, whereas V_s remains fixed. Because i is a *direct current,* we call the ideal battery a *DC* source.

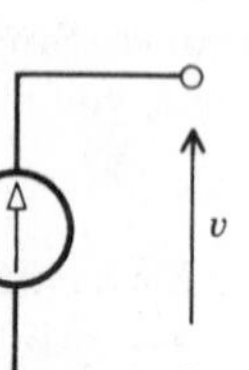

FIGURE 2.2–5
Ideal current source.

Sometimes it proves convenient to use the concept of an *ideal current source,* defined as one whose current i_s is a specified function of time, regardless of the voltage across its terminals. The circuit symbol is given in Fig. 2.2–5. Although less familiar than voltage sources, current sources play an important role as we'll see when we examine certain electronic devices, notably the transistor; but an electric welder also acts more or less like a current source.

We underscore the difference between ideal voltage and current sources by plotting their *i-v* curves in Fig. 2.2–6. At any particular in-

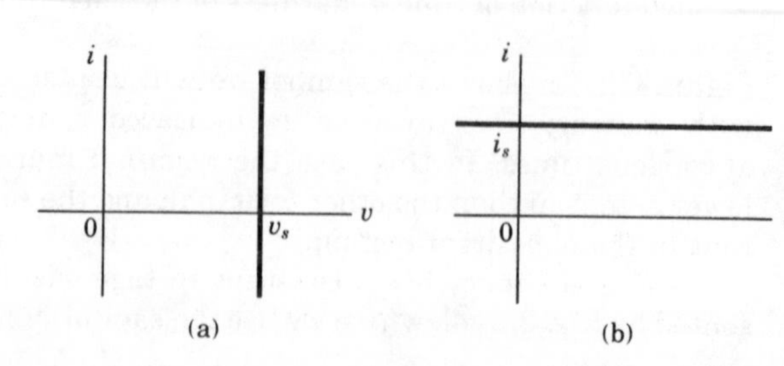

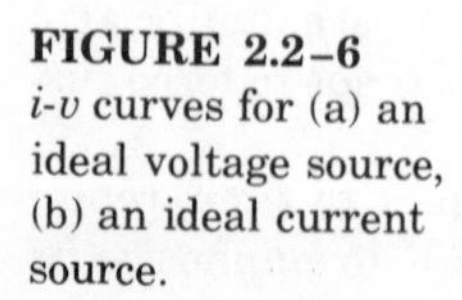

FIGURE 2.2–6
i-v curves for (a) an ideal voltage source, (b) an ideal current source.

stant of time, a voltage source has a specified voltage value v_s but can supply any amount of current, positive or negative, so its i-v curve is a vertical line intersecting the horizontal axis at $v = v_s$. (The curve for an ideal 9-V battery, for instance, intersects at $v = 9.0$ V.) Conversely, a current source has a specified current value independent of the voltage, so its i-v curve is a horizontal line intersecting the vertical axis at $i = i_s$. (Had we plotted v-i curves, the voltage source would have a horizontal line and the current source a vertical line.)

Note that the positive current direction in Fig. 2.2–6 is out of the higher-potential terminal of the source—just the opposite of the curves for passive devices where positive current flows into the higher-potential terminal. This difference reflects the normal mode of operation and allows us to use the i-v curves directly when studying a source-sink combination, as in the example below.

Example 2.2–1
A photodiode

Suppose a photodiode having $I_0 = 0.1\ \mu$A and $I_p = 0$ is driven by an ideal current source with $i_s = 50$ mA, Fig. 2.2–7a. We can calculate the resulting voltage drop across the photodiode by inserting the values for i, I_0, and I_p into its i-v equation and solving for v, as follows:

$$0.05 = 10^{-7}\,(e^{40v} - 1)$$

$$e^{40v} = \frac{0.05}{10^{-7}} + 1 \approx 5 \times 10^5$$

$$v \approx \frac{1}{40} \ln\,(5 \times 10^5) = 0.33 \text{ V}$$

Figure 2.2–7b depicts this solution graphically by plotting the i-v curves for the source and photodiode on the same set of axes; their intersection point gives the value of v, which also equals the voltage rise across the current source.

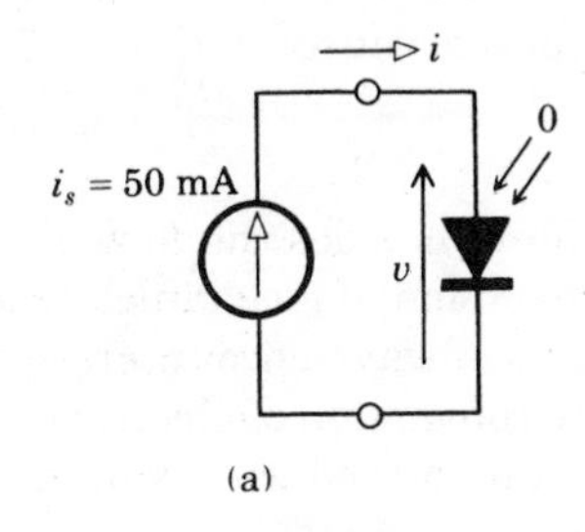

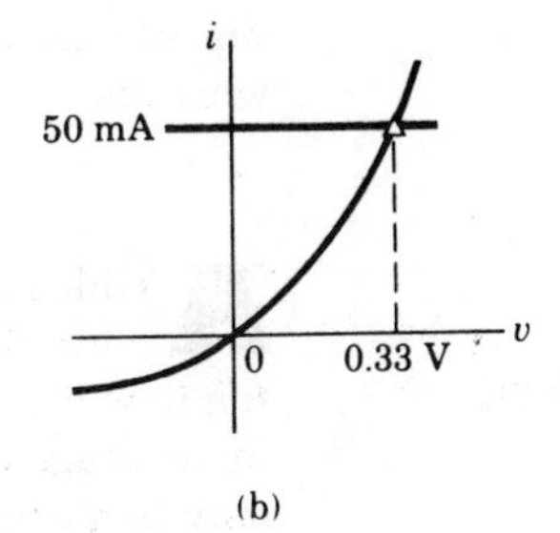

FIGURE 2.2–7 Photodiode circuit with an ideal current source.

Exercise 2.2–1

A device having $i = (0.01\ v)^3$ is connected to a 200-V source. Make a sketch similar to Fig. 2.2–7b, and find the current through the device and the power delivered by the source.

DEVICES AND MODELS

Theoretically, an *ideal* voltage or current source could produce infinite power $p = vi$, for Fig. 2.2–6 implies apparently unlimited values of v or i. But infinite values are physically impossible so ideal sources cannot exist. Why, then, do we bother defining them? The answer to this question is significant, for it relates to one of the most powerful tools in electrical engineering, namely the use of simplified representations or *models* for physical devices.

Models of natural phenomena—expressed as mathematical relations, curves, etc.—are idealizations representing those aspects of the physical characteristics that are pertinent to a particular application. A good model allows one to predict, with reasonable accuracy, how the device will perform under the expected operating conditions. Thus, for a limited range of current, it might be quite acceptable to pretend that a battery acts like an ideal voltage source. By doing so, we can concentrate on significant factors and effects without getting bogged down in the details of a more accurate but very cumbersome description of a battery. Of course, one must always bear in mind the assumptions and limitations of the model, since predictions that go beyond the model's scope are likely to be invalid—for example, that a battery will produce unlimited power.

Virtually every branch of engineering and science involves mathematical models. (As a case in point, Newton's laws of motion are not absolute laws but models that apply only when velocity is small compared to the speed of light.) Electrical engineering, perhaps, does more with models than some other fields. Actually, every circuit diagram is a model, often a very good one, but still a model. From now on, it should go without saying that the various circuit laws and device representations are approximations of physical reality. Where appropriate, the significant limitations of idealized concepts will be discussed.

Exercise 2.2–2

A certain active device is described by the relationship $(5i)^2 + (0.02v)^2 = 1$ for $i \geq 0$ and $v \geq 0$. Sketch the i-v curve and justify the assertion that this device acts like a current source with $i_s = 200$ mA if $v \leq 5$ V, whereas it acts like a voltage source with $v_s = 50$ V if $i \leq 20$ mA.

2.3 RESISTANCE

A kink in a garden hose impedes the flow of water, producing a pressure drop and conversion of mechanical energy to heat. Similarly, the flow of electric current always encounters some *resistance,* resulting in a voltage drop and the conversion of electric energy to heat. Resistance may be desired in a circuit to produce a voltage drop or energy conversion, or it may be an unwanted but unavoidable part of a device or connecting wire. (We'll ignore the very special case of superconductors that have zero resistance.) A *resistor* is a device whose primary electrical characteristic is resistance. The properties of the ideal resistance element are examined here, along with a brief description of electrical conduction in solid materials.

OHM'S LAW

An *ideal* or *linear resistance* is an energy-consuming element described by *Ohm's law*

$$v = Ri \tag{1}$$

which means that voltage is directly proportional to current. The proportionality constant R is the value of the resistance, and the units are *ohms* (Ω). Rewriting Ohm's law as

$$R = \frac{v}{i} \tag{2}$$

yields the unit equation

$$1\ \Omega = 1\ \text{V/A}$$

so resistance is the ratio of voltage to current.

Figure 2.3–1 shows the symbol and i-v curve for resistance. Note carefully that the *slope* of the curve is $1/R$ since, from Ohm's law, $i = v/R$. We interpret this curve as follows: If a voltage v is applied across the terminals of a resistance R, the current through the resistance will equal v/R. Resistance resists the flow of current; and the larger R is, the smaller i will be for a given v. By the way, that interpretation should help you keep track of the three different ways of writing Ohm's law—$v = Ri$, $i = v/R$, and $R = v/i$. Had we plotted voltage versus current, its slope would be R rather than $1/R$, meaning that a voltage $v = Ri$ is produced when a current i flows through a resistance R.

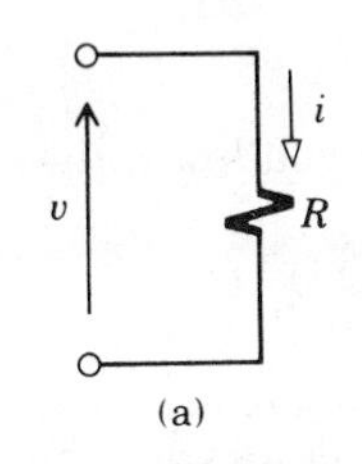

(a)

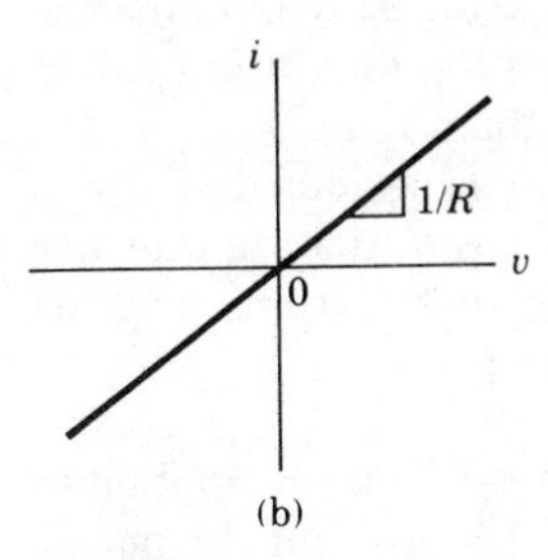

(b)

FIGURE 2.3–1
Ideal resistance and its i-v curve.

For occasional use, it is also convenient to define *conductance* G as the reciprocal of resistance; that is,

$$G \triangleq \frac{1}{R} \tag{3a}$$

so that Ohm's law becomes

$$i = Gv \tag{3b}$$

Conductance is measured in inverse ohms or *siemens* (S) in the SI system of units. However, the earlier and rather droll term *mho* (℧) still dominates in the literature and so will be used throughout this textbook.

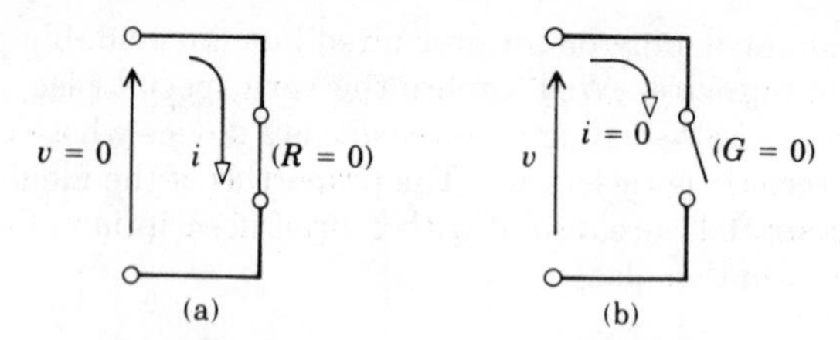

FIGURE 2.3–2 An ON-OFF switch creates: (a) a short circuit; (b) an open circuit.

Referring back to Fig. 2.2–2, we find the flashlight bulb had $i \approx 0.05v$ for small voltages and currents. Under this condition, the bulb can be modeled as a conductance $G = 0.05$ ℧ or resistance $R = 1/G = 20\ \Omega$. Resistance values in typical circuits range from fractions of an ohm to kilohms (kΩ) or even megohms (1 MΩ = 10^6 Ω).

A simple ON-OFF switch also may be modeled in terms of resistance, but with very extreme values. Specifically, in the ON or closed position of Fig. 2.3–2a, the switch creates a *short circuit* or zero-resistance path ($R = 0$); then $v = 0 \times i = 0$ for any value of i. But in the OFF or open position of Fig. 2.3–2b, the switch becomes an *open circuit* with zero conductance ($G = 0$); then $i = 0 \times v = 0$ for any value of v.

POWER DISSIPATION AND OHMIC HEATING

Combining Ohm's law with $p = vi$ gives two equivalent expressions for the power consumed and dissipated by a resistance, namely

$$p = vi = (Ri)i = i^2 R \tag{4a}$$

$$= v(v/R) = v^2/R \tag{4b}$$

Either expression can be used, depending on whether you know the current through the resistance or the voltage across it.

Power dissipation by a resistance element produces heat, a process known as *ohmic heating.* Coils of resistive wire are used precisely for this result in such electrical appliances as ovens, hair dryers, toasters, and so forth. Similarly, an incandescent lightbulb's filament glows when heated by current flow, giving the desired light. Ohmic heating also explains the principle of a simple fuse: when the current reaches the maximum value, heat causes the fuse to melt, thereby "breaking" the circuit.

Unfortunately, however, ohmic heating can cause serious damage to electronic circuits and instrumentation. For this reason, cooling fans are built into some electronic instruments, and large installations such as computer systems require extensive air conditioning.

Example 2.3–1

Figure 2.3–3 shows the circuit and i-v curves for a 9-V battery applied to a 5-kΩ resistance. The current is $i = 9\text{ V}/5\text{ k}\Omega = 1.8$ mA and the power dissipated is $p = (9\text{ V})^2/5\text{ k}\Omega = (1.8\text{ mA})^2 \times 5\text{ k}\Omega = 16.2$ mW. (Note that power comes out directly in milliwatts when we have voltage in volts, resistance in kilohms, and current in milliamps.) Reducing the resistance to 5 Ω gives $i = 1.8$ A and $p = 16.2$ W. In the limit as $R \to 0$, the

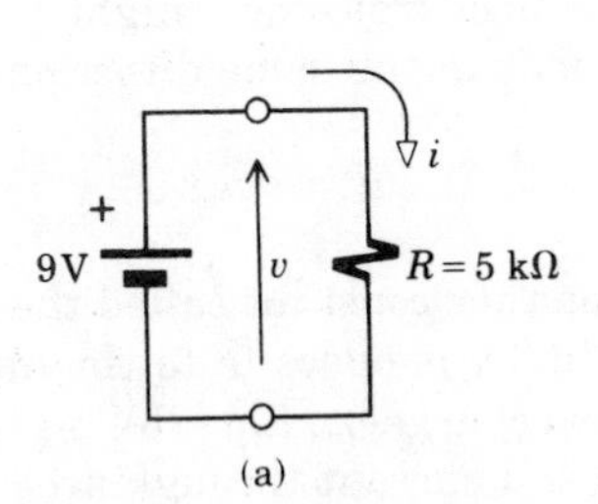

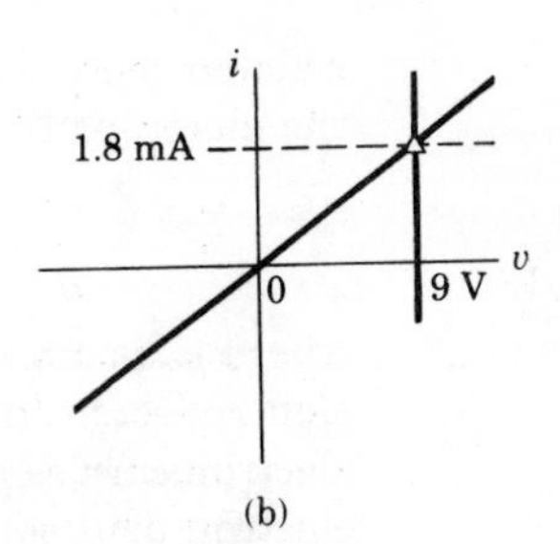

FIGURE 2.3–3
A battery-resistance circuit.

current and power both become infinite, theoretically, because we have assumed an *ideal* voltage source.

Exercise 2.3–1

Find v and p when the battery in Fig. 2.3–3 is replaced by a 3-mA current source.

CONDUCTION AND RESISTIVITY

Ohm's law, as we stated it, focuses on the external characteristics of a resistive element. But the value of the resistance R depends on the material and shape of the element and its temperature. We'll examine these three factors by taking a brief look at what goes on inside a solid conducting material.

A metallic bar or wire is not just a hollow pipe through which charges may be pumped. Rather, it consists of atoms fixed in an orderly crystal lattice, plus a number of free electrons. These electrons have escaped from their parent atoms, leaving them as positively charged ions, as shown in Fig. 2.3–4a. Both the ions and electrons have kinetic energy that manifests itself in random thermal motion. Binding forces constrain the ions to vibrate about their average lattice positions, while the electrons are free to go flying about erratically, colliding here and there with an ion and changing direction. The net charge flow is zero, however, due to the random electron motion in all directions.

Now let an *electric field* ε be applied. The field exerts force on the electrons and they each gain a small velocity component u added to the

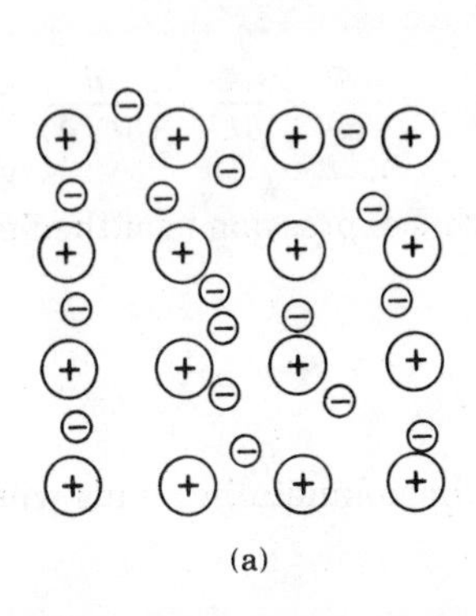

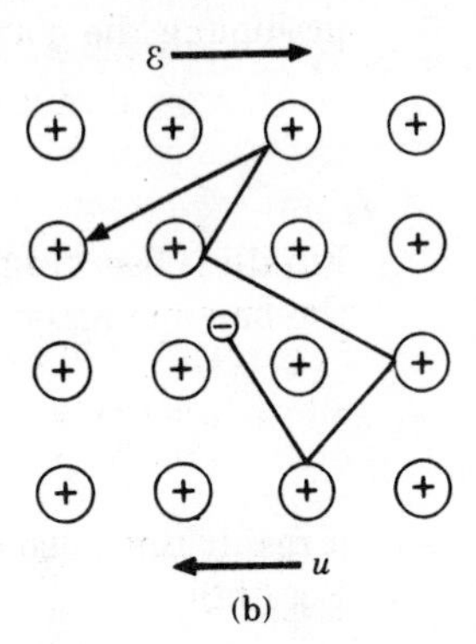

FIGURE 2.3–4
(a) Free electrons and positive ions in a metallic crystal lattice.
(b) Typical electron trajectory with an electric field ε applied.

random motion, so a typical trajectory might look like Fig. 2.3–4b. All the electrons thereby drift in the same direction, with average velocity

$$u = -\mu_e \mathcal{E}$$

where μ_e is a proportionality constant called the *mobility* and the minus sign reflects the fact that u is opposite to the direction of $\mathcal{E}$. Since each electron carries negative charge q_e, Eq. (3b), Section 2.1, tells us that the electron drift constitutes a current through area A given by

$$i = q_e nuA = -|q_e|n(-\mu_e \mathcal{E})A = |q_e|n\mu_e \mathcal{E}A \tag{5}$$

with n being the material's electron density. This current is in the same direction as $\mathcal{E}$ because it consists of negative charges moving in the opposite direction.

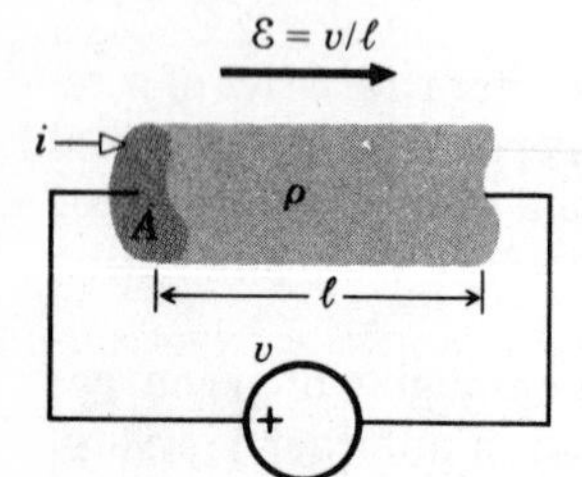

FIGURE 2.3–5
A bar of conducting material with resistivity ρ.

Although it may not appear to be so, Eq. (5) is just another statement of Ohm's law. To bring out the connection, we first define the *resistivity* ρ (rho) as

$$\rho \triangleq \frac{\mathcal{E}A}{i} \tag{6a}$$

$$= \frac{1}{|q_e|n\mu_e} \tag{6b}$$

which depends only on the properties of the material in question and has the units of *ohm-meters* ($\Omega \cdot$ m). (We could also define the material's *conductivity* $\sigma = 1/\rho = |q_e|n\mu_e$.) Next, consider the situation in Fig. 2.3–5, where voltage v has been applied to a uniform bar of conducting material having length ℓ and cross-sectional area A. The voltage gives rise to an electric field directed from higher to lower potential and of value

$$\mathcal{E} = \frac{v}{\ell}$$

producing the current

$$i = \frac{\mathcal{E}A}{\rho} = \frac{vA}{\rho\ell} = \frac{v}{(\rho\ell/A)}$$

Finally, observing that this expression is of the form $i = v/R$, we see that the bar has resistance

$$R = \rho\frac{\ell}{A} \tag{7}$$

a result that also holds for nonmetallic solids with resistivity defined by Eq. (6a).

TABLE 2.3–1 Resistivity of Various Materials at 20°C

Type	*Material*	ρ ($\Omega \cdot$ m)
Conductors	Copper	1.7×10^{-8}
	Aluminum	2.8×10^{-8}
	Nichrome	10^{-6}
	Carbon	3.5×10^{-5}
Semiconductors	Germanium	0.46
	Silicon	2300
Insulators	Rubber	10^{12}
	Polystyrene	10^{15}

Equation (7) reveals that resistance is proportional to resistivity and length, but inversely proportional to area. Consequently, a long thin piece of low-resistivity material might have the same resistance as a short, thick piece of high-resistivity material. By the same reasoning, a wire intended to carry large currents should have low resistivity and large area in order to minimize ohmic heating $p = i^2R = i^2\rho\ell/A$. The physical mechanism of this heating in a metal is suggested by Fig. 2.3–4b, namely energy transferred to the lattice when drifting electrons collide with ions.

Table 2.3–1 lists values of ρ for some representative materials. At the one extreme, metals have exceedingly small resistivities due to the large density n of free electrons, and hence are good electrical *conductors*. To illustrate, the 14-gauge copper wire used in many 15-amp household circuits has a radius of about 0.08 cm, so $A = \pi r^2 \approx 2 \times 10^{-6}$ m² and the resistance of a 10-m length will be

$$R = 1.7 \times 10^{-8} \times \frac{10}{2 \times 10^{-6}} = 0.085\ \Omega$$

corresponding to a maximum drop of 1.3 V at the rated current. At the other extreme are the electrical *insulators* (rubber, plastics, etc.) whose lack of free charge carriers yields resistivities so large that ordinary voltages produce virtually no current flow. In between are the *semiconductors,* about which we'll say much more in Chapter 9.

The tremendous ratio of available resistivities is one reason why electricity is a convenient method for transporting energy from one place to another. Good insulators keep the energy "contained" within the good conductors, which, in turn, waste little power in ohmic heating. For example, the aforementioned copper wire would dissipate a maximum of about 19 W while delivering approximately 1800 W at 120 V. And a larger wire would further reduce the heat loss.

There is, however, one factor not indicated in Table 2.3–1, namely *temperature dependence*. The given values of ρ are at room temperature

(20°C), but resistivity generally increases with temperature for conductors, while it decreases with temperature for insulators. Physically, this difference comes about because high temperature tends to liberate charge carriers in an insulator, whereas the increased thermal vibration of the ions in a metal causes more collisions and reduces the electron mobility μ_e. At extremely high temperatures—as might be encountered in a space mission—all materials act more or less like semiconductors. Obviously, the circuit designer must take special care in such cases. On the other hand, temperature dependence has practical applications in devices such as the resistance-wire thermometer.

Due, in part, to the temperature dependence of ρ, the v-i curve for many resistive elements becomes *nonlinear* at large values of current. An incandescent lamp filament, for instance, has the characteristic previously seen in Fig. 2.2–2, and its resistance at the operating temperature (around 2000°C) is more than 10 times as large as the "cold" resistance. Consequently, when we first energize a lamp, the initial current flow is substantially higher than the normal operating current. Although its duration is short, this large *inrush* current may damage mechanical or electronic switches.

Commercially manufactured resistors are designed to be reasonably linear over their intended operating range. They are usually made of carbon film or very thin wire, and come in a variety of standard resistance values, precision tolerances, and power ratings. Adjustable resistors are also available; they will be discussed in Section 3.1 under the heading "Potentiometers."

Example 2.3–2
A strain gauge

A *strain gauge* consists of many loops of thin resistive wire glued to a flexible backing (see Fig. 2.3–6). It is used to measure the fractional elongation or strain $\Delta\ell/\ell$ of a structural member to which it is attached. Straining the wire makes it somewhat longer and thinner, thereby increasing the resistance a small amount, ΔR.

Specifically, if we assume the volume of the wire remains constant, its area under strained conditions becomes $A' = A\ell/\ell'$ where $\ell' = \ell + \Delta\ell$ is the elongated length. Then, from Eq. (7), the strained resistance R' is

$$\begin{aligned} R' &= \rho\ell'/A' = \rho(\ell + \Delta\ell)^2/A\ell \\ &= (\rho\ell/A)[1 + 2(\Delta\ell/\ell) + (\Delta\ell/\ell)^2] \end{aligned}$$

With $(\Delta\ell/\ell)^2 \ll 1$, we can drop the last term and write $R' \approx R + \Delta R$, where $R = \rho\ell/A$ is the unstrained resistance and

$$\Delta R = 2R(\Delta\ell/\ell)$$

Therefore we can determine mechanical strain by measuring increased resistance.

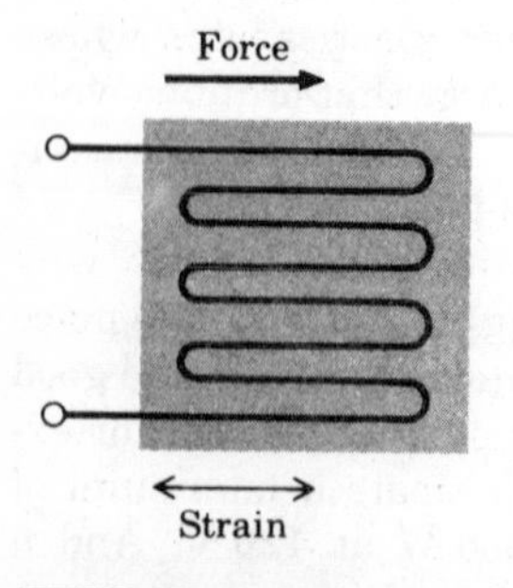

FIGURE 2.3–6
A strain gauge made with resistive wire.

Exercise 2.3–2 Calculate ΔR when $\Delta \ell / \ell = 10^{-3}$ for a strain gauge made from a 100-cm length of nichrome wire having 0.002-cm radius.

LUMPED PARAMETER CIRCUITS

Let's apply what we now know about resistance to the flashlight circuit of Fig. 2.1–1. Because the bulb is a resistance and there is resistance distributed all along the connecting wires, we can redraw the circuit as in Fig. 2.3–7a, which also includes leakage resistance representing the insulation between the bulb's contacts. If we are concerned only with variables measured at the *terminals* of the various elements, as distinguished from those measured in the interior, we can lump the distributed wire resistance at one point and treat the connecting lines as ideal conductors, as in Fig. 2.3–7b. This is called a *lumped parameter model* in that spatially distributed characteristics—resistance in this case—have been concentrated at one point to simplify analysis. Almost all circuit diagrams employ this approach, thereby focusing on the terminal characteristics of the elements rather than their interior behavior.

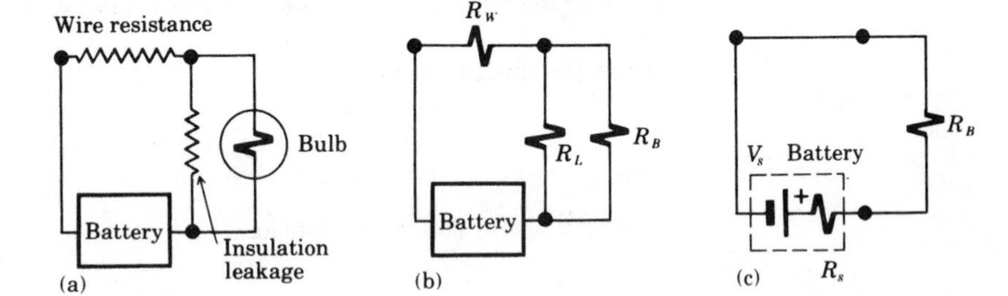

FIGURE 2.3–7
Evolution of a lumped parameter model.

Taking the model one step further, we might ignore the wire resistance entirely since it should be quite small compared to the resistance of the bulb. Similarly, the insulation leakage resistance should be so large that negligible current flows through it. However, the battery also has internal resistance that may cause its terminal behavior to differ significantly from an ideal voltage source. The final diagram then becomes as shown in Fig. 2.3–7c. Hereafter, wire resistance and source resistance will be included when we suspect that they may have an appreciable effect on the circuit's behavior.

2.4 KIRCHHOFF'S LAWS

Figure 2.4–1 illustrates a circuit composed of sources, resistances, and an unspecified element. Circuit analysis is the task of evaluating all voltages and currents, given the source and resistance values and the nature of the other element. Conversely, circuit design involves choosing element values to achieve specified voltages or currents. In either case, one must draw upon Kirchhoff's laws (pronounced Kear-koff) as well as Ohm's law.

Kirchhoff's two laws are *network laws,* that apply to the interconnection of elements rather than to individual elements. Ohm's law, in contrast, is an *element law* that describes a particular element irrespective of how it is connected to other elements. After we state Kirchhoff's laws and examine their implication for common circuit configurations, we'll introduce some simple methods of circuit analysis based on those laws.

KIRCHHOFF'S CURRENT LAW (KCL)

Kirchhoff's current law describes current relations at the *nodes* in a network—meaning any point where two or more elements are connected, such as points a and b in Fig. 2.4–1. An electrical node is analogous to the junction of two or more water pipes. Clearly, the amount of water coming out of a pipe junction must equal the amount going in, and the same holds for current at a node. Physically, because charge must be conserved and no charge accumulates at a node, the charge flowing out exactly equals the charge flowing in. Expressing charge flow in terms of current, KCL states:

The net sum of the currents into any node equals zero.

We write this symbolically as

$$\sum_{\text{node}} i = 0 \tag{1}$$

in which Σ stands for summation.

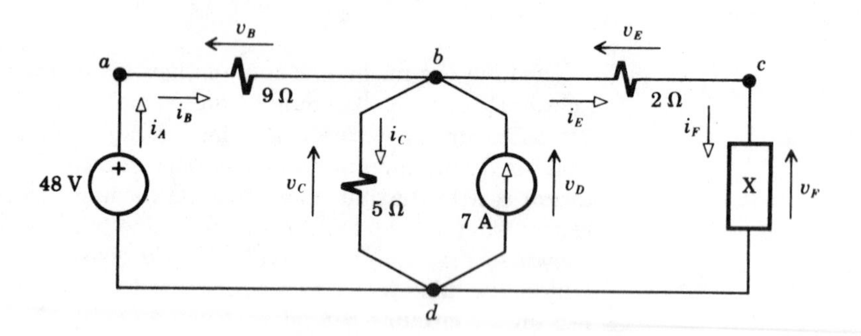

FIGURE 2.4–1

A key factor in KCL is current *direction,* for we recall that the algebraic sign of a current indicates its direction. Accordingly, a positive current flowing out of a node is equivalent to a negative current flowing into that node. To illustrate, in Fig. 2.4–2a, the current i_C and i_E flow out of node b, while i_B and the source current flow in, so KCL yields

$$i_B - i_C + 7\text{A} - i_E = 0$$

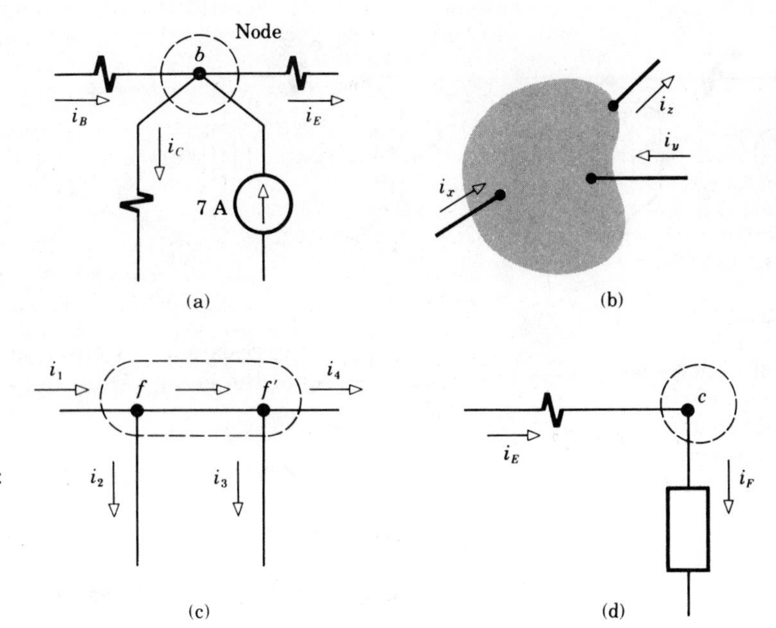

FIGURE 2.4–2 Applications of Kirchhoff's current law: (a) at a node; (b) over a closed surface; (c) for two nodes connected by a conductor; (d) for two elements in series.

Alternatively, we could write

$$i_B + 7\text{A} = i_C + i_E$$

which says that the sum of the currents actually flowing in equals the sum of those actually going out.

Moreover, the same statement holds for any portion of a circuit that could be contained within a closed surface, due to the fact that no circuit element accumulates net charge. Thus, in Fig. 2.4–2b, $i_x + i_y = i_z$ regardless of the details of the circuitry inside the enclosure. Similarly, we can write $i_1 = i_2 + i_3 + i_4$ for Fig. 2.4–2c and not even bother with the current going from f to f'; in fact, these two points constitute just one node since they are directly connected by a perfect conductor.

A particularly simple but significant type of node is one where exactly two elements are connected together as in Fig. 2.4–2d. Clearly, $i_E - i_F = 0$ or $i_E = i_F$, and we see that the current must be the same through both elements. This arrangement is called a *series* connection.

KIRCHHOFF'S VOLTAGE LAW (KVL)

Kirchhoff's voltage law expresses the principle of conservation of energy in terms of the voltages around a *loop,* a loop defined as any closed path in a network. Consider Fig. 2.4–3a which shows one loop from Fig. 2.4–1; if a charge dq goes around this loop, it gains energy from the voltage source

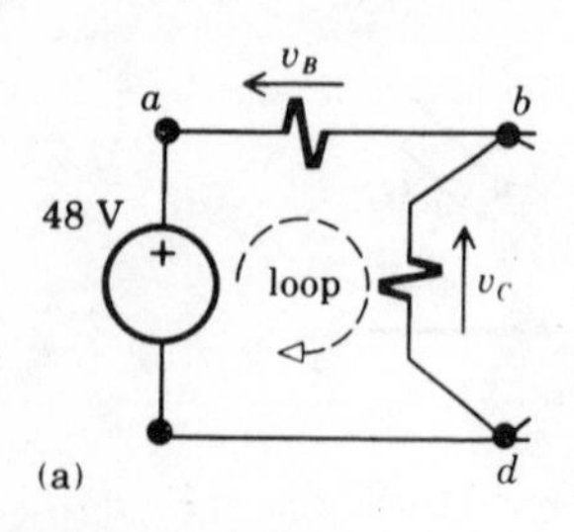

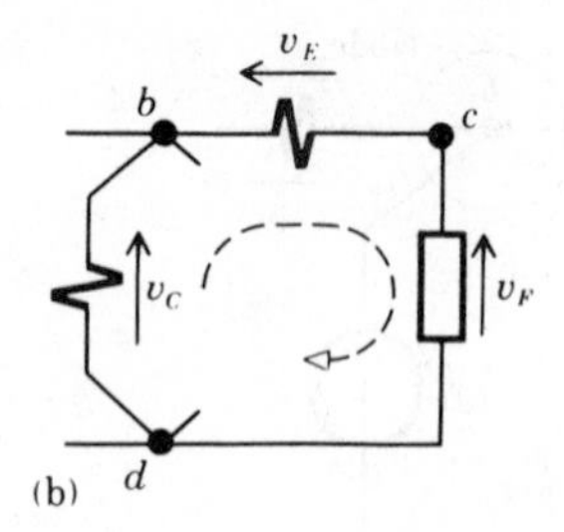

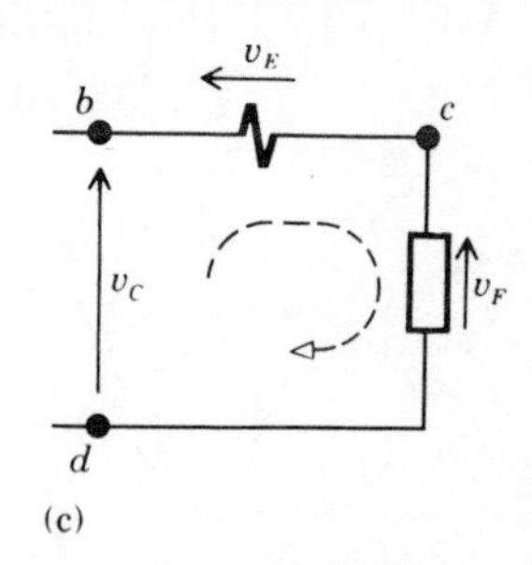

FIGURE 2.4–3
Applications of Kirchhoff's voltage law: (a) around a loop with a source; (b) around a loop without a source; (c) around an electrically unclosed loop.

and delivers energy to the resistances. Conservation of energy requires that the energy loss equal the energy gain, so $v_B\,dq + v_C\,dq = 48 \text{ V} \times dq$ and hence

$$v_B + v_C - 48 \text{ V} = 0$$

In general, Kirchhoff's voltage law states:

The net sum of the voltages around any loop equals zero.

Using summation notation, KVL is written

$$\sum_{\text{loop}} v = 0 \tag{2}$$

in a form analogous to the current law.

Although we introduced KVL with a loop containing a source, the law also applies to loops such as that in Fig. 2.4–3b. There are no sources here, but the arrow for voltage v_C has the opposite direction of the other two voltage arrows as one travels around the loop. Consequently, KVL requires $v_E + v_F - v_C = 0$ or $v_C = v_E + v_F$. A helpful interpretation of this result comes from the observation that, since v_C is the potential difference between points d and b, it must equal the potential difference between d and c, plus the potential difference between c and b. This interpretation further suggests that KVL applies even to loops that do not have electrical closure; thus, the voltage across the open circuit from d to b in Fig. 2.4–3c is still given by $v_C = v_E + v_F$.

One other loop from the original circuit deserves special attention, namely the loop formed by two elements connected together at each end (see Fig. 2.4–4a), which is called a *parallel connection*. KVL gives us $v_D - v_C = 0$ or $v_D = v_C$; hence, the voltage is the same across both elements. Incidentally, such connections are usually drawn as in Fig. 2.4–4b, where the top and bottom nodes have each been replaced by two points, simply to make life easier for the person drawing the circuit.

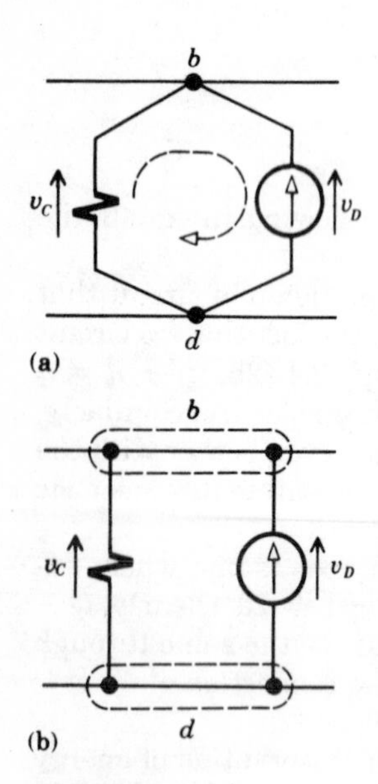

FIGURE 2.4–4
Elements connected in parallel.

Example 2.4–1

Figure 2.4–5 shows a *transistor,* a three-terminal device, connected at nodes B, C, and E. The double-subscript voltage notation shown here is commonly used for electronic devices, with the first subscript identifying the terminal assumed to be at the higher potential. For our present purposes the transistor is taken to have

$$v_{BE} = 0.7 \text{ V} \qquad i_C = 0.9\, i_E$$

This information plus Ohm's and Kirchhoff's laws will allow us to find all other voltages and currents.

First, noting the series connections at E and C, $i_E = 20$ mA, so $i_C = 0.9 \times 20 = 18$ mA and $v_R = 1.5\text{ k}\Omega \times i_C = 27$ V. Next, using KVL around the outer loop gives $v_R + v_{CE} = v_{BE} + 30$ V or $v_{CE} = 0.7 + 30 - 27 = 3.7$ V, from which $v_{CB} = v_{CE} - v_{BE} = 3$ V. Finally, applying KCL to the transistor itself, $i_B + i_C = i_E$ and, therefore, $i_B = i_E - i_C = 2$ mA.

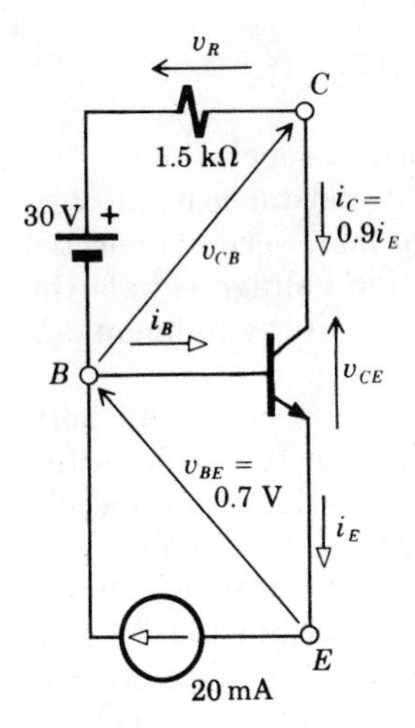

FIGURE 2.4–5
A transistor circuit with double-subscript voltage notation.

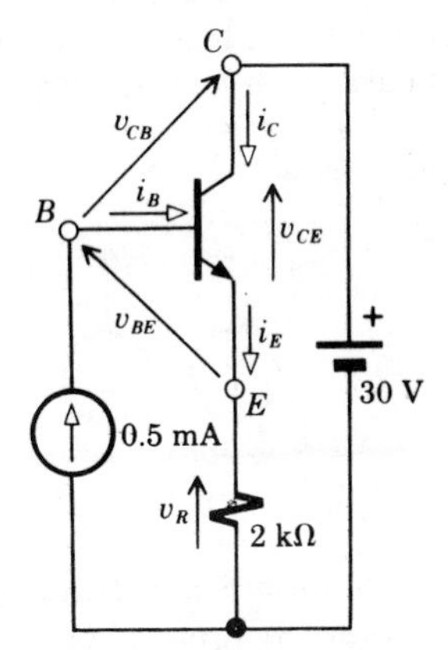

FIGURE 2.4–6
Circuit for Exercise 2.4–1.

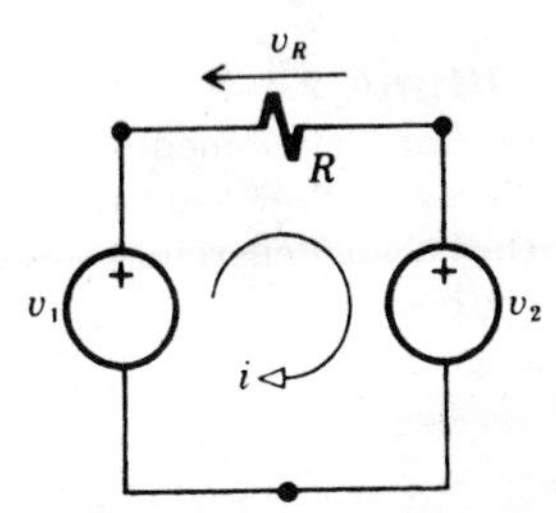

FIGURE 2.4–7
A series circuit with opposing voltage sources.

Exercise 2.4–1

Find the voltages and currents, when the transistor of Example 2.4–1 is used in the circuit of Fig. 2.4–6. Hint: First express i_E in terms of i_B.

SERIES AND PARALLEL CIRCUITS

Circuits consisting entirely of series or parallel connections are usually the easiest to analyze, because there is only one current value in a series circuit and only one voltage value in a parallel circuit. Despite their simplicity, they play an important role in many applications.

The circuit in Fig. 2.4–7 is a *series circuit* in that each of the three nodes connects exactly two elements. Hence, all three elements carry the same current i, indicated by the single current arrow. Given the source voltages v_1 and v_2 and the value of R, it becomes a simple matter to find v_R

from Kirchhoff's voltage law and then solve for i using Ohm's law. Specifically, the voltage across the resistance must satisfy $v_R + v_2 - v_1 = 0$, so

$$v_R = v_1 - v_2 \qquad i = \frac{v_R}{R} = \frac{v_1 - v_2}{R} \tag{3}$$

These seemingly trivial results lead to two important conclusions.

First, the effective source voltage applied to the resistance is the *algebraic sum* of the sources, $v_1 - v_2$. In general, two or more series-connected voltage sources act like one source whose effective voltage equals the algebraic sum of the individual source voltages—as in the common 3-V flashlight with its two 1.5-V batteries.

Second, if v_2 happens to be greater than v_1, then $v_1 - v_2 < 0$ and both v_R and i will be *negative* quantities, meaning that i actually flows the other way. In retrospect, the direction taken for i in the figure was merely a guess based on the assumption that $v_1 > v_2$. Such guesses are not critical for, as we have seen, a "wrong" guess eventually shows up as a negative value. But the voltage polarity across the resistance must be consistent with Ohm's law and the assumed current direction. Therefore, we will always draw voltage and current arrows in *opposite directions* at each resistance. This *passive convention* will also be used for all other two-terminal elements except sources.

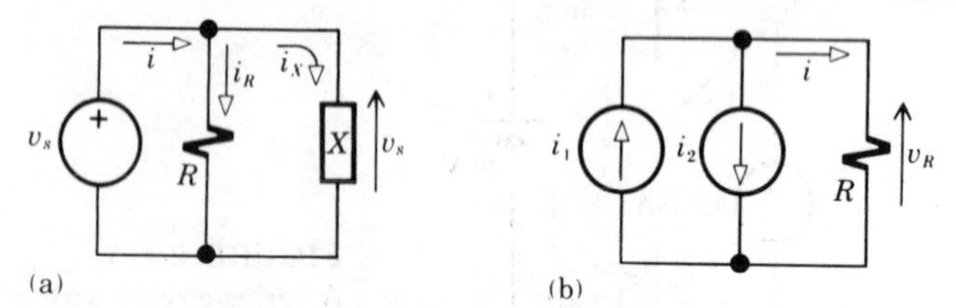

FIGURE 2.4–8 (a) Parallel elements with a voltage source. (b) Parallel current sources.

Now consider a *parallel circuit,* Fig. 2.4–8a. Here, the source voltage appears directly across each of the other elements, whose current directions are taken in accordance with the passive convention. Kirchhoff's current law applied at the upper node shows that the total current from the source will be $i = i_R + i_X$ where $i_R = v_s/R$ while i_X depends on the nature of the unspecified element. Residential AC circuits have this type of structure, with many elements in parallel, and the total current equals the sum of the individual device currents.

If a parallel circuit is driven by one or more current sources, as in Fig. 2.4–8b, then voltage rather than current becomes the unknown that must be found. For the case at hand, KCL and Ohm's law yield

$$i = i_1 - i_2 \qquad v_R = Ri = R(i_1 - i_2) \tag{4}$$

which should be contrasted with the results in Eq. (3).

Exercise 2.4–2

In Fig. 2.4–7, suppose $v_1 = 12$ V and $R = 3\ \Omega$. Redraw the circuit, calculate i, and label all voltage values when the v_2 source is replaced by (a) a closed switch, (b) an open switch, (c) another 3-Ω resistance, (d) a current source with $i_s = 5$ A pointed opposite to i.

Example 2.4–2
Biasing resistors

Turning from analysis to design, suppose two devices are to be powered by a 12-V battery; device A operates at 4 V and 2.0 A while device B requires 5 V and 1.6 A. Connecting them directly in series or parallel with the battery would not work in view of the differing voltage and current requirements. Those differences are overcome with the help of two resistances arranged as in Fig. 2.4–9.

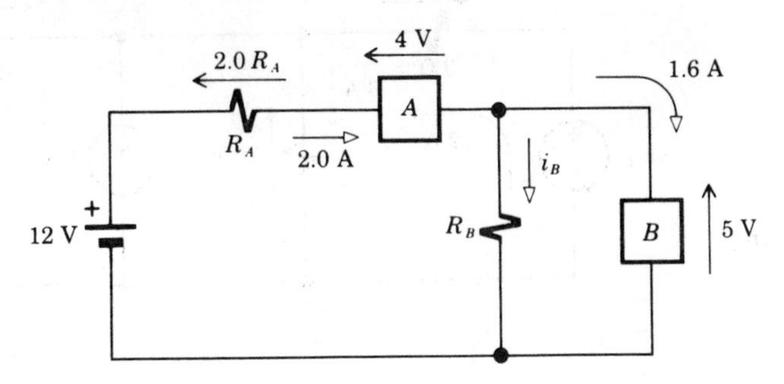

FIGURE 2.4–9
Resistors arranged to bias two devices.

The resistance in series with A provides a voltage drop such that $2.0\ R_A + 4 + 5 = 12$ to satisfy KVL. Thus $R_A = (12 - 9)\ \text{V}/2.0\ \text{A} = 1.5\ \Omega$. The resistance in parallel with B carries a current $i_B = 2.0 - 1.6$ to satisfy KCL. Thus, $R_B = 5\ \text{V}/0.4\ \text{A} = 12.5\ \Omega$. Resistors used in this fashion are called *biasing* resistors.

Exercise 2.4–3

Design a circuit with two biasing resistors so that the devices above can be powered by a 6-V battery. Check your work using the appropriate laws.

BRANCH CURRENTS AND NODE VOLTAGES

Now that we have Kirchhoff's laws in hand, we can begin a complete analysis of a moderately complicated circuit. Specifically, we will determine all the voltages and currents in Fig. 2.4–1, given that the unspecified element may be modeled as an 8-Ω resistance. Two different methods of attack will be illustrated, one involving branch currents, the other involving node voltages.

For the *branch-current method*, the circuit has been redrawn in Fig. 2.4–10a with three simplifications:

- Elements in parallel are shown with the same voltage across them.
- Elements in series are shown with the same current through them.

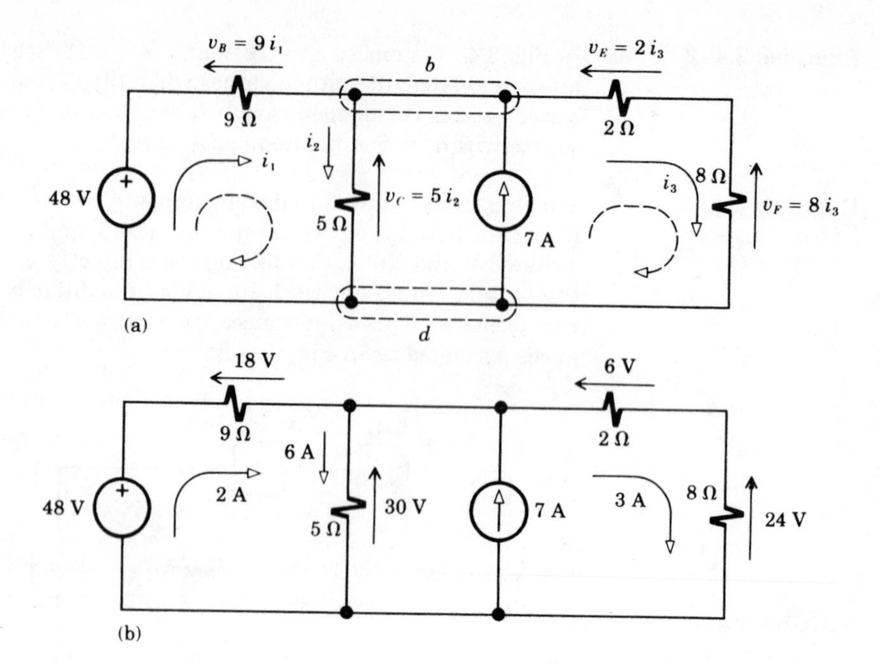

FIGURE 2.4–10 Circuit analysis using branch currents.

- Voltages across resistances are expressed in terms of currents, using Ohm's law and the passive convention.

We have thereby centered attention on three unknowns—labeled i_1, i_2, and i_3—which are the branch currents and have the property that all the unknown voltages are easily found from them.

Evaluating the three branch currents requires *three independent equations*. Invoking KCL at node b gives

$$i_1 + 7 = i_2 + i_3$$

and one is tempted to do the same thing at node d. But, as you can check, KCL at node d yields precisely the same relationship and hence provides no new information. (Expressed in mathematical parlance, the two node equations are not independent.) On the other hand, applying KVL around the left and right loops does give the two additional equations needed; that is,

$$9i_1 + 5i_2 = 48 \qquad 5i_2 = 2i_3 + 8i_3$$

A third loop equation can be written for the outer path, but it contains the same information as the other two.

Solving our three simultaneous equations is a relatively straightforward chore by successive substitution. The KVL equations can be rewritten to get i_1 and i_3 in terms of i_2, so

$$i_1 = \frac{48 - 5i_2}{9} \qquad i_3 = \frac{5i_2}{10}$$

Substituting these into the node equation gives

$$\frac{48 - 5i_2}{9} + 7 = i_2 + \frac{5i_2}{10}$$

from which

$$i_2 = \frac{\frac{48}{9} + 7}{1 + \frac{5}{10} + \frac{5}{9}} = 6 \text{ A}$$

Plugging this back into the previous equations finally yields $i_1 = 2$ A and $i_3 = 3$ A. Figure 2.4–10b shows the circuit with all current and voltage values; and testing loop and node sums quickly checks the accuracy of our results. For instance, taking KVL around the outer loop confirms that $18 + 6 + 24 = 48$, while KCL at node d confirms that $6 + 3 = 2 + 7$.

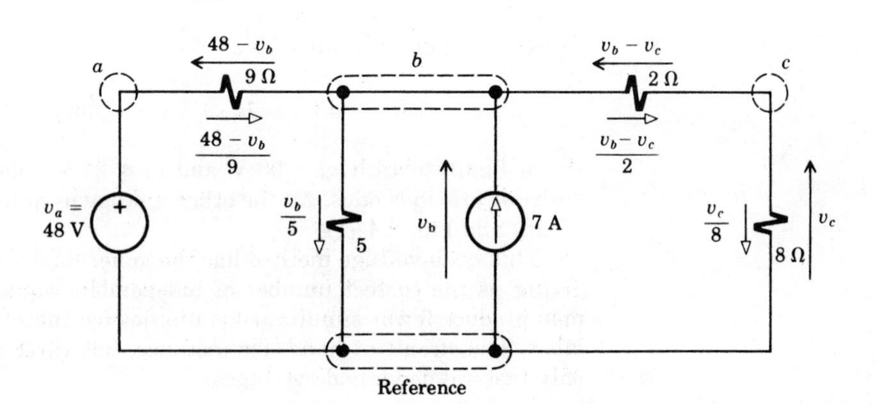

FIGURE 2.4–11
Circuit analysis using node voltages.

For the *node-voltage method,* the circuit is redrawn in Fig. 2.4–11 and labeled entirely in terms of voltages by a three-step process.

- Designate one node as the *reference* and draw voltage arrows to all other nodes. These represent the node voltages, of which v_b and v_c are unknown whereas $v_a = 48$ V due to the voltage source between node a and the reference.
- Invoke KVL to express all remaining voltages as the difference between node voltages.
- Use Ohm's law to write all currents through resistances in terms of node voltages, again following the passive convention.

Thus, for instance, the voltage across the 5-Ω resistance is the node voltage v_b and the current is $v_b/5$; similarly, the voltage across the 2-Ω resistance is $v_b - v_c$ and the current is $(v_b - v_c)/2$.

Having gone through these preliminaries, all that remains is to set down KCL equations at each node having an unknown voltage. Since only v_b and v_c are unknown, we write for node b

$$\frac{48 - v_b}{9} + 7 = \frac{v_b}{5} + \frac{v_b - v_c}{2}$$

and for node c

$$\frac{v_b - v_c}{2} = \frac{v_c}{8}$$

Regrouping these as

$$\left(\frac{1}{9} + \frac{1}{5} + \frac{1}{2}\right) v_b - \frac{1}{2} v_c = \frac{48}{9} + 7 \qquad -\frac{1}{2} v_b + \left(\frac{1}{2} + \frac{1}{8}\right) v_c = 0$$

and clearing of fractions yields

$$73v_b - 45v_c = 1110 \qquad -36v_b + 45v_c = 0$$

so we finally obtain $v_b = 30$ V and $v_c = 24$ V—new results consistent with our previous ones. All the other unknowns quickly follow from these values and Fig. 2.4–11.

This node-voltage method has the general advantage of immediately giving us the correct number of independent equations. In addition, it may produce fewer simultaneous unknowns, thereby reducing algebraic labor. The circuit at hand, for instance, has three branch currents, but only two unknown node voltages.

Example 2.4–3 The circuit in Fig. 2.4–12 has a form commonly occurring in electronics, with currents in milliamps and resistances in kilohms. After identifying the unknown node voltages v_1 and v_2, we can label the currents without

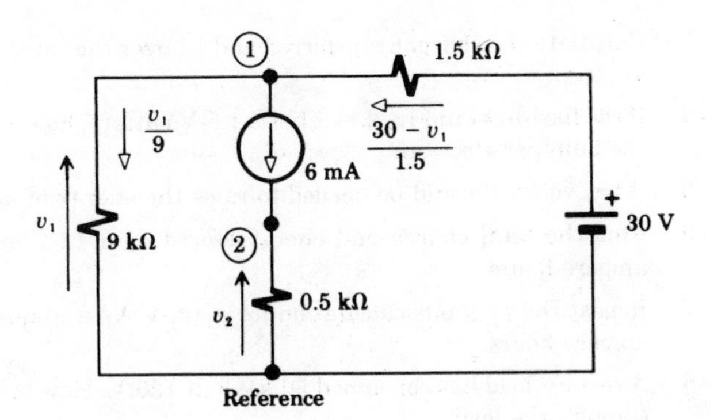

FIGURE 2.4–12 Circuit for Example 2.4–3.

explicitly showing the difference voltages. The KCL equation at node 1 is then

$$\frac{30 - v_1}{1.5\ \text{k}\Omega} = \frac{v_1}{9\ \text{k}\Omega} + 6\ \text{mA}$$

which yields $v_1 = 18$ V. The voltage v_2 did not enter into this calculation because the current source fixes its value at $v_2 = 0.5\ \text{k}\Omega \times 6\ \text{mA} = 3$ V. Since node 1 is at a higher potential than node 2, the current source happens to be consuming power here, namely $p = (v_1 - v_2) \times 6\ \text{mA} = 90$ mW.

Exercise 2.4–4 Repeat the above example using branch-current analysis, and show that $i_1 = v_1/9\ \text{k}\Omega = 2$ mA.

Exercise 2.4–5 Figure 2.4–13 represents a battery-charging circuit. Use the node-voltage method to find all voltages and currents and to show that i_2 is negative.

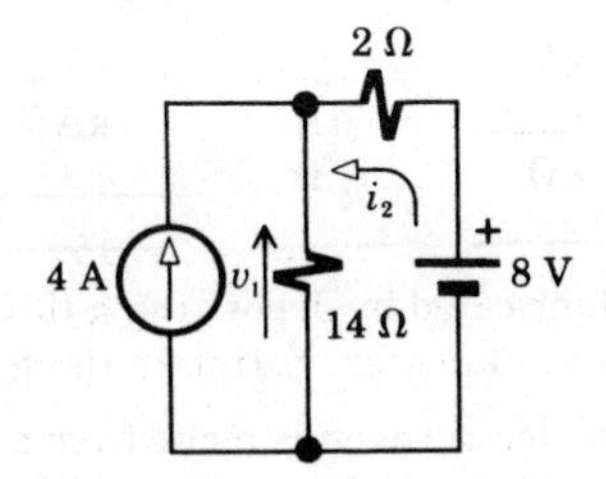

FIGURE 2.4–13 Model of a battery-charging circuit.

PROBLEMS

2.1–1 How many electrons per second pass a given point on a wire that carries $i = 10^{-6}$ A?

2.1–2 Find the average velocity of electrons in a copper wire with $n = 10^{29}$ electrons/m^3 and $A = 2$ mm^2 when $i = 1$ A.

2.1–3 Calculate the charge transferred and i_{av} over the interval $0 \leq t \leq 1$ min if $i(t) = 10^{-3}t$ A.

2.1–4 If the flashlight in Fig. 2.1–1 has a 1.5-V battery, how much energy is delivered to the bulb per electron?

2.1–5 What voltage would be needed to raise the energy of an electron by one joule?

2.1–6 Find the total charge and energy stored by a 12-V auto battery rated for 500 ampere-hours.

2.1–7 Repeat the previous calculation for a 1.5-V AAA alkaline battery rated for 0.75 ampere-hours.

2.1–8 A certain load has consumed 50 kWh at 120 V. How many electrons have passed through the load?

2.1–9 Find i_{av}, assuming the foregoing load was operated continuously for one week.

2.2–1 Repeat Example 2.2–1 with $I_p = 20$ mA.

2.2–2 Repeat Exercise 2.2–1 for a 40-V source connected to a device that has $v = 10^7 i^2$ for $i > 0$ and $v = -10^7 i^2$ for $i < 0$.

2.2–3 Let the photodiode in Fig. 2.2–3 have $I_0 = 0.1\ \mu$A and $I_p = 20$ mA. Justify the following statements:

(a) It acts like a current *sink* if $v \leq -0.1$ V;

(b) It acts like a current source if $0 \leq v \leq +0.1$ V.

2.2–4 Suppose the source in Fig. 2.2–1a has $v = 12/(4 + i^2)$ for $i \geq 0$.

(a) Over what range of current does this source act like a voltage source?

(b) Does it ever act like a current source? Explain your answer.

2.3–1 Let resistance R in Fig. 2.3–3 be adjustable. Show mathematically and graphically that $p \to \infty$ as $R \to 0$.

2.3–2 Example 2.3–1 demonstrates that volts, milliamps, kilohms, and milliwatts constitute a *consistent set of units* that does not require conversion of its magnitude prefixes. Fill in the blanks in the table below so that each column forms another consistent set.

Voltage	V	mV	______	kV	kV	______
Current	mA	______	μA	kA	______	______
Resistance	kΩ	kΩ	Ω	______	______	MΩ
Power	mW	______	______	______	W	μW

2.3–3 A 1-Ω resistance is to be fabricated by depositing a thin film of carbon on a cylinder 9 mm long and 3 mm in diameter. Estimate the film's thickness.

2.3–4 The heating element of an electric oven is made from a 1-mm diameter wire with $\rho = 3 \times 10^{-6}\ \Omega \cdot$ m at the operating temperature. The element dissipates 5 kW at 240 V. How long is the wire?

2.3–5 A 60-W, 120-V lightbulb has a filament 2 cm long.

(a) Find the filament's radius if $\rho = 10^{-6}\ \Omega \cdot$ m at the operating temperature.

(b) Find the inrush current if $\rho = 5 \times 10^{-8}$ at room temperature.

‡ **2.3–6** The resistance of a *fuse* increases with temperature such that $R = R_c(1 + \alpha T)$, where R_c is the "cold" resistance, α is the temperature coefficient, and T the tem-

perature rise above 20°C. The temperature rise is given by $T = kp$ where k is a constant and p the power dissipated by the fuse. Obtain an expression for R in terms of the current i through the fuse, and show that it "blows out" ($R \to \infty$) at $i = 1/\sqrt{\alpha k R_c}$.

2.4–1 Let $v_s = 10$ V and $R = 2$ kΩ in Fig. 2.4–8a, and let the unidentified element be a current source.

(a) Find i if $i_x = 5$ mA.

(b) Find i if $i_x = -5$ mA.

2.4–2 Suppose we know that $i_0 = 5$ A in Fig. P2.4–2. Apply KCL at node d and KVL around loop b-c-d-e to obtain the value of v_s.

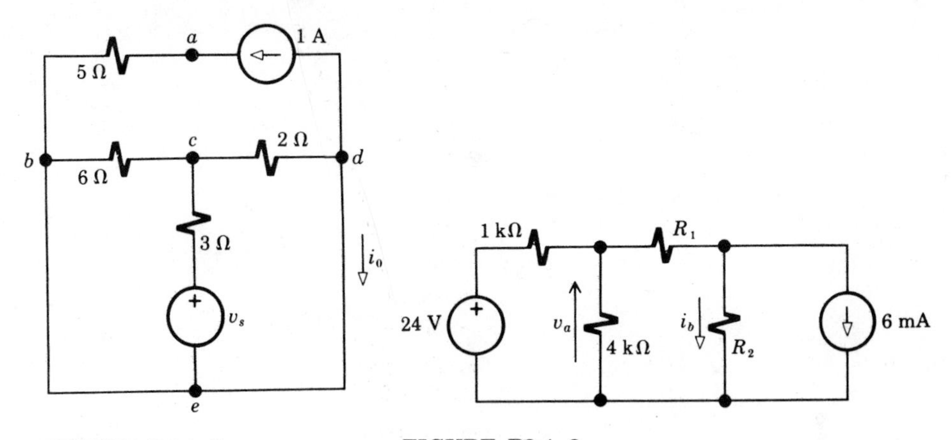

FIGURE P2.4–2 **FIGURE P2.4–3**

2.4–3 If $v_a = 16$ V in Fig. P2.4–3, what is the value of i_b? Note that R_1 and R_2 are unknown. Hint: Consider Fig. 2.4–2b.

2.4–4 Repeat Exercise 2.4–1 with the 2-kΩ resistor relocated in series with the battery.

2.4–5 Modify the circuit of Fig. 2.4–9 for *fail-safe biasing,* in the sense that one device continues to operate properly even when the other device becomes a short circuit or open circuit. Then compare the power drawn from the battery by your circuit and by the original design.

2.4–6 Find all the branch currents in Fig. P2.4–2 when $v_s = 12$ V.

2.4–7 Use the branch-current method to find all the currents in Fig. 3.3–8. You will need two KCL equations and two KVL equations.

2.4–8 Find v_a and v_c relative to node e when $v_s = 12$ V in Fig. P2.4–2.

2.4–9 Use the node-voltage method to find all the unknowns in Fig. P2.4–3 when $R_1 =$ 18 kΩ and $R_2 = 2$ kΩ.

2.4–10 Redraw Fig. 3.3–8 so that the node joining the two 10-kΩ resistors is at the bottom. Take this point as the reference node, and find the remaining two node voltages.

3 RESISTIVE CIRCUITS

This chapter applies our basic circuit concepts to the analysis and design of linear resistive circuits—circuits consisting of ideal sources and resistances. There are two major reasons why resistive circuits deserve further attention. First, they have many practical applications, either in their own right or as approximate models of more complicated circuits. Second, the techniques developed here will be extended in subsequent chapters when we deal with circuits containing nonresistive elements and electronic devices. Accordingly, a firm grasp of resistive circuits serves as the foundation for more exciting topics to come.

We will begin where the last chapter left off, namely with the application of Kirchhoff's laws and Ohm's law to resistive circuits. We will study in detail specific circuit configurations that occur time and again in practice. We then consider source-load circuits in general, and average power from AC sources. Three circuit theorems are also introduced for their value in both analysis and design work. An optional section describes simple DC meters and measurements.

OBJECTIVES

After studying this chapter and working the exercises, you should be able to do each of the following:

- Using the concept of equivalent resistance and divider ratios, find all the voltages and currents in a network comprised of series and parallel resistances (Section 3.1).

- Calculate the voltage, current, and power delivered to a load resistance from a real source, and state the conditions for maximum power transfer (Section 3.2).
- Calculate the average power delivered to a load resistance from an AC source (Section 3.2).
- State Thévenin's theorem and obtain the Thévenin and Norton equivalent circuits for a given one-port network (Sections 3.2 and 3.3).
- Use the superposition theorem to analyze a circuit containing two sources (Section 3.3).
- Draw and explain the basic circuits used for DC meters and null measurements (Section 3.4).†

3.1 SERIES AND PARALLEL RESISTANCE

Consider Fig. 3.1–1 where a battery with internal resistance supplies a "load" network consisting entirely of series and parallel resistances. Completely analyzing this circuit, to the point of finding all voltages and currents, entails considerable work. However, if one is concerned only with a few things—such as the current drain on the battery, the terminal voltage, and total power dissipation—then the concept of equivalent resistance and the handy voltage and current divider ratios can be invoked to expedite those calculations.

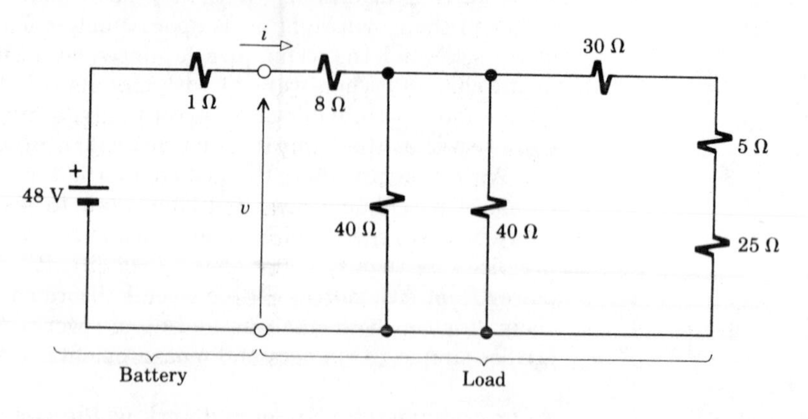

FIGURE 3.1–1 A battery with internal resistance and a resistive load network.

SERIES RESISTANCE AND VOLTAGE DIVIDERS

When two resistances are in series they must have the same current. Applying this fact, plus KVL and Ohm's law to Fig. 3.1–2a, we have

$$v_s = v_1 + v_2 = (R_1 i) + (R_2 i) = (R_1 + R_2)\, i$$

which is of the form $v_s = R_{eq} i$ with

$$R_{eq} = R_1 + R_2 \tag{1}$$

The sum $R_{eq} = R_1 + R_2$ is called the *series equivalent resistance.* It is equivalent in the sense that two resistances in series have the same v-i characteristic as one resistance whose value is the sum of the two. Hence, replacing R_1 and R_2 by a single resistance of value $R_1 + R_2$, Fig. 3.1–2b, produces no changes in the rest of the circuit. While we would seldom make the actual physical replacement, we will often do this mentally for calculations. To illustrate, if $v_s = 12$ V, $R_1 = 5\ \Omega$, and $R_2 = 1\ \Omega$, then $i = 12\text{ V}/(5 + 1)\ \Omega = 2$ A.

Extrapolating Eq. (1) to the case of n resistances in series, the equivalent resistance is the sum

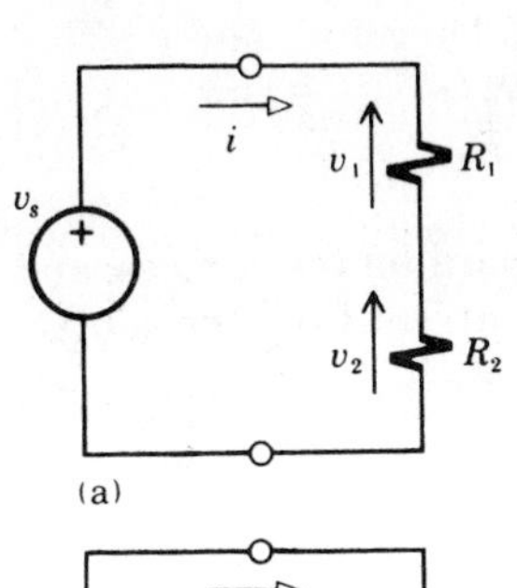

$$R_{eq} = R_1 + R_2 + \cdots + R_n \tag{2}$$

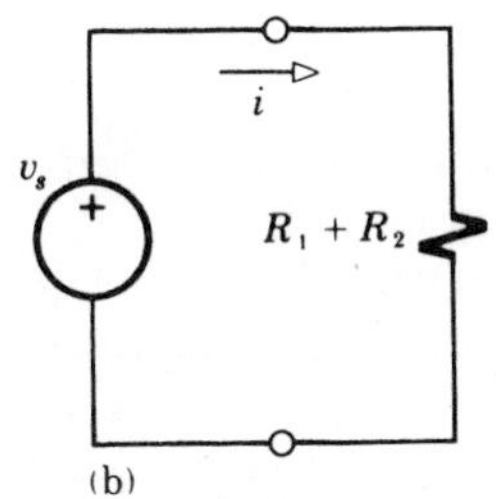

FIGURE 3.1–2
Series equivalent resistance.

The derivation should be obvious.

Now consider the portion of a circuit shown in Fig. 3.1–3 where the voltage v is known (not necessarily as a source voltage). We want to find the voltage v_1 across resistance R_1. This happens to be a common task that keeps popping up in circuit analysis and design. Clearly, $v_1 = R_1 i$ and $i = v/(R_1 + R_2)$, and therefore

$$v_1 = \frac{R_1}{R_1 + R_2} v \tag{3a}$$

which gives us v_1 directly in terms of v without the intermediate calculation of i.

This circuit configuration is called a *voltage divider,* $R_1/(R_1 + R_2)$ being the *voltage-divider ratio.* Similarly, we would use the ratio $R_2/(R_1 + R_2)$ to get v_2. The name comes from the fact that the total voltage v is "divided" between the two resistances and, in fact, $v_1 = v_2 = v/2$ when $R_1 = R_2$. You will find it handy to memorize Eq. (3a) noting carefully that v_1 is proportional to R_1. Also note the approximations

$$v_1 \approx \begin{cases} v & R_1 \gg R_2 \\ \dfrac{R_1}{R_2} v & R_1 \ll R_2 \end{cases} \tag{3b}$$

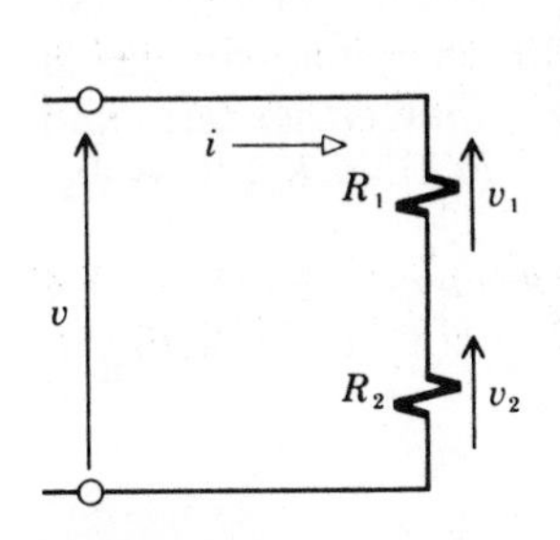

FIGURE 3.1–3
Voltage divider.

so most of the voltage appears across R_1 when $R_1 \gg R_2$, and vice versa.

PARALLEL RESISTANCE AND CURRENT DIVIDERS

When two resistances are in parallel they must have the same voltage drop. Thus, in Fig. 3.1–4a,

$$i = i_1 + i_2 = \frac{v}{R_1} + \frac{v}{R_2} = v\left(\frac{1}{R_1} + \frac{1}{R_2}\right)$$

which has the form $i = v/R_{eq}$ with

$$R_{eq} = \left(\frac{1}{R_1} + \frac{1}{R_2}\right)^{-1} = \frac{R_1 R_2}{R_1 + R_2} \tag{4}$$

Equation (4) is the equivalent resistance for two parallel resistances, and both circuits in Fig. 3.1–4 would draw the same current from the source.

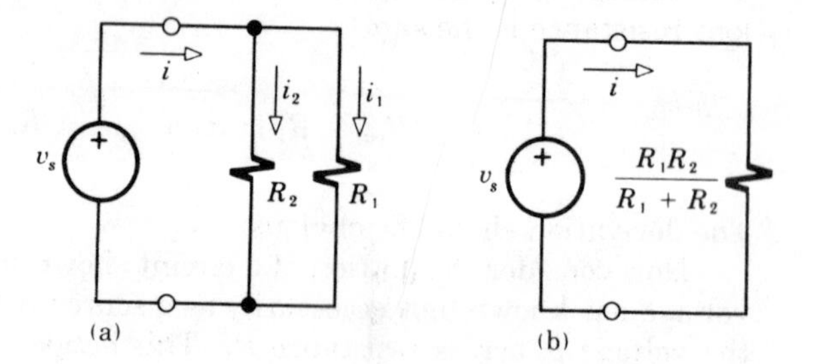

FIGURE 3.1–4 Parallel equivalent resistance.

The right-hand side of the above expression occurs so often in circuit work that we will give it the special notational symbol

$$R_1 \| R_2 \triangleq \frac{R_1 R_2}{R_1 + R_2} \tag{5}$$

where $R_1 \| R_2$ is read as "R_1 in parallel with R_2." Taking $R_1 = 4\ \Omega$ and $R_2 = 6\ \Omega$ for example, the parallel equivalent value is $4\|6 = (4 \times 6)/(4 + 6) = 24/10 = 2.4\ \Omega$.

Observe that the parallel equivalent value is always smaller than either term, and $R_1 \| R_2 = R_1/2$ when $R_2 = R_1$. This should be contrasted with the series case where $R_1 + R_2$ is always larger than either term and $R_1 + R_2 = 2R_1$ when $R_2 = R_1$. Moreover, if $R_1 \gg R_2$ then $R_1 \| R_2 \approx R_2$, whereas $R_1 + R_2 \approx R_1$.

The extension to n parallel resistances is not as obvious as the series case. To derive it we use *conductances* $G_1 = 1/R_1$ and Ohm's law in the form $i_1 = G_1 v$, etc. Then

$$\begin{aligned} i &= i_1 + i_2 + \cdots + i_n \\ &= G_1 v + G_2 v + \cdots + G_n v \end{aligned}$$

which has the form $i = G_{eq} v$ with the equivalent conductance being

$$G_{eq} = G_1 + G_2 + \cdots + G_n \tag{6a}$$

or

$$\frac{1}{R_{eq}} = \frac{1}{R_1} + \frac{1}{R_2} + \cdots + \frac{1}{R_n} \tag{6b}$$

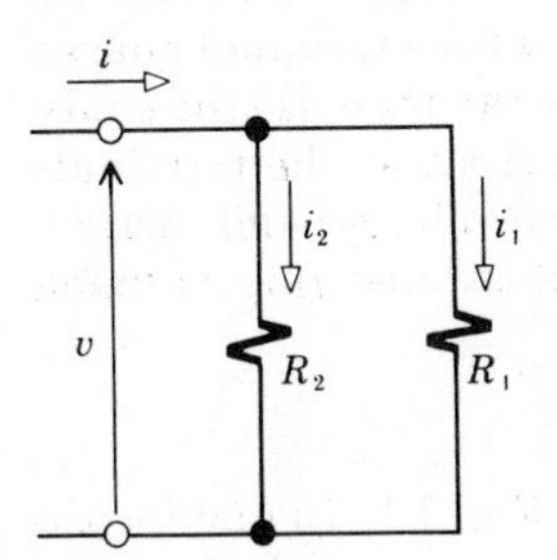

FIGURE 3.1–5
Current divider.

Taking the reciprocal of $1/R_{eq}$ finally gives the parallel equivalent resistance, a routine computation with a calculator. You should prove to yourself that Eq. (6b) leads to Eq. (4) when $n = 2$. (But don't try to use the product-over-sum expression for $n > 2$.)

Just as series resistances form a voltage divider, parallel resistances act as a *current divider*. Specifically, if we know the total current i in Fig. 3.1–5, then the current i_1 through R_1 is given by

$$i_1 = \frac{R_2}{R_1 + R_2} i \tag{7a}$$

$$\approx \begin{cases} i & R_1 \ll R_2 \\ \dfrac{R_2}{R_1} i & R_1 \gg R_2 \end{cases} \tag{7b}$$

These equations should be compared with Eqs. (3a) and (3b), paying special attention to the fact that i_1 is proportional to R_2, so a large R_2 "forces" more current through R_1.

Example 3.1–1
An electric range

The surface cooking units of an electric stove often have two resistance elements and a special switch that connects them to the voltage source individually, in series, or in parallel, as represented by Fig. 3.1–6. So, taking $R_1 < R_2$, there are four possible resistance values: $R_1 \| R_2$, R_1, R_2, and $R_1 + R_2$. Since voltage is constant, $p = v^2/R$ and the lowest "heat" is $p_{min} = v^2/R_{max} = v^2/(R_1 + R_2)$; similarly, $p_{max} = v^2/R_{min} = v^2/(R_1 \| R_2)$. Connecting the elements individually gives the two intermediate "heats" v^2/R_2 and v^2/R_1. Four more heat settings are possible if v has two values, for example, 120 V and 240 V.

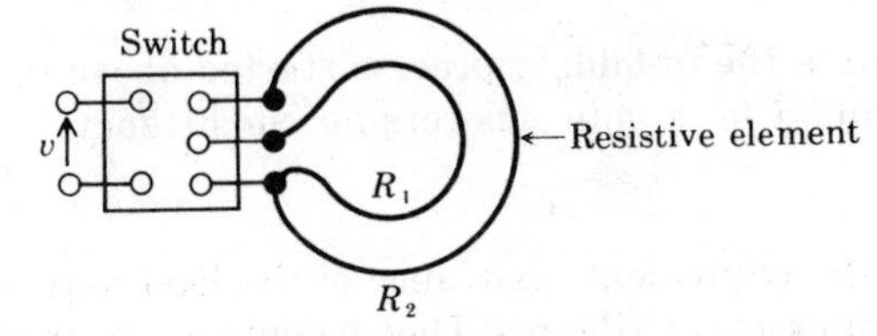

FIGURE 3.1–6
The surface unit of an electric range.

Exercise 3.1–1

Referring to the example above with $v = 120$ V and $R_1 = 60\ \Omega$, find R_2 so that $p_{min} = 80$ W and calculate the other three values of p.

EQUIVALENT RESISTANCE

Now, to tackle problems like the one posed at the start of this section! Suppose we have a network consisting entirely of series and parallel resistances. Repeated application of the above formulas will reduce the entire network to a *single equivalent resistance* whose terminal characteristics are identical to the original network. We can then use the equivalent resistance to calculate the voltage and current at the terminals. Those values in turn can be used, if needed, to find voltages and currents inside the network. The following example illustrates this valuable method.

Example 3.1–2

The load network from Fig. 3.1–1 is redrawn in Fig. 3.1–7a and inspection reveals that the two 40-Ω resistances are in parallel while the 30-Ω, 5-Ω, and 25-Ω resistances are in series. Replacing these by the equivalent values $40 \| 40 = 20\ \Omega$ and $30 + 5 + 25 = 60\ \Omega$ leads to Fig. 3.1–7b. Now we have $20 \| 60 = 15\ \Omega$ in series with 8 Ω, so the equivalent resistance of the network is $8 + 15 = 23\ \Omega$ (see Fig. 3.1–7c).

Putting this equivalent resistance in place of the load network in the original circuit gives the circuit of Fig. 3.1–7d, from which we see that $i = 48\text{ V}/(1 + 23)\ \Omega = 2\text{ A}$ so $p = (2\text{ A})^2 \times (1 + 23)\ \Omega = 96\text{ W}$. Had we been concerned with only the terminal voltage v, we could have derived that directly from the voltage-divider ratio, that is,

$$v = \frac{23\ \Omega}{23\ \Omega + 1\ \Omega} \times 48\text{ V} = 46\text{ V}$$

This result agrees with our value for i since, from KVL, $v = 48\text{ V} - (1\ \Omega \times 2\text{ A}) = 46\text{ V}$.

Moreover, going backwards by "unfolding" some of the equivalent resistances, we get to Fig. 3.1–7e, which has both a voltage divider and a current divider. In particular, the voltage v_1 appears across an equivalent resistance of 15 Ω which is in series with 8 Ω, so $v_1 = (15\ \Omega/23\ \Omega) \times 46\text{ V} = 30\text{ V}$. Similarly, the current through the 60-Ω equivalent resistance is $i_2 = (20\ \Omega/80\ \Omega) \times 2\text{ A} = 1/2\text{ A}$, agreeing with $i_2 = v_1/60\ \Omega$.

Exercise 3.1–2

Continue the unfolding process started above to obtain all voltages and currents. Check your answers for consistency.

Exercise 3.1–3

Find the equivalent resistance of the load network in Fig. 3.1–8, where all values are in kilohms. Then calculate i, p, and the current through the 7-kΩ resistance.

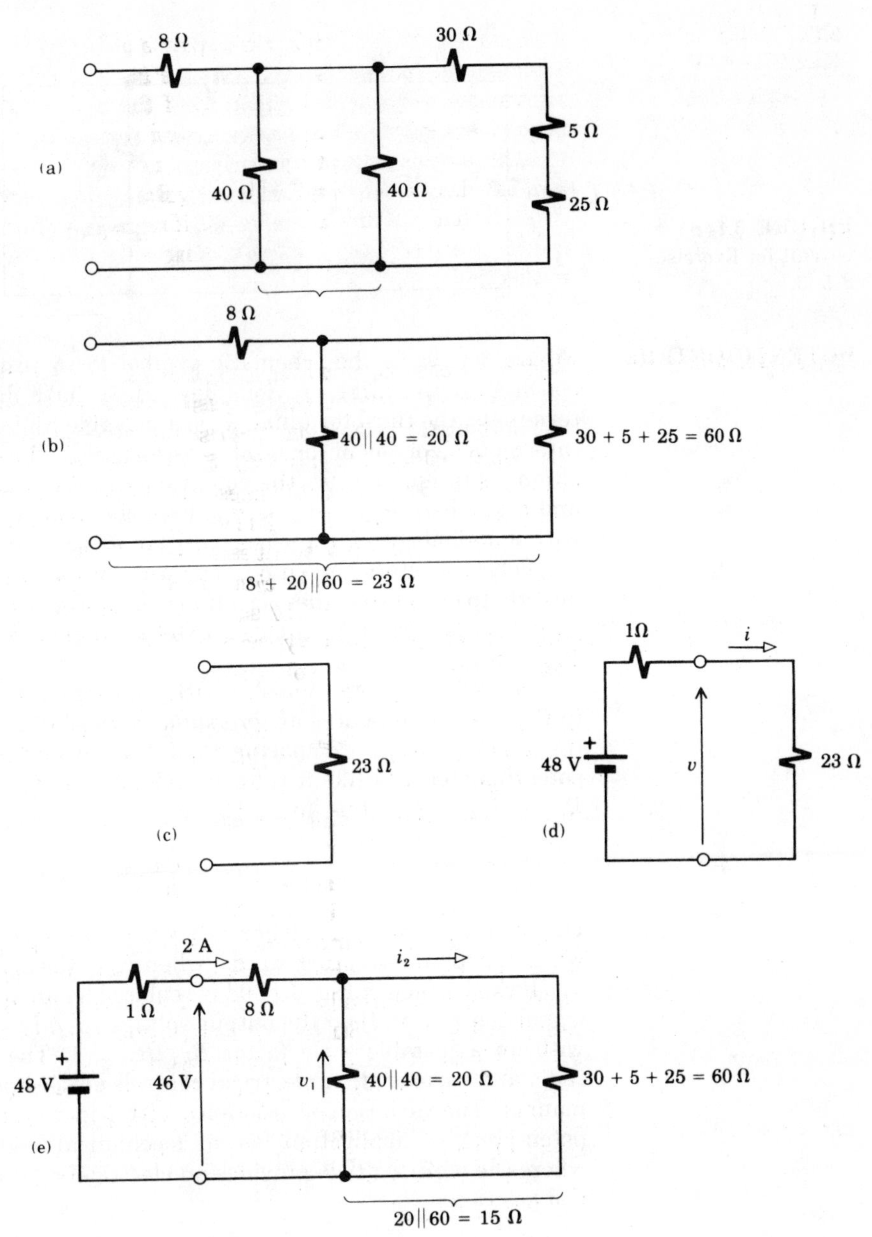

FIGURE 3.1-7
Circuit analysis using series/parallel reduction.

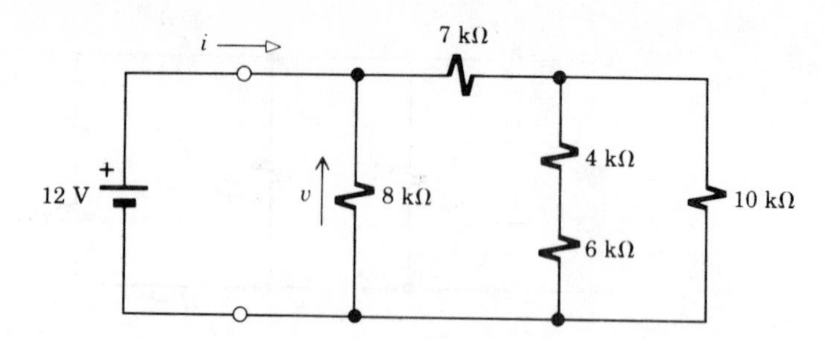

FIGURE 3.1–8
Circuit for Exercise 3.1–3.

POTENTIOMETERS

Figure 3.1–9a is the schematic symbol for a resistive device called a *potentiometer*. Unlike previous devices we have discussed, it has three terminals; the third terminal, w, is a movable contact point or *wiper* that intercepts a portion of the total resistance R_{ab}. Thus, for a fixed wiper position, as in Fig. 3.1–9b, the potentiometer acts like two resistances R_{aw} and R_{wb}, where $R_{aw} + R_{wb} = R_{ab}$. Precision potentiometers are available with dials indicating the wiper position as accurately as 0.1%.

Potentiometers have numerous uses, often with the wiper connected directly to one of the other terminals, as in Fig. 3.1–9c. The short-circuit path around part of R_{ab} yields a two-terminal *adjustable* resistance $R = R_{wb} \leq R_{ab}$.

Combining a potentiometer with a battery or fixed voltage source, as in Fig. 3.1–9d, produces an *adjustable voltage* $0 \leq v_w \leq V_s$ controlled by the wiper position. Comparing Figs. 3.1–9d and 3.1–3 shows that the potentiometer acts like a voltage divider with $R_1 = R_{aw}$ and $R_1 + R_2 = R_{aw} + R_{wb} = R_{ab}$; thus, using Eq. (3a),

$$v_w = \frac{R_{aw}}{R_{ab}} V_s$$

However, this expression holds only when nothing else is connected to the wiper terminal—in other words, when v_w is the *open-circuit* voltage.

If the battery in Fig. 3.1–9d is replaced by an arbitrary time-varying signal source $v(t)$, then the output voltage $v_w(t)$ has the same shape, but with an adjustable scale factor $R_{aw}/R_{ab} \leq 1$. The *volume control* in a radio and various other electrical controls employ potentiometers in this manner. The variation of resistance with wiper position leads to another potentiometer application as a mechanical-to-electrical *transducer*, where the wiper voltage provides an electrical representation of mechanical position.

Example 3.1–3
Angular measurement with a potentiometer

Suppose a mechanical device rotates over an angular range $0° \leq \theta \leq 300°$ and we need an indication of its angular position θ at a point some distance away. This can be done with a circular potentiometer attached to the rotating shaft, plus a battery, connecting wires, and voltmeter (see Fig. 3.1–10). If the potentiometer has $R_{ab} = 10$ kΩ, say, and a linear

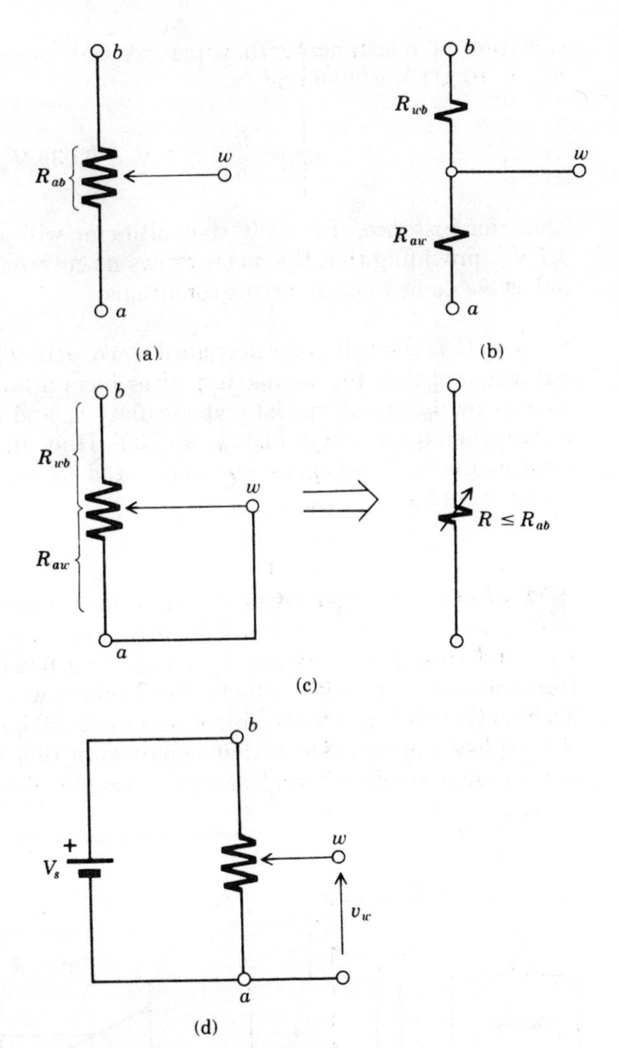

FIGURE 3.1-9
Potentiometer circuits.

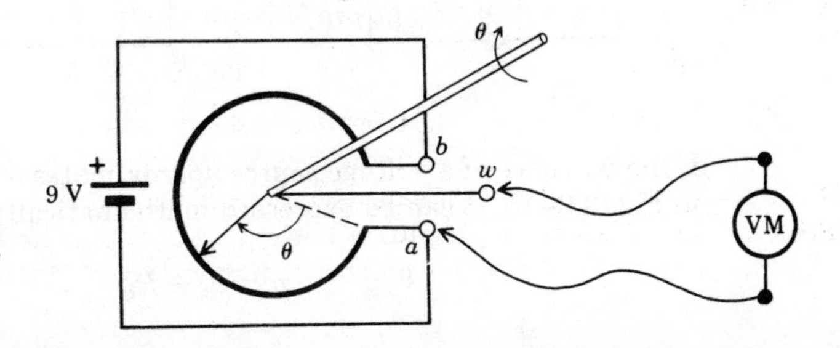

FIGURE 3.1-10
A potentiometer circuit for measuring angular position.

variation of resistance with wiper position over the 300° range, then $R_{aw} = 10\ \text{k}\Omega \times \theta/300°$ and

$$v_w = \frac{R_{aw}}{R_{ab}} \times 9\ \text{V} = 0.03\theta\ \text{V}$$

Thus, for instance, if $\theta = 90°$ the voltmeter will read $v_w = 0.03 \times 90 = 2.7$ V—providing that the meter draws no current from the wiper terminal so we do have open-circuit conditions.

Exercise 3.1–4

Suppose that the voltmeter described above actually acts like a 20-kΩ resistance and that the connecting wires have a total resistance of 1 kΩ. Now draw the circuit model and calculate v_w and the voltage across the meter when (a) $\theta = 300°$ and (b) $\theta = 90°$. Hint: In case (b), consider the total equivalent resistance between a and w, including the meter and wire resistance.

3.2 REAL SOURCES AND POWER TRANSFER

Real or *practical* sources differ from ideal sources by the presence of internal resistance. Consequently, when you connect a source to a load and measure the terminal voltage as a function of current drawn from the source, as in Fig. 3.2–1a, the v-i curve will have a negative slope such as that in Fig. 3.2–1b, due to the increasing internal voltage drop. We explore implications of that behavior in this section, starting with equivalent circuits for real sources. We also discuss AC sources and average power.

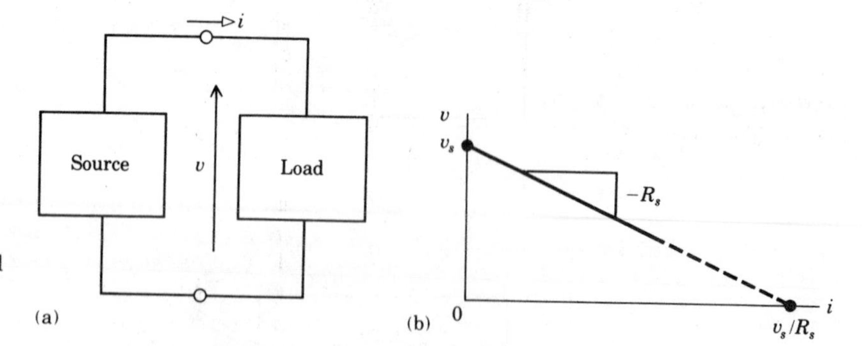

FIGURE 3.2–1 Characteristics of a real voltage source.

EQUIVALENT CIRCUITS FOR REAL SOURCES

If the v-i curve of a voltage source approximates a straight line like that in Fig. 3.2–1b, it can be expressed mathematically as

$$v = v_s - R_s i \tag{1}$$

where v_s is the voltage-axis intercept and $-R_s$ is the slope of the curve, so that the current-axis intercept equals v_s/R_s.

Equation (1) suggests that the practical voltage source may be represented by the equivalent circuit of Fig. 3.2–2a—an *ideal* voltage source v_s in series with a resistance R_s. You should confirm for yourself using KVL that this circuit has $v = v_s - R_s i$ for any value of i into the load. We will call v_s the *source voltage* and R_s the *source resistance*. Actually, it is more descriptive to say that v_s is the *open-circuit voltage* because v equals v_s only when nothing is connected to the terminals so $i = 0$, as in Fig. 3.2–2b. At the other extreme, suppose the terminals are short-circuited with a perfect conductor, as in Fig. 3.2–2c; then $v = 0$ and the resulting *short-circuit current* will be v_s/R_s. (Of course, short-circuit current is something of a fiction, since one would seldom intentionally short-circuit a real source.)

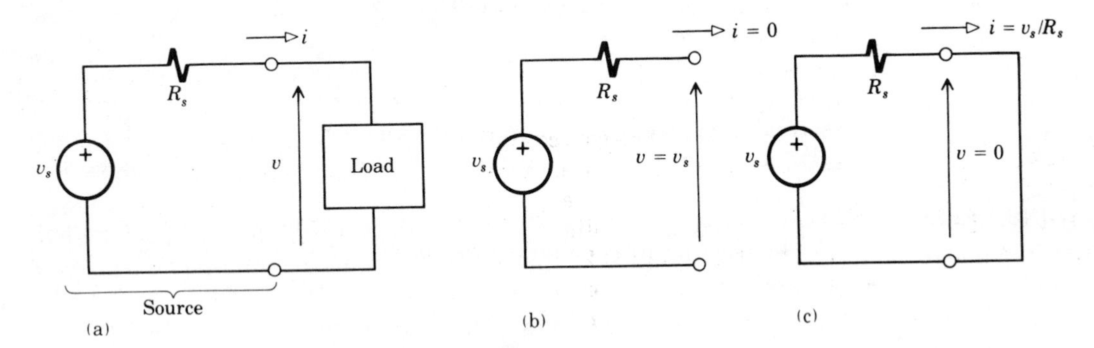

FIGURE 3.2–2
(a) Circuit model for a real voltage source.
(b) Open-circuit voltage.
(c) Short-circuit current.

Turning to practical current sources, if the *i-v* curve is linear (like that of Fig. 3.2–1b with v and i interchanged) then the terminal characteristics are of the form

$$i = i_s - \frac{v}{R_s} \tag{2}$$

In Fig. 3.2–3, we represent this by an ideal current source i_s in parallel with R_s. The *source current* i_s is the short-circuit current, while the open-circuit voltage equals $R_s i_s$, as follow from Eq. (2) or Fig. 3.2–3, with $v = 0$ (short-circuit) or $i = 0$ (open-circuit). Therefore, i_s and $R_s i_s$ are the respective current-axis and voltage-axis intercepts of the *i-v* curve.

These equivalent circuits are *models* for real sources—better models, perhaps, than ideal sources, but still models in the sense of having limited validity. Furthermore, even in their valid ranges, source models are *not* unique as there are other equivalent circuits that could be used.

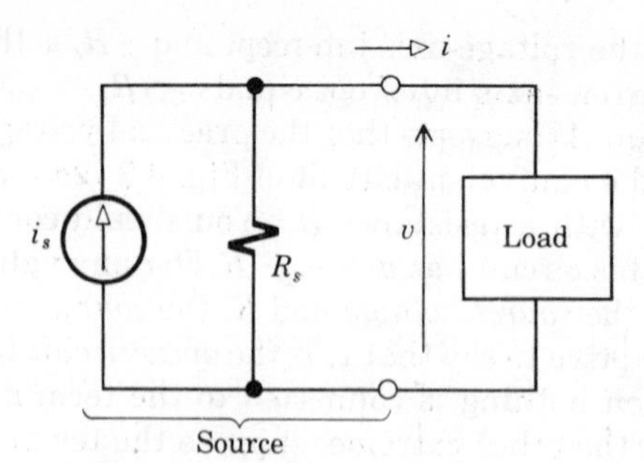

FIGURE 3.2–3
Circuit model for a real current source.

As an important case in point, a voltage source may be represented by the current-source equivalent circuit, or vice versa. This assertion is proved by rewriting Eq. (2) in the form

$$v = R_s i_s - R_s i$$

which is identical to Eq. (1) with $v_s = R_s i_s$. Moreover, R_s has the same interpretation in either circuit, namely the *ratio of open-circuit voltage to short-circuit current.* The functional distinction between real voltage and current sources becomes apparent when we consider the loading effect.

LOADING EFFECT

Let a load resistance R_L be connected to a voltage source, as in Fig. 3.2–4a. The terminal voltage v will be

$$v = v_s - R_s i = \frac{R_L}{R_L + R_s} v_s \tag{3}$$

revealing that v is less than the open-circuit source voltage v_s due to the voltage drop $R_s i$ across the internal source resistance. (A familiar demonstration of that voltage drop occurs when you start an automobile engine with the headlights on; the large current drawn by the starter decreases the voltage across the generator terminals and the headlights momentarily dim.) This loading effect is negligible when $R_s \ll R_L$, and thus $R_s i \ll v_s$.

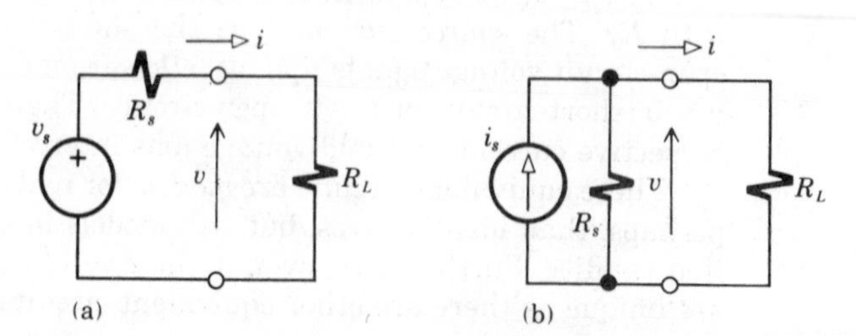

FIGURE 3.2–4
Loading effect due to source resistance.

Similarly, with R_L connected to a current source, as in Fig. 3.2–4b, i_s divides between R_s and R_L, and

$$i = i_s - \frac{v}{R_s} = \frac{R_s}{R_L + R_s} i_s \tag{4}$$

so that the load current is less than the short-circuit source current because of current v/R_s bypassed through the internal source resistance. The loading effect here will be negligible when $R_s \gg R_L$ and, thus, $v/R_s \ll i_s$.

We now see that the key difference between a practical voltage source and a practical current source is the value of R_s relative to the operating conditions. A "good" voltage source—that is, one that acts almost like an ideal voltage source—has a *small* internal resistance so that $v \approx v_s$ over the expected current range. Conversely, a "good" current source has a *large* internal resistance so that $i \approx i_s$ over the expected voltage range. Thus, if R_s is very large, it makes sense to use the current-source equivalent circuit with R_s in parallel, and vice versa when R_s is very small compared to R_L.

Example 3.2–1
Modeling a real source

A variable load resistance is connected to a source and the following two sets of values are measured:

i	v
12 mA	24 V
10 mA	40 V

Since i has less change than v, the current-source model would seem to be appropriate here. Thus, substituting each pair of values into Eq. (2) gives

$$12 \text{ mA} = i_s - 24 \text{ V}/R_s$$
$$10 \text{ mA} = i_s - 40 \text{ V}/R_s$$

which are easily solved to yield $R_s = 8$ kΩ and $i_s = 15$ mA. Equation (4) can then be used to predict v and i for other load values. For instance, with $R_L = 1$ kΩ, $i = (8/9) \times 15 = 13.33$ mA. Such predictions assume a linear i-v curve through the two measured points.

Exercise 3.2–1

Find an appropriate equivalent circuit for a source, given the measured values below, and predict v when $R_L = 200$ Ω.

i	v
0.14 A	11.2 V
0.10 A	12.0 V

Exercise 3.2–2

Electronic circuit designers sometimes use a voltage source plus a large series resistance to approximate an ideal current source. Justify this by finding the smallest value of R in Fig. 3.2–5, such that i decreases by no more than 5% when R_L is increased from 0.5 kΩ to 1 kΩ. You may neglect R_s.

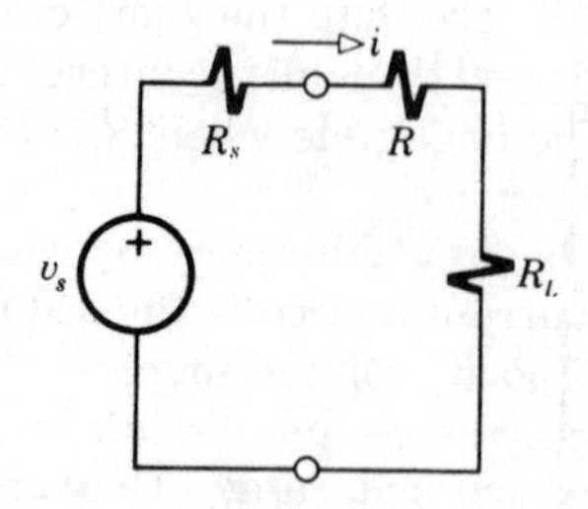

FIGURE 3.2–5 Approximating an ideal current source.

POWER TRANSFER

When a source and load are connected together, power is delivered to the load and dissipated internally in the source. To investigate this situation, we will take the voltage-source model (Fig. 3.2–4a) and compute the delivered power p_L and the internal power dissipation p_s using the fact that $i = v_s/(R_L + R_s)$. Thus,

$$p_L = i^2 R_L = \frac{R_L}{(R_L + R_s)^2} v_s^2$$
$$p_s = i^2 R_s = \frac{R_s}{(R_L + R_s)^2} v_s^2 \qquad (5)$$

A comparison of these expressions reveals that $p_L = (R_L/R_s)p_s$ so the ratio R_L/R_s plays a dominant role here.

Figure 3.2–6 plots p_L and p_s versus R_L/R_s. We see that p_s steadily decreases from its maximum value v_s^2/R_s at $R_L/R_s = 0$, approaching zero as

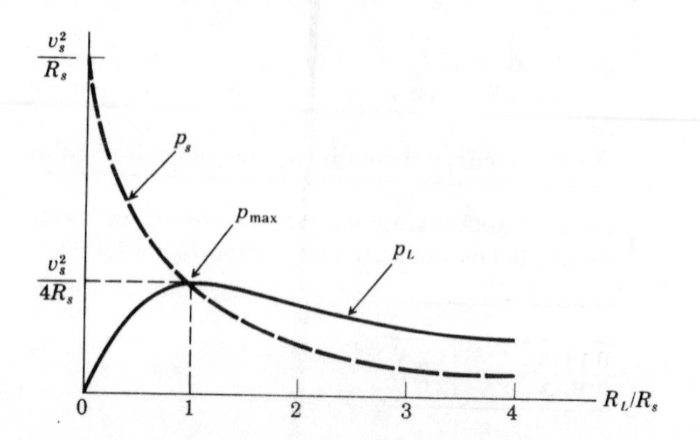

FIGURE 3.2–6 Load power p_L and internal power dissipation p_s plotted versus R_L/R_s.

R_L/R_s increases. Therefore, wasted power dissipation in the source is minimized by making R_L/R_s as large as possible. On the other hand, the curve for p_L has a distinct *maximum* value

$$p_{\max} = \frac{v_s^2}{4R_s} \tag{6}$$

occuring at $R_L/R_s = 1$. (Simple calculus confirms this maximum by solving $dp_L/dR_L = 0$.) Therefore, power transferred to the load is maximized when $R_L = R_s$. The latter observation is formalized in the statement of the *maximum power transfer theorem:*

> Given a source with fixed internal resistance R_s, power transfer is maximum when the load resistance R_L equals R_s.

Under this condition we say that the load is *matched* to the source.

You should carefully observe that maximum power transfer differs from maximum *efficiency*. Efficiency (Eff) is the ratio of load power to total power generated by the source; that is,

$$\text{Eff} = \frac{p_L}{p_L + p_s} \tag{7}$$

usually expressed as a percentage. But p_L equals p_s when the load is matched (see Fig. 3.2–6), so maximum power transfer corresponds to 50% efficiency, with equal power dissipation in source and load. Moreover, the terminal voltage drops to $v = v_s/2$ when $R_L = R_s$. Clearly, electric utility companies would not, nor should not, strive for maximum power transfer. They seek instead maximum efficiency by making p_s as small as possible.

When do we want maximum power transfer? Primarily in those applications where voltage or current signals convey information—communication systems and computers being common examples. In such applications the information-bearing signals account for only a small fraction of the total power consumed by the system, so 50% efficiency is not a serious drawback.

Example 3.2–2 A certain freshly charged automobile battery has $v_s = 12$ V and $R_s = 0.02\ \Omega$. The matched-load power would be $p_{\max} = 12^2/(4 \times 0.02) = 1800$ W when $R_L = 0.02\ \Omega$, $v = 12/2 = 6$ V, and $i = 12/0.04 = 300$ A. More realistically, if $R_L = 1.7\ \Omega$ then $v = (1.7/1.72) \times 12 \approx 11.9$ V, $i = 12/1.72 \approx 7.0$ A, $p_L = vi \approx 83$ W, and Eff $\approx 99\%$. As the battery "runs down," its internal resistance increases, thereby decreasing the terminal voltage and efficiency; for instance, $v = 10.2$ V and Eff $= 85\%$ when $R_s = 0.3\ \Omega$ with $R_L = 1.7\ \Omega$.

Exercise 3.2–3

An electric lawnmower is connected to an outlet with a long extension cord, so the voltage source applied to the mower has $v_s = 120$ V and $R_s = 5\ \Omega$. Electrically, the mower acts like a load resistance whose value depends on the amount of work being done. Draw the equivalent circuit and tabulate the values of i, v, p_L, p_s, and Eff for $R_L = 5$, 15, and 35 Ω.

AC SOURCES AND AVERAGE POWER

Unquestionably, the most abundant electric circuits are those with sinusoidal sources—familiarly known as *alternating-current* or AC circuits. Here we introduce some of the important AC concepts by examining what happens when a sinusoidal source is connected to a resistive load. We will compute the average power and the effective or rms values of AC voltage and current. All of our previous results then apply to AC circuits when expressed in terms of these quantities.

Figure 3.2–7a symbolizes an ideal sinusoidal voltage source whose terminal voltage varies with time as

$$v_s = V_m \cos \omega t \tag{8}$$

The corresponding waveform, plotted in Fig. 3.2–7b, continuously swings up and down between the extremes $+V_m$ and $-V_m$, as follows from the property of the cosine function that $-1 \le \cos \omega t \le 1$ for any value of t.

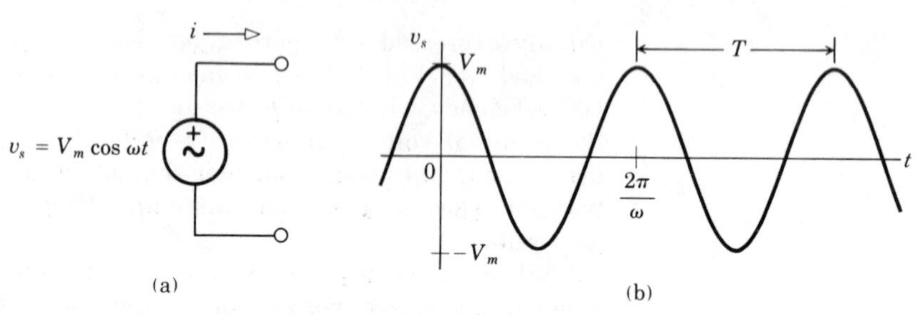

FIGURE 3.2–7 Ideal sinusoidal (AC) voltage source.

We therefore call V_m the *peak value* or *amplitude*, measured in volts. The + sign in Fig. 3.2–7a indicates the terminal at the higher potential, whenever $\cos \omega t$ is positive—which occurs exactly half the time. The rest of the time, $\cos \omega t$ is negative and the + terminal is actually at a lower potential than the other terminal. Because this voltage source tends to produce a sinusoidally alternating current i, we refer to it as an AC source.

The parameter ω in Eq. (8) is the *angular frequency* or oscillation rate in *radians per second* (rad/s), since the angle ωt increases by 2π radians when t increases by $2\pi/\omega$ seconds. The time required for one oscillation is the *period*

$$T = \frac{2\pi}{\omega} \tag{9a}$$

illustrated in Fig. 3.2–7b. Observing that the waveform goes through one complete cycle in one period, and then repeats itself, we define the *cyclical frequency*

$$f = \frac{1}{T} = \frac{\omega}{2\pi} \tag{9b}$$

whose units are cycles per second or *hertz* (Hz).

Angular and cyclical frequency differ by a factor of 2π; that is, $\omega = 2\pi f$. We use ω to save writing 2π in expressions like Eq. (8), but f is more easily measured using an oscilloscope or counter, and most laboratory instruments are calibrated in hertz.

Now suppose an AC voltage source is applied to a resistance R, as in Fig. 3.2–8a. The resulting current will be

$$i = \frac{v_s}{R} = \frac{V_m}{R} \cos \omega t = I_m \cos \omega t \tag{10}$$

which is a sinusoidal waveform, just like Fig. 3.2–7b with peak current $I_m = V_m/R$ amps. The current then flows in the reverse direction exactly half the time, whenever $\cos \omega t$ and v_s are negative.

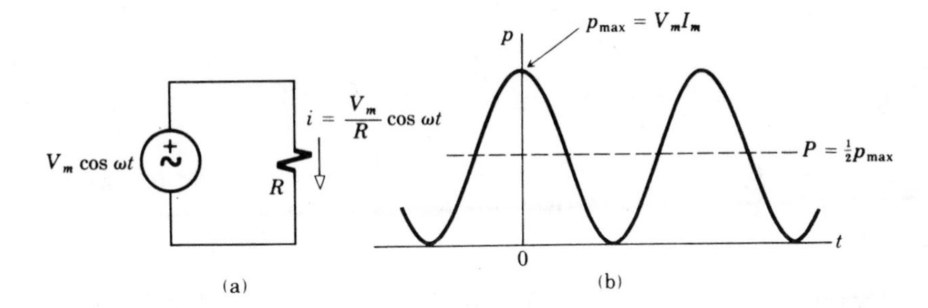

FIGURE 3.2–8
AC current through a resistance, and the instantaneous power dissipation.

Multiplying voltage times current gives the *instantaneous* power

$$p = V_m I_m \cos^2 \omega t = \frac{V_m^2}{R} \cos^2 \omega t = I_m^2 R \cos^2 \omega t \tag{11}$$

as sketched in Fig. 3.2–8b. This waveform oscillates between 0 and $p_{\max} = V_m I_m = V_m^2/R = I_m^2 R$, but never becomes negative, despite the alternations of v_s and i. From the evident symmetry, we conclude that the *average power* P equals one-half the maximum instantaneous power, that is,

$$P = \frac{1}{2} V_m I_m = \frac{1}{2} \frac{V_m^2}{R} = \frac{1}{2} I_m^2 R \tag{12}$$

We interpret P as the average rate at which electrical energy is delivered to the resistance and dissipated as heat. In particular, the total energy delivered in one period is precisely $w = PT$.

AC power calculations are tidied up by defining the *effective values* or *root-mean-square* (rms) *values* of voltage and current

$$V_{\text{rms}} \triangleq \frac{V_m}{\sqrt{2}} \qquad I_{\text{rms}} \triangleq \frac{I_m}{\sqrt{2}} \tag{13a}$$

This eliminates the factor of 1/2 in Eq. (12), since

$$P = V_{\text{rms}} I_{\text{rms}} = \frac{V_{\text{rms}}^2}{R} = I_{\text{rms}}^2 R \tag{13b}$$

which means that an AC voltage or current produces the same average power dissipation in a resistance as would be produced by a DC source with $v_s = V_{\text{rms}}$ or $i_s = I_{\text{rms}}$. In other words, V_{rms} and I_{rms} are the effective values of AC voltage or current insofar as power dissipation is concerned.

The designation "root-mean-square" comes from the fact that V_{rms} equals the square-root of the mean (average) value of v_s^2, defined in general by

$$V_{\text{rms}} = \sqrt{\frac{1}{T}\int_0^T v_s^2 \, dt} \tag{14}$$

This formula, and its equivalent for I_{rms}, gives the effective value of any periodically repeating waveform.

Example 3.2–3

Standard USA residential AC voltage has $V_{\text{rms}} = 120$ V and $f = 60$ Hz. The period is $T = 1/60 = 16.7$ ms, $\omega = 2\pi\, 60 = 377$ rad/s, and the peak value is $V_m = \sqrt{2} \times 120 \approx 170$ V. The actual voltage waveform is $v_s = 170 \cos 377t$ V.

A 60-W lightbulb designed for this voltage has $R = (120 \text{ V})^2/60 \text{ W} = 240\ \Omega$, and the rms current through the bulb will be $I_{\text{rms}} = V_{\text{rms}}/R = 0.5$ A, whereas $I_m = 0.707$ A. If the lightbulb is connected to the source by a long extension cord with 10-Ω wire resistance, the effective voltage across the bulb will be $(240/250) \times 120 = 115.2$ V and it will actually dissipate $115.2^2/240 = 55.3$ W.

Exercise 3.2–4

Find the resistance of a 75-W lightbulb intended for use with the source in Example 3.2–3 (without the extension cord), and compute the resulting peak current.

3.3 CIRCUIT THEOREMS

Equivalence, Thévenin's theorem, and the superposition principle are three important theorems that often simplify the tasks of circuit analysis or design. Although presented here in the context of resistive circuits, these theorems will later be extended to the broader category of circuits containing nonresistive elements and electronic devices.

EQUIVALENCE OF ONE-PORT NETWORKS

A *one-port network,* or "one-port" for short, may contain any number of elements but has just *one pair of terminals* for connection to the world outside. By way of illustration, the left-hand portion of Fig. 3.3-1a qualifies as a one-port, and so does the two-terminal load on the right. The complete circuit is formed, in this instance, by connecting the two one-ports together.

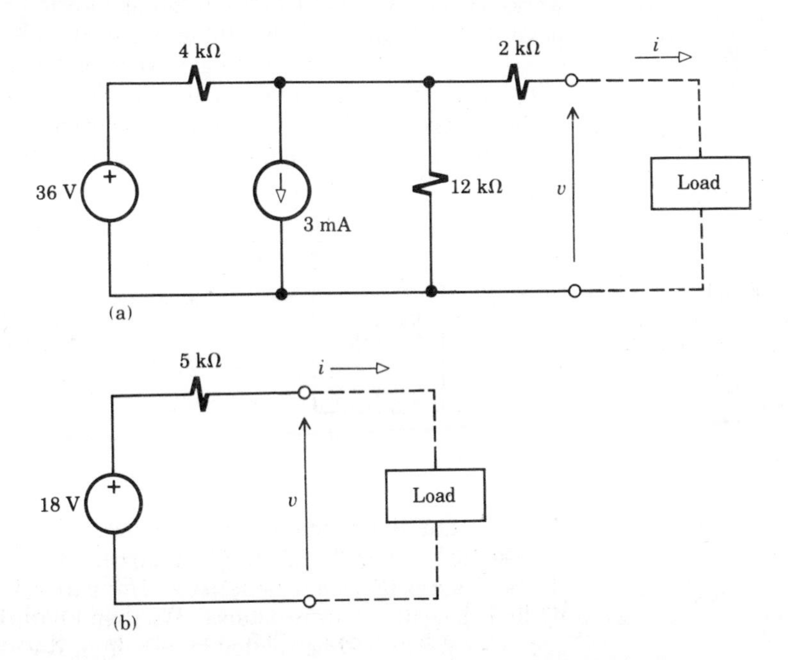

FIGURE 3.3-1 Two equivalent one-port networks.

When we deal with such circuits, equivalence is perhaps the most handy and versatile of the three concepts. It allows us to replace a complicated one-port network with a simpler one for purposes of analysis or design. We have already used this concept in conjunction with equivalent resistance and practical sources. The *equivalence theorem* goes one step further, stating that

> Any two one-port networks are equivalent if they have the same *i-v* characteristics.

In other words, equivalent one-ports always produce the same terminal voltage and current when they are connected to identical external circuitry, regardless of what that circuitry may be.

To underscore the significance of this theorem, we will eventually prove that the two networks in Fig. 3.3–1 have the same *i-v* curves. These one-ports, therefore, are equivalent; and we can use the simpler one to represent the external performance of the other. For instance, suppose we need to calculate the resulting current when a 1-kΩ resistance is connected to the terminals in Fig. 3.3–1a. Referring instead to Fig. 3.3–1b, we immediately see that $i = 18\text{ V}/(5 + 1)\text{ k}\Omega = 3\text{ mA}$.

However, equivalence holds only with respect to the *terminals* behavior, and not to what happens *inside* the networks. Indeed, equivalent networks generally exhibit quite different internal behavior. As a case in point, Fig. 3.3–1a clearly has internal current flow and power dissipation even when $i = 0$ at the terminals, whereas no current flows at all and there is no internal power dissipation in Fig. 3.3–1b when $i = 0$. Accordingly, we use equivalent networks to deal exclusively with terminal characteristics, temporarily ignoring internal conditions.

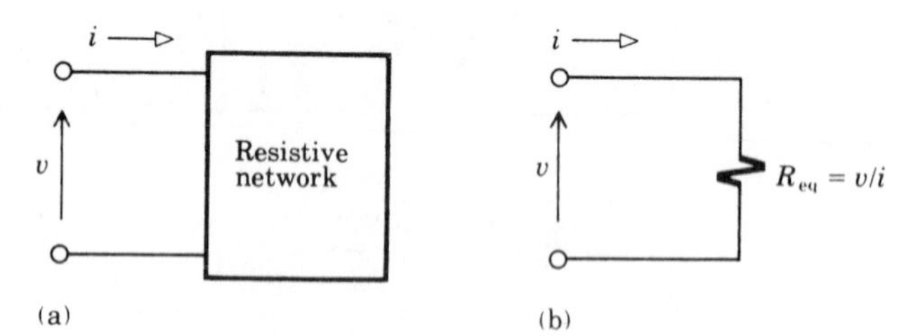

FIGURE 3.3–2
(a) An arbitrary resistive one-port.
(b) Equivalent resistance.

For a familiar example of equivalence, consider the one-port network represented by Fig. 3.3–2a that is composed entirely of resistances. If all the resistances happen to be series and/or parallel connected, we can easily find the equivalent resistance. We then know that the terminal voltage and current will be related by $v = R_{eq}i$. But that same relationship holds for *any* resistive one-port, and leads to the general definition of equivalent resistance as

$$R_{eq} \triangleq \frac{v}{i} \tag{1}$$

corresponding to the circuit model in Fig. 3.3–2b. Equation (1) gives us a formula for computing equivalent resistance that holds even when series/parallel reduction does not work, as we demonstrate in the following example.

Example 3.3–1
Equivalent resistance of a bridge network

Suppose we need to find the equivalent resistance of the *bridge network* in Fig. 3.3–3a, which cannot be simplified by series/parallel reduction. Instead, we mentally apply a source current i and calculate the resulting terminal voltage v to obtain $R_{eq} = v/i$. For this purpose, the circuit is redrawn as Fig. 3.3–3b and labeled with the node voltages v, v_a, and v_b, and the corresponding currents. Using KCL at the upper node then gives

$$i = \frac{v - v_a}{5} + \frac{v - v_b}{4}$$

and, similarly, at nodes a and b,

$$\frac{v - v_a}{5} = \frac{v_a - v_b}{10} + \frac{v_a}{2} \qquad \frac{v - v_b}{4} + \frac{v_a - v_b}{10} = \frac{v_b}{20}$$

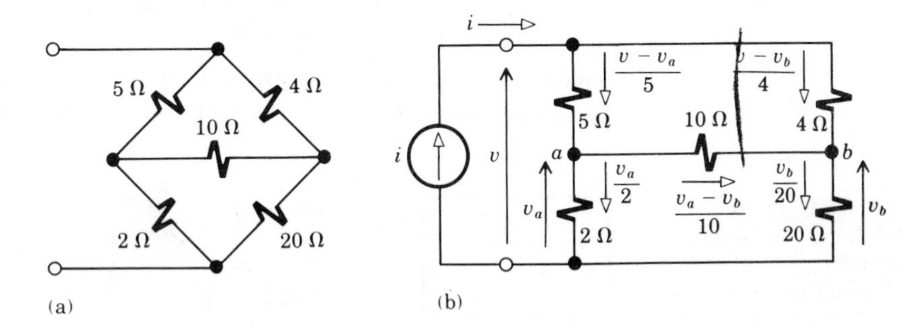

FIGURE 3.3–3 Resistive bridge network.

After combining and rearranging terms, we obtain these three simultaneous equations:

$$\begin{aligned} 0.45v - 0.20v_a - 0.25v_b &= i \\ -0.20v + 0.80v_a - 0.10v_b &= 0 \\ -0.25v - 0.10v_a + 0.40v_b &= 0 \end{aligned}$$

Eliminating v_a and v_b by successive substitution yields $0.205v = i$, so $R_{eq} = v/i = 1/0.205 = 4.88\ \Omega$. The fact that the ratio v/i turns out to be a constant confirms the equivalence theorem.

Exercise 3.3–1

Use Eq. (1) to find R_{eq} when the 10-Ω resistance in Fig. 3.3–3a is replaced by a short circuit. Check your result via series/parallel reduction.

Exercise 3.3–2

Apply the voltage-divider relation to Fig. 3.3–1b to find the value of an external resistance that, when connected to the terminals of Fig. 3.3–1a, results in $v = 12$ V.

THÉVENIN'S THEOREM

Obtaining full advantage of the equivalence theorem requires methods for finding a simple one-port network equivalent to a given, more complicated network. Thévenin's theorem has significance in this regard, for it leads to two basic equivalent structures when the one-port in question contains sources as well as resistances.

Figure 3.3–4a represents an arbitrary one-port consisting of sources and resistances. Let v_{oc} and i_{sc} be the *open-circuit voltage* and *short-circuit current*, respectively, and define the *Thévenin resistance*

$$R_0 \triangleq \frac{v_{oc}}{i_{sc}} \tag{2}$$

(These three parameters differ from v_s, i_s, and R_s used before, in that the network being represented may include more than one source.) *Thévenin's theorem* then states that

> Any one-port composed entirely of sources and linear resistances is equivalent to a voltage source v_{oc} in series with R_o or a current source i_{sc} in parallel with R_o.

The voltage-source circuit (Fig. 3.3–4b) is called the *Thévenin equivalent;* the current-source circuit (Fig. 3.3–4c) is called the *Norton equivalent.*

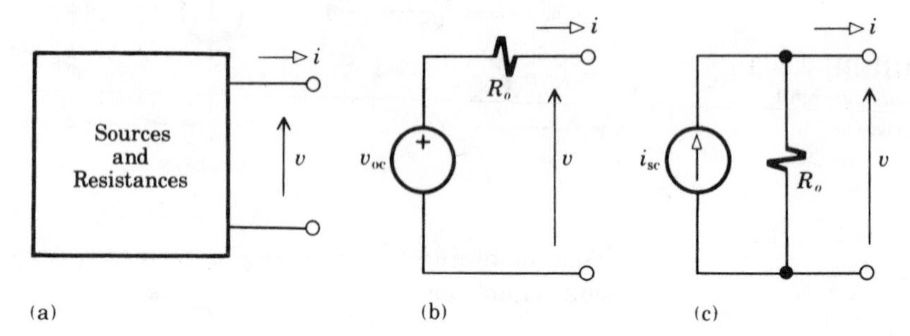

FIGURE 3.3–4 (a) A one-port network containing sources and resistances. (b) Thévenin equivalent circuit. (c) Norton equivalent circuit.

Any two of the three Thévenin parameters fully characterize the terminal behavior of the network, and we obtain an equivalent circuit by determining the open-circuit voltage and short-circuit current. Alternatively, having found either v_{oc} or i_{sc}, we can get R_0 by invoking the following corollary to Thévenin's theorem:

> The Thévenin resistance equals the equivalent resistance seen between a one-port's terminals when all independent sources are suppressed.

Sources are *suppressed* by making the replacements

voltage source → short circuit
current source → open circuit

which ensure no energy flow from the sources. A network with all independent sources suppressed is said to be "dead."

Generally, the corollary provides a simpler way of computing R_o. (It does, however, require special care when there are *dependent* or *controlled* sources present, as encountered in later chapters.) Then, given R_o and either v_{oc} or i_{sc}, we obtain the third parameter via Eq. (2); that is, $i_{sc} = v_{oc}/R_o$ or $v_{oc} = R_o i_{sc}$.

Besides representing complete one-ports, Thévenin's theorem may be applied to portions of a network to simplify intermediate calculations. Moreover, successive conversions back and forth between Thévenin and Norton circuits often saves considerable labor in circuit analysis, especially when there are two or more sources present. As a rule, the Thévenin circuit works best when dealing with a series connection, whereas the Norton circuit is used with parallel connections.

Example 3.3–2
Calculating Thévenin parameters

Let us find the Thévenin parameters for the one-port in Fig. 3.3–5a. Since $i = 0$ under open-circuit conditions (Fig. 3.3–5b), the voltage-divider relation gives $v_{oc} = (10/25) \times 75 = 30$ V. Then, with the termi-

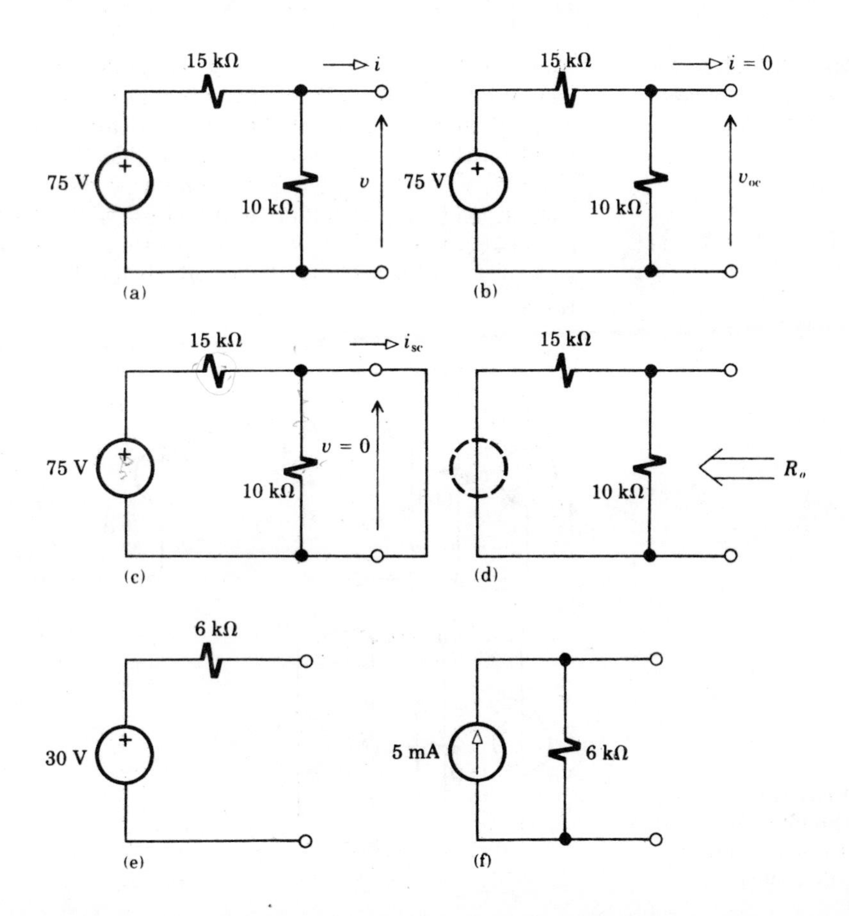

FIGURE 3.3–5
Circuits for Example 3.3–2.

nals short-circuited (Fig. 3.3–5c), no current flows through the 10-kΩ resistance so $i_{sc} = 75/15 = 5$ mA. Hence, $R_o = 30/5 = 6$ kΩ. To verify this result, refer to Fig. 3.3–5d, which shows the "dead" network with the voltage source replaced by a short circuit; the equivalent resistance seen looking in the terminals is $15 \| 10 = 6$ kΩ as expected. Figures 3.3–5e and 3.3–5f are the corresponding Thévenin and Norton circuits.

Exercise 3.3–3

Find the equivalent circuits for Fig. 3.3–6.

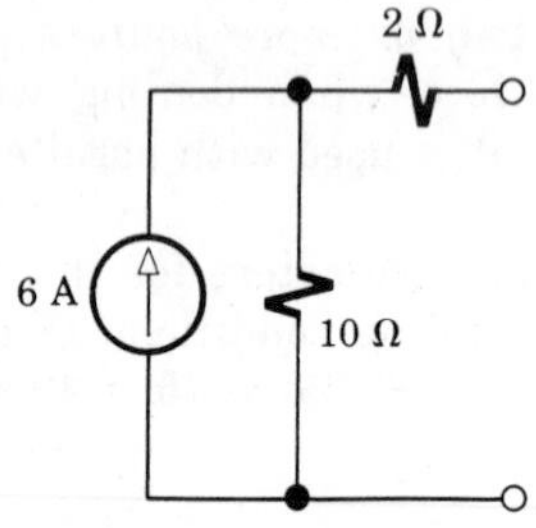

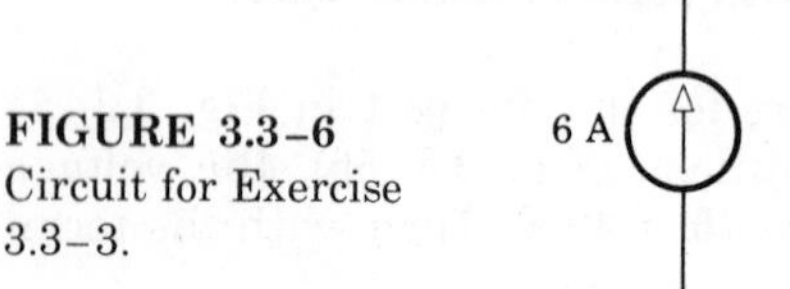

FIGURE 3.3–6
Circuit for Exercise 3.3–3.

Example 3.3–3
Thévenin/Norton conversions

Now consider again the battery-charging circuit from Fig. 2.4–13, repeated here as Fig. 3.3–7a. If you want to compute the charging current i without solving the entire circuit, you can simply "Thévenize" the left-hand portion of the network; this yields Fig. 3.3–7b, from which $i = (56 - 8)$ V$/(14 + 2)$ Ω $= 3$ A. Similarly, "Nortonizing" the right-hand portion leads to Fig. 3.3–7c and facilitates the node-voltage calculation: $v = (14 \| 2)$ Ω $\times (4 + 4)$ A $= 14$ V. Little tricks such as these come easily with practice, especially if you watch for Thévenin and Norton structures.

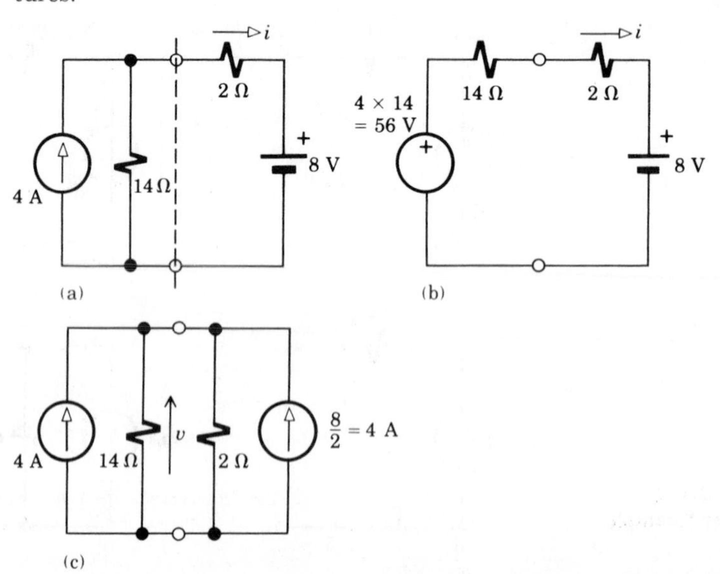

FIGURE 3.3–7
Circuit analysis using Thévenin/Norton conversion.

Exercise 3.3–4 Find the current i in Fig. 3.3–8.

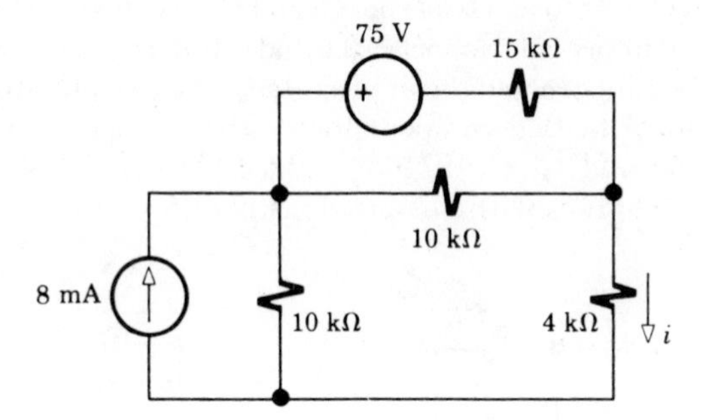

FIGURE 3.3–8 Circuit for Exercise 3.3–4.

SUPERPOSITION AND LINEARITY

Another helpful technique for dealing with networks containing two or more sources stems from the principle of superposition or *superposition theorem,* as follows:

> When a current or voltage in a linear network is the result of several sources acting together, its value is the algebraic sum of the individual contributions from each source acting alone.

Thus, a complicated problem reduces to several easier problems—one for each source—whose answers are added together to get the final result. The contribution from any one source is determined by analyzing the network with all other sources suppressed.

Superposition holds for any type of physical system where cause and effect are linearly related. Formally speaking, let x be the cause and y the corresponding effect related to x by

$$y = f(x) \tag{3a}$$

where $f(x)$ is a function of x. The function is *linear* if replacing x by a sum $x' + x''$ produces

$$y = f(x') + f(x'') \tag{3b}$$

which requires that $f(x' + x'') = f(x') + f(x'')$.

Ohm's law, $v = Ri$, is a linear relationship, for if $i = i' + i''$ then $v = R(i' + i'') = Ri' + Ri''$, and hence the voltage is a linear function of current. We have repeatedly used the phrase *linear resistance* to underscore this fact. As a counter example, the power dissipated by a resistance is *not* a linear function of current, since $p = Ri^2$ and $R(i' + i'')^2 \neq Ri'^2 + Ri''^2$. Superposition therefore applies to voltage and current calculations in a network consisting of linear resistances, but it does not apply to power calculations directly.

The battery-charging circuit of Fig. 3.3–7a provides a good illustration of how to use superposition. It is redrawn in Fig. 3.3–9a with the battery suppressed (short-circuited) and in Fig. 3.3–9b with the current source suppressed (open-circuited). We compute the charging current i by finding its two components i' and i''. Specifically, $i' = (14/16) \times 4 = 3.5$ A and $i'' = -8/16 = -0.5$ A, so $i = i' + i'' = 3.5 - 0.5 = 3.0$ A, which agrees with our previous result.

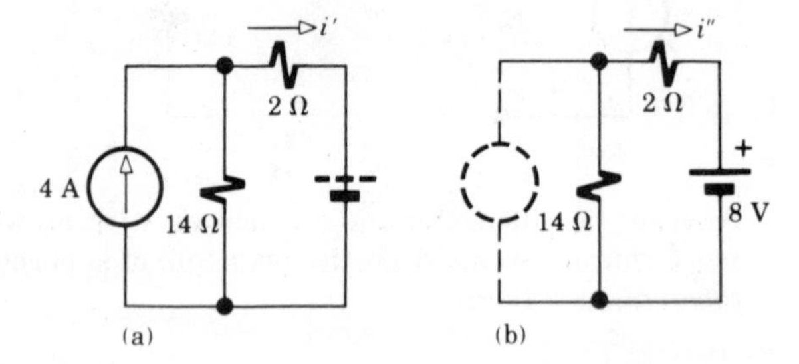

FIGURE 3.3–9
Circuit analysis using superposition.

Exercise 3.3–5

Apply superposition to the problem in Exercise 3.3–4. You may wish to "Thévenize" or "Nortonize" a portion of the circuit as a preliminary step.

Example 3.3–4
A voltage adder

Figure 3.3–10a shows a circuit configuration called a *voltage adder:* v_1 and v_2 are *input* signals while v is the resulting *output.* The drawing is in a kind of short-hand, with all voltages measured with respect to the common bottom terminal identified by a *reference* or *ground* symbol (⏚). Thus, the actual circuit configuration looks like Fig. 3.3–10b.

We apply the superposition principle by mentally suppressing v_2, which puts R_2 directly in parallel with R. (Redraw the circuit with v_2 replaced by a short circuit if you do not visualize this immediately.) The output voltage contribution from v_1 is then

$$v' = \frac{R \| R_2}{R_1 + (R \| R_2)} v_1$$

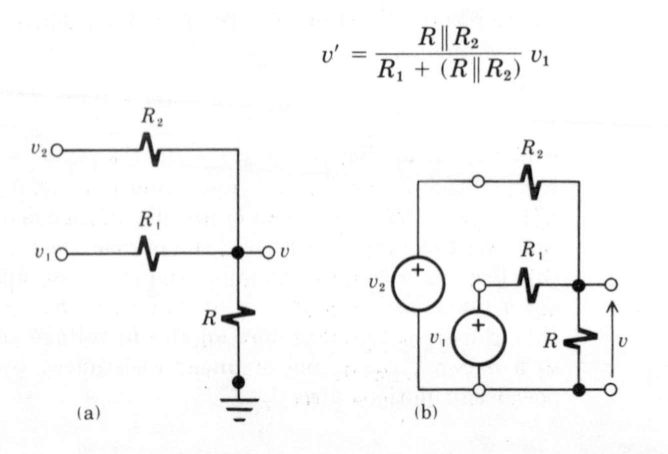

FIGURE 3.3–10
Voltage adder.

By symmetry, the contribution v'' due to v_2 has the same form, but with interchanged subscripts. Therefore, in the common case where $R_2 = R_1$,

$$v = v' + v'' = \frac{R}{2R + R_1}(v_1 + v_2)$$

which justifies the name "voltage adder"; that is, the output voltage v is proportional to $v_1 + v_2$.

Example 3.3–5

We finally return to the network at the beginning of this section (Fig. 3.3–1a) to determine its equivalent circuits by using all the tools developed here.

First, invoking the corollary to Thévenin's theorem, we find R_o by suppressing all sources as in Fig. 3.3–11a. This diagram shows that $R_o = 2 + (4 \| 12) = 5$ kΩ, the resistance value previously given in Fig. 3.3–1b.

Second, we need either v_{oc} or i_{sc}. The former will be easier to find in this case because no current flows through the 2-kΩ resistance under open-circuit conditions, and we can ignore it completely.

Applying superposition to the task, Fig. 3.3–11b immediately gives us $v'_{oc} = (12/16) \times 36 = 27$ V for the voltage source's contribution. As for the contribution from the current source, we start with Fig. 3.3–11c and then Thévenize the portion to the left of the dashed line, being careful with the polarity. Figure 3.3–11d is the result, and gives $v''_{oc} = (12/16) \times (-12) = -9$ V. Hence, $v_{oc} = 27 - 9 = 18$ V, and Fig. 3.3–1b is, indeed,

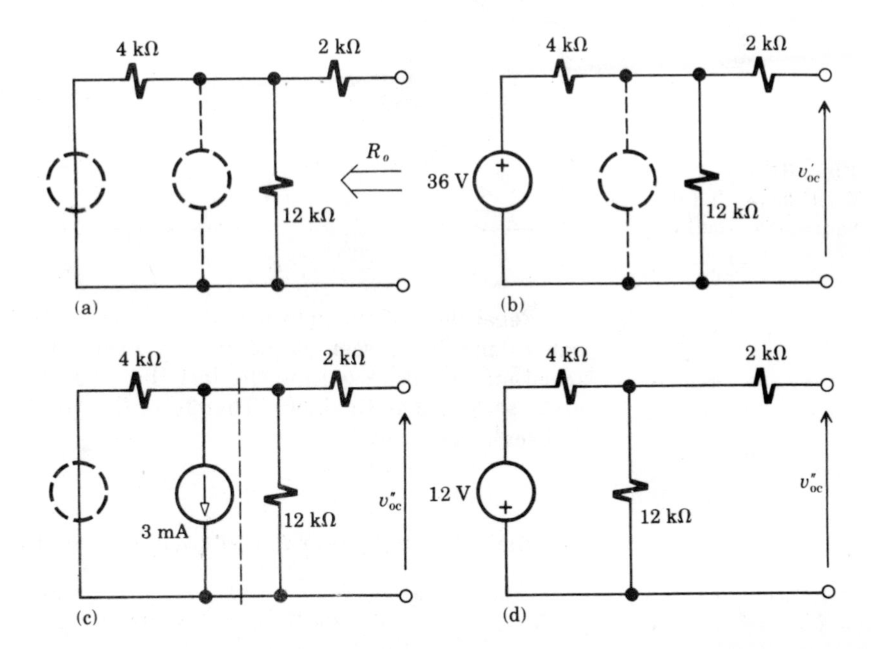

FIGURE 3.3–11
Circuits for Example 3.3–5.

the Thévenin equivalent circuit. The Norton circuit would have $i_{sc} = 18\text{ V}/5\text{ k}\Omega = 3.6\text{ mA}$.

Had there been a resistance directly across the terminals—as in Fig. 3.3–5c, for instance—i_{sc} would be easier to find than v_{oc} because the short circuit bypasses current around that resistance and it can be ignored in the calculation.

3.4 DC METERS AND MEASUREMENTS†

We conclude this chapter with an introduction to DC measuring instruments that can be described in terms of resistive circuits. The capitalized symbols V and I will be used throughout as a reminder that we are dealing with constant or slowly varying voltage and current.

VOLTMETERS AND AMMETERS

Routine measurements of DC current and voltage are usually made with *direct-reading meters* incorporating a *d'Arsonval moving-coil mechanism.* The operating principles of this device will be covered in Chap. 17; here it suffices to say that the meter has an indicating pointer whose angular deflection depends on the amount and direction of the average current I through the meter, as represented in Fig. 3.4–1a. The maximum or *full-scale* deflection corresponds to a constant current I_{fs} flowing in the proper direction, and pointer deflection is proportional to DC current values between zero and I_{fs} with an accuracy of 2–5% of the full-scale value.

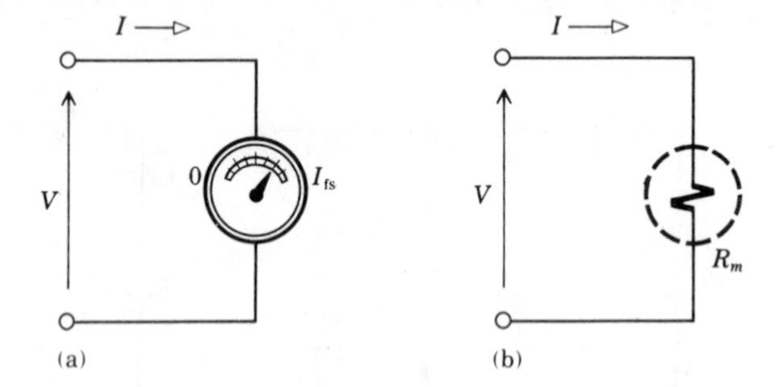

FIGURE 3.4–1 A DC meter and its equivalent circuit.

Regardless of the specific details of the internal configuration, the equivalent circuit seen at the meter's terminals is nothing more than a resistance R_m representing the fact that the meter consumes some energy, as Fig. 3.4–1b shows. Therefore, the voltage required to produce full-scale deflection is

$$V_{fs} = R_m I_{fs} \tag{1}$$

and the maximum power consumption is given by

$$P_{fs} = V_{fs} I_{fs} = I_{fs}^2 R_m = \frac{V_{fs}^2}{R_m}$$

The characteristics of a given meter are fully specified by any two of the four parameters I_{fs}, R_m, V_{fs}, and P_{fs}. Typical values for a low-current meter are $I_{fs} = 50\ \mu A$ and $R_m = 3\ k\Omega$, so $V_{fs} = 150$ mV and $P_{fs} = 7.5\ \mu W$, while a high-current meter might have $I_{fs} = 10$ mA and $R_m = 5\ \Omega$. Low-current meters are said to have greater *sensitivity* because full-scale deflection requires less current.

With the aid of additional resistors and switches, a simple meter can measure voltage or current in various ranges. Figure 3.4–2, for example, shows the arrangement for voltage measurement with two ranges. The unknown voltage to be measured is V_u and the range-setting resistors R_1 and R_2 are called *multipliers*.

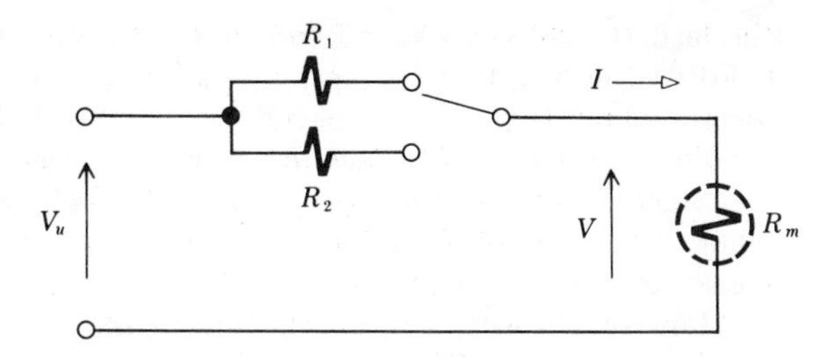

FIGURE 3.4–2
Voltmeter with two ranges.

To analyze this *voltmeter* circuit, assume the switch to be in the upper position, putting R_1 in series with R_m. From the voltage-divider ratio, the voltage across the meter element is $V = R_m V_u/(R_m + R_1)$, so that full-scale deflection corresponds to

$$V_{u_{fs}} = \frac{R_m + R_1}{R_m} V_{fs} = \left(1 + \frac{R_1}{R_m}\right) V_{fs} \tag{2}$$

The same equation applies for the other switch position, with R_2 in place of R_1.

Equation (2) shows that $V_{u_{fs}} \geq V_{fs}$; the multiplier accounts for the difference between the unknown voltage and the meter voltage, as brought out by the alternate expression $V_{u_{fs}} = V_{fs} + R_1 I_{fs}$. To illustrate, take a high-current meter with $I_{fs} = 10$ mA, $R_m = 5\ \Omega$, $R_1 = 95\ \Omega$ and $R_2 = 495\ \Omega$. The voltage ranges then will be $(1 + 95/5) \times V_{fs} = 20 \times 50$ mV $= 1$ V and $(1 + 495/5) \times 50$ mV $= 5$ V.

Putting resistors in parallel with the meter element produces the current-measuring *ammeter* circuit of Fig. 3.4–3. The resistors R_1 and R_2 are known as *shunts,* and full-scale deflection corresponds to $I_{u_{fs}} \geq I_{fs}$ with the shunt carrying the excess current.

The switch in a multirange ammeter must be a special type with a *make-before-break* design. As suggested by the extended contact in Fig. 3.4–3, this switch "makes" the connection to R_2 before it "breaks" contact with R_1, so the meter element always has shunt resistance. Were this not

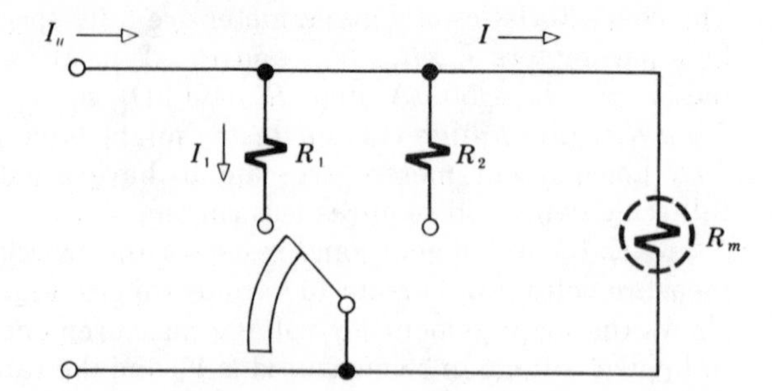

FIGURE 3.4–3
Ammeter with two ranges.

the case, the entire current I_u would flow through the meter with the possibility of destroying the mechanism by excessive ohmic heating. Voltmeters do not require this special switch (Why?), but all meter elements should be protected by a fuse to prevent damage from excessive current.

Since a simple meter can be used for voltage or current measurement, it is possible to construct an instrument called a *multimeter* that measures multiple ranges of voltage and current with one meter element. A rotary switch with several sets of mechanically coupled contacts allows for the proper connection of multipliers and shunts. To illustrate, Fig. 3.4–4 shows the circuit for a simple multimeter that has one voltage range and one current range. The two switches are coupled to move together and the terminal marked "NC" has no connection to it. Redraw the circuit, if necessary, to convince yourself that R_v is a multiplier when the switches are in the upper position while R_a is a shunt when they are in the lower position.

Multimeters are portable and very handy but—like the separate ammeter and voltmeter circuits—they do consume power from the quantity being measured and thereby may significantly alter the value. One must be particularly alert to the possibility of loading effect in voltage measurements. The input resistance of a multirange voltmeter is usually

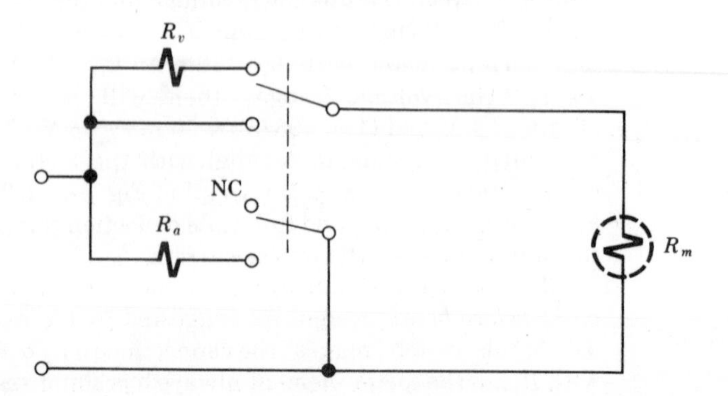

FIGURE 3.4–4
A simple multimeter.

stated in *ohms per volt* of full-scale deflection. For instance, when set on the 5-V scale, a typical 20 kΩ/V meter has an input resistance of 20 kΩ/V × 5 V = 100 kΩ and would draw 50 μA when reading 5 V. The higher the input resistance, the less loading there will be. Other methods for minimizing or eliminating the loading effect will be covered later.

Exercise 3.4–1 Show that the ammeter circuit in Fig. 3.4–3 has

$$I_{u_{fs}} = \left(1 + \frac{R_m}{R_1}\right) I_{fs}$$

Exercise 3.4–2 If the meter in Fig. 3.4–4 has $I_{fs} = 50\ \mu\text{A}$ and $R_m = 3\ \text{k}\Omega$, what values of R_v and R_a yield $V_{u_{fs}} = 30$ V and $I_{u_{fs}} = 2$ A?

OHMMETERS

Figure 3.4–5a diagrams a simplified *ohmmeter* circuit for DC resistance measurement with a d'Arsonval (or equivalent) meter movement. The resistance R is chosen to yield full-scale deflection with the instrument's terminals short-circuited so $R_u = 0$. Thus

$$\frac{V_s}{R + R_m} = I_{fs}$$

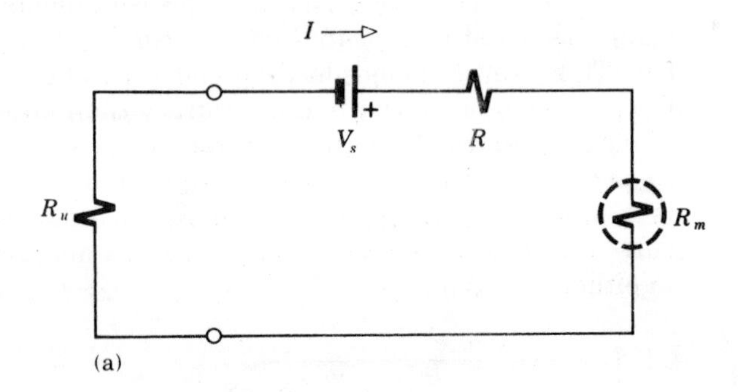

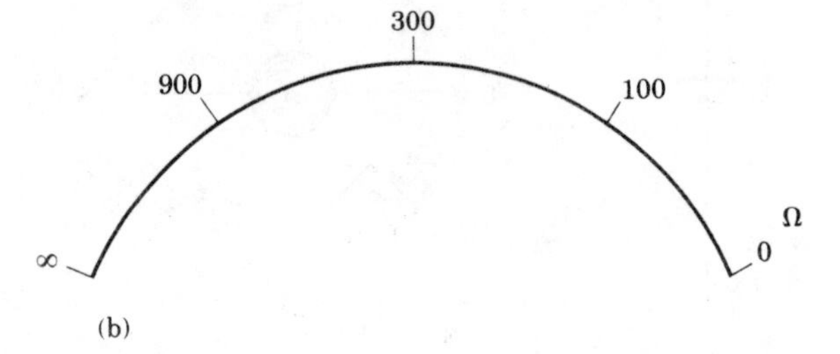

FIGURE 3.4–5 An ohmmeter circuit and its nonlinear scale.

Then, with the unknown resistance R_u in the circuit

$$I = \frac{V_s}{R_u + R + R_m} = \frac{I_{fs}}{1 + R_u/(R + R_m)} \tag{3}$$

which shows that I is inversely proportional to R_u. Specifically, half-scale deflection ($I = 0.5I_{fs}$) corresponds to $R_u = R$, whereas I equals zero when $R_u = \infty$ (open-circuited terminals). Accordingly, the ohms scale is *backwards* and *nonlinear*, as illustrated in Fig. 3.4–5b taking $R + R_m = 300\ \Omega$.

Most multimeters include an ohmmeter circuit similar to Fig. 3.4–5, but in a slightly more complicated form to provide multiple ranges with different half-scale values. A variable resistance, usually labeled "Ohms adjust," is included to compensate for the changing value of V_s as the multimeter's battery ages. A multimeter with scales for volts, ohms, and milliamps is known by the acronym VOM.

Pay careful attention to the fact that the simple ohmmeter can only be used to measure isolated resistors or the equivalent resistance of a connection of resistors. It should not be used to measure the resistance of an electronic component that might be damaged by the sensing current I, nor can it measure the value of a resistance embedded in a network.

BRIDGES AND NULL MEASUREMENTS

Null measurements, made with bridge circuits and related configurations, differ from direct electrical measurements in that the quantity being measured is *compared with* or *balanced against* a *reference* quantity. This strategy avoids loading and other effects of interaction and has the potential for much greater accuracy than direct measurement.

To introduce the null-measurement principle, consider the *Wheatstone bridge* cricuit, Fig. 3.4–6. Here R_a and R_b are standard resistors, R is a precision potentiometer, and R_u an unknown resistance. A low-current *zero-center* meter "bridges" nodes a and b indicating current flow in either direction.

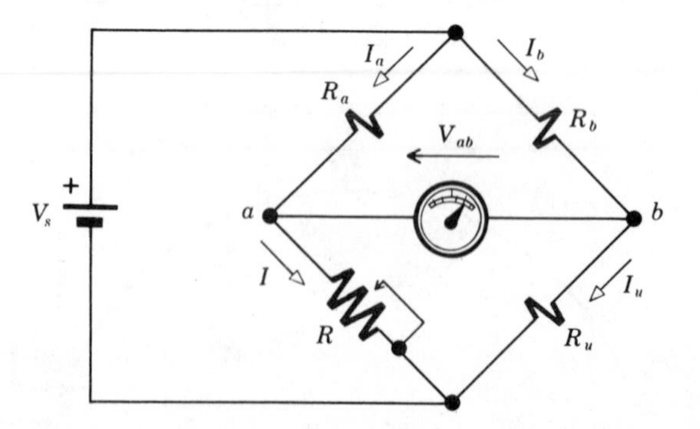

FIGURE 3.4–6
Wheatstone bridge.

To measure R_u, one adjusts R until no current flows either way through the meter. The bridge is then said to be *balanced* and the meter has a *null* reading. The balanced condition implies that $V_{ab} = 0$ (otherwise current would flow through the meter), and that $I_u = I_b$ and $I = I_a$ (from KCL at nodes a and b). Thus, writing KVL equations for the upper and lower triangular loops, we have

$$R_a I_a - R_b I_b = 0 \qquad R_u I_b - R I_a = 0$$

so

$$R_a I_a / R_b I_b = 1 \qquad R_u I_b / R I_a = 1$$

Equating these two expressions and canceling the currents yield

$$R_u = R(R_b/R_a) \tag{4}$$

which gives the unknown resistance in terms of known values. Usually the ratio R_b/R_a equals a power of ten (for example, 10^3 or 10^{-2}), so that the value of R_u can be read from the calibrated dial of the potentiometer. Note that the source voltage V_s does not appear in Eq. (4).

Bridge measurements are inherently more accurate than ohmmeters, which depend on the accuracy of the meter movement. Further, it's a simple matter to modify the bridge circuit to measure small *changes* of resistance for use with transducers such as resistance thermometers and strain gauges.

First-cousin to the Wheatstone bridge is the potentiometric voltage measurement circuit shown in Fig. 3.4–7. This circuit measures an unknown open-circuit source voltage V_u by balancing V_w against it until the meter indicates that $I = 0$. Then, $V_u = V_w = (R_{aw}/R_{ab}) \times V_{\text{ref}}$, where V_{ref} is a known reference voltage. (More sophisticated versions of this circuit use an ordinary voltage source which is calibrated by balancing it against the reference source, thereby reducing undesirable current drain from the reference.) The complete absence of loading under balanced conditions makes this method ideal for measuring sensitive electrochemical processes and the like.

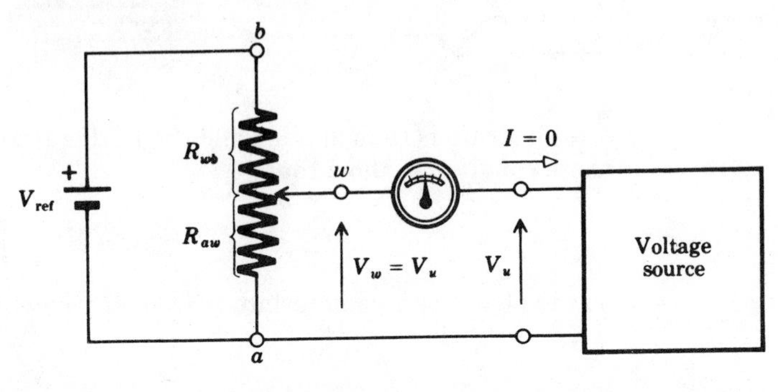

FIGURE 3.4–7
Potentiometric voltage measurement.

Figure 3.4–8 diagrams an *automated* or *self-balancing potentiometer* system wherein the difference voltage $V_u - V_w$ is applied to a voltage amplifier whose output, in turn, drives a small DC motor. If $V_u - V_w \neq 0$, the motor moves the wiper until balance has been achieved. The self-balancing system will continuously follow a *time-varying* voltage, if not too rapid, and is often used in strip-chart and X-Y recorders. This system has the added advantage of drawing negligible current from the voltage source, even when it's in an unbalanced condition.

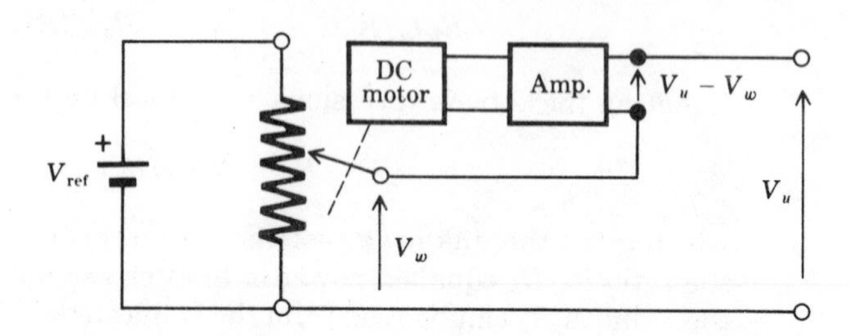

FIGURE 3.4–8 Self-balancing potentiometer.

Exercise 3.4–3

Recall from Section 2.3 that a *strain gauge* has resistance $R_u + \Delta R_u$ where R_u is the unstrained resistance and the change in resistance due to a small strain $\Delta\ell/\ell$ is $\Delta R_u = 2R_u(\Delta\ell/\ell)$. Accordingly, one can measure strain with a Wheatstone bridge that has two potentiometers as in Fig. 3.4–9. The bridge is first balanced using R with $\Delta R = 0$ and no applied

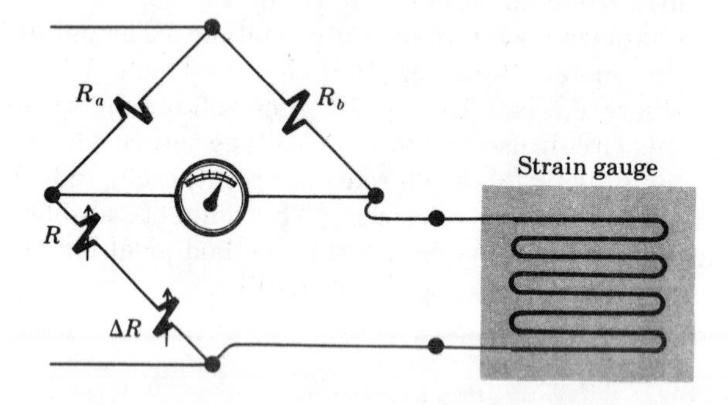

FIGURE 3.4–9 Strain measurement with a Wheatstone bridge.

strain; strain is then applied and the bridge rebalanced using ΔR. Under these conditions, show that

$$\Delta\ell/\ell = \Delta R/2R$$

so the strain can be determined from the potentiometer settings R and ΔR.

PROBLEMS

3.1–1 Generalize Fig. 3.1–3 and Eq. (3a) for the case of n resistors in series.

3.1–2 Show that n equal resistors in parallel yields $R_{eq} = R/n$.

3.1–3 Obtain an expression similar to Eq. (5) for three parallel resistances.

3.1–4 What are the conditions on R_L and R_w such that Fig. 2.3–7b reduces approximately to Fig. 2.3–7c?

3.1–5 Find R_{eq} for Fig. 3.3–3a when the 4-Ω resistance is replaced by
(a) an open circuit (b) a short circuit.

3.1–6 Given four 3-Ω resistors, how would you connect them to obtain the following?
(a) $R_{eq} = 4\ \Omega$ (b) $R_{eq} = 5\ \Omega$

3.1–7 Given five 4-Ω resistors, how would you connect them to obtain the following?
(a) $R_{eq} = 14\ \Omega$ (b) $R_{eq} = 7\ \Omega$

3.1–8 Find all the voltages and currents in Fig. 2.4–10a when the 7-A source is replaced by an open circuit.

3.1–9 Repeat the previous problem with the 48-V source replaced by a short circuit.

3.1–10 Suppose another identical potentiometer is connected across the output terminals in Fig. 3.1–9d and both wipers are set so that $R_{aw} = R_{wb}$. Find the voltage at each wiper terminal in terms of V_s.

‡ **3.1–11** Find the condition on R_1 and R_2 in Fig. 3.1–6 so that $p_{max} = 4\ p_{min}$, and explain why this would not be desireable.

3.2–1 Let $v_s = 10$ V and $i_s = 5$ A in Fig. 3.2–4. Find the condition on R_L so that
(a) $v \geq 0.9\ v_s$ (b) $i \geq 0.9\ i_s$.

3.2–2 Let $v_s = 100$ V and $i_s = 5$ mA in Fig. 3.2–4. Find the condition on R_L so that
(a) $v \geq 0.95\ v_s$ (b) $i \geq 0.95\ i_s$.

3.2–3 A certain source has $v_s = 12$ V and $i_s = 6$ A.
(a) Use the voltage-source model to calculate v, i, p_L, p_s, and Eff when $R_L = 4\ \Omega$.
(b) Repeat, using the current source model.

3.2–4 Repeat the previous calculations with $R_L = 1\ \Omega$.

3.2–5 Use Eqs. (5) and (7) to plot Eff versus R_L/R_s.

‡ **3.2–6** Derive Eq. (6) by solving $dp_L/dR_L = 0$ for R_L.

3.2–7 Let Fig. 3.2–5 represent a source connected to a load by a power transmission line with resistance R. Assuming $R_s \ll R$, show that the efficiency can be expressed as Eff $= 1 - Rp/v_s^2$, where $p = v_s i$ is the total generated power. If p is fixed, what does this result imply for efficient power transmission?

3.2–8 A 60-Ω load is connected to an AC source having $f = 50$ Hz and $V_{rms} = 240$ V. Sketch the resulting current waveform, and find I_{rms} and P.

3.2–9 Repeat the previous problem with $f = 25$ Hz and $V_{rms} = 90$ V.

3.2–10 Sketch the square of the voltage waveforms in Fig. 7.4–1b, d, and e, and find the rms value without resorting to integration.

‡ **3.2–11** Use Eq. (14) to find the rms value of the voltage waveform in Fig. 7.4–1c.

3.3–1 Use Thévenin/Norton conversions to find the Norton equivalent circuit for Fig. P3.3–1.

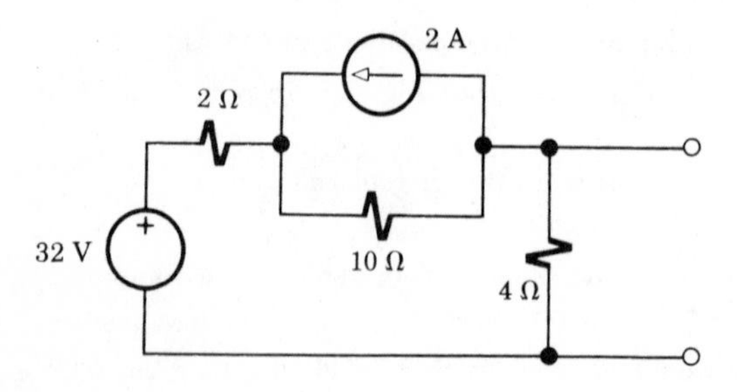

FIGURE P3.3–1

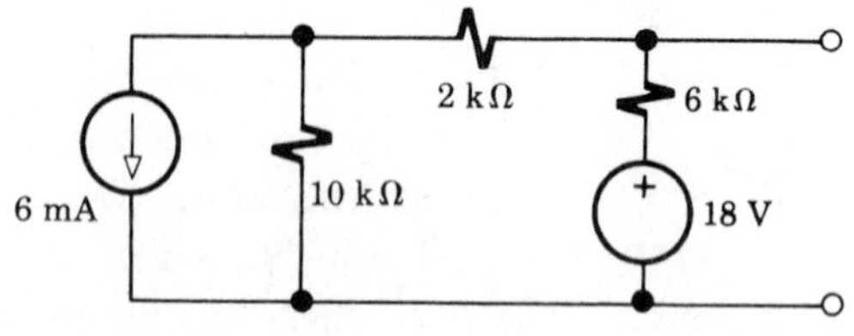

FIGURE P3.3–2

3.3–2 Use Thévenin/Norton conversions to find the Thévenin equivalent circuit for Fig. P3.3–2.

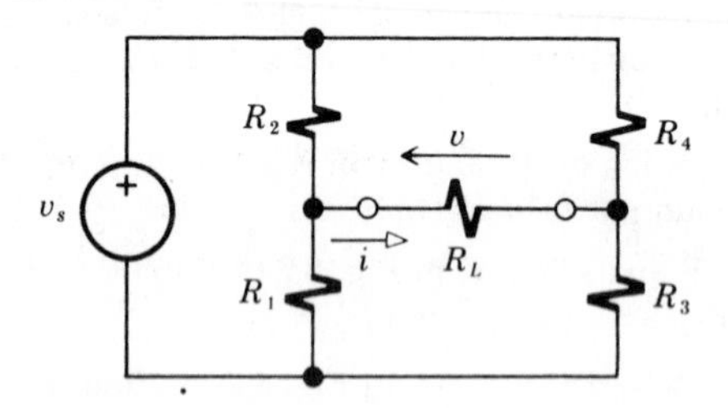

FIGURE P3.3–4

3.3–3 Perform a Norton-to-Thévenin conversion to find i_1 in Fig. 2.4–10a.

3.3–4 Let the bridge circuit in Fig. P3.3–4 have $R_L = \infty$. Show that the open-circuit voltage is

$$v_{oc} = \left(\frac{R_1}{R_1 + R_2} - \frac{R_3}{R_3 + R_4}\right) v_s$$

Then redraw the circuit with $v_s = 0$ to obtain the Thévenin resistance $R_o = (R_1 \| R_2) + (R_3 \| R_4)$.

3.3–5 Use the formulas in the previous problem to calculate $i = i_{sc}$ when Fig. P3.3–4 has $R_L = 0$, $R_1 = R_4 = 3\ \Omega$, $R_2 = R_3 = 6\ \Omega$, and $v_s = 12$ V. Check your results by direct calculations of the currents through R_1 and R_2.

3.3–6 Calculate the maximum power that can be transferred from the source circuit in Fig. 3.3–6, and find the corresponding value of the load resistance.

3.3–7 Let the 9-kΩ resistor in Fig. 2.4–12 be replaced by an adjustable resistance. What resistance value will dissipate the maximum power and what is the value of the maximum power?

3.3–8 Find v_1 in Fig. 2.4–12 using superposition.

3.3–9 Apply superposition to find the current through the 2-kΩ resistance in Fig. P3.3–2 when the terminals are open-circuited.

3.3–10 Use superposition to find v_{oc} and i_{sc} for Fig. P3.3–1. Then compare $R_o = v_{oc}/i_{sc}$ with the equivalent resistance of the dead network.

‡ **3.3–11** Let the ladder network in Fig. 16.1–8 have $n = 3$ sections and input voltages v_0, v_1, and v_2. Show that $v_{oc} = \frac{1}{2}v_2 + \frac{1}{4}v_1 + \frac{1}{8}v_0$ and that $R_o = 3R$. Hint: Start by finding the Thévenin equivalent for everything below the first node, and then move up the ladder.

3.4–1 The ammeter in Fig. 3.4–3 has $R_1 \| R_m = 1\ \Omega$ and is connected in series with a battery V_s and resistor R. What current is measured under the following conditions?

(a) $V_s = 0.6$ V and $R = 20\ \Omega$ (b) $V_s = 6$ V and $R = 200\ \Omega$

3.4–2 Let Fig. 3.1–9d have $V_s = 6$ V and $R_{aw} = R_{wb} = 300\ \Omega$. What value for v_w will be measured by a voltmeter with $R_1 + R_m = 500\ \Omega$?

3.4–3 Design a multimeter circuit similar to that of Fig. 3.4–4 using a meter with $I_{fs} =$ 1 mA and $R_m = 20\ \Omega$ to obtain full-scale ranges of 1 mA, 1 A, 0.2 V, and 20 V.

3.4–4 Repeat the previous design using a meter with $I_{fs} = 100\ \mu$A and $R_m = 2$ kΩ.

3.4–5 The movement in a certain multimeter has $I_{fs} = 100\ \mu$A and $R_m = 200\ \Omega$. Find the input resistance for the 5-V and 10-mA scales.

3.4–6 Suppose the element in Fig. 2.1–6 is a resistance R. The ammeter reads 0.9 mA. The voltmeter has a resistance of 5 kΩ/V and reads 1.8 V on the 2-V scale. What is the value of R?

3.4–7 Figure P3.4–7 is an ohmmeter circuit that has $R'_m = R_z + R_m \gg R$. Resistance R_z is adjusted so that $I = I_{fs}$ when $R_u = 0$. Show that $I \approx I_{fs}/(1 + R_u/R)$. Hence, different half-scale values may be obtained by changing R.

3.4–8 Suppose the bridge in Fig. 3.4–6 is balanced with $R_a = R_b = R = R_u$ and the meter has resistance R_m.

(a) If R_u decreases by a small amount ΔR_u, use the results in Problem 3.3–4 to show that the current through the meter is $I \approx (\Delta R_u/4R)V_s/(R + R_m)$.

(b) Calculate the smallest detectable value of ΔR_u if $R = 1$ kΩ, $V_s = 10$ V, $R_m =$ 2 kΩ, and the minimum measurable current is 2.5 μA.

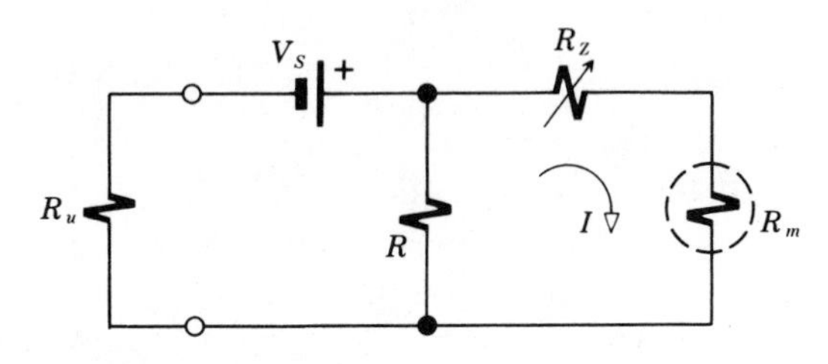

FIGURE P3.4–7

4

TWO-PORTS AND OP-AMPS

A typical electronic system, say a calculator or stereo amplifier, contains hundreds of elements and numerous interconnected circuits. Trying to understand the operation of such a system would be very difficult if we had to deal with each and every component individually, and designing a system part-by-part would be virtually impossible. Obviously, that's not the way it's done. Instead, the designer first works with functional building blocks to obtain a system block diagram with the desired characteristics. Then the details inside the blocks are worked out.

This chapter deals with the concepts, modeling, and analysis of linear building-block networks. These networks may be either *active* or *passive,* depending on whether they contain an active electronic device and its associated energy source. In either case, the distinctive modeling element is the *controlled source,* which we develop first in the context of passive resistive networks. We then use controlled sources to represent the characteristics of ideal transformers and amplifiers, two important new members in our growing family of circuit components. Following a general discussion of amplifier-circuit models, the closing section introduces the handy and versatile operational amplifier or "op-amp." (Later chapters will consider electronic devices and the actual implementation of amplifier circuits.)

OBJECTIVES

After studying this chapter and working the exercises, you should be able to do each of the following:

- Obtain input and output models for a passive two-port network connecting a source and load (Section 4.1).
- Use controlled sources to represent the characteristics of an ideal transformer or amplifier (Sections 4.1 and 4.2).
- Calculate the voltage or current amplification of an amplifier or cascaded amplifiers, including loading effects (Sections 4.2 and 4.3).
- Find the power gain of an amplifier and express it in decibels (Section 4.2).
- Use the concept of the virtual short to analyze and design simple op-amp circuits (Section 4.3).
- Draw the diagrams for a noninverting, inverting, or differential amplifier using an op-amp (Section 4.3).

4.1 TWO-PORTS AND IDEAL TRANSFORMERS

A two-port network—or *two-port*—has four terminals arranged such that one pair forms an *input port* and the other pair forms an *output port.* Depending on its internal elements, the two-port may serve as a matching network, amplifier, or filter, or provide some other function that produces desired effects at the input and output. In this section, we describe techniques for modeling resistive two-ports with the help of our previous theorems, plus the new concept of a controlled source. We also introduce the ideal transformer as a special type of two-port that has important practical uses.

RESISTIVE TWO-PORTS AND CONTROLLED SOURCES

Consider in Fig. 4.1–1a a resistive network connected between a source and a load, a common situation for a two-port. The two-port circuitry, not shown, is assumed to consist exclusively of linear resistances, with no internal sources. Given the two-port's circuit diagram, we might develop expressions relating the four input and output variables, v_1, i_1, v_2, and i_2. Or we might develop separate equivalent-circuit models for the input and output ports, which is the course we pursue here.

First, we look from the source into the input port and note that we see nothing but resistances, including the load resistance R_L. Hence, everything to the right of the input terminals can be modeled by a single *input resistance,* defined in accordance with the equivalence theorem as

$$R_i \triangleq \frac{v_1}{i_1} \tag{1}$$

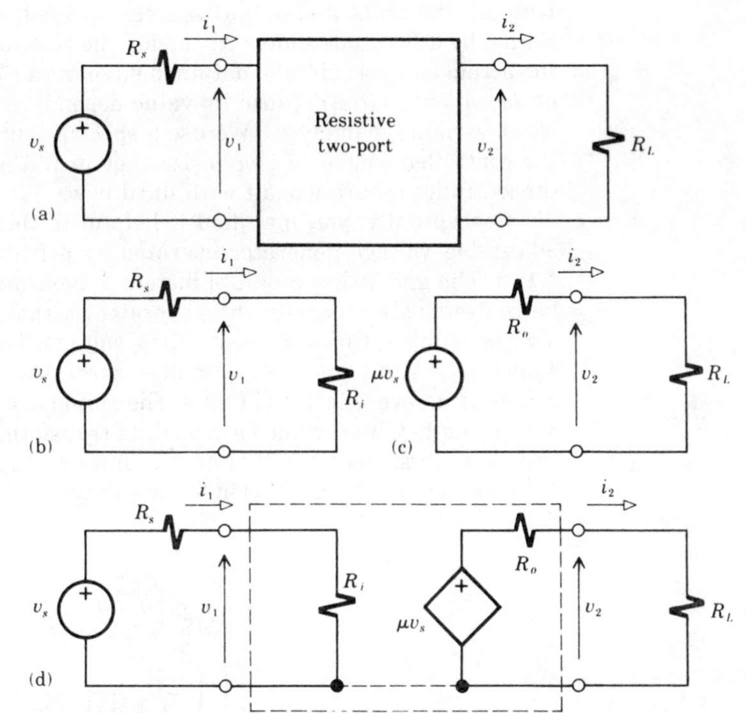

FIGURE 4.1–1
(a) A resistive two-port interfacing a source and load. (b) Input model. (c) Output model. (d) Complete model with a controlled source.

This leads to the equivalent circuit of Fig. 4.1–1b, and brings out the fact that R_i represents the equivalent load on the source.

Next, we look back from R_L into the output port. Since the circuit to the left of the output terminals includes v_s as well as resistances, it must be modeled by a Thévenin or Norton circuit. The Thévenin parameters will be the open-circuit output voltage v_{2-oc}, the short-circuit output current i_{2-sc}, and the *output resistance*

$$R_0 \triangleq \frac{v_{2-oc}}{i_{2-sc}} \tag{2}$$

Clearly, v_{2-oc} and i_{2-sc} will be proportional to v_s, so the Thévenin circuit for the output has the form of Fig. 4.1–1c with $v_{2-oc} = \mu v_s$ where μ (mu) is the proportionality constant. This circuit brings out the fact that μv_s and R_o are the equivalent source parameters as seen from the load resistance. A Norton circuit using i_{2-sc} would convey the same information.

Finally, putting our input and output models together, we have the complete diagram of Fig. 4.1–1d, where the dashed line reflects the possibility that the bottom terminals might be connected and at the same potential level. The diagram has no other direct electrical connection between the input and output sources; instead, the connection is represented by a fictitious source μv_s inside the two-port. (But remember that the actual two-port circuit contained no sources.) We call this a *controlled* or *dependent source* because its value depends on or is controlled by another variable, namely v_s. We use a special diamond-shaped symbol for the controlled source to emphasize that it differs from the *independent* (uncontrolled) sources dealt with until now.

Conceptually, you may find it helpful to think of this source as an adjustable voltage generator operated by a friendly gnome, as in Fig. 4.1–2. The gnome has a pair of meters to measure the two voltages, and he continuously readjusts the generator so that the controlled voltage always equals μ times the controlling voltage. Thus, for instance, if $\mu = \frac{1}{2}$ and $v_s = 10t^3$, our gnome makes sure that the controlled voltage equals $5t^3$ at every instant of time. The notion of a controlled source turns out to be a key ingredient for modeling transformers, transistor devices, and amplifiers; and we will later encounter current sources controlled by a current or voltage in addition to the voltage-controlled voltage source at hand.

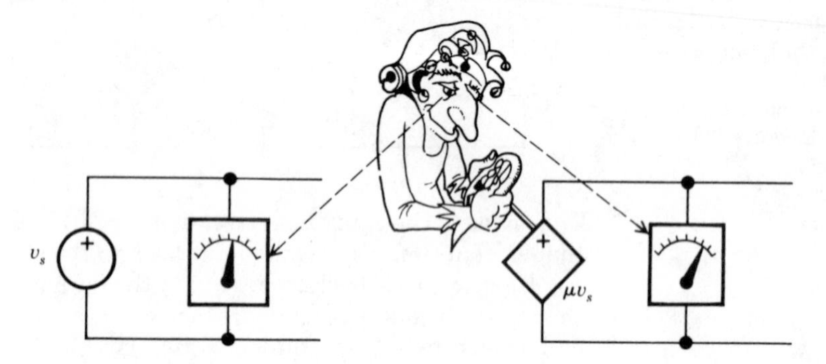

FIGURE 4.1–2 Interpretation of a controlled source.

As a closing comment to this discussion, it must be emphasized that the parameter R_i in Fig. 4.1–1d generally depends on the external load resistance R_L, while μ and R_o generally depend on the external source resistance R_s. This is an obvious disadvantage for it means that different parameter values are required when the same two-port is used in different applications. (An alternative modeling technique involving another set of parameters gets around that problem.) On the other hand, a model similar to that of Fig. 4.1–1d, with fixed parameter values, does apply for certain types of amplifiers and will be explored in the next section.

Example 4.1–1

A certain radio antenna acts like a voltage source with $R_s = 300\ \Omega$ and is to be connected to a radio receiver whose input characteristics are equivalent to a 50-Ω load resistance. A two-port is to be inserted between an-

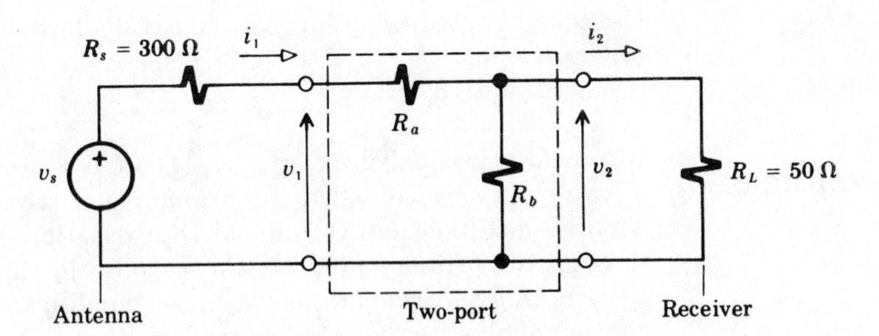

FIGURE 4.1–3
Circuit for Example 4.1–1.

tenna and receiver for purposes of matching R_i to R_s and R_o to R_L. Figure 4.1–3 gives the circuit diagram of a proposed design and we seek to find the corresponding values of R_a and R_b.

Looking into the input port (with R_L connected), we see that

$$R_i = R_a + \frac{R_b R_L}{R_b + R_L}$$

Looking back into the output port (with the source connected) we see that

$$v_{2\text{-oc}} = \frac{R_b}{R_s + R_a + R_b} v_s$$

$$i_{2\text{-sc}} = \frac{v_s}{R_s + R_a}$$

$$R_o = \frac{(R_s + R_a)R_b}{R_s + R_a + R_b}$$

Setting $R_i = R_s = 300$ and $R_o = R_L = 50$ yields two equations with R_a and R_b as unknowns. Both involve the product $R_a R_b$, but they are easily solved to yield

$$R_a = \sqrt{75{,}000} \approx 274\ \Omega \qquad R_b = R_a/5 \approx 55\ \Omega$$

Therefore, from our expression for $v_{2\text{-oc}}$, we obtain $\mu = 55/629$. We can now go back and label Fig. 4.1–1d with these values and also connect the bottom terminals to indicate that they actually are both at the same potential.

Although this two-port satisfies the resistance-matching condition, it does not result in maximum power transfer due to the fact that some of the source power is dissipated in R_a and R_b rather than being delivered to R_L. In fact, more power would be transferred if the two-port were omitted entirely! A better approach involves a transformer for the source-to-load connection.

Exercise 4.1–1

Use the above values in Fig. 4.1–1d to calculate the powers $p_1 = v_1 i_1$ and $p_2 = v_2 i_2$ when $v_s = 10$ V. Then find the power transferred to R_L with the two-port omitted.

IDEAL TRANSFORMERS

A transformer, symbolized by Fig. 4.1–4a, is a two-port device consisting of two coils of wire called the *primary* and *secondary windings.* Both windings act like short circuits for DC excitation, and a constant current through the primary produces no response in the secondary. However, there is *magnetic coupling* between the windings, and a *time-varying* excitation applied to the primary induces a similar time-varying response in the secondary. The underlying physical principles of magnetic induction will be covered in Chapters 17 and 18; here we simply state and apply the characteristics of an ideal transformer—characteristics approximated to greater or lesser degree by many real transformers.

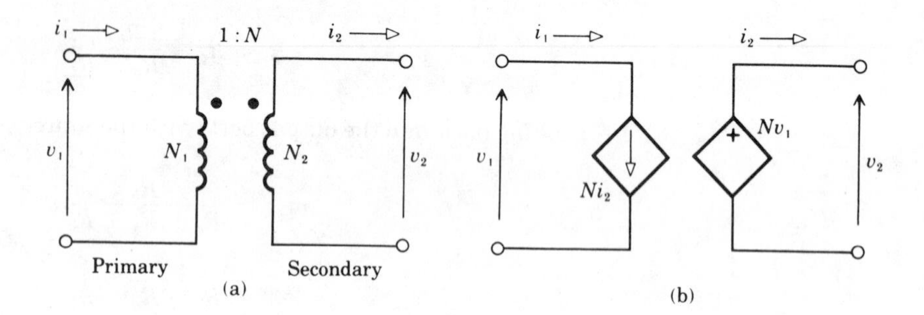

FIGURE 4.1–4
Ideal transformer and its circuit model.

The description of an ideal transformer involves only one parameter, its *turns ratio*

$$N = \frac{N_2}{N_1}$$

where N_1 and N_2 are the number of turns in the primary and secondary windings, respectively. The secondary voltage and current are related to the primary variables by the turns ratio. Specifically, for the polarity dots and reference arrows shown in the figure,

$$v_2 = Nv_1 \qquad i_2 = \frac{i_1}{N} \tag{3}$$

In view of the lack of direct electrical connection, the dots at the ends of the windings are needed to indicate the relative voltage polarities. If the dotted end of the primary has a higher potential than the undotted end, then the same condition holds at the secondary, and vice versa.

Based on Eq. (3), our model for an ideal transformer is as shown in Fig. 4.1–4b. The voltage-controlled voltage source represents the primary-to-secondary coupling $v_2 = Nv_1$, while the current-controlled current source represents the secondary-to-primary coupling $i_1 = Ni_2$—consistent with $i_2 = i_1/N$. This model and Eq. (3) also apply directly to the very common case of AC voltages and currents with the peak values or effective (rms) values in place of the instantaneous values v_1, i_1, and so forth.

If $N > 1$, the output voltage is larger than the input and we have a *step-up transformer;* but note that i_2 will be less than i_1. Conversely, a *step-down transformer* with $N < 1$ has a smaller output voltage but greater current. An ideal transformer with $N \neq 1$ can be either step-up or step-down, depending on which side is taken as the input. Voltage transformation—up or down—is exploited in many transformer applications. If $N = 1$, we have an *isolation transformer* that isolates or decouples the DC potential levels on either side without affecting the time-varying quantities.

For a mechanical analogy to transformer action, consider a rigid pivoted lever (or see-saw) with lengths ratio $\ell = \ell_1/\ell_2$, as in Fig. 4.1–5. If you push down on one end with force f_1 and it moves at velocity u_1, then the upward force and velocity at the other end are

$$f_2 = \ell\, f_1 \qquad u_2 = u_1/\ell$$

identical in form to Eq. (3). Clearly, the mechanical input and output powers $f_1 u_1$ and $f_2 u_2$ must be equal if the pivot is frictionless, and so it is for electrical power in an ideal transformer. In particular,

$$p_2 = v_2 i_2 = N v_1 \frac{i_1}{N} = v_1 i_1 = p_1$$

meaning that an ideal transformer transfers power from input to output without any internal dissipation.

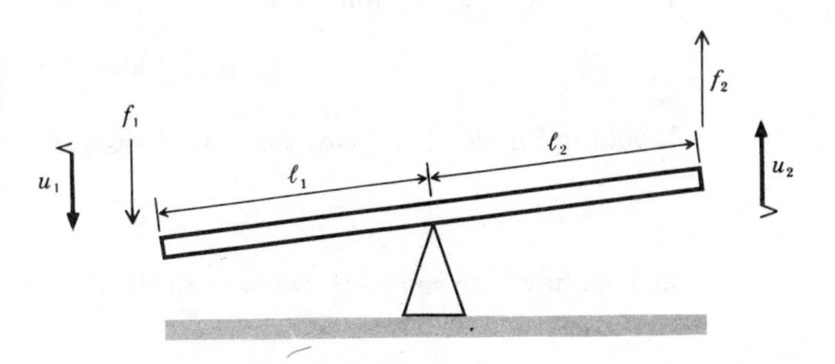

FIGURE 4.1–5
Mechanical analog of a transformer.

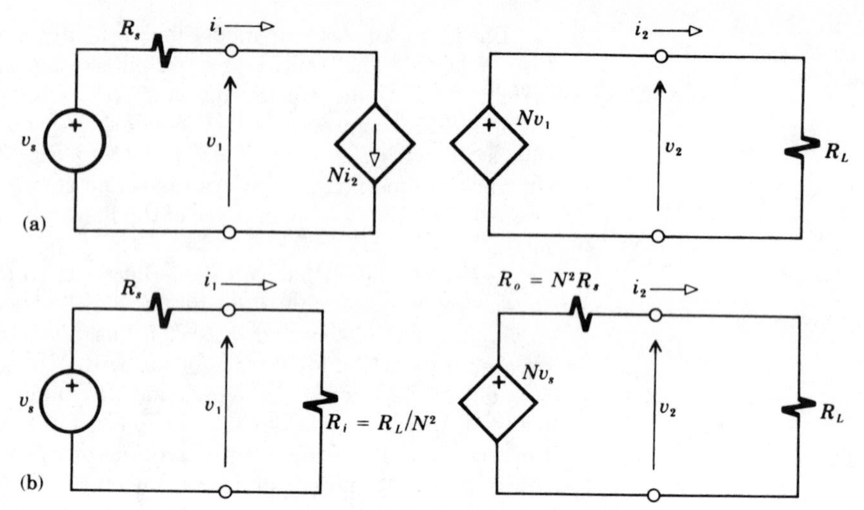

FIGURE 4.1–6
(a) A transformer interfacing a source and load. (b) Equivalent input and output circuits.

Putting this power-transfer property to work, we'll let an ideal transformer be the interface between a source and load, as in Fig. 4.1–6a. The input resistance $R_i = v_1/i_1$ is found by noting that $i_2 = v_2/R_L = Nv_1/R_L$ and $i_1 = Ni_2$. Therefore,

$$R_i = \frac{v_1}{Ni_2} = \frac{v_1}{N(Nv_1/R_L)} = \frac{R_L}{N^2} \tag{4}$$

which is often called the secondary resistance *referred* to the primary. To model the output side, we observe that $i_2 = 0$ under open-circuited conditions, so $i_1 = Ni_2 = 0$, $v_1 = v_s - R_s i_1 = v_s$, and thus

$$v_{2\text{-oc}} = Nv_s \tag{5a}$$

Similarly, $v_2 = 0$ under short-circuited conditions, so $v_1 = v_2/N = 0$, $i_1 = (v_s - v_1)/R_s = v_s/R_s$, and thus

$$i_{2\text{-sc}} = v_s/NR_s \tag{5b}$$

Invoking Eq. (2) then gives the output resistance

$$R_o = N^2R_s \tag{6}$$

and we have the complete two-port model of Fig. 4.1–6b.

If N is such that R_L/N^2 matches R_s, there will be maximum power transferred from source to transformer and thence to the load, for no power is lost in the transformer. Moreover, N^2R_s automatically matches R_L at the output when $R_L/N^2 = R_s$ at the input. The combination of maximum power transfer with resistance matching on both sides makes the transformer a valuable building block in electronic systems with time-varying signals.

Sometimes a transformer has three or more windings, or one winding has additional terminals known as *taps*. For instance, Fig. 4.1–7 shows the diagram and equivalent circuit of an ideal transformer with a tapped secondary. (A segment of the tapped winding may also be wound in the opposite sense, which inverts the voltage polarity. This kind of inversion would be indicated by dotting the other end of the segment.) Such tapped windings permit matching different load resistances, as in the output transformer of an audio amplifier that can drive a 4-Ω, 8-Ω, or 16-Ω loudspeaker. They are also used to provide several different output voltages from one source, as in an electronic power supply or a dual-voltage AC system.

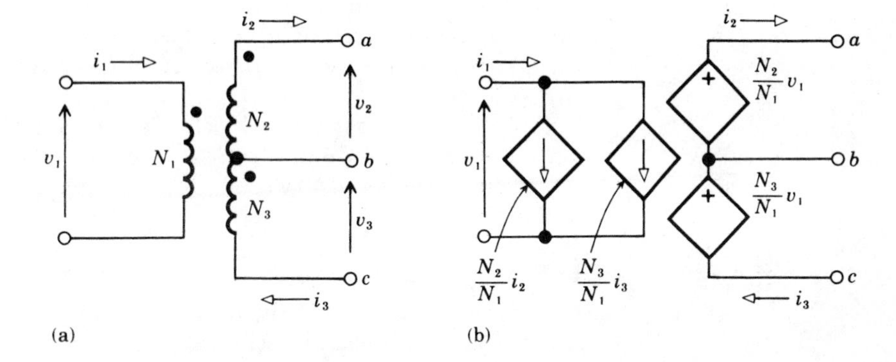

FIGURE 4.1–7 Transformer with a tapped secondary.

Example 4.1–2

Figure 4.1–8 shows an electronic oscillator that generates a 10-V peak sinusoid and has 2.5-kΩ source resistance. A 25-V battery powers the oscillator and is connected to it through the primary of a transformer that also couples the AC voltage to a 100-Ω load resistance. To maximize the AC power transfer, we want $R_i = R_L/N^2 = R_s$ or

$$N = \sqrt{\frac{R_L}{R_s}} = \sqrt{\frac{100\ \Omega}{2500\ \Omega}} = \frac{1}{5}$$

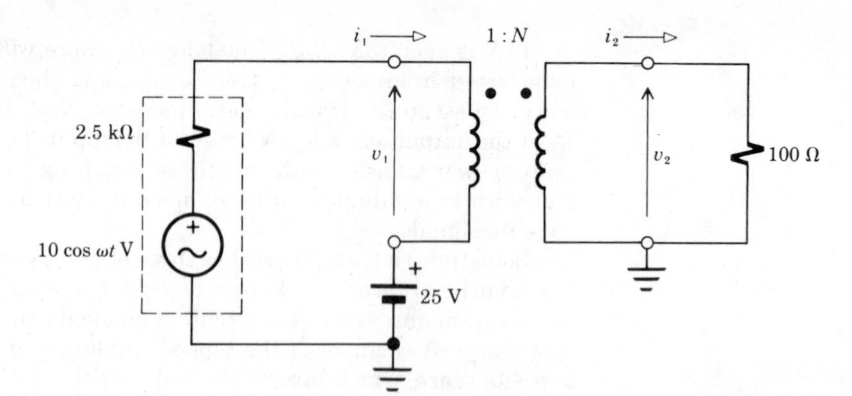

FIGURE 4.1–8
Circuit for Example 4.1–2.

which means a step-down transformer. Thus, since the DC current through the primary is not transformed, $v_2 = N \times 10 \cos \omega t$ V = 2 cos ωt V and $i_2 = v_2/100\ \Omega = 20 \cos \omega t$ mA.

But the primary current consists of a DC and AC component and can be determined using superposition. For the DC component, we imagine the AC voltage source short-circuited and the primary replaced with a short circuit, giving $i_1' = -25$ V/2.5 kΩ = −10 mA. For the AC component, we imagine the battery short-circuited and the primary replaced by $R_i = 2.5$ kΩ, giving $i_1'' = (10 \cos \omega t\ \text{V})/(2.5\ \text{k}\Omega + 2.5\ \text{k}\Omega) = 2 \cos$ ωt mA. Therefore, $i_1 = i_1' + i_1'' = -10$ mA + 2 cos ωt mA, whereas $v_1 = R_i i_1'' = 5 \cos \omega t$ V.

Finally, note that the DC isolation property of the transformer has permitted different ground points on either side. Specifically, the undotted end of the secondary is at ground potential, while the undotted end of the primary is 25 V above ground.

Exercise 4.1–2

In Fig. 4.1–6, let $v_s = 10 \cos \omega t$ V, $R_s = 50\ \Omega$, and $R_L = 200\ \Omega$.

(a) Find v_1, i_1, v_2, i_2, and the average power P through the transformer when N is such that R_i matches R_s.

(b) Repeat the calculations with N such that $R_i = 10R_s$.

Exercise 4.1–3

Let a 120-V (rms) AC source with negligible resistance be applied to the input in Fig. 4.1–7. If $N_1 = 200$, find N_2 and N_3 to provide an effective voltage $V_2 = 180$ V across a 2-kΩ load and $V_3 = 12$ V across a 15-Ω load. Then compute the resulting value of I_1 and the average power through the transformer.

HYBRID PARAMETERS †

For a more general model of a two-port network, independent of the external circuitry, we must work with all four of the terminal variables indicated in Fig. 4.1–9a. By convention, both currents i_2 and i_1 are assumed to flow *into* the two-port since either port might serve as the input for a given application. A complete external description of the two-port is then provided by two equations that relate any two of the terminal variables to the other two.

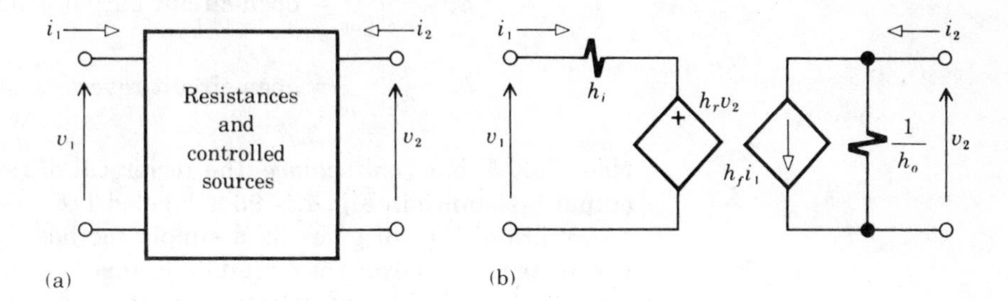

FIGURE 4.1–9 Hybrid-parameter model of a two-port.

If the internal elements of the two-port are restricted to linear resistances and controlled sources, with no *independent* sources, one possible set of equations for the terminal variables is

$$\begin{aligned} v_1 &= h_i i_1 + h_r v_2 \\ i_2 &= h_f i_1 + h_o v_2 \end{aligned} \tag{7}$$

the constants being called the *hybrid parameters* or h parameters. These equations may also be written in the matrix form

$$\begin{bmatrix} v_1 \\ i_2 \end{bmatrix} = \begin{bmatrix} h_{11} & h_{12} \\ h_{21} & h_{22} \end{bmatrix} \begin{bmatrix} i_1 \\ v_2 \end{bmatrix}$$

where $h_{11} = h_i$, $h_{12} = h_r$, etc. Various other sets of parameters for two-port models exist, corresponding to other sets of equations for the terminal variables. Hybrid parameters have the advantage of being particularly suited for models of transistors.

Figure 4.1–9b gives the equivalent circuit or h-parameter model of a two-port based on Eq. (7). This model helps us give a physical interpretation of the h parameters, as follows. First, imagine a short-circuit across the right-hand terminals so that $v_2 = 0$; then $i_2 = h_f i_1$, $v_1 = h_i i_1$, and we can write

$$\begin{aligned} h_i &= \left.\frac{v_1}{i_1}\right|_{v_2=0} = \text{short-circuit input resistance} \\ h_f &= \left.\frac{i_2}{i_1}\right|_{v_2=0} = \text{short-circuit forward current gain} \end{aligned} \tag{8a}$$

(We have used the term "current gain" to describe h_f in anticipation of electronic devices that actually have gain in the sense that $i_2 > i_1$.) Next, imagine an open-circuit at the left-hand terminals so that $i_1 = 0$; then $v_1 = h_r v_2$, $i_2 = h_o v_2$, and we can write

$$h_o = \left.\frac{i_2}{v_2}\right|_{i_1=0} = \text{open-circuit output conductance}$$
$$h_r = \left.\frac{v_1}{v_2}\right|_{i_1=0} = \text{open-circuit reverse voltage gain} \tag{8b}$$

Note that h_o is a *conductance,* the reciprocal of resistance, and that the output resistance in Fig. 4.1–9b is labeled $1/h_o$. Besides the physical interpretation, Eq. (8) gives us a simple method for determining the h-parameter values from the circuit diagram of a two-port by setting $v_2 = 0$ to compute h_i and h_f and setting $i_1 = 0$ to compute h_o and h_r.

Having determined the h parameters, we can investigate the effects of a two-port in a particular application using routine circuit analysis. Consider, for example, the source-load connection in Fig. 4.1–10, where $G_L = 1/R_L$ is the load conductance; thus the equivalent conductance in parallel with the current source is the sum $h_o + G_L$. (Recall that parallel conductances add.) Thus, $v_2 = -h_f i_1/(h_o + G_L)$ and

$$v_1 = h_i i_1 + h_r v_2 = h_i i_1 - h_r \frac{h_f}{h_o + G_L} i_1$$

so the equivalent input resistance is

$$R_i = \frac{v_1}{i_1} = h_i - \frac{h_r h_f}{h_o + G_L}$$

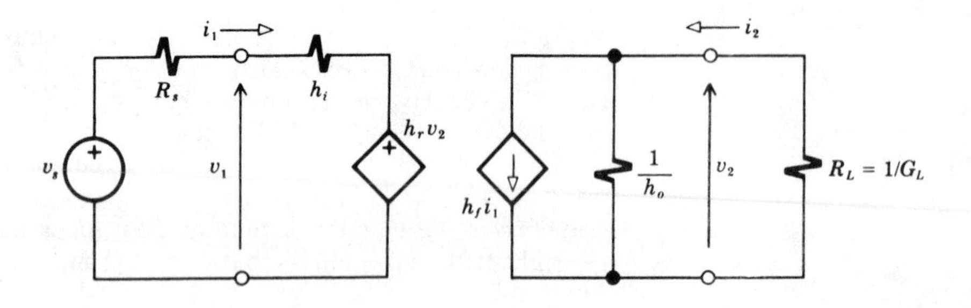

FIGURE 4.1–10

Similarly, one can find $v_{2\text{-oc}}$, $i_{2\text{-sc}}$, and $R_o = v_{2\text{-oc}}/(-i_{2\text{-oc}})$, the minus sign being necessary in view of the reversed direction of i_2.

Exercise 4.1–4 Use Eq. (8) to find the h parameters of the resistive two-port in Fig. 4.1–3.

4.2 AMPLIFIER CONCEPTS

Amplifiers are two-port networks containing electronic devices capable of "enlarging" or "magnifying" the variations of an electrical signal represented by a voltage or current waveform. As such, amplifiers play essential roles in multitudinous applications—transistor radios, TV sets, biomedical instrumentation, industrial process control, and so on, almost without end. This section introduces the general concepts of amplifiers and amplification systems.

Figure 4.2–1a shows a familiar amplification system consisting of an electric guitar, amplifier, and loudspeaker. Figure 4.2–1b represents those same units in a generalized *block diagram*. The *input signal*, denoted $x(t)$, is the voltage or current waveform produced by some *source*—the pickup on the guitar, for instance, or perhaps an electrode sensing brain-wave activity. In any case, we desire an amplified *output signal* $y(t)$ to drive an energy-consuming *load* such as a loudspeaker or strip-chart recorder.

Amplification is achieved with the help of electronic amplifying devices (usually transistors) that have the ability to control a large voltage or current in direct proportion to the smaller input voltage or current. Of course, the large voltage or current must come from something other than the input source, so every amplifier has an associated power supply—for example, a battery or AC-to-DC converter.

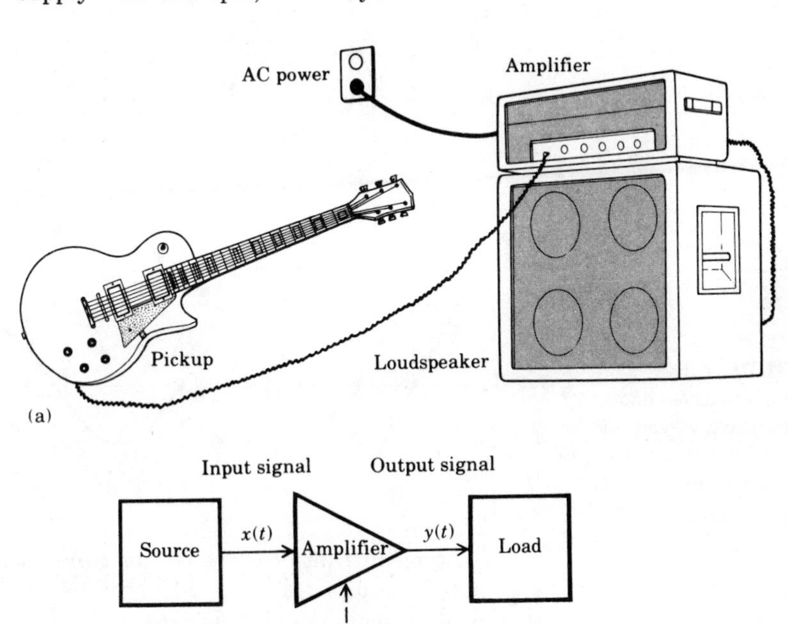

FIGURE 4.2–1
An amplification system and its block diagram.

We are not concerned here with the internal workings of amplifiers and power supplies. Rather, we focus on the input-output relationship, the corresponding equivalent circuits using controlled sources, and the differences between ideal and practical amplifiers.

AMPLIFICATION AND IDEAL AMPLIFIERS

Mathematically, the action of an *ideal amplifier* is expressed by the simple equation

$$y(t) = Ax(t) \tag{1}$$

where A is a constant called the *amplification*. Since $y(t)/x(t) = A$ at any instant of time, plotting y versus x yields a straight line through the origin with *slope* A. Figure 4.2–2a shows this plot, known as the amplifier's *transfer curve,* along with typical waveforms for the case when $A > 1$. We see that the output signal $y(t)$ has the same relative variations as the input signal $x(t)$, but with bigger excursions in the positive and negative directions. Figure 4.2–2b shows the transfer curve and typical waveforms for a *negative* amplification, $A < -1$. The amplifier still produces an enlarged output signal but with a sign inversion or polarity reversal. Many amplifiers have negative amplification, which may or may not be a problem in a particular application.

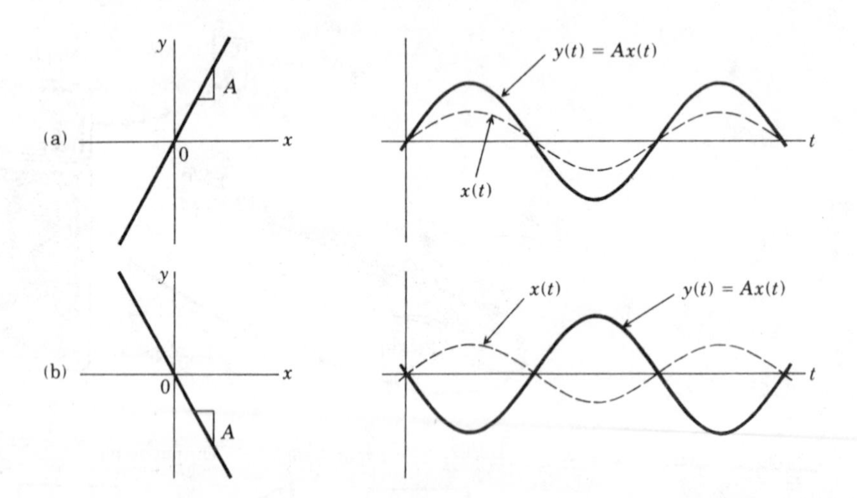

FIGURE 4.2–2 Transfer curves and illustrative waveforms for an amplifier with (a) $A > 1$, (b) $A < -1$.

The typical transfer curve of a *real* (nonideal) amplifier, as given in Fig. 4.2–3, differs from the ideal in two respects: it has a limited straight-line region; and it flattens off towards a constant value for extreme values of x. Physically, the flattening off reflects *saturation* and *cutoff* phenomena in the amplifier because it is unable to produce output

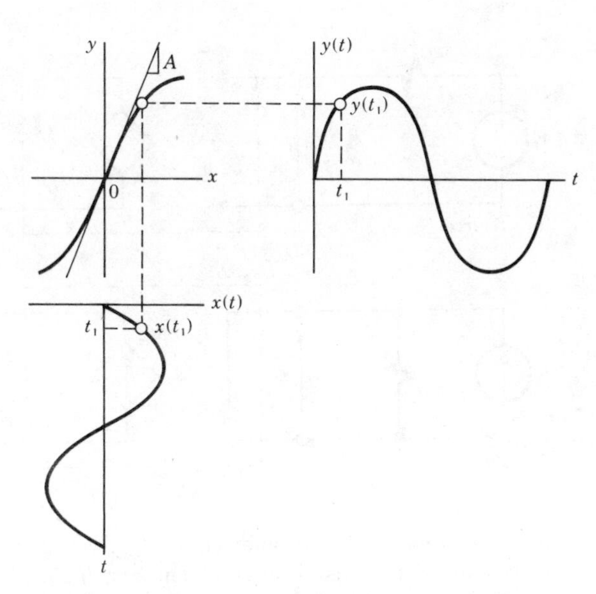

FIGURE 4.2–3
A nonlinear transfer curve and illustrative waveforms.

voltage or current beyond certain limits dictated by the amplifying device and the power supply.

To demonstrate graphically the consequences of a nonlinear transfer curve, we plot the input $x(t)$ with the x-axis turned horizontally, as shown in Fig. 4.2–3. We then project a specific input value such as $x(t_1)$ onto the transfer curve and read off the corresponding output value $y(t_1)$. With the help of a few such points it is possible to sketch the entire output waveform $y(t)$ and compare it with $x(t)$. Clearly, the output does not look exactly like an enlarged version of the input, so $y(t) \neq Ax(t)$. The output thus suffers from *nonlinear distortion* due to the nonlinear shape of the transfer curve, and we say that the amplifier is being *overdriven.*

If, however, the variations of the input are restricted to a small range over which the transfer curve is reasonably linear, then $y(t) \approx Ax(t)$ and a real amplifier approximates ideal amplification. This conclusion forms the theoretical basis of the *small-signal model* for real amplifiers. Unless we state otherwise, we'll assume that the amplifiers we discuss are operating in their linear regions.

Figure 4.2–4a gives the circuit model of an *ideal voltage amplifier,* together with an input voltage source and output load resistance. The amplifier's input terminals lead to an open circuit, so $i_{\text{in}} = 0$ and $v_{\text{in}} = v_s$. A controlled voltage source with $v_a = \mu v_{\text{in}}$ is connected to the output terminals such that $v_{\text{out}} = v_a$. Hence, the source-to-output *voltage amplification* will be

$$A_v = \frac{v_{\text{out}}}{v_s} = \frac{\mu v_{\text{in}}}{v_{\text{in}}} = \mu \tag{2}$$

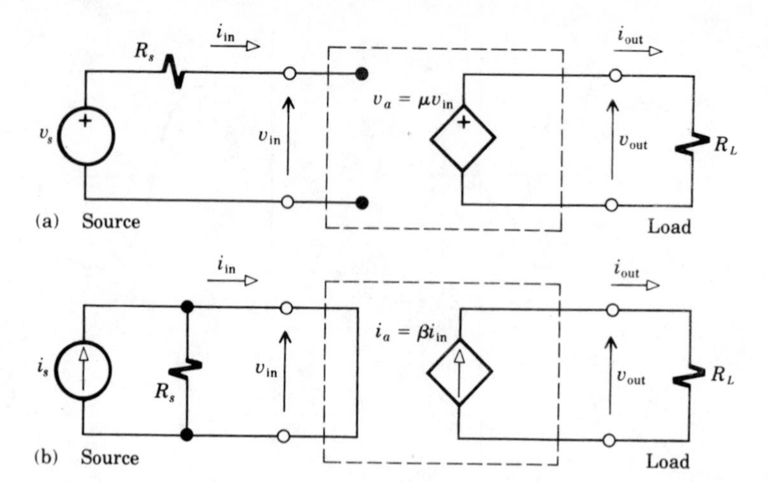

FIGURE 4.2–4
(a) Ideal voltage amplifier. (b) Ideal current amplifier.

and the slope of the transfer curve (v_{out} versus v_s) equals the proportionality constant μ, also known as the *amplification factor*.

Similarly, Fig. 4.2–4b is the model of an *ideal current amplifier* driven by a current source. The short-circuit across the input terminals means that $v_{in} = 0$ and $i_{in} = i_s$, and a controlled current source gives $i_{out} = i_a = \beta i_{in}$. Thus, the source-to-output *current amplification* will be

$$A_i = \frac{i_{out}}{i_s} = \frac{\beta i_{in}}{i_{in}} = \beta \tag{3}$$

so the slope of the current transfer curve equals the amplification factor β (beta).

Note that the performance of these ideal amplifiers is independent of the source and load resistances, R_s and R_L. Moreover, they draw no input power $p_{in} = v_{in} i_{in}$ from the source since either $i_{in} = 0$ or $v_{in} = 0$. The output power $p_{out} = v_{out} i_{out}$ delivered to the load comes entirely from the amplifier's power supply which, by convention, was not included in the circuit model. These ideal characteristics establish a framework for comparing the performance of real amplifiers.

By the way, the symbol μ traces historically to the amplification factor of a triode vacuum tube, while β relates to the model of a bipolar junction transistor. Another type of ideal amplifier is described in the exercise below; its proportionality constant g_m might represent the small-signal characteristics of a field effect transistor or a pentode vacuum tube.

Exercise 4.2–1 Consider the amplifier model in Fig. 4.2–5 where g_m has units of conductance (℧). Show that $A_v = -g_m R_L$.

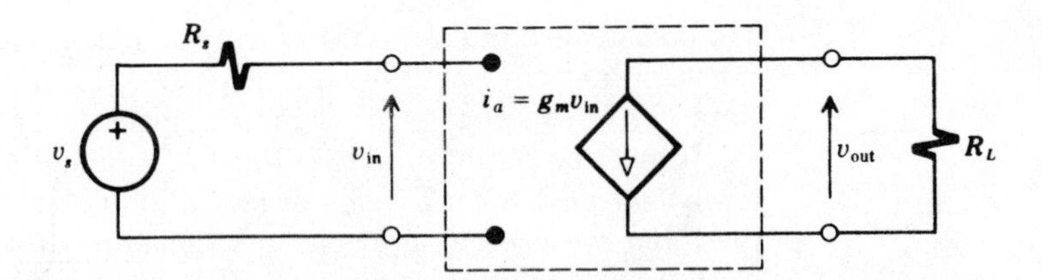

FIGURE 4.2–5
Amplifier model for Exercise 4.2–1.

AMPLIFIER CIRCUIT MODELS AND ANALYSIS

Unlike ideal voltage and current amplifiers, the performance of real amplifiers depends on the source and load circuits. We represent that dependence here by adding resistances to our ideal amplifier models. Then we investigate the implications of the resulting loading effects at input and output.

Figure 4.2–6 gives two possible small-signal models for a real amplifier. Both of these models have an input resistance R_i that equals the equivalent resistance seen looking into the input terminals, that is, $R_i = v_{in}/i_{in}$. It is used in the circuit model to represent the fact that real amplifiers always draw some power from the input source. Both circuits also include an output resistance R_o to represent internal resistance associated with the controlled source.

The Thévenin output circuit, Fig. 4.2–6a, has R_o in series with the controlled voltage source v_a, and we interpret v_a as the open-circuit output voltage. The Norton output circuit, Fig. 4.2–6b has R_o in parallel with the controlled current source i_a, which we correspondingly interpret as the short-circuit output current. The wire connecting the lower input

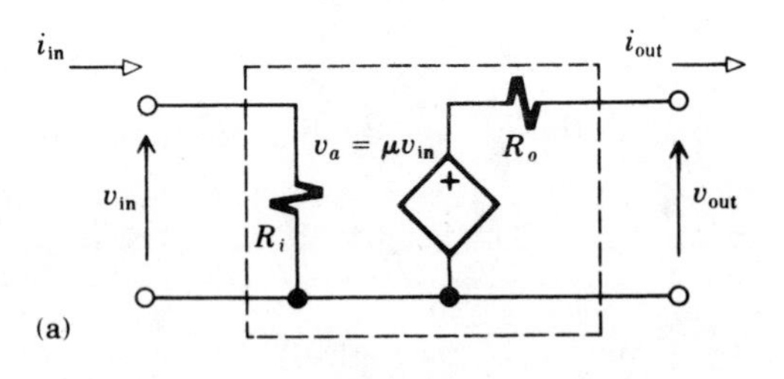

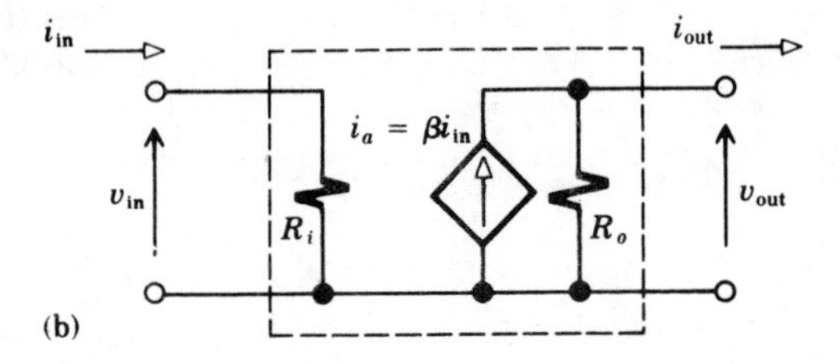

FIGURE 4.2–6
Small-signal models of a real amplifier.

and output terminals carries no current but reflects the fact that these two terminals are at the same potential, usually the common ground point, with respect to small-signal variations.

It must be emphasized that Figs. 4.2–6a and 4.2–6b are different but *equivalent* models for a given practical amplifier. To convert from one to the other we use our previous relationship between the Thévenin and Norton parameters, namely $v_{oc} = R_o i_{sc}$, which becomes $v_a = R_o i_a$ in our present notation. Although these models omit certain other phenomena present in actual amplifiers, they still provide us with valuable approximations for analysis purposes.

The impact of input and output resistance on amplification is brought out by Fig. 4.2–7, where an input voltage source and output load have been connected to the model from Fig. 4.2–6a. Since $v_{in} \neq v_s$ and $v_{out} \neq v_a$ due to loading effect, we must proceed with care when calculating $A_v = v_{out}/v_s$. To this end, we first expand v_{out}/v_s as a "chain" expression

$$A_v = \frac{v_{in}}{v_s} \times \frac{v_a}{v_{in}} \times \frac{v_{out}}{v_a} \tag{4}$$

permitting us to examine each of these voltage ratios.

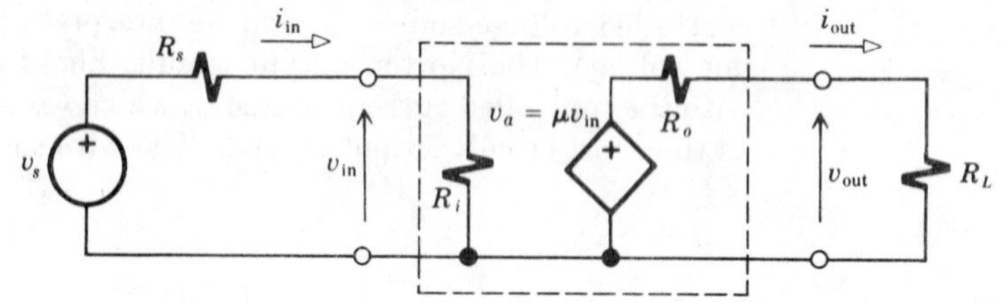

FIGURE 4.2–7
Amplifier with source and load.

Recalling the voltage-divider relation, the first and last ratios are

$$\frac{v_{in}}{v_s} = \frac{R_i}{R_s + R_i} \qquad \frac{v_{out}}{v_a} = \frac{R_L}{R_o + R_L}$$

and, by definition of the controlled source,

$$\frac{v_a}{v_{in}} = \mu$$

Substituting these into Eq. (4) gives

$$A_v = \left(\frac{R_i}{R_s + R_i}\right) \mu \left(\frac{R_L}{R_o + R_L}\right) \tag{5}$$

Clearly, $|A_v| < |\mu|$ and its actual value depends on all four resistances because of loading at the input and output.

Input loading is negligible if the input resistance is large enough; that is, $v_{in} \approx v_s$ if $R_i \gg R_s$. Likewise, $v_{out} \approx v_a$ if the amplifier has small output resistance, $R_o \ll R_L$. Therefore, when

$$R_i \gg R_s \qquad R_o \ll R_L \tag{6a}$$

then

$$A_v \approx \mu \tag{6b}$$

and a practical amplifier closely approximates an ideal voltage amplifier.

If we are concerned with current amplification, we use Norton circuits for both the source and amplifier and write $A_i = i_{out}/i_s = (i_{in}/i_s)(i_a/i_{in})(i_{out}/i_a)$. The details are left as an exercise.

Example 4.2–1

A certain medical transducer has $v_s = 0.5$ V, $R_s = 2$ kΩ, and a maximum current limitation of 0.1 mA. We want to amplify v_s for plotting on a strip-chart recorder having a 5-V scale and 60-Ω input resistance. The proposed amplifier has $R_i = 8$ kΩ, $\mu = 15$, and $R_o = 20$ Ω, giving the complete equivalent circuit in Fig. 4.2–8.

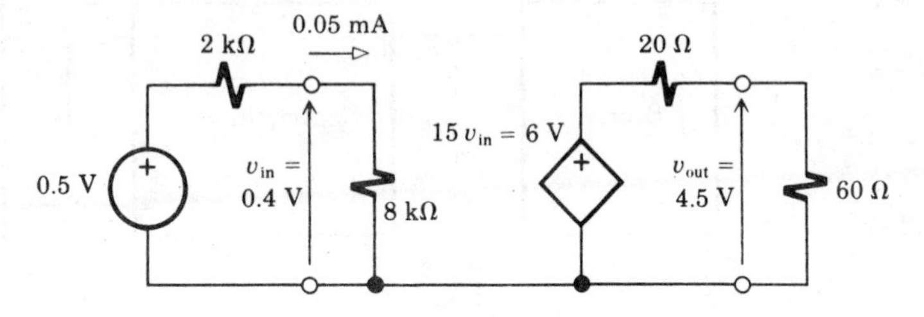

FIGURE 4.2–8 Amplifier circuit for Example 4.2–1.

Substituting values into Eq. (5) gives

$$A_v = \frac{8}{2+8} \times 15 \times \frac{60}{20+60} = 0.8 \times 15 \times 0.75 = 9$$

so $v_{out} = 9\ v_s = 4.5$ V, which is appropriate for the recorder. (Note that input and output loading reduce the amplification by a factor of $0.8 \times 0.75 = 0.6$.) The remaining voltage values are easily computed, and we see that $i_{in} = 0.05$ mA, which falls within the transducer's limitation.

Exercise 4.2–2

By comparing the open-circuit output voltage and short-circuit output current in Fig. 4.2–6, show that the two models in the figure are equivalent if $\mu/\beta = R_o/R_i$.

Exercise 4.2–3

Connect a current source with resistance R_s to the input of Fig. 4.2–6b, and show that

$$A_i = \frac{R_s \beta R_o}{(R_s + R_i)(R_o + R_L)}$$

What would the conditions on the resistances be if $A_i \approx \beta$?

POWER GAIN AND DECIBELS

In addition to voltage and current amplification, we will be interested in the *power gain* of an amplifier. Referring to Fig. 4.2–9, let P_{in} be the *average input power* to the amplifier from the source; that is, P_{in} is the average value of the instantaneous power $p_{\text{in}} = v_{\text{in}} i_{\text{in}}$. Similarly, let P_{out} be the *average output power* from the amplifier to the load. The power gain is then defined as the ratio of output to input power,

$$G \triangleq \frac{P_{\text{out}}}{P_{\text{in}}} \tag{7}$$

and, therefore, $P_{\text{out}} = GP_{\text{in}}$. (Note that G defined here has no relationship to the symbol for conductance.)

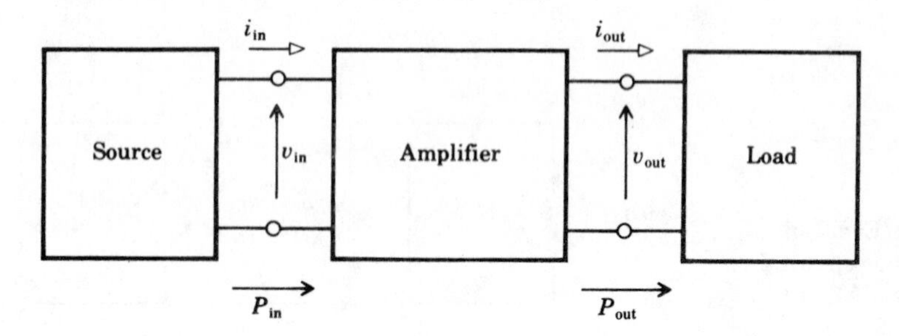

FIGURE 4.2–9

As previously observed, an ideal voltage or current amplifier has $p_{\text{in}} = 0$, so $P_{\text{in}} = 0$, whereas $P_{\text{out}} \neq 0$, meaning that ideal amplifiers have *infinite* power gain. Practical amplifiers always have finite power gain, of course, but the value of G can be very large—so large that a logarithmic measure called the *decibel* (dB) often proves useful. We define the dB power gain to be

$$G_{\text{dB}} \triangleq 10 \log G = 10 \log (P_{\text{out}}/P_{\text{in}}) \tag{8}$$

where log stands for the common logarithm, $\log_{10}$. For example, if a certain amplifier produces $P_{\text{out}} = 5$ W when $P_{\text{in}} = 0.5$ mW, then $G = 5\text{ W}/0.5\text{ mW} = 10^4$ and $G_{\text{dB}} = 10 \log 10^4 = 10 \times 4 = 40$ dB. (We usually drop the dB subscripts when there is no danger of confusion, so $G = 40$ dB clearly means $G_{\text{dB}} = 40$ dB and $G = 10^4$.)

Being a logarithmic unit, the decibel converts powers of 10 to products of 10, for example,

$$G = 10^n \leftrightarrow G_{dB} = 10n \text{ dB}$$

A somewhat strange but logical consequence of this measure is that unit power gain ($G = 1.0 = 10^0$) corresponds to zero dB gain ($G_{dB} = 10 \times 0 = 0$ dB). Furthermore, if $G < 1.0$ then $G_{dB} < 0$ dB. Other dB values are easily computed with the help of a calculator, but you may find it handy to memorize some of the common pairs listed in Table 4.1. Conversion from a dB value back to actual power gain is accomplished via

$$\frac{P_{out}}{P_{in}} = G = 10^{0.1G_{dB}} \tag{9}$$

obtained by inverting Eq. (8).

TABLE 4.1–1 Selected Decibel Values

dB	G	A_v *or* A_i	dB	G	A_v *or* A_i
−10	0.1	0.316	3	2	1.414
−6	0.25	0.5	6	4	2.0
−3	0.5	0.707	10	10	3.16
0	1.0	1.0	20	100	10.0

Strictly speaking, the decibel is defined only for power gain or, equivalently, power ratios. Nonetheless, in practice we often speak of the *voltage* or *current gain* in decibels, defined by

$$A_{dB} \triangleq 10 \log A^2 \tag{10}$$

where A represents A_v or A_i, as the case may be. This definition is based on the fact that power is proportional to the square of voltage or current; and, consequently, power gain is proportional to the square of the voltage or current amplification. Thus, for instance, we can say that A_v equals 40 dB when $A_v = \pm 100 = \pm 10^2$ even if $G \neq 40$ dB. In such cases, a voltage gain of $10n$ dB simply means that the amplification equals $\pm 10^{n/2}$.

Example 4.2–2

Suppose all the voltages and currents in Fig. 4.2–8 are sinusoids with rms values as indicated. Then the average output power is $P_{out} = (4.5 \text{ V})^2/60\ \Omega = 0.338$ W $= 338$ mW, while $P_{in} = (0.4 \text{ V})^2/8\text{ k}\Omega = 0.02$ mW. Thus, the amplifier provides a power gain of $338/0.02 = 16{,}900$, and

$$G_{dB} = 10 \log 16{,}900 = 42.3 \text{ dB}$$

as contrasted with $A_{dB} = 10 \log 9^2 = 19.1$ dB. In view of the relatively small voltage amplification, most of the power gain comes about by "transferring" the signal voltage from a large input resistance to a much smaller load resistance.

Exercise 4.2–4 A current amplifier having $R_i = 2$ kΩ, $\beta = 100$, and $R_o = 5$ kΩ is connected between a 50-Ω load and an AC source with $R_s = 1$ kΩ and $i_s = 3$ mA (rms). Draw the complete equivalent circuit and label it with all rms current values. Then calculate A_i and G in dB.

CASCADED AMPLIFIERS

Frequently, we find it necessary to put two or more amplifiers in *cascade*—the output of the first being applied to the input of the second, and so forth. The amplification produced by the cascade is then proportional to the *product* of the individual amplifications and, as a result, can be much larger than that achieved by a single amplifier.

To demonstrate this cascade multiplication property, consider two cascaded voltage amplifiers with a source and load, as shown in Fig. 4.2–10. Observing that the first output voltage becomes the second input voltage (that is, $v_{12} = v_{\text{out}_1} = v_{\text{in}_2}$) the overall voltage amplification is

$$A_v = \frac{v_{\text{out}}}{v_s} = \frac{v_{\text{in}}}{v_s} \times \frac{v_{a1}}{v_{\text{in}}} \times \frac{v_{12}}{v_{a1}} \times \frac{v_{a2}}{v_{12}} \times \frac{v_{\text{out}}}{v_{a2}}$$

$$= \left(\frac{R_{i1}}{R_s + R_{i1}}\right) \mu_1 \left(\frac{R_{i2}}{R_{o1} + R_{i2}}\right) \mu_2 \left(\frac{R_L}{R_{o2} + R_L}\right)$$

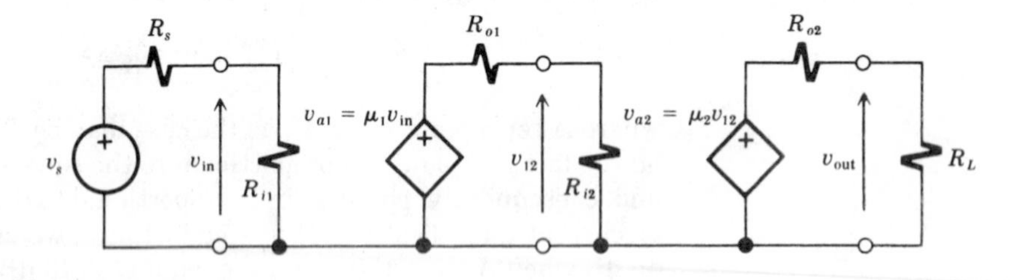

FIGURE 4.2–10 Cascade of two voltage amplifiers.

Minimizing loading effects now requires $R_{o1} \ll R_{i2}$, as well as $R_{i1} \gg R_s$ and $R_{o2} \ll R_L$, in which case

$$A_v \approx \mu_1 \mu_2$$

and we have the product of the individual amplification factors. Similar expressions hold for two cascaded current amplifiers.

When loading is negligible between cascaded amplifiers and at input and final output, the individual amplifications are $A_1 \approx \mu_1$, $A_2 \approx \mu_2$, etc., and n cascaded voltage amplifiers yield

$$A_v \approx A_1 A_2 \cdot \cdot \cdot A_n \tag{11}$$

Accordingly, we could get $A_v = 10^4$ from two amplifiers with $A_1 = A_2 = 100$ or from four amplifiers with $A_1 = A_2 = A_3 = A_4 = 10$, and so forth. We can also correct for unwanted polarity inversion by cascading an even number of inverting amplifiers. Thus if $A_1 = -20$ and $A_2 = -5$, then $A_v = (-20) \times (-5) = +100$.

Exercise 4.2–5

Convert Eq. (11) to decibels, demonstrating that the cascade amplification in dB equals the *sum* of the individual dB values.

4.3 OPERATIONAL AMPLIFIERS

Not too long ago, an electrical engineer who needed an amplifier had to sit down and design it completely, using perhaps a dozen or more individual components. Thankfully, all that has been changed by the modern integrated-circuit operational amplifier, commonly called an "op-amp." Now you can buy an op-amp off the shelf and connect two or three elements to it to create a more reliable, less expensive amplifier than one designed from scratch. The versatile op-amp has thus become a major building-block in today's electronic systems, and helps perform a wide variety of signal-processing operations in addition to amplification.

The name "operational amplifier" actually refers to a large family of general-purpose and special-purpose amplifiers having three characteristics: large input resistance, small output resistance, and very large differential voltage amplification. After describing these characteristics, we'll introduce the valuable concept of the ideal op-amp and use it to explain several op-amp circuits. Other applications are explored in later chapters and in the problems at the end of this chapter.

OP-AMP CHARACTERISTICS AND MODELS

An operational amplifier, represented by Fig. 4.3–1a, is a high-gain voltage amplifier that responds to the *difference* between *two input voltages*. The input terminals, labeled + and −, are called the *noninverting* and *inverting* inputs, respectively. Despite the labels, the applied voltages v_p and v_n may be either polarity with respect to ground. The output v_{out} depends on the *difference voltage*

$$v_d = v_p - v_n$$

across the input terminals. Figure 4.3–1b shows these voltages explicitly and also indicates how the necessary power-supply voltages $+V_{CC}$ and $-V_{CC}$ are connected along with the load resistance.

(a) (b)

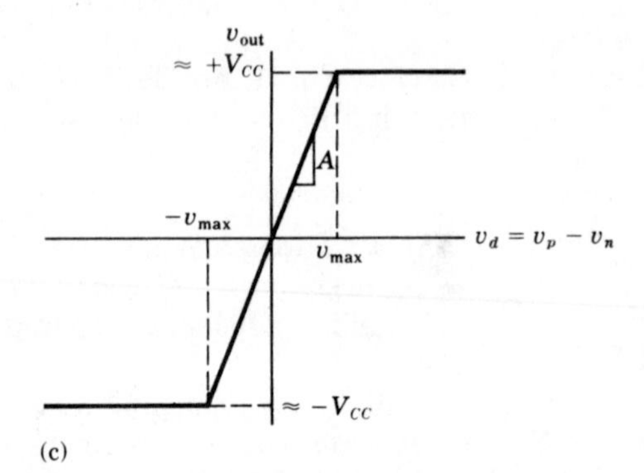

FIGURE 4.3-1
Operational amplifier: (a) symbol; (b) external connections; (c) transfer curve.

The op-amp's transfer curve v_{out} versus v_d, plotted in idealized form in Fig. 4.3-1c, reveals a linear middle region bounded by positive and negative *saturation* on either side. The linear region has slope or amplification A—called the *open-loop differential gain*—such that

$$v_{out} = Av_d = A(v_p - v_n) \tag{1}$$

so $v_{out} = 0$ when $v_p = v_n$. Increasing v_p then causes v_{out} to move in the positive direction, whereas increasing v_n causes v_{out} to move in the negative direction—which explains the "noninverting" and "inverting" designations for the input terminals. Aside from the difference-voltage input, the distinguishing characteristic of an op-amp is its gigantic open-loop gain A, typically 100,000 or more!

When operating in the linear region of the transfer curve, an op-amp may be modeled by the circuit of Fig. 4.3-2. Here, R_i' and R_o' denote the input and output resistance of the op-amp by itself, with representative values of $R_i' \approx 1\ \text{M}\Omega$ and $R_o' \approx 100\ \Omega$. This circuit model holds only over the limited range

$$-v_{max} < v_d < v_{max} \tag{2a}$$

where

$$v_{\max} \approx \frac{V_{CC}}{A} \tag{2b}$$

so that

$$-V_{CC} \lesssim v_{\text{out}} \lesssim V_{CC} \tag{2c}$$

If $|v_d| \geq v_{\max}$, the output saturates at $v_{\text{out}} \approx \pm V_{CC}$ and ceases to respond to further changes of v_p or v_n; we would then have extreme nonlinear distortion rather than linear amplification.

Since the power-supply voltage V_{CC} is usually less than 100 V while $A \geq 10^5$, it follows from Eq. (2) that linear amplification requires the difference voltage to be restricted to $|v_d| \lesssim 10^{-3}$ V = 1 mV. Correspondingly, the currents i_p and i_n will fall in the *nanoamp* region (10^{-9} A), or essentially zero for most purposes. We model this simply by letting $R_i' \to \infty$ in Fig. 4.3–2. On the output side of the circuit, R_o' is normally small enough compared to the load resistance that loading effects are negligible and we can assume $R_o' \to 0$.

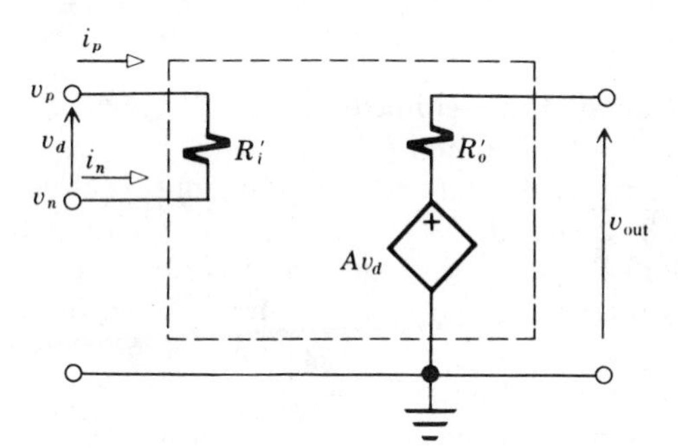

FIGURE 4.3–2
Equivalent circuit of an op-amp.

In view of the restriction on v_d, an op-amp alone has little merit as an amplifier, for it would be driven immediately into saturation by any input signal v_{in} with variations greater than about one millivolt. All linear op-amp circuits therefore have a *feedback* connection that forms a closed loop from the *output* terminal back to the *inverting* terminal such that $|v_d| \ll |v_{\text{in}}|$. This prevents saturation even though $|v_{\text{in}}| > v_{\max}$. The resulting *closed-loop gain* $|v_{\text{out}}/v_{\text{in}}|$ can still be large but is substantially less than the open-loop gain A.

With the feedback connection in place, the input of an op-amp acts as a *virtual short* in the sense that $v_d \approx 0$ (like a short circuit), but $i_p = i_n \approx 0$ (like an open circuit). The concept of a virtual short turns out to be very handy for analyzing and designing op-amp circuits. We formalize it by defining the *ideal op-amp model* shown in Fig. 4.3–3, where the double-headed arrow represents the virtual short. Unlike the model in Fig. 4.3–2, this model does not have an explicit relationship for v_{out} in terms of

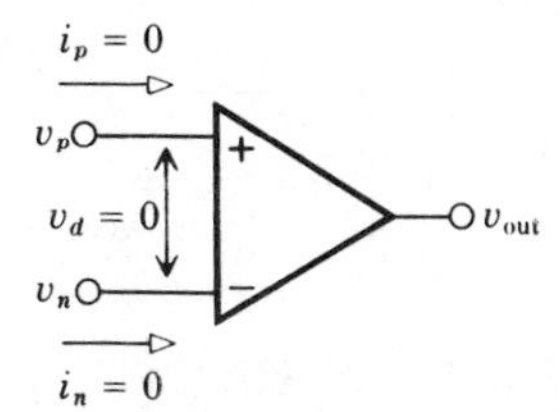

FIGURE 4.3–3
Ideal op-amp with a virtual short.

v_d. Instead, the value of v_{out} must be whatever is required to satisfy the virtual-short conditions $v_d \approx 0$ and $i_p = i_n \approx 0$, independently of the op-amp's gain. In effect, an ideal virtual short implies *infinite* op-amp gain.

Generally speaking, the ideal model yields reasonably accurate results when the closed-loop gain $|v_{out}/v_{in}|$ is small compared to the op-amp's actual open-loop gain A, and when the output voltage stays within the linear region $|v_{out}| < V_{CC}$, corresponding to $|v_d| < v_{max}$.

The following investigations of specific op-amp circuits should clarify the role of the feedback connection, and the use of the ideal model. A later chapter covers the general theory of feedback and its applications beyond op-amp circuits.

NONINVERTING AMPLIFIERS AND VOLTAGE FOLLOWERS

The amplifier circuit diagrammed in Fig. 4.3–4a has v_{in} applied directly to the noninverting terminal, so $v_p = v_{in}$ and $i_{in} = i_p \approx 0$. The feedback resistance R_F and an additional resistance R_1 form a voltage divider across the output such that

$$v_n = \frac{v_{out}}{K} \qquad K = \frac{R_F + R_1}{R_1}$$

with K being the reciprocal of the voltage-divider ratio. This assumes, of course, that $i_n \approx 0$ and $i_1 \approx i_F$.

For a qualitative understanding of this circuit we use the model from Fig. 4.3–2 with $R_i' \to \infty$ and $R_o' \to 0$, yielding the equivalent circuit of Fig. 4.3–4b. Suppose v_{in} goes positive by a few millivolts, causing $v_{out} = Av_d$ to increase towards saturation. But v_n increases in proportion with v_{out} and prevents v_d from increasing too much, since $v_d = v_{in} - v_n$. Thus, thanks to the feedback from v_{out} to v_n, we can end up with $v_d \ll v_{in}$ and linear amplification $v_{out} > v_{in}$. A similar argument applies when v_{in} goes negative.

For a quantitative analysis of Fig. 4.3–4b, we use $v_d = v_{out}/A$ and write

$$v_{in} = v_n + v_d = \frac{v_{out}}{K} + \frac{v_{out}}{A} = \frac{A + K}{KA} v_{out}$$

which gives the closed-loop gain as

$$\frac{v_{out}}{v_{in}} = \frac{KA}{A + K} \tag{3a}$$

Assuming that $A \gg K$, as is usually the case, then $A + K \approx A$ and

$$\frac{v_{out}}{v_{in}} \approx \frac{KA}{A} = K = \frac{R_F + R_1}{R_1} \tag{3b}$$

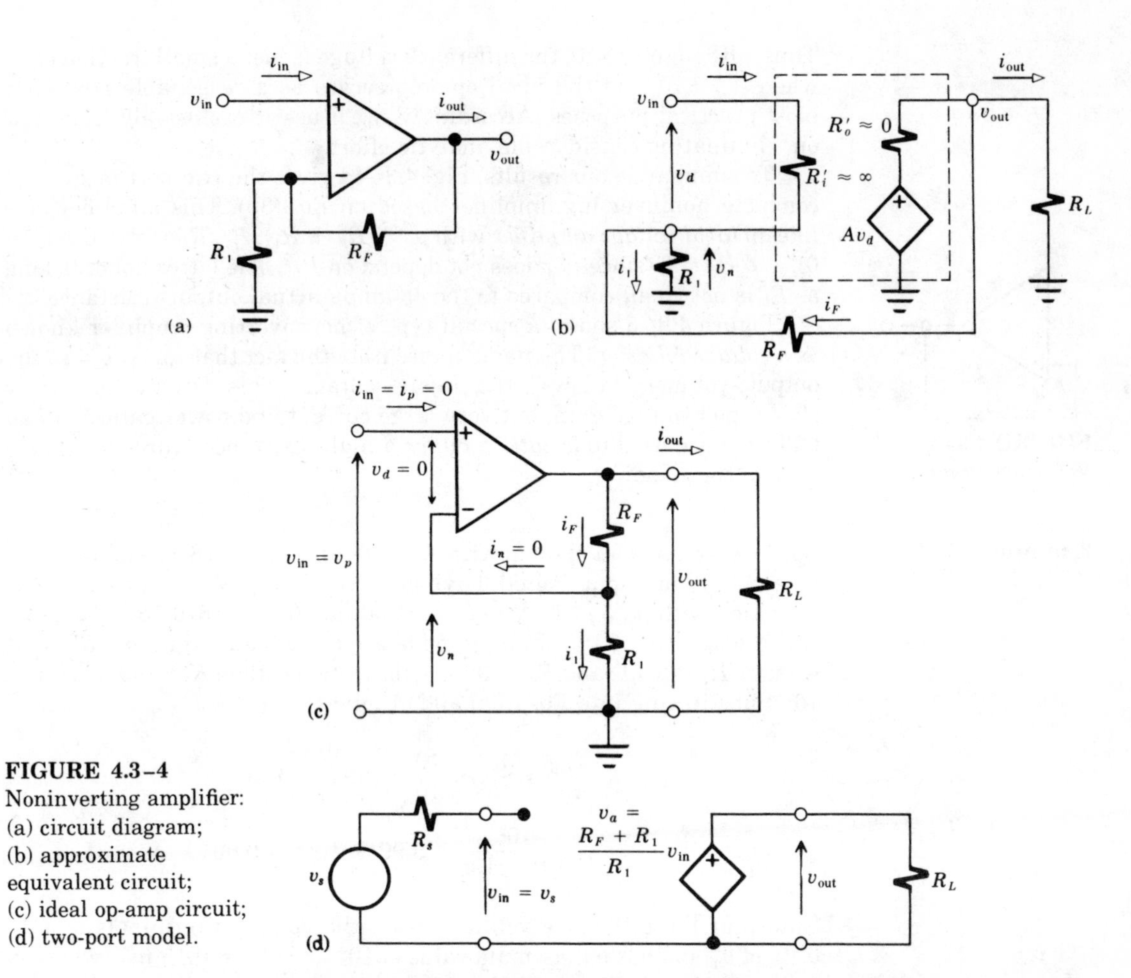

FIGURE 4.3–4
Noninverting amplifier:
(a) circuit diagram;
(b) approximate equivalent circuit;
(c) ideal op-amp circuit;
(d) two-port model.

Significantly, this expression depends on the external resistances, but not on the op-amp's gain A. One can therefore design a noninverting amplifier by simply picking appropriate values for R_F and R_1. (The value of R_F should be small compared to the actual value of R'_i to ensure that $i_F \gg i_n$.) Adjustable gain is easily obtained by using a potentiometer for R_F and R_1.

Now let us repeat the analysis using the *ideal* op-amp model of Fig. 4.3–4c. We still have $v_n = v_{out}/K$, but $v_d = v_{in} - v_n \approx 0$, so $v_{in} \approx v_n = v_{out}/K$ and $v_{out}/v_{in} \approx K$, which agrees with Eq. (3b). Of course v_d cannot be exactly zero (Why not?), and we must go back to Fig. 4.3–4b to find out what its precise value is. Specifically,

$$\frac{v_d}{v_{in}} = \frac{v_d}{v_{out}} \times \frac{v_{out}}{v_{in}} = \frac{1}{A} \times \frac{KA}{A + K} = \frac{K}{A + K} \approx \frac{K}{A} \tag{4}$$

Thus, although $v_d \neq 0$, the difference voltage is just a small fraction of v_{in} when $A \gg K$, and the ideal op-amp serves as a reasonable model for most practical purposes. Accordingly, we'll use it exclusively from now on, eliminating considerable analytic effort.

To summarize our results, Fig. 4.3–4d gives the two-port model of a complete noninverting amplifier based on Eq. (3b). This amplifier acts like an *ideal voltage amplifier* with $\mu \approx (R_F + R_1)/R_1$, $R_i = \infty$ (since $i_{in} \approx 0$), and $R_o = 0$ (since v_{out} does not depend on R_L). The latter holds as long as R_L is not small compared to the op-amp's actual output resistance R_o'.

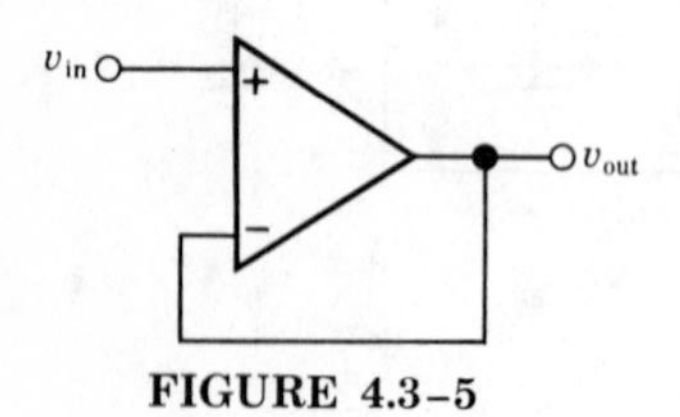

FIGURE 4.3–5
Voltage follower.

Figure 4.3–5 shows a special type of noninverting amplifier known as a *voltage follower*. The name comes from the fact that $v_{out} \approx v_{in}$, so the output voltage "follows" the input voltage. This circuit has unity closed-loop voltage gain, but very large current and power gain. Voltage followers are used to *isolate* or *buffer* a high-resistance source and minimize output loading.

Example 4.3–1

Suppose we have an op-amp with $A = 10^5$ and $V_{CC} = 18$ V, and we want to amplify an input signal having $|v_{in}(t)| \leq 0.15$ V. Ensuring linear operation with $|v_{out}| < 18$ V requires that $|v_{out}/v_{in}| < 18/0.15 = 120$, so a closed-loop gain of 100 would provide a reasonable margin for safety. If we take $R_1 = 1$ kΩ and $R_F = 99$ kΩ, for instance, then $K = (99 + 1)/1 = 10^2$. Substituting into Eqs. (3a) and (4) gives

$$\frac{v_{out}}{v_{in}} = \frac{10^5 \times 10^2}{10^5 + 10^2} = 99.9 \approx 100$$

$$\frac{v_d}{v_{in}} = \frac{10^2}{10^5 + 10^2} = 9.99 \times 10^{-4} \approx 0.001$$

Note that the difference voltage across the op-amp's terminals is only 0.1% of v_{in} and has a maximum value of $10^{-3} \times 0.15 = 0.15$ mV, whereas $|v_{out}|_{max} \approx 100 \times 0.15 = 15 \text{ V} = 10^5\, |v_d|_{max}$.

Exercise 4.3–1

Use the ideal op-amp model to show that a voltage follower has $v_{out}/v_{in} \approx 1$. Then obtain a more exact expression using the model of Fig. 4.3–2 with $R_i' = \infty$ and $R_o' = 0$. Evaluate your result when $A = 10^5$.

INVERTING AND SUMMING AMPLIFIERS

For the *inverting amplifier* diagrammed in Fig. 4.3–6a, we apply v_{in} through R_1 to the inverting terminal and ground the other terminal, making $v_p = 0$. The virtual short now serves as a "virtual ground" so that $v_n \approx 0$, while $i_n \approx 0$ as in Fig. 4.3–6b. Thus, $i_{in} = v_{in}/R_1$, $i_F = -i_{in}$, and

$$v_{out} = v_n + R_F i_F = 0 + R_F(-v_{in}/R_1) = -(R_F/R_1)v_{in}$$

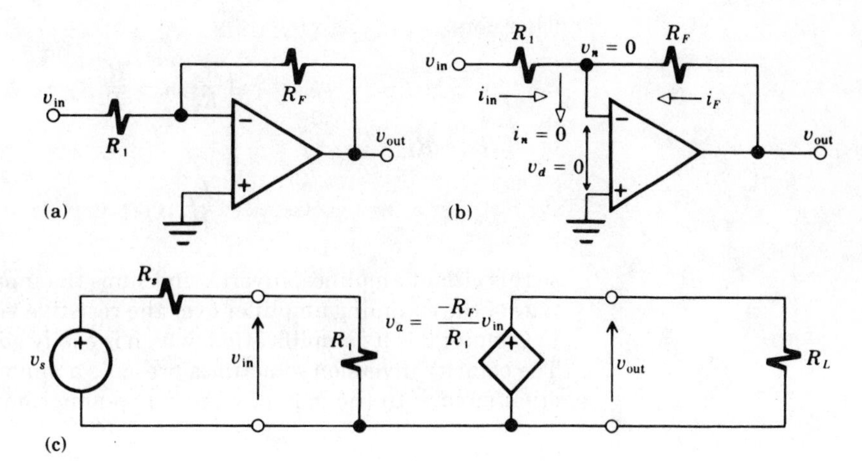

FIGURE 4.3–6
Inverting amplifier:
(a) circuit diagram;
(b) ideal op-amp circuit;
(c) two-port model.

This circuit therefore has a *negative* closed-loop gain

$$\frac{v_{out}}{v_{in}} = -\frac{R_F}{R_1} \tag{5}$$

meaning that v_{in} is enlarged by the factor R_F/R_1 and inverted in polarity at the output. If $R_1 = R_F$, v_{out} equals $-v_{in}$, and we have a unity-gain *inverter*.

Like the noninverting amplifier, the inverting amplifier has $R_o = 0$, but the input resistance is finite, namely

$$R_i = \frac{v_{in}}{i_{in}} = R_1$$

Accordingly, we must give attention to the matter of source loading, using the equivalent circuit diagrammed in Fig. 4.3–6c. If the input source has open-circuit voltage v_s and resistance R_s, then

$$A_v = \frac{v_{out}}{v_s} = \left(\frac{R_1}{R_s + R_1}\right)\left(-\frac{R_F}{R_1}\right) = -\frac{R_F}{R_s + R_1}$$

obtained from Eq. (5), Section 4.2. Whenever possible, we take $R_1 \gg R_s$ to minimize source loading, in which case $A_v \approx -R_F/R_1$.

Adding one or more additional input resistances, as we do in Fig. 4.3–7, changes the inverting amplifier into a *summing amplifier*. Specifically, $v_{out} = R_F i_F$, where

$$i_F = -(i_1 + i_2 + \cdots) = -\left(\frac{v_1}{R_1} + \frac{v_2}{R_2} + \cdots\right)$$

Therefore

$$v_{out} = -\left(\frac{R_F}{R_1} v_1 + \frac{R_F}{R_2} v_2 + \cdots\right) \tag{6a}$$

or, if $R_1 = R_2 = \cdots$,

$$v_{out} = -\frac{R_F}{R_1}(v_1 + v_2 + \cdots) \tag{6b}$$

so this circuit amplifies, inverts, and sums the input voltages. The advantage of the summing amplifier over the resistive voltage adder mentioned in Chapter 3 is its amplification, which is easily adjusted by changing R_F. The polarity inversion sometimes presents a minor annoyance, but can be eliminated with the help of a second op-amp connected as an inverter.

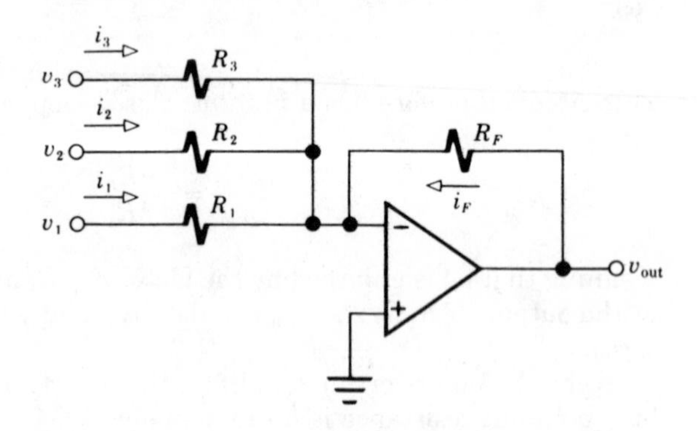

FIGURE 4.3–7 Summing amplifier with three inputs.

Example 4.3–2

Given two signal sources producing $|v_{s_1}(t)| \leq 30$ mV and $|v_{s_2}(t)| \leq 100$ mV, each having $R_s = 20\ \Omega$, we want to design an op-amp system that produces the composite signal $v_{out}(t) = 200\ v_{s_1}(t) - 40\ v_{s_2}(t)$. There are a number of different solutions to this problem, but two op-amps seem to be required in view of the subtraction rather than addition needed to form v_{out}. Figure 4.3–8 gives one possible design, with the first op-amp used as an inverting amplifier and the second as a summing amplifier with inversion. The resistance values were chosen by the following arguments.

Writing the final output as $v_{out} = -40\ [(-5\ v_{s_1}) + v_{s_2}]$ suggests that the inputs to the summing amplifier should be $-5\ v_{s_1}$ and v_{s_2} and that the amplifier should have $R_1 = R_2$ and $R_F/R_1 = 40$. The value of R_2 must be large compared to the 20-Ω source resistance to minimize loading, so we take $R_1 = R_2 = 1\ \text{k}\Omega \gg 20\ \Omega$ and $R_F = 40\ R_1 = 40\ \text{k}\Omega$. The input $-5v_{s_1}$ for the summing amplifier is obtained by applying v_{s_1} to an inverting amplifier with $R_F/R_1 = 5$ and $R_1 \gg R_s$. Again taking $R_1 = 1\ \text{k}\Omega$, we get $R_F = 5\ \text{k}\Omega$ for the inverting amplifier.

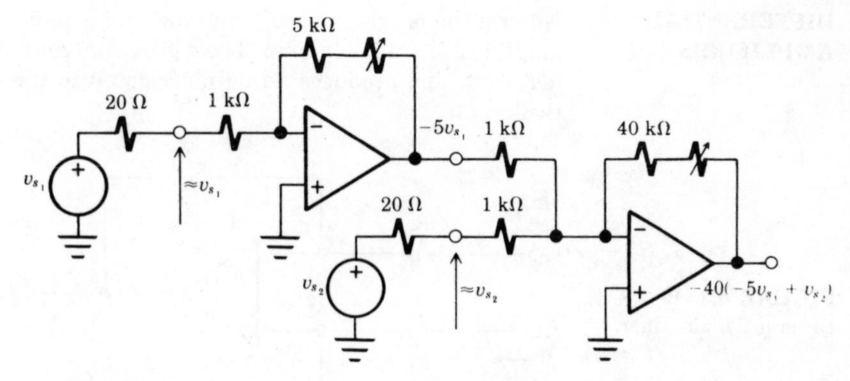

FIGURE 4.3–8 Circuit for Example 4.3–2.

Accounting for the small amount of input loading, the actual voltage gain of the inverting amplifier is $-R_F/(R_s + R_1) = -4.902$. This can be trimmed to the desired value, if necessary, by adding a potentiometer to slightly increase R_F. A trimming "pot" is also shown in the feedback path of the summing amplifier for the same purpose.

A final consideration is the maximum output voltage, which we compute by assuming *worst-case conditions;* that is, v_{s_1} hits its negative maximum at the same time v_{s_2} is at its positive maximum, or vice versa. Then

$$\begin{aligned} |v_{out}|_{max} &= 200\ |v_1|_{max} + 40\ |v_2|_{max} \\ &= 200 \times 30\text{ mV} + 40 \times 100\text{ mV} = 10\text{ V} \end{aligned}$$

Therefore, the second op-amp must have $V_{CC} > 10$ V to ensure operation within the linear range. (What is the equivalent requirement on the first op-amp?)

Exercise 4.3–2 Design a circuit using the op-amp in Example 4.3–1 to produce $v_{out} = -K(3v_{s_1} + v_{s_2})$ with K as large as practical. The signal sources have $|v_{s_1}(t)| \leq 0.1$ V, $R_{s_1} = 200\ \Omega$, $|v_{s_2}(t)| \leq 0.5$ V, and $R_{s_2} = 80\ \Omega$.

Exercise 4.3–3 Show that Fig. 4.3–9 is an inverting *current amplifier* with $A_i = i_{out}/i_s = -(R_F + R_1)/R_1$ independent of R_s and R_L assuming an ideal op-amp.

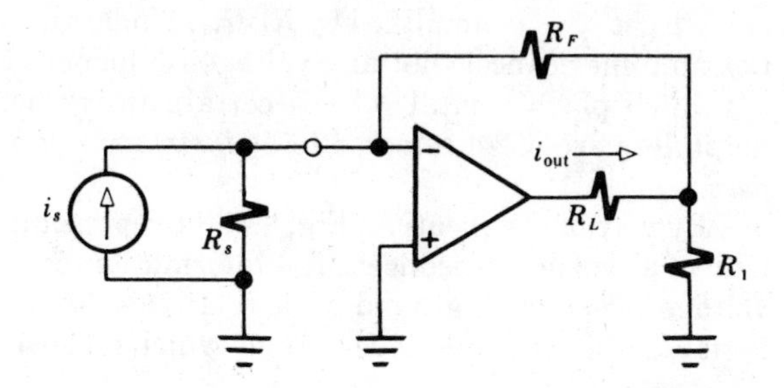

FIGURE 4.3–9 Inverting current amplifier.

DIFFERENTIAL AMPLIFIERS †

None of the previous circuits has directly exploited the difference-voltage amplification of an op-amp. The *differential amplifier* in Fig. 4.3–10 does just that, and produces an output related to the difference between the two inputs.

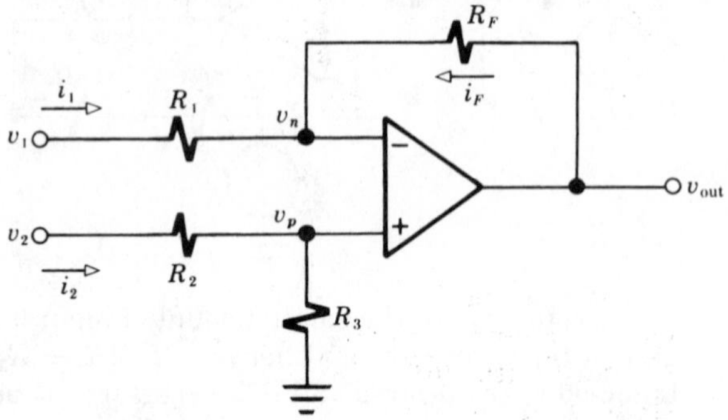

FIGURE 4.3–10
Differential amplifier.

Observing that the upper and lower portions look like inverting and noninverting circuits, respectively, you can quickly confirm that

$$v_{out} = v_n + R_F i_F$$

where

$$v_n = v_p = \frac{R_3}{R_2 + R_3} v_2 \qquad -i_F = i_1 = \frac{v_1 - v_n}{R_1}$$

Algebraic manipulation of these expressions leads to

$$v_{out} = \frac{R_F}{R_1}\left[\frac{1 + (R_1/R_F)}{1 + (R_2/R_3)} v_2 - v_1\right] \tag{7a}$$

If $R_2/R_3 = R_1/R_F$, then

$$v_{out} = \frac{R_F}{R_1}(v_2 - v_1) \tag{7b}$$

and we get $v_2 - v_1$ amplified by R_F/R_1. Under this condition, the differential amplifier cancels out any voltage component common to both v_1 and v_2, which plays a vital role in certain instrumentation systems. Such amplifiers are often referred to as *instrumentation* or *transducer amplifiers*.

By way of clarification, Fig. 4.3–11 represents the voltage characteristic of a typical transducer. The information-bearing signal v_s is a small voltage offset from ground by a relatively large contaminating component v_{cm}, the *common-mode voltage,* which is caused by spurious noise, AC "hum," and so forth. Thus, with respect to ground,

$$v_1 = v_{cm} - (1 - \alpha)v_s \qquad v_2 = v_{cm} + \alpha v_s$$

where $-1 < \alpha < 1$. If we attempt to amplify either v_1 or v_2, the desired signal v_s will be obliterated by the common-mode voltage v_{cm}. But applying these voltages to a differential amplifier with $R_2/R_3 = R_1/R_F$ produces

$$v_{out} = (R_F/R_1)\,[v_{cm} + \alpha v_s - v_{cm} + (1 - \alpha)\,v_s] = (R_F/R_1)\,v_s$$

so the differential amplifier rejects the common-mode voltage and amplifies the desired difference voltage.

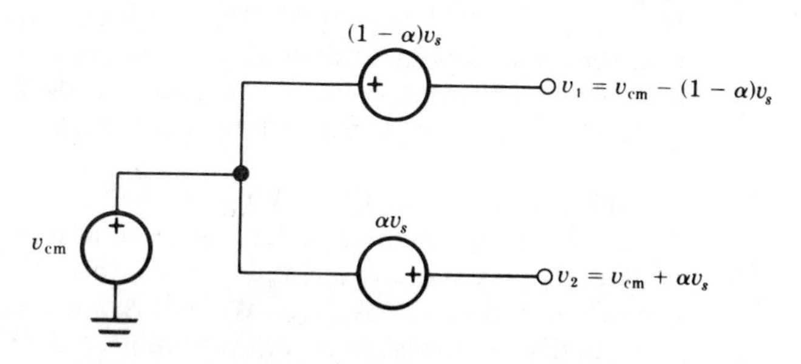

FIGURE 4.3-11 A source with common-mode voltage.

Successful common-mode rejection requires close equality of the resistance ratios, so any gain adjustment involves the tricky business of tracking R_3 along with R_F. More sophisticated differential-amplifier circuits eliminate this problem.

REAL OP-AMP †

Practical op-amps differ from our ideal model in several respects other than finite gain. Although the differences usually are inconsequential, we need to be aware of their possible effects in special applications. The following summary is provided as a guide to interpreting op-amp specification sheets and assessing the need for higher-quality (but more expensive) op-amps in given applications.

Offset voltage The output of a practical op-amp includes a small voltage offset proportional to the closed-loop gain but independent of the input voltage. This is specified in terms of the maximum equivalent *input offset voltage* v_{os}. For example, if $v_{os} = 2$ mV and the circuit has a closed-loop gain of 50, then v_{out} will be offset by an amount not exceeding ± 100 mV (see Problem 4.3-13). Additional external terminals are often provided for the purpose of "nulling out" the offset, but offset *drift* due to temperature variations prevents a perfect null.

Bias and offset currents The internal electronics of an op-amp requires small *bias currents* i_{bp} and i_{bn} that add to i_p and i_n and, consequently, produce unwanted voltages when they flow through external resistance such as R_F. One can easily null out the effect when $i_{bp} = i_{bn}$ (see Problem 4.3–14). But *any* difference $|i_{bp} - i_{bn}|$—called the *input offset current*—creates an additional drifting offset voltage. Since bias and offset currents are typically less than 1 μA, the simplest cure is to keep the external resistances small enough to ignore voltage drops produced by these currents.

Slew rate and frequency response A practical op-amp cannot respond instantaneously to sudden changes of input voltage. The maximum rate of output-voltage change is the op-amp's *slew rate;* for example, an op-amp with a slew rate of 1 V/μs requires at least 20 μs to go from $v_{out} = -10$ V to $+10$ V. This built-in "sluggishness" also means that the op-amp cannot follow sinusoidal voltage variations beyond some *upper frequency limit.* (We defer further considerations of dynamic behavior and frequency response to later chapters.)

Common-mode rejection When $v_p = v_n$, an ideal op-amp has $v_{out} = A(v_p - v_n) = 0$, meaning perfect rejection of any common-mode voltage. However, applying $v_p = v_{cm}$ and $v_n = v_{cm}$ to a practical op-amp results in a small output $v_{ocm} = A_{cm} v_{cm}$. We call A_{cm} the open-loop *common-mode gain,* as distinguished from differential gain A. The *common-mode rejection ratio* (CMRR) expressed in decibels is, then, defined as

$$\text{CMRR} \triangleq 10 \log (A/A_{cm})^2 \tag{8}$$

Op-amps designed for use in differential amplifiers have large values of CMRR, typically 100 dB, to minimize unwanted common-mode output.

Input and output resistance The large but finite input resistance R_i' of a practical op-amp voids our prior assumption that $i_p = i_n = 0$. Likewise, the small but nonzero output resistance R_o' causes v_{out} to depend somewhat on load. As a conservative rule, the corresponding effects are negligible if $R_i' \gg R_s$ and $R_o' \ll R_L$. When source and output loading are critical concerns, we must investigate particular circuits using standard analysis techniques (see, for example, Problem 5.4–8).

PROBLEMS

4.1–1 Repeat the calculations of Example 4.1–1 with $R_s = 8\ \Omega$ and $R_L = 4\ \Omega$.

4.1–2 Show that $R_a = 0$ and $R_b = \infty$ in Example 4.1–1 if $R_s = R_L$.

4.1–3 Design a two-port similar to that of Fig. 4.1–3 to match a 50-Ω source with a 75-Ω load. Calculate the corresponding value of μ in Fig. 4.1–1d.

4.1–4 The two-port in Fig. P4.1–4 has a voltage source v_c controlled by the input voltage v_1. Obtain an expression for $R_i = v_1/i_1$ when a load resistance R_L is connected across the output terminals. Then show that this circuit acts as a *negative resistance* $R_i \approx -R_L$ if $k = 2$ and $R_L \gg R$.

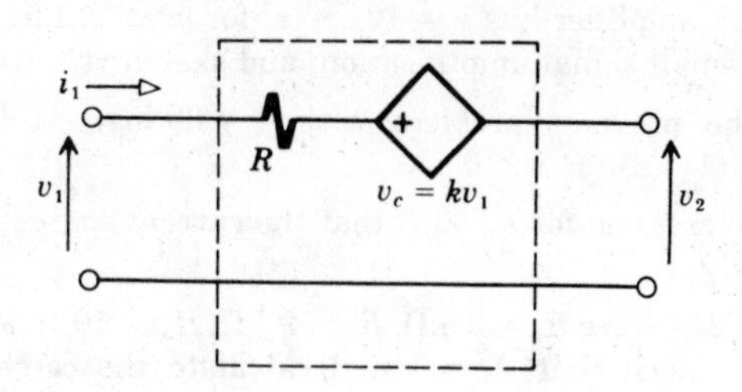

FIGURE P4.1–4

4.1–5 Construct an output model like that of Fig. 4.1–1c for Fig. P4.1–4 with a voltage source v_s and resistance R_s connected to the input.

4.1–6 Find the rms values of all voltages and currents and the average load power in Fig. 4.1–6a when $V_s = 120$ V (rms), $R_s = 1\ \Omega$, $R_L = 36\ \Omega$, and $N = 3$.

4.1–7 Repeat the previous calculations when $R_L = 0.2\ \Omega$, $N = \frac{1}{10}$, and the primary is driven by an AC current source with $I_s = 3$ A (rms) and $R_s = 40\ \Omega$.

4.1–8 Find all the voltages and currents in Fig. P4.1–8 when $R = 20\ \Omega$, $R_L = 2.2\ \Omega$, $N = 10$, and v_s is a time-varying voltage whose instantaneous value equals 120 V. Hint: First find i_2 using appropriate equivalent circuits on each side of R.

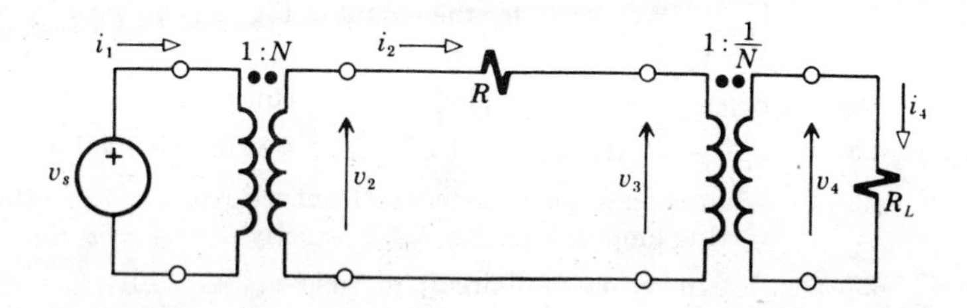

FIGURE P4.1–8

4.1–9 The middle section of Fig. P4.1–8 represents a power transmission line with fixed resistance R. The transformers are for the purpose of improving the efficiency of power transfer to R_L.

(a) Derive an expression for v_4 and show that $v_4 \approx v_s$ if $N \gg \sqrt{R_L/R}$.

(b) Let v_s and $p = v_s i_1$ be fixed quantities. Obtain an expression for Eff, similar to that in Prob. 3.2–7 and find the condition on N such that Eff $\approx 100\%$.

4.1–10 Find the h parameters of the two-port in Fig. 5.4–6a when the three elements are equal resistances R.

4.1–11 Repeat the previous problem for Fig. 5.4–6b.

4.1–12 Find the h parameters for the two-port in Fig. P4.1–4.

4.1–13 Obtain expressions for the Thévenin and Norton parameters as seen looking back from R_L in Fig. 4.1–10.

4.2–1 Suppose an amplifier has $y = 12x - x^3$ for $|x| \leq 2$. Plot the transfer curve; estimate the small-signal amplification; and sketch $y(t)$, when $x(t) = 2 \sin \omega t$.

4.2–2 Repeat the previous problem with $y = 30 \log_{10} (1 + x)$ for $x \geq 0$ and $y = -30 \log_{10} (1 - x)$ for $x < 0$.

4.2–3 Find an expression for g_m such that the current source in Fig. 4.2–6b can be labeled $i_a = g_m v_{\text{in}}$.

4.2–4 Let Fig. 4.2–7 have $R_s = 1$ kΩ, $R_i = 9$ kΩ, $R_o = 10$ Ω, and $R_L = 100$ Ω. Find the value of μ such that $A_v = 1$ and calculate the corresponding value of $A_i = i_{\text{out}}/(v_s/R_s)$.

4.2–5 Obtain a general expression for the power gain of Fig. 4.2–7 and show that $G = \mu^2 R_s/4R_L$ when resistances are matched at input and output.

4.2–6 Calculate A_v and G in dB when a source with $R_s = 1$ kΩ is connected to an amplifier having $R_i = 9$ kΩ, $\beta = 50$, $R_o = 4$ kΩ, and $R_L = 1$ kΩ.

4.2–7 Find G in dB when the amplifier in Problem 4.2–4 has $A_v = 1$.

4.2–8 An amplifier with $R_i = 1$ kΩ has $G = 53$ dB when connected to a 200-Ω load. What are the values of P_{out} and v_{out} if $v_{\text{in}} = 50$ mV?

4.2–9 Consider a cascade of two identical amplifiers like that in Fig. 4.2–6b with $R_i = R_L \ll R_o = R_s$. Obtain expressions for A_i and G.

4.2–10 Referring to Fig. 4.2–10, let $\mu_1 = \mu_2 = -20$, $R_s = R_{i1} = R_{i2} = 5$ kΩ, and $R_{o1} = R_{o2} = 200$ Ω. Find R_L such that $A_v = 100$, and calculate G in dB.

4.3–1 Plot v_{out} versus v_{in} for the circuit of Fig. 4.3–4a with $R_F = 4$ kΩ, $R_1 = 1$ kΩ, and $V_{CC} = 10$ V.

4.3–2 Repeat the previous plot for the circuit of Fig. 4.3–6a.

4.3–3 If $R_s = 50$ Ω and $R_1 = 1$ kΩ in Fig. 4.3–6c, what value of R_F yields $v_{\text{out}} = -10v_s$?

4.3–4 What should the value of the feedback resistor be, so that the output of the inverting amplifier in Fig. 4.3–8 exactly equals $-5v_{s_1}$?

4.3–5 Design an op-amp circuit to yield $v_{\text{out}} = -(10v_1 + v_2 + 0.1v_3)$ when the three input sources have $R_s = 50$ Ω.

4.3–6 Repeat the previous design for $v_{\text{out}} = +(2v_1 + v_2 + 0.5v_3)$.

4.3–7 Figure P4.3–7 is a *voltage-to-current converter* and R_m represents the resistance of an ammeter. Show that i_F in milliamps equals v_{in} in volts, and find v_{out} in terms of v_{in}.

4.3–8 Given a current source i_s with resistance R_s, use an ideal op-amp to build a *current-to-voltage converter* such that v_{out} in volts equals $-i_s$ in microamps.

4.3–9 A resistance R_u is embedded in a network containing no sources. Devise a scheme to measure R_u using an ideal op-amp, voltmeter, reference voltage V_{ref}, and reference resistance R_{ref}.

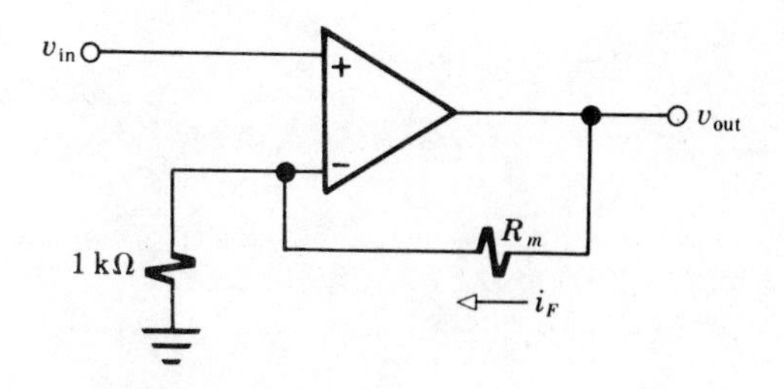

FIGURE P4.3–7

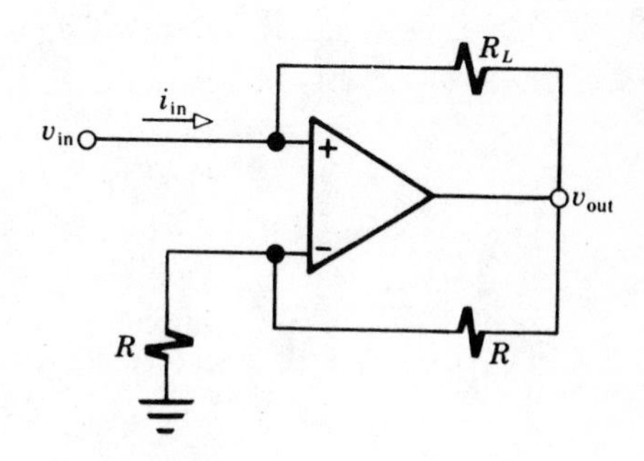

FIGURE P4.3–10

4.3–10 Figure P4.3–10 is a *negative-resistance converter*. Show that $v_{in}/i_{in} = -R_L$.

4.3–11 Investigate the effect of finite op-amp gain on an inverting amplifier, by replacing the ideal op-amp in Fig. 4.3–6b with the model of Fig. 4.3–2 (with $R_i' = \infty$ and $R_o' = 0$). Now show that $v_{out}/v_{in} = -AR_F/[(A + 1)R_1 + R_F]$ and evaluate v_{out}/v_{in} when $A = 10^5$ and $R_F = 10\ R_1$.

4.3–12 Suppose Fig. 4.3–10 has $R_1/R_F = 0.1 = 0.9\ R_2/R_3$. Find v_{out} when the inputs come from Fig. 4.3–11 with $\alpha = \frac{1}{2}$.

4.3–13 When we account for the finite difference gain A and the common-mode gain A_{cm}, the output of a real op-amp is $v_{out} = A(v_p - v_n) + A_{cm}(v_p + v_n)/2$, where $A \gg A_{cm} > 1$. Show that Fig. 4.3–10 then has $v_{out} \approx (v_2 - v_1) + (A_{cm}/2A)(v_2 + v_1)$ when $R_F = R_1 = R_2 = R_3$. Hint: First show that $v_n = (v_1 + v_{out})/2$.

4.3–14 Figure P4.3–14 is the model of a real op-amp with offset voltage v_{os} and bias currents i_{bn} and i_{bp}. Assume that $i_{bn} = i_{bp} = i_b$, and show that an inverting or noninverting amplifier has $v_{out} = (1 + R_F/R_1)v_{os} + R_F i_b$ when $v_{in} = 0$.

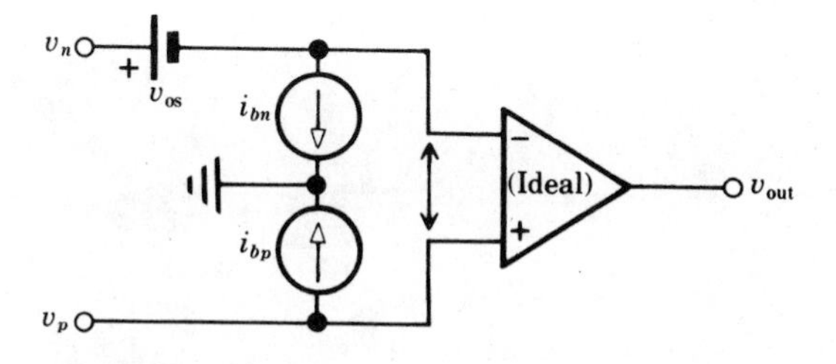

FIGURE P4.3–14

4.3–15 Practical op-amp circuits usually include a resistance R_p in series with the noninverting terminal to compensate for the effects of the bias currents. Using Fig. P4.3–14 with $v_{os} = 0$, obtain an expression for v_{out} when an inverting or noninverting amplifier includes R_p and has $v_{in} = 0$. Then show that $v_{out} = R_F(i_{bn} - i_{bp})$ if $R_p = R_1 \| R_F$.

5

CAPACITANCE, INDUCTANCE, AND IMPEDANCE

C*apacitance* and *inductance,* the new circuit elements introduced in this chapter, have the distinctive ability to absorb energy from a circuit (acting somewhat like a sink), store it temporarily, and return the stored energy to the circuit (acting somewhat like a source). Circuits containing these energy-storage elements have electrical *memory,* in the sense that energy stored at an earlier time may contribute to the present value of a response. Consequently, some significant differences exist between the behavior of these circuits and "memory-less" resistive circuits—especially when time-varying sources are involved.

We will begin with the defining relationships and properties of capacitance and inductance. Then we develop the important concept of *impedance* that allows us to treat energy-storage elements in a fashion similar to the way we treat resistance. An optional section describes systematic methods of network analysis using impedance in loop and node equations.

OBJECTIVES

After studying this chapter and working the exercises, you should be able to do each of the following:

- Write expressions for the voltage-current relationships and for energy stored by capacitance and inductance (Sections 5.1 and 5.2).

- Calculate equivalent capacitance and inductance (Sections 5.1 and 5.2).
- Analyze a circuit with stored energy, assuming DC steady-state conditions (Sections 5.1 and 5.2).
- Write the impedance expressions for the individual circuit elements, and state the conditions under which the impedance concept is applicable (Section 5.3).
- Find the equivalent impedance or admittance of a simple one-port network, and the transfer function of a simple two-port (Section 5.3).
- Obtain the node or loop equations for a network containing resistance, energy-storage elements, and a controlled source (Section 5.4). †

5.1 CAPACITANCE

Capacitance is a circuit element that stores energy in an electric field, but does not dissipate it. Its characteristics are presented here as preparation for our later study of circuits with energy storage.

CAPACITORS AND CAPACITANCE

A capacitor typically consists of two conducting surfaces or plates separated by a dielectric insulation, as in Fig. 5.1–1a, that permits the storage of energy in an electric field between the plates. The dielectric prevents current flow when the applied voltage is *constant* (DC), but a *time-varying* voltage produces a current proportional to the rate of voltage change, namely

$$i = C\frac{dv}{dt} \tag{1}$$

The proportionality constant C is the *capacitance* measured in *farads* (F), having the unit equation

$$1\text{ F} = \frac{1\text{ A}}{1\text{ V/s}} = 1\text{ A}\cdot\text{s/V}$$

since the units of dv/dt are volts per second (V/s). Figure 5.1–1b gives the circuit symbol for capacitance.

Actually, the farad turns out to be a huge quantity, and practical capacitors have values more conveniently expressed in *microfarads* or

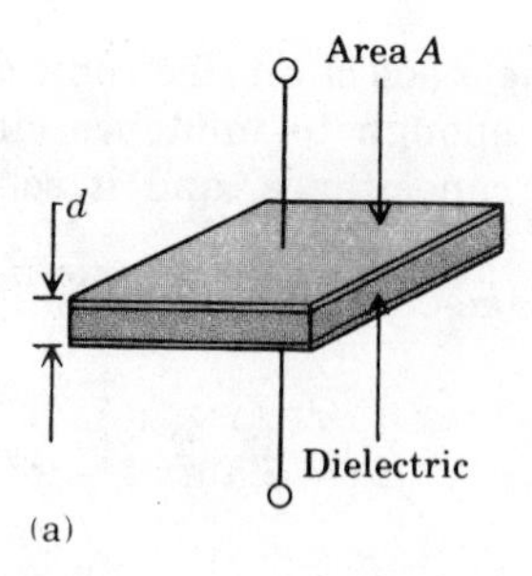

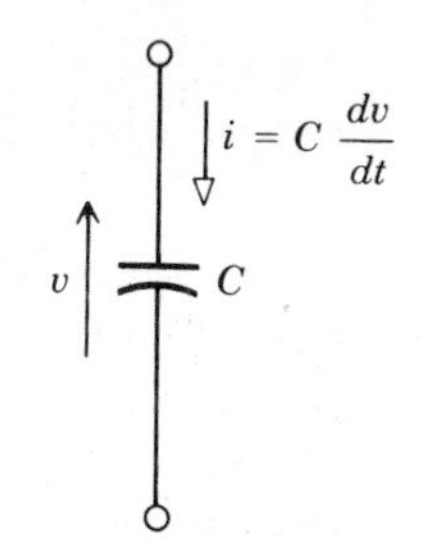

FIGURE 5.1–1
(a) A capacitor and
(b) its circuit symbol.

even *picofarads* ($1\ \mu F = 10^{-6}$ F and 1 pF $= 10^{12}$ F). For example, the capacitance of the parallel-plate structure in Fig. 5.1–1a is

$$C = \frac{\epsilon A}{d} \tag{2}$$

where A is the area of the plates, d their separation, and ϵ the *permittivity* constant of the dielectric. A typical dielectric such as mica has $\epsilon \approx 5 \times 10^{-11}$ F/m, so one-meter by one-meter plates, with one millimeter spacing, corresponds to $C \approx 5 \times 10^{-11} \times 1/10^{-3} = 5 \times 10^{-8}$, or only 0.05 μF.

Commonly available capacitors range from a few picofarads to a few thousand microfarads. The larger values are achieved by rolling layers of aluminum foil into a tubular structure. Such capacitors often have relatively low voltage ratings. *Electrolytic* capacitors are further restricted to voltages of one polarity. Variable capacitors are easily built, by using movable metal plates with air as the dielectric, but their capacitance values are relatively small due to the wider spacing and the lower permittivity $\epsilon_0 = 10^{-9}/36\pi$ F/m. Capacitance also occurs naturally between any two conducting surfaces. When conductors come in close proximity, like wires in a cable or the leads of an electronic device, the resulting capacitance may be large enough to influence circuit behavior. This is called *stray* or *parasitic* capacitance, and it sometimes causes significant problems.

To obtain the voltage-current relationship for capacitance, we rewrite Eq. (1) as

$$dv = \frac{1}{C}\, i\, dt$$

and integrate both sides from $t = -\infty$ to an arbitrary time instant t. This yields

$$v(t) = \frac{1}{C}\int_{-\infty}^{t} i\, dt \tag{3}$$

where the notation $v(t)$ emphasizes that v is a function of time t. Thus, the voltage across a capacitor depends on the entire past history of the current through it, from $t = -\infty$ to the present; in other words, a capacitor "remembers" past values of current. This *memory* capability of capacitance is directly related to its stored energy, as we'll soon see.

A more useful expression for v is obtained by breaking the integral in Eq. (3) into two parts, giving

$$v(t) = V_0 + \frac{1}{C}\int_{0}^{t} i\, dt \qquad t \geq 0 \tag{4a}$$

where $V_0 = \int_{-\infty}^{0} i\, dt$ is the *initial voltage* at $t = 0$, and the second term represents subsequent current flow. For instance, if a constant current $i = I$ is applied at $t = 0$, then

$$v(t) = V_0 + \frac{I}{C}\, t \qquad t \geq 0 \tag{4b}$$

which means that a constant current yields a linearly increasing voltage or *ramp* waveform, like Fig. 5.1–2. If the current stops at time $t = T$, as it does in the figure, the voltage stops increasing but stays constant at $v(t) = V_0 + IT/C$ for $t \geq T$—even though $i = 0$ for $t > T$.

Example 5.1–1

Various *waveform generators* take advantage of the voltage ramp produced across a capacitor by a constant current. As an example of this, the *square-wave* current $i(t)$ plotted in Fig. 5.1–3a generates a *triangle-wave* voltage $v(t)$, Fig. 5.1–3b. We analyze such waveforms in a piecewise fashion, separately considering each successive time interval defined by the points at which $i(t)$ changes value.

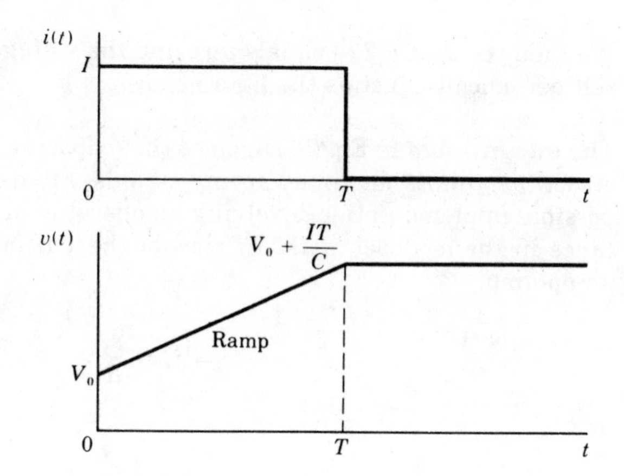

FIGURE 5.1-2
Constant current and voltage ramp.

Specifically, if $V_0 = 0$ then $v(t) = It/C$ over the interval $0 \leq t \leq T$. Therefore, at time $t = T$, $v(T)$ equals IT/C. In the next interval, $i(t) = -I$ so $v(t)$ must decrease linearly starting from IT/C. The formal mathematical expression is

$$v(t) = \frac{IT}{C} - \frac{I}{C}(t - T) \qquad T \leq t \leq 2T$$

(You can check this expression by confirming that it has the correct starting value, $v(T) = IT/C$, and the proper slope, $dv/dt = i/C = -I/C$.)

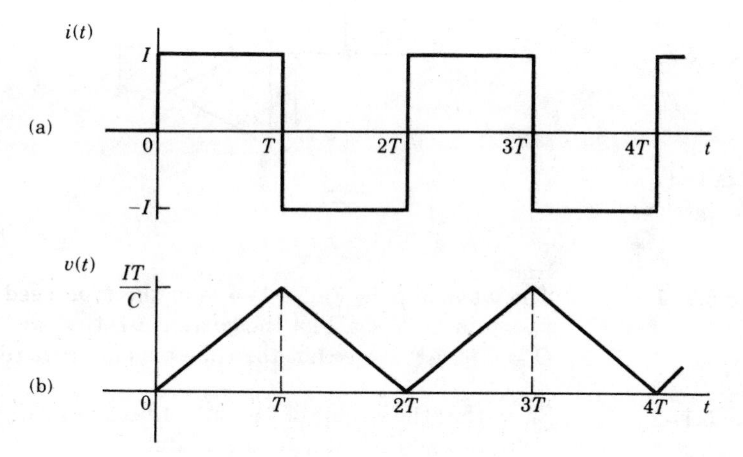

FIGURE 5.1-3
Waveforms for Example 5.1-1.

We then see that $v(2T)$ equals zero and the voltage waveform repeats itself periodically, just as the figure shows.

Example 5.1–2
Op-amp integrator

The integral sign in Eq. (3) suggests that capacitors might be used to construct *integrators* for time-varying signals. Figure 5.1–4 illustrates one possible implementation involving an operational amplifier with capacitance in the feedback path. In view of the virtual short at the input of an op-amp,

$$i_C = i_R = \frac{v_1}{R}$$

and

$$v_2 = -v_C = -\frac{1}{C}\int_{-\infty}^{t} i_C \, dt$$

Substituting for i_C then yields

$$v_2 = -\frac{1}{RC}\int_{-\infty}^{t} v_1 \, dt$$

so the output voltage v_2 is proportional to the integral of the input v_1. This op-amp integrator has applications in signal-processing systems and as the building block of *analog computers*.

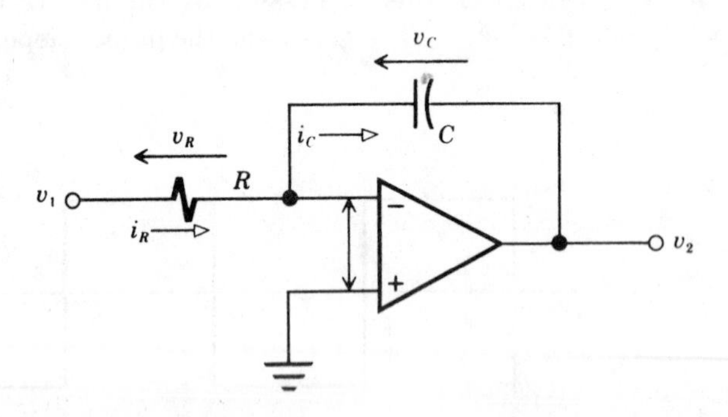

FIGURE 5.1–4
Op-amp integrator.

Exercise 5.1–1

The waveform in Fig. 5.1–5 is of the type used to drive the horizontal sweep in a TV set. Plot the current $i(t)$ that will produce this $v(t)$ across $C = 0.02\ \mu$F. Note that the time axis is in *microseconds* (μs).

Exercise 5.1–2

Show that the circuit of Fig. 5.1–4 becomes a *differentiator* when the resistance and capacitance are interchanged.

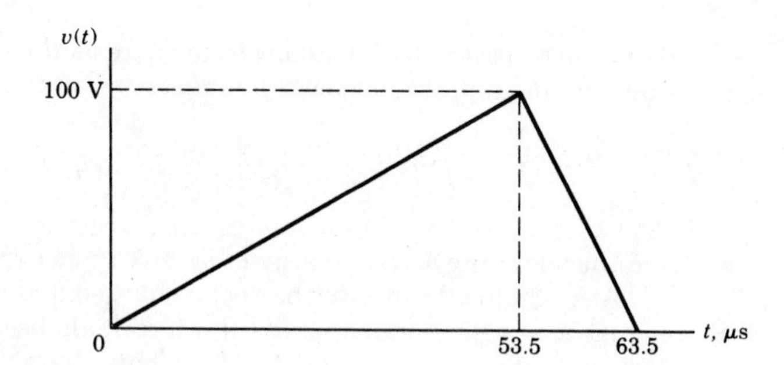

FIGURE 5.1-5
TV sweep waveform.

CHARGE AND ENERGY STORAGE

We can further interpret the capacitive memory effect by recalling that $i = dq/dt$, so the charge q stored by a capacitor at any instant of time is

$$q = \int_{-\infty}^{t} i\, dt$$

Thus, rewriting Eq. (3) in terms of q we have

$$q = Cv \tag{5}$$

which tells us that a one-farad capacitor stores one coulomb of charge per volt. It also tells us that the initial voltage V_0 in Eq. (4) corresponds to an initial charge $q_0 = CV_0$. Therefore, capacitive memory takes the form of *stored charge.*

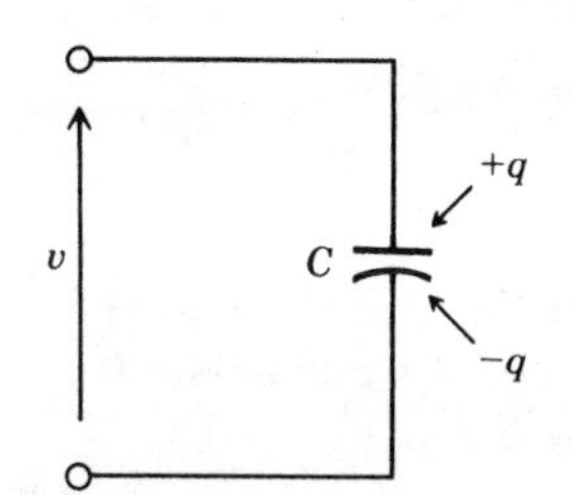

FIGURE 5.1-6
Charge storage in a capacitor.

Now suppose a source is momentarily applied across a capacitor. This produces a charge $+q = Cv$ on the positive terminal plate and an equal but opposite induced charge $-q$ on the other plate (see Fig. 5.1-6). The *work* done by the source to charge the capacitor is

$$w = \frac{1}{2C} q^2 = \tfrac{1}{2} Cv^2 \tag{6}$$

This follows from Eq. (5), the definition $v = dw/dq$, and the fact that $q\, dq = d(\tfrac{1}{2}q^2)$. Thus,

$$dw = v\, dq = \frac{q}{C}\, dq = \frac{1}{C}\, d(\tfrac{1}{2}q^2)$$

so $w = q^2/2C$ and letting $q = Cv$ gives the second expression. The work done by the source also must equal the *energy stored* by the capacitor, for there are no losses here. To show that Eq. (6) *is* the stored energy, we consider the instantaneous power $p = dw/dt = vi$. Since $i = C\,dv/dt$,

$$dw = p\,dt = vi\,dt = vC\frac{dv}{dt}\,dt = Cv\,dv = C\,d(\tfrac{1}{2}v^2)$$

and therefore the stored energy is $w = \frac{1}{2}Cv^2$ as expected.

After an ideal capacitor has been charged and the source removed, a voltage $v = q/C$ remains across the terminals because there is no conducting path for the charge to follow. This charge is stored indefinitely until an energy-consuming device is connected across the terminals. Capacitors are therefore useful in applications such as spot welders, electronic flash lamps, and pulsed lasers requiring large bursts of energy that can be built-up and stored during the relatively long period between bursts.

Ideal capacitors are said to be *lossless* in the sense that there is no internal resistance present to provide a conduction path between the plates. Real capacitors, however, always have some *leakage* that acts like a large parallel resistance through which the capacitor gradually discharges itself. Discharge times for a good quality capacitor may be hours or days.

Example 5.1–3

The voltage and current waveforms illustrated in Example 5.1–1 have been repeated in Fig. 5.1–7, along with the stored energy $w(t) = \frac{1}{2}Cv^2(t)$ and instantaneous power $p(t) = v(t)i(t)$. Observe that $w(t)$ decreases when $p(t) < 0$, meaning that the capacitance is discharging and returning stored energy to the current source. Also note that the average value of $p(t)$ will be zero, reflecting the fact that capacitance does not dissipate power. The waveform for the stored charge $q(t) = Cv(t)$ would have the same shape as the voltage waveform.

Exercise 5.1–3

The voltage across a 500-μF capacitor is $v(t) = 80 \sin 100t$ V. Show that $i(t) = 4 \cos 100t$ A, and sketch the waveforms of $w(t)$, $v(t)$, and $i(t)$ for $0 \leq t \leq 2\pi/100$. (Save your results for use in Exercise 5.2–1.)

EQUIVALENT CAPACITANCE

Although $i = C\,dv/dt$ contains the derivative of voltage and looks quite different from Ohm's law, it is nonetheless a *linear* *i-v* relationship. Proof of linearity follows simply by letting $v = v' + v''$; then

$$i = C\frac{d(v' + v'')}{dt} = C\left(\frac{dv'}{dt} + \frac{dv''}{dt}\right) = C\frac{dv'}{dt} + C\frac{dv''}{dt} = i' + i''$$

Therefore, superposition applies to networks involving capacitors.

A direct consequence of linearity is that two capacitors connected in series or parallel have the same effect as one *equivalent capacitance*,

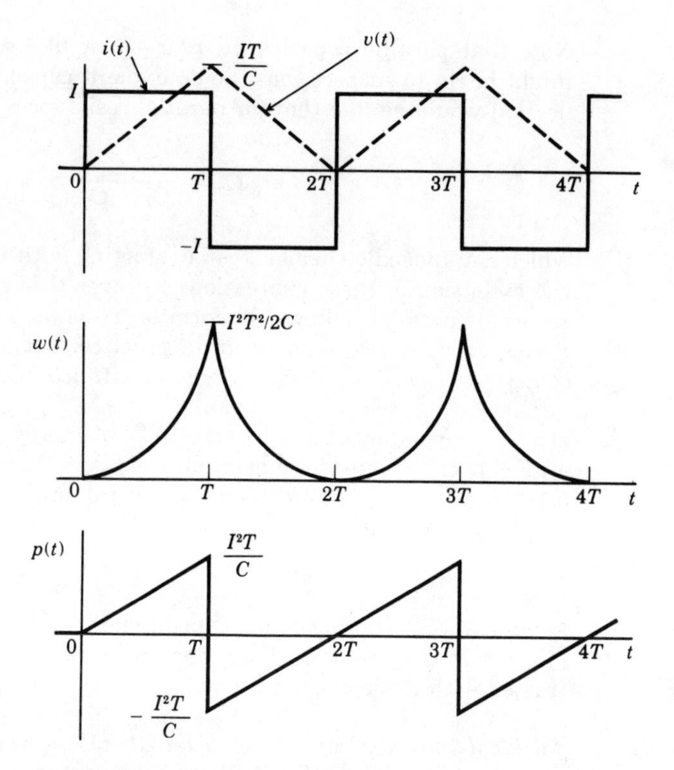

FIGURE 5.1–7
Waveforms for Example 5.1–3.

much like the concept of equivalent resistance. For instance, referring to the parallel connection in Fig. 5.1–8a, we see that

$$i = i_1 + i_2 = C_1 \frac{dv}{dt} + C_2 \frac{dv}{dt} = (C_1 + C_2) \frac{dv}{dt}$$

so the equivalent value for parallel capacitance is

$$C_{\text{eq}} = C_1 + C_2 \tag{7}$$

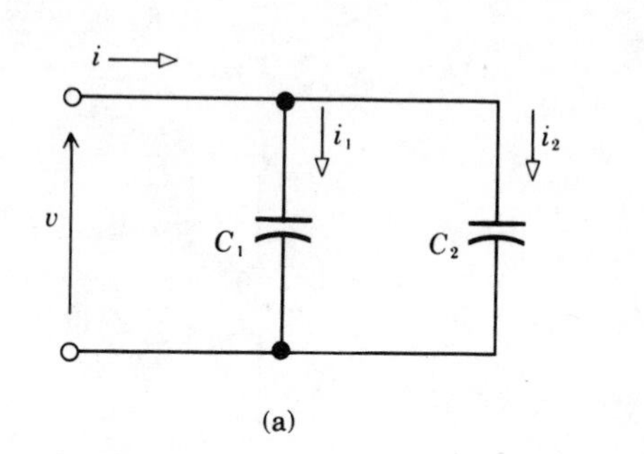

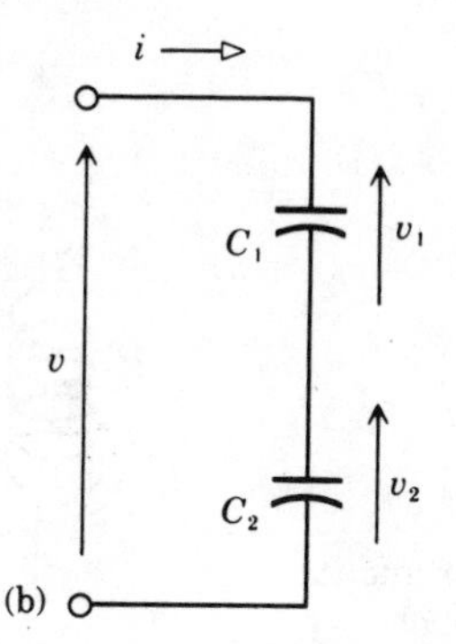

FIGURE 5.1–8
Parallel and series capacitance.

Note that parallel capacitance *adds*—just like series resistances. You might begin to suspect that the equivalent capacitance for a series connection would be like that for parallel resistances, namely

$$C_{eq} = \frac{C_1 C_2}{C_1 + C_2} \tag{8}$$

which can indeed be demonstrated, starting with Fig. 5.1–8b. Moreover, the extension of these expressions to cover three or more capacitors in series (or parallel) follows the form for resistance in parallel (or series). Series capacitors seldom occur in practice, but parallel capacitors are sometimes used to build up a large capacitance "bank" for energy storage.

Exercise 5.1–4

Devise a capacitance bank to store 100 J of energy at 5 kV (5000 V) using 4-μF capacitors rated for a maximum of 3 kV. (Hint: If $C_1 = C_2$, as in Fig. 5.1–8b, then $v_1 = v_2$.) How long will it take to charge this bank from a 2-mA current source?

5.2 INDUCTANCE

Inductance, the other major type of energy-storage element, stores energy in a magnetic field. Its characteristics are presented here and contrasted with those of capacitance.

INDUCTORS AND INDUCTANCE

An *ideal inductor* consists of a length of lossless (resistanceless) wire wound into a coil, like Fig. 5.2–1a. Energy is stored in the magnetic field established around the coil when a current flows through it. A constant current produces no voltage drop across the inductor, but a time-varying current produces a voltage

$$v = L\frac{di}{dt} \tag{1}$$

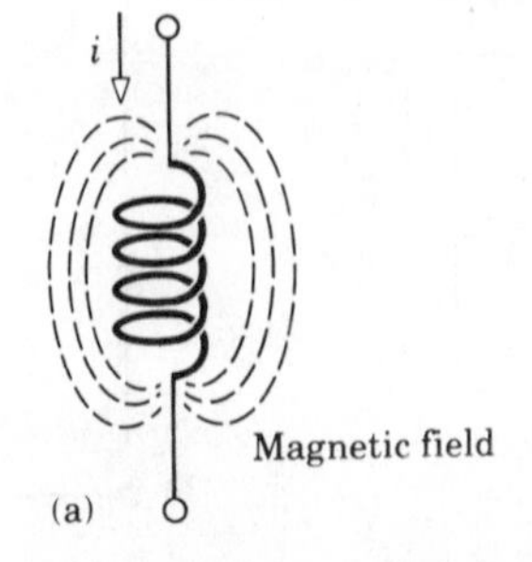

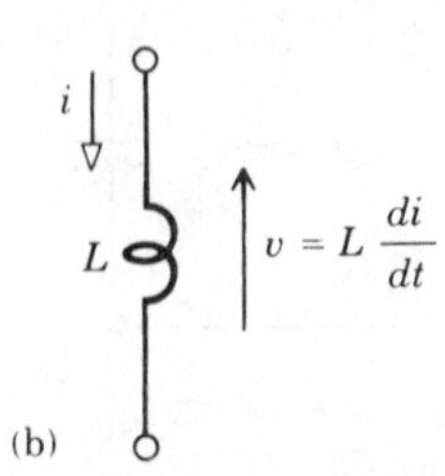

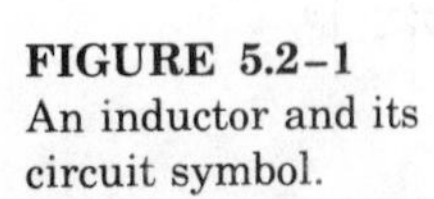

FIGURE 5.2–1
An inductor and its circuit symbol.

with L being the *inductance* measured in *henrys* (H). The unit equation for inductance is

$$1\ \mathrm{H} = 1\ \mathrm{V}\cdot\mathrm{s/A}$$

and Fig. 5.2–1b is the circuit symbol.

Later in Chapter 17 we will show that a cylindrical coil with length ℓ, area A, and N turns of wire has

$$L \approx \mu\frac{N^2A}{\ell}$$

where μ is the *permeability* of the core upon which the coil is wound. Practical values of inductance range from microhenrys to a few henrys, the larger values requiring hundreds of turns and a magnetic (high-μ) core material such as iron. The magnetic core is made movable relative to the coil for a variable inductance. Every circuit also has small amounts of stray inductance due to the fact that current through *any* conductor, even a straight wire, creates a magnetic field around it.

Close examination of the v-i relation in Eq. (1) shows that inductance is the "opposite" of capacitance in that $v = L\,di/dt$ becomes $i = C\,dv/dt$ if i and v are interchanged, and C and L are interchanged. This interchange relationship is called *duality,* and it holds for all the capacitance equations we have previously derived. Thus, replacing C and v by their *duals* (L and i) in $w = \frac{1}{2}Cv^2$, we have

$$w = \tfrac{1}{2}Li^2 \tag{2}$$

for the energy stored by an inductor. Likewise, the dual of the capacitance voltage integral is

$$i(t) = \frac{1}{L}\int_{-\infty}^{t} v\,dt \tag{3a}$$

$$= I_0 + \frac{1}{L}\int_{0}^{t} v\,dt \qquad t \geq 0 \tag{3b}$$

where I_0 is the initial current. We see from Eq. (3) that inductance "remembers" past values of the applied voltage. We also see that a constant voltage $v = V$ applied at $t = 0$ produces a linearly increasing current

$$i(t) = I_0 + \frac{V}{L}t \qquad t \geq 0$$

and we could relabel Fig. 5.1–2 using duality.

Despite the dualism between inductance and capacitance, a real inductor does not hold stored energy as well as a real capacitor. Doing so would require sustaining the current i through a short-circuit across the inductor's terminals, and a real inductor always has some winding resistance in the coil that dissipates the energy rather quickly. Capacitors, therefore, are used almost exclusively for storing energy for any appreciable duration.

Example 5.2–1
Automobile ignition

One commonplace application of inductive energy storage is an automobile ignition system, diagrammed in simplified form in Fig. 5.2–2. The switch (or "points") is closed initially to establish a current $i = Vt/L$ through the inductance associated with the primary winding of a step-up transformer called the ignition coil. Thus, after T seconds have elapsed, there will be energy $w = \frac{1}{2}Li^2 = V^2T^2/2L$ stored in the magnetic field.

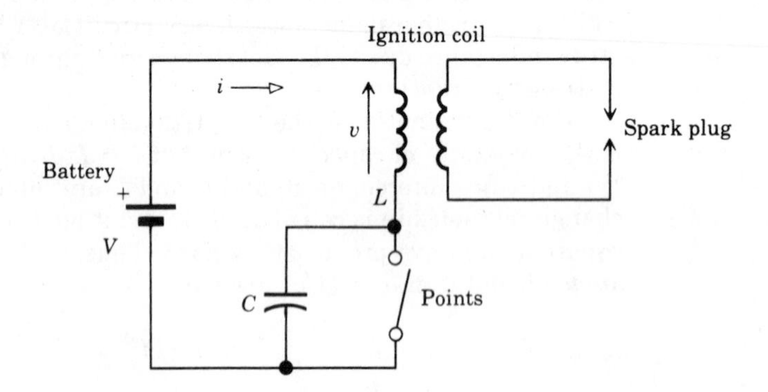

FIGURE 5.2–2
Automobile ignition system.

If the switch is opened, the current rapidly drops to zero and produces a large voltage spike, $v = L\,di/dt$, which is transformed to an even larger spike across the secondary winding. We have now obtained an electrical arc through the gap of the spark plug as the stored energy becomes liberated. A capacitor (known as the "condenser") provides a current path to minimize arcing across the points that would damage the contacts.

Exercise 5.2–1

Suppose a 0.1-H inductor is in series with the capacitor in Exercise 5.1–3, so they both carry the same current. Find and sketch the waveforms of the voltage across the inductor and its stored energy. Then, using your previous results, sketch the total energy stored by the series circuit and apply KVL to find the total voltage across the circuit.

Exercise 5.2–2

Confirm the duality result in Eq. (2) by finding w directly from $p = vi = dw/dt$ with $v = L\,di/dt$.

STEADY-STATE DC BEHAVIOR

Duality applies, as well, to DC behavior, because an inductor has no voltage drop when the current is constant, whereas a capacitor has no current flow when the voltage is constant. Inductance thus acts like a DC *short circuit,* and capacitance acts like an *open circuit* or DC block. Accordingly, when all voltages and currents are constant in a circuit containing energy-storage elements, we can analyze it by making the following mental replacements:

$$\text{Capacitance} \rightarrow \text{Open circuit}$$
$$\text{Inductance} \rightarrow \text{Short circuit}$$

In Fig. 5.2–3, for instance, we easily find that $I_L = 12\text{ V}/8\ \Omega = 1.5\text{ A}$ and $V_C = (6\ \Omega/8\ \Omega) \times 12\text{ V} = 9\text{ V}$.

This analysis, however, holds only when all voltages and currents are constant and, hence, all time derivatives equal zero. When such is the case, we say that the circuit is in the DC *steady-state* condition. In Chapter 8 we will discuss some quantitative measures for determining when a circuit has reached the steady state.

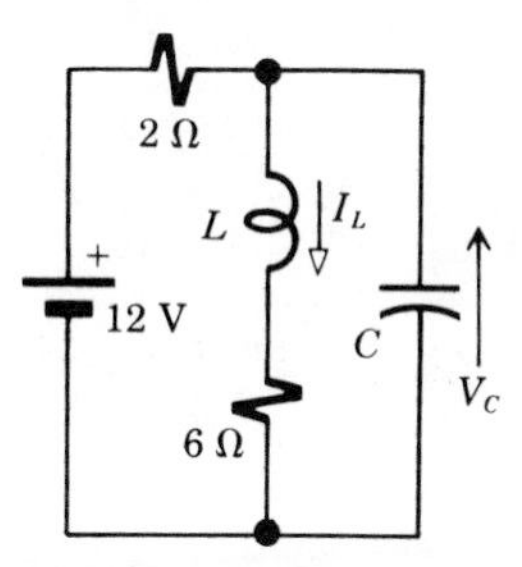

FIGURE 5.2–3 A circuit in the DC steady-state condition.

Exercise 5.2–3

Suppose the battery in Fig. 5.2–3 is replaced by a DC current source with $I_s = 30$ A and $R_s = 16\ \Omega$. Find I_L and V_C, and calculate the total stored energy when $L = 2$ mH and $C = 50\ \mu$F.

EQUIVALENT AND MUTUAL INDUCTANCE

It should be evident that $v = L\, di/dt$ is a *linear v-i* relationship. As a result, superposition applies to networks involving inductors as well as capacitors, and we can speak of *equivalent inductance* for two or more inductors in series or parallel. If two inductors are connected in series, the total voltage drop will be $v_1 + v_2$, where $v_1 = L_1\, di/dt$ and $v_2 = L_2\, di/dt$; therefore, $v = (L_1 + L_2)\, di/dt$, and the equivalent inductance is the sum

$$L_{eq} = L_1 + L_2 \tag{4}$$

One can similarly show for a parallel connection that

$$L_{eq} = \frac{L_1 L_2}{L_1 + L_2} \tag{5}$$

so equivalent inductance obeys exactly the same rules as equivalent resistance.

However, the above formulas assume no interaction between the magnetic fields of the inductances. When magnetic fields are in close proximity, their interaction gives rise to a coupling effect known as *mutual inductance,* symbolized by M in Fig. 5.2–4. Time-varying current through one coil then induces voltage in the other, and the total terminal voltages are

$$v_1 = L_1 \frac{di_1}{dt} \pm M \frac{di_2}{dt} \qquad v_2 = L_2 \frac{di_2}{dt} \pm M \frac{di_1}{dt} \tag{6}$$

The positive signs are taken in Eq. (6) when both currents enter (or leave) the dotted ends of the coils as shown in the figure. This is similar to the dot convention for an ideal transformer. Otherwise, the mutual inductance has the opposite sense, and we take the negative signs in Eq. (6).

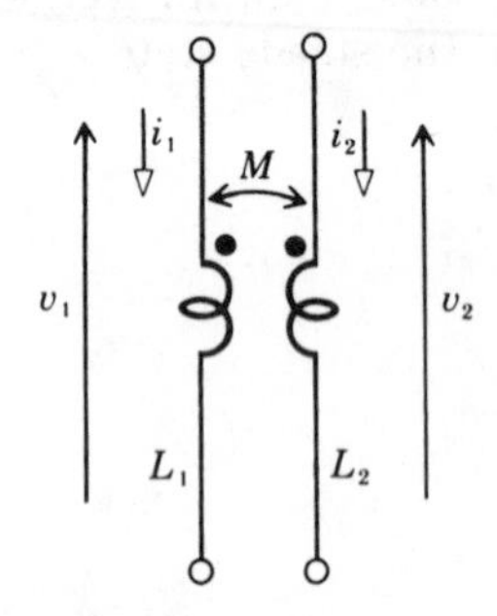

FIGURE 5.2–4
Mutual inductance.

Mutual inductance is the phenomenon underlying the action of a transformer. That relationship is developed in Chapter 18, following the discussions of magnetics and induction in Chapter 17.

Exercise 5.2–4

Suppose the two coils in Fig. 5.2–4 are series connected to form a two-terminal element with $i = i_1 = i_2$ and $v = v_1 + v_2$. Show that the equivalent inductance will be $L_{eq} = L_1 + L_2 + 2M$.

5.3 THE IMPEDANCE CONCEPT

When a circuit contains resistance plus one or more energy-storage elements, its voltage-current relationship becomes a *differential equation* due to the presence of the derivative terms $C\,dv/dt$ or $L\,di/dt$. Such circuits are said to be *dynamic* in the sense that the voltage and current generally have different waveforms—as contrasted with a resistive circuit whose voltage and current waveforms always have the same shape.

Our discussion in this section starts with the formulation of differential equations for dynamic circuits. We then introduce the powerful concept of impedance as a means of simplifying dynamic circuit analysis when the applied source has an exponential time dependence. We also present the related concepts of admittance and transfer functions. These concepts provide us with the tools needed to study AC circuits and transient response in the next three chapters.

DIFFERENTIAL CIRCUIT EQUATIONS

Consider a series RL circuit driven by a time-varying voltage $v(t)$, Fig. 5.3–1a. To find the resulting current waveform $i(t)$, we use Kirchhoff's voltage law

$$v_L + v_R = v$$

together with the element relations

$$v_L = L\frac{di}{dt} \qquad v_R = Ri$$

where the time dependence is implied but not explicitly written, for instance in $v_L = v_L(t)$, etc. Inserting the element equations into KVL yields

$$L\frac{di}{dt} + Ri = v \tag{1}$$

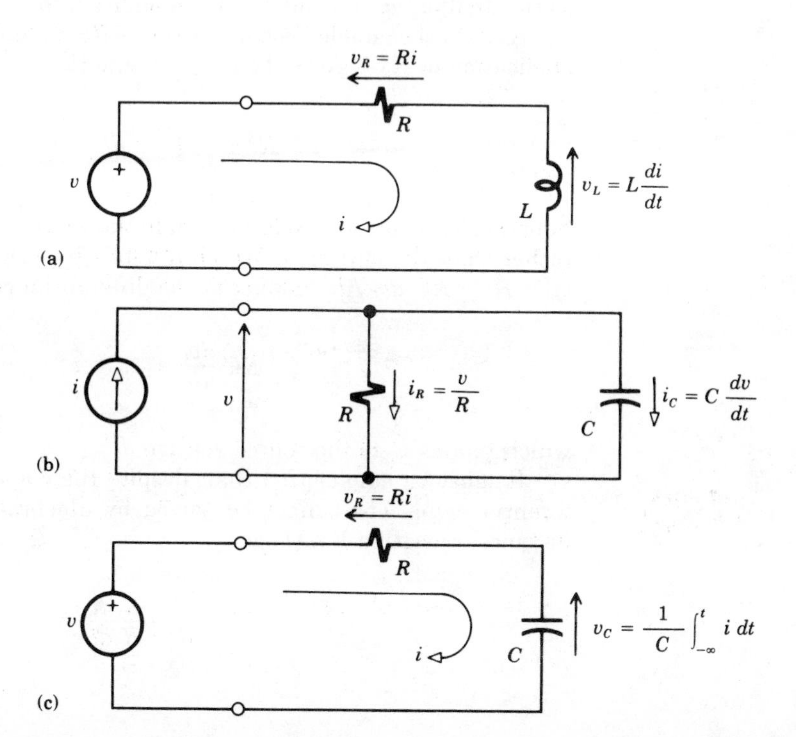

FIGURE 5.3–1 Voltages and currents in first-order circuits.

which provides an indirect rather than a direct relationship between $i(t)$ and $v(t)$.

Equation (1) is called a first-order linear inhomogeneous differential equation—quite a mouthful! It is a *differential* equation because it includes both the unknown current i and its derivative. The designation *first-order* means that only the first derivative di/dt is present, while *inhomogeneous* refers to the nonzero *forcing function* $v = v(t)$ on the right-hand side. It is a *linear* equation in the usual sense that superposition applies. Equations such as this occur time and again when we analyze dynamic circuits.

As a case in point, the voltage across the parallel RC circuit in Fig. 5.3–1b is related to the source current via

$$C\frac{dv}{dt} + \frac{1}{R}v = i \tag{2}$$

which has the same mathematical form as Eq. (1) with v being the unknown and i the forcing function. One obtains Eq. (2) from Kirchhoff's current law $i_C + i_R = i$ plus $i_C = C\,dv/dt$ and $i_R = v/R$. (Incidentally, this circuit may be viewed as the *dual* of the series RL circuit by taking conductance $G = 1/R$ as the dual of resistance and "parallel-connected" as the dual of "series-connected," in addition to the L-C and v-i duality.)

As a third example, look at the series RC circuit of Fig. 5.3–1c. Direct application of KVL gives the *integral* equation

$$Ri + \frac{1}{C}\int_{-\infty}^{t} i\,dt = v$$

Suppose, however, we are interested in the voltage v_C across the capacitor rather than the current i. We eliminate i by writing $i = C\,dv_C/dt$ and $v_R = Ri = RC\,dv_C/dt$, leading to the differential equation

$$RC\frac{dv_C}{dt} + v_C = v \tag{3}$$

which relates v_C to the source voltage v.

It must be emphasized that, despite their apparent simplicity, differential equations cannot be solved by algebraic manipulations. For instance, rewriting Eq. (1) as

$$i = \frac{v}{R} - \frac{L}{R}\frac{di}{dt}$$

gets us no closer to finding i because its unknown derivative di/dt now appears on the right-hand side as part of the "solution." We'll return to this problem after looking at one more circuit equation.

Applying KVL to the series LRC circuit in Fig. 5.3–2 yields an *integral-differential* equation,

$$L\frac{di}{dt} + Ri + \frac{1}{C}\int_{-\infty}^{t} i\,dt = v \tag{4a}$$

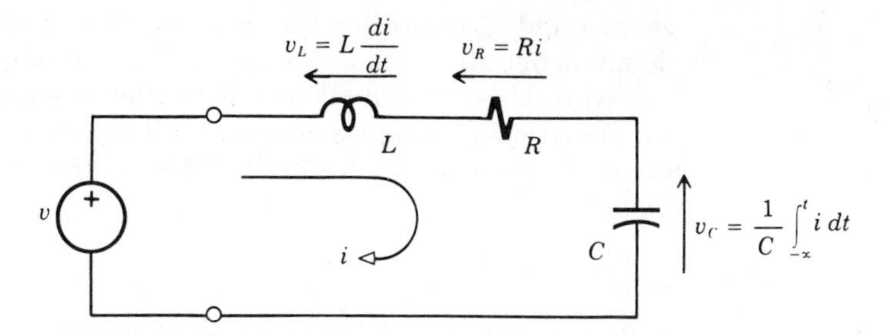

FIGURE 5.3–2
Voltages and current in a second-order circuit.

This can be converted to a differential equation by differentiating both sides with respect to time. The derivative of the sum on the left equals the sum of the derivatives, with

$$\frac{d}{dt}\left(\frac{1}{C}\int_{-\infty}^{t} i\,dt\right) = \frac{1}{C}\frac{d}{dt}\left(\int_{-\infty}^{t} i\,dt\right) = \frac{1}{C}i$$

and

$$\frac{d}{dt}\left(L\frac{di}{dt}\right) = L\frac{d}{dt}\left(\frac{di}{dt}\right) = L\frac{d^2i}{dt^2}$$

which is a *second derivative*—that is, the derivative of the derivative. We thus obtain

$$L\frac{d^2i}{dt^2} + R\frac{di}{dt} + \frac{1}{C}i = \frac{dv}{dt} \tag{4b}$$

whose forcing function is the derivative or slope of the applied voltage.

Equation (4b) is called a *second-order* differential equation, and we refer to Fig. 5.3–2 as a *second-order circuit,* in contrast to the previous *first-order circuits* that were described by first-order differential equations. Second-order equations arise whenever a circuit has two energy-storage elements that cannot be combined into one equivalent inductance or capacitance. And as you might guess, a circuit with n energy-storage elements will generally have an nth-derivative term in its differential equation and would be called an nth-order circuit.

Another way of describing an nth-order circuit is to formulate n *first-order* differential equations with n unknown voltages or currents. For instance, from Fig. 5.3–2 we could write the two equations

$$C\frac{dv_C}{dt} = i \qquad L\frac{di}{dt} = v - Ri - v_C$$

where i and v_C are called the *state variables* because their values completely define the state of the circuit, including the stored energy. This state-variable approach allows one to draw upon sophisticated analytical and numerical techniques for the simultaneous solution of first-order differential equations. Such equations also may be simulated on an analog computer.

Example 5.3–1
Numerical solution

As an illustration of dynamic circuit behavior, let's carry out an approximate numerical solution of a simple differential equation. Suppose a series RL circuit (Fig. 5.3–1a) has $R = 4\ \Omega$, $L = 2$ H, and a constant voltage $v = 40$ V applied at time $t = 0$. Substituting numerical values, Eq. (1) becomes $2\ di/dt + 4i = 40$, which governs the behavior of the current $i(t)$ for $t > 0$. Eventually, we expect the circuit to reach DC steady-state conditions with the inductance acting like a short circuit so $v_L = L\ di/dt = 0$, $di/dt = 0$, and $i = v/R = 10$ A. However, if $i = 0$ initially, the memory effect of inductance prevents an immediate change in the stored energy and there must be some sort of transition into the steady state.

Consider, then, the derivative of the current obtained by rewriting the differential equation as

$$\frac{di}{dt} = \frac{40 - 4i}{2} = 20 - 2i$$

With $i = 0$ at $t = 0$, $i(t)$ has an initial slope of $di/dt = 20$ A/sec and appears to head for $i = 10$ A at $t = 0.5$ sec, as indicated in the plot of Fig. 5.3–3a. But when $i(t)$ increases, we can see from the expression above that its derivative or slope decreases. Thus, for instance, if we approximate $i(t)$ at $t = 0.1$ sec by the straight-line projection $i = 20$ A/sec $\times$ 0.1 sec $= 2$ A, then the new derivative value is $di/dt = 20 - 2 \times 2 =$ 16 A/sec. Using this new slope, the projected value of $i(t)$ at $t = 0.2$ sec will be $i = 2 + 16 \times 0.1 = 3.6$ A, which in turn means that $di/dt =$ $20 - 2 \times 3.6 = 12.8$ A/sec.

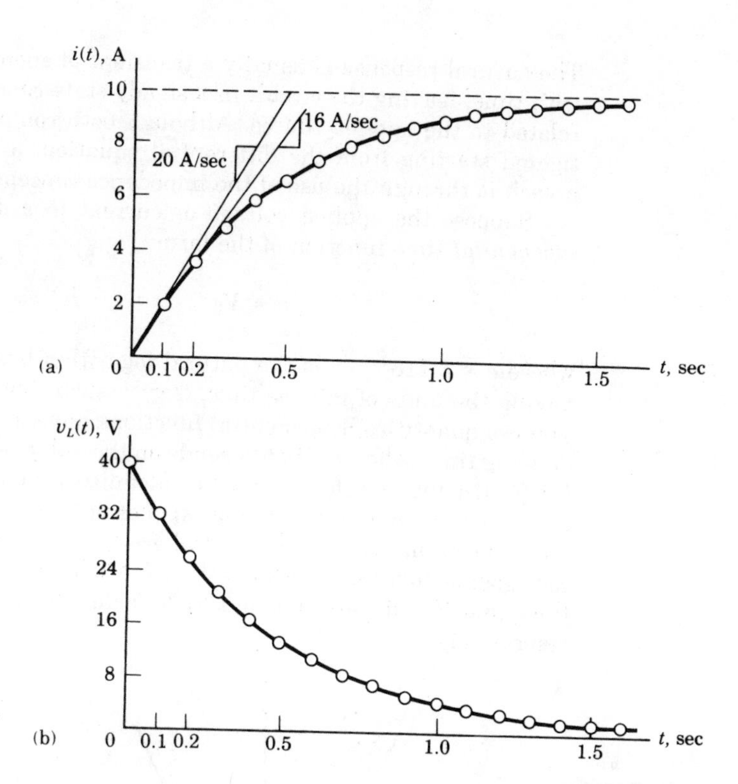

FIGURE 5.3–3
Numerical solution of a differential equation.

Repeating this step-by-step calculation every 0.1 seconds yields the approximate curve of $i(t)$ shown in Fig. 5.3–3a. Figure 5.3–3b plots the resulting voltage across the inductance computed from $v_L = v - Ri$. Both curves exhibit an initial *transient* interval during which the waveforms differ markedly from the constant shape of the applied voltage. But they eventually do approach the steady-state values $i = 10$ A and $v_L = 0$ forced by the constant 20-V source. The transient behavior corresponds to the circuit's readjustment from its initial state to the final steady-state conditions.

Exercise 5.3–1

Let the circuit in Fig. 5.3–1c have $R = 1$ kΩ, $C = 3$ μF, and $v = -20$ V. If $v_C = 10$ V at $t = 0$, use Eq. (3) to evaluate dv_C/dt at $t = 0$. Then calculate the corresponding value of $i = C\, dv_C/dt$, and show that your result agrees with $v = Ri + v_C$.

EXPONENTIAL FUNCTIONS AND IMPEDANCE

The complete response of a dynamic circuit may be viewed as the sum of two components, the steady-state *forced* response and the free or *natural* response. (In mathematical terms, these correspond to the *particular* solution and to the *complementary* solution of the differential equation.)

The natural response is usually a transient phenomenon that dies away with time, leaving the circuit in a steady-state condition that is directly related to the applied source. Although both components can be investigated starting from the differential equation, a more convenient approach is through the use of the impedance concept introduced here.

Suppose the applied voltage or current in a dynamic circuit is an *exponential* time function of the form

$$v = Ve^{st} \qquad i = Ie^{st}$$

where $e = 2.718 \ldots$ is the natural logarithm base and s is a constant having the units of inverse time, $(\text{sec})^{-1}$, such that st and e^{st} are dimensionless quantities. Exponential functions have a decaying, constant, or growing time behavior that depends on the value of s, as sketched in Fig. 5.3–4. (In retrospect, you might recognize that the waveform in Fig. 5.3–3b approximates a decaying exponential.) Note that V or I represents the value of the function at $t = 0$ where $e^{st} = e^0 = 1$. Also note that $s = 0$ corresponds to a DC source $v = V$ or $i = I$. With $s \neq 0$, v and i vary with time, and V and I are constant scale factors measured in volts or amps, respectively.

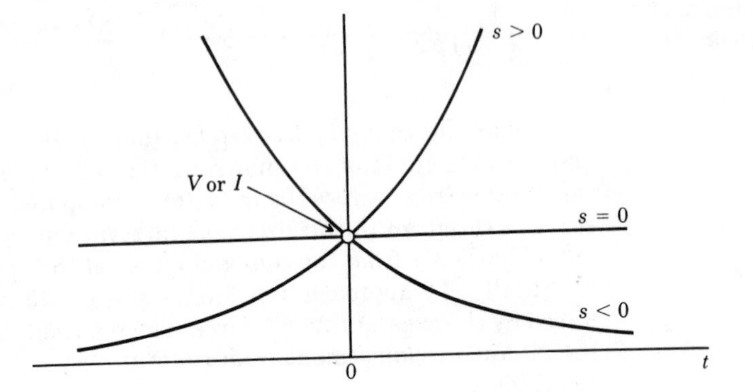

FIGURE 5.3–4
Exponential waveforms $v = Ve^{st}$ or $i = Ie^{st}$.

The key properties of an exponential function pertinent to circuit analysis are, for any constant A,

$$\frac{d}{dt}(Ae^{st}) = sAe^{st} \qquad \int_{-\infty}^{t} Ae^{st}\,dt = \frac{A}{s}e^{st} \tag{5}$$

Thus the derivative or integral of an exponential is *another exponential* having the *same value of* s, but a different *magnitude*. Consequently, if the applied voltage or current has an exponential time variation, *all* forced voltages and currents will be exponential waveforms, and any derivative or integral terms in the circuit equation reduce to algebraic expressions.

To demonstrate these points, take the voltage applied to a series *LRC* circuit (Fig. 5.3–2) to be $v = Ve^{st}$ with given values of V and s. We then assume that $i = Ie^{st}$, where, for the moment, the constant I is unknown. Using Eq. (5) and the element relations, we find the individual voltages are

$$v_L = L\frac{di}{dt} = L(sIe^{st}) = (sL)\,Ie^{st}$$

$$v_R = Ri = RIe^{st}$$

$$v_C = \frac{1}{C}\int_{-\infty}^{t} i\,dt = \frac{1}{C}\left(\frac{I}{s}e^{st}\right) = \left(\frac{1}{sC}\right)Ie^{st}$$

We then define the *element impedances*

$$Z_L \triangleq sL \qquad Z_R \triangleq R \qquad Z_C \triangleq \frac{1}{sC} \tag{6}$$

such that

$$v_L = Z_L Ie^{st} \qquad v_R = Z_R Ie^{st} \qquad v_C = Z_C Ie^{st} \tag{7}$$

Significantly, each of these expressions has the form $v = Zi$, since $i = Ie^{st}$.

But the applied voltage $v = Ve^{st}$ equals the sum $v_L + v_R + v_C$, so

$$Z_L Ie^{st} + Z_R Ie^{st} + Z_C Ie^{st} = Ve^{st}$$

or, upon factoring,

$$(Z_L + Z_R + Z_C)Ie^{st} = Ve^{st}$$

The time variation e^{st} now cancels out on both sides, leaving

$$I = \frac{V}{Z_L + Z_R + Z_C}$$

which expresses I in terms of the source parameters V and s and the element values as contained in the impedances Z_L, Z_R, and Z_C. We have, then, found the unknown constant in the assumed current $i = Ie^{st}$, and Eq. (7) gives the individual voltages—all of them being exponential waveforms.

Observe that the previous analysis did not involve a differential circuit equation because algebraic voltage-current relations in terms of impedances replaced the derivative and integral in Eq. (4a). Additional consideration of the impedance concept is therefore in order.

The impedance $Z_R = R$ associated with a resistance equals the resistance value itself, independent of s, and clearly has the units of *ohms*. The other two impedances depend on the source parameter s as well as the element value, but dimensional analysis shows that the units of $Z_L = sL$ and $Z_C = 1/sC$ also equal ohms. Accordingly, each of the relations in Eq. (7) may be viewed as modified versions of Ohm's law with *impedance replacing resistance*. Putting this another way, impedance "impedes" the flow of current just as resistance "resists" it.

Further confidence in this interpretation is gained by setting $s = 0$, representing a DC source; then $Z_L = 0$ while $Z_C = \infty$, agreeing with our prior conclusions that inductance acts like a DC short circuit and capacitance acts like a DC open circuit. If $s < 0$, representing a decaying exponential, then Z_L and Z_C will have negative values; *negative impedance* simply means that the energy-storage element is returning energy to the circuit rather than storing it.

To summarize our results so far: The impedance concept provides a direct method for finding a dynamic circuit's steady-state response that has been caused by an exponential forcing function. Although that might seem to be a rather limited situation, we will see in Chapters 6 and 8 how impedance relates to steady-state AC circuit analysis and to the study of natural response and transient behavior. Furthermore, advanced mathematical methods permit one to represent virtually any time function with a combination of exponentials. (These methods are called *transforms* because they transform differential equations into algebraic equations.) The remainder of this chapter, therefore, concentrates on impedance per se and the related concepts of admittance and transfer functions.

Example 5.3–2

Let Fig. 5.3–2 have $L = 0.25$ H, $R = 400\ \Omega$, and $C = 100\ \mu$F, and let $v = 30e^{-200t}$ V. Since $s = -200$, $Z_L = -200 \times 0.25 = -50\ \Omega$, and $Z_C = 1/(-200 \times 100 \times 10^{-6}) = -50\ \Omega$. Thus, $I = 30$ V$/(400 - 50 - 50)\ \Omega = 0.1$ A, so $i = 0.1e^{-200t}$ A, $v_R = 40e^{-200t}$ V, and $v_L = v_C = -5e^{-200t}$ V. Note that v_R is actually greater than the applied voltage due to energy being returned from the inductance and capacitance. The steady-state condition therefore requires initial stored energy at $t = -\infty$.

Exercise 5.3–2

Show by substitution that the foregoing result for i satisfies Eq. (4a).

CIRCUIT ANALYSIS USING IMPEDANCE AND ADMITTANCE

Just as we defined the equivalent resistance of a resistive circuit, we can define *equivalent impedance* of a dynamic circuit. Referring to Fig. 5.3–5, let a one-port network consist of linear elements (but no independent

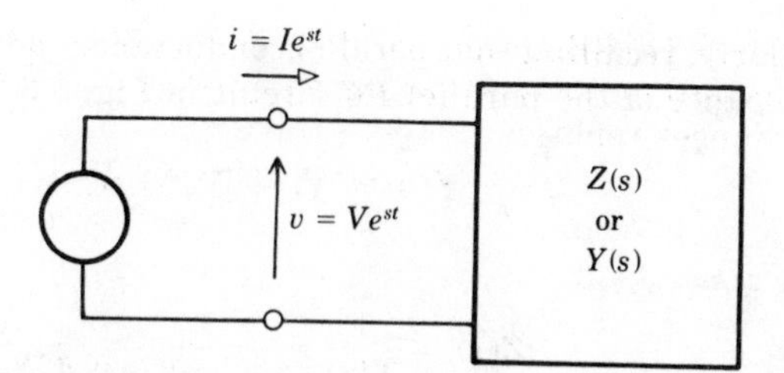

FIGURE 5.3–5
A one-port with exponential excitation.

sources), and let the applied source be an exponential function, either voltage or current. Then both terminal variables v and i will be exponentials in the steady state, and their ratio is the network impedance

$$Z(s) \triangleq \frac{Ve^{st}}{Ie^{st}} \tag{8a}$$

The functional notation $Z(s)$ emphasizes the dependence of impedance on the source parameter s. We also define the *admittance* as the reciprocal of impedance; that is,

$$Y(s) \triangleq \frac{1}{Z(s)} = \frac{Ie^{st}}{Ve^{st}} \tag{8b}$$

whose units are mhos (or siemens), the same as for conductance. Table 5.3–1 lists the impedance and admittance of the individual circuit elements.

TABLE 5.3–1

Element	*Impedance*	*Admittance*
Resistance	R	$G = \dfrac{1}{R}$
Inductance	sL	$\dfrac{1}{sL}$
Capacitance	$\dfrac{1}{sC}$	sC

Equivalent impedance (or admittance) is found from a circuit diagram by combining the individual impedances (or admittances) as if they were resistances (or conductances). For instance, the series *LCR* circuit back in Fig. 5.3–2 has

$$Z(s) = Z_L + Z_R + Z_C = sL + R + \frac{1}{sC} = \frac{s^2LC + sRC + 1}{sC} \tag{9}$$

Similarly, recalling that parallel conductances add, you will find that the admittance of the parallel RC circuit in Fig. 5.3–1b is

$$Y(s) = Y_C + Y_R = sC + G \tag{10a}$$

and

$$Z(s) = \frac{1}{Y(s)} = \frac{1}{sC + G} = \frac{R}{sCR + 1} \tag{10b}$$

(You can also obtain this last result by writing $Z(s) = Z_R \| Z_C$.)

Comparing Eqs. (9) and (10) reveals a significant property of impedance or admittance when written in terms of s as a ratio of polynomials: The highest power of s never exceeds the number of energy-storage elements. Understanding this property, you can quickly check for possible manipulation errors.

Once we know $Z(s)$ or $Y(s)$ for a given network, it becomes a trivial matter to find the steady-state response caused by an exponential forcing function. Specifically, a voltage $v = Ve^{st}$ produces the current $i = Ie^{st}$ with

$$I = \frac{V}{Z(s)} = Y(s)V \tag{11a}$$

while a current $i = Ie^{st}$ produces the voltage $v = Ve^{st}$ with

$$V = \frac{I}{Y(s)} = Z(s)\,I \tag{11b}$$

These equations play the same role in the study of dynamic circuits that Ohm's law did for that of resistive circuits.

If we are concerned with an internal voltage or current, as distinguished from the terminal variables in Fig. 5.3–5, we take the impedance concept one step further and use *impedance in place of resistance* in the various techniques previously developed for resistive circuits:

- Branch-current and node-voltage equations
- Voltage and current dividers
- Thévenin's theorem
- Thévenin/Norton conversion

Furthermore, the *superposition* principle can be invoked to deal with the case of a sum of exponential functions. For instance, if $i = I_1 e^{s_1 t} + I_2 e^{s_2 t}$ is the current through an impedance $Z(s)$, then Eq. (11b) and superposition give the resulting voltage as

$$v = Z(s_1)I_1 e^{s_1 t} + Z(s_2)I_2 e^{s_2 t}$$

The following examples illustrate some of these techniques and help bring out the value of the impedance concept as a labor-saving tool.

Example 5.3–3

Consider the circuit in Fig. 5.3–6. Its impedance is

$$Z(s) = \frac{(Z_L + Z_R)Z_C}{(Z_L + Z_R) + Z_C} = \frac{(sL + R)/sC}{sL + R + 1/sC}$$

so, from Eq. (11b),

$$V = Z(s)I = \frac{sL + R}{s^2LC + sCR + 1} I$$

where the presence of s^2 agrees with the order of the circuit ($n = 2$ energy-storage elements). If $s = 0$, then $Z(0) = R$, and $V = RI$ as expected under DC steady-state conditions. Had we wanted to find i_C, we could have found its scale factor directly from the current-divider ratio

$$I_C = \frac{Z_L + Z_R}{Z_L + Z_R + Z_C} I = \frac{sC(sL + R)}{s^2LC + sCR + 1} I$$

which also equals V/Z_C.

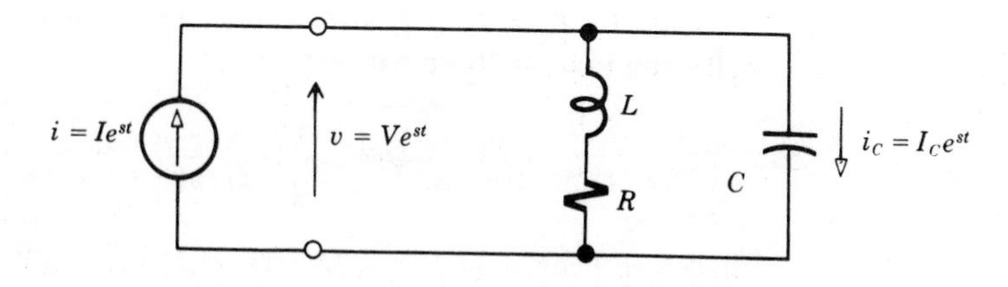

FIGURE 5.3–6 Circuit for Example 5.3–3.

Example 5.3–4 *Miller-effect capacitance*

Figure 5.3–7 gives the circuit diagram and equivalent circuit for an inverting op-amp with a capacitor connecting the input and output terminals. The equivalent circuit comes from Fig. 4.3–6c, which illustrates the fact that $v_{\text{out}} = v_a = -\mu v_{\text{in}}$, where $\mu = R_F/R_1$. We will determine the input admittance $Y(s) = I_{\text{in}}e^{st}/V_{\text{in}}e^{st}$ by assuming exponential time variation for all voltages and currents.

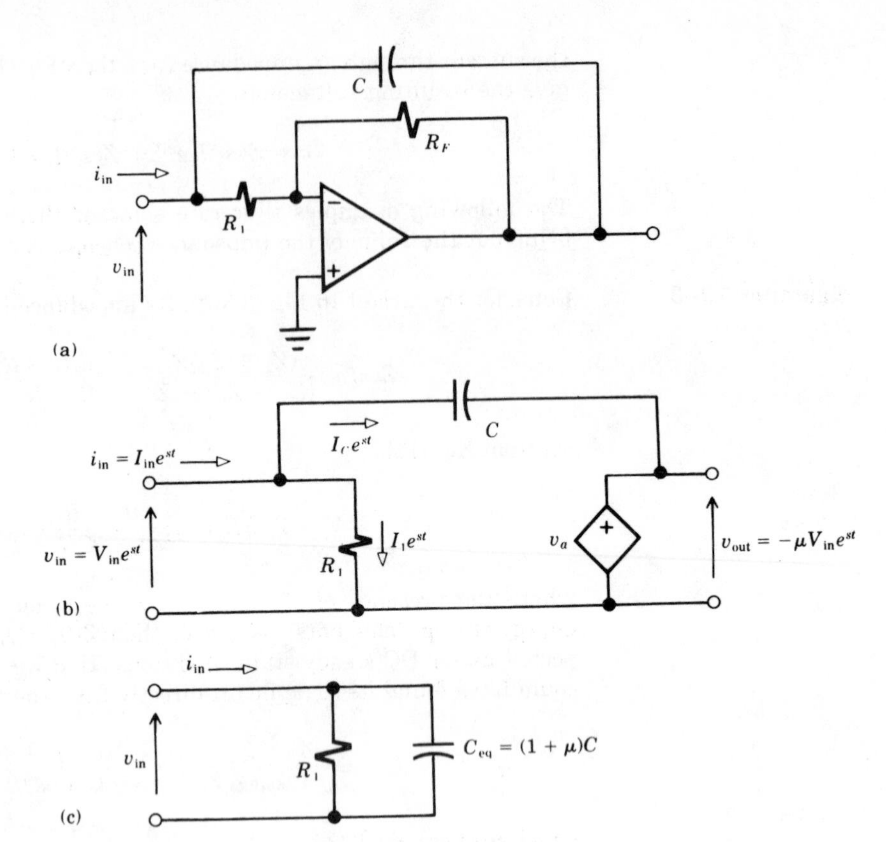

FIGURE 5.3–7
(a) Inverting amplifier with capacitance.
(b) Equivalent circuit; Miller-effect input capacitance.

Clearly, $I_{in} = I_C + I_1$ and $I_1 = V_{in}/R_1$. To find I_C in terms of V_{in}, we write the node-voltage expression

$$I_C = \frac{V_{in} - V_{out}}{Z_C} = \frac{V_{in} - (-\mu V_{in})}{(1/sC)} = s(1 + \mu)CV_{in}$$

Therefore $I_{in}e^{st} = [s(1 + \mu)C + (1/R_1)]\, V_{in}e^{st}$ and

$$Y(s) = s(1 + \mu)C + \frac{1}{R_1}$$

Comparison with Eq. (10a) indicates that the effective input consists of R_1 in parallel with an equivalent capacitance $C_{eq} = (1 + \mu)C$, as in Fig. 5.3–7c. Although $Y(s)$ was derived by assuming exponential functions, the equivalent input circuit holds for *any* time variations of v_{in} and i_{in}.

The apparent multiplication of the capacitance value by $1 + \mu$ is known as the *Miller effect*.

Exercise 5.3–3

Let capacitance C be added in parallel with the resistance in Fig. 5.3–1a. Find $Z(s)$ and obtain an expression for the voltage v_R across the resistance when $v = Ve^{st}$. Check your results with $s = 0$.

TRANSFER FUNCTIONS

Again consider a series *LRC* circuit, but this time arranged as a *two-port* network with the output voltage taken across the resistance (see Fig. 5.3–8). To characterize the input-output relationship, we use the voltage *transfer function*

$$H(s) = \frac{V_{\text{out}}e^{st}}{V_{\text{in}}e^{st}} = \frac{Z_R}{Z_L + Z_C + Z_R} = \frac{sCR}{s^2LC + sRC + 1} \tag{12}$$

Note that $H(s)$ has been written as a ratio of polynomials, with the denominator being the same polynomial that appeared in $Z(s)$, Eq. (9). Given the values of R, C, s, and V_{in}, Eq. (12) could be used to compute V_{out}.

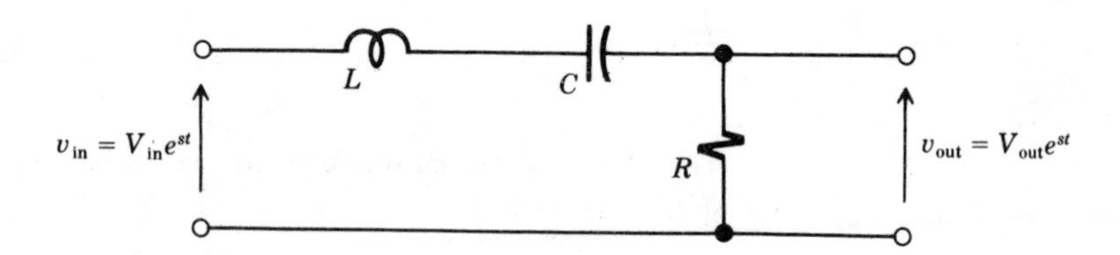

FIGURE 5.3–8

Going from this particular case to an arbitrary linear network, let $x = X_{\text{in}}e^{st}$ denote the applied exponential voltage or current and let $y = Y_{\text{out}}e^{st}$ be any steady-state response of interest, either voltage or current. (Be careful not to confuse the scale factor Y_{out} with our symbol for admittance.) We then make the general definition of a transfer function

$$H(s) \triangleq \frac{Y_{\text{out}}e^{st}}{X_{\text{in}}e^{st}} \tag{13}$$

This definition reduces to our previous definition of impedance or admittance when x and y are taken to be the terminal variables of a one-port network.

Expressions such as Eq. (12) help us study how the output depends on the element values and the source parameter s. Transfer functions are especially valuable for the analysis and design of filters, a topic in Chapter

7, and for the study of feedback and automatic control systems in Chapter 13. In addition, if we want the differential equation that relates the input and output, we can obtain it from $H(s)$ using a method described in Problem 5.3–14.

Example 5.3–5
Compensated probe

Figure 5.3–9 represents a *compensated* 10X *probe* of the type often used with a cathode-ray oscilloscope (CRO), whose input is represented by resistance R. The cable from the CRO to the probe has stray capacitance C in parallel with R. The probe has resistance $R_p = 9R$ and the compensating capacitance C_p is adjusted such that $H(s) = V_{out}e^{st}/V_{in}e^{st}$ is constant, independent of s.

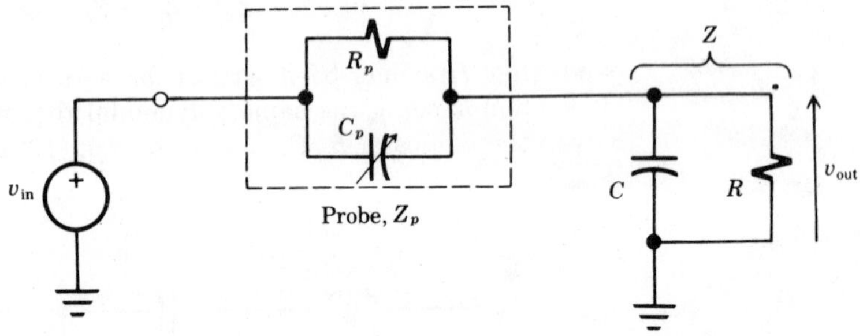

FIGURE 5.3–9
Compensated 10X probe for CRO.

To show that compensation is possible, let $Z(s) = Z_R \| Z_C = R/(sCR + 1)$, and let

$$Z_p(s) = \frac{R_p}{sC_pR_p + 1} = \frac{9R}{sC_p(9R) + 1} = 9\frac{R}{s(9C_p)R + 1}$$

Therefore, if $C_p = C/9$, then $Z_p(s) = 9Z(s)$ and

$$H(s) = \frac{Z(s)}{Z(s) + Z_p(s)} = \frac{Z(s)}{Z(s) + 9Z(s)} = \frac{1}{10}$$

This means that $v_{out} = \frac{1}{10}v_{in}$ for *any* waveform v_{in}, and the waveshape displayed on the CRO looks identical to v_{in}. We call this a "10X" probe because you must multiply the CRO scale by 10 to get the actual values of v_{in}.

Exercise 5.3–4

Obtain an expression for $H(s)$ in Fig. 5.3–6, taking the output as the voltage across L and the input as the applied current.

5.4 NETWORK ANALYSIS METHODS †

Having demonstrated the analytic power of impedance, admittance, and transfer functions, we feel the need of equally powerful methods for obtaining $Z(s)$, $Y(s)$, or $H(s)$ when a network has many branches. The two methods we describe here meet that need by providing compact and convenient ways of directly translating circuit diagrams into network equations. Node analysis is a systematic version of the node-voltage method we have been using all along, while loop analysis involves a new concept of loop currents.

Both methods also lend themselves to *matrix* formulation for numerical solution, this being a virtual necessity if there are many unknowns. Indeed, special network-analysis computer programs have been written based on the concepts of loop currents and node voltages. Intelligent use of these programs therefore requires some familiarity with the underlying theory, so in this section we develop the theory and apply it to illustrative circuits, including circuits with controlled sources. We also derive the wye-delta transformation for three-terminal networks.

NODE ANALYSIS

Consider the circuit of Fig. 5.4–1a whose controlled source presumably represents some electronic device. Taking the bottom node as reference, we see there are two unknown node voltages, v_B and v_{out}, that we need to relate to the source voltage v_{in}. We could then find any other quantity of interest in terms of the node voltages, just as we did with node-voltage equations for resistive circuits. But, due to the presence of energy-storage elements, we must now use impedance in place of resistance when writing the node equations.

To that end, we assume that $v_{\text{in}} = V_{\text{in}} e^{st}$ so all other voltages and currents will be exponential functions. The current-voltage relationship for

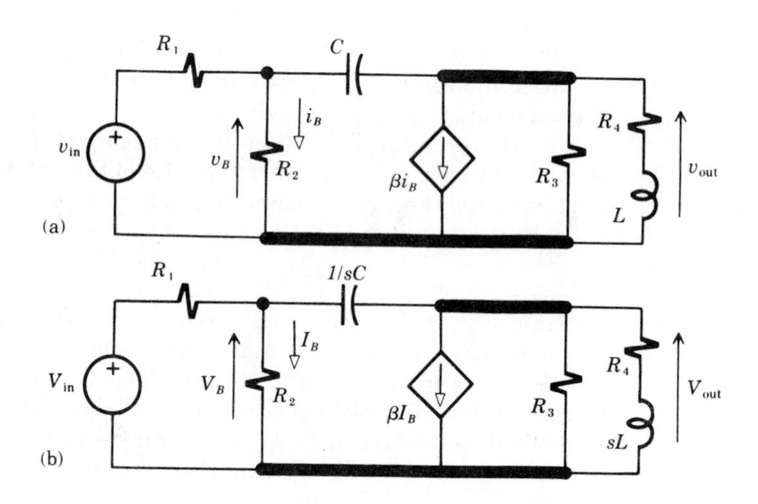

FIGURE 5.4–1
Node analysis.

an arbitrary element then has the form $Ie^{st} = Ve^{st}/Z(s)$, and e^{st} cancels out on both sides. Therefore, we can work entirely with the scale factors and impedances shown on the relabeled circuit diagram of Fig. 5.4–1b. This diagram helps us write the unknown currents for the KCL node equations. For instance, the scale-factor for the current from left to right through the capacitor is $(V_B - V_{out})/(1/sC)$, and the output current through the inductance is $V_{out}/(R_4 + sL)$. The two node equations thus become

$$\frac{V_{in} - V_B}{R_1} = \frac{V_B}{R_2} + \frac{V_B - V_{out}}{(1/sC)}$$

$$\frac{V_B - V_{out}}{(1/sC)} = \beta I_B + \frac{V_{out}}{R_3} + \frac{V_{out}}{R_4 + sL}$$

Upon regrouping we get

$$\left(\frac{1}{R_1} + \frac{1}{R_2} + sC\right) V_B - sCV_{out} = \frac{V_{in}}{R_1}$$

$$-sCV_B + \left(sC + \frac{1}{R_3} + \frac{1}{R_4 + sL}\right) V_{out} = -\beta I_B$$

where all the unknown voltages are on the left and source terms appear on the right.

Now observe that the coefficients of V_B in the first equation and V_{out} in the second are interpreted simply as the sum of all the *admittances* connected to the node in question. Similarly, the coefficient $-sC$ in both equations equals the negative of the admittance connected between the nodes. Furthermore, the right-hand terms equal the source current into the node or, in the case of V_{in}/R, its Norton equivalent. With the help of these interpretations, plus a little practice, you can write down the factored equations immediately.

However, due to the presence of a *controlled* source in this circuit, we still need one more equation to get rid of the unknown control current I_B. Inspecting the circuit shows that I_B is related to the node voltage V_B by

$$I_B = V_B/R_2$$

which can be inserted into the second node equation, thereby yielding two independent equations with two unknowns, V_B and V_{out}. The relationship between a control variable (such as I_B) and one or more of the unknowns is called a *constraint equation.* Generally speaking, constraint equations should be inserted after writing down the factored node equations, since they usually alter the symmetry we have described.

Extrapolating from our example to the case of a circuit having N unknown node voltages (all taken to be positive with respect to the reference node), we get a set of N simultaneous equations of the form

$$
\begin{array}{rcl}
Y_{11}V_1 - Y_{12}V_2 - \cdots - Y_{1N}V_N &=& I_1 \\
-Y_{12}V_1 + Y_{22}V_2 - \cdots - Y_{2N}V_N &=& I_2 \\
\vdots \qquad \vdots \qquad\qquad \vdots && \vdots \\
-Y_{N1}V_1 - Y_{N2}V_2 - \cdots + Y_{NN}V_N &=& I_N
\end{array}
\tag{1a}
$$

where, in general,

Y_{kk} = Sum of all admittances with one terminal at node k
$Y_{kj} = Y_{jk}$ = Sum of admittances connected directly between nodes k and j
V_k = Unknown voltage at node k
I_k = Net equivalent source current into node k

This set of equations becomes more compact in the matrix notation

$$
\begin{bmatrix}
Y_{11} & -Y_{12} & \cdots & -Y_{1N} \\
-Y_{21} & Y_{22} & \cdots & -Y_{2M} \\
\vdots & \vdots & & \vdots \\
-Y_{N1} & -Y_{N2} & \cdots & Y_{NN}
\end{bmatrix}
\begin{bmatrix} V_1 \\ V_2 \\ \vdots \\ V_N \end{bmatrix}
=
\begin{bmatrix} I_1 \\ I_2 \\ \vdots \\ I_N \end{bmatrix}
\tag{1b}
$$

Note that only the main diagonal terms $Y_{11}, Y_{22}, \cdots, Y_{NN}$, are positive, whereas all off-diagonal terms are negative and symmetrical—that is, $-Y_{21} = -Y_{12}$, etc. (This symmetry is destroyed when constraint equations, if any, are inserted.) The source terms $I_1, I_2, \cdots, I_N$ may be positive, negative, or zero.

Example 5.4–1 For an example of node analysis with numerical values, take the circuit of Fig. 5.4–2 which has been labeled in the same manner as Fig. 5.4–1b.

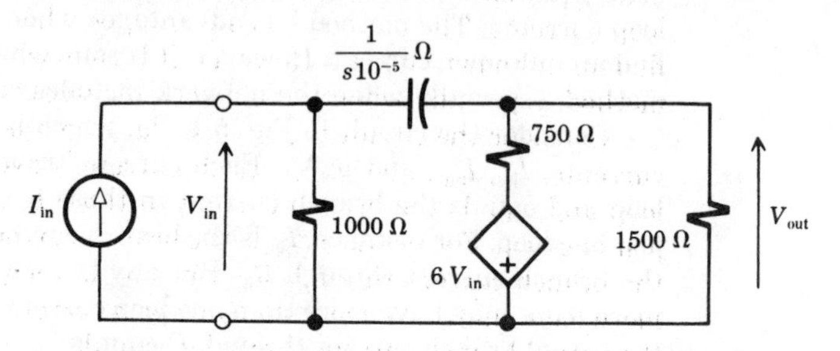

FIGURE 5.4–2 Circuit for Example 5.4–1.

(The 10-μF capacitance has thus become an impedance $1/sC = 1/s10^{-5}$.) Of course, the coefficients in the node equations will be admittances rather than impedances. Therefore, in matrix notation we have

$$\begin{bmatrix} \dfrac{1}{1000} + s10^{-5} & -s10^{-5} \\ -s10^{-5} & s10^{-5} + \dfrac{1}{750} + \dfrac{1}{1500} \end{bmatrix} \begin{bmatrix} V_{in} \\ V_{out} \end{bmatrix} = \begin{bmatrix} I_{in} \\ \dfrac{-6V_{in}}{750} \end{bmatrix}$$

and we do not need an additional constraint equation since the control voltage V_{in} happens to be a node voltage.

Multiplying both equations by 10^5 for convenience gives

$$\begin{aligned} (100 + s)V_{in} - \qquad sV_{out} &= 10^5 I_{in} \\ -sV_{in} + (s + 200)V_{out} &= -800V_{in} \end{aligned}$$

which are easily solved to obtain the transfer function

$$H(s) = \frac{V_{out}}{I_{in}} = 10^3 \frac{s - 800}{11s + 200}$$

It would also be a simple matter to find the input impedance $Z(s) = V_{in}/I_{in}$.

Exercise 5.4–1 Solve the preceding equations for V_{in} in terms of I_{in} to find $Z(s)$. Check your result and the expression for $H(s)$ by setting $s = 0$ and comparing it to a DC steady-state analysis of the circuit.

Exercise 5.4–2 Obtain an expression for $H(s) = V_{out}/I$ when the output voltage is taken across the resistance in Fig. 5.3–6. Hint: Redraw the circuit to bring out the node at the connection of L and R.

LOOP ANALYSIS

The loop method is the dual of the node method. It implicitly satisfies N node equations and produces M loop equations in terms of M unknown loop currents. The method has advantages when $M \leq N$ and we want to find an unknown current. However, it is somewhat trickier than the node method, especially when the network includes current sources.

Consider the circuit in Fig. 5.4–3a, which is labeled with three loop currents, I_{in}, I_{out}, and $g_m V_g$. Each current travels completely around a loop and equals the branch current in those branches that form part of just one loop. For instance, I_{in} is the branch current through R_1, and I_{out} is the branch current through R_3. But any branch in common with two or more loops may have more than one loop current through it; for instance, the actual branch current through C equals $I_{in} - I_{out}$. The source current

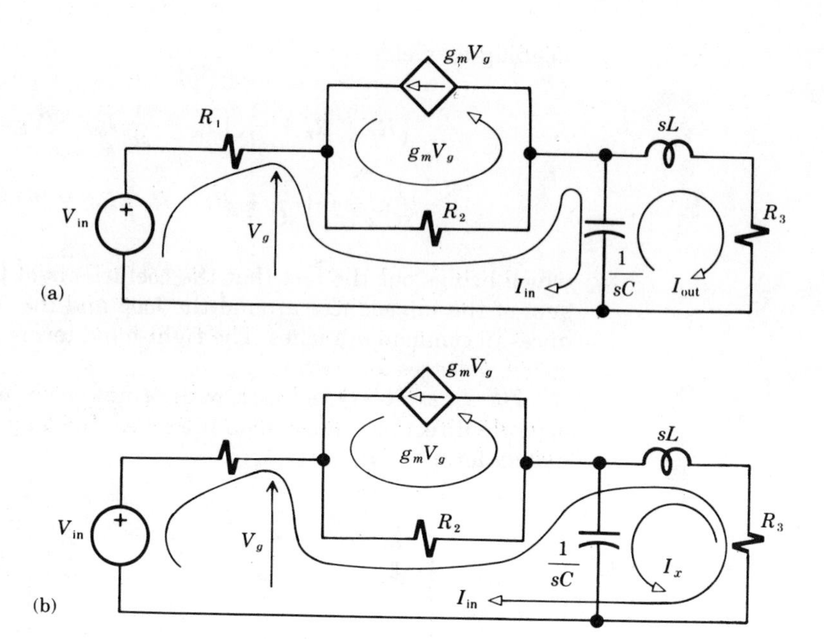

FIGURE 5.4–3
Loop analysis.

$g_m V_g$ is not an independent unknown so we need only $M = 2$ loop equations involving I_{in} and I_{out}—even though the network has three loops.

The selection of loop currents is partly arbitrary, but subject to the restriction that only one loop current may pass through a branch containing a current source. Whenever possible, you should draw unknown loop currents in opposite directions through common branches, and choose them to simplify any constraint equations. The loop currents in Fig. 5.4–3b satisfy these conditions as well, and the constraint equation for the control voltage V_g can be written as

$$V_g = V_{in} - R_1 I_{in}$$

in both cases. But the actual output current in Fig. 5.4–3b is the difference $I_{in} - I_x$ and finding the transfer function $H(s) = I_{out}/V_{in}$ will be easier using the other set of loop currents.

Proceeding with the analysis of Fig. 5.4–3a, we write the KVL loop equations by going around the path of each unknown loop current, summing voltage drops as follows:

$$R_1 I_{in} + R_2(I_{in} + g_m V_g) + \frac{1}{sC}(I_{in} - I_{out}) - V_{in} = 0$$

$$\frac{1}{sC}(I_{out} - I_{in}) + (sL + R_3)I_{out} = 0$$

Regrouping yields

$$\left(R_1 + R_2 + \frac{1}{sC}\right) I_{\text{in}} - \frac{1}{sC} I_{\text{out}} = V_{\text{in}} - R_2 g_m V_g$$

$$-\frac{1}{sC} I_{\text{in}} + \left(\frac{1}{sC} + sL + R_3\right) I_{\text{out}} = 0$$

which brings out the fact that the coefficients of the currents equal the sum of the impedances around the loop and the negative of the impedances in common branches. The right-hand terms are the net equivalent source voltages for each loop.

For an arbitrary network with M unknown loop currents that have opposite directions in common branches, the loop equations take on the matrix form

$$\begin{bmatrix} Z_{11} & -Z_{12} & \cdots & -Z_{1M} \\ -Z_{21} & Z_{22} & \cdots & -Z_{2M} \\ \vdots & \vdots & & \vdots \\ -Z_{M1} & -Z_{M2} & \cdots & Z_{MM} \end{bmatrix} \begin{bmatrix} I_1 \\ I_2 \\ \vdots \\ I_M \end{bmatrix} = \begin{bmatrix} V_1 \\ V_2 \\ \vdots \\ V_M \end{bmatrix} \tag{2}$$

where

Z_{kk} = Sum of all impedances around loop k
$Z_{kj} = Z_{jk}$ = Sum of impedances in the branch common to loops k and j
I_k = Unknown current around loop k
V_k = Net equivalent source voltage around loop k

Comparison of this matrix with the node-equation matrix, Eq. (1b), reveals identical structure and symmetry. Again, symmetry would be destroyed by the insertion of constraint equations.

Example 5.4–2

To underscore the similarity of our two methods, let's redo the bridge-circuit analysis from Example 3.3–1 using matrix equations to find the equivalent impedance of Fig. 5.4–4a when $s = 5000$, so $1/sC = 2\ \Omega$ and $sL = 20\ \Omega$. Since these impedance values equal the corresponding resistances back in Fig. 3.3–3, we know from our prior work that Z_{eq} will have the same value as $R_{\text{eq}} = 4.88\ \Omega$ for the given value of s.

To use node analysis, we apply a current source and identify the three unknown node voltages in Fig. 5.4–4b.

Thus

$$\begin{bmatrix} \frac{1}{5}+\frac{1}{4} & -\frac{1}{5} & -\frac{1}{4} \\ -\frac{1}{5} & \frac{1}{5}+\frac{1}{10}+\frac{1}{2} & -\frac{1}{10} \\ -\frac{1}{4} & -\frac{1}{10} & \frac{1}{4}+\frac{1}{10}+\frac{1}{20} \end{bmatrix} \begin{bmatrix} V \\ V_a \\ V_b \end{bmatrix} = \begin{bmatrix} I \\ 0 \\ 0 \end{bmatrix}$$

Multiplying out gives the same set of equations obtained in Example 3.3–1, and solving for V in terms of I leads to $Z_{eq} = V/I = 4.88\ \Omega$.

To use loop analysis, we apply a voltage source and identify the three unknown loop currents in Fig. 5.4–4c. Thus

$$\begin{bmatrix} 5+2 & -5 & -2 \\ -5 & 5+4+10 & -10 \\ -2 & -10 & 2+10+2 \end{bmatrix} \begin{bmatrix} I \\ I_a \\ I_b \end{bmatrix} = \begin{bmatrix} V \\ 0 \\ 0 \end{bmatrix}$$

Clearly, the solution involves the same amount of work as before because this particular circuit has an equal number of unknown node voltages and unknown loop currents.

FIGURE 5.4–4 Circuits for Example 5.4–2.

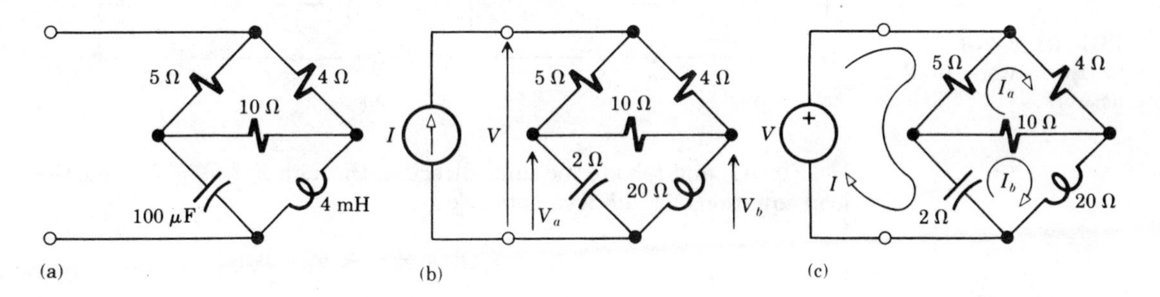

Exercise 5.4–3 Let the 10-Ω resistance in Fig. 5.4–4a be replaced by an open circuit, and let the network be driven by a voltage source $v = 10e^{-5000t}$ having a 6-Ω source resistance. Use loop analysis to find the current through the inductance.

Exercise 5.4–4 Repeat Example 5.3–3 using loop analysis and node analysis.

WYE-DELTA TRANSFORMATIONS

The three-terminal networks in Fig. 5.4–5 are known as *wye* (Y) and *delta* (Δ) networks in view of their configurations. They are also called *tee* (T) and *pi* (Π) networks when arranged as two-ports as shown in Fig. 5.4–6. Occasionally, it's convenient to transform a wye or tee network into an *equivalent* delta or pi network, or vice versa. Let's derive these transformation relations using the loop currents indicated in Fig. 5.4–6.

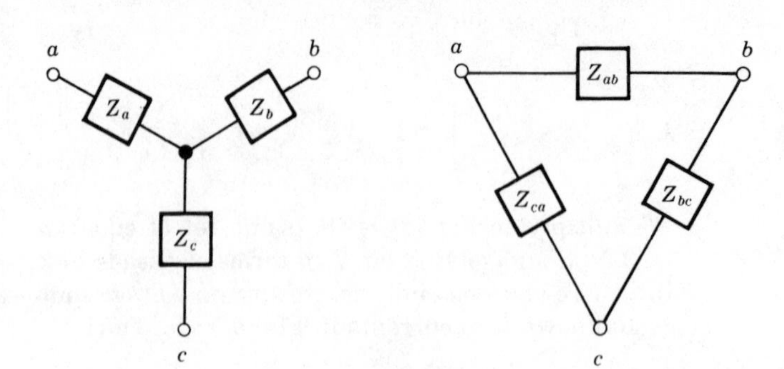

FIGURE 5.4–5
(a) Wye network.
(b) Delta network.

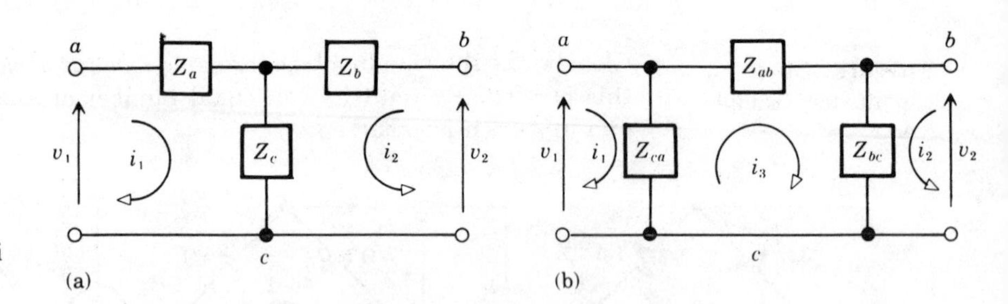

FIGURE 5.4–6
(a) Tee network. (b) Pi network.

Since I_1 and I_2 have the same direction through Z_c in Fig. 5.4–6a, the loop equations for the wye network are

$$
\begin{aligned}
(Z_a + Z_c)I_1 + Z_c I_2 &= V_1 \\
Z_c I_1 + (Z_b + Z_c)I_2 &= V_2
\end{aligned}
\tag{3a}
$$

For the delta network, I_1 and I_3 have opposite directions through Z_{ca}, while I_2 and I_3 have the same direction through Z_{bc}. Thus,

$$
\begin{aligned}
Z_{ca}I_1 + 0 \quad - \quad\quad\quad\quad\quad\quad\quad Z_{ca}I_3 &= V_1 \\
0 + Z_{bc}I_2 + \quad\quad\quad\quad\quad\quad\quad Z_{bc}I_3 &= V_2 \\
-Z_{ca}I_1 + Z_{bc}I_2 + (Z_{ab} + Z_{bc} + Z_{ca})I_3 &= 0
\end{aligned}
$$

Using the third equation to eliminate I_3 from the other two yields

$$
\begin{aligned}
\frac{Z_{ab}Z_{ca} + Z_{ca}Z_{bc}}{Z_{abc}} I_1 + \frac{Z_{ca}Z_{bc}}{Z_{abc}} I_2 &= V_1 \\
\frac{Z_{ca}Z_{bc}}{Z_{abc}} I_1 + \frac{Z_{bc}Z_{ab} + Z_{ca}Z_{bc}}{Z_{abc}} I_2 &= V_2
\end{aligned}
\tag{3b}
$$

where, for convenience, we have defined

$$Z_{abc} = Z_{ab} + Z_{bc} + Z_{ca}$$

Note that the form of Eq. (3b) is comparable to Eq. (3a).

If the two networks are equivalent, the coefficients of the two currents must be equal, and hence

$$Z_a = \frac{Z_{ab}Z_{ca}}{Z_{abc}} \qquad Z_b = \frac{Z_{bc}Z_{ab}}{Z_{abc}} \qquad Z_c = \frac{Z_{ca}Z_{bc}}{Z_{abc}} \tag{4}$$

which is the *delta-to-wye transformation.* Referring to Fig. 5.4–5 then shows that Z_a equals the product of the two delta impedances connected to a (Z_{ab} and Z_{ca}) divided by the sum of the delta impedances. Similar interpretations apply for Z_b and Z_c.

The *wye-to-delta transformation* is obtained by inverting Eq. (4), or by node analysis, as in Fig. 5.4–6. We then get

$$Z_{ab} = \frac{Z^2}{Z_c} \qquad Z_{bc} = \frac{Z^2}{Z_a} \qquad Z_{ca} = \frac{Z^2}{Z_b} \tag{5}$$

in which Z^2 stands for the sum of the pairwise products

$$Z^2 = Z_aZ_b + Z_aZ_c + Z_bZ_c$$

Thus, the delta impedance connecting nodes a and b equals Z^2 divided by the wye impedance at the opposite node (Z_c), and so forth.

The wye-delta transformations are useful in the study of two-port networks and three-phase AC power systems. They can also be used to calculate the equivalent resistance of bridge circuits and related configurations.

Exercise 5.4–5 Replace the lower delta in Fig. 5.4–4b with its wye-equivalent circuit and calculate Z_{eq} by series/parallel reduction.

PROBLEMS

5.1–1 Let capacitance C be connected across the secondary of the transformer in Fig. 4.1–4. Use Eq. (1) to show that $i_1 = N^2C\, dv_1/dt$.

5.1–2 What constant-current value is needed to store $w = 1$ J in an 8-μF capacitance in 0.1 sec?

5.1–3 Sketch $i(t)$, $w(t)$, and $p(t)$ when the voltage across a 1-μF capacitor has the waveform of Fig. P5.1–3.

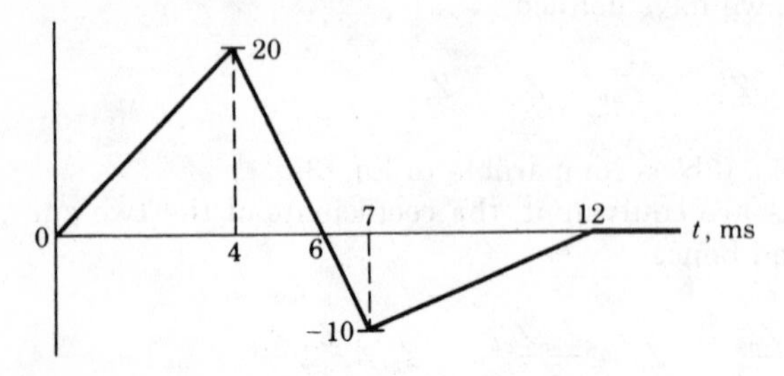

FIGURE P5.1–3

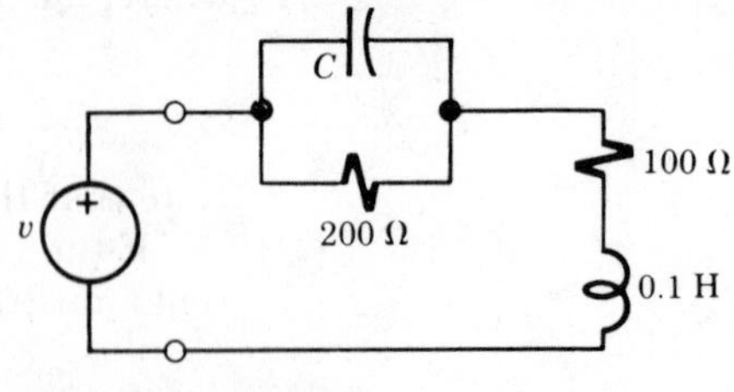

FIGURE P5.2–8

5.1–4 Let Fig. P5.1–3 be the current in milliamps through a 1-μF capacitor. Sketch $v(t)$, $w(t)$, and $p(t)$ assuming $V_0 = 0$.

5.1–5 The electric field strength between the plates in Fig. 5.1–1 is $\mathcal{E} = v/d$, and the dielectric breaks down if $\mathcal{E}$ exceeds a certain value. Calculate the maximum possible stored energy if the capacitor's volume is $Ad = 1\ \text{cm}^3$ and the dielectric is mica, for which $\epsilon \approx 5 \times 10^{-11}$ F/m and $\mathcal{E}_{\max} \approx 2 \times 10^8$ V/m.

5.1–6 Repeat the previous calculation with an air dielectric, for which $\epsilon = 10^{-9}/36\pi$ F/m and $\mathcal{E}_{\max} \approx 3 \times 10^6$ V/m.

5.1–7 Derive Eq. (8) from Fig. 5.1–8b by differentiating $v = v_1 + v_2$ and noting that $dv_1/dt = i/C_1$, etc.

5.1–8 Show that $v_1/v = C_2/(C_1 + C_2)$ in Fig. 5.1–8b.

5.2–1 Show that $v_1 = (L/N^2)\, di_1/dt$ when inductance L is connected across the secondary in Fig. 4.1–4.

5.2–2 A 10-V battery is applied to a 0.5-H inductor with $I_0 = 0$. How long does it take to store $w = 1$ J?

5.2–3 Let Fig. P5.1–3 be the current in amps through a 1-mH inductor. Sketch $v(t)$, $w(t)$, and $p(t)$.

5.2–4 Sketch $i(t)$, $w(t)$, and $p(t)$ when the voltage across a 1-mH inductor with $I_0 = 0$ has the waveform of Fig. P5.1–3.

5.2–5 What value of L in series with the capacitance in Exercise 5.1–3 causes the total voltage across the two elements to be equal to zero?

5.2–6 Suppose the op-amp circuit in Fig. 4.3–6a has R_1 replaced by C and R_F replaced by L. Show that $v_{\text{out}} = -LCd^2v_{\text{in}}/dt^2$.

5.2–7 Let a 0.2-H inductor be in parallel with the capacitor in Exercise 5.1–3. Taking $I_0 = 0$ in Eq. (3b), find the total current into the parallel combination.

5.2–8 Find the DC steady-state voltages and currents in Fig. P5.2–8 when $v = 12$ V. If the energy stored in the capacitor equals the energy stored in the inductor, what is the value of C?

5.2–9 Assume the circuit of Fig. 5.4–4a is in the DC steady-state condition with a 60-V battery across the terminals. Calculate the voltage across the capacitor, the current through the inductor, and the total stored energy.

5.2–10 Repeat Exercise 5.2–4 with the coils connected such that $i = i_1 = -i_2$ and $v = v_1 - v_2$.

5.3–1 Obtain a better approximation for the values of i and v_L at $t = 0.2$ sec in Example 5.3–1 by taking time steps of 0.05 sec starting from $t = 0$.

5.3–2 Use time steps of 0.5 ms to approximate the values of v_C and i at $t = 2.0$ ms in Exercise 5.3–1.

5.3–3 Use impedance analysis to prove Eq. (5), Section 5.2.

5.3–4 Use impedance analysis to confirm the assertion in Problem 5.1–8.

5.3–5 Let an additional resistance R' be in parallel with C in Fig. 5.3–6.

(a) Find $Y(s)$ and $Z(s)$ expressed as a ratio of polynomials, checking your result by setting $s = 0$.

(b) Find i_C when $L = 0.1$ H, $R = R' = 100\ \Omega$, $C = 10\ \mu$F, and $i = 2e^{1000t}$ A.

5.3–6 Repeat the previous problem with R' inserted in series with C rather than in parallel.

5.3–7 Find $Z(s)$ for Fig. P5.2–8 with $C = 5\ \mu$F. Then calculate i when $v = 60 + 30e^{-2000t}$ V.

5.3–8 Repeat the previous problem with C replaced by a 0.2-H inductance.

5.3–9 Find $H(s) = V_{out}/V$ for Fig. P5.2–8 with $C = 5\ \mu$F and the output voltage taken across the RL branch. Evaluate your result at $s = 0$, 1000, and -1000.

5.3–10 Find $H(s) = I_C/I$ for the circuit in Problem 5.3–6 with the element values given in Problem 5.3–5b. Evaluate your result at $s = 0$, 1000, and -1001.

5.3–11 What element should be used for $Z(s)$ in the amplifier circuit model of Fig. P5.3–11 such that $v_{out} = g_m R_L v_{in}$ for any input waveform?

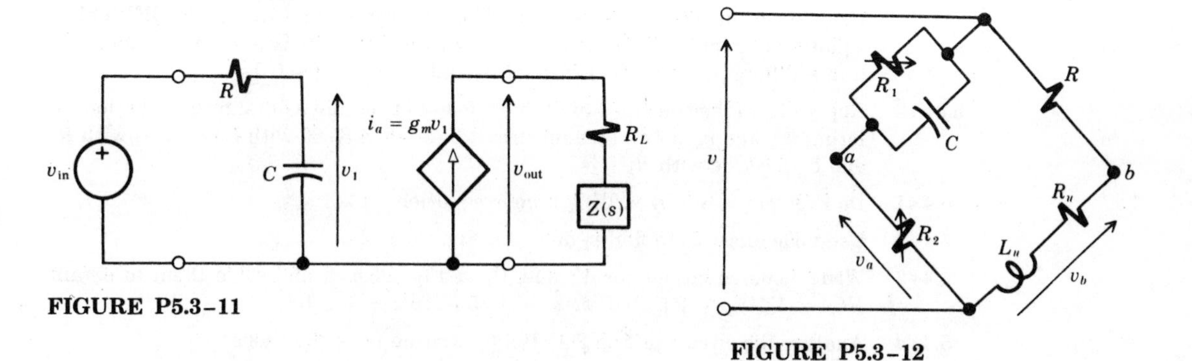

FIGURE P5.3–11

FIGURE P5.3–12

5.3–12 Figure P5.3–12 is a *Maxwell bridge* with adjustable resistances R_1 and R_2 for the purpose of measuring the resistance R_u and inductance L_u of a coil. Write impedance expressions for V_a and V_b to show that the balance condition $V_a - V_b = 0$ corresponds to $R_u = R_2 R/R_1$ and $L_u = CR_2R$.

5.3–13 The *gyrator* in Fig. P5.3–13 is an active two-port network with the idealized properties $i_1 = -v_2/r_2$ and $i_2 = v_1/r_1$, where r_1 and r_2 are constants with the units in ohms.

(a) Find $H(s) = V_2/V_1$ when a load impedance $Z_2(s)$ is connected to the output terminals, and show that $Z(s) = V_1/I_1 = r_1 r_2/Z_2(s)$.

(b) Obtain the equivalent circuit for $Z(s)$ when $Z_2(s)$ consists of R in parallel with C.

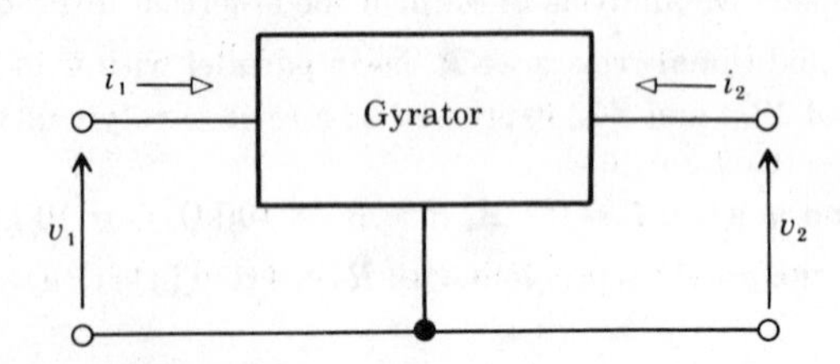

FIGURE P5.3–13

5.3–14 Given $Z(s)$, $Y(s)$, or $H(s)$, we can work backward to obtain the *differential equation* that relates the variables in question. For example, the circuit in Fig. 5.3–2 has

$$Z(s) = \frac{s^2LC + sRC + 1}{sC} = \frac{Ve^{st}}{Ie^{st}} = \frac{v}{i}$$

which can be rewritten as

$$Ls^2i + Rsi + \frac{1}{C}i = sv$$

Comparing this last expression with Eq. (4b), we see that we get the differential equation by replacing s^2i with d^2i/dt^2, si with di/dt, etc. Use this approach to obtain a differential equation relating i_C and i in Example 5.3–3.

5.3–15 Apply the method outlined in Problem 5.3–14 to obtain a differential equation relating v_{out} and v_{in} for an op-amp circuit like Fig. 4.3–6a with L in series with R_1 and C in series with R_F.

5.4–1 Do Problem 5.3–9 by writing a node equation.

5.4–2 Use node analysis to find v_1 and i_1 in Fig. P5.4–2.

5.4–3 Write node equations for V_1 and V_2 in Fig. P5.4–3 and solve them to obtain $H(s) = V_2/V_{in} = 2.5 \times 10^4 s/(s^2 + 1.25 \times 10^5 s + 5 \times 10^9)$.

5.4–4 Analyze the circuit in Fig. 2.4–10a by writing node equations.

5.4–5 Use loop analysis to find i_1 and v_1 in Fig. P5.4–2.

5.4–6 Write loop equations for I_1 and I_2 in Fig. P5.4–6, and show that $H(s) = I_2/I_{in}$ has the same expression as in Problem 5.4–3.

5.4–7 Figure P5.4–7 is called a *twin-tee* circuit. Solve the set of three node equations to obtain $H(s) = (s^2 + a^2)/(s^2 + 4as + a^2)$ where $a = 1/RC$.

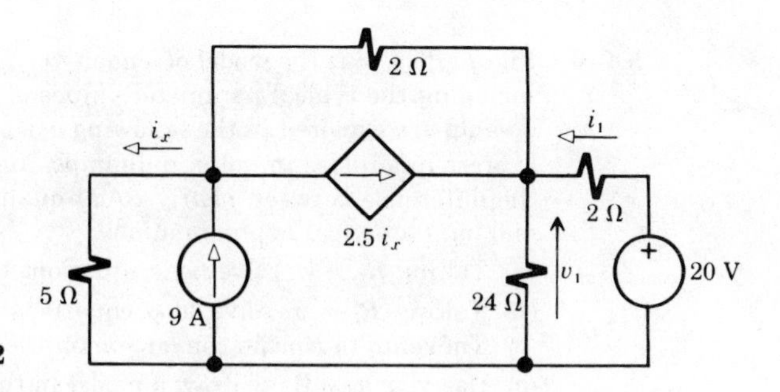

FIGURE P5.4-2

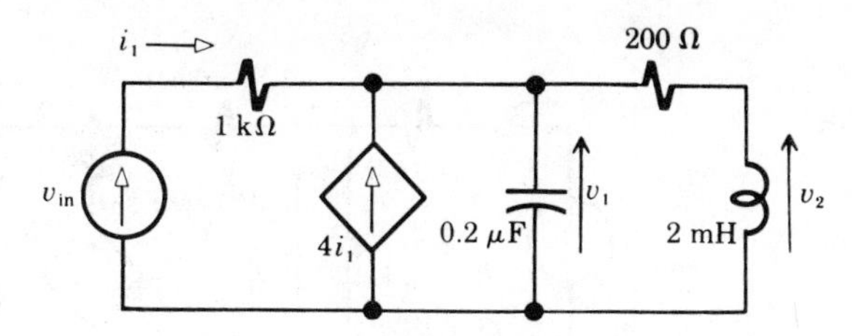

FIGURE P5.4-3

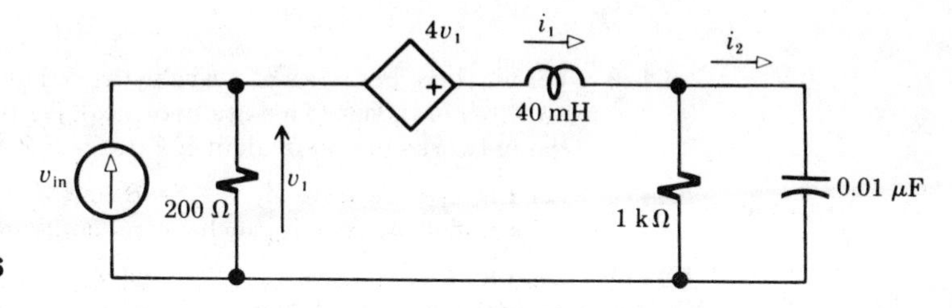

FIGURE P5.4-6

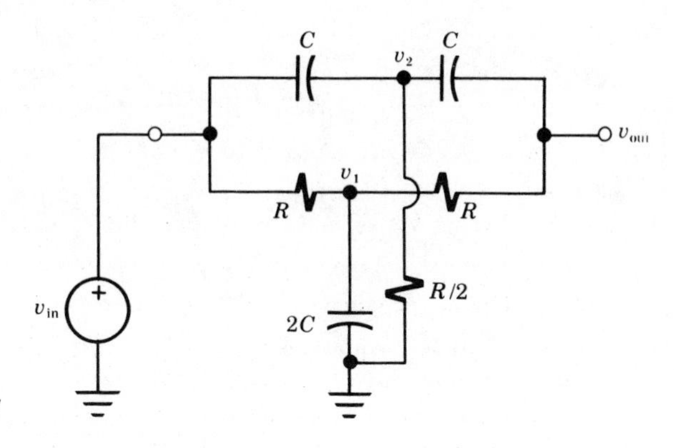

FIGURE P5.4-7

5.4–8 Figure P5.4–8 is the model of a noninverting op-amp circuit like Fig. 4.3–4b, but including the typical parameter values of a real op-amp. The effects of the real op-amp are explored in the following calculations. You will find it convenient to express quantities in volts, milliamps, and kilohms. Some calculations involve the difference between nearly equal quantities, so you should be careful when making numerical approximations.

(a) Taking $R_L = \infty$, solve node equations to evaluate $v_{\text{out}}/v_{\text{in}}$ and $i_{\text{in}}/v_{\text{in}}$.

(b) Taking $R_L = 0$, solve loop equations to evaluate $i_{\text{out}}/v_{\text{in}}$. Hint: Perform a Thévenin-to-Norton conversion on the branch with the controlled source.

(c) Use your results to draw a model in the form of Fig. 4.2–6a, and compare it with Fig. 4.3–4d.

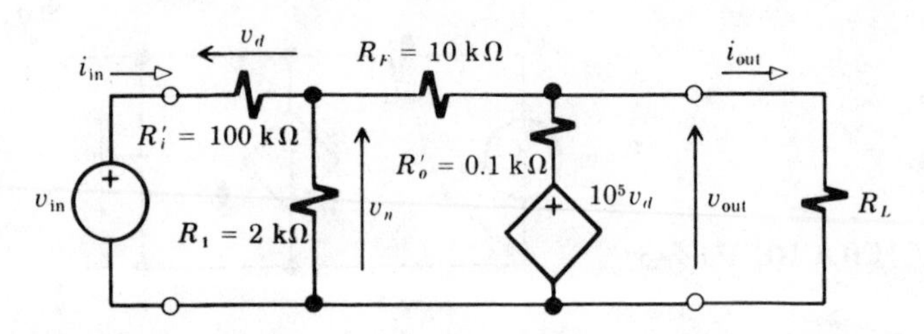

FIGURE P5.4–8

5.4–9 Let all three branches of a wye network have the same impedance $Z_Y(s)$, and let all three branches of a delta network have the same impedance $Z_\Delta(s)$. Show that the networks are equivalent if $Z_\Delta(s) = 3\,Z_Y(s)$.

5.4–10 Let Fig. 5.4–6a have $Z_a = Z_b = R$ and $Z_c = 1/sC$. Draw the equivalent pi network and show that Z_{ab} includes an equivalent inductance.

5.4–11 Let Fig. 5.4–6b have $Z_{ab} = sL$ and $Z_{bc} = Z_{ca} = R$. Draw the equivalent tee network and show that $Y_c = 1/Z_c$ includes an equivalent capacitance.

6 AC CIRCUITS

AC circuits have long been the bread and butter of electrical engineering—in power transmission, consumer products, lighting, and machinery. Further, an understanding of AC circuit response is an essential prerequisite for topics such as communication systems, automatic control, and analog instrumentation. Thus, our work in this chapter has both immediate and subsequent applications.

We begin by defining phasor notation and AC impedance as the keys to efficient AC circuit analysis, with the help of complex-number algebra. These tools are then used to find the voltage-current relationships in circuits that include capacitance and inductance as well as resistance. We also examine average power, resonance, and residential wiring. (Chapter 7 considers the frequency-response characteristics of AC circuits, while Chapter 18 deals with AC power systems.)

OBJECTIVES

After studying this chapter and working the exercises, you should be able to do each of the following:

- Represent a sinusoidal waveform by a phasor, and use phasor addition to evaluate the sum of two or more sinusoids (Section 6.1).

- Calculate the equivalent AC impedance of a one-port network, and find its AC resistance and reactance (Section 6.2).
- Obtain the AC response of a given circuit and determine the average power dissipation (Section 6.3).
- Construct phasor diagrams for the voltages and currents in an RC, RL, or RLC circuits (Section 6.3).
- Evaluate and interpret the resonant frequency and quality factor of a series or parallel RLC circuit (Section 6.3).
- Describe the major features of dual-voltage residential AC circuits (Section 6.4). †

6.1 PHASORS AND COMPLEX NUMBERS

Phasors and complex numbers are somewhat abstract but very important tools for AC circuit analysis. The basic ideas are presented here, following a simple example that will establish a perspective for subsequent sections when we dig into the details of AC circuits.

INTRODUCTION TO AC CIRCUITS

Consider the series RL circuit, Fig. 6.1–1a, driven by an AC current

$$i = I_m \cos \omega t$$

The corresponding *steady-state* (forced) voltages are

$$v_R = Ri = RI_m \cos \omega t$$

$$v_L = L\frac{di}{dt} = LI_m \frac{d}{dt}(\cos \omega t)$$

$$= LI_m\, \omega\, (-\sin \omega t) = \omega LI_m \cos (\omega t + 90°)$$

where we have used the trigonometric relationship $-\sin \omega t = \cos (\omega t + 90°)$. Kirchhoff's voltage law then yields the resulting terminal voltage:

$$v = v_R + v_L = RI_m \cos \omega t + \omega LI_m \cos (\omega t + 90°) \tag{1}$$

Further insight and a simpler expression for v emerges if we represent the two terms on the right-hand side of Eq. (1) as *horizontal projections* of the two *vectors* drawn in Fig. 6.1–1b. One vector has length RI_m and angle ωt with respect to the horizontal axis, so its horizontal projection indeed equals $RI_m \cos \omega t$. The other vector has length ωLI_m and angle $\omega t + 90°$, and its horizontal projection $\omega LI_m \cos (\omega t + 90°)$ happens to be a negative quantity because the drawing assumes a value of t such

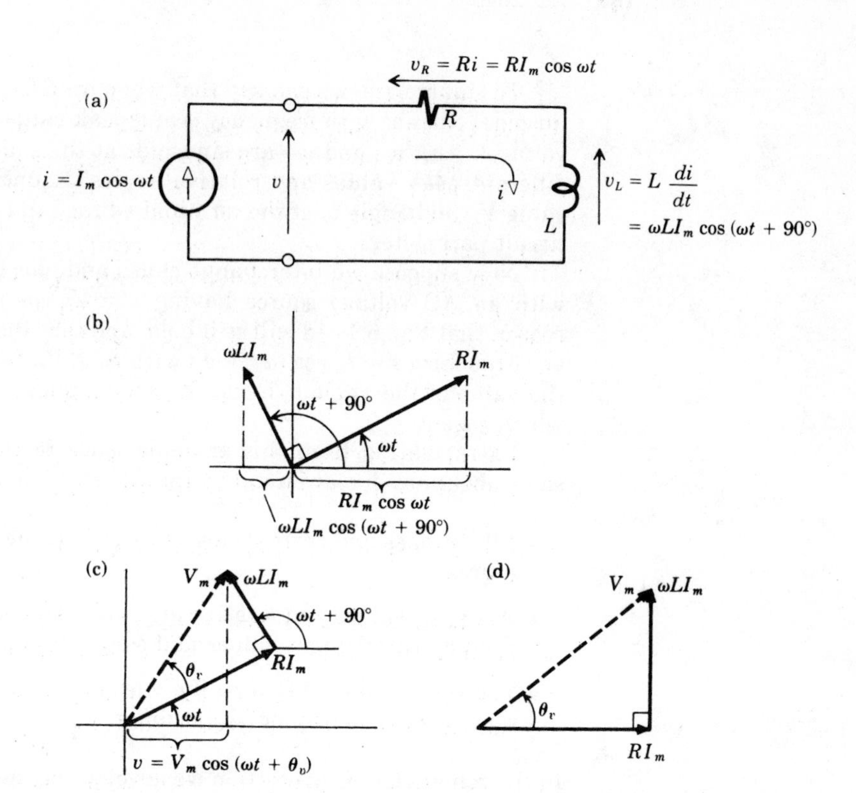

FIGURE 6.1–1

that $0 < \omega t < 90°$. Of course the angles ωt and $\omega t + 90°$ change with time, but the two vectors *always remain perpendicular* to each other.

Taking the algebraic sum of the projections yields the value of v, shown vectorially in Fig. 6.1–1c where the resulting vector sum has a length denoted by V_m and an angle $\omega t + \theta_v$. Thus, we can write

$$v = V_m \cos(\omega t + \theta_v) \tag{2}$$

and our final step requires relating V_m and θ_v to the known parameters I_m, ω, R, and L. For that purpose, the right triangle in Fig. 6.1–1c is redrawn in Fig. 6.1–1d, clearly revealing that

$$V_m^2 = (RI_m)^2 + (\omega LI_m)^2 = [R^2 + (\omega L)^2]\, I_m^2 \qquad \tan\theta_v = \frac{\omega LI_m}{RI_m} = \frac{\omega L}{R}$$

Therefore,

$$V_m = \sqrt{R^2 + (\omega L)^2}\, I_m \qquad \theta_v = \arctan\frac{\omega L}{R} \tag{3}$$

where "arctan" stands for the inverse tangent.

To summarize, we can say that when an RL circuit is driven by a sinusoidal current with frequency ω and peak value I_m, all of the resulting voltages—v_R, v_L, and v—are sinusoids at the same frequency, but have different peak values and relative angles. Equation (3) gives the peak value V_m and angle θ_v of the terminal voltage in terms of the source and circuit parameters.

Now suppose we interchange cause and effect and drive this circuit with an AC voltage source having $v = V_m \cos(\omega t + \theta_v)$. It stands to reason that Fig. 6.1–1d will still hold, and that the resulting current will be of the form $i = I_m \cos(\omega t + \theta_i)$ with $I_m = V_m/\sqrt{R^2 + (\omega L)^2}$. However, the value of the angle θ_i is not obvious—unless $\theta_v = \arctan(\omega L/R)$, in which case $\theta_i = 0$.

Extrapolating from this example leads to three important conclusions about *any* linear circuit in the AC steady-state condition:

- All voltages and currents are sinusoids at the same frequency as the source.
- The peak values and relative angles of the voltages and currents depend on the element values and source frequency.
- The relationships between the various AC waveforms can be represented conveniently by vector diagrams.

In the remainder of this section we develop the mathematical tools of AC circuit analysis based on these conclusions. Specifically, the phasor concept is defined as the vector representation of sinusoids; and the algebra of complex numbers is presented as the means of manipulating phasor expressions.

SINUSOIDS AND PHASORS

In general, any AC voltage waveform can be written as

$$v = V_m \cos(\omega t + \theta_v) \tag{4}$$

which is plotted in Fig. 6.1–2a. Three and only three parameters suffice

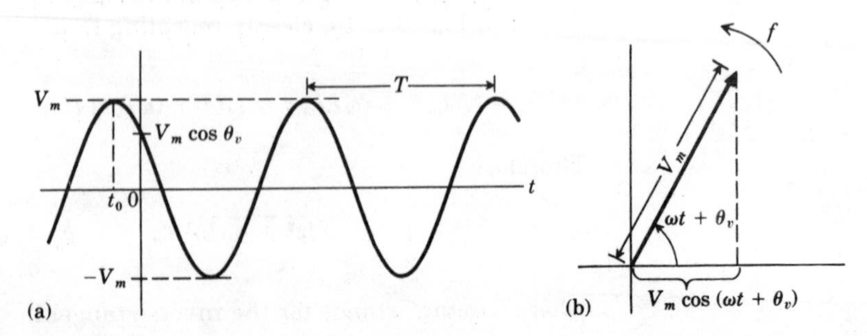

FIGURE 6.1–2
(a) AC waveform.
(b) Rotating-vector representation.

to describe completely such a waveform, namely the peak value or *amplitude* V_m, the *angular frequency* ω, and the *phase angle* θ_v. Those same three parameters are shown in a different but equivalent form in Fig. 6.1–2b, where v is taken as the horizontal projection of a *rotating vector* with length V_m, rotational or cyclical frequency $f = \omega/2\pi$, and angle θ_v at $t = 0$. Identical drawings hold, of course, for an AC current $i = I_m \cos(\omega t + \theta_i)$ with I_m and θ_i replacing V_m and θ_v.

We previously stated in Section 3.2 that angular frequency ω (in rad/s), cyclical frequency f (in hertz), and oscillation period T are related by $\omega = 2\pi f = 2\pi/T$. We also defined the effective or rms value of a sinusoid, but throughout *this* chapter peak values will be used for amplitude and vector length.

The new parameter here is the phase angle θ_v measured in radians or, more commonly, in degrees. Figure 6.1–2b gives a simple interpretation of θ_v as the vector's direction at $t = 0$. However, relative to the actual waveform in Fig. 6.1–2a, θ_v means that the central peak is advanced to the left of the time origin and occurs at time

$$t_0 = -\frac{\theta_v(\text{rad})}{\omega} = -\frac{\theta_v(\text{rad})}{2\pi} T = -\frac{\theta_v(\text{deg})}{360^\circ} T \tag{5}$$

Thus, $v = V_m$ at $t = t_0$, whereas $v = V_m \cos\theta_v$ at $t = 0$. A negative phase angle obviously means a delayed central peak occurring at $t_0 > 0$. Equation (5) is obtained from Eq. (4) by setting $\omega t_0 + \theta_v = 0$ and recalling that 2π rad $= 360^\circ$.

In view of the fact that all AC voltages and currents in a given circuit have the same frequency as the source, differing only in amplitude and phase angle, considerable unnecessary pencil-pushing is eliminated if we represent sinusoidal voltage and current by the short-hand *phasor notation*

$$\underline{V} = V_m \underline{/\theta_v} \qquad \underline{I} = I_m \underline{/\theta_i}$$

Here, $\underline{V}$ stands for a *fixed vector* of length or magnitude V_m at an angle θ_v relative to the horizontal axis, like a snapshot of the rotating vector in Fig. 6.1–2b taken at time $t = 0$. Likewise for the phasor $\underline{I}$.

The term *phasor* designates a vector that represents a sinusoidal time function. Phasor notation allows us to concentrate on the unknown amplitude and angle—V_m and θ_v, or I_m and θ_i—without having to write down the full-blown AC expression at each step of the analysis. Once we find the unknowns, we simply plug their values into $v = V_m \cos(\omega t + \theta_v)$ or $i = I_m \cos(\omega t + \theta_i)$—and this is the only time we need the entire sinusoidal expression.

Besides the symbolic convenience, phasor notation readily lends itself to triangular constructions of the type used in Fig. 6.1–1d to evaluate the sum of two or more sinusoids. Specifically, *Kirchhoff's laws hold in phasor form,* with sinusoids replaced by their phasor representations. For instance, to find the resultant AC current $i = I_m \cos(\omega t + \theta_i)$ in Fig. 6.1–3a, we replace

$$i = I_1 \cos(\omega t + \theta_1) + I_2 \cos(\omega t + \theta_2)$$

by its phasor version

$$\underline{I} = \underline{I}_1 + \underline{I}_2 = I_1 \underline{/\theta_1} + I_2 \underline{/\theta_2}$$

Figure 6.1–3b diagrams this phasor summation. Such phasor manipulations are handled with the help of complex algebra as discussed next.

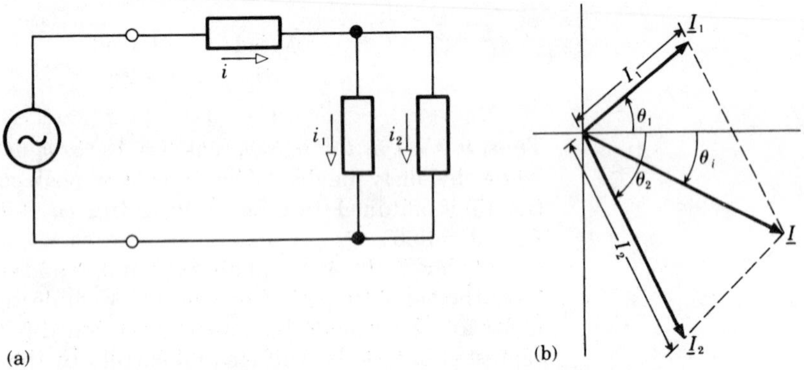

FIGURE 6.1–3
Phasor summation of AC currents.

Example 6.1–1

Suppose the current source in Fig. 6.1–1a has $I_m = 2$ A and $f = 100$ Hz, while $R = 35\ \Omega$ and $L = 0.1$ H. Then $T = 1/100 = 10$ ms, $\omega = 2\pi \times 100 = 628$ rad/s, $\omega L = 62.8$, and inserting values into Eq. (3) gives the amplitude and phase of the terminal voltage as

$$V_m = \sqrt{35^2 + 62.8^2} \times 2 = 144 \text{ V}$$

$$\theta_v = \arctan \frac{62.8}{35} = 60.9°$$

Thus, $\underline{V} = 144 \text{ V} \underline{/60.9°}$ and Eq. (5) tells us that the voltage waveform peaks at $t_0 = -(60.9°/360°) \times 10 \text{ ms} \approx -1.7$ ms.

Exercise 6.1–1

Find the phasor representations of v_R and v_L for the foregoing circuit, and sketch the waveform of v_L.

COMPLEX ALGEBRA

Frequently in engineering analysis we need to represent a quantity that has two distinct but related parts. This can be done using a vector or by using a complex number. The advantages of both methods are combined if a complex number is treated as a vector in a *complex plane.*

The complex plane, Fig. 6.1–4, is a two-dimensional space wherein any point $\underline{A}$ is uniquely specified by two *rectangular coordinates:* its horizontal coordinate or *real part* $\text{Re}[\underline{A}] = A_r$ and its vertical coordinate or *imaginary part* $\text{Im}[\underline{A}] = A_i$. Of course there's nothing "imaginary" about the quantity $\text{Im}[\underline{A}]$; we simply use the designations "real" and "imaginary" to distinguish between the two directional components. Likewise, the *imaginary unit* $j \triangleq \sqrt{-1}$ merely serves as a way of labeling the vertical component of $\underline{A}$ when it is written as the *complex number*

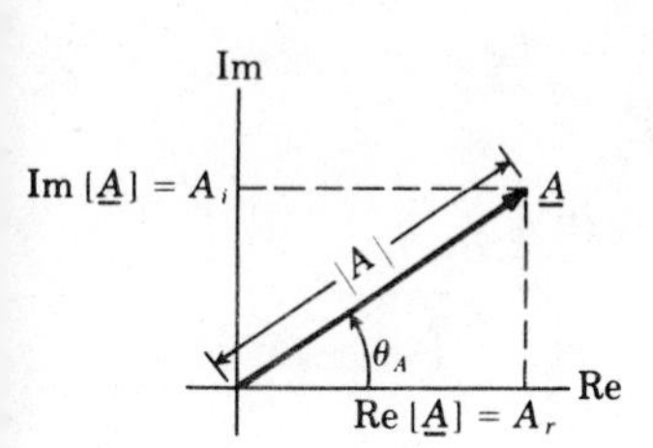

FIGURE 6.1–4 The complex plane and a complex number $\underline{A}$.

$$\underline{A} = \text{Re}[\underline{A}] + j\text{Im}[\underline{A}] = A_r + jA_i \tag{6}$$

Either part of a complex number may be positive, negative, or zero—but it is always a *real* quantity. The quantity $\underline{A}$ is complex only because its imaginary part is multiplied by j.

We also identify the point $\underline{A}$ by its *polar coordinates:* the *magnitude* $|\underline{A}| = |A|$, equal to the distance from the origin to the point, and the *angle* $\measuredangle\underline{A} = \theta_A$, measured counterclockwise relative to the positive real axis. We then write $\underline{A} = |A| \underline{/\theta_A}$, a complex-plane *vector.*

To convert from the polar coordinates to rectangular coordinates, we see in Fig. 6.1–4 that

$$\begin{aligned} A_r &= \text{Re}[\underline{A}] = |A| \cos \theta_A \\ A_i &= \text{Im}[\underline{A}] = |A| \sin \theta_A \end{aligned} \tag{7}$$

and conversely

$$|A| = |\underline{A}| = \sqrt{A_r^2 + A_i^2} \tag{8a}$$

$$\theta_A = \measuredangle\underline{A} = \arctan \frac{A_i}{A_r} \tag{8b}$$

Equation (8b) assumes that $A_r \geq 0$; if $A_r < 0$, the angle can be found from

$$\theta_A = \pm 180° - \arctan\left(\frac{A_i}{-A_r}\right) \tag{8c}$$

Electronic calculators are made-to-order for such computations—especially if the calculator happens to have direct rectangular-to-polar and polar-to-rectangular conversion. (But pity the thousands of engineering students in decades past who sweated over these problems with pencil, paper, and slide rule!)

Euler's theorem formally links complex numbers and vectors via the exponential function. It states that, for any angle ϕ,

$$e^{j\phi} = \cos\phi + j\sin\phi \tag{9}$$

which is a complex number having

$$\mathrm{Re}[e^{j\phi}] = \cos\phi \qquad \mathrm{Im}[e^{j\phi}] = \sin\phi$$

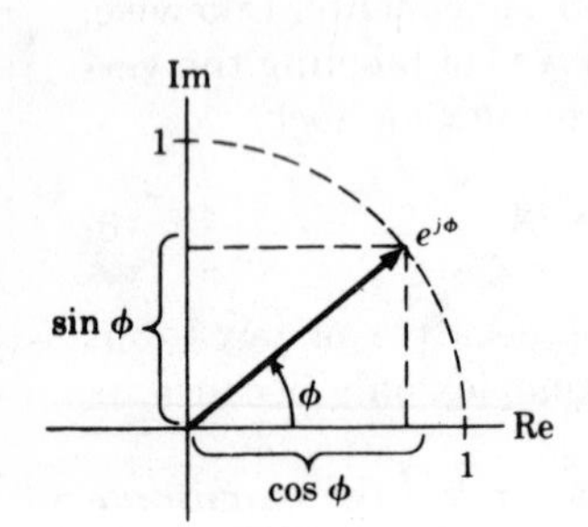

FIGURE 6.1–5
Euler's theorem in the complex plane.

Comparison with Eq. (5) then reveals that

$$|e^{j\phi}| = 1 \qquad \measuredangle e^{j\phi} = \phi$$

and hence

$$e^{j\phi} = 1\underline{/\phi}$$

meaning a *unit-length vector* at angle ϕ, as in Fig. 6.1–5. Therefore, an arbitrary vector with magnitude $|A|$ and angle θ_A can be written as

$$\underline{A} = |A|\, e^{j\theta_A}$$

which is the mathematical expression for the polar form, containing exactly the same information as our informal notation $\underline{A} = |A|\ \underline{/\theta_A}$. Hereafter, we will use both notations.

Addition and *subtraction* in the complex plane obey the usual rules of vector arithmetic. One separately adds or subtracts the horizontal and vertical components. So given the rectangular coordinates of $\underline{A}$ and $\underline{B}$,

$$\begin{aligned}\underline{A} \pm \underline{B} &= (A_r + jA_i) \pm (B_r + jB_i) \\ &= (A_r \pm B_r) + j(A_i \pm B_i)\end{aligned} \tag{10}$$

Multiplication and *division,* however, are more easily done with polar coordinates using the exponential properties $e^{\alpha}e^{\beta} = e^{(\alpha+\beta)}$ and $1/(e^{\alpha}) = e^{-\alpha}$ for any α and β. Specifically,

$$\underline{AB} = (|A|\, e^{j\theta_A})(|B|\, e^{j\theta_B}) = |A|\,|B|\, e^{j(\theta_A+\theta_B)} = |A|\,|B|\ \underline{/\theta_A + \theta_B} \tag{11}$$

and

$$\underline{A}/\underline{B} = \frac{|A|\, e^{j\theta_A}}{|B|\, e^{j\theta_B}} = \frac{|A|}{|B|}\, e^{j(\theta_A-\theta_B)} = \frac{|A|}{|B|}\ \underline{/\theta_A - \theta_B} \tag{12}$$

Note carefully that magnitudes multiply or divide, whereas angles add or subtract.

Equation (11) helps explain how the imaginary unit represents the vertical direction of the complex plane. To see this connection, note that j is a complex number with zero real part and unit imaginary part, so

$$j = 0 + j1 = 1\ \underline{/90^\circ} \tag{13a}$$

Multiplying any vector A by j thus gives

$$j\underline{A} = (1\ \underline{/90^\circ})(|A|\ \underline{/\theta_A}) = |A|\ \underline{/\theta_A + 90^\circ}$$

which is equivalent to rotating it through an angle of 90°. If $\underline{A} = j$ then $j \times j = 1\ \underline{/90^\circ + 90^\circ} = 1\ \underline{/180^\circ}$. Consistency therefore requires the "imaginary" property

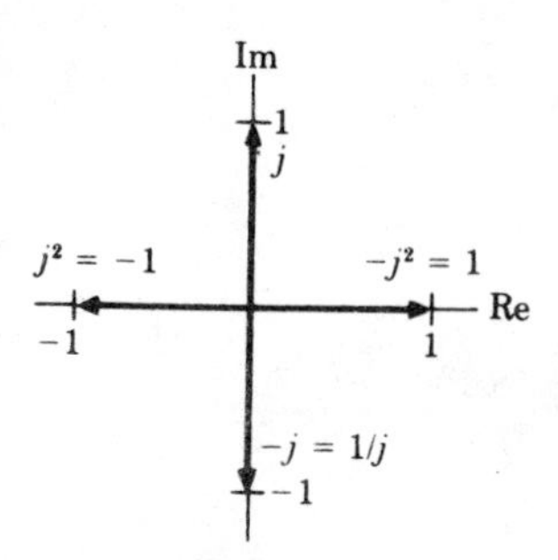

FIGURE 6.1–6 Vector representations of j, j^2, $-j$, and $-j^2$.

$$j^2 = 1\underline{/180^\circ} = -1 \tag{13b}$$

and hence $j = \sqrt{-1}$. A related property is the reciprocal

$$\frac{1}{j} = -j = 1\underline{/-90^\circ} \tag{13c}$$

which is consistent with Eq. (12) when $\underline{A} = 1$ and $\underline{B} = j$. Figure 6.1–6 summarizes these vector characteristics of j.

Multiplication can be carried out in rectangular form if we bear in mind that $j^2 = -1$. Thus

$$\begin{aligned}\underline{AB} &= (A_r + jA_i)(B_r + jB_i)\\ &= A_rB_r + jA_rB_i + jA_iB_r + j^2A_iB_i\\ &= (A_rB_r - A_iB_i) + j(A_rB_i + A_iB_r)\end{aligned}$$

Division in rectangular form is also possible with the help of the *complex conjugate,* which is formed by inverting the sign of the imaginary part of a complex number; specifically,

$$\underline{A}^* \triangleq A_r - jA_i = |A|\underline{/-\theta_A} \tag{14}$$

Multiplying $\underline{A}$ by $\underline{A}^*$ then yields a purely real quantity

$$\underline{AA}^* = |A|^2 = A_r^2 + A_i^2 \tag{15}$$

We apply this property in rectangular division to "rationalize" the denominator, as follows:

$$\underline{A}/\underline{B} = \frac{\underline{A}\underline{B}^*}{\underline{B}\underline{B}^*} = \frac{1}{|\underline{B}|^2}(A_r + jA_i)(B_r - jB_i)$$
$$= \frac{A_rB_r + A_iB_i}{B_r^2 + B_i^2} + j\frac{A_iB_r - A_rB_i}{B_r^2 + B_i^2}$$

Clearly, polar multiplication and division are more simple and more direct if we want the results in polar form.

Example 6.1–2

Suppose we know that $\underline{I}_1 = 9 + j5$ and $\underline{I}_2 = 6 - j13 = 6 + j(-13)$ in Fig. 6.1–3b. Then $\underline{I} = \underline{I}_1 + \underline{I}_2$ has

$$\text{Re}[\underline{I}] = 9 + 6 = 15 \qquad \text{Im}[\underline{I}] = 5 - 13 = -8$$

and hence

$$I_m = |\underline{I}| = \sqrt{15^2 + (-8)^2} = 17$$
$$\theta_i = \measuredangle \underline{I} = \arctan(-8/15) = -28.1°$$

Converting back to rectangular form gives

$$\text{Re}[\underline{I}] = 17 \cos(-28.1°) = 15$$
$$\text{Im}[\underline{I}] = 17 \sin(-28.1°) = -8$$

in agreement with our starting values. Going full circle like this helps detect possible numerical errors. At the very least, you should draw a simple diagram roughly to scale to be sure your results do make sense.

Exercise 6.1–2

Draw the phasor diagram representing $v = 12 \cos(\omega t - 30°) + 20 \cos(\omega t + 45°)$ and find $\underline{V}$ in both rectangular and polar forms.

Exercise 6.1–3

If $\underline{A} = 3 - j4$ find $j\underline{A}$, $\underline{A}\underline{A}$, and $1/\underline{A}$ in rectangular and polar forms.

6.2 AC IMPEDANCE

Having introduced phasor notation and the algebra of complex numbers, we are now prepared to show the role of impedance in AC circuit analysis. We also define the related concepts of AC resistance and reactance, and develop some techniques for AC circuit calculations.

AC IMPEDANCE AND ADMITTANCE

Let Fig. 6.2–1 represent a linear one-port network (containing no independent sources) with an AC excitation at frequency ω. The phasor voltage and current at the input terminals are, in general,

$$\underline{V} = V_m \underline{/\theta_v} \qquad \underline{I} = I_m \underline{/\theta_i}$$

as indicated on the diagram. The network's *AC impedance* is a complex quantity or vector $\underline{Z}$ obtained by setting $s = j\omega$ in $Z(s)$, so that

$$\underline{Z} \triangleq Z(j\omega) \tag{1}$$

The AC impedance relates the terminal phasors via

$$\underline{V} = \underline{Z}\underline{I} \tag{2}$$

which may be called "Ohm's law for AC circuits."

Equation (2) is similar to our impedance relationship for exponential time variations except that we are now dealing with vector quantities, each having a magnitude and angle. This somewhat complicates the calculations although the principles are simple enough. In particular, if $\underline{Z}$ has the polar form

$$\underline{Z} = |Z| \underline{/\theta_z}$$

then Eq. (2) becomes

$$V_m \underline{/\theta_v} = (|Z| \underline{/\theta_z})(I_m \underline{/\theta_i}) = |Z| I_m \underline{/\theta_z + \theta_i}$$

Hence

$$V_m = |Z| I_m \qquad \theta_v = \theta_z + \theta_i \tag{3}$$

and we have separate equations for the magnitudes and angles. The expression $\underline{V} = \underline{Z}\underline{I}$ contains both of these equations in vector notation.

We can also define the AC *admittance* of a network as

$$\underline{Y} \triangleq Y(j\omega) = \frac{1}{Z(j\omega)} \tag{4}$$

Since $\underline{Z} = 1/\underline{Y}$, the magnitudes and angles of the impedance and admittance are related by

$$|Z| = \frac{1}{|Y|} \qquad \sphericalangle \underline{Z} = -\sphericalangle \underline{Y} \tag{5}$$

obtained from Eq. (12), Section 6.1, with $\underline{A} = 1$ and $\underline{B} = \underline{Y}$.

Our next task will be an examination of the characteristics of $\underline{Z}$ (or $\underline{Y}$)

for specific circuits. Before doing so, however, we outline the proof of $\underline{V} = \underline{Z}\underline{I}$, which hinges upon Euler's theorem to represent sinusoidal functions as *complex exponentials* of the form $e^{j\omega t}$. The voltage phasor $\underline{V} = V_m \underline{/\theta_v} = V_m e^{j\theta_v}$ in Fig. 6.2–1 stands for the AC waveform $v = V_m \cos(\omega t + \theta_v)$. But consider, instead, the related complex voltage

$$\begin{aligned} \underline{v} &= V_m \cos(\omega t + \theta_v) + jV_m \sin(\omega t + \theta_v) \\ &= V_m e^{j(\omega t + \theta_v)} = V_m e^{j\theta_v} e^{j\omega t} = \underline{V}e^{j\omega t} \end{aligned}$$

where we have used Eq. (9), Section 6.1, with $\phi = \omega t + \theta_v$. We call $\underline{v} = \underline{V}e^{j\omega t}$ a *rotating phasor* since it is a vector of length V_m and instantaneous angle $\omega t + \theta_v$, just like the rotating vector back in Fig. 6.1–2b. The horizontal projection of $\underline{v}$ then equals the actual AC voltage; that is,

$$\text{Re}[\underline{v}] = \text{Re}[\underline{V}e^{j\omega t}] = V_m \cos(\omega t + \theta_v)$$

Similarly, let

$$\underline{i} = I_m e^{j\theta_i} e^{j\omega t} = \underline{I}e^{j\omega t}$$

so that $\text{Re}[\underline{i}] = I_m \cos(\omega t + \theta_i) = i$.

We have thereby represented the AC voltage and current as the real part of exponential time functions with $s = j\omega$. It then follows from Eq. (11), Section 5.3, that if $\underline{i} = \underline{I}e^{j\omega t}$ and $\underline{v} = \underline{V}e^{j\omega t}$, then

$$\underline{V} = Z(j\omega)\underline{I} = \underline{Z}\underline{I}$$

which holds even when $\underline{V}$, $\underline{I}$, and $s = j\omega$ are complex quantities. For the case of real sinusoids $i = \text{Re}[\underline{i}]$ and $v = \text{Re}[\underline{v}]$, all imaginary parts disappear from the final result but the magnitudes and angles are still related by $\underline{V} = \underline{Z}\underline{I}$.

Example 6.2–1

Let's return to our initial example and consider a series RL circuit with $R = 3\ \Omega$ and $L = 0.1$ H driven by an AC current $i = 2 \cos 40t$. To compute the resulting terminal voltage, we first write the current phasor $\underline{I} = 2 \underline{/0^\circ}$ and note that $\omega = 40$. The circuit's impedance is $Z(s) = R + sL$, so $\underline{Z} = Z(j\omega) = R + j\omega L = 3 + j4$ and conversion to polar form gives $\underline{Z} = 5 \underline{/53.1^\circ}$. Equation (2) then yields

$$\underline{V} = \underline{Z}\underline{I} = (5 \underline{/53.1^\circ})(2 \underline{/0^\circ}) = 10 \underline{/53.1^\circ}$$

Therefore, $v = V_m \cos(\omega t + \theta_v) = 10 \cos(40t + 53.1^\circ)$.

Exercise 6.2–1

Assume a complex current $\underline{i} = 2e^{j40t}$ through the above circuit and find the resulting complex voltage $\underline{v} = R\underline{i} + L\, d\underline{i}/dt$. Then show that Re[$\underline{v}$] equals v as found above using $\underline{V} = \underline{Z}\underline{I}$.

AC RESISTANCE AND REACTANCE

If the network in Fig. 6.2–1 happens to consist of just one element, its AC impedance and admittance are found simply by setting $s = j\omega$ in our prior expressions for $Z(s)$ and $Y(s)$. Specifically, drawing upon Table 5.3–1,

$$\underline{Z}_R = R \qquad \underline{Y}_R = G = \frac{1}{R} \tag{6a}$$

$$\underline{Z}_L = j\omega L = \omega L \,\underline{/90^\circ} \qquad \underline{Y}_L = \frac{1}{j\omega L} = \frac{1}{\omega L}\,\underline{/-90^\circ} \tag{6b}$$

$$\underline{Z}_C = \frac{1}{j\omega C} = \frac{1}{\omega C}\,\underline{/-90^\circ} \qquad \underline{Y}_C = j\omega C = \omega C\,\underline{/90^\circ} \tag{6c}$$

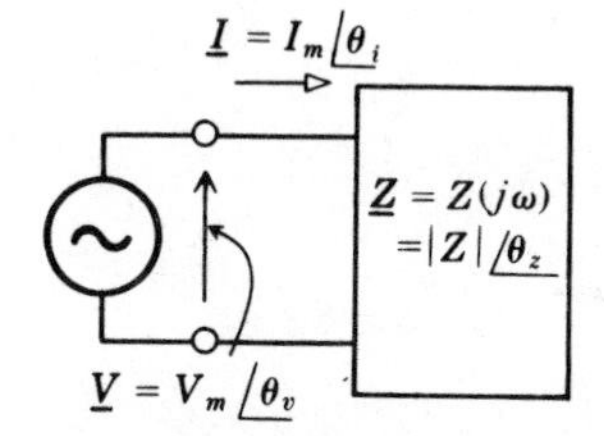

FIGURE 6.2–1
A one-port with AC excitation.

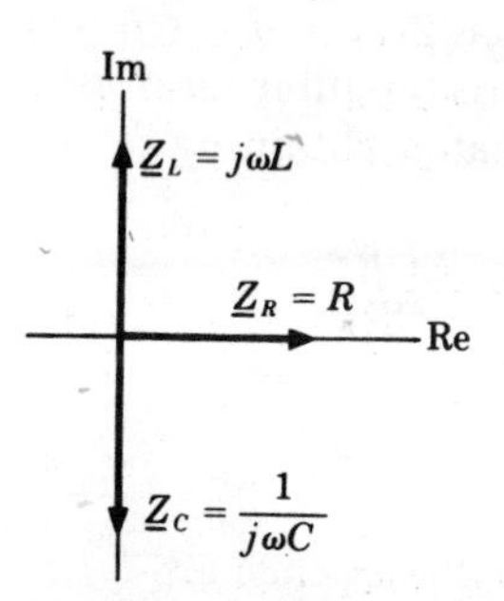

FIGURE 6.2–2
Impedance vectors.

Figure 6.2–2 shows the individual impedance vectors and emphasizes that the energy-storage elements always have an angle of $\pm 90^\circ$ due to the presence of j when we set $s = j\omega$. As before, the magnitude of AC impedance is measured in ohms.

When a network consists of both resistance and energy storage elements, its equivalent AC impedance will be a vector having real (horizontal) and imaginary (vertical) components, written in the general form

$$\underline{Z} = R(\omega) + jX(\omega) \tag{7a}$$

with

$$R(\omega) \triangleq \text{Re}[\underline{Z}] \qquad X(\omega) \triangleq \text{Im}[\underline{Z}] \tag{7b}$$

The real part $R(\omega)$ is called the *AC resistance;* it may or may not involve ω, depending on the particular circuit. The imaginary part $X(\omega)$, called the *reactance,* always involves ω because it reflects the presence of energy storage by inductance and/or capacitance. From Eq. (6), the reactances of inductance and capacitance by themselves are

$$X_L = \text{Im}[\underline{Z}_L] = \omega L \qquad X_C = \text{Im}[\underline{Z}_C] = -\frac{1}{\omega C} \tag{8}$$

which have opposite signs and reciprocal frequency dependence.

Equation (7) expresses $\underline{Z}$ in rectangular form, with horizontal and vertical components as shown in the *impedance diagram* of Fig. 6.2–3. Thus, the corresponding polar coordinates are

$$|Z| = |\underline{Z}| = \sqrt{R^2(\omega) + X^2(\omega)}$$
$$\theta_z = \measuredangle \underline{Z} = \arctan \frac{X(\omega)}{R(\omega)} \tag{9}$$

These conversion formulas are important because we often get $\underline{Z}$ in rectangular form from a circuit diagram, whereas the computations represented by $\underline{V} = \underline{Z}\underline{I}$ require $\underline{Z}$ in polar form.

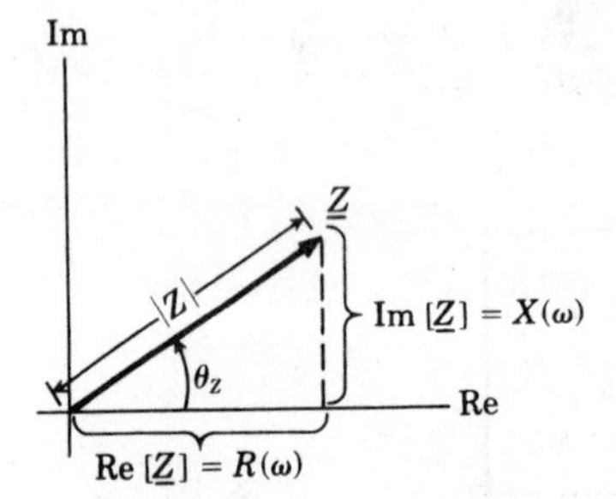

FIGURE 6.2–3 Impedance diagram showing AC resistance and reactance.

Example 6.2–2 *Impedance of a parallel RC circuit*

In Section 5.3, we found that a parallel RC circuit has $Z(s) = R/(sCR + 1)$ so $\underline{Z} = Z(j\omega) = R/(j\omega CR + 1)$. We put this into rectangular form using the complex conjugate to rationalize the denominator, obtaining

$$\underline{Z} = \frac{R}{1 + j\omega CR} \times \frac{1 - j\omega CR}{1 - j\omega CR} = \frac{R - j\omega CR^2}{1 + (\omega CR)^2}$$
$$= \frac{R}{1 + (\omega CR)^2} + j\frac{-\omega CR^2}{1 + (\omega CR)^2}$$

By comparison with Eq. (7), the AC resistance and reactance are

$$R(\omega) = \frac{R}{1 + (\omega CR)^2} \qquad X(\omega) = -\frac{\omega CR^2}{1 + (\omega CR)^2}$$

The *negative* value of $X(\omega)$ is characteristic of *capacitive reactance.*

Exercise 6.2–2

Obtain expressions for $R(\omega)$ and $X(\omega)$ for a parallel RL circuit, starting from $Z(s)$, and confirm that $X(\omega) > 0$.

AC CIRCUIT CALCULATIONS

At last we put to work the tools that have been developed for AC circuit calculations. In general one proceeds as follows:

- Identify the frequency ω of the source and its amplitude and phase (I_m and θ_i or V_m and θ_v).
- Find the impedance at the source frequency, and convert to polar form $\underline{Z} = |Z| \angle \theta_z$.
- Apply Ohm's law for AC circuits, $\underline{V} = \underline{Z}\underline{I}$, to obtain the phasor response $\underline{V} = V_m \angle \theta_v$ or $\underline{I} = I_m \angle \theta_i$.
- Substitute the phasor parameters into the corresponding AC waveform $v(t)$ or $i(t)$.

Additional intermediate steps may involve a voltage-divider ratio, Thévenin's theorem, etc., with AC impedance in place of resistance.

For the impedance calculation per se, we obviously could start with $Z(s)$ as we did in the previous example. This approach yields a rectangular expression that brings out the dependence on the source frequency ω, but conversion to polar form often involves a lot of algebra. An alternative approach having particular value for numerical computations starts with the AC impedance of the individual elements; these are then combined in the usual manner with the help of the rules for vector addition, multiplication, etc. Sometimes it is easier to find the equivalent admittance and get the impedance from $\underline{Z} = 1/\underline{Y}$. The following example illustrates these techniques.

Example 6.2–3

Assume the circuit in Fig. 6.2–4a has $R_1 = 80\ \Omega$, $L = 0.14$ H, $C = 2.5\ \mu$F, and $R_2 = 800\ \Omega$. If the source voltage is $v(t) = 60 \cos 1000t$, then $\omega = 1000$ rad/s and the impedances of the energy-storage elements are

$$\underline{Z}_L = j1000 \times 0.14 = j140\ \Omega = 140\ \Omega \angle 90^\circ$$

$$\underline{Z}_C = \frac{1}{j1000 \times 2.5 \times 10^{-6}} = -j400\ \Omega = 400\ \Omega \angle -90^\circ$$

while, of course, $\underline{Z}_{R_1} = 80\ \Omega$ and $\underline{Z}_{R_2} = 800\ \Omega$. For purposes of AC circuit analysis, the diagram becomes as shown in Fig. 6.2–4b. (You should always make a diagram like this at the start of such problems.)

The calculation of the total equivalent impedance $\underline{Z}$ is best carried out in three steps. First, the series RL section has equivalent impedance

$$\underline{Z}_a = \underline{Z}_{R_1} + \underline{Z}_L = 80\ \Omega + j140\ \Omega$$

Next, recalling that parallel admittances add directly, the RC section has

$$\underline{Y}_b = \frac{1}{800} + \frac{1}{-j400} = (1.25 + j2.5) \times 10^{-3} = 2.795 \times 10^{-3} \angle 63.4^\circ$$

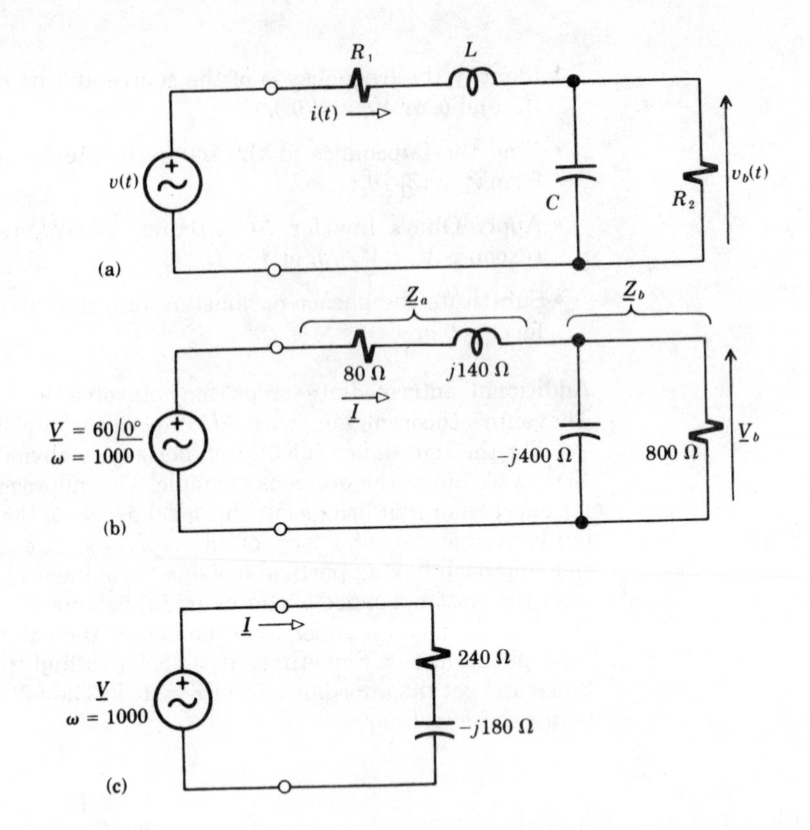

FIGURE 6.2–4 Circuits for Example 6.2–3.

and

$$\underline{Z}_b = 1/\underline{Y}_b = 358\ \Omega\ \underline{/-63.4^\circ} = 160\ \Omega - j320\ \Omega$$

Note that we obtain $\underline{Z}_b$ in both polar and rectangular form to prepare for the final step, namely adding the real and imaginary parts of $\underline{Z}_a$ and $\underline{Z}_b$ to get

$$\begin{aligned}\underline{Z} = \underline{Z}_a + \underline{Z}_b &= (80 + 160) + j(140 - 320) \\ &= 240\ \Omega - j180\ \Omega = 300\ \Omega\ \underline{/-36.9^\circ}\end{aligned}$$

The equivalent AC resistance and reactance of the network when $\omega = 1000$ are thus

$$R(\omega) = \text{Re}[\underline{Z}] = 240\ \Omega$$

$$X(\omega) = \text{Im}[\underline{Z}] = -180\ \Omega$$

The AC resistance in this case is less than the DC resistance (which equals 80 + 800 = 880 Ω), and the reactance is a negative quantity. As far as the source is concerned, it "sees" the equivalent circuit of Fig. 6.2–4c, which includes a capacitance (and no inductance) because of the sign of reactance. This circuit model holds *only* for $\omega = 1000$.

The resultant current from the source is now easily found with Ohm's law for AC circuits. Specifically,

$$\underline{I} = \underline{V}/\underline{Z} = \frac{60 \text{ V } \underline{/0^\circ}}{300\ \Omega\ \underline{/-36.9^\circ}} = 0.2 \text{ A } \underline{/36.9^\circ}$$

and the corresponding AC waveform is

$$i(t) = 0.2 \cos(1000t + 36.9^\circ) \text{ A}$$

Had we been concerned only with the voltage $v_b(t)$ across the RC section, we could have omitted the calculation of $i(t)$ and gone immediately to the voltage-divider expression

$$\underline{V}_b = \frac{\underline{Z}_b}{\underline{Z}_a + \underline{Z}_b}\underline{V} = \frac{358\ \underline{/-63.4^\circ}}{300\ \underline{/-36.9^\circ}}\,60\ \underline{/0^\circ} = 71.6 \text{ V } \underline{/-26.5^\circ}$$

from which

$$v_b(t) = 71.6 \cos(1000t - 26.5^\circ)$$

The surprising result that $v_b(t)$ has a larger amplitude (71.6 V) than the source voltage merely reflects the fact that $\underline{Z}_b$ has a larger magnitude than $\underline{Z}_a + \underline{Z}_b$ due to partial cancelation of the reactances. The related phenomenon of *resonance* will be examined in the next section.

Exercise 6.2–3

Repeat the preceding calculations with $\omega = 2000$ and the positions of the capacitance and inductance interchanged.

6.3 AC CIRCUIT ANALYSIS

This section goes into further details of the behavior of circuits under AC steady-state conditions, with particular emphasis on instantaneous and average power, phasor diagrams, and resonance.

INSTANTANEOUS AND AVERAGE POWER

When driven by $v(t) = 60 \cos 1000t$, the circuit in Example 6.2–3 was found to draw $i(t) = 0.2 \cos(1000t + 36.9^\circ)$. Figure 6.3–1 plots these two waveforms along with their product $p(t) = v(t)i(t)$, which equals the *instantaneous power* delivered by the source. The oscillatory shape of $p(t)$ means that the source alternately supplies power to the circuit, when $p(t) > 0$, and receives power from the circuit, when $p(t) < 0$. The plot also

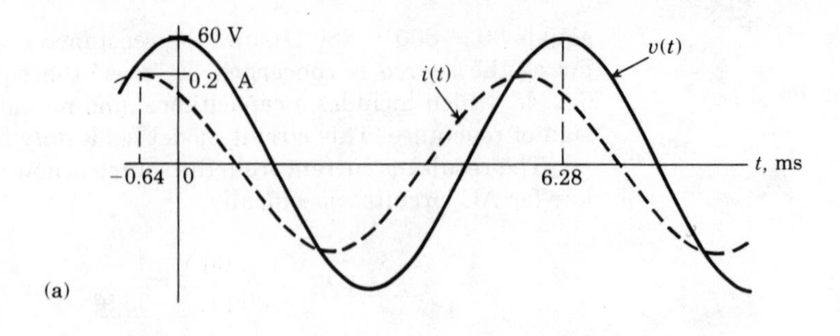

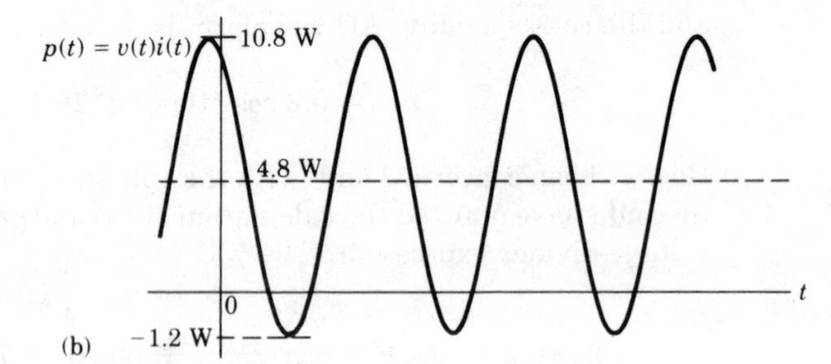

FIGURE 6.3–1
AC voltage, current, and power waveforms.

shows that the positive peak is bigger than the negative peak so that, on the average, the source delivers more power than it receives.

For a general proof of these properties, let the terminal voltage and current of an arbitrary one-port be

$$v = V_m \cos(\omega t + \theta_v) \qquad \text{and} \qquad i = I_m \cos(\omega t + \theta_i)$$

then

$$\begin{aligned} p = vi &= V_m \cos(\omega t + \theta_v)\, I_m \cos(\omega t + \theta_i) \\ &= \tfrac{1}{2} V_m I_m \cos(\theta_v - \theta_i) + \tfrac{1}{2} V_m I_m \cos(2\omega t + \theta_v + \theta_i) \end{aligned}$$

where we have invoked the trigonometric expansion for the product of cosines, $\cos\alpha \cos\beta = \frac{1}{2}\cos(\alpha - \beta) + \frac{1}{2}\cos(\alpha + \beta)$. The first term of the expression for p, being a constant, represents the *average power* P delivered to the one-port. The second term reflects the presence of energy storage, but does not contribute to P.

To express average power dissipation in terms of the impedance, we recall that $V_m = |Z|\, I_m$ and $\theta_v = \theta_i + \theta_z$ or $\theta_v - \theta_i = \theta_z$. Thus,

$$P = \tfrac{1}{2} V_m I_m \cos(\theta_v - \theta_i) = \tfrac{1}{2} \frac{V_m^2}{|Z|} \cos\theta_z = \tfrac{1}{2} I_m^2\, |Z| \cos\theta_z \tag{1a}$$

Alternatively, since $|Z| \cos \theta_z = \text{Re}[\underline{Z}] = R(\omega)$,

$$P = \tfrac{1}{2} I_m^2\, R(\omega) \tag{1b}$$

which serves as a further interpretation of the AC resistance. If the network contains only energy-storage elements, its AC resistance equals zero and $P = 0$, consistent with the fact that ideal inductance and capacitance never dissipate power.

Example 6.3–1

We previously found that $\underline{Z} = 300\ \Omega\ \underline{/-36.9^\circ}$ in Fig. 6.3–2. Hence, applying Eq. (1a) with $V_m = 60$ V, the total average power delivered by the source is

$$P = \frac{1}{2} \frac{(60\text{ V})^2}{300\ \Omega} \cos(-36.9^\circ) = 4.8\text{ W}$$

as represented by the dashed line in Fig. 6.3–1b. Equation (1b) gives the same result with $I_m = 0.2$ A and $R(\omega) = 240\ \Omega$. This power is shared entirely by the two resistances.

In particular, the amplitude of the current through the 80-Ω resistance is $I_{R_1} = |\underline{I}| = 0.2$ A so

$$P_{R_1} = \tfrac{1}{2} I_{R_1}^2\, R_1 = 1.6\text{ W}$$

and the remaining 3.2 W must be dissipated in the 800-Ω resistance. Checking out this conclusion, we compute $|\underline{I}_{R_2}| = |\underline{V}_b|/R_2 = 0.0895$ A and $P_{R_2} = \frac{1}{2}(0.0895\text{ A})^2 \times 800\ \Omega = 3.2$ W as expected. But note that $P \neq \frac{1}{2} I_m^2 (R_1 + R_2)$, since $R(\omega) \neq R_1 + R_2$.

Exercise 6.3–1

Use Eqs. (1a) and (1b) to compute P in Fig. 6.3–2 when R_1 and L are replaced by a short circuit so $\underline{Z} = \underline{Z}_b$ given in Example 6.2–3. Then find the current through C and sketch $p_C(t) = v(t)i_C(t)$ to confirm that it has zero average value.

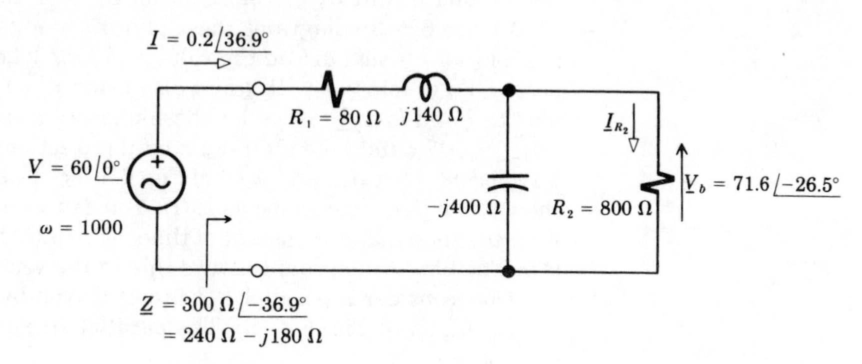

FIGURE 6.3–2
Circuit for Example 6.3–1.

PHASOR DIAGRAMS

Phasor diagrams provide an informative picture of the voltage and current relationships in an AC circuit. As a simple but important example, consider the series RC circuit in Fig. 6.3–3a with a voltage source having $\underline{V} = V_m \,\underline{/0^\circ}$. (We normally take zero phase angle for an applied source.) The impedance is

$$\underline{Z} = \underline{Z}_R + \underline{Z}_C = R + \frac{1}{j\omega C} = R - j\frac{1}{\omega C}$$

which has

$$|Z| = \sqrt{R^2 + \left(\frac{1}{\omega C}\right)^2} = \frac{1}{\omega C}\sqrt{1 + (\omega CR)^2}$$

$$\theta_z = \arctan\frac{(-1/\omega C)}{R} = -\arctan\frac{1}{\omega CR}$$

where, for θ_z, we have used $\arctan(-\phi) = -\arctan\phi$. The magnitude and angle of the resulting current phasor are given by

$$I_m = \frac{V_m}{|Z|} = \frac{\omega C}{\sqrt{1 + (\omega CR)^2}}V_m \qquad \qquad (2)$$

$$\theta_i = \theta_v - \theta_z = +\arctan\frac{1}{\omega CR}$$

and the phasor voltages across R and C are

$$\underline{V}_R = \underline{Z}_R\underline{I} = RI_m \,\underline{/\theta_i} \qquad \underline{V}_C = \underline{Z}_C\underline{I} = \frac{I_m}{\omega C}\,\underline{/\theta_i - 90^\circ}$$

the 90° phase shift in $\underline{V}_C$ coming from $\underline{Z}_C = (1/\omega C)\,\underline{/-90^\circ}$.

Figure 6.3–3b diagrams these phasors along with $\underline{V}$ and $\underline{I}$. The current phasor $\underline{I}$ is said to *lead* the voltage phasor $\underline{V}$ because $\theta_i > \theta_v = 0$. The phasor $\underline{V}_R$ is colinear with $\underline{I}$ and also leads $\underline{V}$, whereas $\underline{V}_C$ is perpendicular to $\underline{I}$ and *lags* $\underline{V}$. If we let these phasors rotate counterclockwise at rate $f = \omega/2\pi$ and take their horizontal projections, we get the actual AC waveforms $v(t)$, $v_R(t)$, and $v_C(t)$ of Fig. 6.3–3c. The current waveform, not shown, has the same shape as $v_R(t)$. You can visually check that $v_R(t) + v_C(t)$ equals $v(t)$ at any instant of time, as required by Kirchhoff's voltage law. The phasor diagram conveys this in the vector sum $\underline{V}_R + \underline{V}_C = \underline{V}$.

Now consider a parallel RC circuit driven by a current source with $\underline{I} = I_m \,\underline{/0^\circ}$, as in Fig. 6.3–4a. The parallel structure suggests using the

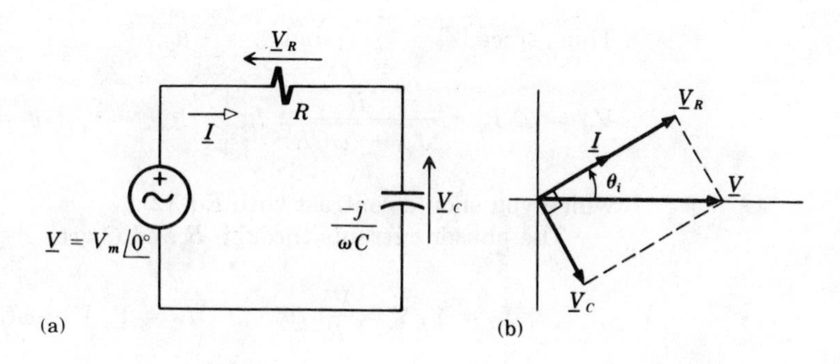

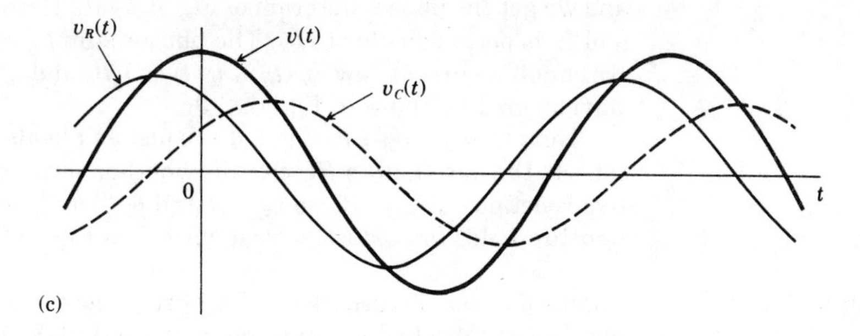

FIGURE 6.3–3
(a) Series RC circuit.
(b) Phasor diagram.
(c) Waveforms.

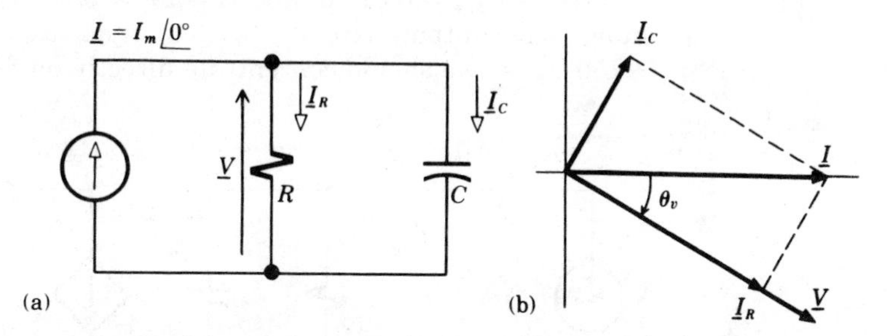

FIGURE 6.3–4
(a) Parallel RC circuit;
(b) Phasor diagram.

equivalent AC admittance

$$\underline{Y} = \underline{Y}_R + \underline{Y}_C = \frac{1}{R} + j\omega C$$

which has

$$|Y| = \left(\frac{1}{R}\right)^2 + (\omega C)^2 = \frac{1}{R}\sqrt{1 + (\omega CR)^2}$$

$$\theta_y = \arctan\frac{\omega C}{1/R} = \arctan(\omega CR)$$

Thus, since $|Z| = 1/|Y|$ and $\theta_z = -\theta_y$,

$$V_m = |Z|\,I_m = \frac{R}{\sqrt{1 + (\omega CR)^2}}\,I_m \qquad \theta_v = \theta_z + \theta_i = -\arctan(\omega CR) \quad (3)$$

which you should contrast with Eq. (2).

The phasor currents through R and C are

$$\underline{I}_R = \underline{Y}_R\,\underline{V} = \frac{V_m}{R}\,\underline{/\theta_v} \qquad \underline{I}_C = \underline{Y}_C\,\underline{V} = \omega C V_m\,\underline{/\theta_v + 90^\circ}$$

and we get the phasor diagram of Fig. 6.3–4b. Here, $\underline{I}_R$ is colinear with $\underline{V}$, and $\underline{I}_C$ is perpendicular to $\underline{I}_R$. The phasor sum $\underline{I}_R + \underline{I}_C = \underline{I}$ corresponds to Kirchhoff's current law, $i_R(t) + i_C(t) = i(t)$, and one could easily sketch waveforms like those in Fig. 6.3–3c.

Note that $\underline{V}$ lags $\underline{I}$ in Fig. 6.3–4b just as $\underline{I}$ leads $\underline{V}$ in Fig. 6.3–3b. This reflects the fact that an RC circuit, whether series or parallel, has capacitive reactance $X(\omega) < 0$, so $\theta_z < 0$ and $\theta_i > \theta_v$. The opposite phasor relationship holds for inductive reactance; see Fig. 6.1–1d for an example.

Example 6.3–2

Figure 6.3–5a represents the high-frequency model of a certain amplifier. We want to find $v_{\text{out}}(t)$ when $v_s(t)$ is a 0.01-V (10-mV) sinusoid at f = 50 MHz, so $\underline{V}_s = 0.01\,\underline{/0^\circ}$ and $\omega = 2\pi \times 50 \times 10^6 = 10^8\pi$ rad/s. Examining the output circuit reveals that $v_{\text{out}} = 300\ \Omega \times (-2.5\ v_{\text{in}}) = -750\ v_{\text{in}}$, so we should concentrate directly on finding v_{in}.

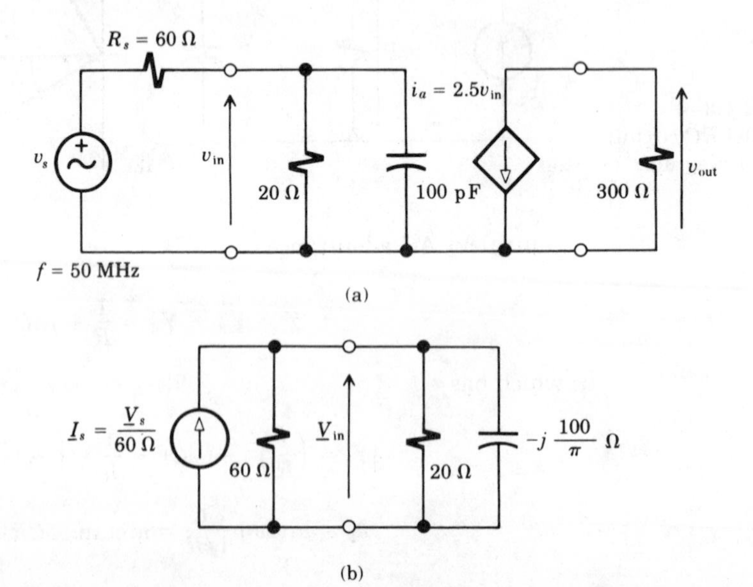

FIGURE 6.3–5
Circuits for Example 6.3–2.

To that end, we convert the source to its Norton equivalent and redraw the input circuit labeled with phasors and impedances (see Fig. 6.3–5b). The equivalent resistance is $R = 60\ \Omega \| 20\ \Omega = 15\ \Omega$, and $\underline{Z}_C = -j/\omega C = -j100/\pi\ \Omega$. Inserting values in Eq. (3) gives

$$|\underline{V}_{\text{in}}| = \frac{15}{\sqrt{1 + (15\pi/100)^2}} \times \frac{0.01}{60} = 0.00226\ \text{V}$$

$$\sphericalangle \underline{V}_{\text{in}} = -\arctan(15\pi/100) = -25.2°$$

Therefore,

$$v_{\text{out}}(t) = -750\ v_{\text{in}}(t) = -1.7 \cos(10^8 \pi t - 25.2°)$$

so this amplifier has a voltage gain of $-1.7/0.01 = -170$, plus a *phase shift* of $-25.2°$ at the frequency in question.

Exercise 6.3–2

Construct phasor diagrams for a series RL circuit and a parallel RL circuit, and show that inductive reactance results in $\underline{V}$ leading $\underline{I}$ in both cases.

RESONANCE

Inductance and capacitance have opposite AC properties in two respects: inductive reactance $X_L = \omega L$ is positive and proportional to frequency, while capacitive reactance $X_C = -1/\omega C$ is negative and inversely proportional to frequency. A circuit containing both types of energy-storage elements exhibits a distinctive behavior owing to the joint effect of these properties. Depending on the excitation frequency, the inductance or capacitance may dominate, or the two reactances may cancel out each other and produce the condition known as resonance. Here we will examine the properties of simple series and parallel resonant circuits; the discussion is continued in the next chapter under the heading of tuned circuits.

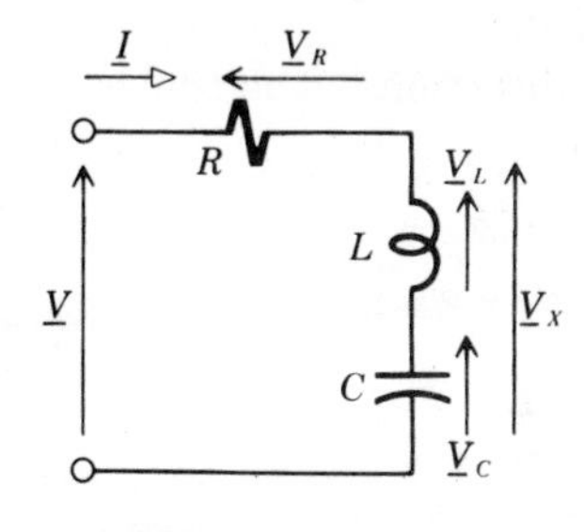

FIGURE 6.3–6
Series resonant circuit.

Consider the series RLC circuit in Fig. 6.3–6. Its total impedance is the sum

$$\underline{Z} = R + jX_L + jX_C = R + j\omega L - \frac{j}{\omega C} = R + j\left(\omega L - \frac{1}{\omega C}\right) \tag{4}$$

having the *net reactance*

$$X = X_L + X_C = \omega L - \frac{1}{\omega C}$$

Clearly, if $\omega L > 1/\omega C$ then $X > 0$ and the circuit is *inductive* because it has positive reactance like a series RL circuit. Conversely, if $\omega L < 1/\omega C$ then $X < 0$ and the circuit is *capacitive*. The borderline between these two cases occurs when $\omega L = 1/\omega C$, called the *resonance* condition. The *resonant frequency* ω_0 satisfies $\omega_0 L = 1/\omega_0 C$, that is,

$$\omega_0^2 = \frac{1}{LC} \qquad \omega_0 = \frac{1}{\sqrt{LC}} \tag{5}$$

The circuit appears to be purely *resistive* at ω_0 since $X = 0$ and $\underline{Z} = R$.

The corresponding phasor diagrams provide further insight about these three conditions. Suppose a circuit is driven by an AC voltage source with $\underline{V} = V_m \,/0°$. The resulting current phasor $\underline{I} = \underline{V}/\underline{Z}$ will have

$$I_m = \frac{V_m}{\sqrt{R^2 + X^2}} = \frac{V_m}{\sqrt{R^2 + \left(\omega L - \dfrac{1}{\omega C}\right)^2}}$$
$$\theta_i = -\arctan\frac{X}{R} = -\arctan\left(\frac{\omega L}{R} - \frac{1}{\omega CR}\right) \tag{6a}$$

and the individual phasor voltage drops will be

$$\underline{V}_R = RI_m \,\underline{/\theta_i} \qquad \underline{V}_L = \omega L I_m \,\underline{/\theta_i + 90°} \qquad \underline{V}_C = \frac{I_m}{\omega C}\,\underline{/\theta_i - 90°} \tag{6b}$$

Since $\underline{V}_L$ and $\underline{V}_C$ are colinear but have opposite directions, it is convenient to introduce the net reactive voltage phasor

$$\underline{V}_X = \underline{V}_L + \underline{V}_C$$

Kirchhoff's voltage law in phasor form then becomes

$$\underline{V}_R + \underline{V}_X = \underline{V}$$

which will help our phasor constructions.

Figure 6.3–7a shows the phasor diagram below resonance (that is, $\omega < \omega_0$), where $\underline{V}_X$ is in the same direction as $\underline{V}_C$ so $\underline{I}$ leads $\underline{V}$ and the circuit is capacitive (see Fig. 6.3–3b for comparison). The conditions exactly at resonance are shown in Fig. 6.3–7b; here $\underline{V}_L$ and $\underline{V}_C$ cancel each other leaving $\underline{V}_X = 0$, so $\underline{V}_R = \underline{V}$ and $\underline{I}$ is in phase with $\underline{V}$—just as if the circuit consisted only of resistance. Finally, Fig. 6.3–7c gives the relationships above resonance, where $\underline{V}_X$ is in the same direction as $\underline{V}_L$, so $\underline{I}$ lags $\underline{V}$, and the circuit is inductive.

The series resonance phenomenon has two additional characteristics deserving mention. First, the average power dissipation is *maximum* at

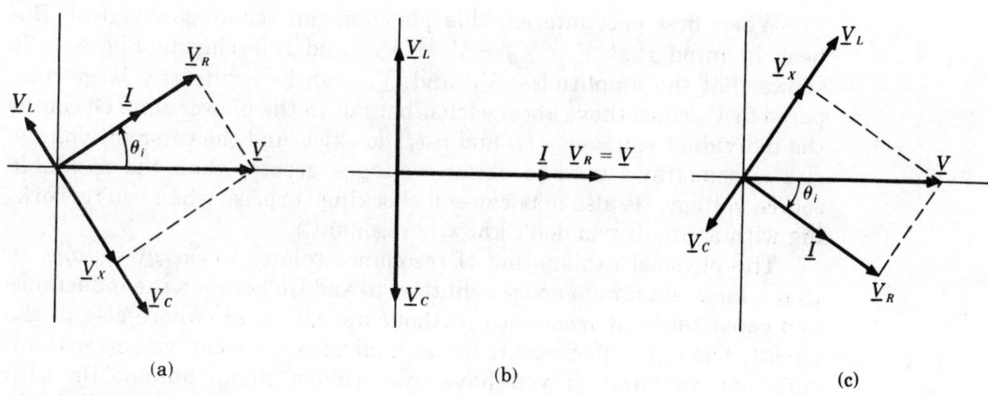

FIGURE 6.3–7
Phasor diagrams:
(a) below resonance, $\omega < \omega_0$;
(b) at resonance, $\omega = \omega_0$; (c) above resonance, $\omega > \omega_0$.

ω_0, and

$$P = \frac{1}{2}\frac{V_m^2}{R} = \frac{V_{\text{rms}}^2}{R}$$

since the impedance has its minimum value, $\underline{Z} = R$.

Second, although the net reactive voltage $\underline{V}_X$ has zero amplitude at resonance, the individual amplitudes $|\underline{V}_L|$ and $|\underline{V}_C|$ are quite large—possibly even greater than the source amplitude V_m! Specifically, using Eq. (6b) with $\omega = \omega_0$, the voltage amplitudes across the inductance and capacitance are

$$|\underline{V}_L| = \omega_0 L I_m = \omega_0 L \frac{V_m}{R} = \frac{1}{R}\sqrt{\frac{L}{C}}\,V_m$$

$$|\underline{V}_C| = \frac{I_m}{\omega_0 C} = \frac{1}{\omega_0 C}\frac{V_m}{R} = \frac{1}{R}\sqrt{\frac{L}{C}}\,V_m$$

where we have substituted $\omega_0 = 1/\sqrt{LC}$. Defining the *quality factor* of a series resonant circuit as

$$Q \triangleq \frac{\omega_0 L}{R} = \frac{1}{\omega_0 CR} = \frac{1}{R}\sqrt{\frac{L}{C}} \tag{7}$$

we have at resonance

$$|\underline{V}_L| = |\underline{V}_C| = QV_m$$

These amplitudes will be greater than V_m if $Q > 1$, a phenomenon known as the *resonant voltage rise*.

When first encountered, this phenomenon seems paradoxical. But bear in mind that $\underline{V} = \underline{V}_R + \underline{V}_L + \underline{V}_C$, and referring to Fig. 6.3–7b shows that the amplitudes $|\underline{V}_L|$ and $|\underline{V}_C|$ can be arbitrarily large compared to V_m since they cancel each other out in the phasor sum. Of course the individual voltages $v_L(t)$ and $v_C(t)$ do exist, and the resonant voltage rise is sometimes used to obtain voltages greater than the available source voltage. (It also may cause a shocking surprise when you're working with a circuit you don't know is resonant.)

The physical explanation of resonance relates to *energy storage,* in that a large amount of energy shuttles to and fro between the inductance and capacitance at resonance, without appearing anywhere else in the circuit. The same effect occurs in mechanical suspension systems with insufficient damping. If you have ever ridden in an automobile with worn-out shock absorbers you probably experienced an unpleasant throbbing at some particular speed—a mechanical resonance condition with energy oscillating between kinetic energy in the vibrating chassis and potential energy in the springs. Replacing the shock absorbers increases the damping and decreases the mechanical Q, which in turn, decreases the vibration amplitude.

We've dealt first with series resonance because its properties are probably simpler to grasp than parallel resonance. But we can now apply the series results to the parallel RLC circuit driven by a current source, Fig. 6.3–8, if we use the admittance

$$\underline{Y} = \underline{Y}_R + \underline{Y}_C + \underline{Y}_L = \frac{1}{R} + j\left(\omega C - \frac{1}{\omega L}\right) \tag{8}$$

Comparison with Eq. (4) shows that $\underline{Y}$ has the same form as $\underline{Z}$ for a series RLC circuit with L and C interchanged and R replaced by $1/R$—another example of *duality*. Moreover, the current and voltage phasors are related by $\underline{V} = \underline{I}/\underline{Y}$, which is the dual of $\underline{I} = \underline{V}/\underline{Z}$. Thus our previous equations hold for parallel resonance when $\underline{I}$ and $\underline{V}$ are interchanged, along with the modifications above.

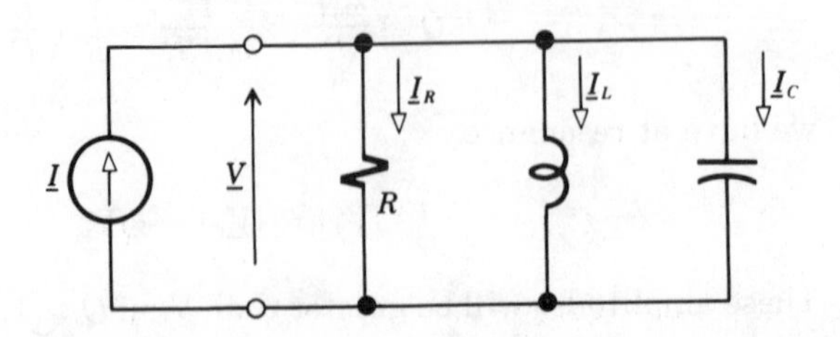

FIGURE 6.3–8
Parallel resonant circuit.

For instance, a parallel *RLC* circuit has a resonant *current* rise in the sense that, at $\omega = \omega_0$,

$$|\underline{I}_L| = |\underline{I}_C| = QI_m$$

where the quality factor now is given by the dual of Eq. (7), namely

$$Q = \omega_0 CR = \frac{R}{\omega_0 L} = R\sqrt{\frac{C}{L}} \tag{9}$$

Note that a high-Q parallel circuit has large resistance, whereas a high-Q series circuit has small resistance. The resonant frequency itself, however, remains the same since interchanging L and C in Eq. (5) still yields

$$\omega_0 = \frac{1}{\sqrt{LC}}$$

You can easily show from Eq. (8) that the imaginary part of $\underline{Y}$ equals zero when $\omega = \omega_0$.

Example 6.3–3
Resonant power transfer

The reactance-cancellation effect at resonance has practical application in AC power transmission when it is desired to maximize power transfer to a load having a series reactive component. For example, suppose we wish to deliver as much power as possible to the 25-Ω load resistance in Fig. 6.3–9, given a 60-Hz source with $V_{\text{rms}} = 120$ V. The total circuit can be made resonant at the source frequency, that is, at $\omega_0 = 2\pi \times 60 = 377$, by adding a series capacitance

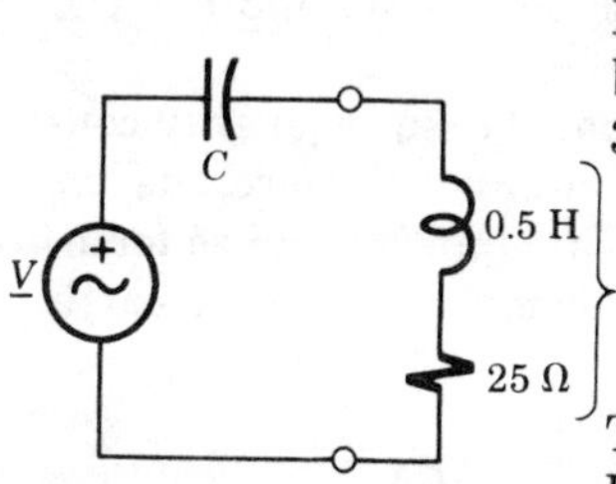

FIGURE 6.3–9
Circuit for Example 6.3–3.

$$C = \frac{1}{\omega_0^2 L} = \frac{1}{377^2 \times 0.5} = 14.1\ \mu\text{F}$$

Then $P = V_{\text{rms}}^2/R = 576$ W, compared to 9.9 W without the capacitance. However, resonance brings along a voltage rise by the factor

$$Q = \frac{\omega_0 L}{R} = \frac{377 \times 0.5}{25} = 7.54$$

so $|\underline{V}_C| = |\underline{V}_L| = Q \times \sqrt{2}\ V_{\text{rms}} = 1280$ V and the voltage amplitude across the RL load is $\sqrt{V_L^2 + V_R^2} = 1291$ V, which may be damaging to the elements.

Exercise 6.3–3

Find ω_0 and Q for a series circuit with $R = 50\ \Omega$, $L = 4$ mH, and $C = 0.1\ \mu$F. Then draw the phasor diagram when the circuit is driven at resonance by a *current* source having $I_m = 3$ A.

Exercise 6.3–4

A parallel resonant circuit in an AM radio has $\omega_0 = 2\pi \times 10^6$, $Q = 100$, and $R = 5\ \text{k}\Omega$. What are the values of C and L?

6.4 RESIDENTIAL CIRCUITS AND WIRING †

No discussion of AC circuits would be complete without at least some coverage of the wiring arrangements that distribute electric power to the outlets of a typical home. This section, therefore, describes several of the major features of residential circuits. As we will be concerned only with voltage and current magnitudes, capital letters will be used throughout to denote rms values.

A word of warning before we begin: This material is intended for illustrative purposes and does not qualify you to be an electrician!

DUAL-VOLTAGE SERVICE ENTRANCE

Most homes today have a *dual-voltage* AC supply provided by a three-line entrance cable like that of Fig. 6.4–1a. One line, called the *neutral,* is connected to an *earth ground.* The other two lines, labeled B and R, are "hot" in the sense that they have 170-V peak sinusoidal potential variations relative to ground potential, with one sinusoid inverted in polarity compared to the other. The two sources shown actually correspond to the secondaries of a center-tapped transformer, like that in Fig. 4.1–7, whose primary is connected to the power line at a utility pole.

Formally, we write $v_B = 170 \cos \omega t$ and $v_R = -v_B = -170 \cos \omega t$, with rms values $V_B = V_R = 170/\sqrt{2} = 120$ V. The line-to-line voltage is then $v_{BR} = v_B - v_R = 340 \cos \omega t$, as shown in Fig. 6.4–1b, and has the rms value $V_{BR} = 340/\sqrt{2} = 240$ V.

After passing through the electric meter that measures energy consumption, the entrance cable terminates at the *main panel.* Here, the hot lines connect to individual circuits for lighting, appliances, and so forth, while the neutral connects to a *busbar* and thence to the local earth ground.

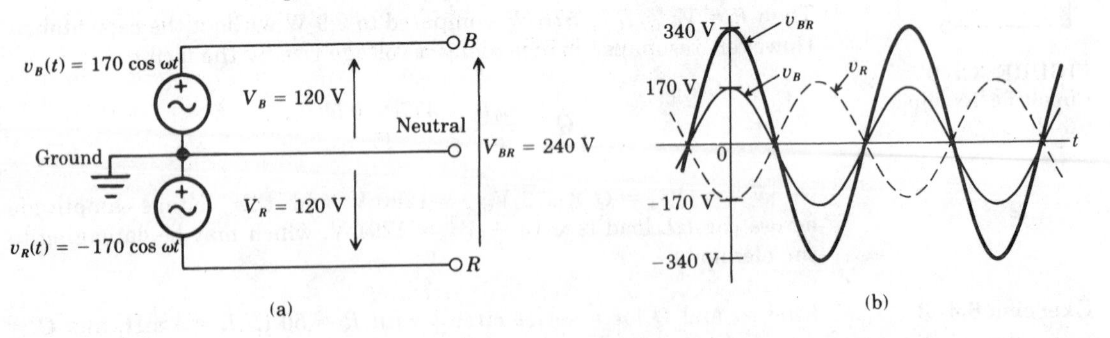

FIGURE 6.4–1 Dual-voltage AC supply.

Figure 6.4–2 diagrams a main panel with *breakers* serving the joint role of disconnecting switches and overcurrent protection. The modern

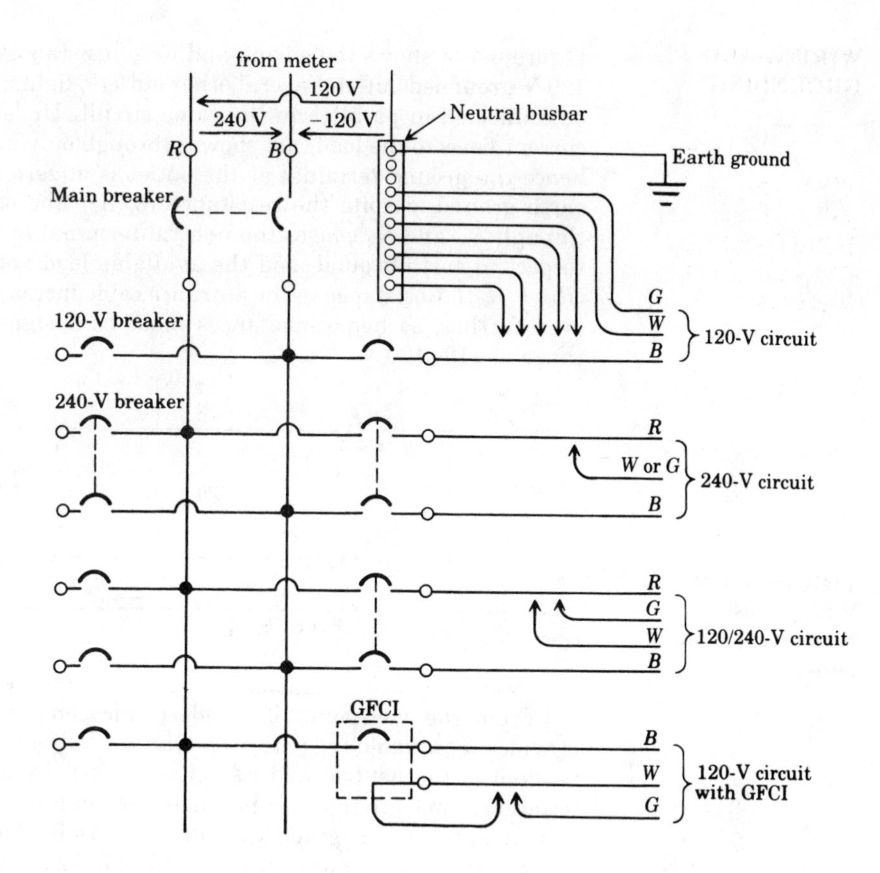

FIGURE 6.4–2 Main panel with breakers and typical circuit connections.

thermal-magnetic breaker is a spring-loaded switch having a bimetallic element that opens under small but continuous current overload, and a magnetic coil that trips the switch instantly under heavy overloads. The breaker labeled GFCI has additional features described later. Older designs would have separate switches and fuses instead of breakers.

The figure also shows four different types of circuits going out of the panel. Each circuit has a minimum of three wires, and they have been labeled in accordance with standard insulation color codes as follows:

Hot = *B* (black) or *R* (red)
Neutral = *W* (white)
Ground = *G* (green) or uninsulated

Every outgoing hot wire must connect to a breaker, whereas every neutral wire and ground wire must be tied directly to earth ground at the neutral busbar. The functional difference between neutral and ground wires will be brought out by considering circuit wiring.

WIRING AND GROUNDING

Figure 6.4–3 shows the wiring and wire resistances from the panel to a 120-V grounded outlet. Several other outlets, lights, etc., would typically be connected in parallel on the same circuit. Under normal conditions, current flows to the load (not shown) through only hot and neutral wires; hence the ground terminal at the outlet is at zero volts with respect to earth ground, despite the resistance R_G. On the other hand, current I through R_W and R_B causes the neutral terminal to be at $R_W I$ volts with respect to earth ground, and the available load voltage becomes $120 - (R_W + R_B)I$. Resistance in the entrance cable increases this loading effect even further, so home appliances must be designed to operate over a range of 110–120 V.

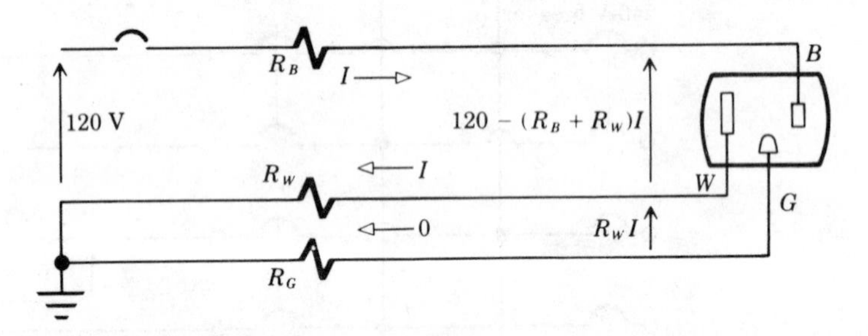

FIGURE 6.4–3 Wiring resistance between panel and outlet.

From the viewpoint of an electronics engineer, the ground wire provides a valuable *reference potential* for voltage measurements, independent of the neutral and its voltage offset. In addition, electrical interference and ground-loop problems are minimized by connecting all instruments to one ground point. But vastly more important is the ground wire's role in protecting human life against electrical shock.

Table 6.4–1 lists effects of various levels of 60-Hz AC current on the human body. The 100–300 mA range turns out to be the most dangerous. Larger currents induce a temporary heart contraction that actually protects it from fatal damage. As the table implies, the amount of current rather than voltage is the key factor in electrical shock; voltage enters the picture when we take account of body resistance, which ranges from around 500 kΩ down to 1 kΩ, depending on whether the skin is dry or wet. Thus, a person with wet skin risks electrocution from AC voltages as low as 100 V.

TABLE 6.4–1 Effects of AC Electrical Shock

Current	*Effects*
1–5 mA	Threshold of sensation.
10–20 mA	Involuntary muscle contractions ("can't-let-go").
20–100 mA	Pain, breathing difficulties.
100–300 mA	Ventricular fibrilation, POSSIBLE DEATH.
>300 mA	Respiratory paralysis, burns, unconsciousness.

Now suppose you touch the metal frame of an ungrounded appliance that has an internal wiring fault between the frame and the hot line as in Fig. 6.4–4a. Your body then provides a possible conducting path to ground, and you may experience a serious shock, especially if you're standing on a damp concrete floor or other grounded surface so that the current passes through your chest. The circuit breaker offers no help in this situation, for it is designed to protect the circuit against excess currents of 15 A or more.

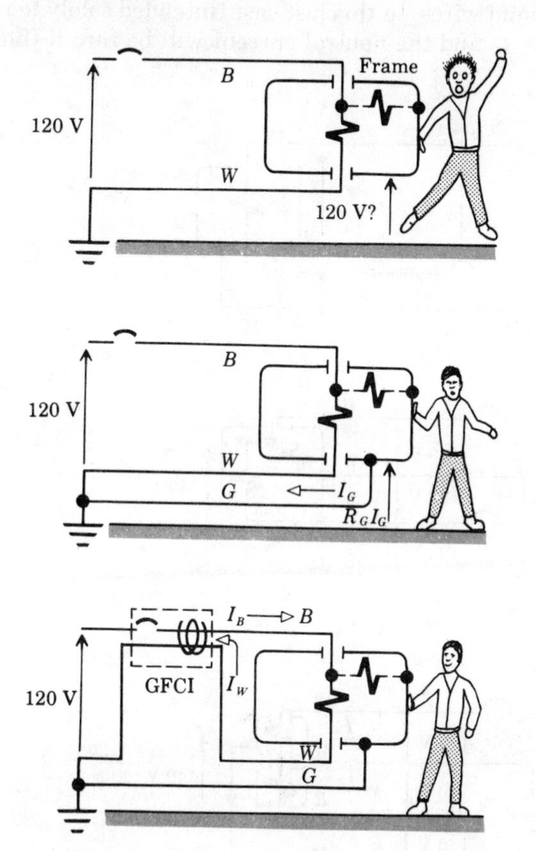

FIGURE 6.4–4 (a) Ungrounded appliance with wiring fault. (b) Grounded appliance with wiring fault. (c) Appliance connected to ground-fault circuit interrupter.

An internal connection from the frame to neutral significantly improves the situation, but even better is the grounded frame shown in Fig. 6.4–4b. This arrangement keeps the frame at ground potential in absence of a wiring fault or, at worst, a few volts from ground if a fault results in current through the ground wire. The best possible shock protection is afforded by the *ground-fault circuit interrupter* (GFCI) shown in Fig. 6.4–4c. The GFCI has a sensing coil around the hot and neutral wires, and any ground-fault current—through you *or* the ground wire—that results in an imbalance $|I_B - I_W| > 5$ mA induces a current in the sensing coil and opens the circuit. The GFCI may be located at an outlet or it may be part of a circuit breaker at the main panel, as in Fig. 6.4–2.

Figure 6.4–5 diagrams three wiring patterns involving 240 V: (a) a 240-V load, such as an electric heating unit; (b) a dual-voltage load, such as an electric range with 120-V and 240-V elements; (c) a so-called "three-wire circuit" with two 120-V loads sharing common neutral and ground wires. In this last case (intended solely to reduce wire costs), $I_W = I_B - I_R$ and the neutral current will be zero if the loads are equal.

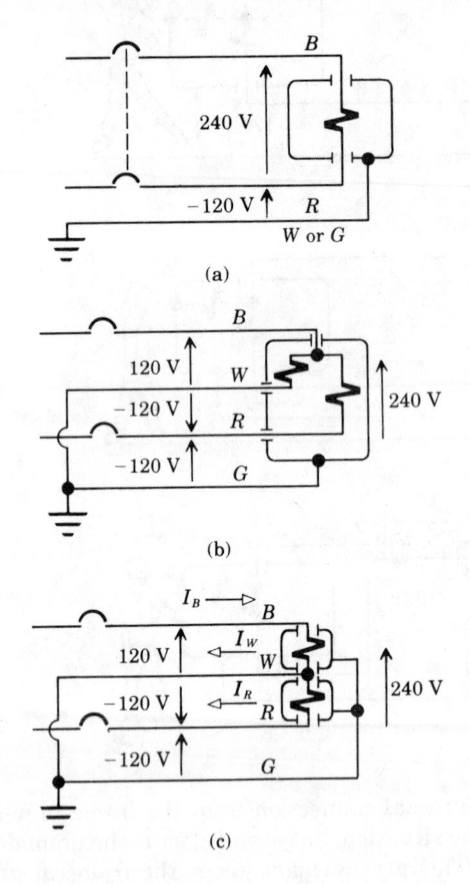

FIGURE 6.4–5
(a) 240-V load.
(b) Dual-voltage load.
(c) Three-wire circuit for two 120-V loads.

Finally, Fig. 6.4–6 shows how a light or some other device can be controlled independently from two different locations using single-pole double-throw (SPDT) switches commonly known as "three-way" switches. The hot wire is switched between two "travelers" at the first switch and from the travelers to the light at the second; therefore, we have a complete circuit only when both switches are either up or down, and flipping either switch opens the circuit. Control at three or more locations is also possible using "four-way" switches that interchange the travelers. In any case, neutral and ground wires are never switched.

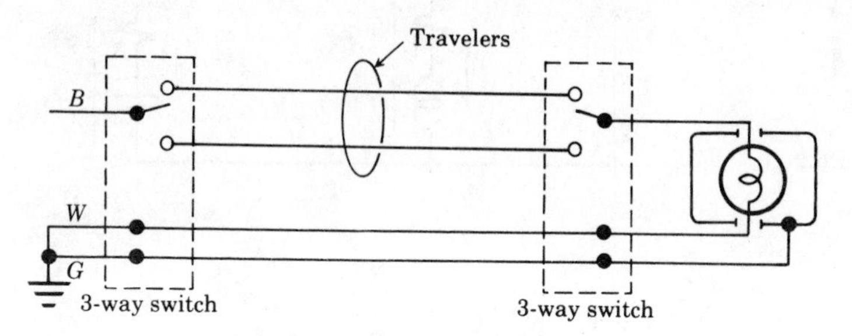

FIGURE 6.4–6 Connection of three-way switches.

Exercise 6.4–1 Why must the two hot wires in Fig. 6.4–5c come from the two different hot lines? Hint: Consider the neutral current.

Exercise 6.4–2 Redraw Fig. 6.4–6 with the light in the middle and the wires from the panel arriving at the light's location.

PROBLEMS

6.1–1 Use Eq. (5) and Fig. 6.1–2a to show that $\cos(\omega t - 90°) = \sin \omega t$.

6.1–2 Repeat the previous problem for $\cos(\omega t + 90°) = -\sin \omega t$.

6.1–3 Evaluate $\underline{A} + \underline{B}$, $\underline{AB}$, and $\underline{A}/\underline{B}$ in rectangular and polar form when $\underline{A} = 5 - j12$ and $\underline{B} = 15\ \underline{/53.1°}$.

6.1–4 Repeat the previous calculations with $\underline{A} = 25\ \underline{/16.3°}$ and $\underline{B} = -(24 + j18)$.

6.1–5 Taking $\underline{A} = A_r + jA_i = |A|\ \underline{/\theta_A}$, reduce each of the following: $\underline{A} + \underline{A}^*$, $\underline{A} - \underline{A}^*$, $1/\underline{A}^*$, and $\underline{A}^*/\underline{A}$.

6.1–6 Carry out the division $(a + jb)/(c + jd)$ in rectangular and polar forms to show that $\arctan[(bc - ad)/(ac + bd)] = \arctan(b/a) - \arctan(d/c)$. Simplify this relation for the case where $a = 0$ and check it trigonometrically.

6.1–7 Look up the series expansions for e^x, $\cos \phi$, and $\sin \phi$, and use them to justify Eq. (9).

6.1–8 Draw the phasor diagram and express $v_1 + v_2$ and $v_1 - v_2$ in the form of Eq. (4) when $v_1 = 10 \cos(\omega t + 135°)$ and $v_2 = 10 \cos(\omega t + 45°)$.

6.1–9 Repeat the previous problem with $v_1 = \sqrt{8} \cos(\omega t - 45°)$ and $v_2 = 2 \sin \omega t$. Hint: See Problem 6.1–1.

6.2–1 The circuit in Fig. P6.2–1 has $R_1 = 6\ \Omega$, $L = 2$ mH, $R_2 = 0$, and $C = 25\ \mu$F, and

is driven by a current source with $\underline{I} = 2\text{ A }\underline{/0°}$ and $\omega = 4000$.
(a) Show that $\underline{Z} = 5\sqrt{10}\ \Omega\ \underline{/-18.5°}$ and obtain an expression for $v(t)$.
(b) Find i_1 and i_2, and use a phasor sum to confirm that $i_1 + i_2 = i$.

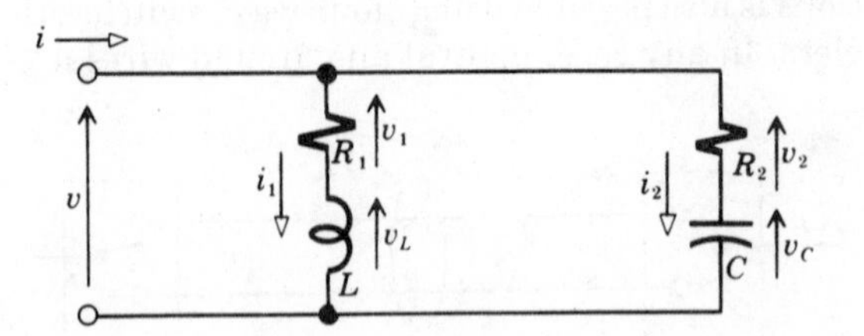

FIGURE P6.2–1

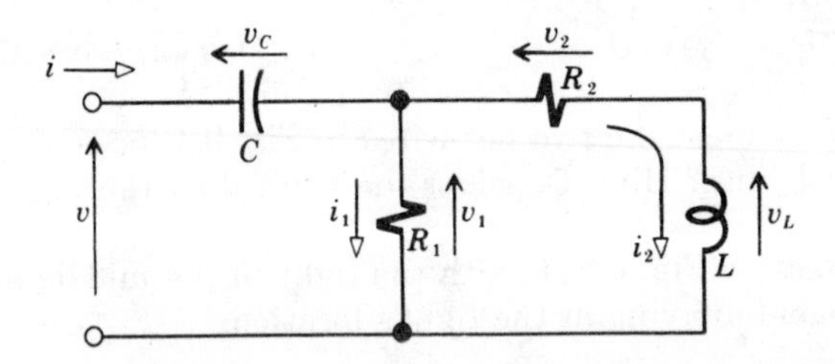

FIGURE P6.2–3

6.2–2 Repeat the previous problem with $R_1 = 0$, $R_2 = 6\ \Omega$, and $\omega = 5000$, in which case $\underline{Z} = 5\sqrt{10}\ \Omega\ \underline{/+18.5°}$.

6.2–3 The circuit in Fig. P6.2–3 has $C = 20\ \mu\text{F}$, $R_1 = 80\ \Omega$, $R_2 = 0$, and $L = 12$ mH, and is driven by a voltage source with $\underline{V} = 100\text{ V }\underline{/0°}$ and $\omega = 5000$.
(a) Show that $\underline{Z} = (28.8 + j28.4)\ \Omega$ and obtain an expression for $i(t)$.
(b) Find v_C and v_1 and use a phasor sum to confirm that $v_C + v_1 = v$.

6.2–4 Repeat the previous problem with $C = 10\ \mu\text{F}$, $R_1 = 25\ \Omega$, $R_2 = 7\ \Omega$, and $L = 4.8$ mH, in which case $\underline{Z} = (12.5 - j10.625)\ \Omega$. (The results of this problem are needed in Problem 6.3–1.)

6.2–5 Figure P6.2–5 represents a source circuit with $\underline{V}_s = 260\text{ mV }\underline{/0°}$, $\omega = 10^6$, and $L = 0$.

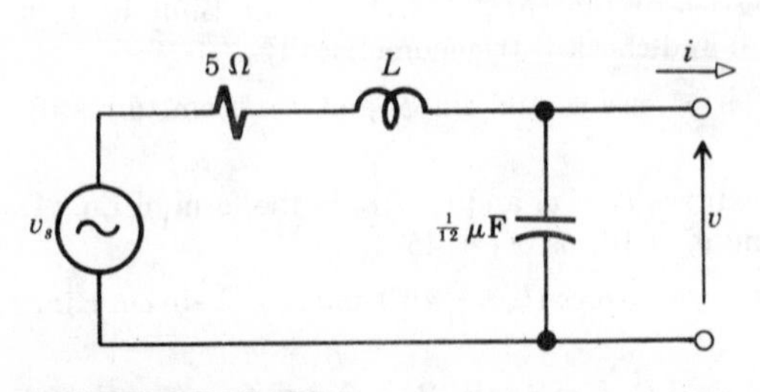

FIGURE P6.2–5

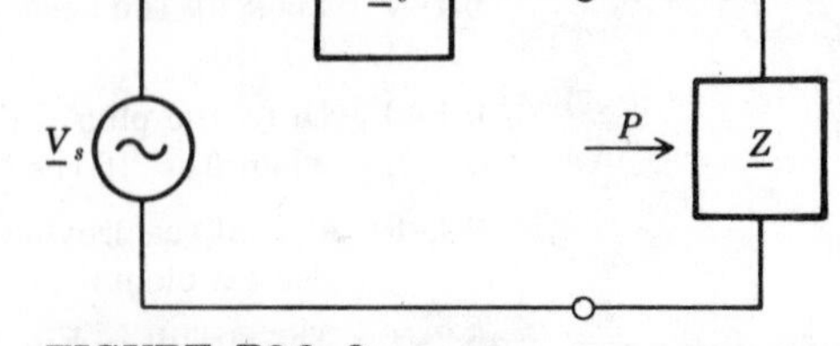

FIGURE P6.3–3

(a) Obtain the Thévenin equivalent circuit by calculating the phasors $\underline{V}_{oc}$ and $\underline{I}_{sc}$ and $\underline{Z}_o = \underline{V}_{oc}/\underline{I}_{sc}$. Check the latter by direct calculation of the equivalent impedance when $\underline{V}_s = 0$.

(b) Find the output current i when a 10-Ω resistance is connected to the terminals.

6.2–6 Repeat the previous problem including $L = 24\ \mu$H.

6.2–7 If Fig. 6.2–4a has $R_1 = 100\ \Omega$, $L = 50$ mH, $C = 2\ \mu$F, and $R_2 = 500\ \Omega$, for what value of ω (besides $\omega = 0$) is $X(\omega) = 0$? Find the corresponding value of $R(\omega)$.

6.2–8 Repeat the previous problem with $R_1 = 160\ \Omega$ relocated to be in parallel with L.

6.2–9 Let Fig. P6.2–3 have $R_1 = 11\ \Omega$, $R_2 = 7\ \Omega$, and $L = 2.4$ mH. What value of C results in $\angle \underline{Z} = 0°$ at $\omega = 10^4$? (The results of this problem are needed in Problem 6.3–9.)

6.2–10 Let Fig. P6.2–1 have $R_2 = 100\ \Omega$ and $C = 5\ \mu$F. If $|\underline{I}_2| = |\underline{I}_1|$ at $\omega = 1000$ and $|\underline{I}_2| = 2|\underline{I}_1|$ at $\omega = 2000$, what are the values of R_1 and L?

6.2–11 Let Fig. 7.2–3a have $R = 100\ \Omega$, $L = 50$ mH, and $C = 10\ \mu$F. Find two values of ω such that $|\underline{V}_{out}| = |\underline{V}_{in}|/\sqrt{2}$. Hint: Write $\underline{Z} = R + jX$ and show that the condition requires $X^2 = R^2$ or, equivalently, $\pm X = R$.

6.3–1 Use Eq. (1) to calculate P for the circuit in Problem 6.2–4. Check your results by summing the power dissipated in each resistance.

6.3–2 Repeat the previous problem for the circuit in Fig. P6.2–1 with $R_1 = 20\ \Omega$, $\omega L = 15\ \Omega$, $R_2 = 7\ \Omega$, $1/\omega C = 24\ \Omega$, and $I_m = 3$ A.

6.3–3 Figure P6.3–3 represents an AC voltage source with fixed source impedance $\underline{Z}_s = R_s + jX_s$. It can be shown that *maximum power transfer* to the load occurs when $\underline{Z} = \underline{Z}_s^*$.

(a) If $\underline{V}_s = 170$ V at $\omega = 2\pi \times 60$ and $\underline{Z}_s = 5\ \Omega\ \angle 45°$, what two series elements should make up the load for maximum power transfer, and what is the corresponding value of the load power P?

(b) Find P if $\underline{Z}$ equals R_s rather than $\underline{Z}_s^*$.

6.3–4 Repeat the previous problem with $\underline{V}_s = 10$ V at $\omega = 5000$ and $\underline{Z}_s = 50\ \Omega\ \angle -60°$.

6.3–5 Suppose $\underline{Z} = R$ in Fig. P6.3–3. Obtain an expression for the power P dissipated by R and solve $dP/dR = 0$ to show that maximum power transfer in this case occurs when $R = |\underline{Z}_s|$.

6.3–6 Let Fig. P6.2–1 have $R_2 = 0$ and $\omega L = 1/\omega C = R_1$ at some value of ω. Without making detailed calculations, construct a phasor diagram showing all voltage and current phasors when $\angle \underline{I} = 0°$.

6.3–7 Repeat the previous problem with $R_1 = 0$ and $\omega L = 1/\omega C = R_2$.

6.3–8 Let Fig. P6.2–1 have $\omega L/R_1 = 1/\omega C R_2$ so $|\underline{V}_C| = |\underline{V}_L|$. Taking $\angle \underline{V} = 0°$, construct a phasor diagram showing all voltage phasors and confirm that $\angle(\underline{V}_L - \underline{V}_C) = 90°$.

6.3–9 Use the results of Prob. 6.2–9 to construct a phasor diagram showing all the voltage phasors when $\angle \underline{V} = 0°$.

6.3–10 Suppose the winding resistance of a coil is given by $R_w = 100\sqrt{L}\ \Omega$ where L is the coil's inductance in henrys. Let a capacitor be connected to the coil to form a series-resonant circuit with $R = R_w$.

(a) Calculate R_w, ω_0, and Q if $L = 40$ mH and $C = 25\ \mu$F.

(b) Find the values of C, L, and R_w if $\omega_0 = 10^6$ and $Q = 500$.

6.3–11 Find the amplitude of the voltage across the capacitance and across the coil when the circuit in Prob. 6.3–10b is driven at resonance with $V_m = 50$ V.

6.3–12 A parallel resonant circuit is to have $Q = 100$ using $C = 100\ \mu$F. Find the values of L and R when

(a) $\omega_0 = 10^4$ (b) $\omega_0 = 10$.

Comment on the feasibility of these two circuits.

6.3–13 The coil in Problem 6.3–10 is to be used for the inductance of a parallel resonant circuit having $\omega_0 = 10^4$ and $Q = 100$. Find values for L, C, and R such that R_w has negligible effect when $\omega \approx \omega_0$.

6.3–14 A general definition of quality factor that applies to any resonant phenomenon is $Q = 2\pi W_s/W_d$, where W_s is the average stored energy and $W_d = PT$ is the energy dissipated per period. Show that a series resonant circuit with $\omega = \omega_0$ has $W_s = w_C + w_L = V_m^2 L/2R^2$ and, therefore, $2\pi W_s/W_d = \omega_0 L/R$.

6.3–15 Repeat the previous analysis for a parallel resonant circuit, in which case $w_C + w_L = I_m^2 RC/2$ and $2\pi W_s/W_d = \omega_0 CR$.

6.3–16 Let Fig. P6.2–1 have $R_1 = 40\ \Omega$, $L = 10$ mH, $C = 1\ \mu$F, and $R_2 = 0$.

(a) Determine the resonant frequency by finding the value of ω at which $\sphericalangle \underline{Y} = 0°$. Compare with the value of $1/\sqrt{LC}$.

(b) Calculate $\underline{Z}$, $|\underline{I}_1|/|\underline{I}|$, and $|\underline{I}_2|/|\underline{I}|$ at resonance. Compare these current ratios with $Q_{par} = R\sqrt{C/L}$ and $Q_{ser} = 1/Q_{par}$. (These results will be needed in Prob. 6.3–18.)

6.3–17 Repeat the previous problem with $R_1 = 0$ and $R_2 = 40\ \Omega$.

6.3–18 Use the definition from Prob. 6.3–14 to calculate Q of the circuit in Problem 6.3–16. Hint: Although $w_C + w_L$ is not constant, W_s is easily computed when you draw on the fact that the average value of $\cos^2(\omega t + \theta)$ equals $\frac{1}{2}$ for any θ.

FREQUENCY RESPONSE AND FILTERS

Chapter 6 dealt with AC circuits operating at fixed source frequencies, so we paid little attention to frequency dependence. But the signals in an electronic system generally contain many different frequency components, and the response to any one component depends on its frequency, as well as on the circuit parameters. A frequency-selective network, then, may function like a *filter,* in the sense that it produces a larger response at some frequencies than at others.

This chapter develops the concepts of frequency response and filtering, starting with simple lowpass and highpass filters. We will also examine bandpass filters, tuned circuits, and filters that employ active devices to achieve both amplification and filtering. Optional sections introduce the Bode plot, a graphic means of displaying frequency-response characteristics, and periodic steady-state analysis using the Fourier series.

OBJECTIVES

After studying this chapter and working the exercises, you should be able to do each of the following:

- Calculate the cutoff frequency of a first-order lowpass or highpass filter and sketch its frequency response (Section 7.1).

- Design a simple lowpass, highpass, or bandpass filter using an op-amp (Sections 7.1 and 7.2).
- Calculate the center frequency and bandwidth of a tuned circuit and sketch its frequency response (Section 7.2).
- Construct the Bode plot of a transfer function that consists of a product of first-order functions (Section 7.3). †
- Interpret the Fourier-series expansion of a periodic waveform, and obtain an approximation for the resulting steady-state response when the waveform is applied to a frequency-selective network (Section 7.4). †

7.1 LOWPASS AND HIGHPASS FILTERS

A *first-order* circuit has been defined to contain only one reactive element, either capacitance or inductance. Such circuits exhibit a lowpass or highpass frequency response. Those characteristics are studied here, along with simple filters.

RL AND *RC* FREQUENCY RESPONSE

Suppose a variable-frequency source $v = V_m \cos \omega t$ drives a series RL circuit, Fig. 7.1–1a. The resulting current will be $i = I_m \cos(\omega t + \theta_i)$ with

$$I_m = \frac{V_m}{\sqrt{R^2 + (\omega L)^2}} \qquad \theta_i = -\arctan \frac{\omega L}{R} \tag{1}$$

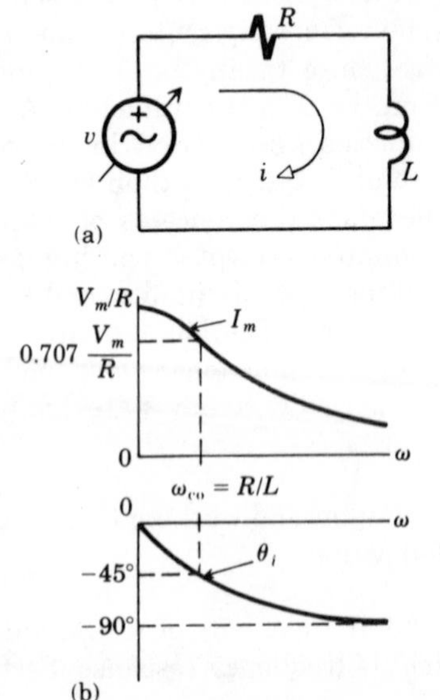

FIGURE 7.1–1
(a) *RL* circuit.
(b) Frequency-response curves.

If all other parameters are kept fixed, I_m and θ_i vary with ω as shown in Fig. 7.1–1b when the source frequency goes from low to high values. At the low-frequency end where $\omega \to 0$,

$$I_m = \frac{V_m}{\sqrt{R^2 + 0}} = \frac{V_m}{R} \qquad \theta_i = -\arctan 0 = 0°$$

This low-frequency response agrees with our prior conclusion that inductance acts like a *DC short circuit,* since an AC source with $\omega = 0$ would actually be a DC source $v = V_m \times \cos 0 = V_m$. At the other end of the frequency axis where $\omega \to \infty$ we see that

$$I_m = 0 \qquad \theta_i = -\arctan \infty = -90°$$

meaning that the inductance effectively becomes an *open-circuit* or "AC choke" at high frequencies. The low- and high-frequency behavior of an inductance are compactly summarized by its impedance $\underline{Z}_L = j\omega L$.

An easily computed intermediate point on the curves in Fig. 7.1–1b occurs where $\omega = R/L$ so $\omega L = R$ and

$$\begin{aligned} I_m &= \frac{V_m}{\sqrt{R^2 + R^2}} = \frac{V_m}{\sqrt{2}R} = 0.707\frac{V_m}{R} \\ \theta_i &= -\arctan 1 = -45° \end{aligned} \tag{2a}$$

This point serves as a rough boundary between the two extremes, and we define the so-called *cutoff frequency*

$$\omega_{\text{co}} = \frac{R}{L} \tag{2b}$$

The cutoff frequency is also known as the *half-power* point in view of the fact that $P = V_m^2/4R$ at ω_{co}, precisely one-half the maximum possible AC power dissipation $P = V_m^2/2R$ when $\omega \to 0$. (The DC power dissipation also equals $V_m^2/2R$ if the source has $v = V_{\text{rms}} = V_m/\sqrt{2}$.)

If capacitance replaces the inductance in Fig. 7.1–1a we find just the opposite frequency response. Specifically, since $\underline{Z} = R + 1/j\omega C$,

$$I_m = \frac{V_m}{\sqrt{R^2 + (1/\omega C)^2}} = \begin{cases} 0 & \omega \to 0 \\ V_m/R & \omega \to \infty \end{cases}$$

and

$$\theta_i = \arctan\left(\frac{1}{\omega CR}\right) = \begin{cases} +90° & \omega \to 0 \\ 0° & \omega \to \infty \end{cases}$$

Therefore, the capacitance goes from a DC *open circuit* to a high-frequency *short circuit,* again as summarized by $\underline{Z}_C = 1/j\omega C$. The cutoff frequency is

$$\omega_{co} = \frac{1}{RC} \tag{3}$$

at which $I_m = V_m/\sqrt{2}\,R$ and $\theta_i = +45°$.

Parallel RL and RC circuits and other first-order circuits exhibit similar frequency response. Second-order circuits (with two reactive elements) will be discussed in Section 7.2.

Exercise 7.1–1

Suppose $R = 200\ \Omega$, $C = 5\ \mu$F, and $I_m = 0.1$ A in the parallel circuit of Fig. 6.3–4. Calculate or approximate $\underline{V}$ at $\omega = 10$, 10^3, and 10^5, and show that $|\underline{I}_R| = |\underline{I}_C|$ when $\omega = 1/RC$.

RL AND RC FILTERS

Because of the shape of I_m versus ω in Fig. 7.1–1b, we say that a series RL circuit functions as a *lowpass filter* in the sense that a low-frequency applied voltage produces much more current than a high-frequency voltage source. Thus, if the applied signal consists of many different frequency components, the circuit "passes" all low frequencies, $\omega \ll \omega_{co}$, but rejects or "filters out" all high frequencies, $\omega \gg \omega_{co}$.

We formalize this concept by arranging the circuit as the *two-port* network in Fig. 7.1–2, with input voltage phasor $\underline{V}_{in} = V_{in}\ \underline{/\theta_{in}}$ and output $\underline{V}_{out} = V_{out}\ \underline{/\theta_{out}}$. We then introduce the *AC voltage transfer function* as the ratio of the phasors at an arbitrary frequency $f = \omega/2\pi$, that is

$$\underline{H}(f) \triangleq \frac{\underline{V}_{out}}{\underline{V}_{in}} = \frac{V_{out}}{V_{in}}\ \underline{/\theta_{out} - \theta_{in}} \tag{4}$$

which corresponds to letting $s = j2\pi f$ in $H(s)$ as defined by Eq. (13), Section 5.3. *Cyclical* frequency f is used here, rather than ω, to reflect the common practice of measuring frequency in hertz. In general, $\underline{H}(f)$ is a complex quantity and can be written in the polar form

$$\underline{H}(f) = |H(f)|\ \underline{/\theta(f)} \tag{5a}$$

where, by definition,

$$|H(f)| = \frac{V_{out}}{V_{in}} \qquad\qquad \theta(f) = \theta_{out} - \theta_{in} \tag{5b}$$

called the *amplitude ratio* and *phase shift,* respectively. To clarify the meaning of these terms, note that any AC input voltage $v_{in}(t) = v_{in}$ cos

$(2\pi ft + \theta_{\text{in}})$ produces a steady-state AC output voltage with

$$V_{\text{out}} = |H(f)|\, V_{\text{in}} \qquad \theta_{\text{out}} = \theta(f) + \theta_{\text{in}}$$

and

$$v_{\text{out}} = V_{\text{out}} \cos\,(2\pi ft + \theta_{\text{out}})$$

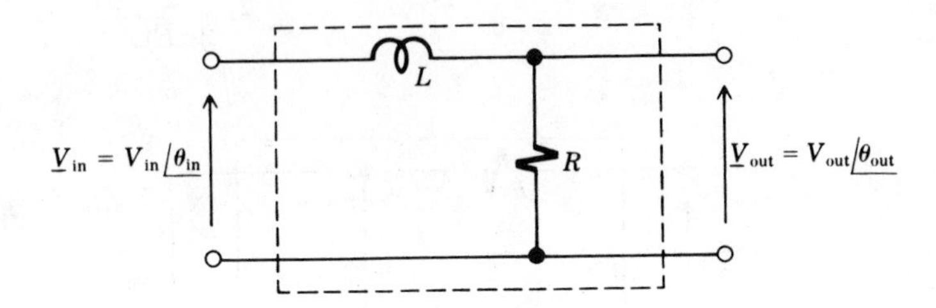

FIGURE 7.1–2
RL lowpass filter.

The transfer function for Fig. 7.1–2 is easily found from the voltage-divider ratio

$$\underline{V}_{\text{out}} = \frac{R}{R + j\omega L}\,\underline{V}_{\text{in}} = \frac{1}{1 + j(\omega L/R)}\,\underline{V}_{\text{in}}$$

Hence, letting $\omega = 2\pi f$ and inserting the cyclical cutoff frequency

$$f_{\text{co}} = \frac{\omega_{\text{co}}}{2\pi} = \frac{R}{2\pi L} \tag{6}$$

we obtain

$$\underline{H}(f) = \frac{1}{1 + j(2\pi fL/R)} = \frac{1}{1 + j(f/f_{\text{co}})} \tag{7a}$$

which has

$$|H(f)| = \frac{1}{\sqrt{1 + (f/f_{\text{co}})^2}} \qquad \theta(f) = -\arctan \frac{f}{f_{\text{co}}} \tag{7b}$$

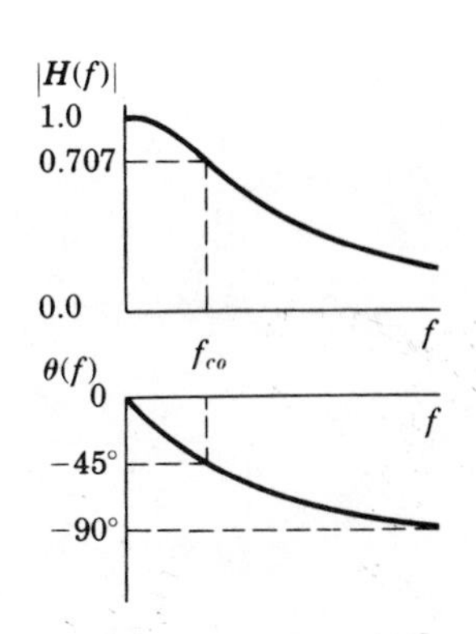

FIGURE 7.1–3
Amplitude ratio and phase shift of a lowpass filter.

The amplitude ratio and phase shift are plotted in Fig. 7.1–3 and have the same shape as our previous curves I_m and θ_i versus ω (see Fig. 7.1–1). This *RL* two-port is, in fact, a lowpass filter because any input components at $f \ll f_{\text{co}}$ appear at the output with amplitude and phase virtually unchanged, while components at $f \gg f_{\text{co}}$ are "choked" by the inductance and appear greatly reduced in amplitude at the output. The cutoff frequency or half-power point is defined such that $|H(f_{\text{co}})| =$

$1/\sqrt{2} = 0.707$ and roughly divides the *passband* ($f < f_{co}$) from the *stopband* ($f > f_{co}$).

An *RC* circuit also acts as a lowpass filter when arranged as in Fig. 7.1–4, because the capacitance becomes a short circuit as $f \to \infty$ and "shorts out" any high-frequency voltage across the output. The transfer function is identical to Eq. (7) with

$$f_{co} = \frac{1}{2\pi RC} \tag{8}$$

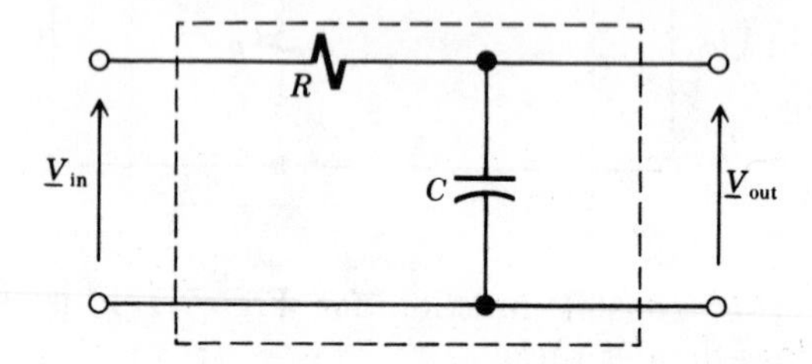

FIGURE 7.1–4
RC lowpass filter.

You might suspect that interchanging resistance and reactance in Fig. 7.1–2 would produce a *highpass* filter. And, indeed, that is true. In particular, for Fig. 7.1–5a

$$\underline{H}(f) = \frac{j\omega L}{R + j\omega L} = \frac{j(2\pi fL/R)}{1 + j(2\pi fL/R)} = \frac{j(f/f_{co})}{1 + j(f/f_{co})} \tag{9a}$$

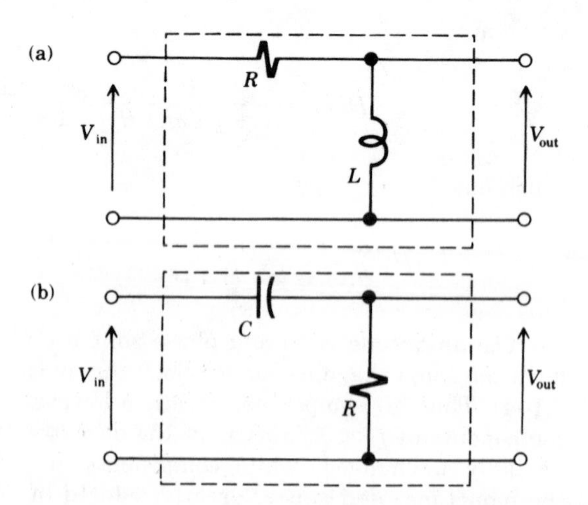

FIGURE 7.1–5
Highpass filters.

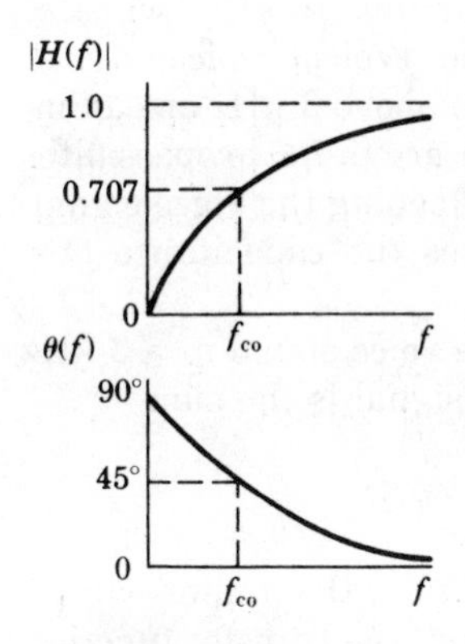

FIGURE 7.1–6
Amplitude ratio and phase shift of a highpass filter.

and conversion to polar form yields

$$
\begin{aligned}
|H(f)| &= \frac{(f/f_{co})}{\sqrt{1 + (f/f_{co})^2}} \\
\theta(f) &= 90^\circ - \arctan(f/f_{co})
\end{aligned}
\tag{9b}
$$

as plotted in Fig. 7.1–6. The amplitude-ratio curve confirms that this circuit blocks low frequencies and passes high frequencies, $f \gg f_{co}$, as does the RC circuit of Fig. 7.1–5b.

It is frequently more informative to express amplitude ratios $|H(f)|$ in *decibels,* using the definition

$$H_{\rm dB} \triangleq 10 \log |H(f)|^2$$

We then plot $H_{\rm dB}$ versus frequency on a logarithmic axis. Such graphs are known as *Bode plots* (pronounced Bo-dee). They can be quickly sketched using the techniques in Section 7.3. For reference purposes, Fig. 7.1–7 shows the Bode plots of our lowpass and highpass filter functions. Note that $H_{\rm dB} \approx -3$ dB at $f = f_{co}$; consequently, the cutoff frequency is sometimes called the "3-dB frequency." The negative dB values here merely reflect the fact that $|H(f)| \leq 1$—that is, these filters do not amplify.

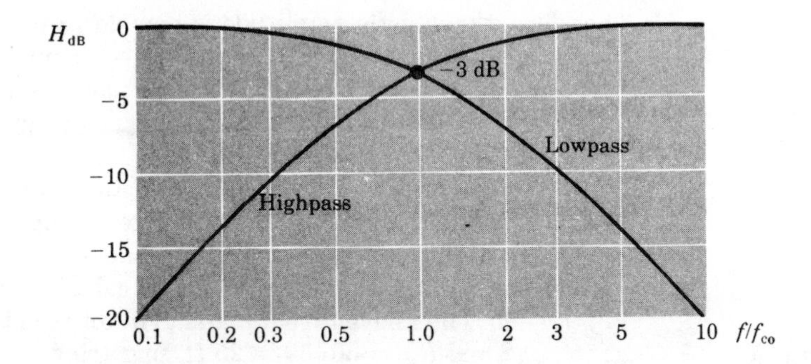

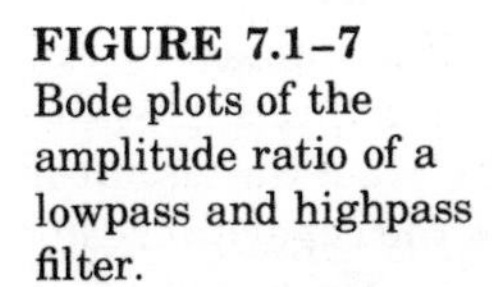

FIGURE 7.1–7
Bode plots of the amplitude ratio of a lowpass and highpass filter.

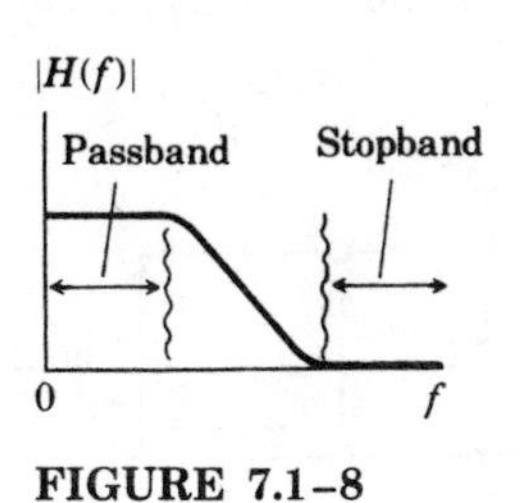

FIGURE 7.1–8

Despite the similarity of RL and RC circuits, virtually all simple filter designs employ capacitance rather than inductance due to the greater availability and lower cost of capacitors and the unwanted resistance associated with inductors. More sophisticated filters, using two or more reactive elements, have a "squarer" frequency response of the type sketched in Fig. 7.1–8, where $|H(f)|$ is essentially constant or "flat" over the passband and there is a relatively narrow transition between the passband and stopband.

Example 7.1–1

A home intercom system has developed an annoying 10-kHz "whistle" whose amplitude is about ten-percent of the typical voice signal. Knowing that voice frequency components much above 3 kHz are unimportant for intelligibility, the owner decides to insert an RC lowpass filter with $f_{co} = 4$ kHz to get rid of the whistle while keeping the voice signal. He chooses $R = 2$ kΩ, arbitrarily, and computes the capacitance $C = 1/2\pi R f_{co} \approx 0.02$ μF.

To analyze the strategy here, let's model the voice signal as a 3-kHz sinusoid with 5-V amplitude, so the total input signal is the sum

$$v_{in}(t) = 5 \cos 2\pi f_1 t + 0.5 \cos 2\pi f_2 t$$

where $f_1 = 3$ kHz and $f_2 = 10$ kHz. By superposition, the output will be the sum of the individual outputs determined from the transfer function given in Eq. (7) with $f_1/f_{co} = 3\text{ kHz}/4\text{ kHz} = 0.75$ and $f_2/f_{co} = 10\text{ kHz}/4\text{ kHz} = 2.5$, respectively. Inserting numerical values yields

$$\underline{H}(f_1) = 0.80 \underline{/-36.9^\circ} \qquad \underline{H}(f_2) = 0.371 \underline{/-68.2^\circ}$$

from which

$$v_{out}(t) = 4.0 \cos (2\pi f_1 t - 36.9^\circ) + 0.19 \cos (2\pi f_2 t - 68.2^\circ)$$

Therefore, the whistle amplitude is now down to roughly five-percent of the signal's slightly reduced amplitude—not much improvement, but a first-order filter can do no better in this particular case.

After installing the filter, the owner is dismayed to find virtually no output signal at all, voice or whistle. What's been overlooked is the *loading effect* due to a 50-Ω source resistance and its matched load resistance, the equivalent circuit with filter being as diagrammed in Fig. 7.1–9a. However, the owner can still achieve the desired filtering by discarding the 2-kΩ resistance entirely and calculating C from the circuit in Fig. 7.1–9b. The Thévenin equivalent circuit seen by the capacitance has $v_{oc} = v_{in}/2$ and $R_{eq} = 50 \| 50 = 25$ Ω, so getting $f_{co} = 4$ kHz requires $C = 1/(2\pi \times 25\ \Omega \times 4 \times 10^3\text{ Hz}) \approx 1.6$ μF. The output is then $v_{out} \approx v_{in}/2$ at $f \ll f_{co}$ due to the voltage divider.

FIGURE 7.1–9 Circuits for Example 7.1–1.

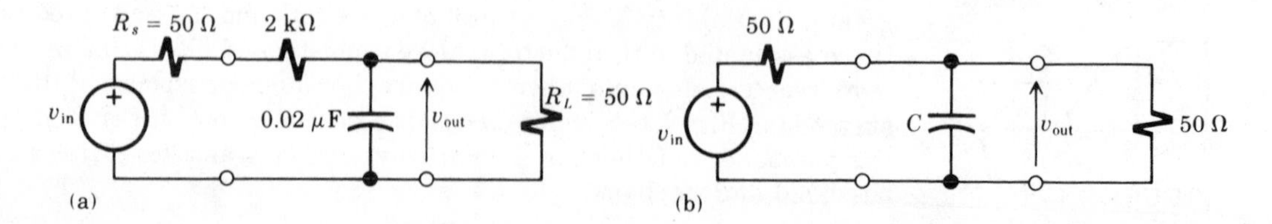

Exercise 7.1–2

A certain telemetry signal consists of two individual signals: $v_a(t)$, with frequencies below 10 kHz, and $v_b(t)$, with frequencies above 50 kHz. The circuit in Fig. 7.1–10 has been proposed as a means of separating the two signals. Determine appropriate values for R_a and R_b. Then calculate the actual output amplitudes when $v(t)$ is a 30-kHz sinusoid with 10-V amplitude.

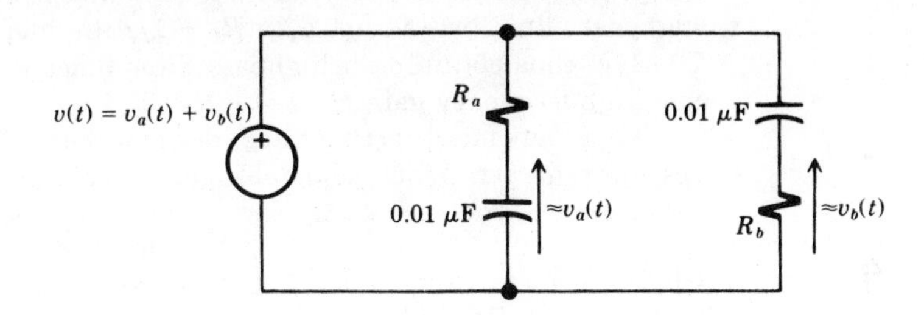

FIGURE 7.1–10 Circuit for Exercise 7.1–2.

ACTIVE FILTERS

Loading effects and other shortcomings of passive filters may be overcome by including an amplifying device—transistor or op-amp—to make an *active* filter. We'll briefly introduce these techniques using the ideal op-amp circuit in Fig. 7.1–11a. If the impedances were resistances R_1 and R_F, you would recognize this as a simple inverting amplifier with gain $v_{\text{out}}/v_{\text{in}} = -R_F/R_1$. Inserting impedances in place of resistances immediately gives us the transfer function

$$\underline{H}(f) = \frac{\underline{V}_{\text{out}}}{\underline{V}_{\text{in}}} = \frac{-\underline{Z}_F}{\underline{Z}_1} \tag{10}$$

where, in general, both $\underline{Z}_F$ and $\underline{Z}_1$ vary with frequency.

To build an active lowpass filter, we take $\underline{Z}_1 = R_1$ and $\underline{Z}_{F_F} = \underline{Z}_{R_F} \| \underline{Z}_{C_F}$, Fig. 7.1–11b. Then

$$\underline{Z}_F = \frac{(1/j\omega C_F)R_F}{(1/j\omega C_F) + R_F} = \frac{R_F}{1 + j2\pi f R_F C_F}$$

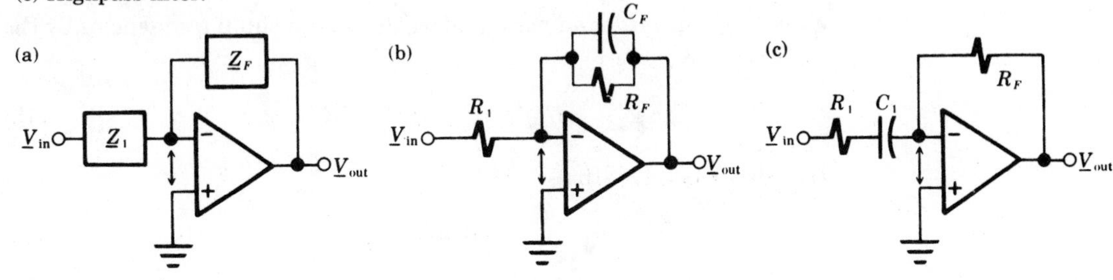

FIGURE 7.1–11 (a) Inverting amplifier with impedances. (b) Lowpass filter. (c) Highpass filter.

and

$$\underline{H}(f) = -\frac{R_F}{R_1}\frac{1}{1 + j(f/f_{co})} \qquad f_{co} = \frac{1}{2\pi R_F C_F}$$

which is a first-order lowpass function with low-frequency gain $\underline{H}(0) = -R_F/R_1$. The input impedance equals R_1, whose value should be large compared to any source impedance to minimize loading effect. We build a highpass filter by taking $\underline{Z}_1 = R_1 + 1/j\omega C_1$ and $\underline{Z}_F = R_F$ as in Fig. 7.1–11c, thus obtaining a highpass filter function with $f_{co} = 1/2\pi R_1 C_1$ and high-frequency gain $\underline{H}(\infty) = -R_F/R_1$.

More advanced active-filter designs have "squarer" frequency-response characteristics, adjustable gain, and large input impedance (see, for example, Problem 7.1–15).

Exercise 7.1–3

Choose element values such that the op-amp highpass filter has $\underline{H}(\infty) = -10$, $f_{co} = 50$ Hz, and an input impedance no less than 1 kΩ.

7.2 BANDPASS FILTERS AND TUNED CIRCUITS

Bandpass filters include at least two reactive elements arranged to yield a frequency-response characteristic that combines lowpass and highpass filtering. Such filters then have a passband that falls between the lower and upper cutoff frequencies. (Sometimes the bandpass effect comes from unavoidable rather than intentional reactances, as in many electronic amplifier circuits.) Tuned circuits are bandpass filters that exploit the resonance phenomenon to achieve a very narrow passband. These circuits play important roles in communication electronics—radio, radar, TV, and so forth.

BANDPASS FILTERS

A simple bandpass filter might have the characteristics sketched in Fig. 7.2–1, where f_ℓ and f_u stand for the *lower* and *upper cutoff frequencies*, respectively. The amplitude ratio is relatively "flat" over the passband and has the value

$$|H(f)| \approx |K| \qquad f_\ell < f < f_u \tag{1a}$$

We call $|K|$ the *midband gain*, and we define the cutoff frequencies by the property that

$$|H(f_\ell)| = |H(f_u)| = |K|/\sqrt{2} \tag{1b}$$

The filter's bandwidth is

$$B = f_u - f_\ell \tag{1c}$$

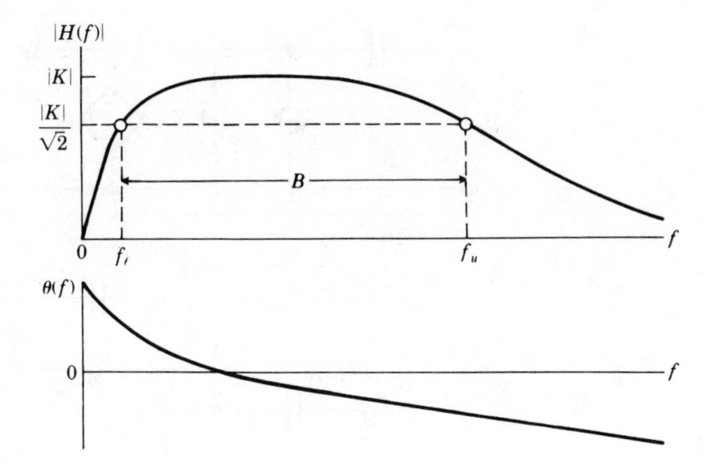

FIGURE 7.2–1 Amplitude ratio and phase shift of a bandpass filter.

and we say we have a *wideband* filter when $f_u \gg f_\ell$ so $B \approx f_u$. The transfer function of a wideband filter is expressed mathematically by

$$\underline{H}(f) = K \frac{j(f/f_\ell)}{1 + j(f/f_\ell)} \frac{1}{1 + j(f/f_u)} \tag{2}$$

which, if $K = 1$, is simply the product of first-order highpass and lowpass functions with cutoff frequencies $f_\ell \ll f_u$. Note that the *highpass* function accounts for the *lower* cutoff frequency f_ℓ, and vice versa.

It's tempting to conclude from Eq. (2) that you could build a bandpass filter by just connecting together a highpass and a lowpass filter. But that conclusion overlooks the *interaction* between the two circuits. To eliminate interaction and, simultaneously, provide midband amplification, you would have to insert a voltage amplifier between the two sections as diagrammed in Fig. 7.2–2a. This configuration is easily analyzed using the chain expansion $\underline{H}(f) = \underline{V}_{\text{out}}/\underline{V}_{\text{in}} = (\underline{V}_1/\underline{V}_{\text{in}})(\underline{V}_a/\underline{V}_1)(\underline{V}_{\text{out}}/\underline{V}_a)$.

If the amplifier has $R_i \gg R_\ell$ and $R_o \ll R_u$, the transfer function will take the form of Eq. (2) with $K = \mu$, and

$$f_\ell = \frac{1}{2\pi R_\ell C_\ell} \qquad f_u = \frac{1}{2\pi R_u C_u} \tag{3}$$

Equations (2) and (3) also hold for the op-amp bandpass filter in Fig. 7.2–2b, whose midband gain and polarity inversion corresponds to $K = -R_u/R_\ell$.

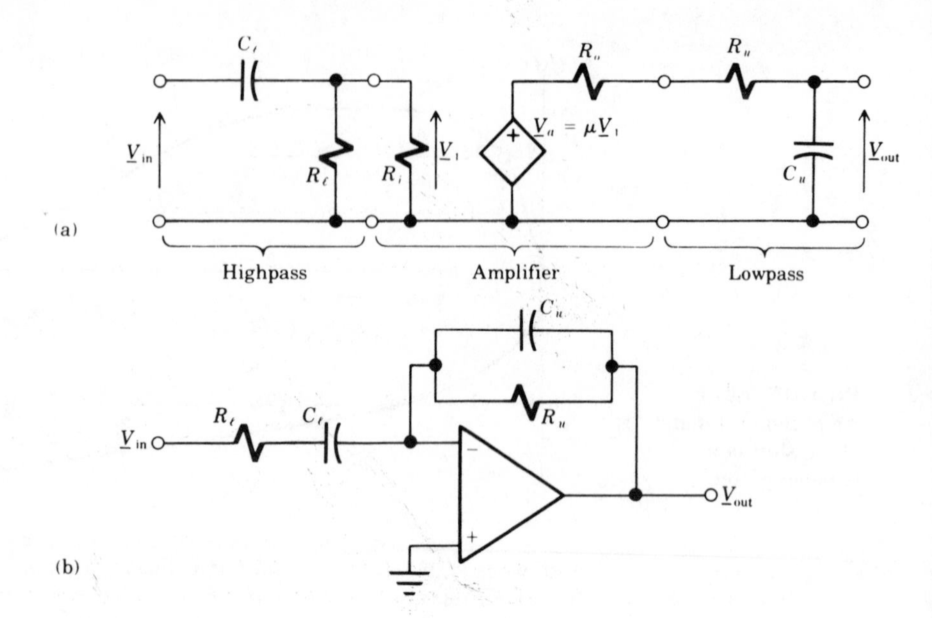

FIGURE 7.2–2 (a) Active bandpass filter. (b) Op-amp implementation.

Example 7.2–1

Suppose that, along with the 10-kHz whistle, the intercom system in Example 7.1–1 also suffers from a low-frequency 60-Hz "hum" and an inadequate signal level. We can cure all these problems in one fell swoop with an op-amp bandpass filter, using element values chosen according to the following line of reasoning.

Since the source has $R_s = 50\ \Omega$ we take $R_\ell = 1\ \text{k}\Omega \gg R_s$ to minimize input loading. We then want C_ℓ, such that the lower cutoff frequency from the highpass function falls above 60 Hz to reduce the hum, but below the significant voice frequencies. An appropriate compromise frequency might be 200 Hz, obtained with $C_\ell \approx 0.08\ \mu\text{F}$. The midband voltage gain between 200 Hz and 4 kHz equals $-R_u/R_\ell$, and we can increase the signal level by a factor of three, say, if $R_u = 3\ \text{k}\Omega$. This, in turn, leads to $C_u \approx 0.01\ \mu\text{F}$ for the lowpass function to have an upper cutoff of 4 kHz.

Exercise 7.2–1

If the values of f_ℓ and f_u in Eq. (2) do not satisfy the wideband condition $f_\ell \ll f_u$, then $|H(f)|$ will be curved rather than flat over the passband and the actual cutoff frequencies differ from f_ℓ and f_u. Demonstrate this by calculating $|H(f)|$ at $f = 10$, 20, 30, and 40, taking $K = 1$, $f_\ell = 10$, and $f_u = 40$ in Eq. (2).

Exercise 7.2–2

Use Eq. (10), Section 7.1, to derive $\underline{H}(f)$ for the op-amp bandpass filter.

TUNED CIRCUITS

Many applications call for a *narrowband* filter whose passband is centered at some frequency f_0 and whose bandwidth B is small compared to the center frequency. This type of characteristic can be implemented with a high-Q resonant circuit "tuned" to the desired center frequency. Consider, for instance, the tuned circuit in Fig. 7.2–3a. Qualitatively, we see that the inductance shorts out low-frequency voltage components while the capacitance shorts out high frequencies; consequently, the filter passes only those frequencies in the vicinity of resonance where the LC section has maximum impedance.

To emphasize that we are dealing with *parallel* resonance, we perform a Thévenin-to-Norton conversion and redraw the circuit per Fig. 7.2–3b. Taking the input and output to be the indicated currents then, relative to Fig. 7.2–3a, $\underline{I}_{in} = \underline{V}_{in}/R$ and $\underline{V}_{out} = R\underline{I}_{out}$, so $\underline{I}_{out}/\underline{I}_{in} = \underline{V}_{out}/\underline{V}_{in}$ and the two circuits have identical transfer functions. In practice, the parallel configuration driven by a current source often occurs as the model of a tuned amplifier.

For a quantitative analysis of Fig. 7.2–3b, we use the parallel admittance $\underline{Y} = \underline{I}_{in}/\underline{V}_{out}$ from Eq. (8), Section 6.3, and write $\underline{I}_{out} = \underline{V}_{out}/R = (\underline{I}_{in}/\underline{Y})/R = \underline{I}_{in}/R\underline{Y}$. Thus

$$\underline{H}(f) = \frac{\underline{I}_{out}}{\underline{I}_{in}} = \frac{1}{R\underline{Y}} = \frac{1}{1 + j(\omega CR - R/\omega L)}$$

which can be put in a more useful form by introducing the resonant frequency and quality factor

$$f_0 = \frac{\omega_0}{2\pi} = \frac{1}{2\pi\sqrt{LC}} \qquad \text{(4a)}$$

$$Q = \omega_0 CR = \frac{R}{\omega_0 L} \qquad \text{(4b)}$$

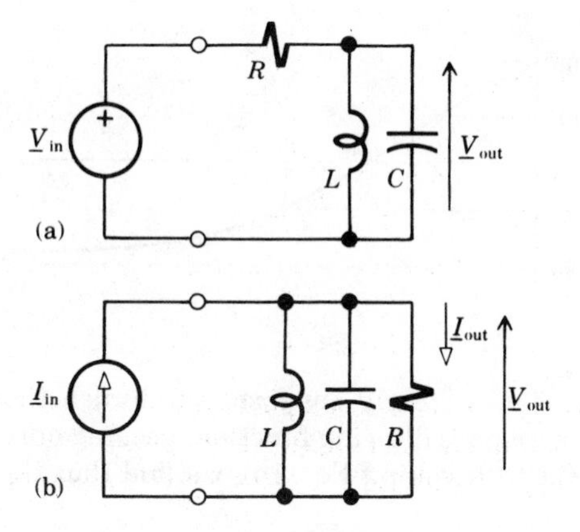

FIGURE 7.2–3
Tuned circuits.

so that

$$\omega CR = \omega_0 CR \frac{\omega}{\omega_0} = Q \frac{f}{f_0}$$

$$\frac{R}{\omega L} = \frac{R}{\omega_0 L} \frac{\omega_0}{\omega} = Q \frac{f_0}{f}$$

Making these substitutions yields

$$\underline{H}(f) = \frac{1}{1 + jQ\left(\frac{f}{f_0} - \frac{f_0}{f}\right)} \tag{5}$$

with

$$|H(f)| = \frac{1}{\sqrt{1 + Q^2\left(\frac{f}{f_0} - \frac{f_0}{f}\right)^2}}$$

$$\theta(f) = -\arctan Q\left(\frac{f}{f_0} - \frac{f_0}{f}\right)$$

which are plotted in Fig. 7.2–4. The midband "gain" is $|H(f_0)| = 1$.

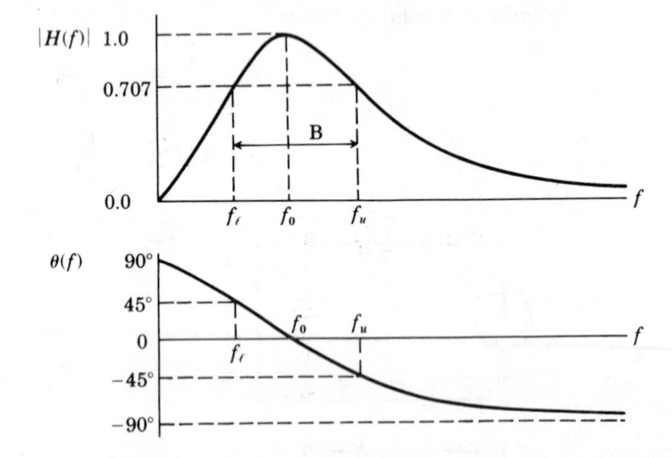

FIGURE 7.2–4 Amplitude ratio and phase shift of a tuned circuit.

This figure clearly supports our hunch that the tuned circuit performs a bandpass filtering function, passing only those frequencies in the vicinity of f_0. Invoking Eq. (1b), we find that the cutoff frequencies must satisfy

$$Q\left(\frac{f}{f_0} - \frac{f_0}{f}\right) = \pm 1$$

which leads to a quadratic equation whose solutions are

$$f_\ell = f_0 \sqrt{1 + \left(\frac{1}{2Q}\right)^2} - \frac{f_0}{2Q} \qquad f_u = f_0 \sqrt{1 + \left(\frac{1}{2Q}\right)^2} + \frac{f_0}{2Q} \tag{6a}$$

Therefore, the filter's bandwidth is

$$B = f_u - f_\ell = \frac{f_0}{Q} \tag{6b}$$

and a high-Q filter has a narrow bandwidth. When $Q \gg 1$, we can use the approximations

$$f_\ell \approx f_0 - \frac{f_0}{2Q} = f_0 - \frac{B}{2}$$

$$f_u \approx f_0 + \frac{f_0}{2Q} = f_0 + \frac{B}{2}$$

derived from the above expressions with $(1/2Q)^2 \ll 1$. In this very common case, the amplitude-ratio curve is essentially symmetrical around f_0, illustrated by Fig. 7.2–5.

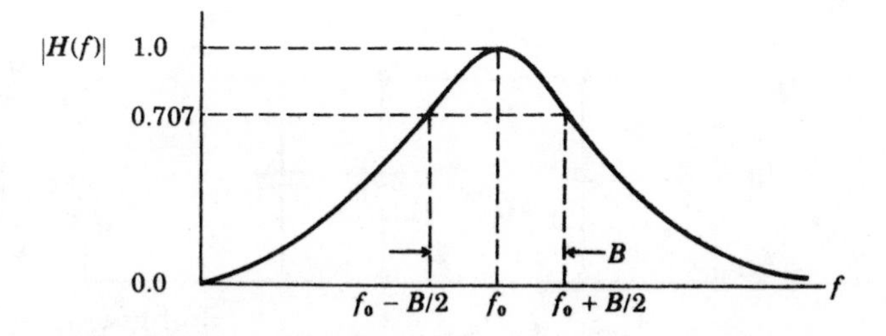

FIGURE 7.2–5 Amplitude ratio of a high-Q filter.

Finally, we should take account of the inevitable series resistance associated with a real inductor, which leads to the more realistic tuned-circuit model in Fig. 7.2–6. The admittance of this circuit is

$$\underline{Y} = \frac{1}{R_1} + j\omega C + \frac{1}{R_2 + j\omega L} = \frac{1}{R_1} + \frac{j\omega CR_2 - \omega^2 LC + 1}{R_2 + j\omega L}$$

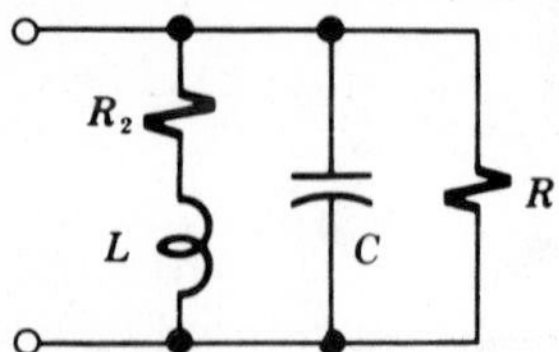

FIGURE 7.2–6
A practical tuned circuit.

However, if $R_2 \ll \omega_0 L$ then $R_2 + j\omega L \approx j\omega L$ for the frequencies of interest and

$$\underline{Y} \approx \frac{1}{R_1} + \frac{CR_2}{L} + j\omega C + \frac{1}{j\omega L}$$

which has the same form as that of a parallel *RLC* circuit with

$$\frac{1}{R} = \frac{1}{R_1} + \frac{CR_2}{L} \tag{7}$$

Our previous results therefore apply to Fig. 7.2–6 by computing R from Eq. (7).

Example 7.2–2
Tuned amplifier

Figure 7.2–7a represents the output stage of a tuned amplifier with transformer coupling to a 50-Ω load. The amplifier is to operate over a range of 100 ± 2.5 kHz, delivering maximum signal power to the load. We need to find C, L, and the turns ratio N to suit the specifications, and also determine $\underline{V}_{out}$ when $\underline{I}_s = 10\text{ mA} \angle 0°$ and $f = 100$ kHz.

Referring the load resistance to the primary side of the transformer gives the parallel-resonant circuit in Fig. 7.2–7b. For maximum power transfer the reactances should cancel out—which they will at resonance—and the referred load resistance should equal the source resistance. Thus, we want $50\ \Omega/N^2 = 3.2\text{ k}\Omega$ or $N = \sqrt{50/3200} = \frac{1}{8}$, corresponding to a step-down transformer ($N < 1$). The equivalent parallel resistance then is $R = 3.2\text{ k}\Omega/2 = 1.6\text{ k}\Omega$.

From the frequency specifications we take $f_0 = 100$ kHz and $B = 2 \times 2.5 = 5$ kHz, so $Q = \frac{100}{5} = 20$. Hence, inserting values in Eq. (4b) yields $C = Q/\omega_0 R \approx 0.02\ \mu\text{F}$ and $L = R/\omega_0 Q \approx 130\ \mu\text{H}$. Finally, since

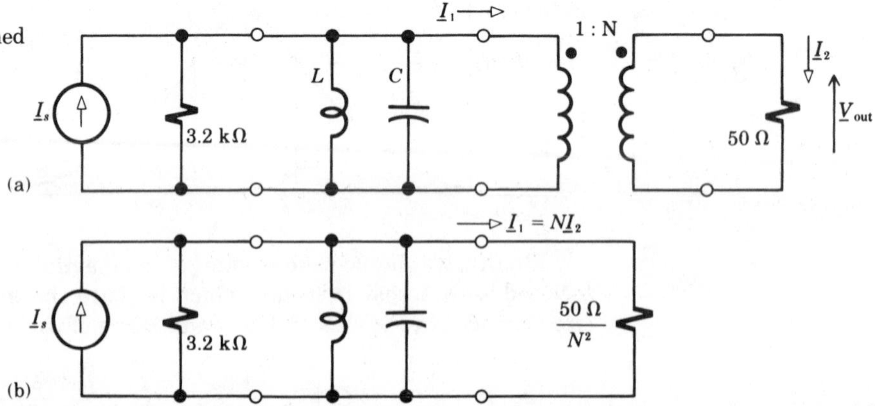

FIGURE 7.2–7
Output stage of a tuned amplifier.

$\underline{I}_C + \underline{I}_L = 0$ at resonance, the current-divider ratio gives $\underline{I}_1 = N\underline{I}_2 = \underline{I}_s/2$; therefore, $\underline{V}_{out} = 50\ \Omega \times (\underline{I}_s/2N) = 2\ \text{V}\ \underline{/0°}$.

Exercise 7.2–3

The intermediate-frequency amplifier in an FM tuner has $f_0 = 10.7$ MHz and $B = 250$ kHz. Find Q, R, and C for the corresponding tuned circuit when $L = 10\ \mu$H.

Exercise 7.2–4

Suppose the tuned circuit of Exercise 7.2–3 actually has the form of Fig. 7.2–6 with $R_2 = 5\ \Omega$. Use Eq. (7) to find the value of R_1.

7.3 BODE PLOTS †

The transfer function of a two-port network actually represents two related functions, the amplitude ratio or gain $|H(f)|$ and the phase shift $\measuredangle \underline{H}(f)$. It often proves convenient to express gain in decibels and to plot the dB gain and phase versus frequency on a logarithmic axis, which constitutes a Bode plot. Such plots are valuable for studying the frequency-response characteristics of amplifiers, filters, and linear systems in general. They can be rapidly sketched by taking advantage of simple asymptotic behavior, rather than by making extensive numerical calculations. In describing and illustrating the techniques for sketching these plots, we use cyclical frequency f throughout; but the method remains the same with angular frequency $\omega = 2\pi f$ for the independent variable—as normally used in systems engineering.

FACTORED TRANSFER FUNCTIONS

Given a transfer function $\underline{H}(f)$, the starting point for constructing its Bode plot is to write the factored expression

$$\underline{H}(f) = K\,\underline{H}_1(f)\,\underline{H}_2(f) \cdots \tag{1}$$

where K is a real constant (possibly negative) and $\underline{H}_1(f)$, $\underline{H}_2(f)$, . . . , are simple functions with known Bode plots. The overall gain is then the product

$$|H(f)| = |K|\,|H_1(f)|\,|H_2(f)| \cdots$$

which becomes a *sum* when converted to decibels. Specifically, using the short-hand notation

$$H_{dB} \triangleq |H(f)|_{dB} = 10 \log |H(f)|^2 = 20 \log |H(f)| \tag{2}$$

we have the overall dB gain

$$H_{dB} = K_{dB} + H_{1_{dB}} + H_{2_{dB}} + \cdots \tag{3a}$$

with $K_{\text{dB}} = 20 \log |K|$, $H_{1_{\text{dB}}} = 20 \log |H_1(f)|$, etc. Moreover, since the angle of a product of complex numbers equals the sum of the individual angles, the overall phase $\theta = \measuredangle \underline{H}(f)$ will be

$$\theta = \theta_K + \theta_1 + \theta_2 + \cdots \tag{3b}$$

where $\theta_1 = \measuredangle \underline{H}_1(f)$, etc., and $\theta_K = 0°$ or $\pm 180°$ depending upon whether K is positive or negative.

Equations (3a) and (3b) bring out the underlying strategy here. For if you know the Bode plots of the individual factors, you simply *add them together* to get the complete Bode plot. That addition property holds directly for the phase shifts, while dB conversion makes it possible to add the gains. We can now focus attention on the individual factors that might make up a typical transfer function per Eq. (1).

FIRST-ORDER FUNCTIONS

Most of the factored functions of interest are first order, consisting of constants and f but of no higher powers of f. To begin with, consider the linear function

$$\underline{H}_a(f) = j\frac{f}{F} = \frac{f}{F} \underline{/90°} \tag{4a}$$

in which F stands for an arbitrary constant. The gain and phase are

$$H_{a_{\text{dB}}} = 20 \log \left(\frac{f}{F}\right) \qquad \theta_a = 90°$$

as plotted in Fig. 7.3–1a. The gain curve is a straight line passing through 0 dB at $f = F$ where $\underline{H}_a(F) = 1$. It has a slope of $+20$ dB per *decade*, meaning that $H_{a_{\text{dB}}}$ increases by 20 dB when the frequency increases by a factor of 10—one decade along the logarithmic frequency axis. This slope is also equivalent to about $+6$ dB per *octave*, when the frequency increases by a factor of 2.

Next, take the case of the *highpass* function

$$\underline{H}_b(f) = \frac{j(f/F)}{1 + j(f/F)} = \frac{jf}{F + jf} \tag{4b}$$

The Bode plot can be constructed by taking advantage of the low-frequency and high-frequency approximations

$$\underline{H}_b(f) \approx \frac{jf}{F} = \underline{H}_a(f) \qquad f < 0.1F$$

$$\underline{H}_b(f) \approx \frac{jf}{jf} = 1 \underline{/0°} \qquad f > 10F$$

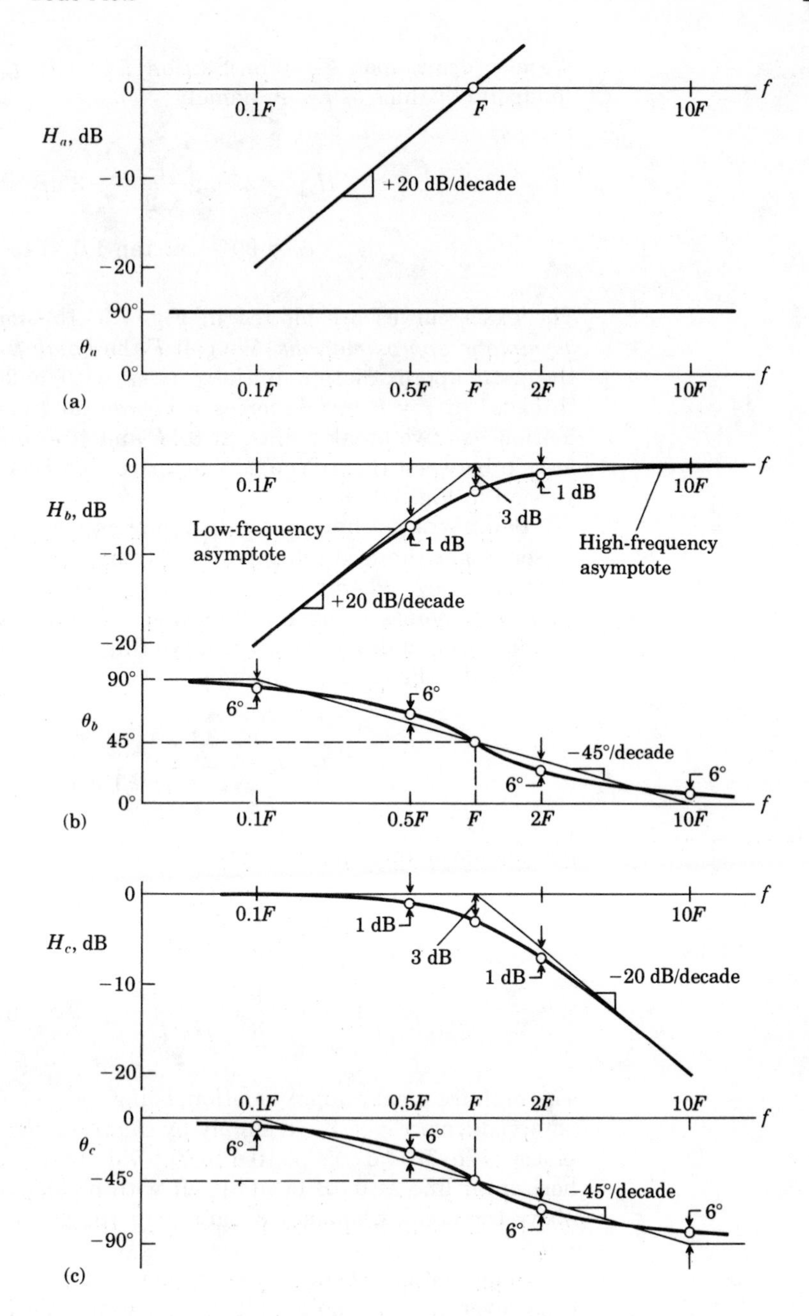

FIGURE 7.3–1
Bode plots of first-order functions:
(a) $\underline{H}_a(f) = jf/F$;
(b) $\underline{H}_b(f) = jf/(F + jf)$;
(c) $\underline{H}_c(f) = F/(F + jf)$.

We also draw upon Eq. (9b), Section 7.1, with $f_{co} = F$ to compute the "midpoint" values at $f = F$, namely

$$H_{b_{dB}} = 20 \log \frac{1}{\sqrt{2}} \approx -3 \text{ dB}$$

$$\theta_b = 90° - \arctan 1.0 = 45°$$

The exact curves are plotted in Fig. 7.3–1b along with straight-line *asymptotic approximations*. We call F the *break frequency* here because the gain approximation initially rises with a 20-dB slope and then "breaks" at $f = F$ and becomes a horizontal line. The phase approximation has two break points, at $0.1F$ and $10F$, and a slope of $-45°$ per decade between them. You can compare this Bode plot with the linear version in Fig. 7.1–6.

Being straight lines, the gain and phase asymptotes are easy to draw on semilogarithmic graph paper. And frequently they alone provide sufficient accuracy. When more precise values are required, you merely plot additional points using the indicated correction terms relative to the asymptotes and draw a smooth curve through them.

Finally, the *lowpass* function

$$\underline{H}_c(f) = \frac{1}{1 + j(f/F)} = \frac{F}{F + jf} \tag{4c}$$

has the asymptotic approximations

$$\underline{H}_c(f) \approx \frac{F}{F} = 1 \underline{/0°} \qquad f < 0.1F$$

$$\underline{H}_c(f) \approx \frac{F}{jf} = \frac{1}{\underline{H}_a(f)} \qquad f > 10F$$

The high-frequency approximation, being the *reciprocal of* $\underline{H}_a(f)$, is then obtained from Fig. 7.3–1a simply by *changing the signs* of the gain and phase. (Why?) Thus, as plotted in Fig. 7.3–1c, the gain curve starts as a horizontal line at 0 dB but falls off with a slope of -20 dB per decade above the break frequency F. Similarly, the phase curve goes from 0° to $-90°$.

Combinations of the three functions in Fig. 7.3–1, plus their reciprocals, cover an amazingly wide variety of practical cases. Two examples should help demonstrate this point and illustrate the technique of using Eqs. (1) and (3).

Example 7.3–1 Consider the function

$$\underline{H}(f) = \frac{j12{,}000f}{(200 + jf)(4000 + jf)} = 3\frac{jf}{200 + jf}\frac{4000}{4000 + jf}$$

$$= 3\frac{j(f/200)}{1 + j(f/200)}\frac{1}{1 + j(f/4000)}$$

which we recognize as a *bandpass* function with $K = 3$, $f_\ell = 200$ and $f_u = 4000$. (Such manipulations may take a few trials to get a suitable form.) We construct the Bode plot by graphically adding the asymptotes of highpass and lowpass functions that have break frequencies of 200 and 4000, respectively. Then, accounting for K, we shift the entire gain curve up by $K_{\text{dB}} = 20 \log 3 \approx 9.5$ dB. Had K been a negative quantity, the phase curve would also be shifted up or down to include $\theta_K = \pm 180°$. (Incidentally, with $K = -3$, our results would completely correspond to the op-amp filter in Example 7.2–1.)

Figure 7.3–2 shows the individual asymptotic approximations from Figs. 7.3–1b and 7.3–1c, the sums with K_{dB} added to the gain curve, and the final smooth plots obtained using the correction terms. Observe that the gain correction terms do not "overlap" because the break frequencies are more than a decade apart—consistent with the characteristics of a wideband filter. However, the phase corrections do have some overlap; thus, for instance, the total phase correction at $f = 400$ consists of $-6°$ from the highpass function at $f = 2F = 2f_\ell$ plus another $-6°$ from the lowpass function at $f = F/10 = f_u/10$.

If the gain or phase correction terms are important, and/or the break frequencies are less than a decade apart, you may find it more accurate to graphically add the individual smooth curves rather than the asymptotes.

Example 7.3–2 Suppose we are given a transfer function in the form

$$H(s) = \frac{158 \times 10^5\,(s + 251)}{s(s + 1260)(s + 12{,}600)}$$

Setting $s = j\omega = j2\pi f$, dividing numerator and denominator by $(2\pi)^3$, and rewriting yields

$$\underline{H}(f) \approx \frac{4 \times 10^5\,(jf + 40)}{jf(jf + 200)(jf + 2000)}$$

$$= \frac{40 + jf}{jf}\,\frac{200}{200 + jf}\,\frac{2000}{2000 + jf}$$

$$= K\underline{H}_1(f)\,\underline{H}_2(f)\,\underline{H}_3(f)$$

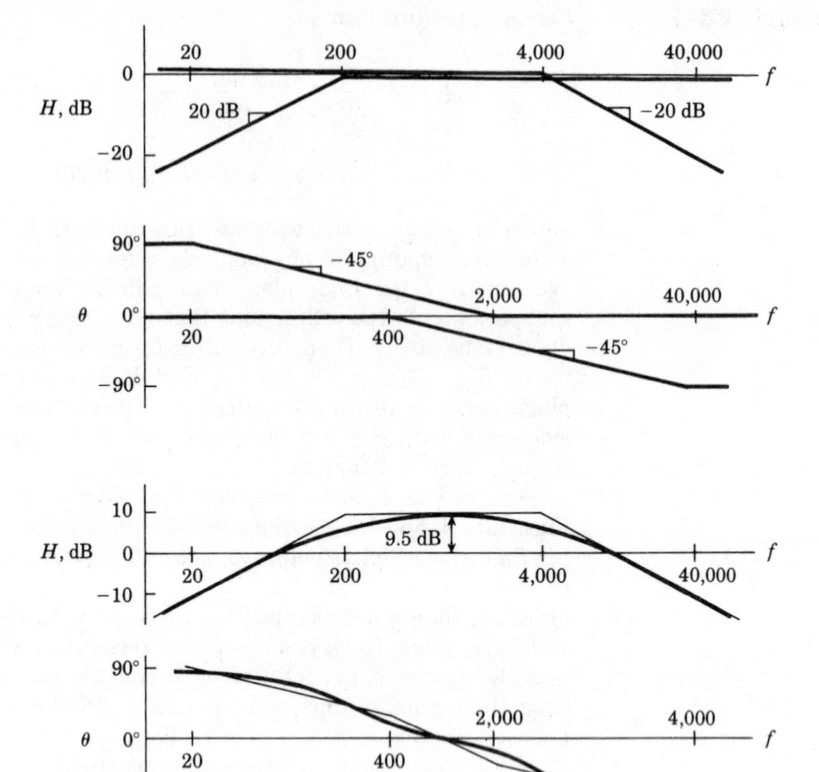

FIGURE 7.3–2
Bode plot for Example 7.3–1: (a) asymptotes; (b) final curves with correction terms.

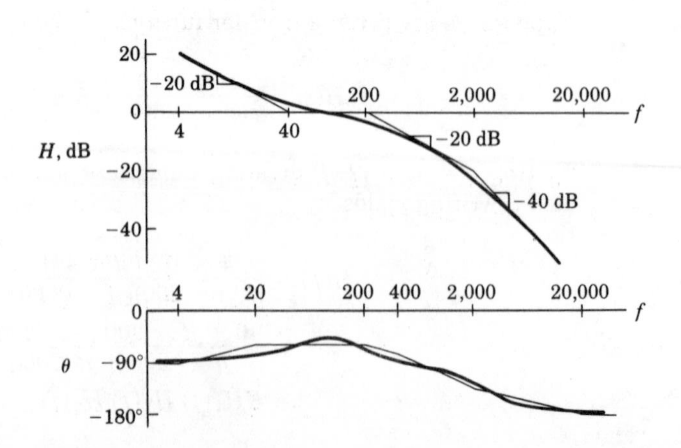

FIGURE 7.3–3
Bode plot for Example 7.3–2.

We see that $K = 1$, $\underline{H}_1(f)$ is the reciprocal of a highpass function with $F = 40$, and $\underline{H}_2(f)$ and $\underline{H}_3(f)$ are lowpass functions with $F = 200$ and 2000. Note that the factors have been arranged in order of increasing break frequency, which helps in the construction of the Bode plot.

The complete Bode plot is drawn in Fig. 7.3–3. The individual asymptotes have been omitted, but they come from Fig. 7.3–1b (with the signs reversed) and from Fig. 7.3–1c. Due to the joint effects of $\underline{H}_2(f)$ and $\underline{H}_3(f)$, the high-frequency gain has a slope of -40 dB per decade and the phase approaches $-180°$.

Exercise 7.3–1

Draw the Bode plot of $\underline{H}(f) = j1000f/(100 + jf)^2$ and determine the maximum value of H_{dB}. Hint: There are two break frequencies at $f = 100$.

Exercise 7.3–2

Use the reciprocal of the lowpass function to obtain the gain and phase asymptotes for

$$\underline{H}(f) = \frac{-4000(50 + jf)}{(200 + jf)(1000 + jf)}$$

Then estimate the lower and upper cutoff frequencies of the passband.

SECOND-ORDER FUNCTIONS

Second-order functions involve f^2 as well as f, and usually occur in a form equivalent to

$$\underline{H}(f) = \frac{1}{1 + j2\zeta(f/f_0) - (f/f_0)^2} \tag{5}$$

where the parameter ζ (zeta) is called the *damping ratio.* If $\zeta > 1$, we have an *overdamped* case and $\underline{H}(f)$ can be rewritten as a product of two first-order lowpass functions. If $\zeta < 1$, we have an *underdamped* or *resonant* case with break frequency f_0 and asymptotes determined from the approximations

$$H_{\text{dB}} \approx \begin{cases} 0 & f < 0.1\, f_0 \\ -40 \log\left(\dfrac{f}{f_0}\right) & f > 10\, f_0 \end{cases}$$

$$\theta \approx \begin{cases} 0° & f < 0.1\, f_0 \\ -180° & f > 10\, f_0 \end{cases}$$

However, the exact behavior in the vicinity of f_0 depends critically upon the damping ratio.

Figure 7.3–4 shows the Bode plot of Eq. (5) for selected values of ζ. The asymptotes are clearly poor approximations near f_0 when $\zeta < 0.5$, and the exact curves do not lend themselves to simple correction terms. Consequently, one may need to resort to a computer program for the detailed calculations if accuracy is at all important.

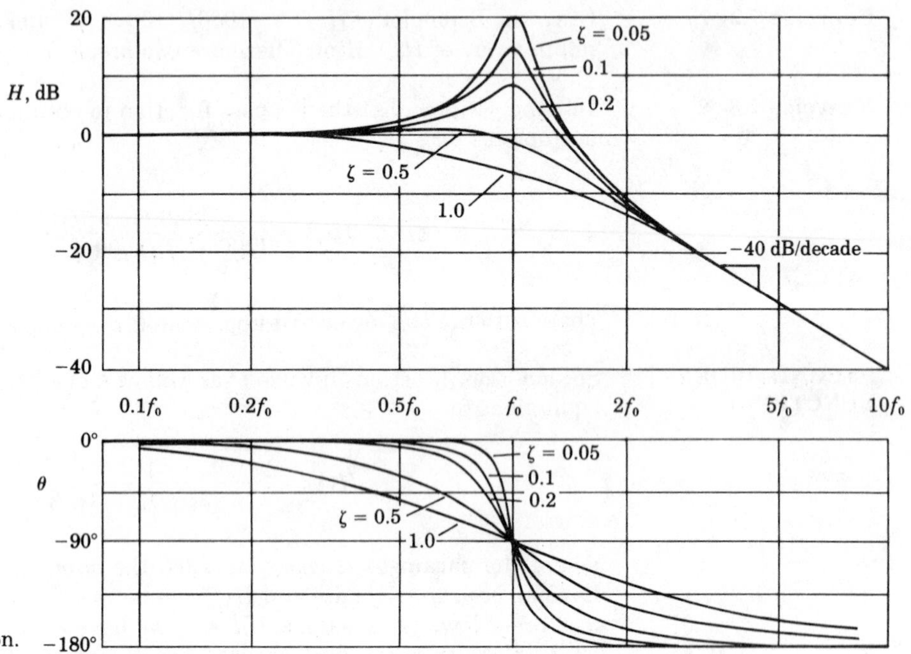

FIGURE 7.3–4
Bode plot of a second-order function.

Exercise 7.3–3

Show that the tuned-circuit transfer function in Eq. (5), Section 7.2, can be written as

$$\underline{H}(f) = \frac{j2\zeta(f/f_0)}{1 + j2\zeta(f/f_0) - (f/f_0)^2}$$

with $\zeta = \frac{1}{2}Q$. Then use Figs. 7.3–1a and 7.3–4 to sketch the Bode plot when $Q = 5$.

7.4 FOURIER-SERIES ANALYSIS †

When a linear network is driven in the steady state by a nonsinusoidal but periodic waveform, the response can be calculated (or at least approximated) through the use of Fourier-series analysis—a means of expanding periodic waveforms as a sum of sinusoids at harmonically related frequencies. We can then invoke superposition and the frequency-response characteristics of the network to determine the resulting periodic steady-state response, also expressed as a sum of sinusoids. Later, in Chapter 14 we extend these concepts in a general discussion of spectral analysis.

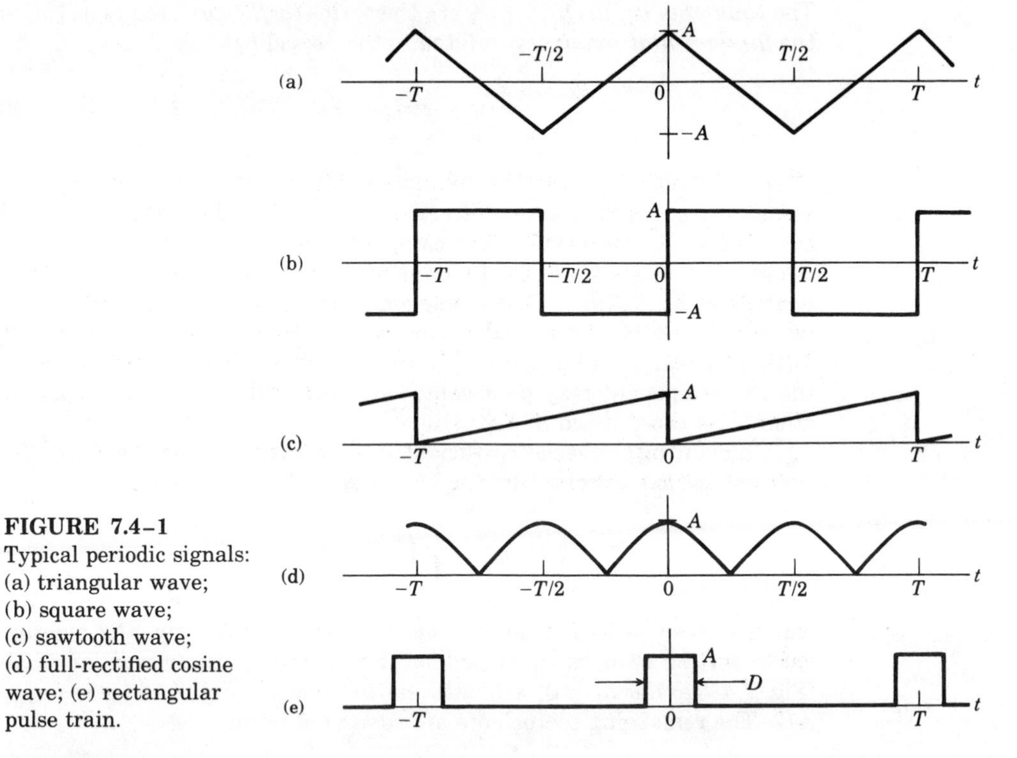

FIGURE 7.4–1 Typical periodic signals: (a) triangular wave; (b) square wave; (c) sawtooth wave; (d) full-rectified cosine wave; (e) rectangular pulse train.

PERIODIC WAVEFORMS AND FOURIER SERIES

The waveforms in Fig. 7.4–1 represents typical signals that occur in various electronic circuits. Although clearly *not* sinusoidal, each signal is *periodic* in that it repeats itself every T seconds. Mathematically, we write this property as

$$x(t + T) = x(t)$$

with T being the period of repetition.

The *Fourier-series theorem* states that almost any such waveform can be decomposed into an infinite series

$$\begin{aligned} x(t) &= a_0 + (a_1 \cos \omega_0 t + b_1 \sin \omega_0 t) \\ &\quad + (a_2 \cos 2\omega_0 t + b_2 \sin 2\omega_0 t) \\ &\quad + \cdots \end{aligned} \tag{1a}$$

$$= a_0 + \sum_{n=1}^{\infty} (a_n \cos n\omega_0 t + b_n \sin n\omega_0 t) \tag{1b}$$

The constants $a_0, a_1, b_1, \ldots$, are the series *coefficients* and ω_0 is called the *fundamental frequency*, related to the period by

$$\omega_0 = 2\pi f_0 = \frac{2\pi}{T} \tag{2}$$

(Be careful not to confuse the fundamental frequency of a signal with the resonant frequency of a circuit.) According to Eq. (1), a periodic signal contains a DC (zero-frequency) component a_0, a sinusoidal component $a_1 \cos \omega_0 t + b_1 \sin \omega_0 t$ at the fundamental frequency f_0, and similar components at $2f_0, 3f_0, \ldots$ These integer multiples of f_0 are known as *harmonics*, $2f_0$ being the second harmonic, $3f_0$ the third harmonic, and so forth. Depending on the specific waveform, some of the harmonics and/or the DC component may be absent, but there will never be components other than those listed in Eq. (1).

Calculating the series coefficients involves integrating the waveform over one period. Specifically, the DC component is found from

$$a_0 = \frac{1}{T} \int_0^T x(t)\, dt \tag{3}$$

which is seen to be the *average value* of $x(t)$. Frequently, the average value will be obvious by inspection; for instance, the triangular wave (Fig. 7.4–1a) has $a_0 = 0$, while the sawtooth wave (Fig. 7.4–1c) has $a_0 = A/2$. The remaining coefficients are computed from

$$a_n = \frac{2}{T} \int_0^T x(t) \cos n\omega_0 t\, dt \tag{4a}$$

$$b_n = \frac{2}{T} \int_0^T x(t) \sin n\omega_0 t\, dt \tag{4b}$$

We have presented these formulas primarily for the sake of completeness, since our concern here is the use and interpretation of the Fourier series rather than integration techniques. To help facilitate matters, Table 7.4–1 lists coefficient expressions for the waveforms in Fig. 7.4–1.

TABLE 7.4–1 Fourier-Series Coefficients for the Waveforms in Fig. 7.4–1

Waveform	*Symmetry*	a_0	a_n or b_n
(a) Triangular wave	Even and half-wave	0	$a_n = \begin{cases} \dfrac{8A}{\pi^2 n^2} & n = 1, 3, 5, \ldots \\ 0 & n = 2, 4, 6, \ldots \end{cases}$
(b) Square wave	Odd and half-wave	0	$b_n = \begin{cases} \dfrac{4A}{\pi n} & n = 1, 3, 5, \ldots \\ 0 & n = 2, 4, 6, \ldots \end{cases}$
(c) Sawtooth wave	Odd, except for a_0	$\dfrac{A}{2}$	$b_n = -\dfrac{A}{\pi n}$
(d) Full-rectified cosine wave	Even	$\dfrac{2A}{\pi}$	$a_n = \begin{cases} 0 & n = 1, 3, 5, \ldots \\ \dfrac{4A}{\pi(n^2 - 1)} & n = 2, 6, 10, \ldots \\ \dfrac{-4A}{\pi(n^2 - 1)} & n = 4, 8, 12, \ldots \end{cases}$
(e) Rectangular pulse train	Even	$\dfrac{AD}{T}$	$a_n = \dfrac{2A}{\pi n} \sin \dfrac{\pi D n}{T}$

(More extensive tables are given in mathematics handbooks.) Additional entries can be generated by performing simple operations on waveforms with known coefficients.

As the table indicates, various types of waveform *symmetry* cause some of the series coefficients to be zero. In particular, an *even* function with the property

$$x(-t) = x(t) \tag{5a}$$

has symmetry about the vertical axis, illustrated by Fig. 7.4–1a. Due to this symmetry,

$$b_n = 0 \qquad n = 1, 2, 3, \ldots \tag{5b}$$

meaning that the Fourier series contains only even (cosine) functions and, possibly, a constant term. Conversely, an *odd* function with the property

$$x(-t) = -x(t) \tag{6a}$$

has symmetry about the origin, illustrated by Fig. 7.4–1b. Then

$$a_n = 0 \qquad n = 0, 1, 2, \ldots \tag{6b}$$

and the series contains only odd (sine) functions. The sawtooth waveform (Fig. 7.4–1c) would have odd symmetry if we subtracted the constant $a_0 = A/2$, so its series has $a_n = 0$ for $n = 1, 2, \ldots$

A function may also have *half-wave symmetry,* defined by

$$x\left(t + \frac{T}{2}\right) = -x(t) \tag{7a}$$

which means that the second half of each period looks like the first half turned upside down, as in Figs. 7.4–1a and 7.4–1b. For such waveforms

$$a_n = b_n = 0 \qquad n = 2, 4, 6, \ldots \tag{7b}$$

and the Fourier series consists only of *odd harmonics.* On the other hand, the series for the full-rectified cosine (Fig. 7.4–1d) contains only even harmonics—not because of symmetry but because the actual repetition period is $T/2$ rather than T.

You should review Table 7.4–1 in light of these symmetry effects. And while doing that, also observe that the magnitudes of a_n or b_n generally decrease as n increases; in other words, the higher harmonic components tend to be smaller than the lower ones. (Some exceptions to this property will be encountered in Chapter 14.) Furthermore, the coefficients of a relatively "smooth" waveform such as the triangular wave decrease more rapidly than those of a "jumpy" or discontinuous waveform such as the square wave. We conclude, then, that discontinuous signals have more significant high-frequency content than continuous signals, other factors being equal.

Example 7.4–1
Half-rectified cosine wave

To illustrate how we generate new entries for Table 7.4–1, consider the *half-rectified cosine wave* in Fig. 7.4–2a. This waveform may be viewed as the sum of the two waveforms in Fig. 7.4–2b, where $x'(t) = A \cos \frac{2\pi}{T} t$ and $x''(t)$ is a full-rectified cosine (Fig. 7.4–1d). From Table 7.4–1, the Fourier-series expansion of $x''(t)$ is

$$x''(t) = \frac{2A}{\pi} + \frac{4A}{3\pi} \cos 2\omega_0 t - \frac{4A}{15\pi} \cos 4\omega_0 t + \cdots$$

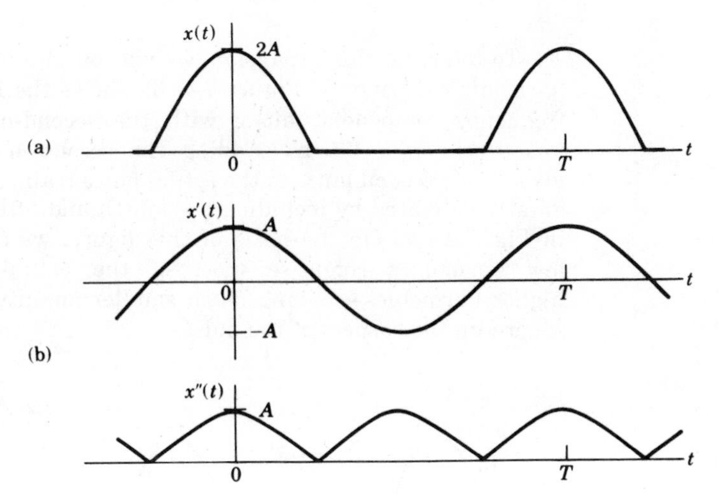

FIGURE 7.4–2 (a) Half-rectified cosine wave. (b) Decomposition into cosine wave plus full-rectified wave.

Adding $x'(t)$ then yields

$$x(t) = x'(t) + x''(t)$$
$$= \frac{2A}{\pi} + A \cos \omega_0 t + \frac{4A}{3\pi} \cos 2\omega_0 t - \frac{4A}{15\pi} \cos 4\omega_0 t + \cdots$$

Thus, in comparison with Eq. (1a), a half-rectified cosine with twice the amplitude of a full-rectified cosine has the same Fourier-series expansion, plus a fundamental term with $a_1 = A$.

If the half-rectified wave had amplitude A rather than $2A$, we would divide all the coefficients by 2, so $a_0 = A/\pi$, $a_1 = A/2$, etc. Such rectified waveforms occur in electronic AC-to-DC power supplies.

Example 7.4–2

The rectangular pulse train in Fig. 7.4–1e consists of pulses of height A and duration D. The ratio D/T is called the duty cycle, since the pulse is "on" for D seconds of each period and "off" the remaining time. Pulse trains are used for timing purposes and to represent digital information. If a particular pulse train has $A = 18$, $D = \frac{1}{3}$ ms, and $T = 1$ ms, then $f_0 = 1/T = 1$ kHz, $a_0 = A/3 = 6$, and Table 7.4–1 gives $a_n = (36/\pi\, n) \sin(\pi n/3)$. Thus

$$x(t) = 6 + 9.92 \cos \omega_0 t + 4.96 \cos 2\omega_0 t$$
$$- 2.48 \cos 4\omega_0 t - 1.98 \cos 5\omega_0 t + \cdots$$

so the harmonics at $3f_0, 6f_0, \ldots$, are missing, and the harmonics at $4f_0$, $5f_0$, $10f_0$, $11f_0, \ldots$ have inverted amplitudes—all due to the factor $\sin(\pi n/3)$.

To interpret this expansion, we will build-up the waveform from its harmonic components. Figure 7.4–3a shows the DC plus fundamental-frequency component, along with the second-harmonic term, which has period $\frac{1}{2}f_0 = 0.5$ ms. Adding the second harmonic, Fig. 7.4–3b, gives a fair resemblance to the actual pulse train, and the resemblance is greatly enhanced by including the fourth and fifth harmonics as plotted in Fig. 7.4–3c. On the basis of this figure, we conclude that the first few harmonics comprise most of the actual waveform, and all higher harmonics—having much smaller amplitudes—merely serve to square-up the corners of the pulses.

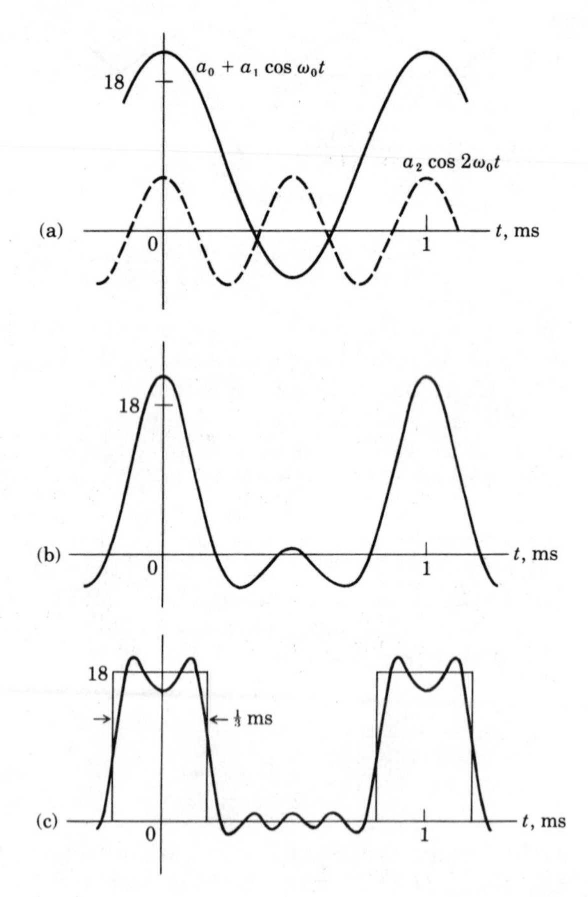

FIGURE 7.4–3 Reconstruction of a pulse train: (a) DC component plus first harmonic; (b) sum of first three components; (c) sum of first six components.

Exercise 7.4–1 The even-symmetry square wave in Fig. 7.4–4 can be "generated" by subtracting a constant from a rectangular pulse train with $D/T = \frac{1}{2}$. Use this approach to show that all coefficients are zero except

$$a_n = \begin{cases} \dfrac{4B}{\pi n} & n = 1, 5, 9, \ldots \\ -\dfrac{4B}{\pi n} & n = 3, 7, 11, \ldots \end{cases}$$

Then let $B = A$ and compare the Fourier expansions of the odd and even square waves.

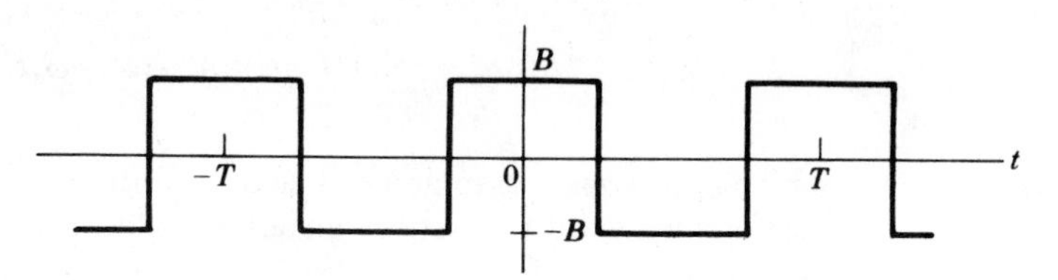

FIGURE 7.4–4 Even-symmetry square wave.

Exercise 7.4–2

Sketch an approximation of the triangular wave (Fig. 7.4–1a) using the first two nonzero terms of its Fourier series. Then add the next term and compare your results with the actual waveform. You will find it convenient to assume $A = 225\ \pi^2/8$ and $T = 60$.

PERIODIC STEADY-STATE RESPONSE

Suppose a periodic signal is applied to the input of a linear two-port network. Under steady-state conditions, each sinusoidal component of the input produces a sinusoidal output term whose amplitude and phase can be calculated from the input parameters and the network's transfer function. Therefore, by superposition, the periodic steady-state response equals the sum of the individual output components.

As a preliminary step to carrying out the foregoing process, we first convert the "cosine-plus-sine" version of the Fourier series to the "amplitude-and-phase" expansion

$$x(t) = a_0 + \sum_{n=1}^{\infty} A_n \cos(n\omega_0 t + \phi_n) \tag{8}$$

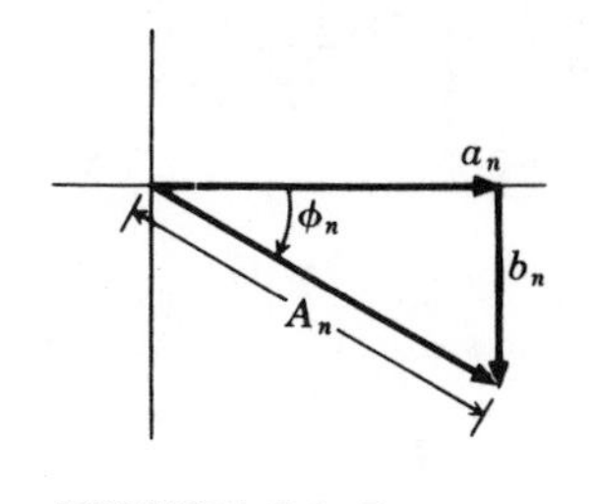

FIGURE 7.4–5 Phasor diagram of Fourier coefficients.

We obtain this form from Eq. (1) by writing $b_n \sin n\omega_0 t = b_n \cos(n\omega_0 t - 90°)$ so the phasor diagram of Fig. 7.4–5 yields

$$a_n \cos n\omega_0 t + b_n \sin n\omega_0 t = A_n \cos(n\omega_0 t + \phi_n)$$

with

$$A_n = \sqrt{a_n^2 + b_n^2} \qquad \phi_n = -\arctan \frac{b_n}{a_n} \tag{9}$$

The advantage of Eq. (8) is that each harmonic component can be represented by a *phasor* of length A_n and angle ϕ_n computed from a_n and b_n via Eq. (9). If $b_n = 0$ (corresponding to even symmetry) then $A_n = a_n$ and $\phi_n = 0°$ or $\pm 180°$ depending on the sign of a_n. Similarly, if $a_n = 0$ (odd symmetry), then $A_n = b_n$ and $\phi_n = \pm 90°$.

Now let the network in question have the transfer function $\underline{H}(f) = |H(f)| \underline{/\theta(f)}$. An input component at frequency nf_0 produces an output phasor of length $|H(nf_0)|\, A_n$ and angle $\phi_n + \theta(nf_0)$. The total output is then

$$y(t) = \underline{H}(0)a_0 + \sum_{n=1}^{\infty} |H(nf_0)|\, A_n \cos\,[n\omega_0 t + \phi_n + \theta(nf_0)] \qquad (10)$$

which expresses $y(t)$ in general as an infinite sum of sinusoids—not very useful if we want the actual output waveform! However, there are many practical applications where the purpose of the network is to remove all but one or two of the harmonic components. In such cases, Eq. (10) tells us how to compute the dominant output terms and how to estimate the size of the remaining unwanted but much smaller components.

Example 7.4–3
Harmonic generator

Figure 7.4–6a represents a harmonic generator intended to produce a 3-MHz sinusoidal output starting from a square wave with $A = 5$ and $f_0 = 1/T = 1$ MHz. The square wave is applied to a tuned-circuit filter having $Q = 10$ and resonant frequency $f_{res} = 3$ MHz. Thus, from Eq. (5), Section 7.2,

$$\underline{H}(nf_0) = \left[1 + jQ\left(\frac{nf_0}{f_{res}} - \frac{f_{res}}{nf_0}\right)\right]^{-1} = \left[1 + j10\left(\frac{n}{3} - \frac{3}{n}\right)\right]^{-1}$$

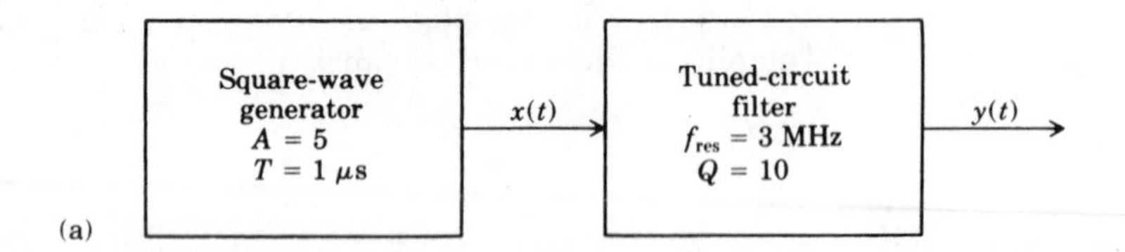

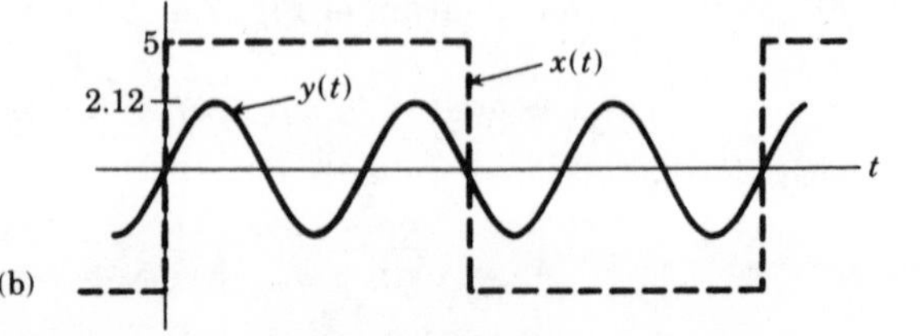

FIGURE 7.4–6
Harmonic generator: (a) block diagram; (b) waveforms.

Although the filter has been tuned to the desired third-harmonic component, it will pass, to some extent, the other harmonics.

We investigate the shape of $y(t)$ by assuming the square wave has odd symmetry and tabulating the numerical values of the first three nonzero input and output terms:

n	A_n	ϕ_n	$\lvert H(nf_0)\rvert A_n$	$\phi_n + \theta(nf_0)$
1	6.37	−90°	0.12	− 1°
3	2.12	−90°	2.12	− 90°
5	1.27	−90°	0.06	−177°

Since the third harmonic does, indeed, have a much larger amplitude than any other output component, we can write

$$y(t) \approx 2.12 \cos (3\omega_0 t - 90°)$$

This waveform and the input waveform are drawn in Fig. 7.4–6b. Schemes such as these are used to overcome the high-frequency limitations of some electronic generators.

Exercise 7.4–3 Obtain a three-term approximation for $y(t)$ when a sawtooth wave with $A = 2$ is applied to a first-order lowpass filter having $f_{co} = f_0$.

PROBLEMS

7.1–1 Let $\underline{V} = 1 \,\underline{/0°}$ in Fig. 6.3–3b. Draw the phasors $\underline{V}_R$ and $\underline{V}_C$ for $\omega = 1/RC$, $\omega < 1/RC$, and $\omega > 1/RC$; and infer the path (or *locus*) of the tip of each of these phasors as ω goes from zero to infinity. Compare your results with Figs. 7.1–3 and 7.1–6.

7.1–2 Repeat the previous problem for $\underline{I}_R$ and $\underline{I}_C$ in Fig. 6.3–4 with $\underline{I} = 1 \,\underline{/0°}$.

7.1–3 Suppose resistance R' is put in parallel with C in Fig. 7.1–4. Show that the transfer function has the form $\underline{H}(f) = K/[1 + j(Kf/f_{co})]$ with $K < 1$. Taking $R' = 4R$, find f/f_{co} such that $|H(f)| = K/\sqrt{2}$, and sketch $|H(f)|$ versus f.

7.1–4 Repeat the previous problem for R' in parallel with L in Fig. 7.1–5a, in which case $\underline{H}(f) = jK(f/f_{co})/[K + j(f/f_{co})]$.

7.1–5 Suppose resistance R' is in parallel with C in Fig. 7.1–5a. Show that $\underline{H}(f) = [K + j(f/f_{co})]/[1 + j(f/f_{co})]$ where $K < 1$ and $f_{co} > 1/2\pi RC$. If $R' \gg R$, what are the major differences in the frequency response here as compared to that of Fig. 7.1–6?

‡ **7.1–6** Confirm that Fig. P7.1–6 has $\underline{H}(f) = 1/[1 - (f/f_{co})^2 + j3(f/f_{co})]$, where $f_{co} = 1/2\pi RC$. Explain why $\underline{H}(f) \neq 1/[1 + j(f/f_{co})]^2$. Find f/f_{co} such that $|H(f)| = 1/\sqrt{2}$, and calculate the corresponding value of $\theta(f)$.

7.1–7 A second-order *Butterworth filter* has $\underline{H}(f) = 1/[1 - (f/f_{co})^2 + j\sqrt{2}(f/f_{co})]$ which gives a "squarer" lowpass response than a first-order filter. Sketch $|H(f)|$ versus

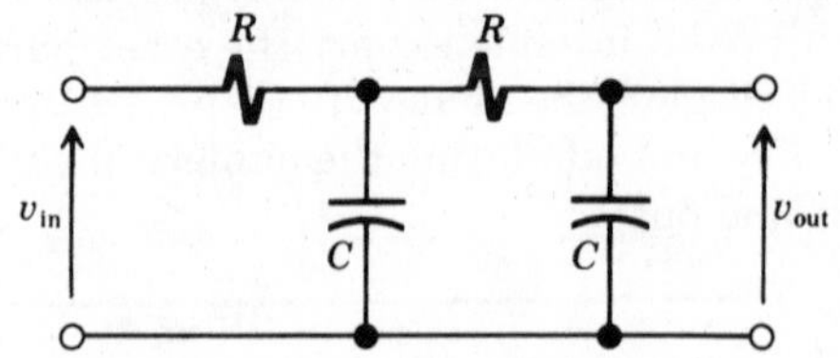

FIGURE P7.1–6

f by calculating values at $f/f_{co} = 0.5, 0.75, 1, 1.25, 1.5$, and 2; and compare your drawing with Fig. 7.1–3.

7.1–8 The *lattice network* in Fig. P7.1–8 is readily analyzed by using voltage-divider ratios when the circuit is redrawn as shown. Find $\underline{H}(f)$ when $\underline{Z}_1 = R$ and $\underline{Z}_2 = 1/j\omega C$. Then justify the name *all-pass filter* by sketching $|H(f)|$ and $\theta(f)$.

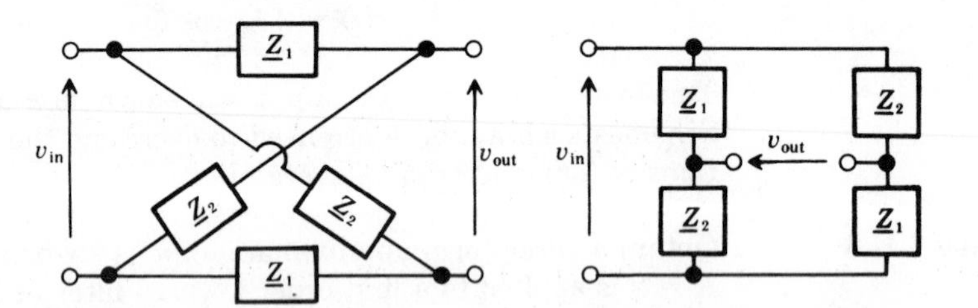

FIGURE P7.1–8

7.1–9 Repeat the previous problem with $\underline{Z}_1 = 1/j\omega C$ and $\underline{Z}_2 = R$.

7.1–10 Let the compensated probe in Fig. 5.3–9 have $R_p = 9R$ and $C_p = kC$. Find $\underline{H}(f)$ in terms of $f_{co} = 1/2\pi RC$ and confirm that $H(f) = 1$ when $k = \frac{1}{9}$. What is the probe's effect if k is so small that $9k(f/f_{co}) \ll 1$ or k is so large that $9k(f/f_{co}) \gg 1$ for all frequencies of interest?

‡ **7.1–11** Obtain the overall transfer function for a cascade of two op-amp filters like that in Fig. 7.1–11b, where each has $R_F = R_1$ and the same values of f_{co}. Find f/f_{co} such that $|H(f)| = 1/\sqrt{2}$, and calculate the corresponding value of $\theta(f)$. Also evaluate $|H(f)|$ and $\theta(f)$ at $f = f_{co}$ and $f \to \infty$.

7.1–12 Let the active lowpass filter in Fig. 7.1–11b have a finite-gain op-amp. Use the results of Prob. 4.3–11 with $\underline{Z}_F$ replacing R_F to obtain $\underline{H}(f)$, and show that $f_{co} = [1 + R_F/(A+1)R_1]/2\pi R_F C_1$.

7.1–13 Repeat the previous problem for the highpass filter in Fig. 7.1–11c, in which case $f_{co} = 1/2\pi R_1 C_1 \, [1 + R_F/(A+1)R_1]$.

7.1–14 Let resistance R_o be in parallel with the controlled current source in the amplifier circuit of Fig. P5.3–11, and let $Z(s)$ be an inductance L.

(a) Obtain the transfer function in the form

$$\underline{H}(f) = K \frac{1 + j(f/f_2)}{[1 + j(f/f_1)][1 + j(f/f_3)]}$$

(b) Sketch $|H(f)|$ when $R_o = 4R_L$ and $L = R_L/RC$, and compare with a sketch of $|H(f)|$ when $L = 0$ to determine the purpose of L.

‡ **7.1–15** Show that the active filter in Fig. P7.1–15 has $\underline{H}(f)$ as given in Prob. 7.1–7, with $f_{co} = 1/2\pi\sqrt{2}\,RC$. Hint: First show that $\underline{V}_a = (1 + j\omega RC)\,\underline{V}_{out}$, and then write a node-voltage equation.

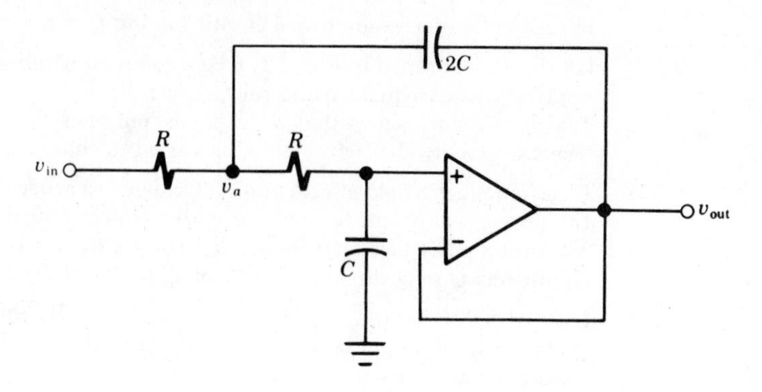

FIGURE P7.1–15

7.2–1 Carry out all the details between Eqs. (5) and (6).

7.2–2 Let $f_1 < f_0$ and $f_2 > f_0$ be two frequencies at which a tuned circuit has the same amplitude ratio; that is, $|H(f_1)| = |H(f_2)|$. Show that $f_1 f_2 = f_0^2$ or, equivalently, $f_1/f_0 = f_0/f_2$, which means that $|H(f)|$ has *geometric symmetry*.

7.2–3 Sketch and compare $|H(f)|$ versus f/f_0 for a tuned circuit with $Q = 1$ and $Q = 10$ by calculating $|H(f)|$ when $f/f_0 = \frac{1}{2}, \frac{3}{4}, \frac{9}{10}, \frac{19}{20}, 1, \frac{20}{19}, \frac{10}{9}, \frac{4}{3}$, and 2. Use the geometric-symmetry property (Problem 7.2–2) to avoid duplicate calculations.

7.2–4 Derive from Eqs. (5) and (6) the handy approximation

$$\underline{H}(f) \approx \frac{1}{1 + j2Q(f - f_0)/f_0} = \frac{1}{1 + j2(f - f_0)/B}$$

which holds when $|f - f_0| \ll f_0$. Hint: Let $f = f_0(1 + \delta)$, where $\delta \ll 1$ so that $(1 + \delta)^{-1} \approx 1 - \delta$.

7.2–5 Figure P7.2–5 represents a *double-tuned amplifier* with $\underline{H}(f) = \underline{I}_{out}/\underline{I}_s = -g_m R_1 \underline{H}_1(f)\underline{H}_2(f)$ where $\underline{H}_1(f)$ and $\underline{H}_2(f)$ are the transfer functions of the tuned circuits at the input and output. Derive the foregoing expression for $\underline{H}(f)$. Then use the approximation from Problem 7.2–4 for $\underline{H}_1(f)$ and $\underline{H}_2(f)$ and let $|H(f)|/|H(f_0)| = 1/\sqrt{2}$ to find the overall bandwidth B in terms of f_0 and Q when both tuned circuits have the same values of f_0 and Q.

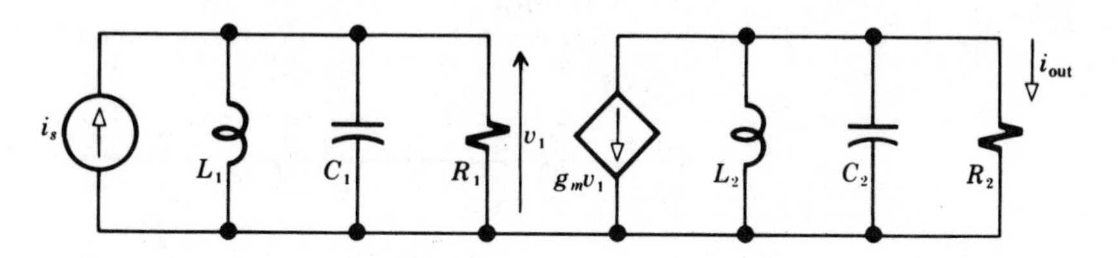

FIGURE P7.2–5

‡ **7.2–6** The amplifier in Fig. P7.2–5 is said to be *stagger-tuned* if both circuits have the same bandwidth, say $B_1 = B_2 = 2\Delta$, but are tuned to the opposite sides of the

center frequency f_c so that $f_{0_1} = f_c + \Delta$ and $f_{0_2} = f_c - \Delta$. Use the approximation of Prob. 7.2–4 for $\underline{H}_1(f)$ and $\underline{H}_2(f)$ to calculate $|H(f)|$ at $|f - f_c| = 0, \pm\Delta, \pm\sqrt{2}\Delta$, and $\pm 2\Delta$. Then sketch $|H(f)|/|H(f_c)|$ versus f and compare your results with the amplitude ratio of a single tuned circuit having $f_0 = f_c$ and $B = 2\sqrt{2}\Delta$.

7.2–7 Let the R_2–L branch in Fig. 7.2–6 be a coil with winding resistance R_2 which, for relatively low frequencies, is related to L by $R_2 = \alpha\sqrt{L}$ where α is a constant. Taking $R_1 = \infty$, show that B depends only on the coil. Hence, we have a "constant-bandwidth" circuit if C is varied to change f_0.

7.2–8 Electromagnetic-field effects at high frequencies cause the resistance of the coil in the previous problem to increase with frequency such that the coil alone has a constant quality factor $Q_c = \omega_0 L/R_2$. Taking $R_1 = \infty$ in Fig. 7.2–6, show that the circuit has $Q = Q_c$, independently of f_0, so $B = f_0/Q_c$.

7.2–9 Let Fig. 7.2–6 have $L = 10\ \mu$H and $R_2 = 0.5\ \Omega$. By considering the value of R_1, explain why it would be impossible to obtain $B = 1$ kHz with $f_0 = 2$ MHz. (The results are to be used in the next problem.)

7.2–10 Devise a scheme using a *negative-resistance converter* (Problem 4.3–10) to meet the specifications in Problem 7.2–9.

7.2–11 A low-frequency vibration analyzer requires a tuned circuit like Fig. 7.2–3b, with $f_0 = 2$ Hz and $B = 0.5$ Hz.

(a) Taking $R = 5$ kΩ, find L and C and explain why this circuit would be unrealistic.

(b) Repeat with $R = 5\ \Omega$.

‡ **7.2–12** Suppose we have available a *gyrator* (as in Problem 5.3–13) with $r_1 = r_2 = 1$ kΩ. Using two capacitors but no inductor, design a circuit to meet the requirements of Problem 7.2–11 with $R = 2$ kΩ.

7.2–13 Consider a filter consisting of a *series RLC* circuit with the output voltage taken across the capacitor.

(a) Find $H(f) = \underline{V}_{\text{out}}/\underline{V}_{\text{in}}$ in terms of $f_0 = 1/2\pi\sqrt{LC}$ and $Q_{\text{ser}} = (1/R)\sqrt{L/C}$.

(b) Evaluate $|H(f)|$ at $f = 0$, f_0, and ∞, and sketch $|H(f)|/|H(f_0)|$ when $Q_{\text{ser}} \gg 1$. Compare your results with Fig. 7.2–5.

(c) For what value of Q_{ser} does this circuit become a Butterworth lowpass filter as described in Prob. 7.1–7?

‡ **7.2–14** Find $\underline{H}(f) = \underline{V}_{\text{out}}/\underline{V}_{\text{in}}$ for the circuit in Fig. P7.2–14. Then show that if $C_2 \ll C_1$, $\underline{H}(f)$ approximates the bandpass function in Eq. (2) with $K = 1$, $f_\ell = 1/2\pi R_1 C_1$, and $f_u = 1/2\pi R_2 C_2$.

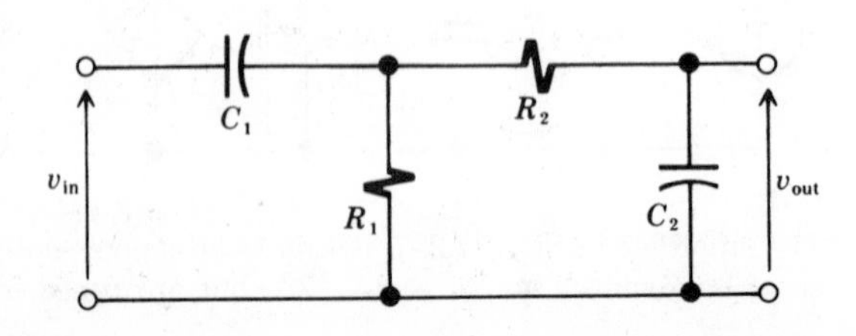

FIGURE P7.2–14

7.3–1 Find the asymptotes and construct the Bode diagram for $\underline{H}(f) = (F - jf)/F$. Compare your results with Fig. 7.3–1c.

7.3–2 Repeat the previous problem for $\underline{H}(f) = (F - jf)/jf$, and compare the results with Fig. 7.3–1b.

7.3–3 Construct the Bode diagram, including asymptotic correction terms, for $\underline{H}(f) = (40 + jf)/4(10 + jf)$.

7.3–4 Repeat the previous problem for $\underline{H}(f) = 20(10 + jf)/jf(40 + jf)$.

7.3–5 Construct the asymptotic Bode diagram for $\underline{H}(f) = 10^6/[(10 + jf)(200 + jf)(1000 + jf)]$ and estimate H_{dB} when $\theta = -180°$.

7.3–6 Repeat the previous problem for $\underline{H}(f) = -10^6 jf/[(10 + jf)(200 + jf)(1000 + jf)]$.

7.3–7 Consider the Butterworth transfer function in Problem 7.1–7. Use appropriate approximations plus a few calculated values of $\underline{H}(f)$ to construct the asymptotes and sketch the Bode diagram.

‡ **7.3–8** If $\zeta > 1$, the denominator of Eq. (5) can be factored as $1 + j2\zeta(f/f_0) - (f/f_0)^2 = [1 + j(f/F_1)][1 + j(f/F_2)]$, where $F_1 = (\zeta - \sqrt{\zeta^2 - 1})f_0$ and $F_2 = (\zeta + \sqrt{\zeta^2 - 1})f_0$. Confirm this factorization by expanding the right-hand side and solving for F_1 and F_2.

7.3–9 If the circuit in Fig. P7.2–14 has $R_1 = R_2$ and $C_1 = C_2$, then

$$\underline{H}(f) = j(f/f_0)/[1 + j3(f/f_0) - (f/f_0)^2]$$

where $f_0 = 1/2\pi R_1 C_1$. Use the factorization in Problem 7.3–8 to plot H_{dB} and estimate its maximum value.

7.3–10 Plot H_{dB} for $\underline{H}(f) = -f^2/(100 + j2f - f^2)$.

7.3–11 Plot H_{dB} for a *notch filter* having $\underline{H}(f) = (100 + jf - f^2)/(10 + jf)^2$.

7.4–1 Evaluate a_0 and classify the symmetry of each of the waveforms in Fig. P7.4–1.

7.4–2 Construct the waveform in Fig. P7.4–1a by combining a constant with a triangular wave; then find the Fourier-series coefficients.

7.4–3 Construct the waveform in Fig. P7.4–1b by combining a square wave with a sawtooth wave; find the Fourier-series coefficients.

7.4–4 Construct the waveform in Fig. 7.4–1c by combining a square wave with a triangular wave; find the Fourier-series coefficients.

7.4–5 Let $x'(t)$ be an even square wave and $x''(t)$ be an odd square wave with the same amplitude. Sketch $x(t) = x'(t) + x''(t)$, and write out its Fourier-series expansion through the fifth harmonic.

7.4–6 Let $x'(t)$ be a sawtooth wave with $A = 4$ and $x''(t)$ be a triangular wave with $A = 1$. Sketch $x(t) = x'(t) + x''(t) - 1$ and write out its Fourier-series expansion through the third harmonic.

7.4–7 The *rms value* of a periodic waveform $x(t)$ is related to its Fourier coefficients by

$$x_{rms} = \sqrt{a_0^2 + \tfrac{1}{2}\sum_{n=1}^{\infty}(a_n^2 + b_n^2)}$$

Use this formula with $1 \le n \le 4$ to estimate the rms value of a full-rectified cosine wave, and compare your result with the exact value $x_{rms} = A/\sqrt{2}$.

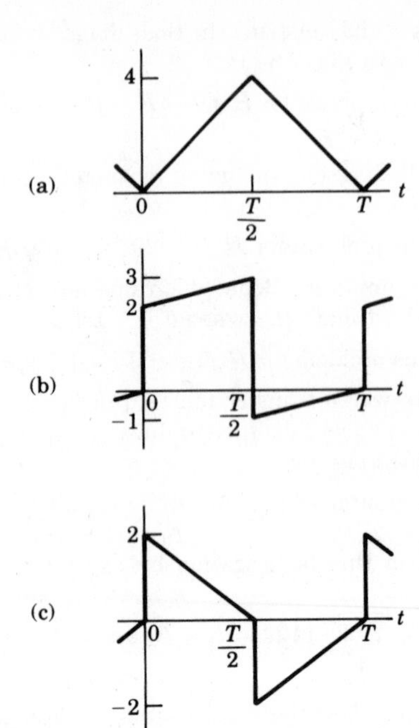

FIGURE P7.4–1

7.4–8 Use the formula in Problem 7.4–7 with $1 \leq n \leq 5$ to estimate the rms value of a triangular wave. Compare your result with the rms value of the first harmonic alone.

7.4–9 A triangular wave is applied to a first-order lowpass filter. Find f_{co}/f_0 such that, at the output, the third-harmonic amplitude equals 5% of the first-harmonic amplitude. Then compute the corresponding percentage for the fifth harmonic.

7.4–10 A square wave is applied to a first-order lowpass filter. Find f_{co}/f_0 such that, at the output, the third-harmonic amplitude equals 12.5% of the first-harmonic amplitude. Then compute the corresponding percentage for the fifth harmonic.

‡ **7.4–11** Show that it would be impossible, in Problem 7.4–10, to make the third-harmonic amplitude equal to 10% of the first-harmonic amplitude.

7.4–12 Obtain an approximation for $y(t)$ when a full-rectified cosine wave with $A = \pi$ is applied to a first-order lowpass filter having $f_{co} = f_0$.

7.4–13 Repeat the previous problem with a half-rectified cosine wave.

7.4–14 Repeat Exercise 7.4–3 with a Butterworth filter having $f_{co} = f_0$ (as in Problem 7.1–7).

8 TRANSIENT RESPONSE

Previous chapters have developed methods for finding the steady-state or forced response of a circuit. Here we are concerned with the *transient response* that occurs before a circuit reaches steady-state conditions. The study of this behavior is important, on the one hand, because transients often become limiting factors in the performance of an electrical or electronic system. On the other hand, they may be put to various practical uses, as in timing circuits and waveform generators.

To divide this topic into manageable pieces, we first examine the *natural response* of first-order *RL* and *RC* circuits that have initial stored energy but are not subject to applied excitation. We then combine the expressions for the natural and forced responses to obtain the complete transient behavior of a first-order circuit. Finally, we consider the natural response and transient behavior of circuits with two energy-storage elements.

OBJECTIVES

After studying this chapter and working the exercises, you should be able to do each of the following:

- Calculate the time constant of an *RC* or *RL* circuit, and use the continuity conditions to obtain the natural response (Section 8.1).

- Sketch the DC transient response, step response, and pulse response of a first-order circuit, and determine the time at which the response has a specified value (Section 8.2).
- State the relationship between natural behavior and impedance or admittance (Sections 8.1 and 8.3).
- Find the characteristic polynomial for a second-order circuit, and determine if it is overdamped, underdamped, or critically damped (Section 8.3). †
- Write the expression for the complete response of a first-order or second-order circuit, evaluate the initial-condition constants, and identify the transient response (Sections 8.2 and 8.3). †

8.1 NATURAL RESPONSE OF FIRST-ORDER CIRCUITS

Natural response refers to the voltages and currents in a circuit caused by stored energy, as distinguished from the forced response caused by an applied source. In particular, if a circuit has initial stored energy and the source is suddenly removed, the memory effect of energy storage produces voltages and currents that persist in the circuit until all of the initial energy has been used up. We will see that the natural response in any first-order circuit (RC or RL) always has a decaying exponential waveform whose properties depend on the initial stored energy and the circuit's impedance $Z(s)$.

NATURAL RESPONSE OF RC CIRCUITS

Consider the typical situation diagrammed in Fig. 8.1–1a. The switch

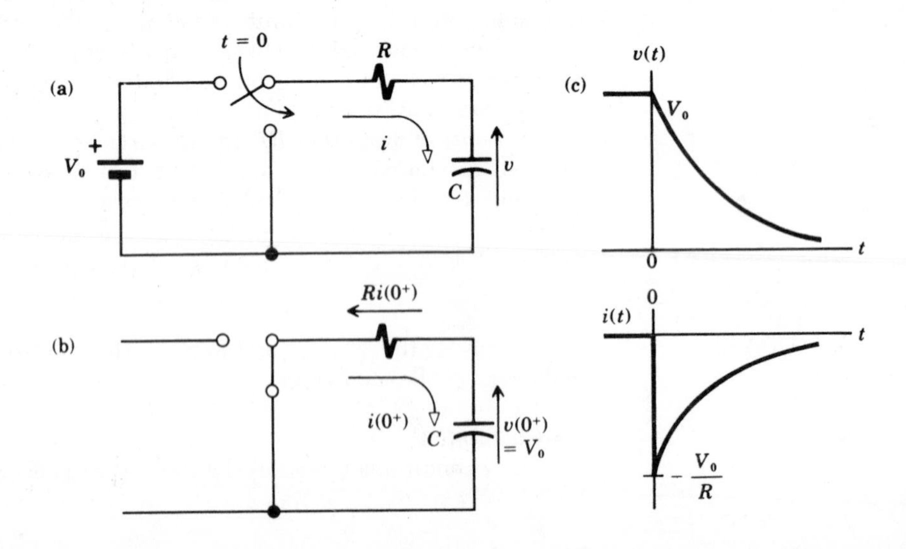

FIGURE 8.1–1
(a) A switched RC circuit. (b) Initial conditions at $t = 0^+$. (c) Natural-response waveforms.

has been in the upper position for a long time, charging the capacitor to V_0 volts; the charged capacitor acts like a DC block, so $i = 0$. The switch is then turned to the lower position, removing the source. For convenience we take $t = 0$ as the switching time and let $t = 0^+$ stand for the instant immediately after switching, at which time

$$v(0^+) = V_0 \tag{1}$$

because the stored energy $w = \frac{1}{2}Cv^2$ cannot change value instantaneously.

We prove Eq. (1) by setting $t = 0^+$ in the integral expression Eq. (4a), Section 5.1,

$$v(0^+) = V_0 + \frac{1}{C}\int_0^{0^+} i\,dt$$

whose second term is vanishingly small for any *finite* current i. Besides being physically impossible, *infinite* current would produce infinite voltage across the resistance and violate KVL. We are thus led to the important and general conclusion that when a circuit branch contains resistance and capacitance, the capacitor's voltage never makes an abrupt or discontinuous change. This is called the *continuity condition for voltage across a capacitance,* and is similar to the condition of a moving mass whose momentum tends to prevent sudden velocity changes.

But current through a capacitance may exhibit a jump or *discontinuity,* and does so in this circuit at $t = 0$. This can be seen from the equivalent circuit at $t = 0^+$, Fig. 8.1–1b, where $Ri(0^+) + v(0^+) = 0$, so

$$i(0^+) = \frac{-v(0^+)}{R} = \frac{-V_0}{R}$$

The fact that $i(0^+)$ is *negative* means, of course, that the current is flowing in a direction opposite to the reference arrow and that the capacitor is beginning to discharge, dissipating its stored energy as ohmic heating in the resistance. Eventually the capacitor discharges completely as $t \to \infty$, leaving the circuit with

$$i(\infty) = 0 \qquad v(\infty) = 0$$

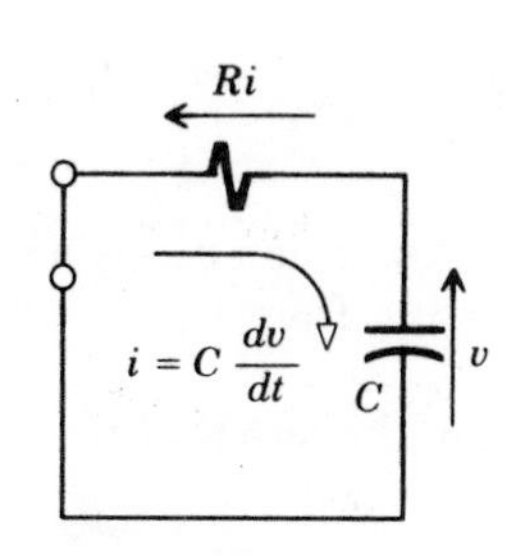

FIGURE 8.1–2
Circuit condition for $t > 0$.

Having determined the initial and final value of the natural responses, we can see why the waveforms $v(t)$ and $i(t)$ must look something like Fig. 8.1–1c.

We next seek the exact shape of those curves by formulating the circuit's differential equation. For that purpose, Fig. 8.1–2 represents the circuit at any time $t > 0$ after switching, and Kirchhoff's voltage law requires that

$$Ri + v = 0$$

whose right-hand side equals zero because there is no applied source after the switch has been turned. Substituting $i = C\,dv/dt$ then gives

$$RC\frac{dv}{dt} + v = 0 \tag{2}$$

which governs the natural behavior of $v(t)$ between its initial and final value. We call this a *homogeneous* differential equation, meaning that there is no forcing function.

Before proceeding to the solution, we find it convenient to introduce the *time constant* τ (tau) defined by

$$\tau \triangleq RC \tag{3}$$

which has the units of time. Inserting τ and rewriting Eq. (2) as

$$\frac{dv}{dt} = -\frac{1}{\tau}v$$

reveals that v must be *proportional to its own derivative*. The mathematical function with this property is the decaying exponential $Ae^{-t/\tau}$, since

$$\frac{d}{dt}(Ae^{-t/\tau}) = -\frac{1}{\tau}Ae^{-t/\tau}$$

Taking $A = V_0$ then yields the correct initial value at $t = 0^+$ in our solution

$$v(t) = V_0e^{-t/\tau} \qquad t \geq 0 \tag{4}$$

The stipulation $t \geq 0$ emphasizes the point that this expression describes the natural response of a discharging capacitor starting from $v = V_0$ at $t = 0$.

As for the natural behavior of the current i, we use the fact that $Ri = -v$ (see Fig. 8.1–2), so

$$i(t) = -\frac{1}{R}v(t) = -\frac{V_0}{R}e^{-t/\tau} \qquad t > 0 \tag{5}$$

confirming our previous result that $i(0^+) = -V_0/R$. The power dissipation in the resistance is then $p = Ri^2$, and it can be shown that the total energy dissipated exactly equals the initial stored energy.

Our results for the simple RC circuit easily generalize to circuits containing two or more resistances, providing we take care when calculating the initial voltage and the equivalent series resistance in the discharge mode. Take Fig. 8.1–3a for example. With the switch in the upper position, we can redraw the circuit as in Fig. 8.1–3b and the initial voltage will be the Thévenin equivalent voltage as seen looking back from the capacitor—namely $V_0 = R_2 V_s/(R_s + R_1 + R_2)$. With the switch in the lower position we get the circuit of Fig. 8.1–3c, which shows that the capacitor will discharge into the equivalent resistance $R_1 \| R_2$ and hence $\tau = (R_1 \| R_2)C$.

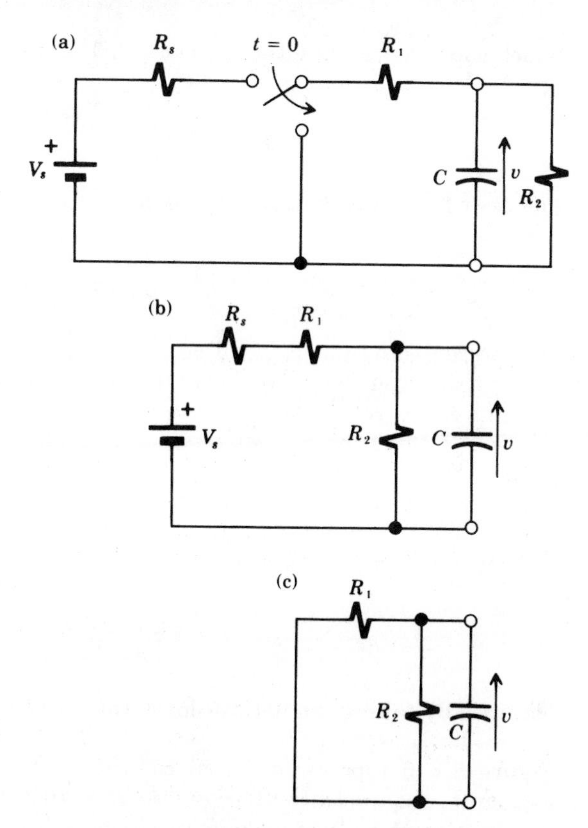

FIGURE 8.1–3
(a) Switched circuit.
(b) Equivalent circuit for $t < 0$. (c) Equivalent circuit for $t > 0$.

Now let's return to the natural-response waveforms. Figure 8.1–4 shows a plot of the general decaying exponential function $y(t) = Y_0 e^{-t/\tau}$,

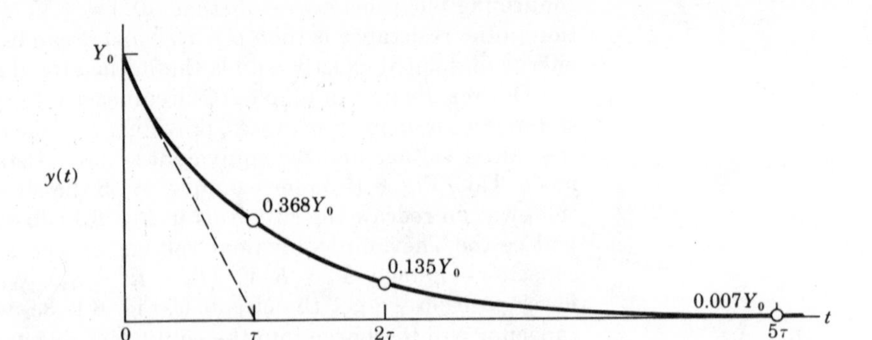

FIGURE 8.1–4
The decaying exponential $y(t) = Y_o e^{-t/\tau}$

which could represent either $v(t)$ or $i(t)$. The *initial slope* is

$$\frac{dy}{dt}(0) = -\frac{Y_0}{\tau} \tag{6a}$$

and near $t = 0$, the decay is approximately *linear*, that is,

$$y(t) \approx Y_0\left(1 - \frac{t}{\tau}\right) \qquad t \leq \frac{\tau}{10} \tag{6b}$$

as represented by the dashed line.

The time constant τ provides a measure of the rate of decay. In particular, at $t = \tau$, $y(t)$ has decayed to about 37% of its initial value since $y(\tau) = Y_0 e^{-1} = 0.368\ Y_0$. After two time constants have elapsed, $y(t) = y(2\tau) = Y_0 e^{-2} = 0.135\ Y_0$. Although $y(t)$ never completely reaches zero in finite time, it becomes negligibly small after five time constants when $y(5\tau) = 0.007\ Y_0$—less than one percent of the initial value. If we need to find the value of t corresponding to a specified value of $y(t)$, we use

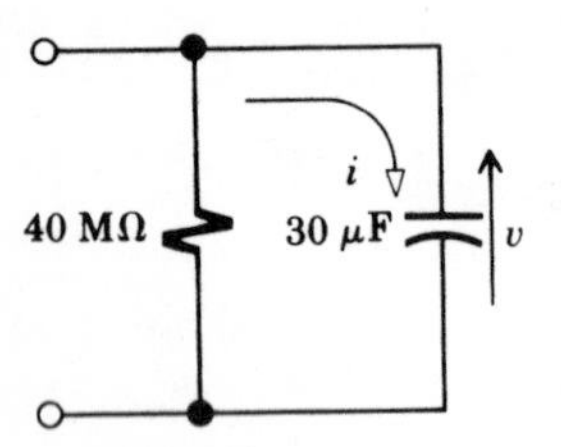

FIGURE 8.1–5
Capacitor with internal leakage resistance.

$$t = -\tau \ln \frac{y(t)}{Y_0} \tag{6c}$$

obtained by taking the natural logarithm of $y(t)$.

Example 8.1–1

Figure 8.1–5 represents a real capacitor with internal leakage. The capacitor was initially charged to $V_0 = 1000$ V and now discharges slowly through its own leakage resistance with time constant $\tau = RC = 40\ \text{M}\Omega \times 30\ \mu\text{F} = 1{,}200$ sec $= 20$ min. One hour later when $t/\tau = 3$, the voltage will have decayed to $v = 1000e^{-3} = 49.8$ V. But the leakage current is very small—$i(0^+) = -1000\ \text{V}/40\ \text{M}\Omega = -25\ \mu\text{A}$—and at $t = 60$ sec the voltage has changed little, since from Eq. (6b), $v \approx 1000 \times$

$(1 - 60/1200) = 950$ V. Incidentally, these values are typical of a power-supply capacitor in a TV set, which explains why you might get a nasty "zap" from a set that's been turned off for as much as several minutes.

Exercise 8.1–1

A 10-μF capacitor is required to hold at least 50% of its initial charge for 30 seconds. Estimate the minimum acceptable leakage resistance and sketch $v(t)$ and $i(t)$ taking $V_0 = 100$ V.

NATURAL RESPONSE OF *RL* CIRCUITS

Now consider a switched *RL* circuit with a constant current source, as in Fig. 8.1–6. Assuming the switch has been open for a long time prior to $t = 0$, the inductance acts like a short circuit with $v = 0$ and $i = I_0$. After the switch closes, R and L form a series loop without the current source, but

$$i(0^+) = I_0 \tag{7}$$

because the stored energy $w = \frac{1}{2}Li^2$ cannot change value instantaneously, and *current* through an *RL* branch must have *continuity*. It then follows that

$$v(0^+) = -Ri(0^+) = -RI_0$$

so $v(t)$ has a jump discontinuity at $t = 0$.

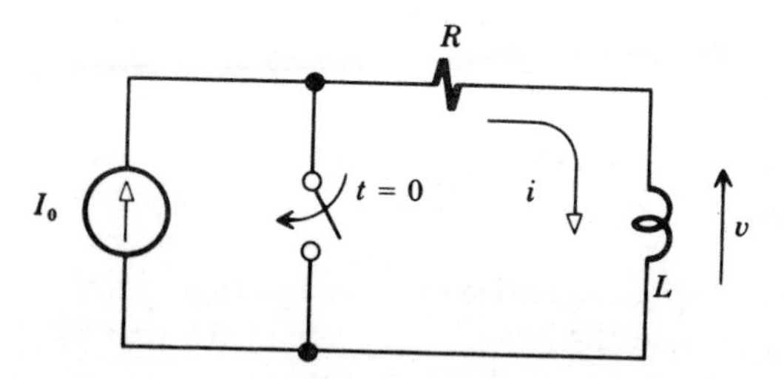

FIGURE 8.1–6
Switched RL circuit.

These conclusions are the duals of those for an *RC* circuit, where voltage across the capacitance is continuous and the current discontinuous. Carrying the dual notion further suggests that the curves for $i(t)$ and $v(t)$ must look like the opposite of the *RC* curves—that is, of the interchange current and voltage curves back in Fig. 8.1–1c. Specifically,

$$\begin{aligned} i(t) &= I_0 e^{-t/\tau} \qquad t \geq 0 \\ v(t) &= -RI_0 e^{-t/\tau} \qquad t > 0 \end{aligned} \tag{8}$$

with the time constant now being

$$\tau \triangleq \frac{L}{R} \tag{9}$$

Again, these results hold for more complicated RL circuits, including those with voltage sources; all you have to do is find the appropriate initial values and equivalent resistance.

Example 8.1–2

If the switch in Fig. 8.1–7 has been closed for a long time, DC steady-state analysis gives $i = 20/25 \times 18\ \text{V}/(5 + 5\|20)\ \Omega = 1.6$ A for $t < 0$. The continuity condition then requires $I_0 = i(0^+) = 1.6$ A; so, for $t \geq 0$, $i(t) = 1.6\, e^{-t/\tau}$ where $\tau = 0.2\ \text{H}/25\ \Omega = 8$ ms because the resistance in series with the inductance is $R = 20 + 5 = 25\ \Omega$ after the switch opens. Due to the larger resistance and the continuity of current, this circuit produces a negative voltage "spike" with peak value $-RI_0 = -40$ V—more than twice the battery voltage. If we want to know when the voltage has decayed to -4 V, for instance, the application of Eq. (6c) yields $t = -8\ \text{ms} \times \ln(-4/-40) = 18.4$ ms.

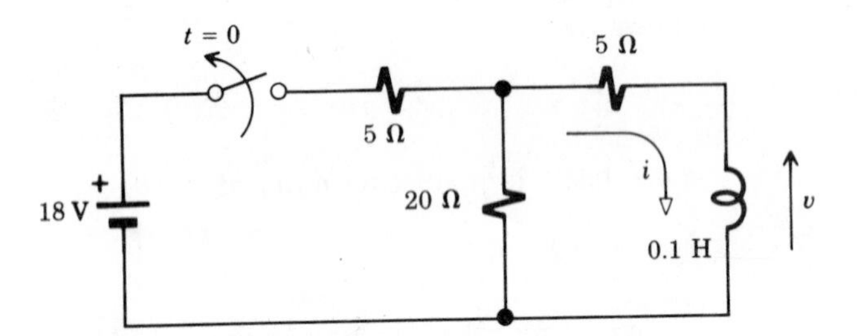

FIGURE 8.1–7 Circuit for Example 8.1–2.

Exercise 8.1–2

Confirm by substitution that $i(t)$ in Eq. (8) satisfies the differential circuit equation, $L\, di/dt + Ri = 0$.

NATURAL BEHAVIOR AND IMPEDANCE

You may have noticed in passing that the *time constant* of an RC or RL circuit and the *cutoff frequency* of the corresponding filter are inversely proportional to each other; that is

$$f_{co} = \frac{1}{2\pi\tau} \tag{10}$$

This expression suggests a connection between natural response and forced response. A more general relationship emerges when we consider the circuit's impedance $Z(s)$.

To develop that relationship, consider a one-port network having stored energy at $t = 0$ when the terminals are suddenly short-circuited as in Fig. 8.1–8a. We expect a natural-response current of the form $i = I_0 e^{st}$, which would ordinarily produce a voltage $v = Z(s) I_0 e^{st}$ across the terminals. But the short circuit requires that $v = 0$, so either $Z(s) = 0$ or $I_0 e^{st} = 0$. Since $I_0 e^{st}$ reflects the dissipation of stored energy, we are left with the conclusion that

$$Z(s) = 0 \tag{11}$$

for the short-circuit natural behavior.

Physically, zero impedance permits natural current flow without an applied voltage, and the value of s in $i = I_0 e^{st}$ must be such that $Z(s) = 0$. We call that value the *root* of $Z(s)$. Now let's turn this argument around: We determine the time constant for the *short-circuit natural response* of a first-order circuit by finding the root of $Z(s)$. For example, the impedance of a series RC circuit is $Z(s) = R + 1/sC$, which equals zero when $s = -1/RC = -1/\tau$.

Next, consider Fig. 8.1–8b, which represents a network that has initial stored energy at $t = 0$ when the terminals are open-circuited. We expect a natural-behavior voltage of the form $v = V_0 e^{st}$ while $i = Y(s) V_0 e^{st} = 0$. Hence,

$$Y(s) = 0 \tag{12}$$

and the root of the admittance describes the *open-circuit natural response*.

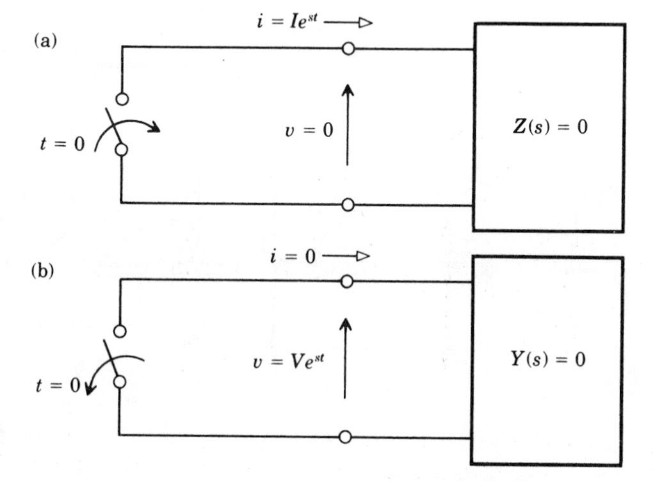

FIGURE 8.1–8
(a) Short-circuit natural behavior, $Z(s) = 0$.
(b) Open-circuit natural behavior, $Y(s) = 0$.

Equations (11) and (12) have special value for analyzing higher-order circuits with two or more energy-storage elements, in which case there will be more than one possible root of the impedance or admittance.

Exercise 8.1–3

Find the roots of $Z(s)$ and $Y(s)$ for the network to the right of the switch in Fig. 8.1–7. Justify your results by redrawing the network and calculating the time constants under short-circuit and open-circuit conditions.

8.2 FIRST-ORDER TRANSIENTS

When a source applied to an RC or RL circuit makes an abrupt change at some particular time, stored energy within the circuit cannot change value instantaneously. As a result, the various currents and voltages undergo a transitional readjustment process that constitutes the transient response. Our investigation of this response begins with a general treatment of DC transients in RC and RL circuits, including the important step response and pulse response that occur in digital electronics and switching circuits. We then consider the complete response produced by an arbitrary input.

DC TRANSIENT RESPONSE

Consider Fig. 8.2–1a, where the applied voltage changes discontinuously from V_A to V_B at some instant of time, say $t = 0$. Based on our previous study of RC circuits we know that the initial voltage across the capacitor will be

$$V_0 = v(0^+) = \frac{R_2}{R_1 + R_2} V_A$$

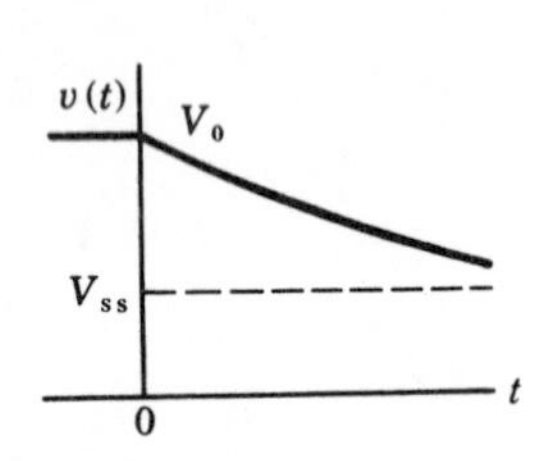

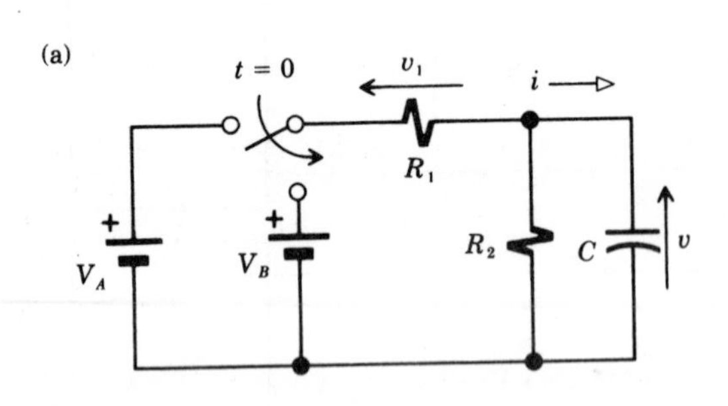

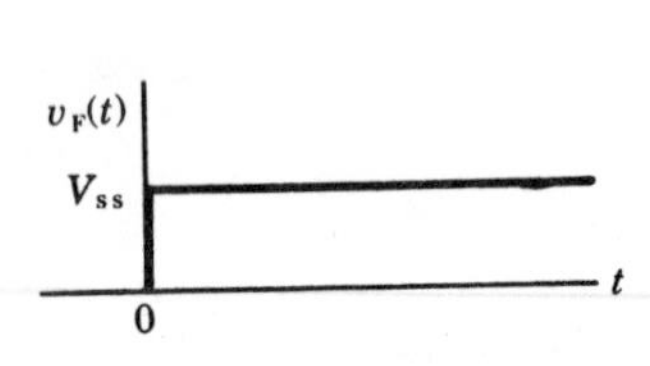

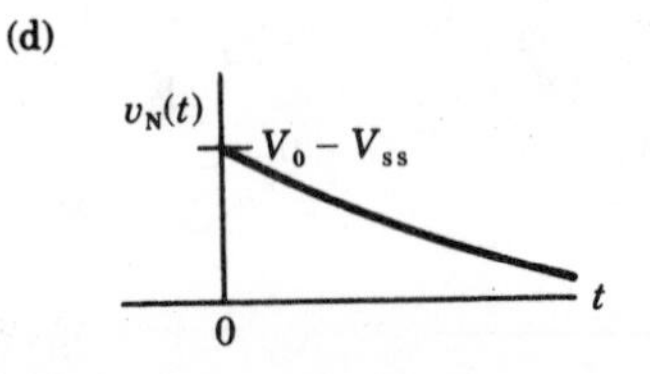

FIGURE 8.2–1
(a) RC circuit with switched voltage.
(b) Transient response.
(c) Forced response.
(d) Natural response.

and it seems reasonable to infer the final steady-state value

$$V_{ss} = v(\infty) = \frac{R_2}{R_1 + R_2} V_B$$

approached as $t \to \infty$. Figure 8.2–1b shows the expected transient response of $v(t)$ taking $V_{ss} < V_0$. The voltage is continuous at $t = 0$ (from the continuity condition) and then decays towards its final value.

An important interpretation of transient behavior comes from the superposition principle by viewing $v(t)$ as the sum of two terms for $t > 0$. One term is the *forced* response $v_F(t) = V_{ss}$, as in Fig. 8.2–1c, which is caused by the applied voltage V_B; the other term is the *natural* response $v_N(t)$, as in Fig. 8.1–1d, which is caused by the *change* of stored energy. The *complete* transient response is then the sum $v(t) = v_N(t) + v_F(t)$.

To carry out an analysis of Fig. 8.2–1a, we first find the Thévenin equivalent of everything to the left of the capacitance, which yields the circuit model of Fig. 8.2–2 where $R = R_1 \| R_2$ and

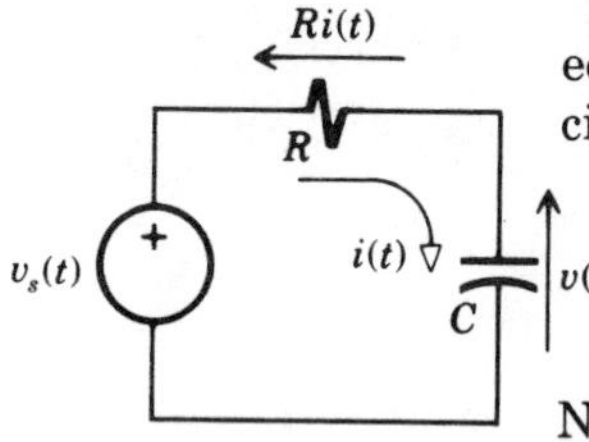

FIGURE 8.2–2

$$v_s(t) = \begin{cases} V_0 & t < 0 \\ V_{ss} & t > 0 \end{cases} \qquad (1)$$

Note that this model applies to any circuit that has an arbitrary number of resistors plus one capacitor. In general, then, R represents the Thévenin equivalent resistance seen from the terminals of the capacitance, and V_0 and V_{ss} are the Thévenin source voltages before and after the switching time.

The natural response for $t > 0$ has the form

$$v_N(t) = Ae^{-t/\tau}$$

where $\tau = RC$ while A is a constant to be determined from the initial conditions. Adding the forced response $v_F(t) = V_{ss}$ to $v_n(t)$ yields the complete response

$$v(t) = Ae^{-t/\tau} + V_{ss} \qquad t \geq 0$$

and we are ready to evaluate A by setting $t = 0^+$. Thus

$$v(0^+) = A + V_{ss} = V_0$$

so $A = V_0 - V_{ss}$ and we finally have

$$v(t) = (V_0 - V_{ss})\, e^{-t/\tau} + V_{ss} \qquad (2a)$$

$$= V_0 e^{-t/\tau} + V_{ss}(1 - e^{-t/\tau}) \qquad (2b)$$

which is the voltage transient response of an RC circuit for $t \geq 0$. The corresponding transient current is easily found from $i = (V_{ss} - v)/R$ for $t > 0$. If $V_{ss} = 0$—meaning no forced term—Eq. (2) reduces to our previous result for the natural response alone. In any case, the time interval from $t = 0$ to about $t = 5\tau$ comprises the transitional readjustment between the initial and final values.

If inductance replaces capacitance in Fig. 8.2–2 the transient-response current has the same form as Eq. (2) with $\tau = L/R$, V_0 replaced by $I_0 = V_0/R$, and V_{ss} replaced by $I_{ss} = V_{ss}/R$. We need not pursue our discussion of RL transients any further, since the results would parallel those of the more common RC circuit.

Example 8.2–1

Go back to Fig. 8.2–1a, and suppose the element values are $V_A = -12$ V, $V_B = +20$ V, $R_1 = 4$ kΩ, $R_2 = 12$ kΩ, and $C = 100$ μF. Then

$$V_0 = \frac{12}{16}(-12) = -9 \text{ V} \qquad V_{ss} = \frac{12}{16}20 = +15 \text{ V}$$

$$R = 4 \parallel 12 = 3 \text{ k}\Omega \qquad \tau = 3 \text{ k}\Omega \times 100\ \mu\text{F} = 0.3 \text{ sec}$$

and Eq. (2) gives

$$\begin{aligned} v(t) &= -24e^{-t/0.3} + 15 \\ &= -9e^{-t/0.3} + 15(1 - e^{-t/0.3}) \end{aligned}$$

as shown in Fig. 8.2–3a. The capacitance, then, charges up to its new

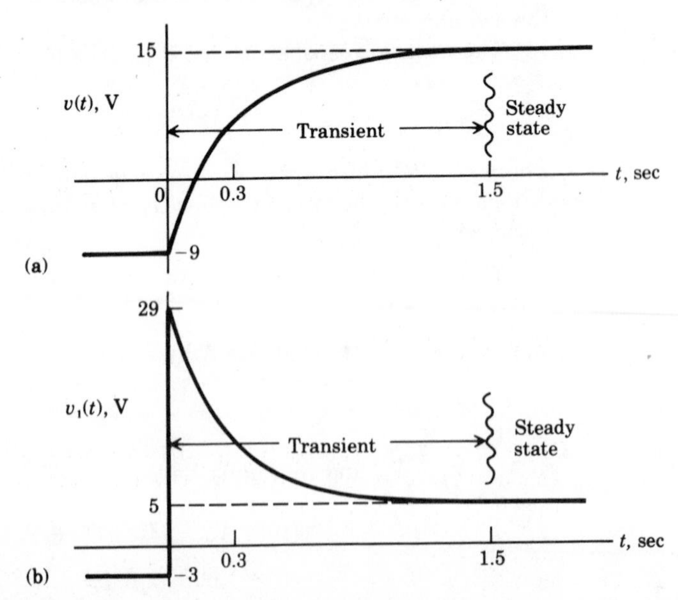

FIGURE 8.2–3
Waveforms for Example 8.2–1.

value of 15 V in about $5\tau = 1.5$ sec, and the maximum charging current is $[15 - (-9)]$ V/3 kΩ = 8 mA at $t = 0^+$.

Figure 8.2–3b shows the transient voltage $v_1(t)$ across resistance R_1. This waveform is obtained from Fig. 8.2–1a, where we find that $v_1(t) = V_A - v(t)$ for $t < 0$ while $v_1(t) = V_B - v(t)$ for $t > 0$. Observe that $v_1(t)$ must make a jump of $V_B - V_A = 32$ V at $t = 0$, since $v(t)$ has continuity.

Exercise 8.2–1

Find the sketch $i(t)$, $v_1(t)$, and $v(t)$ when a 0.3-H inductance replaces the capacitance in the previous example. Hint: Note that $v = v_s - Ri$, as in Fig. 8.2–2.

STEP RESPONSE

A special but important transient waveform is the *step response,* which is the behavior of a circuit that has no initial stored evergy at the moment a *constant* source is suddenly applied. We model this case by writing Eq. (1) as

$$v_s(t) = \begin{cases} 0 & t < 0 \\ V_s & t > 0 \end{cases} \tag{3}$$

representing an input voltage step of "height" V_s. Letting $V_0 = 0$ and $V_{ss} = V_s$ in Eq. (2b), we obtain the step response of an RC circuit

$$v(t) = V_s(1 - e^{-t/\tau}) \qquad t \geq 0 \tag{4}$$

as plotted in Fig. 8.2–4.

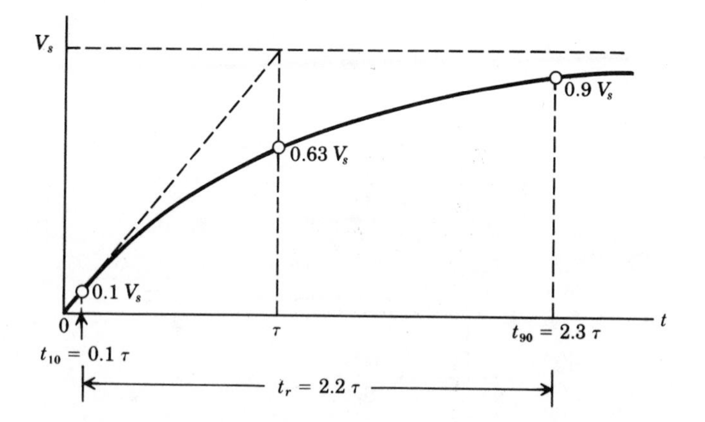

FIGURE 8.2–4
Step response of an RC circuit showing the rise time t_r.

From our previous study of the exponential function, we know that $e^{-t/\tau}$ decreases almost linearly near $t = 0$ so $(1 - e^{-t/\tau})$ begins to *rise* in a nearly linear fashion and

$$v(t) \approx \frac{V_s}{\tau} t \qquad t \leq \frac{\tau}{10} \tag{5}$$

which is a good approximation up to the 10-percent time $t_{10} \approx 0.1\tau$, where $v(t_{10}) = 1.0\ V_s$. The rise continues less rapidly, passing through the 63-percent point at $t = \tau$ where $v(\tau) = V_s(1 - e^{-1}) \approx 0.63\ V_s$, and is essentially over at the 90-percent time $t_{90} \approx 2.3\tau$, where $v(t_{90}) = V_s(1 - e^{-2.3}) = 0.9\ V_s$. The time interval between t_{10} and t_{90} is termed the *rise time*

$$t_r \triangleq t_{90} - t_{10} \approx 2.2\tau$$

which, like the time constant τ, is another measure of the "speediness" of the step response.

Rise-time considerations play a major role in the design of high-speed digital electronic circuits. They also figure in more prosaic circumstances such as the simple charging of a capacitor. To clarify this point, observe from Fig. 8.2–4 that v never reaches the full value V_s exactly, but that there will be no measurable difference between v and V_s if the source has been applied for a time interval large compared to t_r. That's what we imply in statements such as "The switch has been closed for a *long* time."

Exercise 8.2–2

A 100-V source with 5 Ω internal resistance is connected to a 20-μF capacitance at $t = 0$. Evaluate t_r and use your calculator to find v at $t = 1.0$ ms.

PULSE RESPONSE

A transient waveform having particular importance in digital electronics is the response to a *rectangular pulse* of duration D plotted in Fig. 8.2–5a.

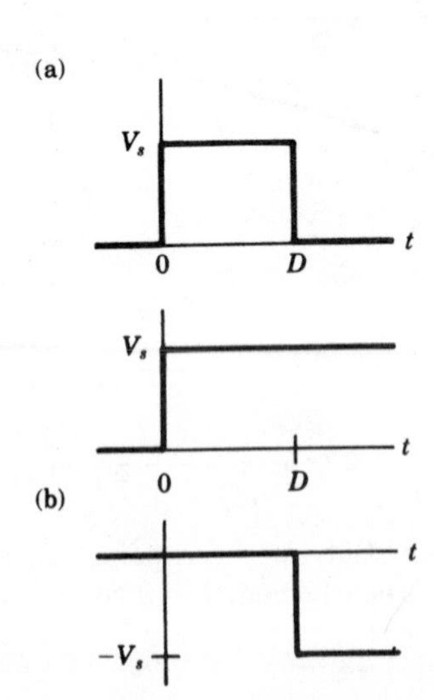

FIGURE 8.2–5
(a) Rectangular pulse.
(b) Decomposition into two step functions.

For the purpose of analysis, we will view this pulse as the *sum of two step functions*—an upward step at $t = 0$ followed by a downward step at $t = D$, as in Fig. 8.2–5b. Invoking superposition now yields a resulting pulse response that is the sum of two step responses—the second step response *delayed* D seconds from the first.

Before proceeding further, we need to take a short digression on the subject of *time delay*. Delaying any time function, say $x(t)$, by D units shifts the entire function to the right along the time axis and produces a new time function written as $x(t - D)$. Conversely, $x(t + D)$ is the time-advanced version of $x(t)$. In short, replacing t with $t \mp D$ shifts a time function to the right by $\pm D$ units. Let's now apply superposition and the time-delay concept to calculate the pulse response of an RC circuit.

If there is no initial stored energy at $t = 0$, the pulse response starts out exactly like the step response

$$v'(t) = V_s(1 - e^{-t/\tau}) \qquad t \geq 0$$

which holds until the pulse ends at $t = D$. But we have modeled the end of the pulse as a delayed step of height $-V_s$, and the response to that term alone is found from Eq. (4) by replacing V_s with $-V_s$ and substituting $t - D$ for t to account for the delay; that is,

$$v''(t) = -V_s[1 - e^{-(t-D)/\tau}] \qquad t \geq D$$

The total pulse response is then $v(t) = v'(t) + v''(t)$, in which $v''(t) = 0$ for $t < D$. Our final result, therefore, must be written as two equations

$$v(t) = V_s(1 - e^{-t/\tau}) \qquad 0 \leq t \leq D \tag{6a}$$

and

$$\begin{aligned} v(t) &= V_s(1 - e^{-t/\tau}) - V_s[1 - e^{-(t-D)/\tau}] \\ &= V_s[e^{-(t-D)/\tau} - e^{-t/\tau}] \qquad t > D \end{aligned} \tag{6b}$$

Equation (6a) describes the transient during the applied pulse, and Eq. (6b) the transient after the pulse has ended. Setting $t = D$ in either of these gives the value of the response at the end of the pulse, that is,

$$v(D) = V_s(1 - e^{-D/\tau}) \tag{7}$$

Equation (6b), then, can be expressed in terms of $v(D)$ as

$$v(t) = v(D)e^{-(t-D)/\tau} \qquad t \geq D \tag{8}$$

which represents a decaying *natural response* starting at $t = D$ with initial value $V_0 = v(D)$.

Figure 8.2–6 shows the pulse response for three different values of the time constant τ relative to the pulse duration D. When $\tau \ll D$ the response rises quickly since $t_r = 2.2\tau \ll D$; then it essentially flattens off at the applied pulse height and decays quickly from $v(D) \approx V_s$. When $\tau \approx D$ the rise and decay are less rapid, and $v(D)$ must be computed from Eq. (7); for instance, $v(D) = V_s(1 - e^{-1}) = 0.63\ V_s$ if $\tau = D$. When $\tau \gg D$ the rise approximates a linear ramp and $v(D) \approx V_s D/\tau \ll V_s$.

(a)

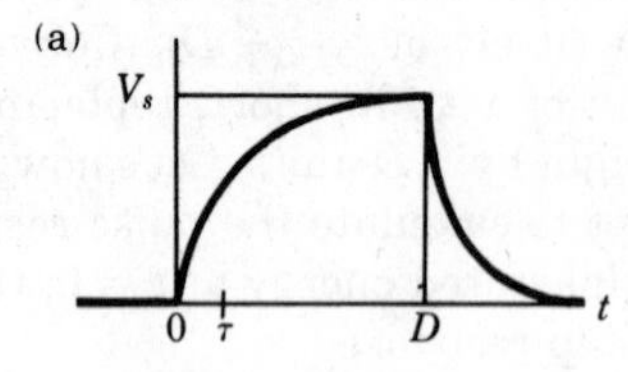

(b)

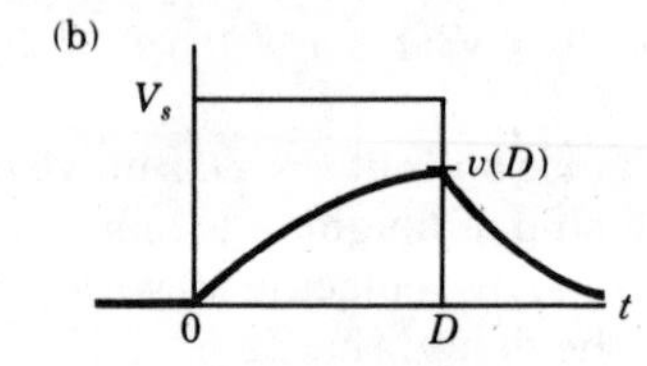

(c)

FIGURE 8.2–6
Pulse response of an RC circuit: (a) $\tau \ll D$; (b) $\tau \approx D$; (e) $\tau \gg D$.

We therefore reach the important conclusion that the shape of the pulse response of an *RC* (or *RL*) circuit depends critically on the relative values of τ and D. Nonetheless, the decaying "tail" at the end of the pulse always has the shape of the natural response and falls off at the same speed that the step response rises. Hence, the *rise* time t_r also serves as a measure of the *fall time*. A reasonably faithful reproduction of a rectangular pulse shape, as in Fig. 8.2–6a, requires $\tau \le D/5$ so that $v(D) \ge 0.99\ V_s$.

Another important conclusion here relates to the *frequency response* of the circuit. In particular, the waveforms in Fig. 8.2–6 correspond to the pulse response of a *lowpass filter*, that is to the voltage across the capacitance in an *RC* circuit or across the inductance in an *RL* circuit. Since the filter's cutoff frequency is $f_{co}\ 1/2\pi\tau$, the reproducibility condition $\tau \le D/5$ becomes

$$f_{co} \ge \frac{1}{D} \tag{9}$$

Thus, a small pulse duration requires a large cutoff frequency.

Exercise 8.2–3

The circuit in Fig. 8.2–2 becomes a *highpass* filter if we take the output as the voltage across the resistance, namely $v_1(t) = Ri(t) = v_s(t) - v(t)$. Use Fig. 8.2–6 to sketch the pulse response of $v_1(t)$ for three values of τ. From your sketches, justify the assertion that a highpass filter should have $f_{co} \leq 1/30D$ for good pulse reproduction.

COMPLETE RESPONSE

Finally, we outline the method for determining the complete response of a first-order circuit—a circuit possibly containing initial stored energy—when an arbitrary excitation is applied at $t = 0$. Let $x(t)$ and $y(t)$ denote the applied input and the response of interest, either of which may be a voltage or a current. We proceed with the solution as follows:

- Evaluate the time constant τ from the root of $Z(s)$ if $x(t)$ is a voltage source or $Y(s)$ if $x(t)$ is a current source.
- Write the natural response in the form

$$y_N(t) = Ae^{-t/\tau}$$

  with A being an unknown constant.
- Find the forced response $y_F(t)$ by assuming temporarily that $x(t)$ has been applied for all time $t > -\infty$.
- Write the complete response as

$$y(t) = y_F(t) + y_N(t) \qquad t \geq 0$$

  and evaluate the constant A from the initial condition on $y(t)$ at $t = 0^+$.

Since $y_N(t)$ dies away with time, $y(t)$ approaches $y_F(t)$ as $t \to \infty$. In other words, the complete response eventually reaches the *steady-state condition* $y(t) = y_F(t)$ after an initial transient interval. It is this observation that justifies calculating $y_F(t)$ via steady-state analysis. Thus, depending on $x(t)$, $y_F(t)$ will be the steady-state DC response, AC response, exponential response, and so forth. The mechanics of the method are illustrated in the example below.

Example 8.2–2

Figure 8.2–7a shows an AC voltage source applied to an RL circuit at $t = 0$. There is no initial stored energy, so $i(0^+) = 0$. We want to find $i(t)$ for $t \geq 0$. The impedance is $Z(s) = 12 + 0.1s$ which equals zero when $s = -120$. Hence, the natural response is $i_N(t) = Ae^{-120t}$. For the forced response we have $\underline{V} = 170 \angle -90°$, $\omega = 377$, and $Z(j\omega) = 12 + j37.7 =$

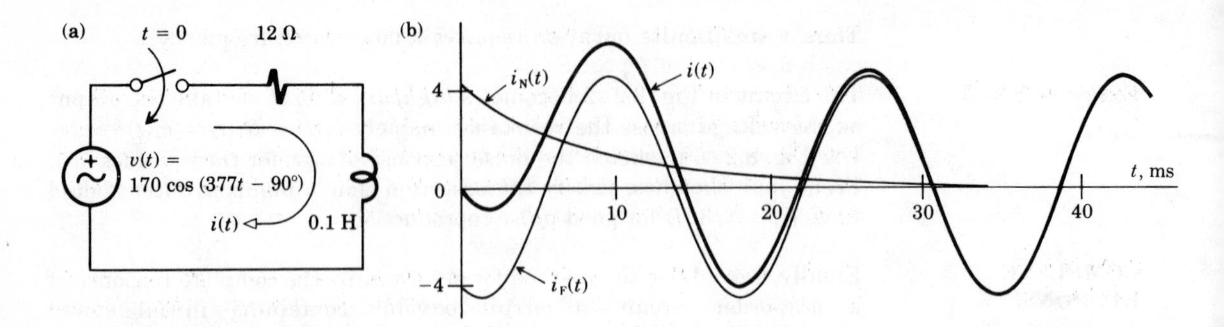

FIGURE 8.2–7
Circuit and waveforms for Example 8.2–2.

.40 $\underline{/72^\circ}$, which gives $i_F(t) = (170/40)\cos(377t - 90^\circ - 72^\circ)$. Therefore,

$$i(t) = 4.25\cos(377t - 162^\circ) + Ae^{-120t}$$

$$i(0^+) = 4.25\cos(-162^\circ) + A = 0$$

Solving the latter yields $A = 4.04$, and our final result is

$$i(t) = 4.25\cos(377t - 162^\circ) + 4.04e^{-120t}$$

for $t \geq 0$.

The plot of this waveform in Fig. 8.2–7b reveals a transient interval lasting about 40 ms, after which $i(t)$ becomes a steady-state AC current. The transient duration could have been predicted from the fact that $i_N(t)$ decays with time constant $\tau = \frac{1}{120} \approx 8$ ms and is essentially zero for $t \geq 5\tau \approx 40$ ms.

Note that the natural-response component appears here even though there was no initial stored energy. The reason is that $i_N(t)$ is needed to satisfy the continuity condition. Had there been an initial current of just the right value, namely $i(0) = -4.04$, we would have obtained $A = 0$ and there would be no transient in the complete response.

Exercise 8.2–4 Repeat the above example with L replaced by $C = 250\ \mu$F and $\underline{V} = 170\ \underline{/0^\circ}$. Remember that the voltage across the capacitance must be continuous.

8.3 SECOND-ORDER TRANSIENTS †

When a circuit contains two or more energy-storage elements, its transient response may include time functions other than the decaying exponentials that characterized first-order circuits. Of particular interest are circuits with at least one capacitance and one inductance, which may exhibit an oscillatory or underdamped natural behavior. We will examine the various possibilities by focusing on transients in second-order circuits. The extension to higher-order circuits is conceptually straightforward, but often involves laborious conputations. Fortunately, much can be learned from the study of second-order circuits. Moreover, many higher-order circuits have a dominant second-order behavior that is of primary concern.

NATURAL RESPONSE

A simple introduction to second-order transients is provided by the circuit in Fig. 8.3–1, where the switch is opened at $t = 0$. We want to find the open-circuit natural response $v_N(t)$ for $t > 0$. Steady-state DC analysis of the circuit with the switch closed shows that $v_C = V_0$ and $i_L = V_0/R_1 = I_0$ for $t < 0$. The continuity of voltage across a capacitance and current through an inductance then require

$$v_C(0^+) = V_0 \qquad i_L(0^+) = I_0$$

immediately after the switch has been opened. Furthermore, closer inspection of the diagram reveals that the capacitance discharges entirely through R_1, while the energy stored by the inductance must dissipate entirely in R_2. We conclude, then, that $v_N(t)$ will be the sum of *two* decaying exponentials, namely

$$v_N(t) = V_0 e^{-t/\tau_1} + R_2(-I_0)e^{-t/\tau_2}$$

with $\tau_1 = R_1C$ and $\tau_2 = L/R_2$.

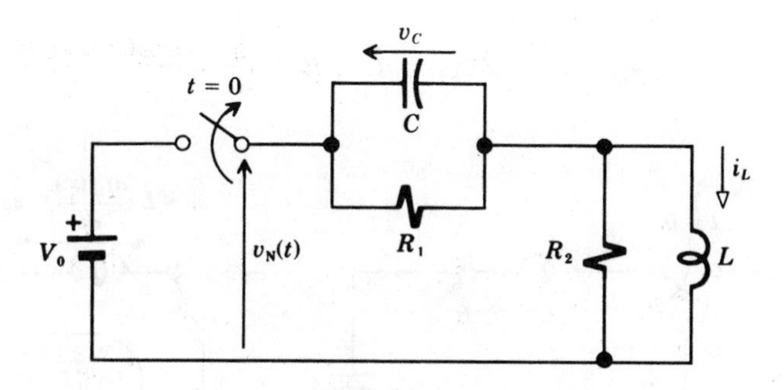

FIGURE 8.3–1

The existence of two exponential terms is confirmed mathematically by showing that the admittance $Y(s)$ has *two roots*. (Recall that open-circuit natural behavior corresponds to $Y(s) = 0$.) Since impedance calculation happens to be easier for the circuit at hand, we start with

$$Z(s) = \left(R_1 \| \frac{1}{sC}\right) + (R_2 \| sL) = \frac{R_1}{sCR_1 + 1} + \frac{sLR_2}{sL + R_2}$$

from which

$$Y(s) = \frac{1}{Z(s)} = \frac{(sCR_1 + 1)(sL + R_2)}{(sL + R_2)R_1 + (sCR_1 + 1)sLR_2}$$

Clearly, $Y(s)$ equals zero when its numerator equals zero, so we write

$$(sCR_1 + 1)(sL + R_2) = 0$$

and there are indeed two roots, say $s = p_1$ and $s = p_2$, given by

$$p_1 = -\frac{1}{R_1 C} = -\frac{1}{\tau_1} \qquad p_2 = -\frac{R_2}{L} = -\frac{1}{\tau_2}$$

You should check that $Y(s) = 0$ when $s = p_1$ or p_2.

Now consider the series RLC circuit in Fig. 8.3–2a with $v_C = V_0$ and $i_L = 0$ immediately after the terminals are short-circuited at $t = 0$. Presumably, the natural-response current will be of the form

$$i_N(t) = A_1 e^{p_1 t} + A_2 e^{p_2 t} \tag{1}$$

where the constants A_1 and A_2 incorporate the initial conditions, and p_1 and p_2 are the roots of $Z(s)$, since we are dealing with short-circuit behavior. We find these roots from

$$Z(s) = sL + R + \frac{1}{sC} = \frac{s^2LC + sCR + 1}{sC} = 0$$

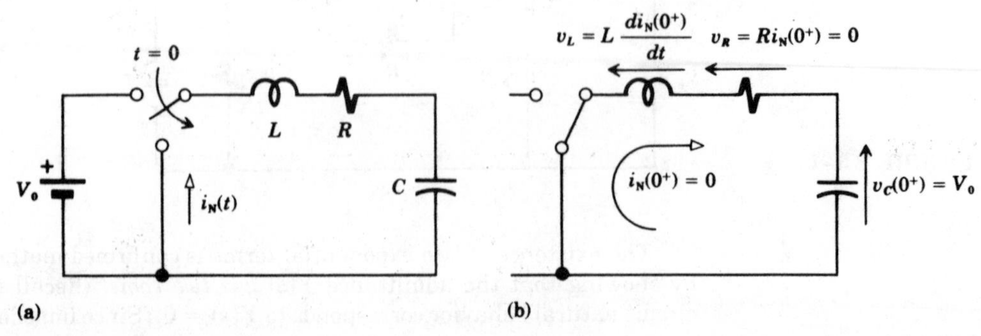

FIGURE 8.3–2 (a) A switched RLC circuit. (b) Initial conditions.

or, upon dividing the numerator by LC,

$$s^2 + \frac{R}{L}s + \frac{1}{LC} = 0 \tag{2}$$

This quadratic equation does not factor directly, but application of the quadratic formula yields the roots

$$p_1 = -\alpha + \sqrt{\alpha^2 - \omega_0^2} \qquad p_2 = -\alpha - \sqrt{\alpha^2 - \omega_0^2} \tag{3a}$$

where

$$\alpha = R/2L \qquad \omega_0 = 1/\sqrt{LC} \tag{3b}$$

We recognize ω_0 as the resonant frequency, while α is called the *damping coefficient.*

Each of the exponential terms in $i_N(t)$ thus depends on all three elements. This reflects the fact that, unlike the situation in Fig. 8.3–1, the natural-response current passes through all three elements. Consequently, the energy-storage elements interact with each other during the natural response. We also see this interaction in the constants A_1 and A_2.

To evaluate the constants, we observe that $i_N = i_L$ for $t > 0$ and $i_L(0^+) = 0$. Hence, setting $t = 0^+$ in Eq. (1),

$$i_N(0^+) = A_1 + A_2 = 0 \tag{4a}$$

but we still need another equation to solve for A_1 and A_2. That other equation comes from $v_C(0^+) = V_0$ and the initial-condition diagram of Fig. 8.3–2b, which shows that $v_L(0^+) = L\,[di_N/dt](0^+) = -v_R(0^+) - v_C(0^+) = -V_0$. We thus have a condition on the initial *slope* of $i_N(t)$, namely $[di_N/dt](0^+) = -V_0/L$. Differentiating Eq. (1) then gives

$$\frac{di_N}{dt}(0^+) = A_1p_1 + A_2p_2 = -\frac{V_0}{L} \tag{4b}$$

and simultaneous solution of Eqs. (4a) and (4b) will yield values for A_1 and A_2. Again, unlike the situation in Fig. 8.3–1, each of the two constants depends on both initial conditions.

Generalizing our results, the natural-response voltage or current in a second-order circuit takes the form

$$y_N(t) = A_1e^{p_1t} + A_2e^{p_2t} \tag{5}$$

where p_1 and p_2 are the roots of a quadratic *characteristic equation.* The characteristic equation may be written for convenience as

$$s^2 + 2\alpha s + \omega_0^2 = 0 \tag{6}$$

where the parameter ω_0 may or may not relate to a resonance phenomenon, depending upon the value of α. The left-hand side of Eq. (6) is called the *characteristic polynomial,* and its two roots are expressed compactly by

$$p_1, p_2 = -\alpha \pm \sqrt{\alpha^2 - \omega_0^2} \tag{7}$$

The constants A_1 and A_2 may then be found from the initial-value equations

$$A_1 + A_2 = y_N(0^+) \qquad A_1 p_1 + A_2 p_2 = \frac{dy_N}{dt}(0^+) \tag{8}$$

with $y_N(0^+)$ and $[dy_N/dt](0^+)$ satisfying the continuity conditions.

By the same reasoning, the natural response of an nth-order circuit is

$$y_N(t) = A_1 e^{p_1 t} + A_2 e^{p_2 t} + \cdots + A_n e^{p_n t}$$

where $p_1, p_2, \ldots, p_n$ are the roots of a characteristic polynomial involving powers of s through s^n. Each of the n energy-storage elements will have a continuity condition, and the constants $A_1, A_2, \ldots, A_n$ are evaluated by solving n simultaneous initial-value equations for $y_N(0^+)$, $[dy_N/dt](0^+)$, $[d^2y_N/dt^2](0^+)$, etc. Such computations usually require computer-aided numerical methods when $n \geq 3$.

Example 8.3–1

Let the circuit in Fig. 8.3–2 have $R = 100\ \Omega$, $L = 0.4$ H, $C = 250\ \mu$F, and $V_0 = 12$ V. The characteristic equation becomes

$$s^2 + 250s + 10^4 = 0$$

whose roots are

$$p_1 = -50 \qquad p_2 = -200$$

corresponding to decaying exponentials with time constants $\tau_1 = -1/p_1 = \frac{1}{50} = 20$ ms and $\tau_2 = -1/p_2 = \frac{1}{200} = 5$ ms. We then evaluate A_1 and A_2 by noting that $A_2 = -A_1$, per Eq. (4a), and substituting in Eq. (4b):

$$A_1(-50) + (-A_1)(-200) = \frac{-12}{0.4}$$

Thus, $A_1 = -\frac{30}{150} = -0.2$ A, $A_2 = +0.2$ A, and

$$i_N(t) = -0.2e^{-50t} + 0.2e^{-200t} \qquad t \geq 0$$

as plotted in Fig. 8.3–3.

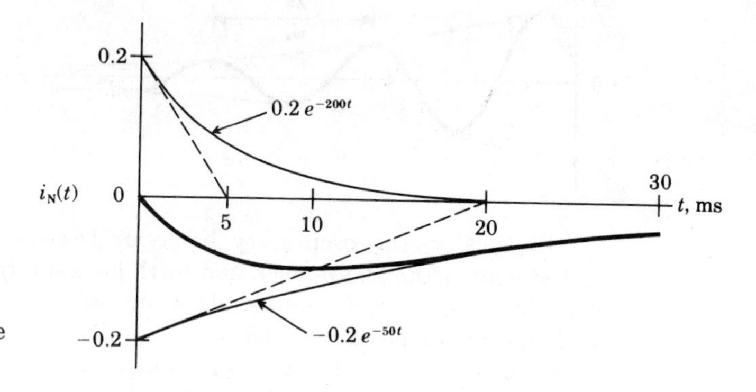

FIGURE 8.3–3
Waveforms for Example 8.3–1.

This waveform looks like the combination of two first-order waveforms, except that the time constants differ slightly from the individual time constants $L/R = 4$ ms $= \frac{1}{250}$ and $RC = 25$ ms $= \frac{1}{40}$. The negative value of $i_N(t)$ means, of course, that the capacitance is discharging through R and L. The zero initial value preserves continuity of current through the inductance, and the negative initial slope equals $-V_0/L$.

Exercise 8.3–1

Suppose the terminals in Fig. 8.3–1 are short-circuited instead of open-circuited. Show that the roots of the characteristic polynomial are given by Eq. (7) with $\omega_0 = 1/\sqrt{LC}$ and $\alpha = 1/2RC$ where $R = R_1 \| R_2$.

UNDERDAMPED NATURAL BEHAVIOR

So far we have assumed that the roots p_1 and p_2 are unequal negative quantities, meaning that $\alpha > \omega_0$ in Eq. (7). This will always be the case for nonresonant circuits. But the element values in a resonant circuit may result in $\alpha < \omega_0$, so the roots will be *complex conjugates* with negative real parts, namely,

$$p_1, p_2 = -\alpha \pm j\omega_N \tag{9a}$$

where

$$\omega_N \triangleq \sqrt{\omega_0^2 - \alpha^2} \tag{9b}$$

The natural response then takes the form

$$y_N(t) = Ae^{-\alpha t} \cos(\omega_N t + \theta) \tag{10}$$

with A and θ the initial-condition constants. We describe this as *underdamped* behavior because $y_N(t)$ oscillates at the *natural frequency* $\omega_N \leq \omega_0$ with a decaying amplitude or *envelope* $Ae^{-\alpha t}$ that has the time constant $\tau = 1/\alpha$, Fig. 8.3–4.

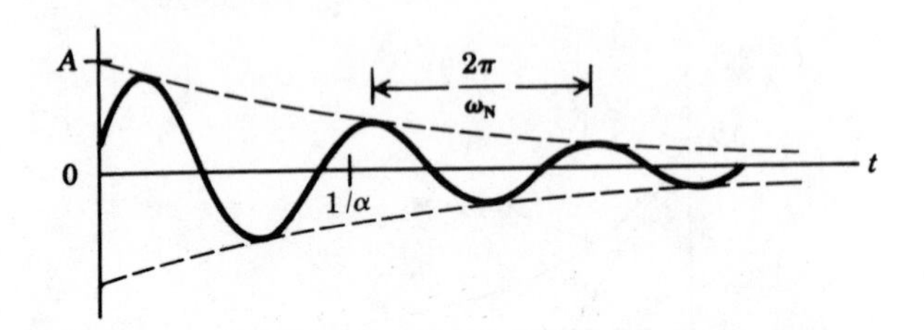

FIGURE 8.3–4
Underdamped natural behavior.

Physically, the oscillatory behavior results from an exchange of stored energy that flows back and forth between the capacitance and inductance. If the circuit were lossless, the oscillations would go on forever without decay since $\alpha = 0$ and $\tau \to \infty$. Oscillations do not occur when $\alpha > \omega_0$ because energy is dissipated in the resistance at a faster rate than it can be exchanged; we say that those circuits are *overdamped.*

The extraordinary assertion that complex roots correspond to an oscillatory time function can be proved using Euler's theorem plus the implication in Eq. (8) that when p_1 and p_2 are complex conjugates, so are the constants A_1 and A_2. We therefore write their polar versions as $\underline{A}_1 = |\underline{A}_1| e^{j\theta}$ and $\underline{A}_2 = \underline{A}_1^* = |\underline{A}_1| e^{-j\theta}$ and insert them along with p_1 and p_2 into the exponential expression, as follows:

$$\begin{aligned} y_N(t) &= \underline{A}_1 e^{p_1 t} + \underline{A}_2 e^{p_2 t} \\ &= |\underline{A}_1| e^{j\theta}\, e^{(-\alpha + j\omega_N)t} + |\underline{A}_1| e^{-j\theta}\, e^{(-\alpha - j\omega_N)t} \\ &= |\underline{A}_1| e^{-\alpha t}[e^{j(\omega_N t+\theta)} + e^{-j(\omega_N t+\theta)}] \end{aligned}$$

But for any ϕ, $e^{j\phi} = \cos\phi + j\sin\phi$, while $e^{-j\phi} = \cos(-\phi) + j\sin(-\phi) = \cos\phi - j\sin\phi$; so

$$e^{j\phi} + e^{-j\phi} = 2\cos\phi$$

Therefore, letting $\phi = \omega_N t + \theta$,

$$y_N(t) = 2\,|\underline{A}_1|\, e^{-\alpha t} \cos(\omega_N t + \theta)$$

which is identical to Eq. (10) with

$$A = 2\,|\underline{A}_1| \qquad \theta = \measuredangle \underline{A}_1 \tag{11}$$

Note that all the imaginary parts have disappeared, as they must for $y_N(t)$ to be a real function of time.

Example 8.3–2

Suppose the circuit in Fig. 8.3–2 has $R = 100\ \Omega$, $L = 0.4$ H, and $C = 16\ \mu$F, so $\omega_0 = 1/\sqrt{LC} \approx 395$ while $\alpha = R/2L = 125$. Since $\alpha < \omega_0$, the circuit is underdamped and the natural response oscillates at the natural frequency $\omega_N = \sqrt{395^2 - 125^2} = 375$ rad/s. The oscillation envelope decays with time constant $1/\alpha = 8$ ms.

Exercise 8.3–2

Repeat the calculations of Example 8.3–1 with the resistance reduced to 64 Ω. In particular, show that $\underline{A}_1 = j/4$ and $\underline{A}_2 = \underline{A}_1^*$.

CRITICAL DAMPING

Critical damping is the dividing line between overdamped and underdamped behavior. It occurs when the element values are adjusted precisely such that $\alpha = \omega_0$; then $p_1 = p_2 = -\alpha$, and we say that the roots are *repeated*.

Mathematicians have shown that repeated roots give rise to a natural behavior of the form te^{pt}, along with e^{pt}. Thus, with critical damping, the natural response becomes

$$y_N(t) = A_1 e^{-\alpha t} + A_2 te^{-\alpha t} \tag{12}$$

whose first term is just a decaying exponential. The second term, however, starts with a linear rise, goes through a maximum value of $A_2/\alpha e = 0.368\ A_2/\alpha$ at $t = 1/\alpha$, and then decays as $t \to \infty$ (see Fig. 8.3–5). The constants are evaluated from

$$A_1 = y_N(0^+) \qquad -\alpha A_1 + A_2 = \frac{dy_N}{dt}(0^+) \tag{13}$$

instead of from Eq. (8).

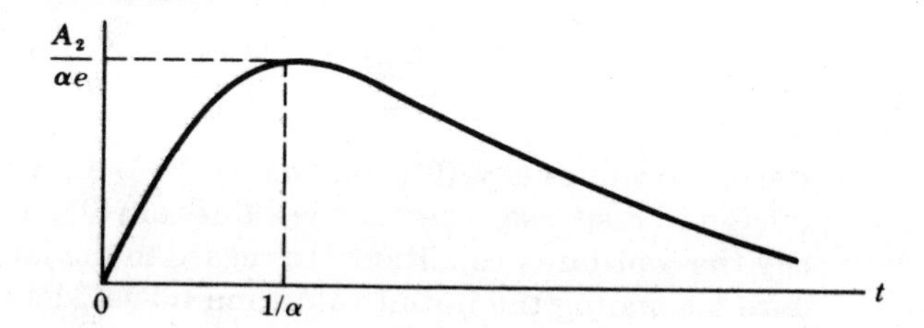

FIGURE 8.3–5 The critically damped component $A_2 te^{-\alpha t}$.

Example 8.3–3

A series RLC is critically damped when $R/2L = 1/\sqrt{LC}$ or $R = 2\sqrt{L/C}$. If we assume that all other values are those in Example 8.3–1, and if $R = 2\sqrt{0.4\ \text{H}/250\ \mu\text{F}} = 80\ \Omega$, then $\alpha = \omega_0 = 100$ and Eq. (13) gives

$$A_1 = 0 \qquad -100A_1 + A_2 = -30$$

since $i_N(0^+) = 0$ and $[di_N/dt](0^+) = -12/0.4 = -30$. Thus

$$i_N(t) = -30te^{-100t}$$

which has only one term (because $A_1 = 0$) and which will, therefore, look like Fig. 8.3–5.

Exercise 8.3–3

Obtain the characteristic equation of a parallel *RLC* circuit from $Y(s) = 0$. Then show that a critically damped series or parallel *RLC* circuit has a quality factor of $Q = \frac{1}{2}$.

COMPLETE RESPONSE

The complete transient response of a second-order circuit, like that of a first-order circuit, consists of the forced response plus the natural response, namely.

$$y(t) = y_F(t) + y_N(t)$$

If the applied input $x(t)$ is a constant, an exponential, or a sinusoid, the forced response is easily found by using the *transfer function* $H(s)$ and the steady-state analysis techniques described in Chaps. 5 and 6. The transfer function also has significance relative to the natural response, for if we write it as a ratio of polynomials, its *denominator* will be identical to the *characteristic polynomial*. Therefore, solving

$$\frac{1}{H(s)} = 0 \tag{14}$$

yields the roots p_1 and p_2 for $y_N(t)$, and we need not bother with $Z(s)$ or $Y(s)$.

The two constants in $y_N(t)$ are evaluated using Eq. (8) or (13) with

$$y_N(0^+) = y(0^+) - y_F(0^+) \tag{15}$$

$$\frac{dy_N}{dt}(0^+) = \frac{dy}{dt}(0^+) - \frac{dy_F}{dt}(0^+)$$

Here, the values of $y_F(0^+)$ and $[dy_F/dt](0^+)$ are determined directly from the forced response, whereas the values of $y(0^+)$ and $[dy/dt](0^+)$ must satisfy the continuity conditions. In regard to the latter, a systematic procedure for finding the initial value and initial slope of $y(t)$ is as follows:

First, label the circuit diagram with *symbols* (not values) for all voltages and currents at $t = 0^+$. Second, write down the initial input value $x(0^+)$, and use the continuity conditions to evaluate initial capacitor voltages and inductor currents, say

$$v_C(0^+) = V_0 \qquad i_L(0^+) = I_0 \tag{16a}$$

Third, apply Kirchhoff's laws and Ohm's law to find all other voltages and currents at $t = 0^+$. Fourth, calculate dx/dt at $t = 0^+$ and use the initial capacitor currents and inductor voltages to evaluate

$$\frac{dv_C}{dt}(0^+) = \frac{i_C(0^+)}{C} \qquad \frac{di_L}{dt}(0^+) = \frac{v_L(0^+)}{L} \tag{16b}$$

Finally, calculate all other initial slopes by noting that if $v_R = Ri_R$ then $dv_R/dt = R\, di_R/dt$; similarly, if Kirchhoff's law requires $v_1 = v_2 + v_3$, then $dv_1/dt = dv_2/dt + dv_3/dt$—and likewise for any sum of currents. This procedure is easier than it sounds, as our next example shows.

Example 8.3–4
Step response of an RLC circuit

In Section 8.2 we found the voltage *step response* of a first-order circuit to be $v(t) = V_s(1 - e^{-t/\tau})$. Let's derive the corresponding expression for the *RLC* circuit in Fig. 8.3–6a with input voltage $v_s(t) = V_s$ for $t > 0$ and no initial stored energy. The transfer function is

$$H(s) = \frac{1/sC}{sL + R + \dfrac{1}{sC}} = \frac{1/LC}{s^2 + 2\left(\dfrac{R}{2L}\right)s + \dfrac{1}{LC}}$$

whose denominator is, indeed, the characteristic polynomial in Eq. (2), with the roots given by Eq. (3). The forced response due to the constant input V_s is $v_F(t) = H(0)V_s = V_s$, and is easily checked by DC steady-state

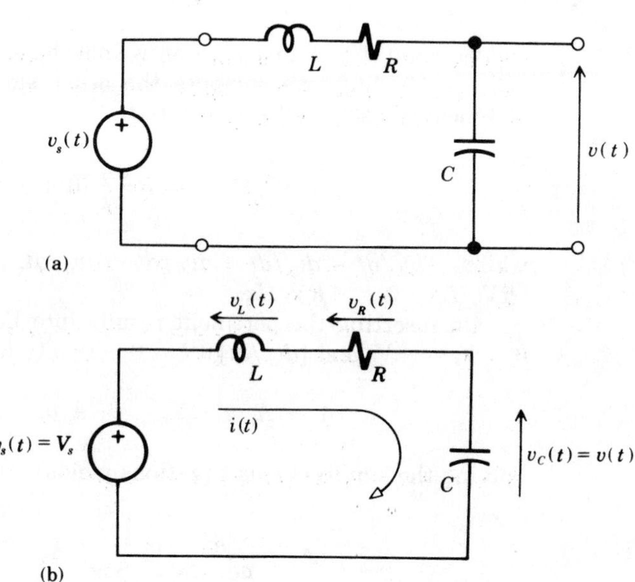

FIGURE 8.3–6

analysis. Thus, adding $v_F(t)$ and $v_N(t)$ yields

$$v(t) = V_s + A_1 e^{p_1 t} + A_2 e^{p_2 t}$$

if we assume distinct (nonrepeated) roots, $p_1 \neq p_2$.

To determine the constants A_1 and A_2, we'll use Eqs. (8) and (15) with $y(t) = v(t)$ and $y_F(t) = v_F(t) = V_s$, so $y_F(0^+) = V_s$ and $[dy_F/dt](0^+) = 0$. However, we first must find the initial value and initial slope of $v(t)$ that, being the voltage across the capacitor, can also be denoted by $v_C(t)$.

Taking the first step in the systematic procedure, Fig. 8.3–6b shows all the voltages and currents of interest. The initial input value is $v_s(0^+) = V_s$, and the absence of initial stored energy plus the continuity conditions tell us that $v_C(0^+) = 0$ and $i(0^+) = i_L(0^+) = 0$. Since $i_R(t) = i(t)$, $v_R(0^+) = Ri(0^+)$ and, from KVL,

$$v_L(t) = v_s(t) - v_R(t) - v_C(t)$$

so

$$v_L(0^+) = V_s - 0 - 0 = V_s$$

which finishes the third step. As for the initial slopes, we know that $[dv_s/dt](0^+) = 0$; applying Eq. (16b) then gives

$$\frac{dv_C}{dt}(0^+) = \frac{0}{C} = 0 \qquad \frac{di_L}{dt}(0^+) = \frac{V_s}{L}$$

and we could stop at this point for we now have the needed initial slope of $v(t) = v_C(t)$. But let's compute the other slopes for the sake of completeness. Clearly, with $i_R(t) = i_L(t)$,

$$\frac{dv_R}{dt}(0^+) = R\frac{di_L}{dt}(0^+) = \frac{RV_s}{L}$$

while $dv_L/dt = dv_s/dt - dv_R/dt - dv_C/dt$, so $[dv_L/dt](0^+) = 0 - (RV_s/L) - 0 = -RV_s/L$.

On inserting the pertinent results into Eq. (15), we have $v_N(0^+) = 0 - V_s = -V_s$ and $[dv_N/dt](0^+) = 0 - 0 = 0$. Equation (8) then becomes

$$A_1 + A_2 = -V_s \qquad A_1 p_1 + A_2 p_2 = 0$$

Solving the simultaneous equations yields

$$A_1 = \frac{p_2}{p_1 - p_2} V_s \qquad A_2 = -\frac{p_1}{p_2} A_1$$

which hold for an underdamped or overdamped circuit, but not for critical damping ($p_1 = p_2$).

Taking $p_1 = -50$ and $p_2 = -200$ (as in Example 8.3–1), we get $A_1 = -\frac{4}{3}V_s$ and $A_2 = \frac{1}{3}V_s$ and the overdamped step response

$$v(t) = V_s(1 - \tfrac{4}{3}e^{-50t} + \tfrac{1}{3}e^{-200t})$$

sketched in Fig. 8.3–7a. In contrast with the step response of a first-order circuit (Fig. 8.2–4), $v(t)$ does not begin to rise immediately but has zero initial slope—directly attributable to the two continuity conditions. Consequently, the rise time for an overdamped RLC circuit is longer than that of a first-order circuit with $\tau = -1/p_1$.

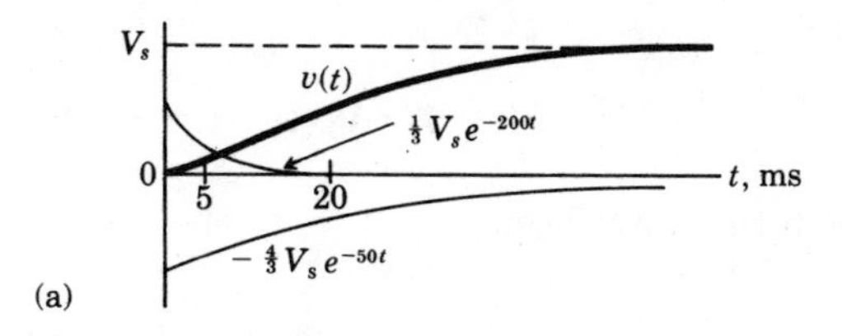

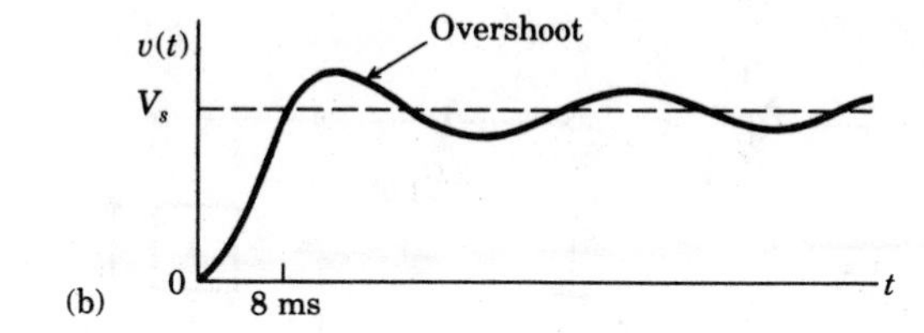

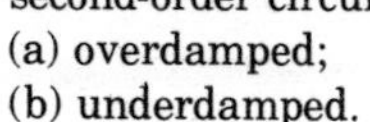
FIGURE 8.3–7 Step response of a second-order circuit: (a) overdamped; (b) underdamped.

Turning to the underdamped case $p_1 = -125 + j375$ and $p_2 = -125 - j375$ (as in Example 8.3–2), we get the complex constant

$$\underline{A}_1 = \frac{-125 - j375}{2j375}V_s = (-1 + j\tfrac{1}{3})\frac{V_s}{2}$$

and you can check that $\underline{A}_2 = \underline{A}_1^*$. We then write $v_N(t)$ in the form of Eq. (10) with

$$A = 2\,|\underline{A}_1| = 1.054\ V_s \qquad \theta = \measuredangle \underline{A}_1 = 161.6^\circ$$

so

$$v(t) = V_s[1 - 1.054e^{-125t}\cos(375t + 161.6^\circ)]$$

The underdamped step response therefore has an oscillatory transient, as sketched in Fig. 8.3–7b. A comparison of this with the overdamped response reveals two significant differences—a faster rise time and an *overshoot,* in the sense that the transient portion of $v(t)$ has peaks that exceed the steady-state value.

A compromise between the overdamped and underdamped responses is the response that is critically damped. Such a response has the fastest possible rise time, but without overshoot and oscillations. By the way, critical damping is just what you want in an automobile's suspension system, which approximates the same dynamic behavior as that of an *RLC* circuit, with the mass and springs acting like energy-storage elements and the shock absorbers providing the resistance.

Exercise 8.3–4 Use Eq. (13) to show that the critically-damped step response of an *RLC* circuit is

$$v(t) = V_s(1 - e^{-\alpha t} - \alpha t e^{-\alpha t})$$

Sketch this waveform.

PROBLEMS

8.1–1 The switch in Fig. P8.1–1 has been in the upper position for a long time. Find $v(t)$ and $i(t)$ for $t > 0$ if X is a 10-μF capacitor.

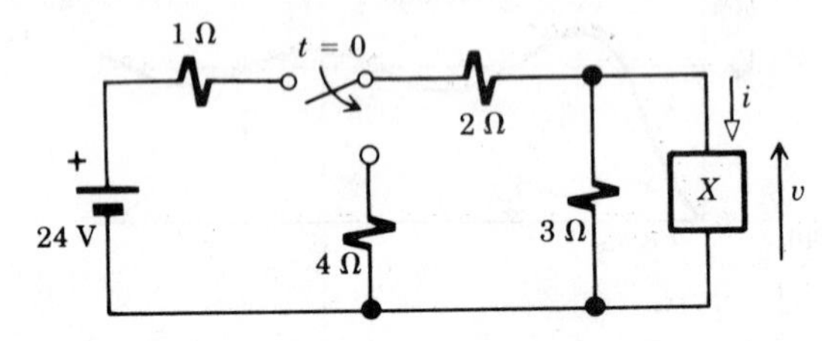

FIGURE P8.1–1

8.1–2 The switch in Fig. P8.1–2 has been in the upper position for a long time. Find $v(t)$, $i(t)$, and $v_1(t)$ for $t > 0$ if X is a 0.5-μF capacitor.

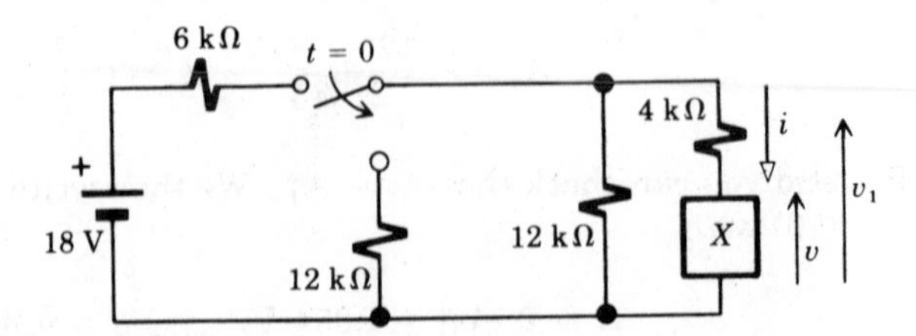

FIGURE P8.1–2

8.1–3 Repeat Problem 8.1–1 taking X as a 0.6-H inductor.

8.1–4 Repeat Problem 8.1–2 taking X as a 50-mH inductor.

8.1–5 Given a 12-V battery with $R_s = 0.4\ \Omega$, design a circuit using a switch, resistor, and inductor to produce a 3000 V voltage spike with $\tau = 5\ \mu s$.

8.1–6 Repeat the previous design with a capacitor instead of the inductor to produce a 60-A current spike.

8.1–7 Calculate the percentage error in Eq. (6b) when $t = \tau/20$ and $\tau/10$.

8.1–8 Consider the decaying exponential $y(t) = Y_0 e^{-t/\tau}$. Let $y(t)$ have the value $y(t_0)$ at $t = t_0$, and show that you can write $y(t) = y(t_0)e^{-(t-t_0)/\tau}$. Then express dy/dt in terms of $y(t_0)$.

‡ **8.1–9** As $t \to \infty$, the total charge that passes to the left through R in Fig. 8.1–2 must equal the initial charge stored by the capacitor. Use this fact to prove that $\int_0^\infty e^{-t/\tau}\,dt = \tau$.

‡ **8.1–10** The total energy dissipated by the resistance in Fig. 8.1–2 is $w = \int_0^\infty p\,dt$ where $p = Ri^2$. Use a modification of the integral relationship in the previous problem to show that w equals the initial energy stored by the capacitor.

8.2–1 Let the circuit in Fig. P6.2–3 have $R_1 = 10\ k\Omega$ and $R_2 = 3\ k\Omega$. Use the continuity conditions to find the values of all voltages and currents at $t = 0^+$ if $v = 18$ V for $t < 0$ and $v = -12$ V for $t > 0$. Also find the final values as $t \to \infty$. (Your results will be needed in Problem 8.3–8.)

8.2–2 Repeat the previous calculations for the circuit in Fig. P8.2–2. (The results will be needed in Problem 8.3–7.)

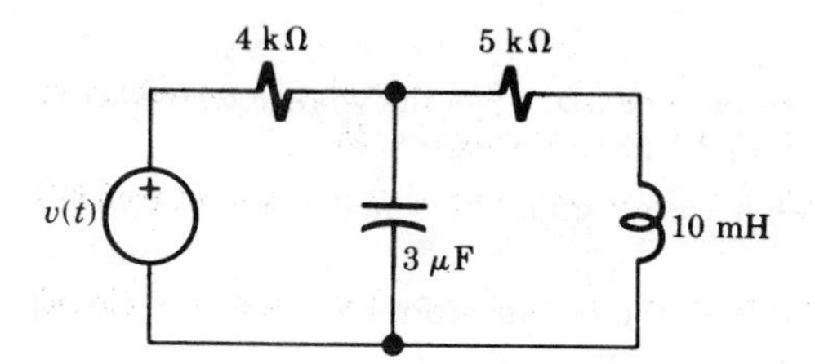

FIGURE P8.2–2

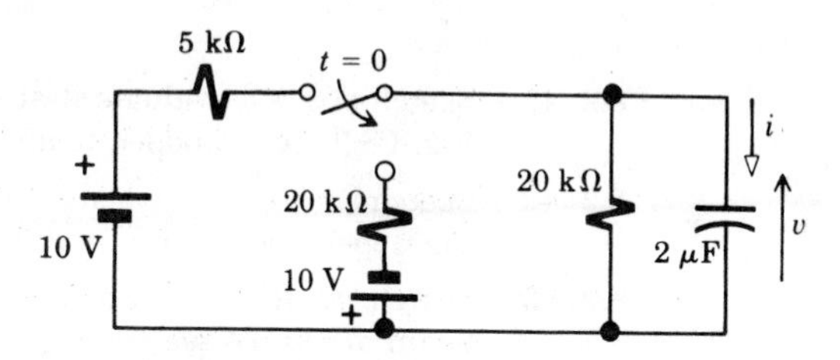

FIGURE P8.2–4

8.2–3 Carry out the *exact* solution of Example 5.3–1, assuming $v = 0$ for $t < 0$.

8.2–4 The switch in Fig. P8.2–4 has been in the upper position for a long time. Find and sketch $v(t)$ and $i(t)$ for $t > 0$, and calculate the time at which $v = 0$. (These results are needed in Problem 8.2–12.)

8.2–5 The switch in Fig. P8.2–5 has been closed for a long time. Find and sketch $i(t)$ and $v(t)$ for $t > 0$, and calculate the time at which $i = 0$. (These results are needed in Problem 8.2–13.)

8.2–6 Find and sketch the step response of the circuit in Fig. P7.1–8 when $\underline{Z}_1$ is a resistance R and $\underline{Z}_2$ a capacitance C.

8.2–7 Let $v_{out} = v_L - v_C = v - v_1 - v_C$ in Fig. P6.2–1; now find and sketch the step response when $R_1 = 100\ \Omega$, $R_2 = 1\ k\Omega$, $L = 0.1$ H, and $C = 1\ \mu F$.

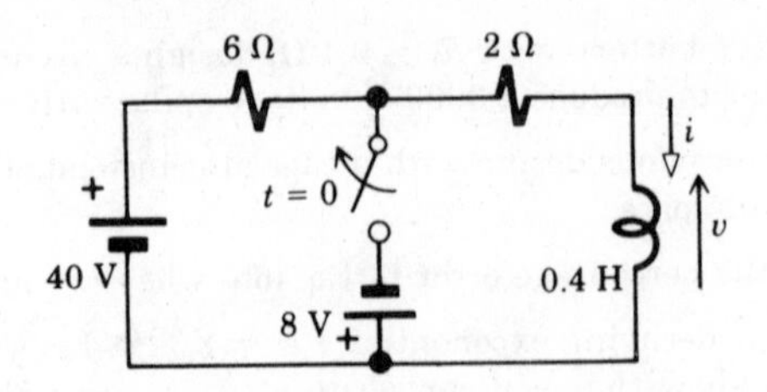

FIGURE P8.2–5

8.2–8 Repeat the previous problem with $C = 2\ \mu F$.

8.2–9 Let $C = 0.5\ \mu F$ replace R_L in Fig. P3.3–4. Find the current through C when $R_1 = R_4 = 6\ k\Omega$, $R_2 = R_3 = 3\ k\Omega$, and v_s is an 18-V step at $t = 0$.

8.2–10 Let the voltage waveform in Fig. P8.2–10 be applied to an RC lowpass filter, and suppose you want $v(2D) \leq \frac{1}{2}v(D)$ so that the spacing between pulses is evident at the output. Show that this condition requires $f_{co} \geq 1/9D$.

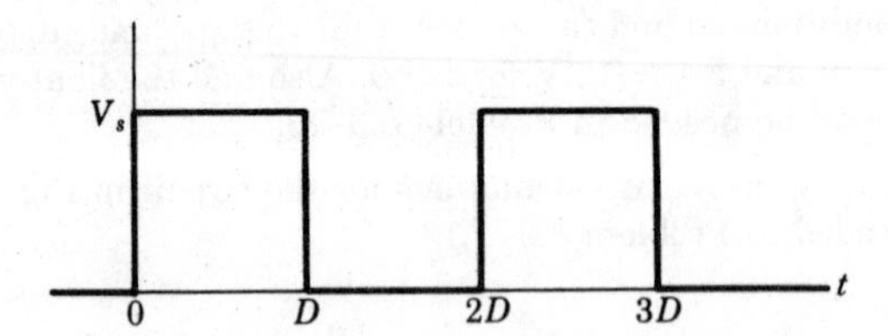

FIGURE P8.2–10

8.2–11 Sketch $v(t)$ and evaluate it at $t = D$, $2D$, and $3D$, when the voltage waveform in Fig. P8.2–10 is applied to an RC lowpass filter having $\tau = D$.

‡ **8.2–12** Sketch $v(t)$ for $t > 0$ when the switch in the circuit of Problem 8.2–4 is returned to the upper position at $t = 30$ ms.

‡ **8.2–13** Sketch $i(t)$ for $t > 0$ when the switch in the circuit of Problem 8.2–5 is closed again at $t = 0.4$ sec.

8.2–14 Suppose the applied voltage in Example 8.2–2 had been $v(t) = 170 \cos (377t - 18°)$. Show that $i(t) = i_F(t)$.

8.2–15 Find and sketch $i(t)$ when $v(t) = 30e^{-60t}$ V in Example 8.2–2.

8.2–16 Find and sketch the complete response $v(t)$ when the circuit in Fig. 8.2–2 has $R = 5\ k\Omega$, $C = 1\ \mu F$, and $v_s(t) = 40e^{-1000t}$ V for $t > 0$, and there is no initial stored energy.

8.2–17 Repeat the previous problem with $v_s(t) = 40(1 - e^{-1000t})$ V.

8.3–1 Obtain expressions for the roots of $Z(s)$ and $Y(s)$ for the circuit in Fig. P6.2–1 unless you find an unfactored quadratic, in which case simply write the expressions for α and ω_0.

8.3–2 Repeat the previous problem for Fig. P6.2–3.

8.3–3 Write the expression for $i_N(t)$ for the circuit in Fig. P5.2–8 with $C = 50\ \mu F$.

8.3–4 Write the expression for $v_N(t)$ for the circuit in Fig. 7.2–6 with $C = 1\ \mu F$, $R_1 = 1\ k\Omega$, $L = 40$ mH, and $R_2 = 600\ \Omega$.

8.3–5 Repeat Problem 8.3–3 with $C = 5\ \mu F$.

8.3–6 Repeat Problem 8.3–4 with $L = 400$ mH.

8.3–7 Evaluate the initial slopes of all voltages and currents in the circuit of Problem 8.2–2.

8.3–8 Evaluate the initial slopes of all voltages and currents in the circuit of Problem 8.2–1, taking $C = 0.02\ \mu F$ and $L = 0.1$ H.

8.3–9 Find and sketch the step response of the circuit in Fig. 8.3–6 when $R = 30\ \Omega$, $L = 0.01$ H, $C = 50\ \mu F$, and the output is taken as the voltage across the resistance.

8.3–10 Repeat the previous problem with the output taken as the voltage across the inductance.

8.3–11 Repeat the previous problem using the element values in Example 8.3–4, still taking the output as the voltage across the inductance.

‡ **8.3–12** Find the step response of the circuit in Problem 7.1–6. Express your answer in terms of V_s and $\tau = RC$.

‡ **8.3–13** When the energy-storage elements of a second-order network are both capacitors or both inductors, the natural response must be overdamped. Confirm this assertion for the circuit in Fig. P7.2–14 by obtaining the characteristic polynomial of $Z(s)$ and showing that $\alpha^2 - \omega_0^2 > 0$.

PART II

Large-scale integrated circuits contain over 1000 components on a semiconductor chip smaller than a postage stamp. Hundreds of identical chips are fabricated simultaneously on silicon wafers. (Courtesy of Texas Instruments)

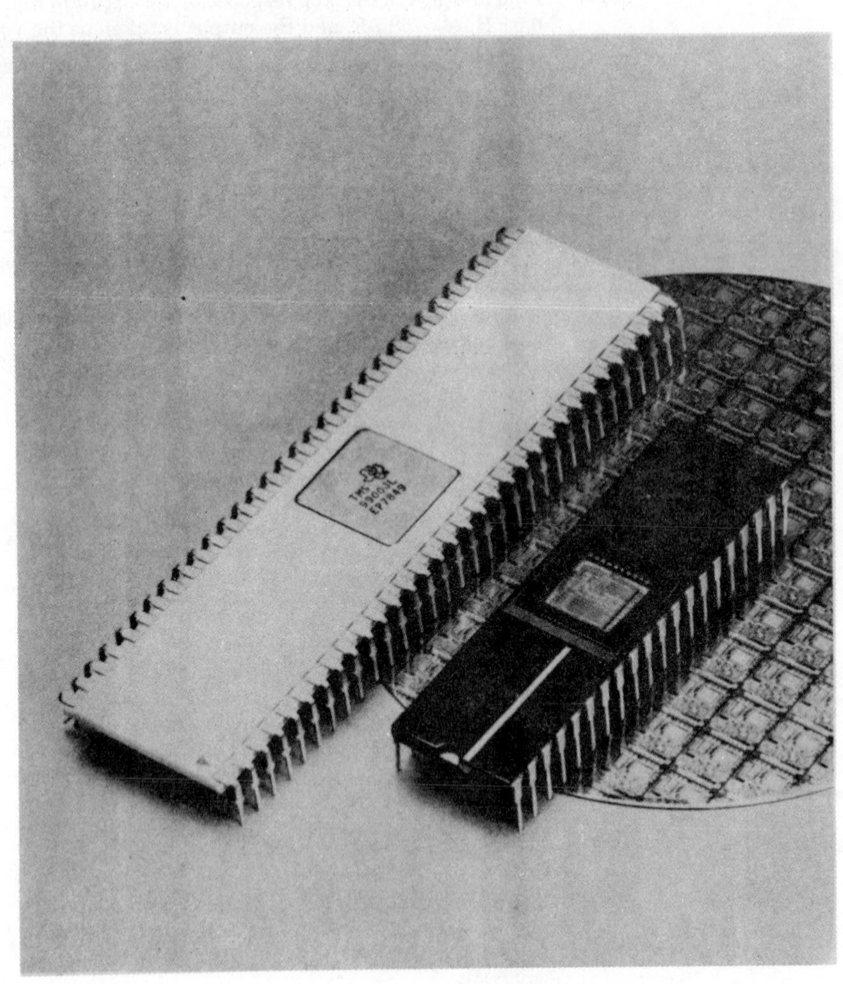

ELECTRONIC DEVICES AND CIRCUITS

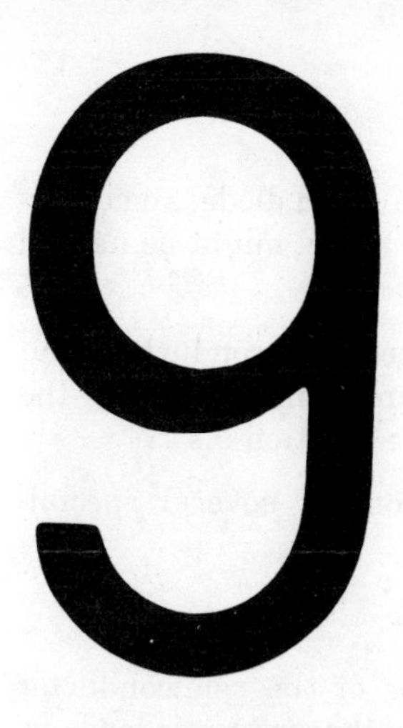

DIODES AND SEMICONDUCTORS

This chapter begins our study of electronic devices and circuits by examining the class of elements called diodes—two-terminal elements that, like most electronic devices, have nonlinear electrical behavior. Nonlinearity complicates circuit analysis and requires new methods of attack. But the richness and variety of practical applications more than compensates for the somewhat increased analytic difficulty.

We start with a description of the external characteristics of a semiconductor diode, and then develop special methods for dealing with nonlinear circuit elements. One method involves an idealized model that can be augmented to represent the behavior of real diodes in various operating states. The chapter closes with a discussion of the physical phenomena underlying diode behavior, and a description of various special-purpose devices.

OBJECTIVES

After studying this chapter and working the exercises, you should be able to:

- Sketch the *i-v* curves for a semiconductor diode, ideal diode, and Zener diode (Sections 9.1 and 9.2).
- Use a load line to find the operating point of a nonlinear element, given its *i-v* curve (Section 9.1).

- Develop a simple piecewise-linear model for a given diode, and determine under what conditions a more complex model might be needed (Section 9.2).
- Calculate the resistivity of a pure or doped semiconductor, and identify the current components in a *pn* junction, explaining the operating principles of a semiconductor diode (Section 9.3). †
- Describe the characteristics and applications of several special-purpose devices (Section 9.4). †

9.1 DIODE CONCEPTS

This section describes the external behavior of the semiconductor diode in terms of its *i-v* curve. The *nonlinear* characteristic leads us to two new methods for nonlinear circuit analysis: a graphical or numerical approach based on the concept of a load line, and a piecewise-linear method based on the concept of the ideal diode.

THE IDEAL DIODE

The gross behavior of a diode is like an electrical *one-way valve:* it conducts current much more readily in one direction, called the *forward direction,* than it does in the *reverse direction.* Most semiconductor diodes conduct so well in the forward direction that they almost seem to be short circuits, while so little current flows in the reverse direction that they seem to be open circuits. We can put this another way: A diode acts like a *switch* that closes for current flow in the forward direction, but opens up to prevent current flow in the reverse direction.

The *ideal diode* symbolized by Fig. 9.1–1a is a fictitious device having precisely these characteristics. With a negative applied voltage, the diode is OFF (like an open switch) and no current flows as in Fig. 9.1–1b. With a slightly positive applied voltage, say $v = 0^+$, the diode is ON (like a closed switch) and any amount of current may flow in the forward direction as in Fig. 9.1–1c. The value of i, in this latter case, is dictated by the external circuitry; and one should usually put some resistance in series with the diode to prevent excessive current.

We summarize these characteristics of the ideal diode by plotting its *i-v* curve in Fig. 9.1–1d, which consists of two straight lines corresponding to the two states ON and OFF. Note that the arrow-like device symbol indicates the allowed direction of current flow $i \geq 0$ when $v = 0^+$, which we hereafter take as $v = 0$. Also notice that the short-circuit condition of the ON state makes it impossible, theoretically, to apply any positive voltage $v > 0$ to the diode. Despite its obviously far-fetched behavior, the ideal diode turns out to be a reasonable model for real diodes under certain conditions—as you'll see when we examine the characteristics of real diodes.

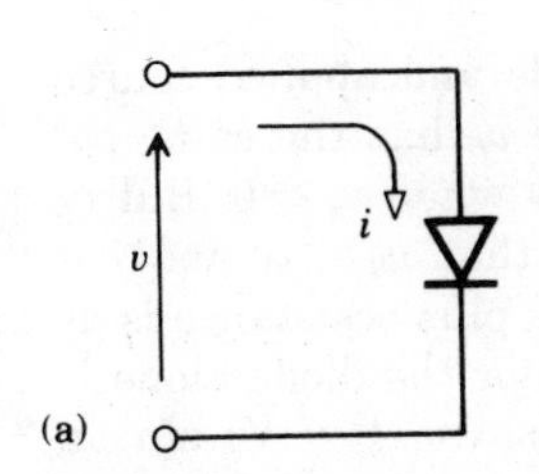

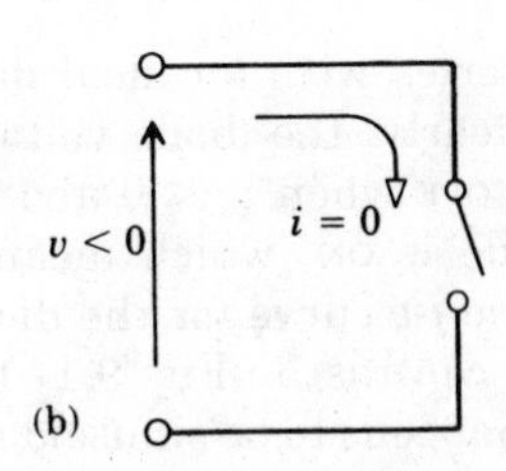

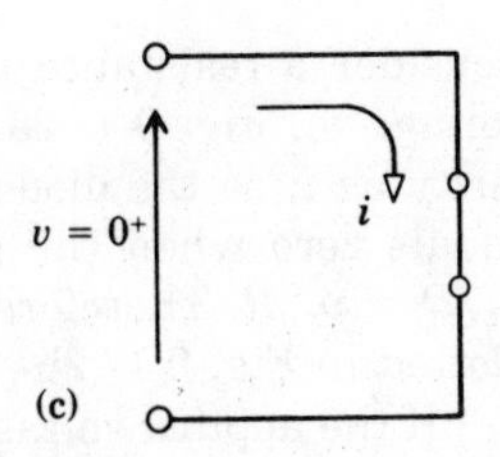

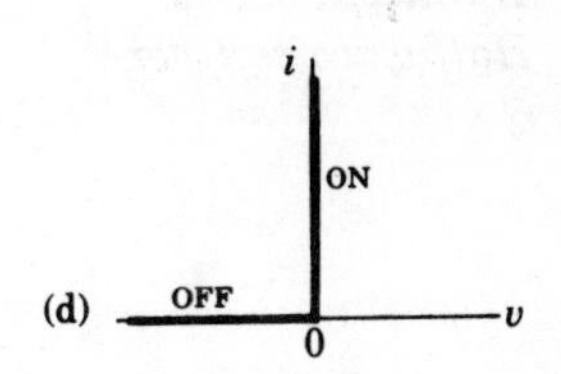

FIGURE 9.1–1
Ideal diode, equivalent circuits for the OFF and ON states, and the *i-v* curve.

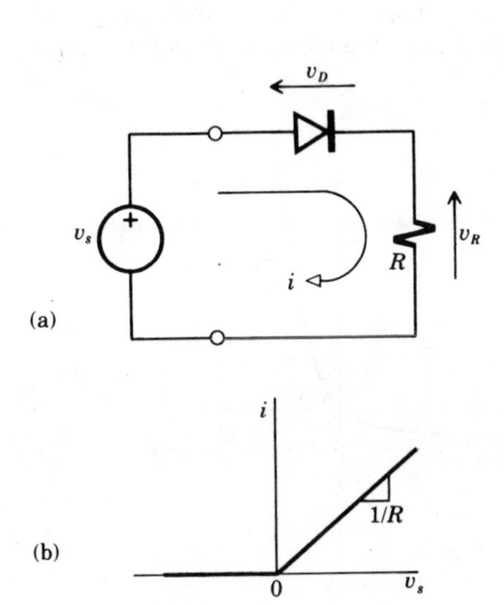

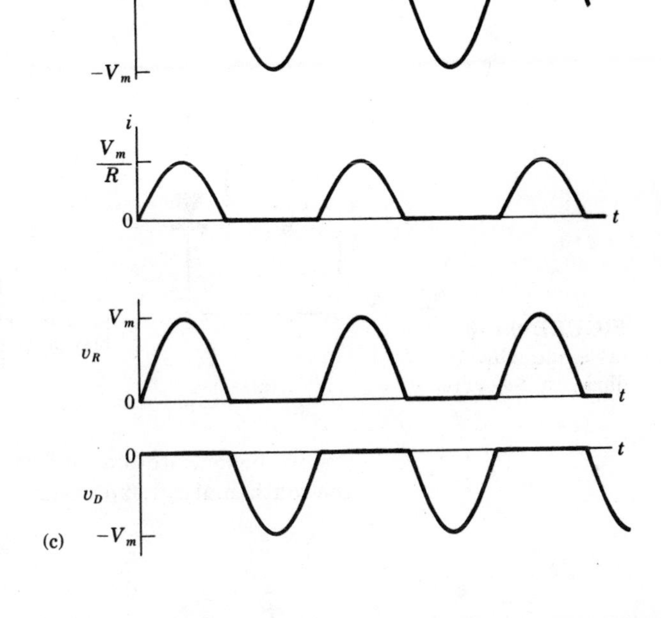

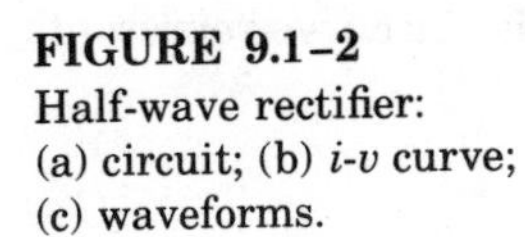
FIGURE 9.1–2
Half-wave rectifier:
(a) circuit; (b) *i-v* curve;
(c) waveforms.

Example 9.1–1
Half-wave rectifier

Consider a resistance in series with an ideal diode and applied source voltage v_s, Fig. 9.1–2a. Clearly, the diode voltage v_D has the same polarity as v_s so the diode is OFF when $v_s < 0$ and ON when $v_s > 0$. But v_D equals zero when the diode is ON, which means that $v_R = v_s$ and $i = v_R/R = v_s/R$. Therefore, the *i-v* curve for the diode plus resistance is as plotted in Fig. 9.1–2b—in contrast to Fig. 9.1–1d for the diode alone.

If the applied voltage happens to be sinusoidal, say $v_s(t) = V_m \sin \omega t$, current flows only during the half cycles when $v_s(t) > 0$ so $v_D(t) = 0$, $v_R(t) = v_s(t)$, and $i(t) = (V_m/R) \sin \omega t$. Otherwise, $i(t)$ equals zero during the negative half cycles of $v_s(t)$, so $v_R(t) = 0$ and $v_D(t) = v(t)$. The various waveforms are then as shown in Fig. 9.1–2c. We call this circuit a *half-wave rectifier* because it produces unidirectional or nonalternating current from an AC source. Rectification is the first step in the important task of AC-to-DC conversion for electronic power supplies, etc.

Exercise 9.1–1

Repeat the above example with the diode's terminals reversed.

SEMICONDUCTOR DIODES

Figure 9.1–3 gives the circuit symbol and *i-v* curve for a typical semiconductor diode. The *i-v* curve has three distinct regions marked *forward bias, reverse bias,* and *reverse breakdown,* depending on the value of *v*. Under normal operating conditions we keep the diode away from the reverse-breakdown region, and further discussion of that region is deferred to the next section.

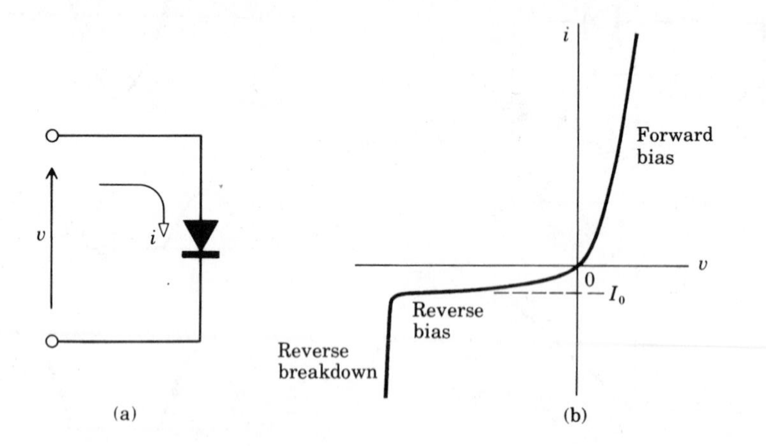

FIGURE 9.1–3 (a) Semiconductor diode. (b) *i-v* curve.

In the forward-bias and reverse-bias regions, the *i-v* curve represents the mathematical expression

$$i = I_0 (e^{\lambda v} - 1) \tag{1a}$$

in which

$$\lambda = \frac{|q_e|}{\eta kT} = \frac{1.17 \times 10^4}{\eta T} \tag{1b}$$

where $|q_e|$ is the magnitude of the electronic charge, k is Boltzmann's constant, and T is the temperature in degrees kelvin. The factor η (eta) in Eq. (1b) depends on the type of semiconductor used; but, for most purposes, can be taken to have unity value. Letting $\eta = 1$ and $T = 300$ K, which corresponds to *room temperature,* we get

$$\lambda \approx 40 \text{ V}^{-1} = \frac{1}{25 \text{ mV}}$$

and

$$i = I_0(e^{40v} - 1) \tag{2}$$

which is the same expression given for the photodiode in Sec. 2.2 when $I_p = 0$ (no incident light). The constant I_0 is called the *reverse saturation current* since $i \approx -I_0$ when $\lambda v \leq -5$ or $v \leq -0.1$ V at room temperature. The value of I_0 depends on the particular diode and ranges from a few picoamps (10^{-12} A) to a fraction of a milliamp.

The curve in Fig. 9.1–3 certainly does not look very much like that of an ideal diode—even if we ignore reverse breakdown—because the current and voltage scales were chosen to bring out the difference. Suppose, however, we put some numbers into Eq. (2) and replot it on different scales. Taking $I_0 = 1$ nA $= 10^{-9}$ A, for instance, yields Fig. 9.1–4a which still has the shape of Fig. 9.1–3 but which shows the truly minute values of voltage and current. With larger voltages and currents included, that same curve appears as shown in Fig. 9.1–4b, where we no longer see the reverse current due to the fact that $I_0 \ll 10\ \mu$A. Going even further in this direction yields Fig. 9.1–4c which now looks much more like an ideal diode although it still represents $i = I_0(e^{40v} - 1)$.

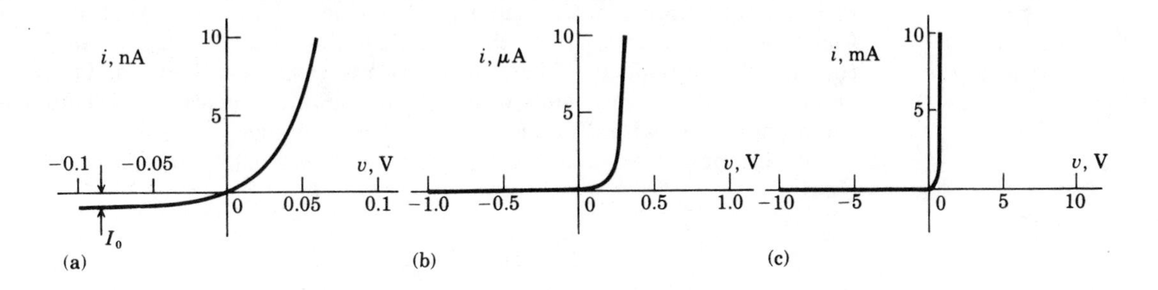

FIGURE 9.1–4 Semiconductor diode curves plotted on different scales (omitting reverse breakdown).

We have through this process arrived at a significant conclusion: When the voltages in a diode circuit exceed a few volts and the currents are very large compared to the reverse saturation current, the diode's behavior differs only slightly from that of an ideal diode. The impact of this conclusion will be brought home when we examine a simple diode circuit and contrast the amount of labor required for its exact analysis with that required for an approximate analysis using the ideal diode instead of a real diode.

NONLINEAR ELEMENTS AND LOAD-LINE ANALYSIS

Consider the apparently trivial circuit in Fig. 9.1–5 where the unidentified element could be a real diode or any other *nonlinear* device. By this we mean that the *i-v* curve is not a straight line through the origin. Mathematically, the *i-v* equation for a nonlinear element has the form

$$i = f(v) \tag{3a}$$

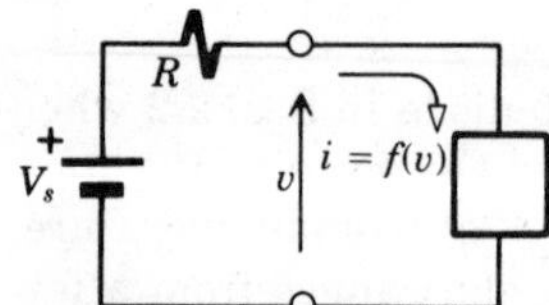

FIGURE 9.1–5 Circuit with a nonlinear device.

with $f(v)$ being a nonlinear function of v in the sense that

$$f(v' + v'') \neq f(v') + f(v'') \tag{3b}$$

when $v = v' + v''$. Real diodes clearly fall in this category, judging from the curvature of Fig. 9.1–3b. An obvious consequence of nonlinearity is that *superposition does not hold;* less apparent is the complication it brings to circuit analysis.

Despite the presence of a nonlinear element, Kirchhoff's laws still apply to Fig. 9.1–5 for they describe connection relationships and are independent of the types of elements. Furthermore, linear resistance continues to obey Ohm's law. Therefore, applying KVL and Ohm's law to the left-hand portion of the circuit gives

$$i = \frac{V_s - v}{R} \tag{4}$$

called the *load-line equation.* This equation describes the battery-plus-resistance, regardless of the load element connected in series with them. Conversely, $i = f(v)$ describes the nonlinear device itself, regardless of the rest of the circuit. Both equations have two variables, i and v; and the values of i and v in the complete circuit must simultaneously satisfy both equations—otherwise KVL or KCL would be violated.

Accordingly, we equate the right-hand sides of Eq. (3a) and Eq. (4) to obtain

$$\frac{V_s - v}{R} = f(v) \tag{5}$$

whose solution $v = v_Q$ is the *operating voltage* for the particular device in this particular circuit. The corresponding operating current will be $i_Q = f(v_Q) = (V - v_Q)/R$. Because Eq. (5) is a nonlinear equation, we must use numerical or graphical methods to obtain the values of v_Q and i_Q.

To carry out the *graphical solution* we start with a plot of $i = f(v)$—the device's *i-v* curve—and superimpose the load line on the same set of axes, as illustrated in Fig. 9.1–6. The load line is easily drawn by connecting the points representing the *open-circuit voltage* V_s (at $i = 0$) and the *short-circuit current* V_s/R (at $v = 0$). The *intersection* of the load line and device curve is then the *operating point* Q, since it is the point where $f(v)$ equals $(V_s - v)/R$ and thus is the solution of Eq. (5). The graphical method has merit because it works even when the device is described by an experimental *i-v* curve instead of a theoretically derived functional expression. On the other hand, graphical analysis may require some rather tedious plotting and has limited accuracy.

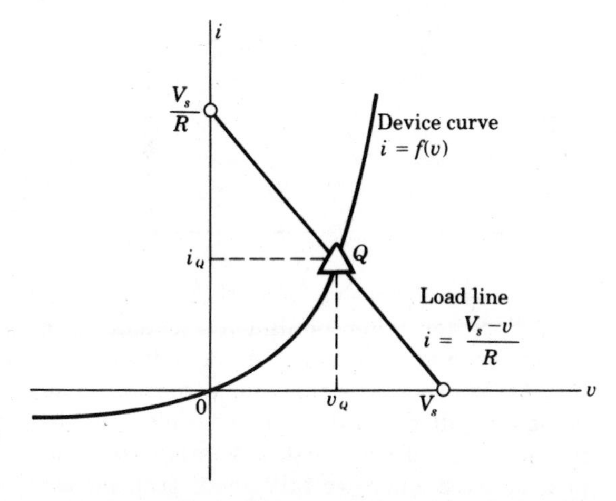

FIGURE 9.1–6
Load-line analysis showing operating point Q.

A more accurate, more modern approach would be a *numerical solution* arrived at by calculator or digital computer. To set up the computation, we rewrite Eq. (5) in the form

$$f(v) - \frac{V_s - v}{R} = 0$$

and use a *root-finding algorithm* such as bisection or Newton's method. Of course, we must have an appropriate expression for $f(v)$.

By way of example, suppose the device in question is a real diode with $I_0 = 1\text{ nA} = 10^{-6}\text{ mA}$ and suppose that $V_s = 10\text{ V}$ and $R = 2\text{ k}\Omega$ so

$V_s/R = 5$ mA. Equation (5) then becomes

$$5 \text{ mA} - \frac{v}{2 \text{ k}\Omega} = 10^{-6} \text{ mA } (e^{40v} - 1)$$

whose numerical solution yields $v_Q = 0.38464$ V, from which $i_Q = 5 - (0.38464/2) = 4.80768$ mA. Alternatively, drawing the load line on Fig. 9.1–4c gives $v_Q \approx 0.4$ V and $i_Q \approx 4.8$ mA.

But what do we get if we assume an *ideal* diode in this circuit? We get a much simpler problem whose results are not very far from the mark. Specifically, from the diagram in Fig. 9.1–7 we conclude that current flows in the forward direction, so $v_Q = 0$ V and $i_Q = 10$ V/2 kΩ = 5 mA. The latter is off by about four percent, and the small but nonzero voltage drop across the real diode may or may not be significant, depending on the application.

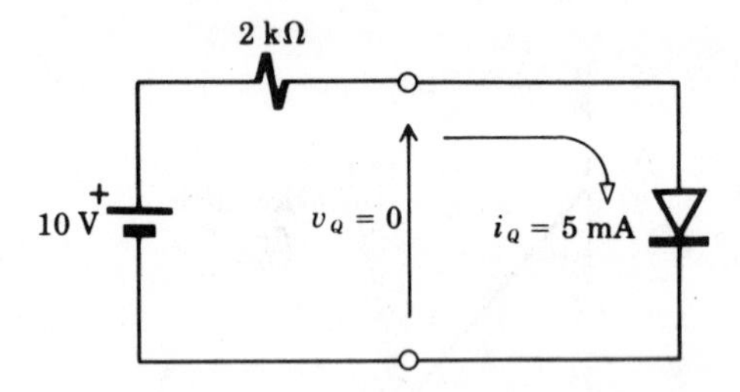

FIGURE 9.1–7 Circuit with an ideal diode.

In summary, we've found that analyzing even a simple circuit with a real diode (or other nonlinear element) is a messy problem, at best, whereas the assumption of an ideal diode leads to quick but less accurate results. We therefore turn our attention to developing improved models using ideal diodes and other linear elements to represent more accurately the behavior of real diodes. The load-line concept will reappear in the next chapter when we talk about transistors; this concept also helps in the study of waveform distortion caused by nonlinearity, as illustrated in the following example.

Example 9.1–2

Suppose the battery in Fig. 9.1–5 is replaced by a sinusoidal source $v_s(t) = 0.1 \sin \omega t$ V, and let the rest of the circuit be a 10-MΩ resistance and a semiconductor diode with $I_0 = 1$ nA. We can find the time-varying current $i(t)$ by drawing several load lines corresponding to different values of $v_s(t)$. Four such load lines are shown in Fig. 9.1–8a, representing $v_s/R = \pm 10$ nA when $v_s = \pm 0.1$ V (that is, when $\sin \omega t = \pm 1$) and $v_s/R = \pm 5$ nA when $\sin \omega t = \pm 0.5$ so $v_s = \pm 0.05$ V. Reading off the values of i at the intersection points and plotting them versus time gives the waveform $i(t)$ in Fig. 9.1–8b. Figure 9.1–8b also includes a sinusoid with

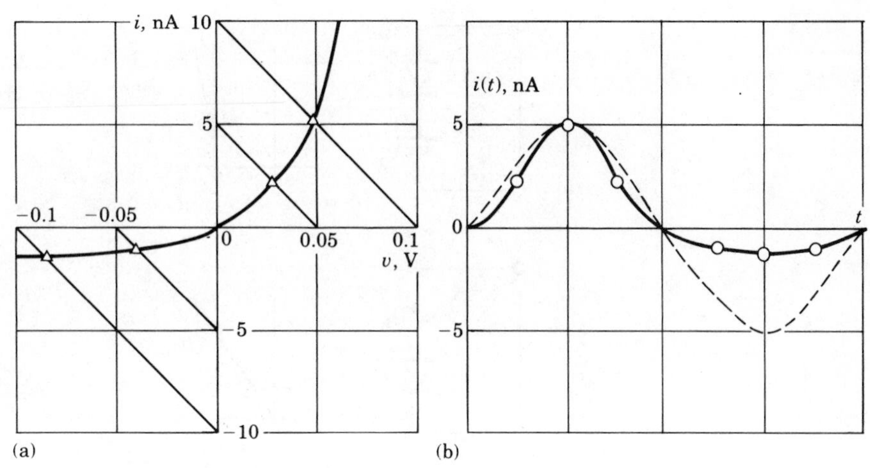

FIGURE 9.1–8
Load lines and waveform for Example 9.1–2.

the same positive peak as $i(t)$, which brings out the fact that the current has neither a sinusoidal shape nor a half-wave rectified shape like that of Fig. 9.1–2c.

Exercise 9.1–2

The *i-v* relationship for a *vacuum-tube diode* is given approximately by the "$\frac{3}{2}$ power law,"

$$i = \begin{cases} Kv^{3/2} & v \geq 0 \\ 0 & v < 0 \end{cases}$$

Plot this curve for $-100 \text{ V} \leq V \leq 100 \text{ V}$, taking $K = 10^{-3}$, and find v_Q and i_Q when $V_s = 100$ V and $R = 100\ \Omega$. Then sketch $i(t)$ when $v_s(t) = 100 \sin \omega t$ V. (Save your plot for use in Exercise 9.2–1.)

9.2 DIODE STATES AND MODELS

This section develops circuit models that represent the behavior of a real diode in various operating states. The basic strategy of doing so is to approximate the smooth but nonlinear *i-v* curve by a *piecewise-linear* curve consisting of straight line segments. Each straight line segment then corresponds to a linear circuit, which simplifies the analysis or design process. Linearization also allows us to invoke superposition for the purpose of deriving a small-signal model.

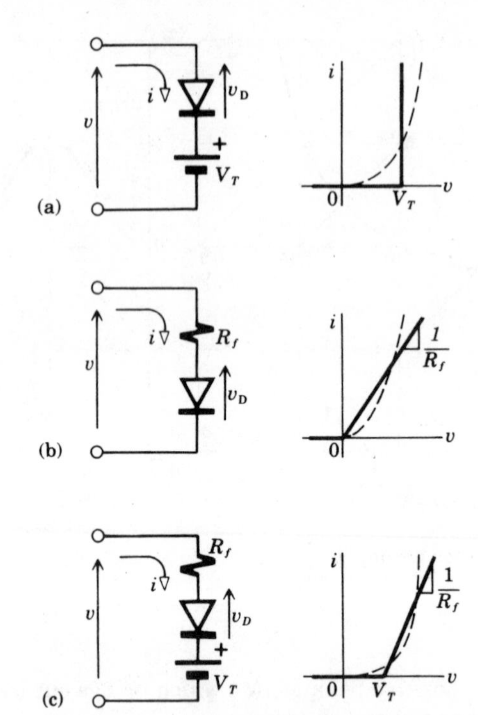

FIGURE 9.2–1 Forward-biased diode models: (a) with threshold voltage V_T; (b) with forward resistance R_f; (c) with threshold voltage and forward resistance.

FORWARD-BIAS MODEL

The most apparent difference between a real and ideal diode is the non-zero voltage drop when a real diode conducts in the forward direction. Figure 9.2–1 shows three different ways of modeling this feature using an ideal diode. The dashed curves represent the actual diode in question. The linearized *i*-*v* curves follow directly from the circuit models because the ideal diode acts like a switch and its voltage v_D can only be negative (when $i = 0$) or zero (when $i > 0$).

The battery in Fig. 9.2–1a has polarity such that $v_D = v - V_T$. The ideal diode is, therefore, OFF when $v < V_T$. The diode turns ON when $v = V_T$, at which time any amount of current may then flow: the *i*-*v* curve, then, looks like that of an ideal diode offset to the right by V_T volts. Similarly, the resistance in Fig. 9.2–1b has no effect when $v < 0$ for $v_D < 0$ and $i = 0$; but with $v > 0$, the ideal diode is ON, $v_D = 0$, and $i = v/R_f$. This *i*-*v* curve, therefore, has slope $1/R_f$ for $v > 0$. The parameters V_T and R_f are known as the *threshold voltage* and *forward resistance,* respectively.

Combining forward resistance and threshold voltage in one circuit leads to Fig. 9.2–1c as a more accurate model of a real diode and as a circuit of practical interest in its own right. An easy way to analyze such circuits is to locate the *break points.* These are the coordinates of the *i*-*v* curve where a sudden change in slope occurs because an ideal diode has changed state (OFF or ON). At a break point, both the current and voltage in an ideal diode equal zero. In the present case, zero current means zero voltage drop across R_f. Simultaneously, zero diode voltage requires that $v = V_T$. The break point is therefore at $i = 0$, $v = V_T$. Above this point, any voltage increment $v - V_T$ produces a current increment $(v - V_T)/R_f$. On the other hand, voltages less than V_T produce no current at all, since the diode is OFF.

The combination of an ideal diode, forward resistance, and threshold voltage yields a quite satisfactory model for most real diodes operating in the forward-bias state. This model also holds for the reverse-bias state when the reverse saturation current I_0 is negligibly small, as is often true in practice. The value of V_T roughly corresponds to the "knee" of the actual i-v curve; and for a typical silicon diode, one usually takes

$$V_T \approx 0.7 \text{ V} \tag{1}$$

The value of the forward resistance R_f generally falls in the range 5–50 Ω. This small resistance may also be ignored in many practical circuits, thereby reducing our model to a simple diode plus battery, as in Fig. 9.2–1a.

Example 9.2–1

Consider a half-wave rectifier (as in Fig. 9.1–2a), which has peak AC source voltage $V_m = 3$ V, load resistance $R = 100\ \Omega$, and a real diode modeled by Fig. 9.2–1c with $V_T = 0.7$ V and $R_f = 10\ \Omega$. When the voltage v across the diode is greater than V_T, the ideal diode in the model is ON, and we have the equivalent circuit of Fig. 9.2–2a. Thus

$$v_R = \frac{100\ \Omega}{110\ \Omega}(v_s - 0.7 \text{ V})$$

whose maximum value is 2.1 V when $v_s = 3$ V. The condition $v \geq 0.7$ V corresponds to $v_s \geq 0.7$ V, obtained by solving the circuit for v_s with the break-point values $v = 0.7$ and $i = 0$.

When $v_s < 0.7$ V, our model predicts zero current flow, since the ideal diode will be OFF. The real diode would actually have $|i| \leq I_0$ in the reverse-bias state. If $I_0 = 0.1\ \mu$A, for instance, then $|v_R| \leq 10^{-5}$ V, and we are justified in ignoring the reverse current.

Figure 9.2–2b shows the resulting waveform $v_R(t)$ along with $v_s(t)$. Comparing with Fig. 9.1–2c, we see that the real diode reduces the maximum value of $v_R(t)$ and increases its OFF time. With a larger AC voltage, say $V_m > 10$ V, those differences would be inconsequential.

Exercise 9.2–1

Suppose the vacuum-tube diode in Exercise 9.1–2 is operated in a circuit where $40 \leq v \leq 80$ V. Find values for R_f and V_T so that the diode model in Fig. 9.2–1c yields a reasonable approximation of the i-v curve over the

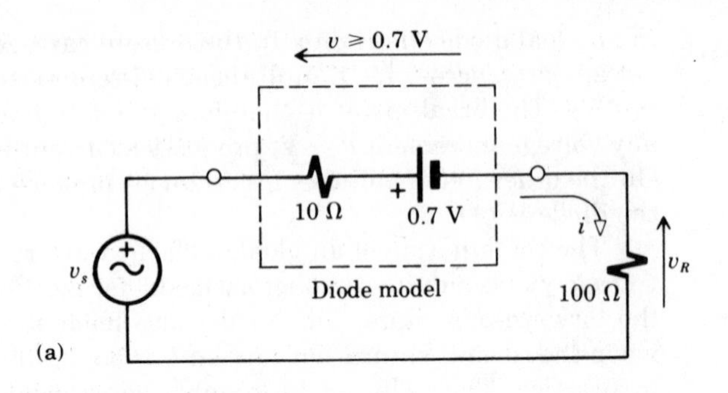

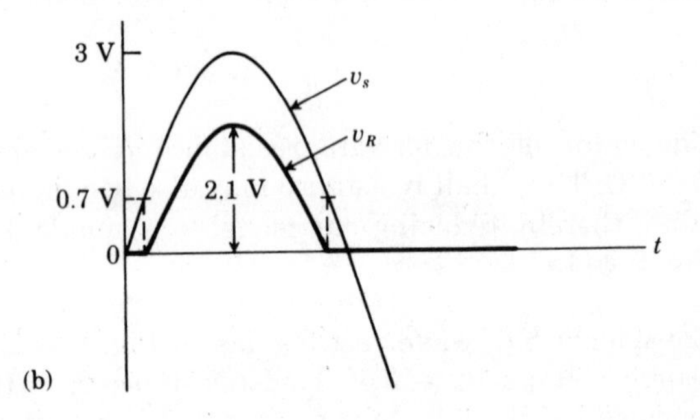

FIGURE 9.2–2
Circuit and waveforms for Example 9.2–1.

operating voltage range. Then use your model to estimate the value of i when v = 60 V.

Exercise 9.2–2

The circuit in Fig. 9.2–3 is one way of modeling the reverse-bias state and the forward-bias state together. Show that the break point of the piecewise i-v curve is at $i = -I_0$, $v = V_T - R_f I_0$. Then sketch and label the curve, taking $V_T > R_f I_0$.

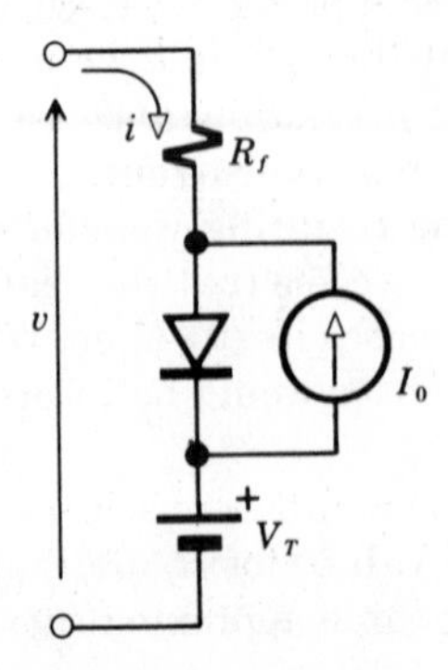

FIGURE 9.2–3
Combined forward and reverse bias model.

REVERSE BREAKDOWN AND ZENER DIODES

Applying too much reverse voltage to any real diode causes it to break down and conduct heavily in the reverse direction (see Fig. 9.1–3). The voltage across the diode then stays nearly constant over a wide current range. *Zener diodes* are special-purpose devices designed expressly to utilize this breakdown effect for the purpose of establishing a reference voltage.

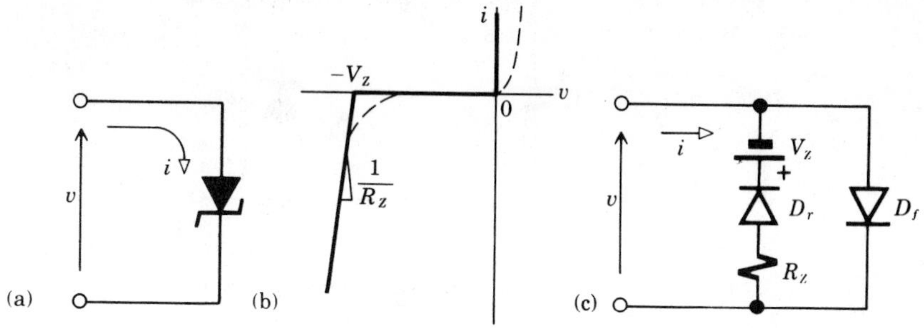

FIGURE 9.2–4
Zener diode: (a) symbol; (b) linearized *i-v* curve; (c) circuit model.

The device symbol, linearized *i-v* curve, and circuit model are given in Fig. 9.2–4. The *i-v* curve indicates that a Zener diode approximates an ideal diode for $v > -V_Z$, called the Zener voltage. But when the reverse bias exceeds V_Z (that is, $v < -V_Z$), the diode conducts in the reverse direction and acts like a small reverse resistance R_Z in series with a battery V_Z. The two branches of the circuit model reflect these two operating characteristics. With appropriate values of R_Z and V_Z, this model could represent any real diode in its reverse-breakdown region.

Figure 9.2–5a diagrams a typical Zener-diode circuit. You should watch out for the fact that the Zener is upside down here compared to Fig. 9.2–4, the inverted connection being the key to such circuits for it puts the diode in its breakdown region if $V_s > V_Z$. Assuming the latter holds, diode D_f in our model is OFF while diode D_r is ON, and we have the simplified equivalent circuit of Fig. 9.2–5b. Straightforward analysis then shows that

$$v_{\text{out}} \approx V_Z \tag{2a}$$

providing that

$$V_Z < V_s \ll \left(\frac{R}{R_Z} + 2\right) V_Z \tag{2b}$$

Under these conditions the Zener establishes a *fixed reference voltage* $v_{out} \approx V_Z$ regardless of the value of V_s. Zener diodes are available with values of V_Z from about 2 V to 200 V with tolerances of $\pm 5\%$ or better.

Exercise 9.2–3

Calculate v_{out} in Fig. 9.2–5 when $V_s = 12$ V, $R = 500\ \Omega$, $R_Z = 20\ \Omega$, and $V_Z = 9$ V. Repeat with $V_s = 18$ V.

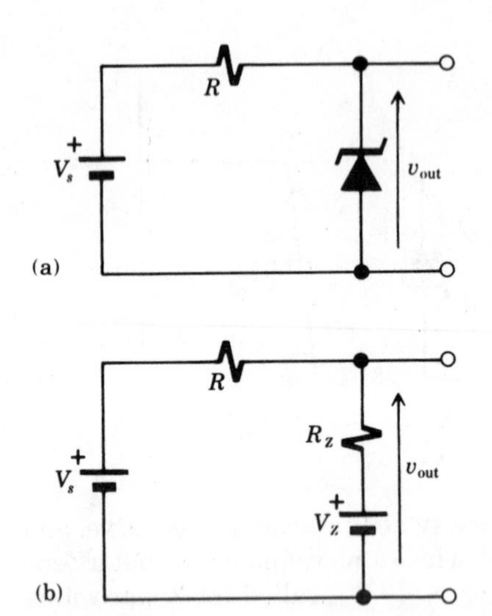

FIGURE 9.2–5
(a) Zener diode circuit.
(b) Equivalent circuit.

SMALL-SIGNAL MODEL

Another type of diode model relates to circuits like Fig. 9.2–6a where the applied voltage consists of a DC component V_{DC} plus a small time-varying signal $v_s(t)$. The DC voltage forward biases the diode at some operating point Q, while $v_s(t)$ produces small voltage and current variations such that

$$v = V_Q + v_d(t) \qquad i = I_Q + i_d(t)$$

Clearly, V_Q and I_Q are the operating-point values due to V_{DC} alone—that is, when $v_s(t) = 0$—found from the usual load-line analysis in Fig. 9.2–6b. Therefore, if we linearize the diode's i-v curve at the operating point, we can invoke the superposition principle and develop a *small-signal model* relating the variations $v_d(t)$ and $i_d(t)$ due to $v_s(t)$.

For this purpose, draw a straight line of slope $1/r_d$ tangent to the curve at Q, as shown. We call r_d the *dynamic* or *small-signal resistance* of the diode, in view of its property that

$$i_d \approx \frac{1}{r_d} v_d$$

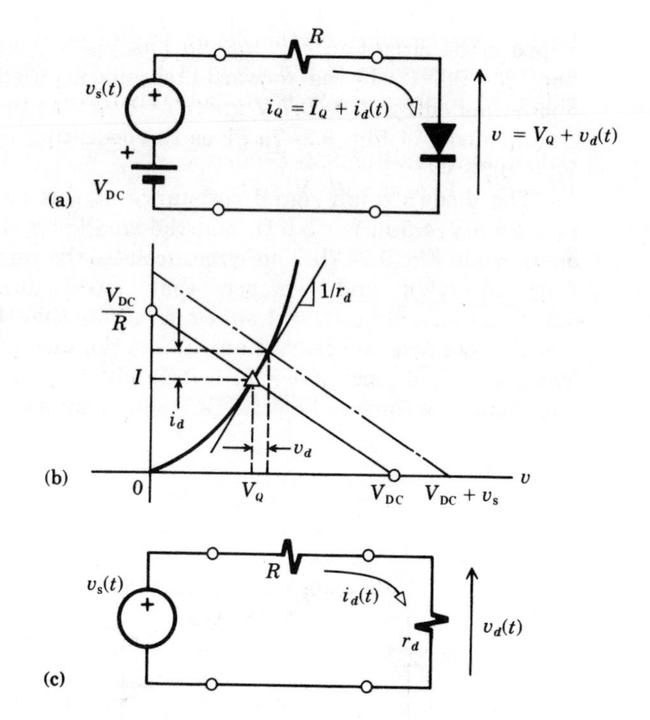

FIGURE 9.2–6 (a) Diode circuit with DC bias and small signal. (b) Load-line analysis. (c) Small-signal equivalent circuit.

for small variations about the operating point. Accordingly, Fig. 9.2–6c becomes the small-signal equivalent circuit, the diode simply being replaced by its small-signal resistance.

Be careful not to confuse r_d with the resistance R_f in the forward-bias model (Fig. 9.2–1). The forward-bias model approximates the diode's *entire i-v* curve, whereas the small-signal resistance represents the behavior around a specific *operating point* and, in fact, the value of r_d depends on the point in question. To bring out that dependence, we start with the definition of the slope of the tangent line at Q, namely

$$\frac{1}{r_d} \triangleq \left.\frac{di}{dv}\right|_{v=V_Q}$$

which means that we let $v = V_Q$ after we have differentiated. Since $i = I_0(e^{40v} - 1)$ at room temperature, $di/dv = 40I_0e^{40v}$ and

$$r_d \approx \frac{25\ mV}{I_Q} \tag{3}$$

assuming $V_Q \geq 0.05$ V so that $I_Q \approx I_0 e^{40V_Q}$. A more complete small-signal model would include a parallel capacitance, discussed in Section 9.4.

Example 9.2–2

Suppose the circuit in Fig. 9.2–6a has $V_{DC} = 3$ V, $v_s(t) = 0.1 \sin \omega t$ V, and $R = 500\ \Omega$. If the forward-bias characteristics of the diode are approximated by $V_T = 0.7$ V and $R_f = 10\ \Omega$ (as in Fig. 9.2–2a), the DC circuit model of Fig. 9.2–7a gives the operating point $I_Q \approx 4.5$ mA and $V_Q \approx 0.75$ V.

The diode's small-signal resistance at this operating point will be $r_d = 25\ \text{mV}/4.5\ \text{mA} = 5.6\ \Omega$, and the small-signal equivalent circuit is as shown in Fig. 9.2–7b. This circuit relates the small variations $i_d(t)$ and $v_d(t)$ to $v_s(t)$, and we see that $i_d(t) = (0.1 \sin \omega t\ \text{V})/505.6\ \Omega \approx 0.2 \sin \omega t$ mA and $v_d(t) \approx 1 \sin \omega t$ mV. Note that these sinusoids are *not* rectified because the DC voltage biases the diode to be ON all the time. Thus, the total current $i = I_Q + i_d(t)$ will consist of a small sinusoidal variation superimposed on the DC component as in Fig. 9.2–7c.

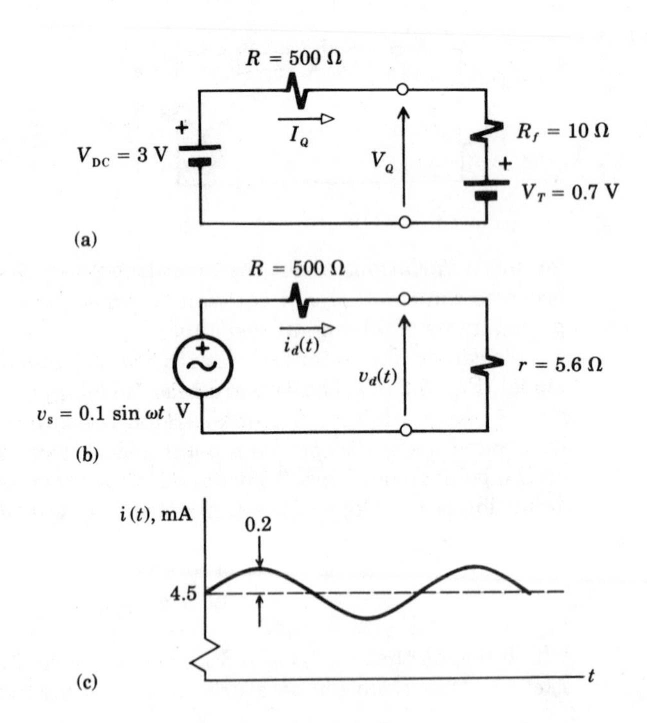

FIGURE 9.2–7 Circuits and waveform for Example 9.2–2.

Exercise 9.2–4

The small-signal equivalent circuit in Fig. 9.2–6c predicts that $i_d(t) = v_s(t)/(R + r_d)$. Confirm this prediction directly from Fig. 9.2–6a using the facts that $V_Q = V_{DC} - RI_Q$ and $v_d(t) \approx r_d i_d(t)$.

9.3 SEMICONDUCTORS AND *pn* JUNCTIONS †

Having described and modeled the *i-v* characteristics of a semiconductor diode, we should give some consideration to the underlying physical phenomena that account for the external behavior. This section takes a brief look at what goes on inside a piece of current-carrying semiconductor and describes how diodes are created by forming a semiconductor junction.

SEMICONDUCTORS

Semiconductors are materials that, in pure crystalline form, have no free electrons at absolute zero temperature. The important ones come from the fourth column of the periodic table of the elements—namely germanium and silicon—and crystallize in a lattice where each atom has four nearest neighbors and shares its four valence electrons with them, one electron to each neighbor. Although the actual lattice structure is three dimensional, the two-dimensional representation of Fig. 9.3–1 very nicely shows an approximate cross section of the crystal. The shared electrons constitute covalent bonds, represented by pairs of lines. The label "+4" on the atoms means a charge of $+4\,|q_e|$ without the valence electrons. The net electric charge is, of course, zero.

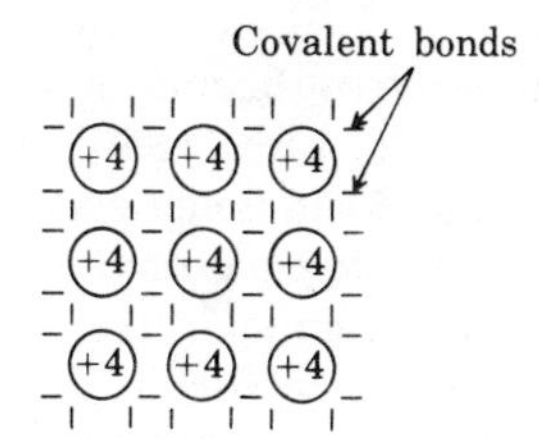

FIGURE 9.3–1 Semiconductor crystal lattice with covalent bonds.

As the temperature increases, the atoms vibrate more and more energetically about their average positions. The valence electrons also gain energy. At room temperature, some electrons with higher than normal kinetic energy manage to break loose completely from their bonds and fly off as *free electrons*. Conduction may now occur thanks to the presence of this somewhat limited number of free electrons.

There is also another equally important but less obvious mechanism for conduction. When a free electron moves off, it leaves a single electron instead of a pair orbiting around the two parents. The net charge in that vicinity is now positive. Suppose that the electron marked a in Fig. 9.3–2 has become a free electron and disappears from the picture. With an overall electric field $\mathcal{E}$ applied, all orbits are distorted and bulge slightly to the right. Suddenly, electron b leaves its job of bonding the horizontal pair and replaces the vacancy at a. Since b is now a vacant position, the

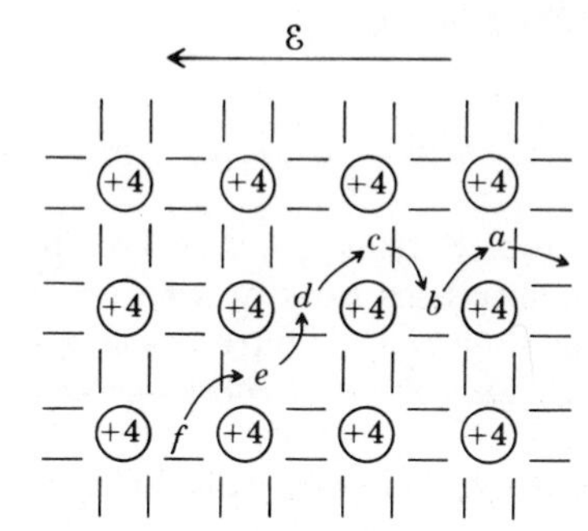

FIGURE 9.3–2 Transfer of bound electrons from left to right—equivalent of a positive hole moving from right to left.

electron at c quickly moves to fill it. Then d fills the c vacancy, and so on. It's easy to see that, on the average, electrons are moving from left to right, even though they are not free in the usual sense.

It's more convenient, however, to think of a missing orbital electron as a positive charge at its former location rather than as an absence of a negative charge. We call these fictitious positive charges *holes*, for that's what they really are; and they play a major role in the electrical properties of semiconductors. To clarify, observe from Fig. 9.3–2 that the hole moves from right to left, in the general direction of the field $\mathcal{E}$, whereas the escaped negatively-charged electrons go off to the right. Therefore, both holes and electrons contribute to conduction, although the holes have less mobility due to the covalent bonding.

With or without an applied field, holes exhibit random thermal motion and, from time to time, a hole and an electron in the same vicinity will *recombine*—the electron becoming bound again, and the hole disappearing. But other electron-hole pairs are continually being *generated* by thermal ionizations, and equilibrium conditions require equal rates of generation and recombination. For a pure or *intrinsic* semiconductor, the equilibrium density of free electrons is called the *intrinsic concentration*, denoted by n_i and expressed in electrons per unit volume (m^{-3}). The equilibrium hole concentration must also equal n_i because generation and recombination always involve *electron-hole pairs*. The value of n_i for silicon at moderate temperatures is given by

$$n_i = 5 \times 10^{21} T^{3/2} e^{-6380/T} \tag{1}$$

or about 1.5×10^{16} m^{-3} when $T = 300$ K. Since silicon has roughly 10^{28} atoms/m^3, we conclude that only a minute fraction of the atoms will be ionized at any one time at room temperature. (Incidentally, silicon happens to be the most common element used in semiconductor devices.)

Although the hole and electron concentrations are equal, the two types of mobile charges do not contribute equally to current flow because the hole mobility μ_p is less than the electron mobility μ_e. Thus, we write the intrinsic resistivity as

$$\rho_i = \frac{1}{|q_e|\, n_i \mu_e + |q_e|\, n_i \mu_p} = \frac{1}{|q_e|\, n_i (\mu_e + \mu_p)} \tag{2}$$

by extension of Eq. (6b), Section 2.3. Referring to Eq. (1) then reveals that n_i increases and ρ_i decreases rapidly with increasing temperature. As a result, the resistance of a piece of pure silicon near room temperature has a temperature coefficient of about -8% per kelvin.

Example 9.3–1

Near room temperature, the mobilities for silicon are

$$\mu_e = 0.13 \qquad \mu_p = 0.05$$

in SI units. Let's find the resistance from end to end of a bar of length ℓ = 10 mm and area A = 1 mm^2. From Eq. (2), the resistivity is $\rho_i = [1.6 \times 10^{-19} \times 1.5 \times 10^{16}\ (0.13 + 0.05)]^{-1} = 2300\ \Omega \cdot \text{m}$ at $T = 300$ K where $n_i = 1.5 \times 10^{16}$. Therefore, $R = \rho_i \ell / A = 23$ MΩ. If the temperature increases to 310 K, Eq. (1) gives $n_i = 3.15 \times 10^{16}$, so $\rho_i = 1102\ \Omega \cdot \text{m}$ and the resistance drops to about 11 MΩ.

Exercise 9.3–1

Repeat the previous calculations for $T = 290$ K.

n-TYPE AND *p*-TYPE DOPING

Most semiconductor devices are made by adding controlled amounts of impurities to pure crystalline silicon, producing a *doped* semiconductor. Doping elements from the fifth column of the periodic table (phosphorus, arsenic, antimony) are called *donors* because they donate their fifth valence electron to the pool of free electrons. The resulting *n-type* semiconductor conducts mostly by electron flow. Doping elements from the third column (aluminum, gallium, indium) are called *acceptors* because they accept a free electron to make up a full set of covalent bonds. The resulting *p-type* semiconductor conducts mostly by the motion of holes.

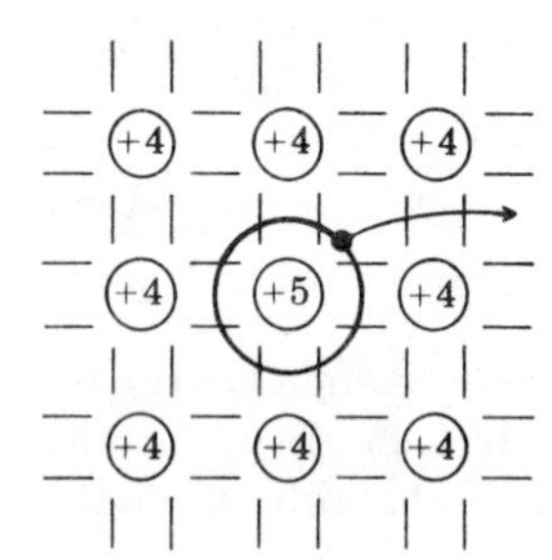

FIGURE 9.3–3
Silicon crystal with *n*-type doping by donor atoms.

To explain how doping works, consider the region in an *n*-type crystal where a donor atom has replaced the silicon atom, as shown at the center of Fig. 9.3–3. Since only four of the five valence electrons are needed as covalent bonds for the four adjacent atoms, the fifth electron orbiting the donor atom requires so little additional energy to escape that at any reasonable temperature it goes off as a free electron, leaving a net charge of +1 at the location of the impurity. Typical *n*-type semiconductors at ordinary temperatures have electron concentrations about equal to the impurity concentration, and electrons from thermal ionization contribute relatively little to the total. The resistivity is therefore appreciably less temperature-sensitive than that of an intrinsic semiconductor.

Usually, we make the donor concentration much larger than n_i, so the electron concentration in an *n*-type semiconductor has an enhanced value $n \gg n_i$. This concentration also increases the probability of electron-hole recombinations, thus decreasing the hole concentration to $p \ll n_i$. Accordingly, we say that electrons are the *majority carriers*, while holes are the *minority carriers*. However, it can be shown that the concentration product remains constant at

$$np = n_i^2 \tag{3}$$

because the generation and recombination rates must still be equal.

A similar situation exists for acceptor doping. In this case an impurity atom, while electrically neutral, would have only three valence electrons to share with four neighbors. The result is a vacancy or hole. It takes very little thermal energy for the hole to move off, leaving a fixed negative charge (the fourth and borrowed electron) in the vicinity of the impurity atom. Conduction is, in this instance, almost entirely by holes, which are now the majority carriers; and their concentration is about the same as that of the impurity atoms. We thus have $p \gg n_i$ and $n \ll n_i$, but with $np = n_i^2$ as before.

Whether n-type or p-type, a relatively small amount of doping produces a drastic change in the concentration of majority and minority carriers. The resistivity becomes

$$\rho = \frac{1}{|q_e|(n\mu_e + p\mu_p)} \tag{4}$$

which, as a rule, will be much less than ρ_i. The following example provides a numerical illustration.

Example 9.3–2 Suppose pure silicon at $T = 300$ K has been doped with a mere 50 parts per billion of acceptor impurities, meaning 50 acceptor atoms for every 10^9 silicon atoms. Since the crystal has 10^{28} atoms per cubic meter, the hole concentration produced by the acceptors will be $p = 50 \times 10^{-9} \times 10^{28} = 5 \times 10^{20}$, compared to $n_i = 1.5 \times 10^{16}$, while $n = n_i^2/p = 4.5 \times 10^{11} \ll p$. Equation (4) now yields $\rho \approx (|q_e|p\mu_p)^{-1} = 0.25\ \Omega \cdot \text{m}$—about 0.01% of the intrinsic resistivity.

Exercise 9.3–2 What donor concentration, in parts per billion, should be added to the bar in Example 9.3–1 to obtain $R = 2400\ \Omega$? Calculate $p\mu_p/n\mu_e$ for this doping to confirm the assertion the holes play a negligible role in conduction.

pn JUNCTIONS

There are several methods for producing a semiconductor crystal that has an abrupt change from a predominantly donor or n-type region to a predominantly acceptor or p-type region. The boundary between the two regions is called a *pn junction*. When terminals are attached to the p and n sides of the crystal, we have a *semiconductor diode*.

Figure 9.3–4 is a simplified representation of a pn junction, with

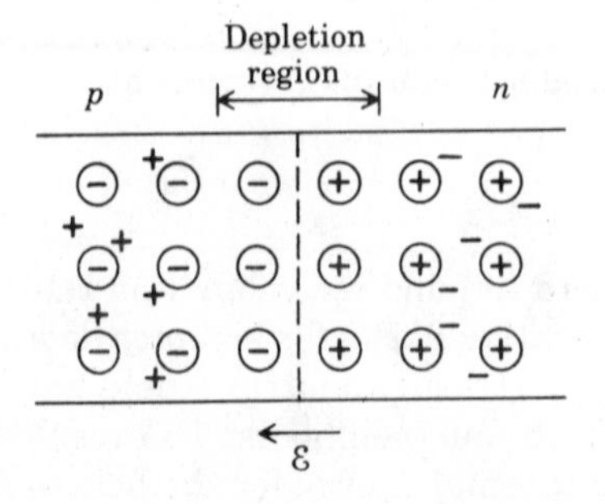

FIGURE 9.3–4
pn junction showing depletion region.

p-type material to the left of the dashed line and *n*-type material to the right. The circles stand for impurity atoms only. The scale of the figure is quite different from that of previous figures, for we have omitted the millions of silicon atoms fixed in the lattice between the impurity atoms. The impurity atoms, of course, are actually distributed at random.

The donor atoms in the *n*-type region are positively ionized, having donated an electron, while the acceptor atoms in the *p*-type region are negatively ionized, having accepted an electron. The resulting free electrons and holes move about randomly, but the net charge equals zero in any area of reasonable size—except in the immediate vicinity of the junction. Here there is a *depletion region* almost totally devoid of mobile charges, because any holes moving from left to right across the junction have recombined with electrons; likewise for electrons moving across the junction from right to left. Thus, the depletion region has an electrostatic field $\mathcal{E}$ pointing from the immobile positive ions on the *n* side to the immobile negative ions on the *p* side. We infer quite correctly from the presence of this electric field that there is an effective potential difference between the *p* and *n* sides of the crystal.

Now let's assume that the right-hand part of the *n*-type region is connected to a metal wire by a *nonrectifying* or *ohmic* contact, and the left-hand part of the *p* region is similarly connected to the other end of the same wire. If all parts are at the same temperature, we would expect no net current flow, since the system has no energy source. Nevertheless, the internal potentials of the wire, the *p*-type material, and the *n*-type material are all different, as we show in Fig. 9.3–5, where the potential of the metal has been chosen arbitrarily. The potential difference $V_0 = V_n - V_p$ is called the *contact potential* of the junction, and typically equals a few tenths of a volt.

Although no current flows through the connecting wire that short-circuits a *pn* junction, there are actually *two opposing currents* within the diode itself. Figure 9.3–6 depicts these currents, labeled I_0 and I_D, and their constituent carriers. A description of how these currents originate will help explain the *i-v* characteristics of a semiconductor diode with an external applied voltage.

The current I_0 originates in or near the depletion region where thermal ionization of the silicon atoms continually generates new hole-electron pairs, just as in an intrinsic semiconductor. The internal electric field $\mathcal{E}$ separates the holes and electrons and sweeps them in opposite directions, producing equivalent positive charge transfer from the *n* side to the *p* side. We call I_0 the *drift current* because it consists of charge carriers that drift under the action of an electric field, similar to charged motion in a metal. From our study of intrinsic semiconductors, we expect I_0 to be quite small at room temperature, but to increase rapidly if temperature goes up. Specifically, if $I_0 = I_0(T_1)$ at temperature T_1, then at a

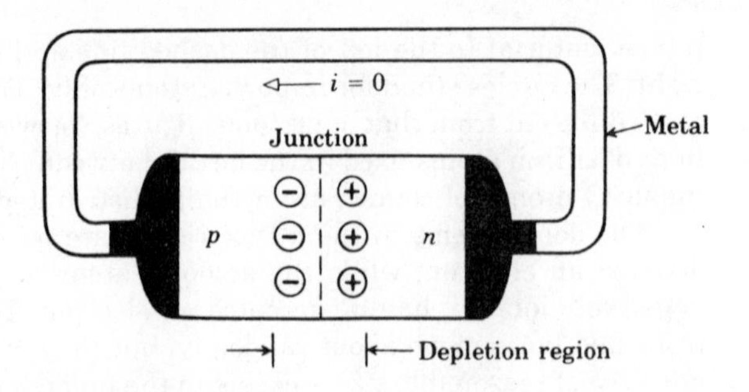

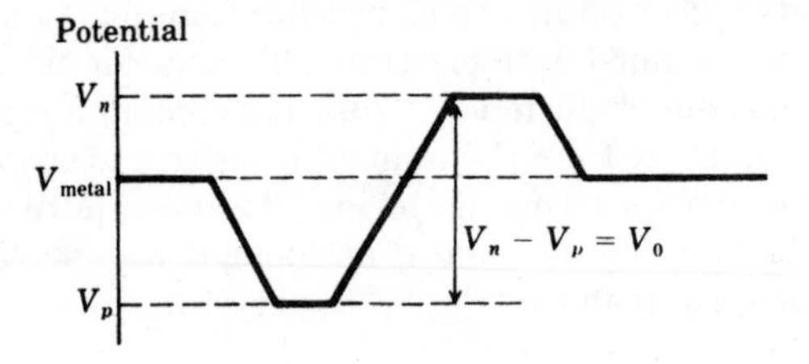

FIGURE 9.3–5
Short-circuited *pn* junction with contact potential V_0.

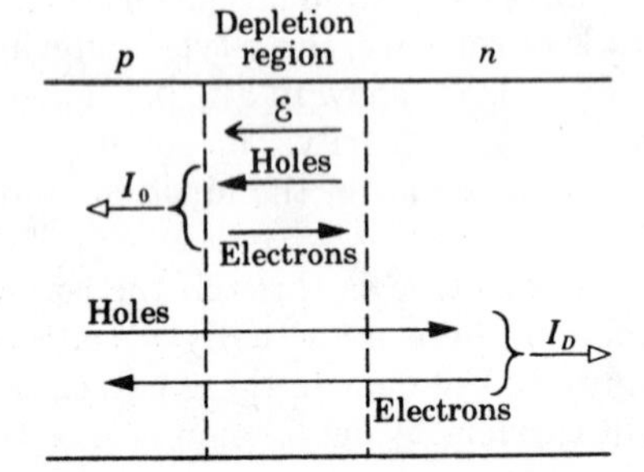

FIGURE 9.3–6
Current components in a *pn* junction.

higher temperature T_2,

$$I_0(T_2) \approx I_0(T_1) \times 2^{(T_2 - T_1)/10} \tag{5}$$

which means that the drift current approximately doubles for every ten-degree increase of temperature.

The current I_D also consists of electrons and holes. However, it originates outside the depletion region and travels *against* the electric field, unlike ordinary currents. To understand this unusual process, recall that electrons and holes have thermal motion with randomly directed velocities. Consequently, a few of the holes on the p side will be injected into the depletion region with sufficient velocity to overcome the potential barrier $V_n - V_p$ and enter the n-type material—where they are *minority* carriers and quickly recombine with majority-carrier electrons. The average time before recombination is known as the minority carrier *lifetime.* Likewise, some electrons from the n side will cross the depletion region and recombine in the p-type material. We call I_D the *injection* or *diffusion current.* It overcomes the potential barrier by virtue of the *diffusion process* that always exists between regions of higher and lower concentrations of thermal particles. Familiar examples of diffusion in action are the evaporation of water and the odor of gasoline, both due to the natural movement of particles to regions of lower concentration.

For the equilibrium condition back in Fig. 9.3–5, with $V_n - V_p = V_0$, the value of V_0 must be such that the diffusion current exactly equals the drift current and $i = I_D - I_0 = 0$. But I_D is highly sensitive to the potential barrier, and an external applied voltage upsets the equilibrium and produces a net current—which we know will be much greater in one direction than the other. Let's see how this works in terms of the two current components.

Consider first the situation in Fig. 9.3–7a, where an external source raises the potential of the n side of the diode relative to the p side. This

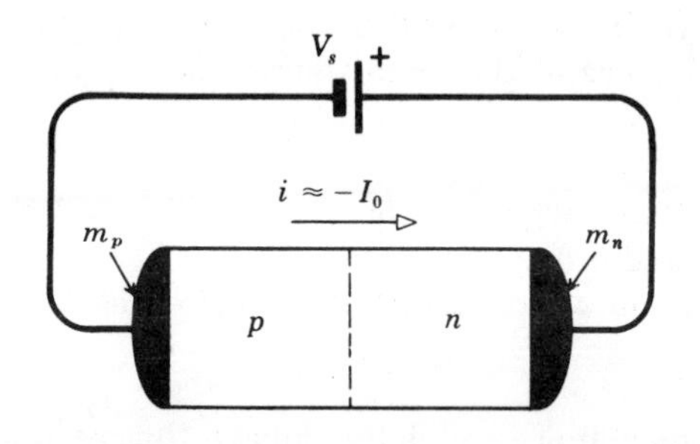

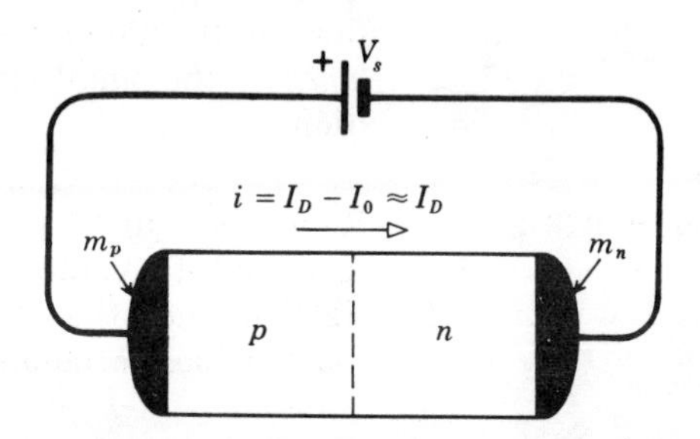

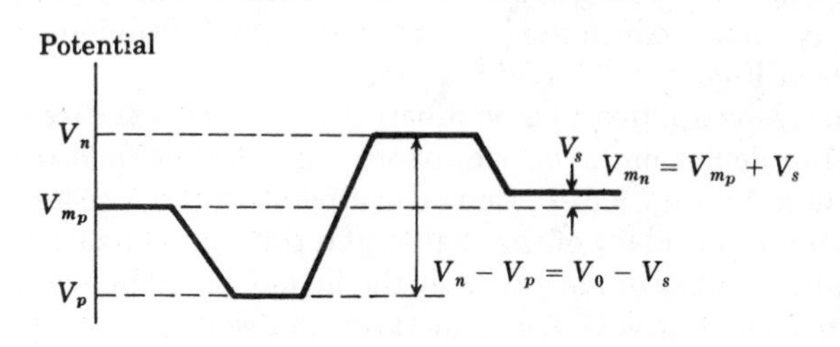

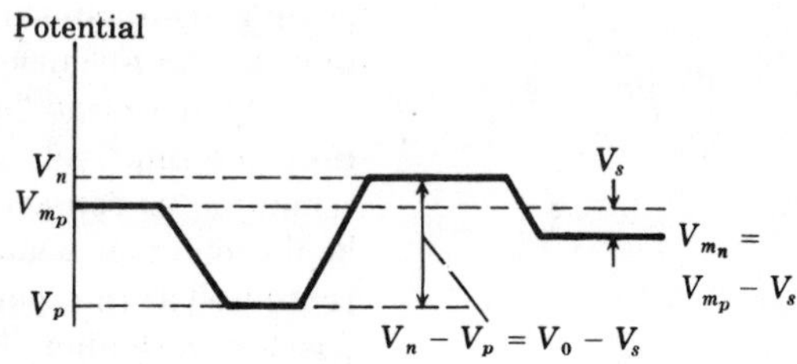

FIGURE 9.3–7
(a) Reverse-biased *pn* junction. (b) Forward-biased *pn* junction.

voltage adds to the contact potential, increases the diffusion barrier, and effectively stops I_D if V_s exceeds a few tenths of a volt. The remaining current is $i \approx -I_0$, temperature dependent but relatively small. We say that the junction is *reverse biased,* and we recognize I_0 as the *reverse saturation current.*

Next, let the applied voltage raise the potential of the p side as in Fig. 9.3–7b. This voltage has little effect on I_0, but it lowers the potential barrier and allows the diffusion current to increase markedly. The junction is now *forward biased,* and diffusion accounts for the major portion of the net current $i = I_D - I_0 > 0$.

For an arbitrary applied voltage v, it can be shown that the diffusion current has an exponential dependence of the form $I_D = Ke^{\lambda v}$, so

$$i = I_D - I_0 = Ke^{\lambda v} - I_0$$

with λ as given in Eq. (1), Section 9.1. But $i = 0$ when $v = 0$ and $e^{\lambda v} = 1$; therefore the constant K must equal I_0 and

$$i = I_0 e^{\lambda v} - I_0 = I_0(e^{\lambda v} - 1)$$

This is the i-v relationship for a semiconductor diode that we started with in Section 9.1. Of course, the forward current in an actual diode can't increase exponentially to arbitrarily large values. Instead, it becomes limited by the ohmic resistance of the bulk material in the p and n regions *and* by the small effective resistance of the metal-semiconductor junctions.

Exercise 9.3–3

Consider a diode that has $i = 10\ \mu\text{A}$ when $v = 50$ mV and $T = 293$ K.
(a) Find the values of the drift and diffusion currents.
(b) What are the values of I_0, I_D, and i if the temperature increases 20 K but v is held constant?

REVERSE BREAKDOWN

Finally, we examine the phenomenon of reverse breakdown that occurs in a semiconductor diode with a large reverse bias. Either of two breakdown mechanisms may then produce the large reverse current previously seen in the i-v curves of Figs. 9.1–3b and 9.2–4b.

When a large voltage is applied to a pn junction in the reverse direction, the relatively thin depletion layer acquires a high electric field intensity $\mathcal{E}$, as represented by Fig. 9.3–8. Consider a point in the stressed depletion layer near the p boundary of the region. If a thermal ionization is created at that point, the electron is forced by the field toward the more positive n region. In the relatively short distance between collisions, when the field is high, this electron can gain enough kinetic energy of motion to cause a new ionization at the next collision. At that point we have

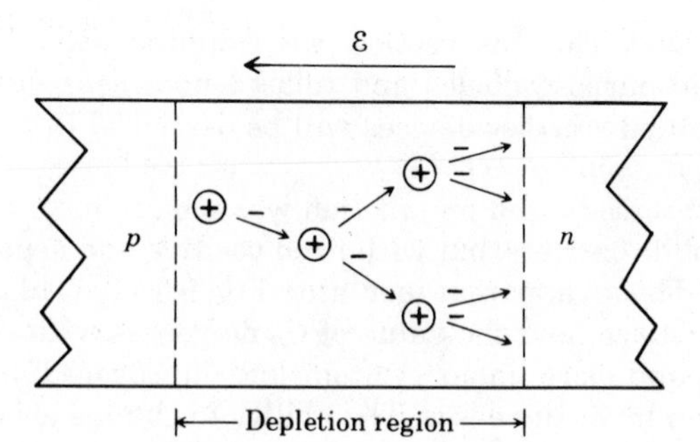

FIGURE 9.3–8 Avalanche breakdown caused by collisions between ions and accelerated free electrons.

the new electron and the original one, both ready to accelerate in the positive direction. If, after the next collision, they each produce a new hole-electron pair, there are now four electrons. At the next collision there could be eight, then sixteen, and so forth in geometric progression. All the holes, of course, travel in the opposite direction of the electrons, adding to the total current. This mechanism, usually called *avalanche breakdown,* causes a rapid increase of current when the reverse voltage exceeds a critical value.

An identical result—the rapid increase of current when the reverse voltage exceeds a critical value—occurs in diodes that have narrow depletion layers. In this case, however, no avalanche mechanism is required. As the voltage rises, the field intensity becomes high enough to cause valence electrons to break their bonds, producing large numbers of charge carriers. This mechanism is called *Zener breakdown* but is generally limited to diodes that break down at voltages less than about 6 to 8 V. The general name of *Zener diode* is commonly applied, however, regardless of the mechanism involved, to any diode designed to be used in the breakdown region.

You should not infer from the term "breakdown" that there is direct damage to the diode. But the current-voltage product represents power dissipation that does heat the junction, and a diode *is* damaged if the temperature remains high enough for a long enough time to cause impurity atoms to diffuse across the junction, degrading it.

9.4 SPECIAL-PURPOSE DEVICES †

Although junction diodes behave, in general, like electrical one-way valves, the fabricators have a number of parameters under their control that can alter a diode's characteristics. Variations in such things as impurity concentrations and depletion layer thickness may result in diodes that are less than optimum for some purposes, but serve others very well. In this section we examine some interesting and useful special-purpose diodes and related nonlinear devices. (SCRs and other controlled rectifier devices will be described in Section 11.4.)

JUNCTION CAPACITANCE AND VARACTORS

The *reverse*-biased *pn* junction was seen to have a depletion layer devoid of mobile carriers, but with fixed positive and negative charges (ions) separated by an apparent insulator, Fig. 9.4–1a. This fulfills the definition of *capacitance*, and the value of C_R decreases with reverse voltage because more and more impurity atoms are "uncovered" and the width of the depletion layer increases. The ability to change a capacitance with applied voltage is a handy one used to advantage in applications such as the automatic frequency control (AFC) circuit of an FM radio. Diodes fabricated especially to enhance this property are known as *varactors* or *varicaps*, symbolized in Fig. 9.4–1b. The reverse capacitance is given by

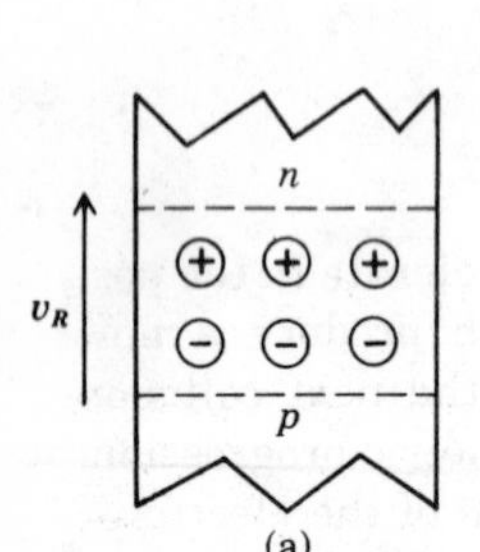

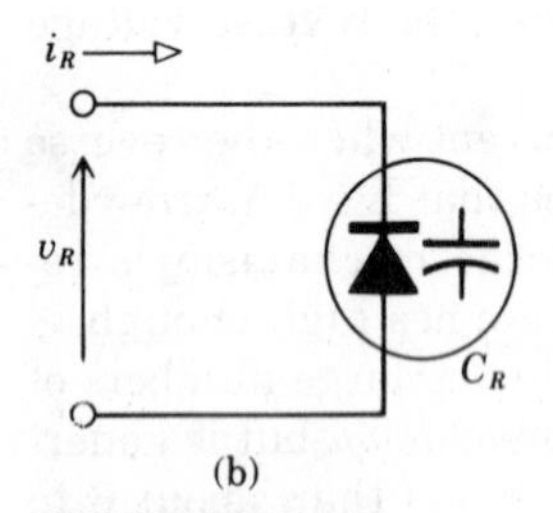

FIGURE 9.4–1
(a) Uncovered ions create capacitance in a reverse-biased junction.
(b) Circuit symbol of a varactor.

$$C_R \approx C_C + \frac{C_0}{\sqrt{1 + 2v_R}}$$

where C_C and C_0 are constants. Small variations of v_R produce the reverse current

$$i_R = C_R \frac{dv_R}{dt}$$

in addition to the constant component I_0. Reverse capacitance, also known as *transition* capacitance, has nominal values in the range of 5–500 pF.

Another type of capacitance is caused by the high concentration of minority carriers in the depletion region of a *forward*-biased diode. This is called *diffusion capacitance* C_D; its value is proportional to the DC forward current, and may be much greater than C_R. Including C_D, the small-signal model of a diode becomes as shown in Fig. 9.4–2 and has a built-in time constant

$$\tau = r_d C_D$$

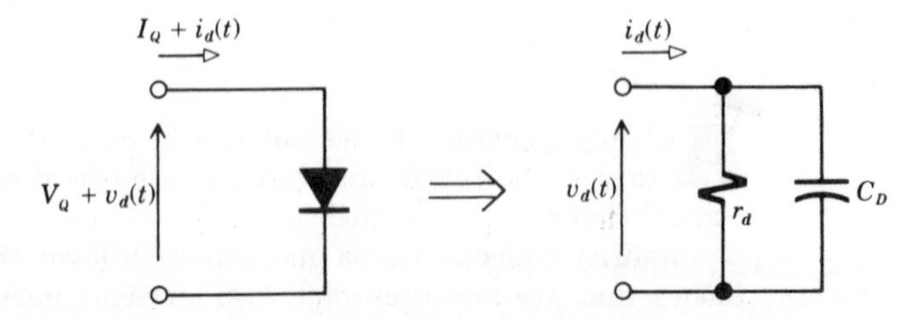

FIGURE 9.4–2
Diffusion capacitance added to the small-signal model of a forward-biased diode.

where τ is the average lifetime of minority carriers, ranging from a few nanoseconds to hundreds of microseconds.

OHMIC CONTACTS AND SCHOTTKY DIODES

Attaching the leads to a diode requires formation of a *nonrectifying* or *ohmic* junction between the metal and the semiconductor. This is usually accomplished by increasing the impurity concentration to obtain a high-conductivity region adjacent to the metallic contact. Such heavily doped regions would be labeled with the symbols n^+ or p^+.

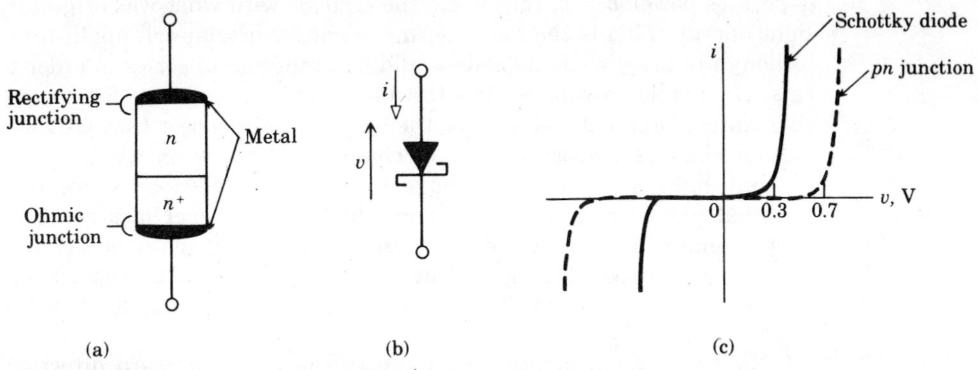

FIGURE 9.4–3 Schottky diode: (a) construction; (b) symbol; (c) *i-v* curve.

It is also possible to construct a *rectifying* metal-to-semiconductor junction instead of a *pn* junction. When this is done using *n*-type material, a positive voltage on the metal with respect to the semiconductor causes electrons to be injected into the metal with such high velocity that they are sometimes called *hot carriers.* Since the current flows in the form of majority carriers only, the effective capacitance is very low, as is the forward voltage drop. Such devices, known as *Schottky diodes,* have advantages in high-speed applications.

Figure 9.4–3 shows the structure, symbol, and *i-v* curve of a Schottky diode. We see that the metal-to-semiconductor junction results in smaller values of forward threshold and reverse breakdown voltage, when compared to a *pn* junction.

PHOTO DEVICES

Light-sensitive and light-emitting devices may be grouped under the general heading of photo devices. Some of the major ones are described here.

It is possible to fabricate a *pn* junction very close to the surface of a crystal. An ohmic contact may then be made by evaporation, for instance, with a surface material so thin that it is transparent to light. When light impinges upon this device, many photons begin to interact with the valence electrons of atoms in the depletion region and generate hole-electron pairs by photoionization. The amount of reverse saturation current, $I_0 + I_p$, now depends on the number of incident photons per second, or light intensity (as well as on the temperature), as illustrated in Fig. 9.4–4. Diodes designed to be used as light sensors are called *photodiodes*. Those intended to produce electric power from light are called *solar cells*.

The load line labeled A in Fig. 9.4–4b is drawn, as customary, for a fixed resistance and a positive supply voltage. This arrangement produces very little change in voltage or current for varying light levels. The load line labeled B is for the same resistance value, but the supply voltage is zero, so the line goes through the origin. Now the illuminated diode generates power $p = vi$ that heats the resistor with what was originally light energy. This is the basic arrangement for a solar-cell application, although in practice many cells would be connected in series in order to raise the available voltage. The third load line, labeled C, is for a negative value of applied voltage. We then have a light sensor that produces voltage changes proportional to the changes of light intensity.

But light sensors do not have to be diodes. *Photocells* are two-terminal photoconductive devices in which a thin layer of a semiconducting material (usually cadmium sulphide) produces many new charge carriers when exposed to light. Thanks to these additional carriers, a photocell's resistance under illumination drops to a small fraction of the "dark" resistance.

Now consider a junction diode operating in the forward direction: Large numbers of electrons are injected into the p region where they recombine with holes; and large numbers of holes are injected into the n region where they recombine with electrons. The recombination process

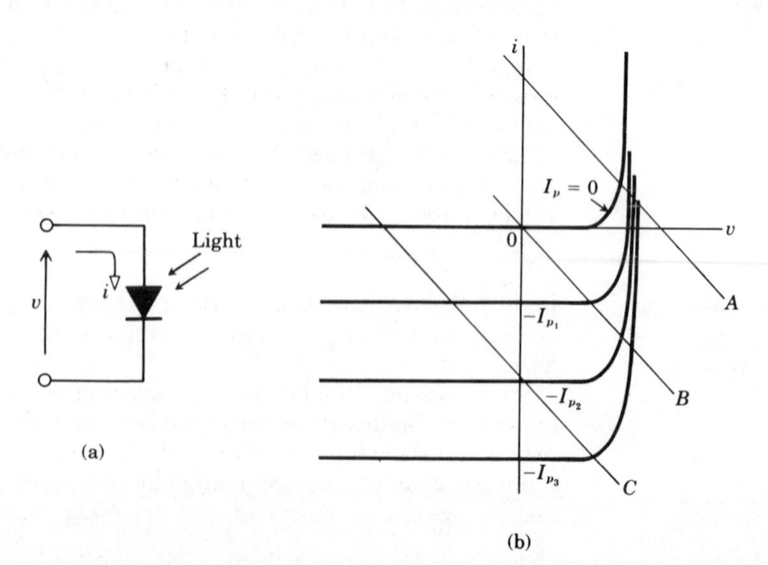

FIGURE 9.4–4
Photodiode curves with load lines for diode operation A, solar cell B, and light sensor C.

in both cases releases energy, which is not surprising since an external energy source is required to produce hole-electron pairs. Under favorable circumstances, the energy released by a recombination may be in the form of a light photon that has an optically clear path out of the diode. Such diodes act like photocells in reverse, producing light when electrically energized. These are called *light emitting diodes,* or *LEDs*. LEDs come in a half dozen colors, have forward voltage drops of one to two volts, and—unlike most other light sources—change their intensity very rapidly in response to sudden changes in diode current.

A familiar application is the seven-segment display used in pocket calculators, each segment consisting of one or several LEDs, as in Fig. 9.4–5a. Another application is the *optoisolator,* Fig. 9.4–5b, which contains an LED whose light output directly illuminates a photodiode inside an opaque package. The result is a four-terminal device in which neither of the two input terminals is connected to an output terminal. This ability to isolate one circuit from another is very useful, but is generally limited to ON/OFF signals. The variations of a continuous signal would be distorted by the nonlinearity of the optoisolator.

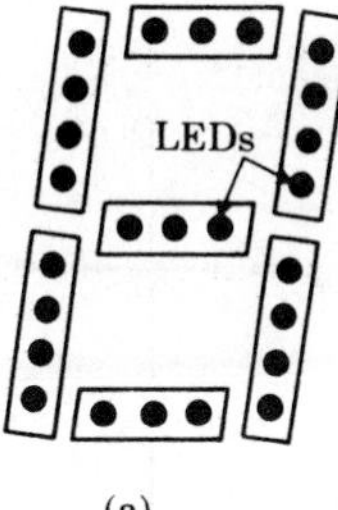

(a)

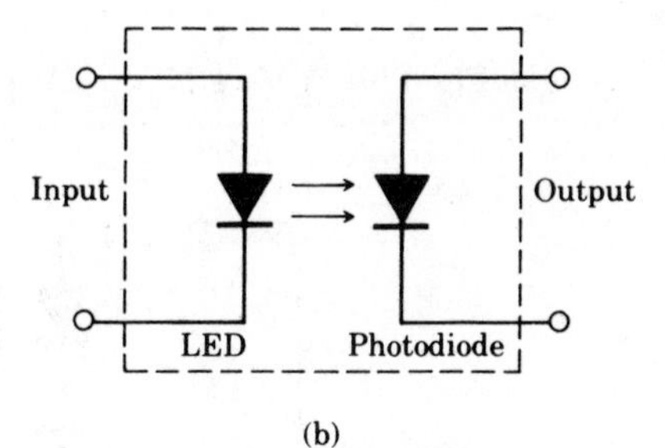

(b)

FIGURE 9.4–5
Light emitting diodes in: (a) seven-segment display, (b) optoisolator.

THERMISTORS AND VARISTORS

A *thermistor* is a two-terminal device made of polycrystalline semiconductor material instead of a single crystal. The polycrystalline material produces many new hole-electron pairs when the temperature rises, so the device exhibits a negative temperature coefficient of resistance. Figure 9.4–6 shows the thermistor symbol and a typical curve of resistance versus temperature. A thermistor may be put in one branch of a bridge circuit for the purpose of sensing temperature variations.

Another useful polycrystalline device is the *varistor* (Fig. 9.4–7), whose resistance is quite large until the applied voltage exceeds a critical value for either polarity. The device then breaks down and conducts heavily, acting something like a Zener diode. Varistors are used to prevent damage to other elements by the sudden high-voltage spikes that sometimes occur on power and communication lines.

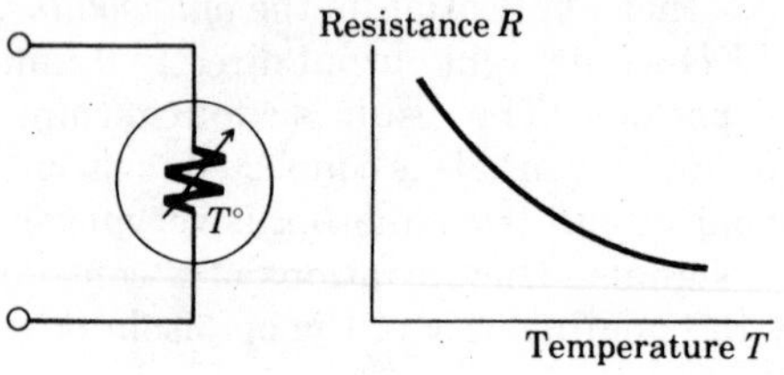

Resistance R

$T°$

Temperature T

FIGURE 9.4–6
Thermistor symbol and resistance curve.

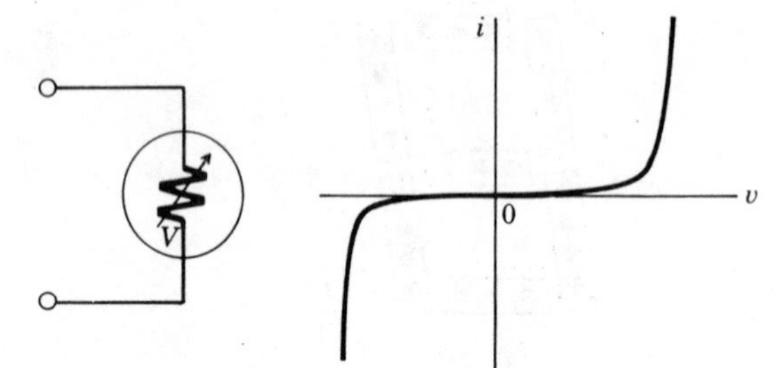

FIGURE 9.4–7
Varistor symbol and *i-v* curve.

PROBLEMS

9.1–1 The identical diodes in Fig. P9.1–1 have $I_0 = 1\ \mu$A. Find i_1, i_2, and i when $v =$ 50 mV.

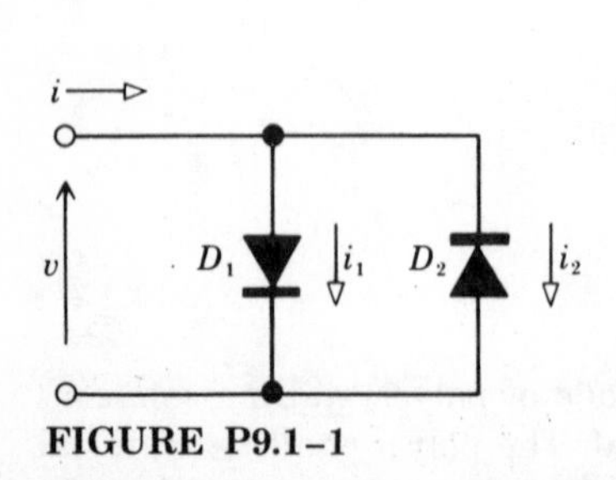

FIGURE P9.1–1

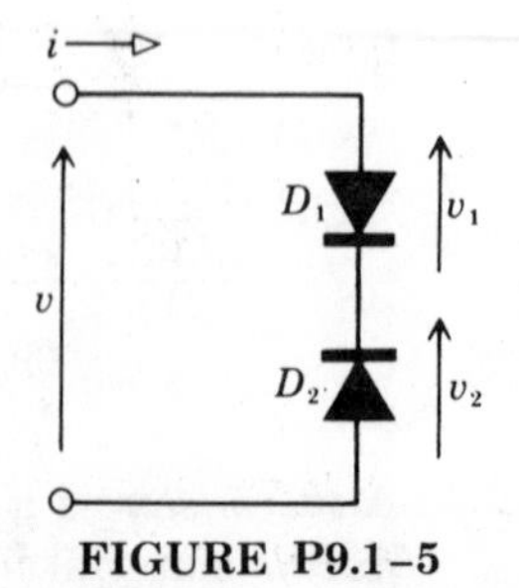

FIGURE P9.1–5

9.1–2 Find v, i_2, and i when the circuit in the previous problem has $i_1 = 5\ \mu A$.

9.1–3 A certain semiconductor diode has $i = 1$ mA when $v = v_1 \gg 25$ mV. Assume $i = 10$ mA when $v = v_2$, and evaluate the voltage change $v_2 - v_1$.

9.1–4 Repeat the previous calculation with $i = 100$ mA when $v = v_2$.

9.1–5 The diodes in Fig. P9.1–5 are identical and have $I_0 = 1\ \mu A$. Find i, v_1, and v_2 when $v = 5$ V.

9.1–6 Find i_1, i_2, and v for the circuit in Problem 9.1–1 when $i = 1$ mA.

9.1–7 Suppose the nonlinear element in Fig. 9.1–5 has $i = 5 \times 10^{-6}\, v^2$ for $v \geq 0$. Solve analytically for v_Q and i_Q when $V_s = 400$ V and $R = 1\ k\Omega$.

9.1–8 Repeat the previous problem with a nonlinear element having $i = 0.03\sqrt{v}$.

9.1–9 Repeat Example 9.1–2 when $v_s(t)$ has the waveform in Fig. 7.4–1b with $A = 0.1$ V.

9.1–10 Repeat Example 9.1–2 when $v_s(t)$ has the waveform in Fig. 7.4–1a with $A = 0.1$ V.

9.1–11 Figure P9.1–11 shows a photodiode (Fig. 2.2–3) with $I_p > 0$ serving as the source applied to an ordinary diode. Make a sketch similar to Fig. 9.1–6 for this situation, and indicate the operating point. Hint: Note that the photodiode current here has a direction opposite that of Fig. 2.2–3.

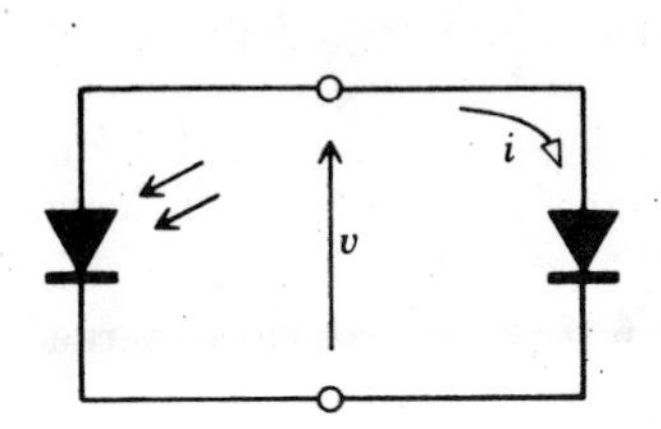

FIGURE P9.1–11

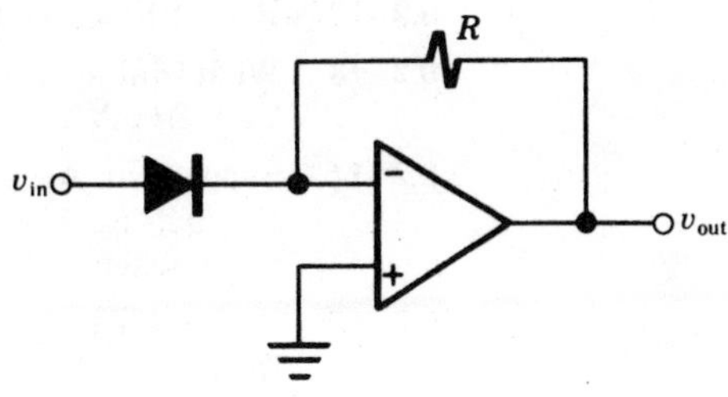

FIGURE P9.1–13

9.1–12 Repeat the previous problem with the photodiode inverted.

9.1–13 The op-amp in Fig. P9.1–13 is ideal, and $v_{in} \gg 25$ mV.
(a) Show that this circuit acts as an *exponential amplifier*, with $v_{out} \approx -RI_0 e^{40v_{in}}$.
(b) Show that interchanging the diode and resistor creates a *logarithmic amplifier*, with $v_{out} \approx -0.025 \ln (v_{in}/RI_0)$.

9.1–14 Devise an op-amp system whose output voltage is proportional to the *product* of two positive input voltages v_1 and v_2. You may draw on the results of the previous problem, together with the fact that $e^{(\ln x + \ln y)} = xy$.

9.2–1 The specification sheet for a certain diode states that $i = 300$ mA when $v = 1.6$ V. Assuming $V_T = 0.7$ V, find the value for R_f in Fig. 9.2–1c. Then calculate i when this diode is put in the circuit of Fig. 9.1–2a with $v_s = 5$ V and $R = 15\ \Omega$.

9.2–2 Repeat the previous problem with a diode stated to have $i = 20$ mA when $v = 1.2$ V.

9.2–3 Plot the i-v curve for Fig. P9.1–1, taking both diodes to have $V_T = 0.7$ V and $R_f = 10\ \Omega$. Then sketch $i(t)$ when $v(t) = 3 \sin \omega t$ V.

9.2–4 Repeat the previous problem with $V_T = 0.3$ V and $R_f = 20\ \Omega$ for diode D_2.

9.2–5 As an alternative in Fig. 9.2–3, the reverse current of a diode could be modeled by putting a large reverse resistance R_r in parallel with the circuit Fig. 9.2–1c. Draw the corresponding i-v curve, assuming $R_r \gg R_f$.

9.2–6 Suppose the diodes in Fig. P9.1–5 are identical Zener diodes, with $V_Z = 9$ V and $R_Z = 20\ \Omega$. Plot the i-v curve.

9.2–7 Repeat the previous problem taking D_1 to have $V_Z = 6$ V and $R_Z = 10\ \Omega$.

9.2–8 Let $V_s = V_Z + \Delta V$ in Fig. 9.2–5, and show that $v_{\text{out}} = V_Z + \Delta V/(1 + R/R_Z)$. What is the value of V_s, assuming $R = 9R_Z$ and $v_{\text{out}} = 1.2\ V_Z$?

9.2–9 Let $R_Z = R/9$ in Fig. 9.2–5a. Obtain the Thévenin equivalent circuits for the two cases $0 < V_s < V_Z$ and $V_s > V_Z$.

9.2–10 Suppose a load is connected to the terminals of Fig. 9.2–5a and draws a variable current $0 \le i_{\text{out}} \le 0.5$ A. Plot v_{out} versus i_{out} when $V_s = 30$ V, $R = 20\ \Omega$, $V_Z = 25$ V, and $R_Z = 5\ \Omega$. Hint: Consider the two end points and the Zener's break point.

9.2–11 Find the value of R for the circuit in the previous problem such that the Zener's break point occurs at $i_{\text{out}} = 0.5$ A. Then plot v_{out} versus i_{out}.

9.2–12 Repeat Example 9.2–2 with $R = 100\ \Omega$.

9.2–13 What value of V_{DC} should be used in Example 9.2–2 such that $v_d(t) = 2 \sin \omega t$ mV?

9.2–14 Suppose an AC source $v_s(t) = 5 \sin \omega t$ V is added in series with the battery in Fig. 9.2–5a. Use a method like that used in Example 9.2–2 to find and sketch $v_{\text{out}}(t)$ when $V_s = 30$ V, $R = 20\ \Omega$, $V_Z = 25$ V, and $R_Z = 5\ \Omega$.

9.2–15 Repeat the previous problem with $V_s = 25$ V, so that the Zener will be OFF part of the time.

9.3–1 If we let $\rho(T)$ represent the resistivity at an arbitrary temperature T, the temperature coefficient of resistance at temperature T_1 can be approximated by calculating $100\ [\rho(T_1 + \Delta T/2) - \rho(T_1 - \Delta T/2)]/\Delta T\ \rho(T_1)$, which gives the result in percent per degree kelvin. Use this expression and the results from Example 9.3–1 and Exercise 9.3–1 to find the temperature coefficient for silicon at $T_1 = 300$ K.

9.3–2 Repeat the previous calculation for $T_1 = 310$ K.

9.3–3 A bar of n-type silicon in an integrated circuit is 5-μm thick and 20-μm wide. If $n = 2 \times 10^{23}$, what should the length of the bar be to obtain $R = 1$ kΩ?

9.3–4 Repeat the previous problem with p-type silicon having $p = 2 \times 10^{23}$.

9.3–5 Consider a doped semiconductor having a donor concentration N_d and an acceptor concentration N_a. Since charge neutrality requires that $n + N_a = p + N_d$, show that the electron concentration is found by solving the quadratic expression $n^2 + (N_a - N_d)n - n_i^2 = 0$. Then confirm that $n \approx N_d$ if $N_d \gg n_i^2$ and $N_a \ll N_d$.

9.3–6 Use the expression in the previous problem to find n and p when $N_a = 10^{18}$ and $N_d = 0.1 \times 10^{18}$. Confirm, in this case, that $p \approx N_a$.

9.3–7 Figure P9.3–7 represents a block of n-type semiconductor carrying current i in the direction shown. If a magnetic field B is applied perpendicular to i, the electrons are forced downward and the *Hall-effect voltage* $v_H = Bi/|q_e|nw$ can be measured across the block. (For a p-type semiconductor, the voltage is $v_H = -Bi/|q_e|pw$, since the holes travel in the opposite direction and will also be forced downward.) Devise a method for measuring the mobility μ_e from this experiment.

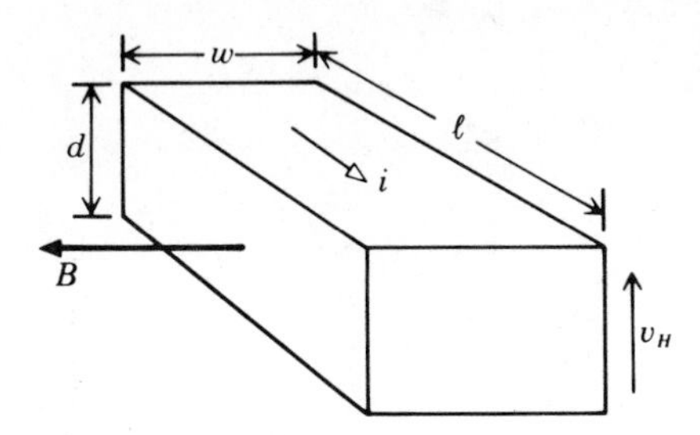

FIGURE P9.3–7

9.3–8 If the diode in Fig. 9.3–5 has hole concentration p_p on the p side and electron concentration n_n on the n side, the contact potential at the junction is given by

$$V_0 = \frac{T}{1.17 \times 10^4} \ln \left(\frac{p_p n_n}{n_i^2} \right)$$

and the width of the depletion region is

$$W = \sqrt{\frac{2\epsilon V_0}{q_e} \left(\frac{1}{p_p} + \frac{1}{n_n} \right)}$$

where $\epsilon \approx 10^{-10}$ F/m. Calculate V_0 and W when $p_p = n_n = 1000\ n_i$.

9.3–9 Repeat the previous problem with $p_p = 10^4\ n_i$ and $n_n = 100\ n_i$. Then use the fact that the net charge in the depletion region must equal zero to calculate the widths on each side of the junction.

9.3–10 Suppose i is held constant in Exercise 9.3–3. What is the value of v at the higher temperature?

9.3–11 Two identical diodes are connected in parallel, conducting in the same direction. Diode D_1 is at $T = 300$ K, while D_2 is at $T = 320$ K. Find the individual currents when $v = 0.25$ V and the total current is 1 mA.

9.3–12 A diode having $I_0 = 1\ \mu$A at $T = 300$ K is connected to a 1-mA current source. Obtain an expression for v in terms of T.

10

TRANSISTORS AND INTEGRATED CIRCUITS

The transistor family embraces a diverse and growing collection of semiconductor devices with literally thousands of different members. Despite this multiplicity, all transistors are *three-terminal* elements that have the distinctive ability to *control* a voltage or current in accordance with the value of another voltage or current. There are two major types of transistors: *field-effect transistors* (FETs), which are voltage controlled, and *bipolar junction transistors* (BJTs), which are inherently current controlled.

Depending upon the circuit configuration and application, a given FET or BJT may operate in either a linear or nonlinear mode. As the active device in electronic amplifiers and instrumentation systems, the transistor is operated within its linear regime and acts like a *controlled source*. When operated at the extremes of its nonlinear mode, a transistor effectively becomes a *switch* for use in digital computers or other electronic switching systems.

In this chapter we introduce the external characteristics and circuit models of transistors, and discuss the underlying physical phenomena in FETs and BJTs. We also describe semiconductor device fabrication and the technology of integrated circuits.

OBJECTIVES

After studying this chapter and working the exercises, you should be able to do each of the following:

- Identify the symbols for FETs and BJTs; sketch typical sets of characteristic curves; and label the operating regions (Sections 10.1 and 10.2).
- Use load-line analysis or a large-signal model to find the operating point of a transistor circuit (Sections 10.1 and 10.2).
- Plot the transfer curves and waveforms for a simple transistor amplifier or switching circuit (Sections 10.1 and 10.2).
- Draw the small-signal model of a FET or BJT (Sections 10.1 and 10.2).
- Sketch the internal structure of a FET or BJT, and explain the conduction mechanisms (Section 10.3). †
- Describe the processes used to fabricate an integrated circuit (Section 10.4). †

10.1 FIELD EFFECT TRANSISTORS

The family tree of FETs charted in Fig. 10.1–1 has three main branches and six individual devices, each with its own circuit symbol. All of these devices owe their behavior to a *field effect:* Voltage applied to the *gate* (G) creates an electric field that controls the flow of charges in a semiconductor *channel* from the *source* (S) to the *drain* (D). Consequently, all FETs exhibit similar operating characteristics.

We start this section by examining the characteristics and circuit models of one device, the n-channel enhancement MOSFET, and summa-

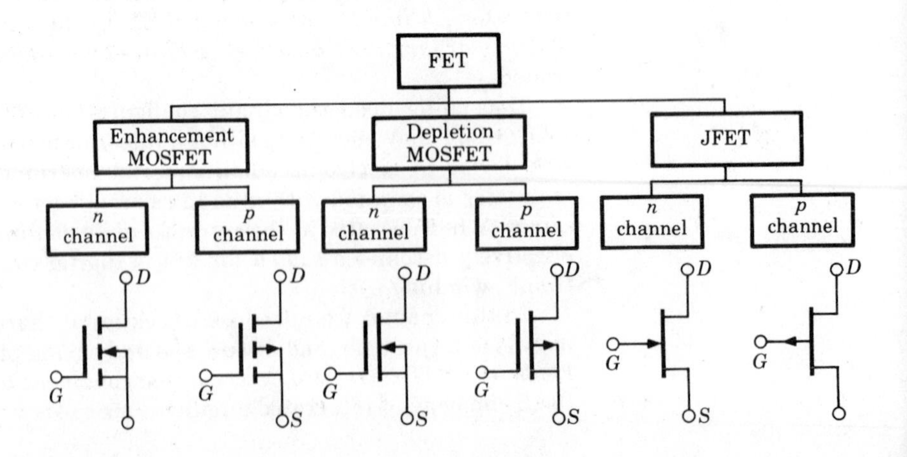

FIGURE 10.1–1
FET family tree.

rize the differences in the behavior of its p-channel sibling. Our emphasis here is upon the external i-v relationships and how they lead to amplification and switching action. The internal physical mechanisms will be discussed in Section 10.3, along with the characteristics of depletion MOSFETs and JFETs.

ENHANCEMENT MOSFET CHARACTERISTICS

Consider the n-channel enhancement MOSFET in Fig. 10.1–2. The prefix MOS means that the device consists of a metal-oxide-semiconductor "sandwich." The oxide layer insulates the gate from the channel, as represented by the gap in the symbol, so virtually no current i_G flows through the gate terminal. Nor does appreciable channel current flow when the gate-to-source voltage v_{GS} equals zero. The broken channel line in Fig. 10.1–2 symbolizes this "normally OFF" property. Applying a positive value of v_{GS} *enhances* the conduction of an n-type channel. *Negative* charges then flow from source to drain, corresponding to the indicated drain current i_D and a positive value for the drain-to-source voltage v_{DS}.

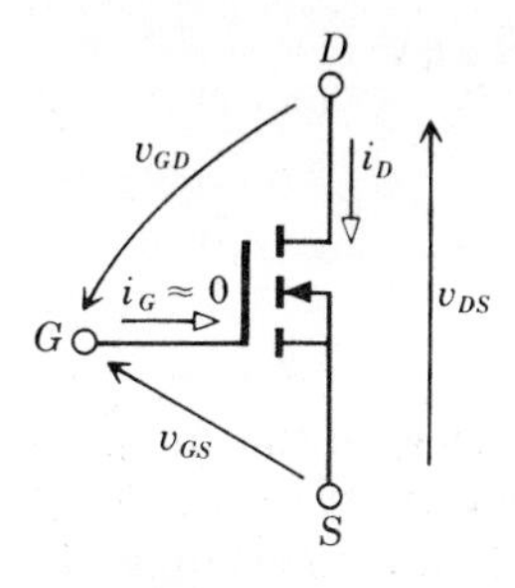

FIGURE 10.1–2
n-channel enhancement MOSFET.

As there are *two* independent voltages associated with a MOSFET, we must learn how the drain current depends on each of them. (Mathematically speaking, i_D is a function of two variables, v_{GS} and v_{DS}.) To examine this dependence, suppose we first hold v_{GS} constant at some value V_{GS} and apply an adjustable voltage for v_{DS}, as in Fig. 10.1–3a. The resulting plot of i_D versus v_{DS} theoretically looks like Fig. 10.1–3b. The curve has a steep initial rise, goes through a rounded "knee," and flattens off horizontally. (The curve of an actual device might be more rounded and not exactly horizontal.) The device eventually breaks down at $v_{DS} = BV_{DS}$ where i_D increases precipitously. Repeating our measurements for different values of v_{GS} yields a *set of characteristic curves,* as in Fig. 10.1–4, which omits the breakdown region.

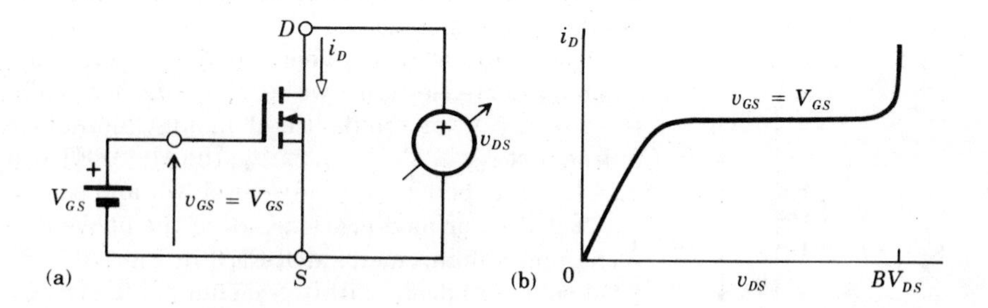

FIGURE 10.1–3

Given values for any two of the three variables—i_D, v_{DS}, and v_{GS}—we can determine the third from these curves, although interpolation is required for intermediate values of v_{GS}. To illustrate: If $v_{DS} = 6$ V and $i_D = 5$ mA, then we estimate that $v_{GS} \approx 4.25$ V, since the point in question falls about midway between the curves for $v_{GS} = 4$ V and $v_{GS} = 4.5$ V.

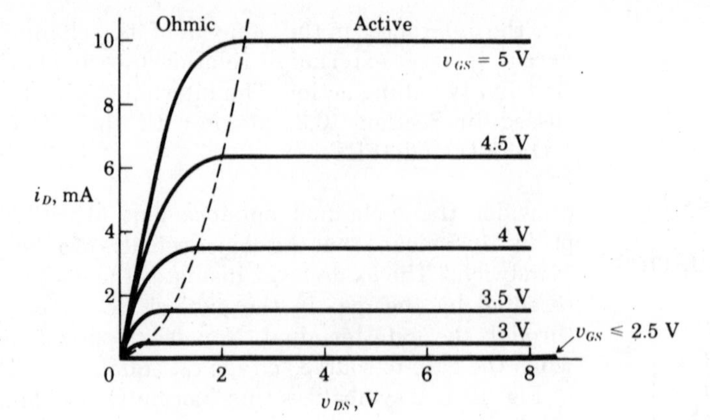

FIGURE 10.1–4 Characteristic curves of an *n*-channel enhancement MOSFET.

Contrast this set of curves for a three-terminal device with the single *i-v* curve that describes a diode or other two-terminal device (Fig. 9.1–3, for instance), where the value of one variable immediately determines the value of the other variable.

Each curve in Fig. 10.1–4 has the same shape, but the height of the knee depends upon the gate voltage. Consequently, in the region to the right of the knees, the MOSFET acts like a *voltage-controlled current source,* in the sense that i_D stays relatively constant at a value determined by v_{GS} rather than increasing with v_{DS}. We will refer to this region as the *active* operating region, instead of the more proper, but cumbersome, terms *beyond-pinchoff* or *current-saturation* region, both of which are found in the literature. To the left of the knees is the *ohmic* operating region (also called the *below-pinchoff* region) where i_D increases with v_{DS}, similar to a resistance.

The value of i_D anywhere in the active or ohmic regions decreases when v_{GS} decreases until we reach a *threshold voltage* V_T, which happens to equal 2.5 V for the device at hand. Channel conduction therefore requires that $v_{GS} > V_T$. If $v_{GS} \leq V_T$, the MOSFET is in the *cutoff* region and acts like an open circuit between drain and source, so $i_D \approx 0$.

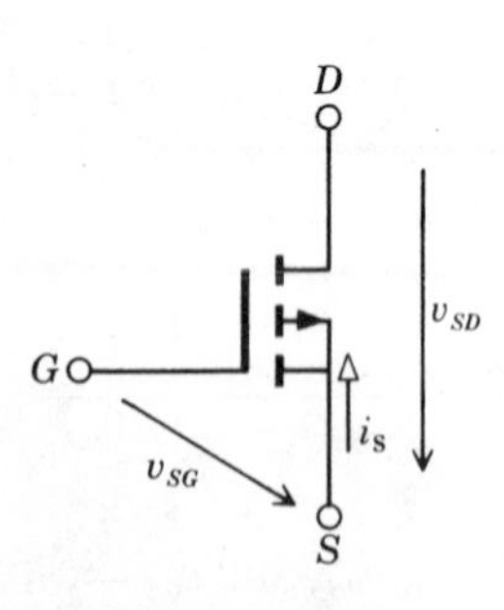

FIGURE 10.1–5 *p*-channel enhancement MOSFET.

With minor modifications, all of the above description applies to the *p*-channel enhancement MOSFET in Fig. 10.1–5. *Positive* charges flow from source to drain within a *p*-channel device, constituting the current $i_S = -i_D$. This current flows when $v_{SG} = -v_{GS} > V_T$ and $v_{SD} = -v_{DS} > 0$. The characteristic curves then look like those in Fig. 10.1–4 with i_S, v_{SD}, and v_{SG} replacing i_D, v_{DS}, and v_{GS}, respectively. We say that *n*-channel and *p*-channel devices are *complementary,* having the same characteristics but opposite current direction and voltage polarities.

TRANSFER CURVES, SWITCHES, AND AMPLIFIERS

The practical implications of a MOSFET's characteristics are brought out more clearly when we look at the input-output curves for the simple circuit in Fig. 10.1–6. The input is a variable voltage applied at the gate, while the output circuit consists of a supply battery V_{DD} and resistance R_D. Note carefully that the "output" current i_D actually goes into the drain terminal; this current direction results in *negative voltage amplification,* an inherent property of many electronic amplifiers. We want to find the variation of output current i_D and voltage v_{DS} caused by varying the input voltage v_{GS}. To do that, because of the MOSFET's nonlinear drain-to-source behavior, we must use the graphical *load-line* approach previously introduced to handle diode circuits.

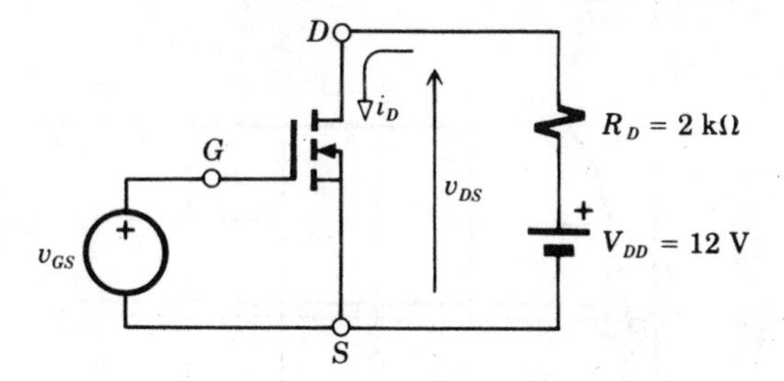

FIGURE 10.1–6
A MOSFET circuit.

Recall from Section 9.1 that a load line represents the *i-v* relation at the terminals of the load circuit. For the case at hand, then, the load-line equation obtained from an inspection of Fig. 10.1–6 will be

$$i_D = \frac{V_{DD} - v_{DS}}{R_D} = \frac{12\ \text{V} - v_{DS}}{2\ \text{k}\Omega}$$

The load line itself is easily constructed by connecting the short-circuit point ($i_D = V_{DD}/R_D = 6$ mA at $v_{DS} = 0$) to the open-circuit point ($v_{DS} = V_{DD} = 12$ V at $i_D = 0$). Figure 10.1–7a shows this load line superimposed on the characteristic curves. As before, the intersection of the load line and the device curve defines the operating point Q.

The only new factor here is the dependence of Q on the control variable, and we must use a different curve for each different value of v_{GS}. Thus, if $v_{GS} = 4$ V the Q point is at $i_D = 3.6$ mA and $v_{DS} = 12$ V − (2 kΩ × 3.6 mA) = 4.8 V. If v_{GS} changes to 3.5 V, the Q point moves down the load line to $i_D = 1.6$ mA and $v_{DS} = 8.8$ V. Plotting Q-point values for different values of gate voltage then gives the *transfer curves* i_D and v_{DS} as functions of v_{GS} in Fig. 10.1–7b.

The transfer curves reveal that our circuit has three more-or-less distinct operating states, as follows:

1. The *cutoff state* is characterized by minimum current and maximum voltage,

$$i_D \approx 0 \qquad v_{DS} \approx V_{DD}$$

corresponding to the MOSFET's cutoff condition $v_{GS} \leq V_T$.

2. The *saturation state* is characterized by large current and small voltage,

$$i_D \approx V_{DD}/R_D \qquad v_{DS} \ll V_{DD}$$

corresponding to the MOSFET's ohmic region where it approximates a small resistance.

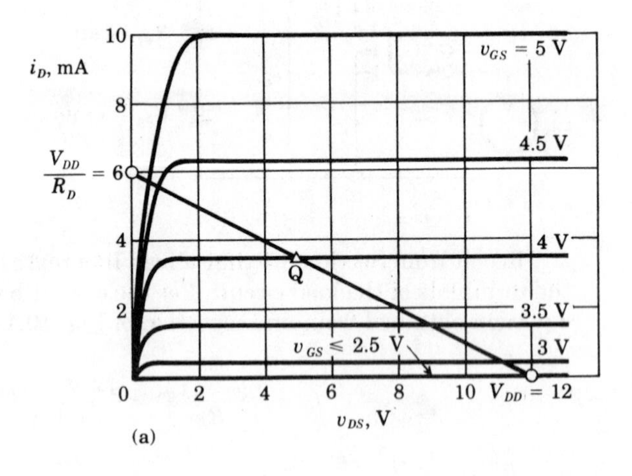

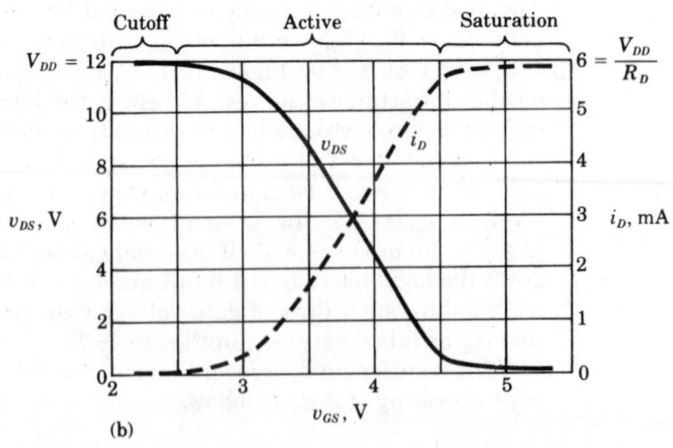

FIGURE 10.1–7
Load line and transfer curves of a MOSFET circuit.

3. The *active state,* between the extremes of cutoff and saturation, is characterized by i_D increasing with gate voltage while v_{DS} decreases. This state corresponds to the MOSFET operating as a voltage-controlled current source.

Observe that the values of i_D and v_{DS} in the cutoff and saturated states depend primarily on the load-circuit elements, and that changing v_{GS} by a few volts drives the circuit from cutoff to saturation, or vice versa. Also observe that the transfer curves are roughly linear near the middle of the active state so that small variations of v_{GS} around the mid-point will produce proportional variations of i_D and v_{GS}. These observations happen to be the keys to transistor switching circuits and amplifiers, as demonstrated in the following examples.

Example 10.1–1
MOSFET switching circuit

Let the applied gate voltage in Fig. 10.1–6 be a sinusoid $v_{GS}(t) = 6 \sin \omega t$ V, as sketched in Fig. 10.1–8a. Figure 10.1–8b shows the resulting output waveforms $i_D(t)$ and $v_{DS}(t)$ obtained with the help of the

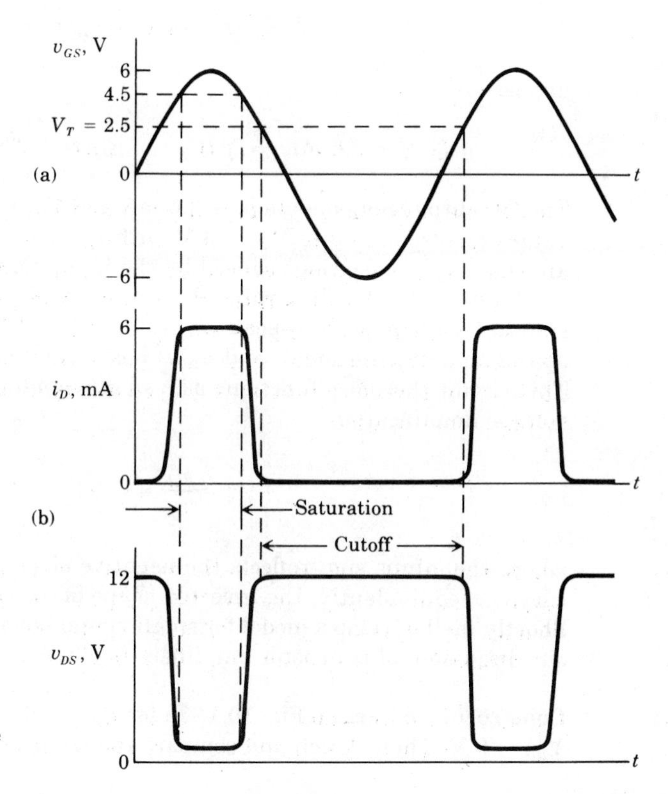

FIGURE 10.1–8
Waveforms for Example 10.1–1.

transfer curves and the graphical projection method from Fig. 4.2–3. The distorted shape of these waveforms comes about because the circuit is cut off ($i_D \approx 0$, $v_{DS} \approx 12$ V) whenever $v_{GS}(t) \leq 2.5$ V and saturated ($i_D \approx$ 6 mA, $v_{DS} \approx 1$ V) whenever $v_{GS}(t) > 4.5$ V.

In view of the severe output distortion, we certainly don't have a useful amplifier here. Instead, we have a clipping or waveshaping circuit or, more fundamentally, a *switching circuit* that changes suddenly from saturation (ON) to cutoff (OFF) when the gate voltage changes by about 2.0 V. The MOSFET thereby turns power ON and OFF to the load resistance R_D, and does so without drawing power from the controlling input. (Chapter 11 goes into the details of transistor switching circuits and their applications.)

Example 10.1–2
MOSFET amplifier

For relatively undistorted amplification, a MOSFET circuit must be restricted to *small variations* of voltage and current around a *DC operating point in the active region*. This condition is implemented in Fig. 10.1–9a by adding a biasing battery $V_{GS} = 4$ V to the small input signal $v_{gs}(t)$ such that

$$v_{GS}(t) = 4 \text{ V} + v_{gs}(t)$$

and so

$$i_D(t) = 3.6 \text{ mA} + i_d(t) \qquad v_{DS}(t) = 4.8 \text{ V} + v_{ds}(t)$$

The DC output components $I_D = 3.6$ mA and $V_{DS} = 4.8$ V are the Q-point values corresponding to $V_{GS} = 4$ V with $v_{gs}(t) = 0$, while $i_d(t)$ and $v_{ds}(t)$ are the output variations caused by the input signal $v_{gs}(t)$.

Figure 10.1–9b illustrates these waveforms, showing $v_{gs}(t)$ to be a sinusoid with a peak-to-peak swing of 0.5 V. The output signals are approximately sinusoidal, and $v_{ds}(t)$ has a peak-to-peak swing of 4.8 V. This circuit therefore functions as a *small-signal amplifier* and provides voltage amplification

$$A_v = \frac{v_{ds}}{v_{gs}} \approx -9.6$$

where the minus sign reflects the negative slope of the voltage transfer curve or, equivalently, the inverted shape of $v_{ds}(t)$ as compared to $v_{gs}(t)$. Shortly we'll develop a model for small-signal behavior in preparation for our discussion of transistor amplifiers in Chapter 12.

Exercise 10.1–1

Construct load lines on Fig. 10.1–7a for $R_D = 1.5$ kΩ with $V_{DD} = 12$ V and $V_{DD} = 6$ V. Then sketch and compare the transfer curves.

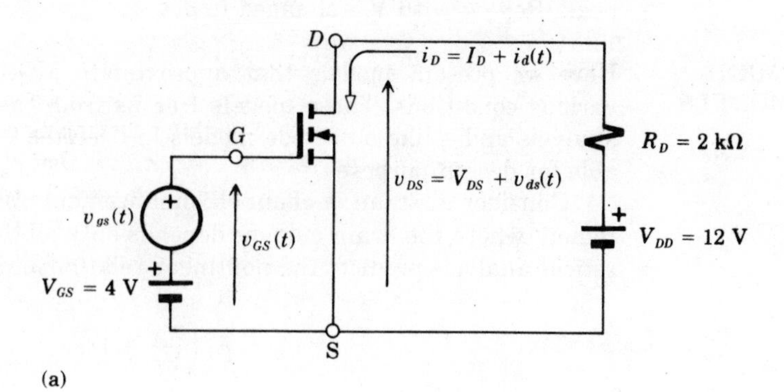

(a)

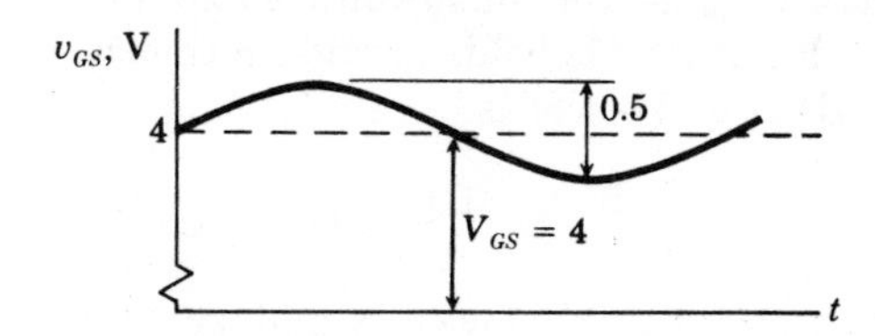

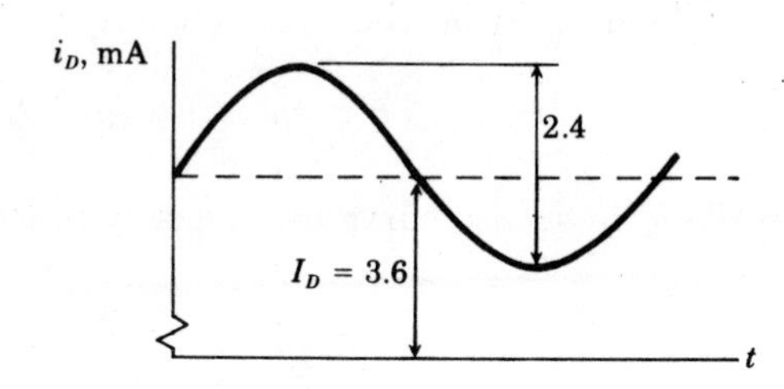

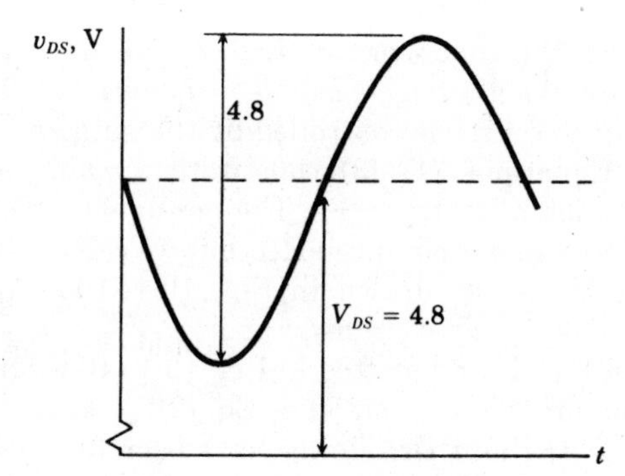

(b)

FIGURE 10.1-9
Small-signal amplifier:
(a) circuit;
(b) waveforms.

Exercise 10.1–2 Sketch $i_D(t)$ and $v_{DS}(t)$ for Fig. 10.1–9a when $v_{gs}(t)$ swings 2 V peak to peak. Repeat with V_{GS} changed to 3 V.

ENHANCEMENT MOSFET MODELS

Here we present models that approximate MOSFET behavior under various conditions. These models free us from the tedium of graphical analysis and—like our diode models in Chapter 9—prove to be invaluable for design purposes.

Consider first an *n*-channel enhancement MOSFET in its active region, where the drain current depends only on the gate voltage. Theoretical analysis predicts the nonlinear relationship

$$i_D = K\left(\frac{v_{GS}}{V_T} - 1\right)^2 \tag{1}$$

where V_T is the threshold voltage and K equals the value of i_D when $v_{GS} = 2V_T$. Equation (1) holds providing that v_{DS} is below the breakdown voltage BV_{DS} and satisfies

$$v_{DS} > v_{GS} - V_T \tag{2}$$

which defines the lower boundary of the active region. As for the ohmic-region behavior, theory predicts that if

$$0 \le v_{DS} \le \tfrac{1}{4}(v_{GS} - V_T) \tag{3a}$$

then the i_D-versus-v_{DS} curve has a nearly constant slope of $1/R_{DS}$, where

$$R_{DS} = \frac{V_T/2K}{(v_{GS}/V_T) - 1} \tag{3b}$$

Under the conditions of Eq. (3), we have $i_D \approx v_{DS}/R_{DS}$, the MOSFET acting like a *voltage-variable resistance* (VVR) whose resistance R_{DS} from drain to source is controlled by the gate voltage.

Equations (1)–(3) constitute a *mathematical model* of MOSFET behavior when $v_{GS} > V_T$. The nonlinear nature of these expressions precludes a linear circuit model, but we can construct "universal" characteristic curves as drawn in Fig. 10.1–10a. Another universal MOSFET curve is the *active transfer characteristic*, i_D versus v_{GS}, obtained directly from Eq. (1) and plotted in Fig. 10.1–10b. This plot displays the transfer behavior when v_{DS} satisfies Eq. (2).

Note that these curves have been fully labeled in terms of the device parameters K and V_T, and therefore represent *any* *n*-channel enhancement MOSFET. (They also apply to *p*-channel devices with the previously noted change of variables.) Accordingly, device manufacturers need not

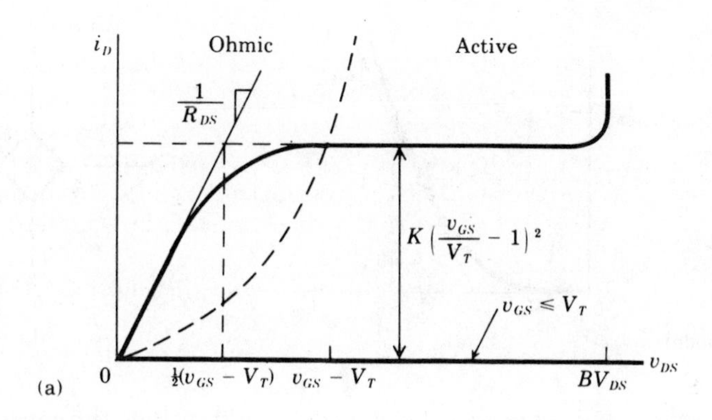

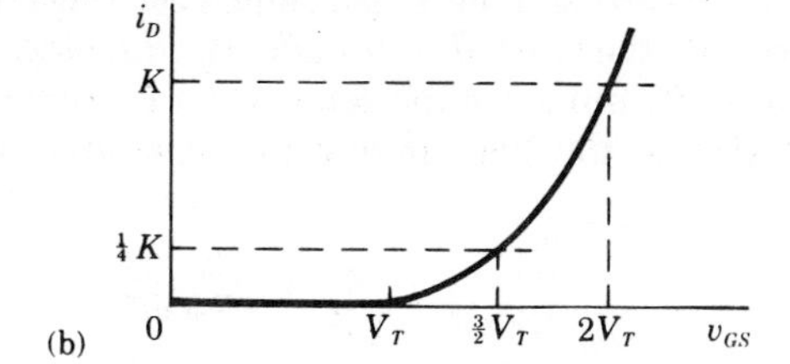

FIGURE 10.1–10
(a) Universal enhancement MOSFET characteristic curves.
(b) Active transfer characteristic.

supply actual curves for each particular model. And they usually don't. Instead, they merely list the parameter values or equivalent data on the specification sheet. Typical MOSFETs have $K \approx$ 1–100 mA and $V_T \approx$ 1–6 V, with a breakdown voltage BV_{DS} in the range of 20–50 V.

For a linear circuit model of the MOSFET, we assume the same small-signal conditions as in Example 10.1–2. Specifically, let

$$v_{GS}(t) = V_{GS} + v_{gs}(t)$$

$$i_D(t) = I_D + i_d(t)$$

so there are small variations, $v_{gs}(t)$ and $i_d(t)$, around a DC operating point ($v_{GS} = V_{GS}$, $i_D = I_D$) in the active region. In Figure 10.1–11a we interpret this condition, using the active transfer characteristic.

If the input variations are small enough, we can replace the transfer characteristic by a straight line tangent to it at Q. Then

$$i_d(t) \approx g_m v_{gs}(t) \tag{4}$$

where g_m stands for the *slope* of the tangent line. Equation (4) justifies the *small-signal model* of Fig. 10.1–11b, which represents the MOSFET as a voltage-controlled current source with an open circuit at the gate termi-

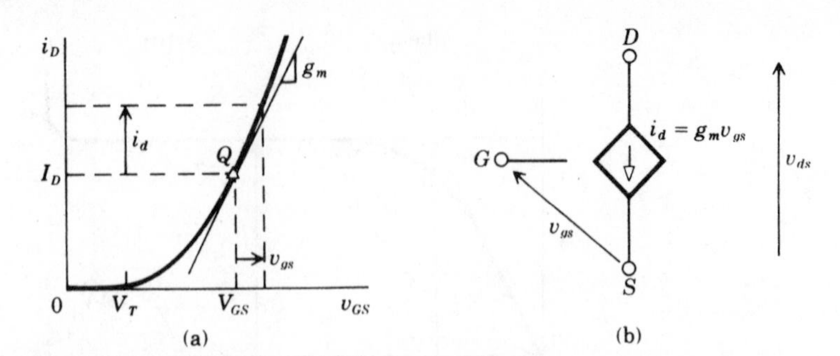

FIGURE 10.1-11
(a) Small-signal analysis.
(b) Small-signal model.

nal. (A more refined model would include the capacitance associated with the gate structure and, perhaps, some resistance.)

We call g_m the *mutual* or *transfer conductance,* or more simply *transconductance;* its units are the same as for conductance, namely mhos (℧). Mathematically, the slope in question is defined by

$$g_m \triangleq \left.\frac{di_D}{dv_{GS}}\right|_Q \tag{5}$$

the subscript Q indicating that the derivative must be evaluated at the operating point. Differentiating Eq. (1) and setting $v_{GS} = V_{GS}$ then yields

$$g_m = g_{mo}\left(\frac{V_{GS}}{V_T} - 1\right) \tag{6a}$$

with

$$g_{mo} = \frac{2K}{V_T} \tag{6b}$$

which equals the slope at $v_{GS} = 2V_T$.

Example 10.1-3

Comparing Fig. 10.1-10a with the characteristic curves previously given in Fig. 10.1-7a, we see that that particular MOSFET has $V_T = 2.5$ V and $K \approx 10$ mA since $i_D \approx 10$ mA when $v_{GS} = 5\text{ V} = 2V_T$. The breakdown region was not shown; but, presumably, BV_{DS} exceeds 12 V.

If we desire this MOSFET to operate in the active state with $i_D = 3.6$ mA, for instance, the value of v_{GS} can now be found analytically without using the characteristic curves. Inserting numbers into Eq. (1) and rewriting it as

$$\left(\frac{v_{GS}}{2.5\text{ V}} - 1\right) = \sqrt{\frac{3.6\text{ mA}}{10\text{ mA}}} = \pm 0.6$$

we get

$$v_{GS} = 2.5\text{ V }(1 \pm 0.6) = 4.0\text{ V} \qquad \text{or} \qquad 1.0\text{ V}$$

The correct value must be 4.0 V, since 1.0 V is less than V_T and would put the MOSFET in cutoff with $i_D = 0$. Equation (2) then requires that $v_{DS} > 4 - 2.5 = 1.5$ V for active-state operation. These results are consistent with the Q point determined by the graphical load-line method.

To analyze the small-signal behavior of the amplifier circuit in Fig. 10.1–9, we use Eq. (6) to obtain $g_{mo} = 2 \times 10$ mA/2.5 V = 8 m℧ and $g_m = 8\,[(4/2.5) - 1] = 4.8$ m℧ with $V_{GS} = 4$ V. Since Fig. 10.1–11b represents variations with respect to the DC operating point, and since $v_{ds}(t) = -R_D i_d(t)$, our complete *small-signal equivalent circuit* becomes as diagrammed in Fig. 10.1–12. This circuit predicts that $v_{ds} = -R_D i_d = -R_D(g_m v_{gs}) = -9.6 v_{gs}$, so

$$A_v = \frac{v_{ds}}{v_{gs}} = -R_D g_m = -2\ \text{k}\Omega \times 4.8\ \text{m℧} = -9.6$$

agreeing with our previous result.

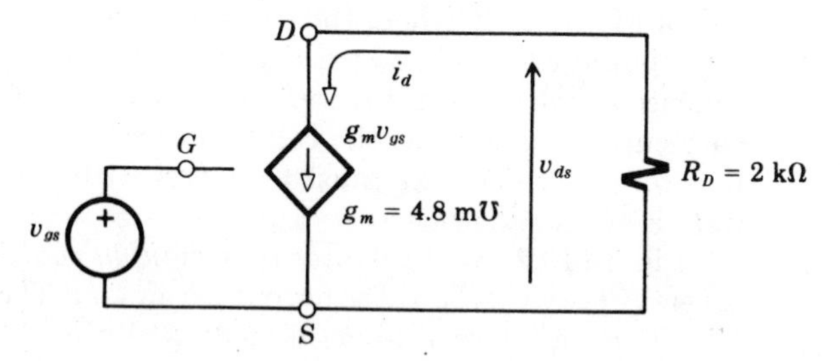

FIGURE 10.1–12 Small-signal equivalent circuit for Example 10.1–3.

Exercise 10.1–3

If a MOSFET has $K = 16$ mA and $V_T = 4$ V, what values of v_{GS} and v_{DS} give $i_D = 9$ mA in the active region?

Exercise 10.1–4

Let the above MOSFET be put in the circuit of Fig. 10.1–9 with $R_D = 3$ kΩ and $V_{DD} = 12$ V. Find V_{GS} such that $A_v = -6$, and calculate the resulting values of I_D and V_{DS}.

10.2 BIPOLAR JUNCTION TRANSISTORS

The family of bipolar junction transistors has just two branches, the *npn* and *pnp* BJT, whose circuit symbols are shown in Fig. 10.2–1. Both devices contain a *semiconductor junction* between the *base* (B) and the *emitter* (E). This junction conducts like a diode in the direction indicated by the arrowhead, and the amount of base current controls the current at the *collector* (C). The term *bipolar* reflects the fact that the base current consists of positive and negative charges, as distinguished from the unipolar charge flow in a field-effect transistor.

You've probably guessed by now that the *npn* and *pnp* BJTs are *complementary* devices. We may therefore concentrate on the *npn* BJT, which has many similarities to the *n*-channel enhancement MOSFET. As we've

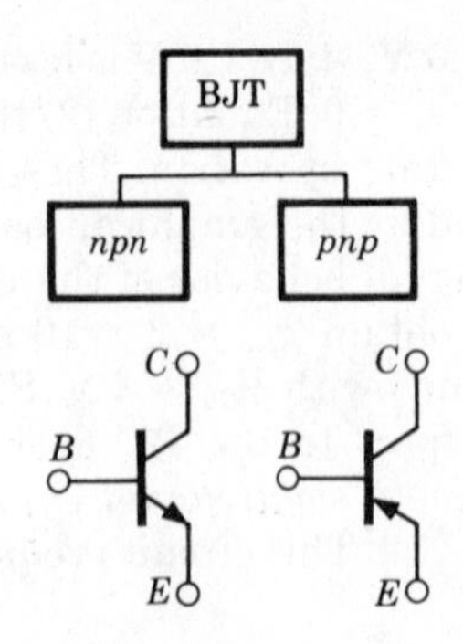

FIGURE 10.2–1
BJT family tree.

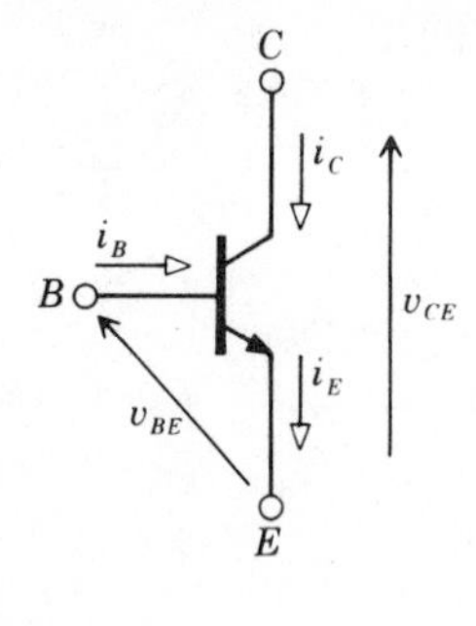

FIGURE 10.2–2
npn transistor in the common-emitter configuration.

done previously, we'll first consider external characteristics and circuit models and then take up device physics in the next section.

COMMON-EMITTER CHARACTERISTICS

The majority of BJT applications call for the *common-emitter configuration* of Fig. 10.2–2, where the emitter serves as the common terminal for the input at the base and the output at the collector. The normal voltage polarities and current directions for an *npn* BJT are as shown, and the base current i_B controls the collector current i_C and the emitter current $i_E = i_C + i_B$. Hence, we must now deal with *two* currents, i_B and i_C, as well as two voltages.

Figure 10.2–3a illustrates typical *input characteristics* in the form of i_B versus v_{BE} with v_{CE} as the output parameter. These curves take the general shape of a forward-biased diode with threshold voltage $V_T \approx 0.7$ V, and they show only a slight dependence on v_{CE}. Consequently, we can reasonably approximate the input using an ideal diode plus V_T, independent of the output. This approximation greatly simplifies the analysis and design of BJT circuits.

Turning to the *output characteristics* in Fig. 10.2–3b shows somewhat idealized curves of i_C versus v_{CE} for fixed values of i_B (actual device

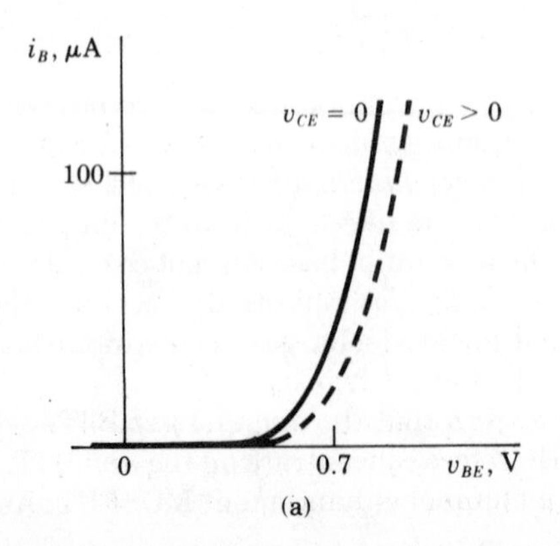

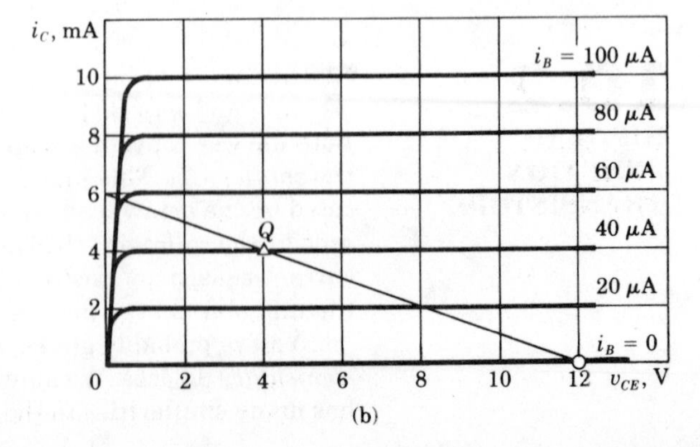

FIGURE 10.2–3
Common-emitter characteristics:
(a) input; (b) output.

curves would have less uniform vertical spacing and would tend to rise a little with increasing v_{CE} until breakdown occurs at BV_{CE}). These characteristics resemble FET curves and, like a FET, the BJT has an *active region* where it functions as a *controlled source*. But now the output current i_C is controlled by current i_B, rather than a voltage.

The "knees" that form the left-hand boundary of the active region occur at lower voltages than those of a FET, and there is a more uniform vertical spacing between curves. The region labeled *saturation* corresponds to the ohmic region of a FET. The BJT is in *cutoff* with $i_C \approx 0$ if $i_B = 0$, which is analogous to the FET cutoff condition.

Example 10.2–1

The similarities and differences noted in the previous discussion become more evident when we consider the circuit in Fig. 10.2–4a. The output side has the same element values as our previous FET circuit (see Fig. 10.1–6), while the input includes a current-limiting resistance R_B in series with the voltage source v_B. If $v_B > 0.7$ V, the base is forward-biased and

$$i_B \approx \frac{v_B - 0.7 \text{ V}}{R_B}$$

Otherwise, $i_B \approx 0$ for $v_B < 0.7$ V.

Again, we want to find transfer curves that display the dependence of output current i_C and voltage v_{CE} on the input variable v_B. Proceeding as before, we superimpose the load line for the output circuit on the characteristic curves and locate operating points for various values of i_B. These, in turn, lead to the transfer curves plotted in Fig. 10.2–4b.

Comparing the BJT transfer curves with those of a FET (Fig. 10.1–7b) shows that both circuits have an active operating state between the extremes of cutoff and saturation. We conclude, then, that the BJT circuit, as well as the FET, can function like an electronic switch or a

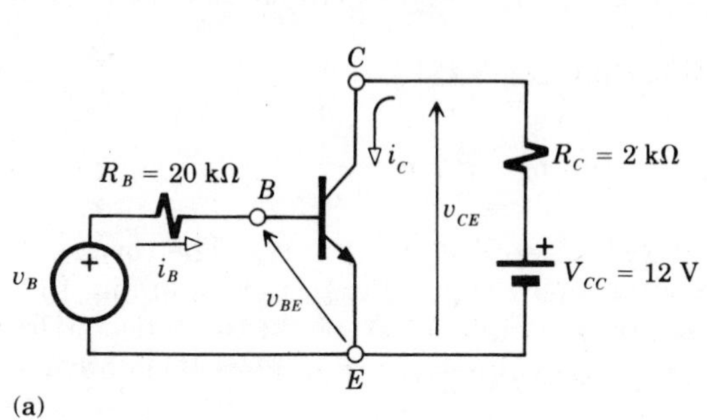

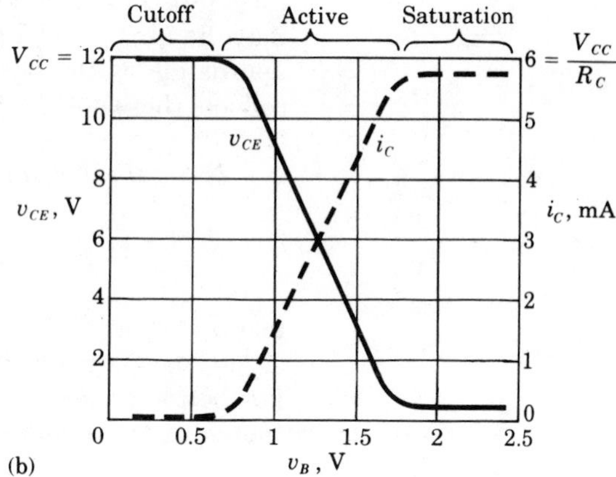

FIGURE 10.2–4 Circuit and transfer curves for Example 10.2–1.

small-signal amplifier. Indeed, the BJT has better amplification characteristics in view of the more linear shape of the transfer curves throughout its active region. For the same reason, a BJT amplifier can produce larger undistorted voltage and current variations, thereby delivering higher output signal power. (The recently developed power MOSFETs, however, show great promise for large-signal applications.) When used as a switching device, a BJT acts more nearly like a short circuit in the saturated or ON state, where v_{CE} is typically less than one volt.

On the other hand, a BJT circuit would require some power $p_{\text{in}} = v_{BE}\, i_B$ from the input source, whereas a FET amplifier has $p_{\text{in}} \approx 0$, since $i_G \approx 0$. An additional consideration favoring FETs is the fact that they require less space on an integrated-circuit chip, a topic we pursue in Section 10.4.

Exercise 10.2–1

Estimate the small-signal voltage amplification A_v and current amplification A_i for Fig. 10.2–4 with $v_B(t) = 1.3\text{ V} + 0.2 \sin \omega t\text{ V}$.

BJT CIRCUIT MODELS

The nearly linear transfer curves in Fig. 10.2–4b suggest the possibility of constructing a *linear* circuit model for a BJT with *large* signal variations—something that could not be done for a FET. Actually, we'll develop three large-signal models, corresponding, respectively to the active, cutoff, and saturation regions of the common-emitter characteristics. Then we'll consider the small-signal model.

A common-emitter BJT circuit in the *active state* must have $i_B > 0$ which, from Fig. 10.2–3a, requires that $v_{BE} \approx V_T$. Referring to Fig. 10.2–3b then shows us that the collector current varies in almost direct proportion to the base current, so we write

$$i_C \approx \beta i_B \tag{1}$$

The proportionality constant β represents the DC *current gain*, and typically falls in the range of about 20 to 100. (Special *supergain* transistors may have $\beta > 1000$.) Equation (1) holds, providing $v_{CE} > v_{BE}$, which approximates the lower boundary of the active region. Figure 10.2–5a collects these observations in an approximate circuit model for the active state.

The *saturated state* is characterized by

$$i_C < \beta i_B \tag{2}$$

and occurs when $i_B > 0$ and $v_{BE} \approx V_T$ but $v_{CE} < v_{BE}$. The corresponding circuit model, neglecting the small value of v_{CE}, is given in Fig. 10.2–5b. For better accuracy, we could include a battery in the collector lead to represent the nonzero saturated value $v_{CE} = V_{\text{sat}}$. Most BJTs have $V_{\text{sat}} \approx$

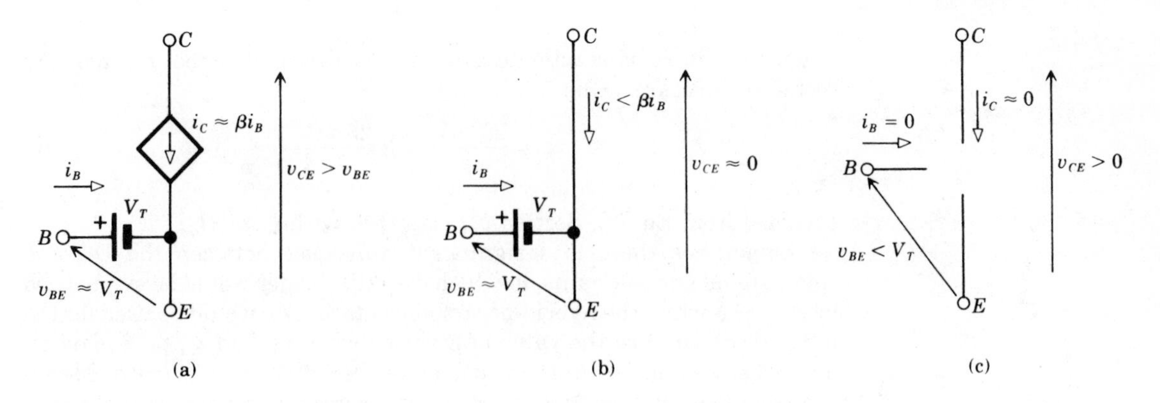

FIGURE 10.2–5
BJT large-signal models: (a) active; (b) saturated; (c) cutoff.

0.2 V, which is often inconsequential compared to the other circuit voltages.

The *cutoff state* occurs when $v_{BE} < V_T$, $i_B = 0$, and $v_{CE} > 0$, in which case

$$i_C \approx I_{CEO} \tag{3}$$

where we have introduced the *collector cutoff current* I_{CEO}. At room temperature, this current is usually less than a few microamps and could not be distinguished from zero on the characteristic curves in Fig. 10.2–3b. To the extent that I_{CEO} may be ignored, a BJT in cutoff is simply a three-way open circuit with $i_C = i_B = 0$ as in Fig. 10.2–5c.

Because the collector cutoff current increases exponentially with temperature, it might become an important consideration if the temperature rises. To include I_{CEO} in Fig. 10.2–5c, we merely complete the collector-to-emitter connection with a current source producing $i_C = I_{CEO}$. That same source could also be put in parallel with the controlled source in Fig. 10.2–5a for a more refined model of the active state.

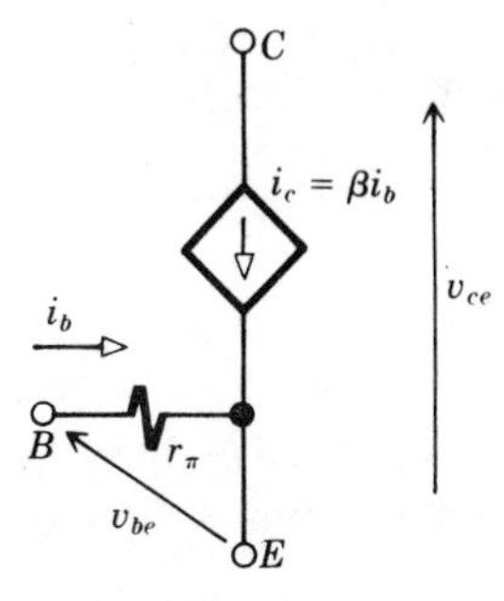

FIGURE 10.2–6
Common-emitter small-signal model.

The rough but handy large-signal models in Fig. 10.2–5 facilitate the study of BJT switching circuits that switch between ON (saturated) and OFF (cutoff). For a BJT amplifier circuit, we would use the large-signal active-state model to determine the DC operating point, and then invoke a small-signal model for the signal variations around the operating point.

The common-emitter *small-signal* model in Fig. 10.2–6 is derived by assuming active-region operation and letting $i_B(t) = I_B + i_b(t)$, $i_C(t) = I_C + i_c(t)$, etc. It then follows from Eq. (1) that $i_c(t) \approx \beta i_b(t)$, meaning that the *small-signal current gain* approximately equals the DC current gain and that the output circuit simply consists of a current-controlled current source. For the input side of the model we take the approach we used in Section 9.2 to determine the *small-signal resistance* of the base-to-emitter diode characteristic. This resistance, denoted r_π, has the room-temperature value

$$r_\pi \approx \frac{25 \text{ mV}}{I_B} = \beta \frac{25 \text{ mV}}{I_C} \tag{4}$$

obtained from Eq. (3), Section 9.2, together with $I_C \approx \beta I_B$.

Sometimes there is a significant difference between the DC and small-signal current gains, in which case the former would be symbolized by β_0 or β_{DC} while the hybrid-parameter notation h_{fe} would be used for the latter. Furthermore the value of β varies with v_{CE} and i_C, producing an upward slope and nonuniform spacing in the active-region curves. Manufacturers usually specify a typical range or minimum value of β at a specific operating point for each particular model. It then stands to reason that BJT circuit designs should not depend critically upon the value of β. By the same token, our approximate large- and small-signal models will be sufficiently accurate for most practical work. If desired, the active-region slope can be represented by a resistance in parallel with the controlled source in Fig. 10.2–5a or 10.2–6.

Example 10.2–2

Given the amplifier circuit of Fig. 10.2–7a with specified values for R_C, V_{CC}, and the BJT parameters β and V_T, we want to find the values for R_B and V_B in order to obtain $A_v = v_{ce}/v_b = -20$. Another matter of concern is the limitation on $v_b(t)$, such that the circuit stays within the active state, thus preventing serious distortion of the output signal variations.

We start by noting that $i_{C_{\max}} = V_{CC}/R_C = 5$ mA, which suggests that we take the Q point at $I_C = \frac{1}{2} i_{C_{\max}} = 2.5$ mA to allow for positive and negative signal swings. The other DC values then are $V_{CE} = V_{CC} - R_C I_C = 12.5$ V and $I_B \approx I_C/\beta = 0.025$ mA; and Eq. (4) yields $r_\pi \approx 25 \text{ mV}/I_B = 1 \text{ k}\Omega$.

To obtain the value of R_B, we replace the BJT by its small-signal model and drop the Q-point quantities, as shown in Fig. 10.2–7b. The resulting small-signal relationships are $v_{ce} = -5 \text{ k}\Omega \times 100\, i_b$ and $i_b = v_b/(R_B + 1 \text{ k}\Omega)$, so

$$A_v = \frac{v_{ce}}{v_b} = -\frac{100 \times 5 \text{ k}\Omega}{R_B + 1 \text{ k}\Omega} = -20$$

whose solution gives $R_B = 24 \text{ k}\Omega$.

To obtain the value of the DC bias voltage V_B, we replace the BJT by its active-state large-signal model and drop the small-signal terms, as shown in Fig. 10.2–7c. Thus,

$$I_B = \frac{V_B - 0.7 \text{ V}}{24 \text{ k}\Omega} = 0.025 \text{ mA}$$

whose solution gives $V_B = 1.3$ V.

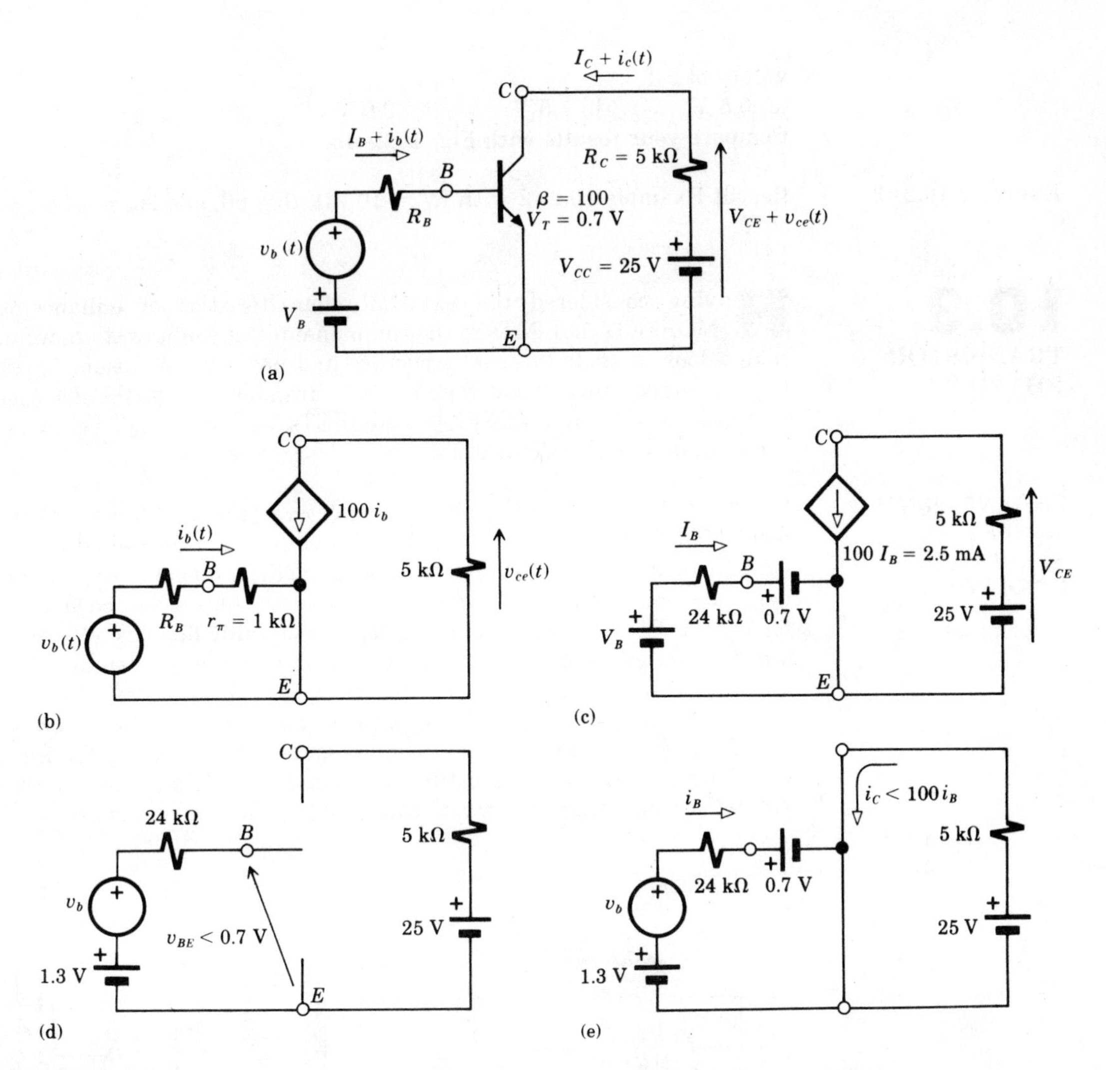

FIGURE 10.2–7
Circuits for Example 10.2–2.

Finally, to obtain the limitations on $v_b(t)$, we consider the large-signal variations that would drive the circuit into cutoff or saturation. The cutoff condition of Fig. 10.2–7d requires $v_b + 1.3\text{ V} < 0.7\text{ V}$ or $v_b < -0.6$ V. The saturated condition of Fig. 10.2–7e has $i_B = (v_b + 1.3\text{ V} - 0.7\text{ V})/24\text{ k}\Omega$ and requires $i_C = 5\text{ mA} < 100\ i_B$, which means that $v_b > +0.6$ V. These conditions will be prevented, and the circuit will stay in the active state if we impose the limitation $-0.6 \le v_b(t) \le +0.6$ V.

Exercise 10.2–2 Assuming that the BJT in Fig. 10.2–4a has $\beta = 100$ and that $V_T = 0.7$ V, use the large-signal models to find i_B, i_C, and v_{CE} for each of the following

values of v_B:
(a) 0.5 V, (b) 1.5 V, (c) 2.0 V.
Compare your results with Fig. 10.2–4b.

Exercise 10.2–3 Repeat Example 10.2–2 with $R_C = 10\ \text{k}\Omega$, $\beta = 60$, and $A_v = -30$.

10.3 TRANSISTOR PHYSICS †

Having considered the external characteristics of enhancement MOSFETs and BJTs in the common-emitter configuration, we now take a look at their internal structure and physical operation. In addition, to round out our coverage of the transistor family, the discussion will include depletion MOSFETs and JFETs, as well as BJT behavior in the common-base configuration.

ENHANCEMENT MOSFET STRUCTURE AND OPERATION

Figure 10.3–1a depicts the physical structure of an *n*-channel enhancement MOSFET. The device is built on a piece of silicon, called the *substrate,* which has moderate *p*-type doping. Heavily doped n^+-type regions near each end provide high-conductivity contacts for the source and drain terminals. The gate terminal connects to a metallic film (usually aluminum) insulated from the substrate by a thin oxide layer (usually silicon dioxide). The resistance of the oxide layer is so large—$10^{12}\ \Omega$ or more—that current through the gate is entirely negligible. (Thus, a MOSFET is also termed an *insulated-gate* FET or IGFET.) Figure 10.3–1b shows the same structure reoriented to correspond to the circuit symbol, which has the substrate connected to the source terminal.

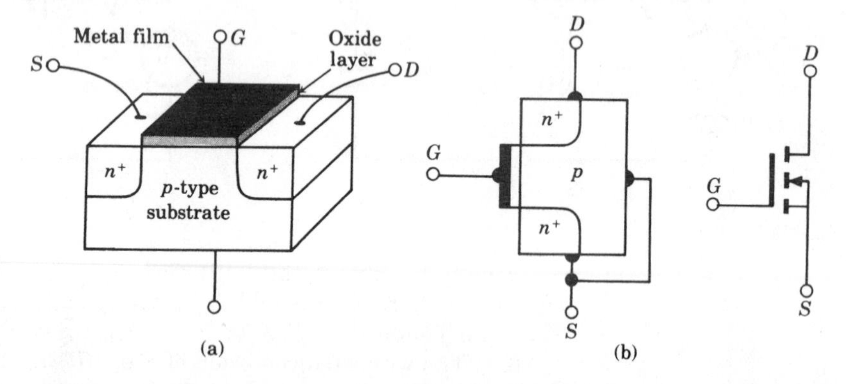

FIGURE 10.3–1 Structure of an *n*-channel enhancement MOSFET.

If the gate voltage v_{GS} equals zero, no appreciable drain current flows when $v_{DS} > 0$ because of the *reverse-biased pn junction* at the drain end. (There would be a reverse-biased junction at the source end if $v_{DS} < 0$.) However, a positive voltage applied to the gate produces an electric field $\mathcal{E}$ across the substrate that pushes holes to the right and pulls electrons to the left toward the gate, as shown in Fig. 10.3–2a. A stronger field results in the situation of Fig. 10.3–2b where the region along the gate has become an *n-type channel,* due to attracted mobile electrons, and a *depletion region* devoid of mobile carriers insulates the channel from the rest of the substrate.

Now current i_D is possible when $v_{DS} > 0$, and it consists almost entirely of majority-carrier electrons that flow upward through the channel from source to drain. The electric field has thereby *enhanced* conduction. No portion of i_D goes through the substrate because it forms a reverse-biased *pn* junction with the channel. The arrowhead on the circuit symbol points in the forward direction of this junction.

The gate-threshold voltage V_T is the minimum value of v_{GS} required to form the channel. If $v_{GS} \leq V_T$ the MOSFET is cutoff, with $i_D \approx 0$. Increasing v_{GS} above V_T enhances conduction by increasing the channel width w, and i_D increases. For small positive values of v_{DS} the channel appears to be a resistance R_{DS} whose value is inversely proportional to w. This explains the shape of the characteristic curves in the ohmic region.

For larger values of v_{DS}, the gate-to-drain voltage $v_{GD} = v_{GS} - v_{DS}$ is significantly reduced, and the field near the drain end will be less than that near the source. In fact, if $v_{DS} > v_{GS} - V_T$, then $v_{GD} < V_T$ and one might expect the channel to be "pinched off" at the top. But complete pinchoff ($w \to 0$) cannot happen because the potential difference v_{DS} requires nonzero current through the channel. Instead, the channel becomes "pinched down" as in Fig. 10.3–2c, and i_D remains constant rather than increasing with v_{DS}. This behavior corresponds to the MOSFET's active region where it functions as a current source controlled by the gate voltage.

Excessively large v_{DS} products *avalanche breakdown* between the drain and substrate, and i_D abruptly rises. To avoid possible damage, the device should be operated below the breakdown voltage BV_{DS}. The gate voltage has a similar upper limit. Moreover, since the gate acts as one

FIGURE 10.3–2
(a) Electric field in an enhancement MOSFET.
(b) Formation of *n*-type channel when $v_{GS} > V_T$.
(c) Channel pinched down when $v_{GD} < V_T$.

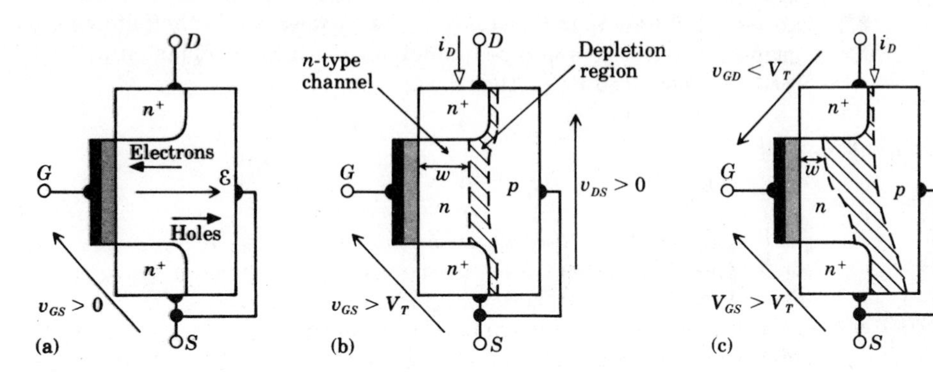

plate of a nearly leakage-free capacitor, stray electrostatic charges may build up to the point that the oxide layer is ruptured and permanently destroyed by high-voltage breakdown. For this reason, one must be extremely careful when handling MOSFETs. For the same reason, manufacturers of integrated circuits often build in *diode protection circuits* connected to MOSFET gates.

The foregoing description qualitatively justifies the characteristic curves and *i-v* relationships presented in Section 10.1 for the *n*-channel device. The structure of the *p-channel* enhancement MOSFET differs from Fig. 10.3–1 only in its end regions, which are p^+-type, and its substrate, which is *n*-type. As a result, *negative* gate voltage creates a *p*-type channel with mobile holes that flow from source to drain, constituting $i_D < 0$ when $v_{DS} < 0$. (The arrowhead on the circuit symbol back in Fig. 10.1–1 now points from the *p*-type channel to the *n*-type substrate.) The complementary behavior of a *p*-channel MOSFET follows directly from the interchange of *n*-type and *p*-type doping.

If the construction of the MOSFET is physically symmetrical, as is often the case, the channel conducts equally well in either direction. Thus, a *p*-channel MOSFET could have $i_D > 0$, like an *n*-channel device, and vice versa. The only operational difference would then be the polarity of the gate voltage. However, the greater mobility of electrons favors *n*-channel MOSFETs for many applications. Special-purpose CMOS devices combine complementary *n*-channel and *p*-channel MOSFETs in one unit.

DEPLETION MOSFETS AND JFETS

The enhancement MOSFET may be described as a "normally OFF" device, in the sense that $i_D \approx 0$ when $v_{GS} = 0$. On the other hand, depletion MOSFETs and JFETs are "normally ON," with significant channel conductance and drain current when $v_{GS} = 0$. The unbroken channel lines on the circuit symbols in Fig. 10.3–3 reflect this property. A negative voltage applied to the gate *depletes* the channel of mobile electrons, assuming an *n*-type channel, and decreases i_D as we describe in the following paragraphs. The less common *p*-channel devices have complementary characteristics, and need no further discussion.

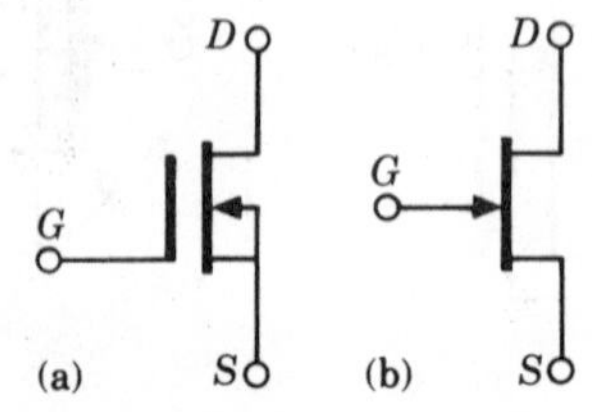

FIGURE 10.3–3 Depletion MOSFET and JFET (*n*-channel).

We build a depletion MOSFET by adding a lightly doped n-type channel to the enhancement MOSFET structure, as in Fig. 10.3–4a. This channel exists and permits conduction even when $v_{GS} = 0$. Applying a negative gate voltage produces an electric field that attracts holes toward the gate and pushes electrons away. Electron-hole recombinations then create *two depletion regions* as shown in Fig. 10.3–4b—one of which reduces the effective channel width w, thereby decreasing i_D for a given value of v_{DS}. The depletion MOSFET therefore has an ohmic region of operation where it acts like a voltage-variable resistance for small values of v_{DS}.

Eventually, the depletion region completely blocks the channel, as in Fig. 10.3–4c; and drain current is "pinched off." This condition corresponds to the cutoff state and occurs when $v_{GS} \leq V_P$, with V_P being a negative quantity known as the *pinchoff voltage*.

With $v_{GS} > V_P$, the channel becomes "pinched down" near the drain end as v_{DS} increases, as shown in Fig. 10.3–4d, and i_D remains relatively constant. This corresponds, of course, to the active state where the device functions as a controlled current source. Theoretically, drain current and gate voltage in the active region are related by

$$i_D = I_{DSS}\left(1 - \frac{v_{GS}}{V_P}\right)^2 \tag{1}$$

where I_{DSS} equals the value of i_D when $v_{GS} = 0$. The lower boundary of the active region follows from $v_{GD} < V_P$ or

$$v_{DS} > v_{GS} - V_P \tag{2}$$

Comparing these expressions with Eqs. (1) and (2) in Section 10.1, shows that the parameters I_{DSS} and V_P are analogous to the enhancement MOSFET parameters K and V_T.

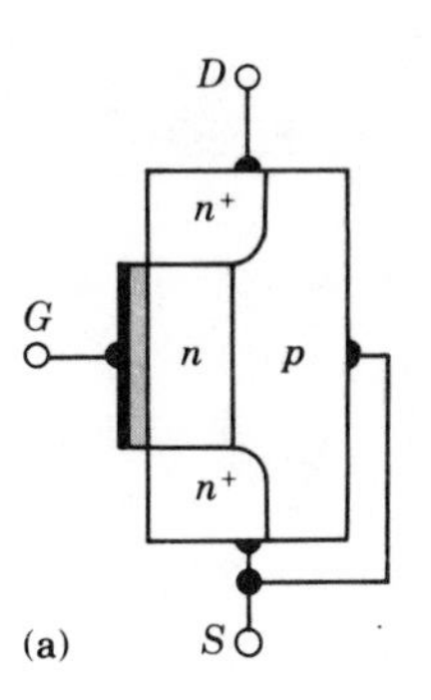

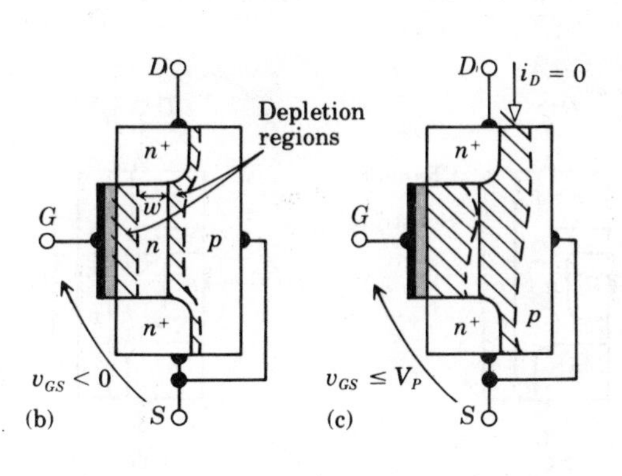

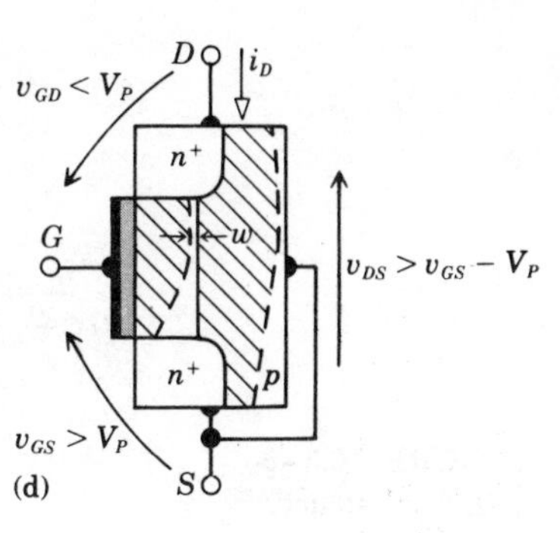

FIGURE 10.3–4 Depletion MOSFET: (a) structure; (b) depletion regions; (c) channel pinched off when $v_{GS} \leq V_P$; (d) channel pinched down when $v_{GS} > V_P$.

The characteristic curves and active transfer characteristic plotted in Fig. 10.3–5 further emphasize the similarity between depletion and enhancement MOSFETs. In fact, the channel in a depletion MOSFET becomes *enhanced* if $v_{GS} > 0$ so that electrons are drawn from the substrate into the channel. The dashed lines in the figure represent this enhancement mode.

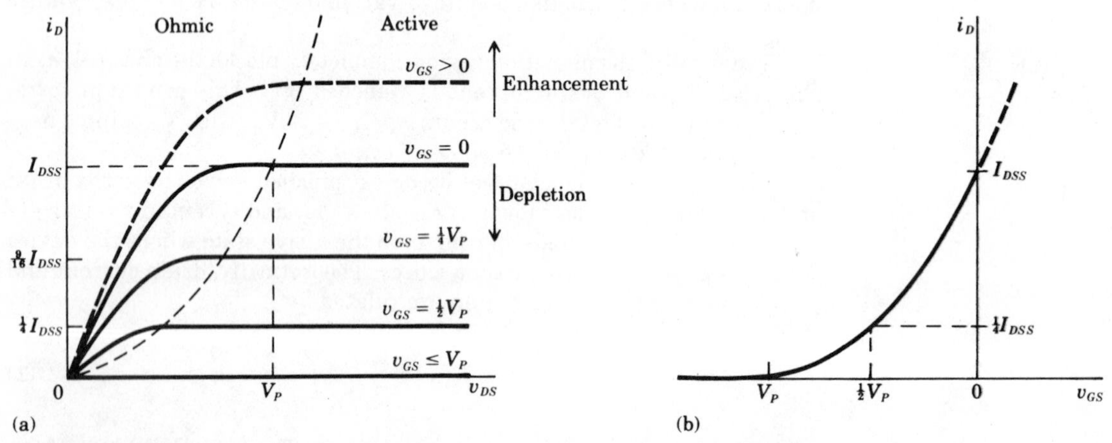

FIGURE 10.3–5 Characteristic curves and active transfer characteristic of depletion MOSFET (or JFET).

A *junction* FET (or JFET) has a *pn* junction rather than an insulating layer between the gate terminal and the channel. The junction is formed with a heavily doped p^+-type region in the *n*-type channel, as in Fig. 10.3–6a. (The *p*-type substrate, if present, plays no role other than structural.) The forward direction of this junction is from gate to channel, as indicated by the arrowhead on the circuit symbol in Fig. 10.3–3. Normal operating conditions call for a negative gate voltage to *reverse-bias* the junction, in which case there will be a *reverse gate current* $i_G \approx -I_{GSS}$ typ-

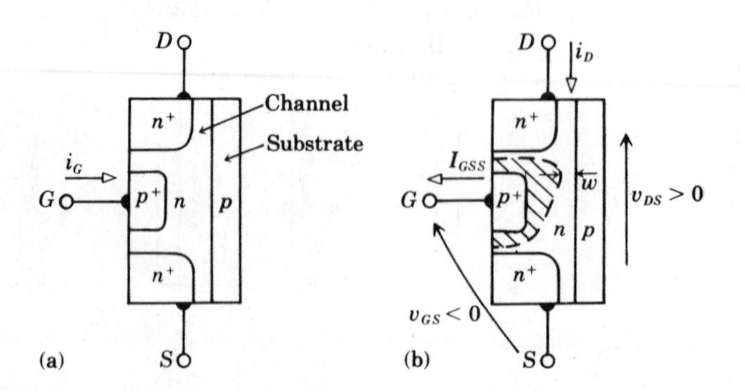

FIGURE 10.3–6 JFET: (a) structure; (b) depletion region.

ically around 1 nA—a small current but one much larger than that in a MOSFET.

Because of the reverse-biased junction, a depletion region with its accompanying electric field penetrates the channel and reduces the effective conducting width w (see Fig. 10.3–6b). Depending on the values of v_{GS} and v_{DS}, the channel may become pinched off or pinched down, just as in a depletion MOSFET. Thus, our previous device parameters (V_P and I_{DSS}), expressions, and curves also describe n-channel JFET behavior, with the additional restriction that $v_{GS} \leq 0$ V. (Actually, we could allow positive gate voltage up to the junction's threshold of about 0.7 V.)

Finally, the *small-signal model* of a depletion MOSFET or JFET is identical to Fig. 10.1–11b. The transconductance is given in terms of I_{DSS} and V_P by

$$g_m = g_{mo}\left(1 - \frac{V_{GS}}{V_P}\right) \tag{3a}$$

with

$$g_{mo} = \frac{-2I_{DSS}}{V_P} \tag{3b}$$

which has a positive value, since V_P is negative. Because v_{GS} is restricted to negative values, g_{mo} represents the *maximum* transconductance of a JFET.

Exercise 10.3–1 Given an n-channel depletion MOSFET with $I_{DSS} = 10$ mA and $V_P = -5$ V, explain how you could use a battery to make it act like an n-channel enhancement MOSFET with $K = 10$ mA and $V_T = 5$ V.

Exercise 10.3–2 Repeat Exercise 10.1–4 for a JFET with $I_{DSS} = 8$ mA and $V_P = -4$ V.

BJT STRUCTURE AND OPERATION

Figure 10.3–7a illustrates the structure of an *npn* transistor. The base is a thin layer of lightly doped p-type material sandwiched between the n-type collector and the heavily doped n^+-type emitter. The device therefore consists of *two pn junctions,* as brought out more clearly by Fig. 10.3–7b. Current always crosses at least one junction, and involves both minority and majority carriers. The circuit symbol in Fig. 10.3–7c identifies the emitter terminal with an arrowhead pointing in the forward direction of the emitter junction. A *pnp* transistor has reversed n-type and p-type regions, resulting in complementary characteristics.

A casual glance at the structure suggests that a BJT could be modeled by two diodes connected back to back at the base terminal. In fact, the current flow across either junction alone does follow the familiar *i-v* characteristic of a semiconductor diode. However, the two-diode model

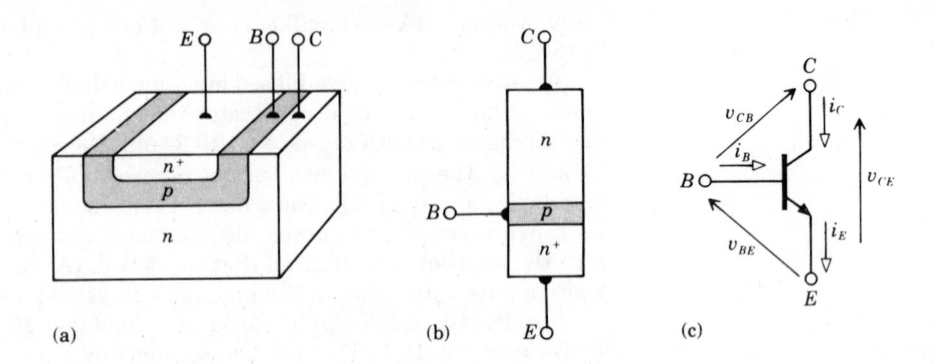

FIGURE 10.3–7 *npn* transistor structure.

fails to account for interaction within the shared base region when current crosses both junctions. And this is the condition in which the BJT acts as a controlled current source.

To start with a simplified case, let's suppose the collector junction is short circuited ($v_{CB} = 0$) and the emitter junction is forward biased ($v_{BE} > 0$), as in Fig. 10.3–8a. Although we might expect $i_C = 0$ and $i_E = i_B$, we actually find that $i_C \gg i_B$ and that $i_E \approx i_C$. These results are explained when we consider the current components shown in Fig. 10.3–8b.

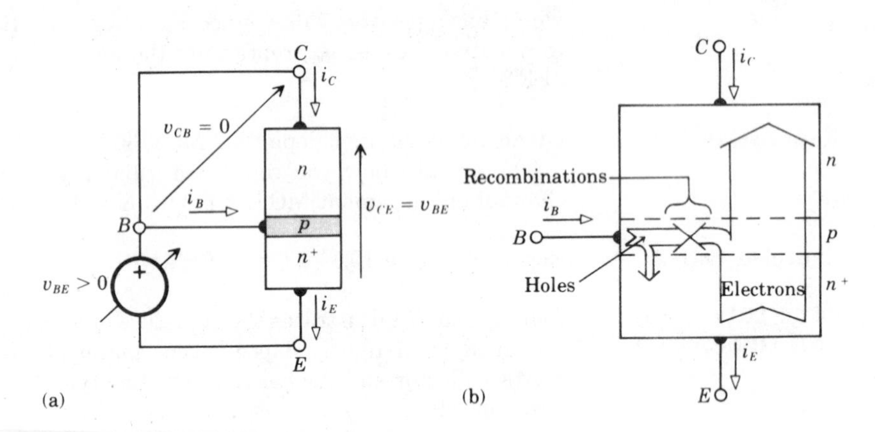

FIGURE 10.3–8

The emitter current consists primarily of electrons (majority carriers) that diffuse across the forward-biased junction into the base region. Holes likewise diffuse in the opposite direction, from base to emitter, but they amount to just a tiny portion of i_E in view of the relative base and emitter doping levels. Because the base has low conductivity and is very thin—usually less than 0.01 mm—most of the "emitted" or "injected" electrons travel completely across and become "collected," thereby constituting the collector current. (Electron flow from emitter to collector means $i_C > 0$, which is consistent with $v_{CE} > 0$.) Letting α stand for the collected fraction of emitted electrons, we write

$$i_C = \alpha i_E \tag{4a}$$

The electrons remaining in the base recombine with holes, constituting the base current

$$i_B = (1 - \alpha)\, i_E \tag{4b}$$

which omits the negligible base-to-emitter hole current.

Solving the foregoing expressions for i_E and i_C in terms of i_B yields $i_E = i_B/(1 - \alpha)$ and $i_C = \alpha i_B/(1 - \alpha)$. Therefore

$$i_C = \beta i_B \qquad i_E = (\beta + 1)\, i_B \tag{5}$$

where we have defined the current gain by

$$\beta \triangleq \frac{\alpha}{1 - \alpha} \tag{6}$$

Equation (5) clearly reveals that a small base current controls the much larger collector and emitter currents—provided that $\beta \gg 1$, which in turn requires that $\alpha \approx 1$. For instance, if 98% of the emitter electrons reach the collector, then $\alpha = 0.98$ and $\beta = 0.98/0.02 = 49$, so $i_C = 49\, i_B$ and $i_E = 50\, i_B \approx i_C$. Increasing the collection factor about 0.5% to $\alpha = 0.985$ yields $\beta \approx 66$, a 34% increase of the current gain.

The heavy emitter doping and large surface area of the collector in Fig. 10.3–7a are designed specifically for large current gain. Typically, α falls in the range 0.95–0.99, which corresponds to β from about 20 to 100. Operating a transistor in the *reverse* mode, with current pulling from collector and emitter, produces much smaller current gain. (Consequently, an FET would be used for most applications that require bidirectional currents.)

Now let the collector-to-base voltage take on a nonzero value, subject to the condition that $v_{CB} > -v_{BE}$ so $v_{CE} > 0$. The emitter-to-collector electron current will not change significantly, but the collector and base currents will include an additional term due to majority and minority carriers crossing only the collector junction. We denote this component by i_{CO} and investigate it using the circuit of Fig. 10.3–9a, where $i_E = 0$, $i_C = i_{CO}$, and $i_B = -i_{CO}$. Since *positive* v_{CB} *reverse* biases the collector junction, the resulting plot of i_{CO} versus v_{CB} in Fig. 10.3–9b has the same shape as a diode *i-v* curve (Fig. 9.1–3) with the axes rotated 180° to reflect the reversed current direction and voltage polarity.

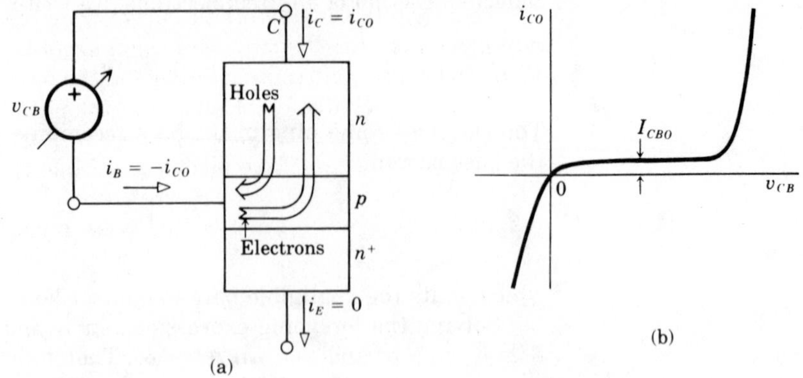

FIGURE 10.3–9 (a) (b)

The *reverse saturation current* I_{CBO} has a value around 10 nA for a silicon BJT at room temperature. However, like the drift current in a semiconductor diode, the value of I_{CBO} approximately doubles for every 10° increase of temperature, as expressed by Eq. (5), Section 9.3. (Germanium BJTs have much larger I_{CBO} and are seldom used due to the resulting problems of thermal instability.)

For the more general case of $v_{BE} > 0$ and $v_{CE} > -v_{BE}$, we sum the current components from the two previous cases to obtain the total collector current

$$i_C = \alpha i_E + i_{CO} \tag{7a}$$

where

$$i_{CO} = I_{CBO}\,(1 - e^{-\lambda v_{CB}}) \tag{7b}$$

The total base current is therefore $i_B = (1 - \alpha)i_E - i_{CO}$, so $i_E = (i_B + i_{CO})/(1 - \alpha)$ and

$$i_C = \frac{\alpha}{1-\alpha}\, i_B + \left(\frac{\alpha}{1-\alpha} + 1\right) i_{CO} = \beta i_B + (\beta + 1)\, i_{CO} \tag{8}$$

which gives i_C in terms of i_B and i_{CO}. The active region of operation has $\lambda v_{CB} \gg 1$, so $i_{CO} \approx I_{CBO}$ and

$$i_C = \alpha i_E + I_{CBO} = \beta i_B + I_{CEO}$$

where

$$I_{CEO} = (\beta + 1)\, I_{CBO} \tag{9}$$

Consequently, the collector cutoff current I_{CEO} will be much larger than the reverse saturation current I_{CBO}.

To obtain the characteristic curves from the above relationships, we will start with the *common-base configuration* shown in Fig. 10.3–10a. Here, the emitter-current direction and the base-emitter voltage polarity have been reversed to emphasize that the emitter is the input terminal. The input characteristics are sketched in Fig. 10.3–10b in terms of I_E versus v_{BE}. We again have diode curves like the common-emitter input (Fig. 10.2–3a) save that i_E includes i_C and is much larger than i_B. The output characteristics, i_C versus v_{CB}, are sketched in Fig. 10.3–10c. These curves come directly from Eq. (7) by adding αi_E to i_{CO} in Fig. 10.3–9b.

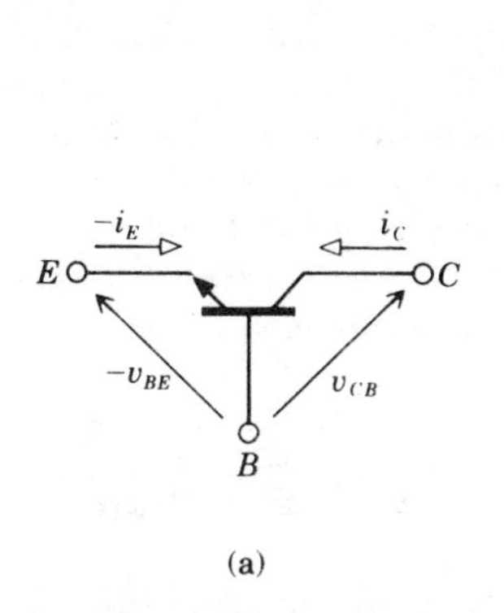

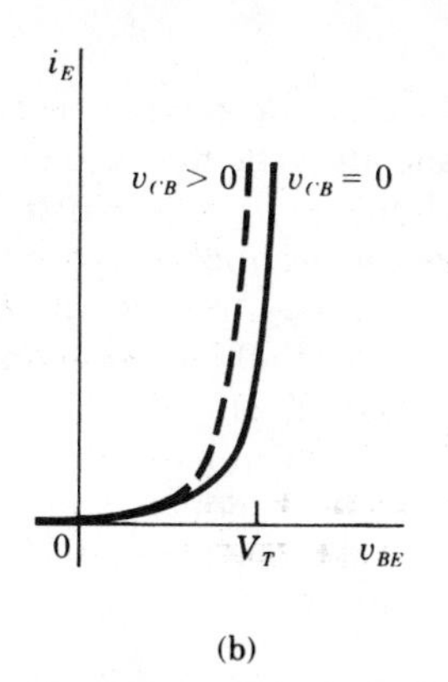

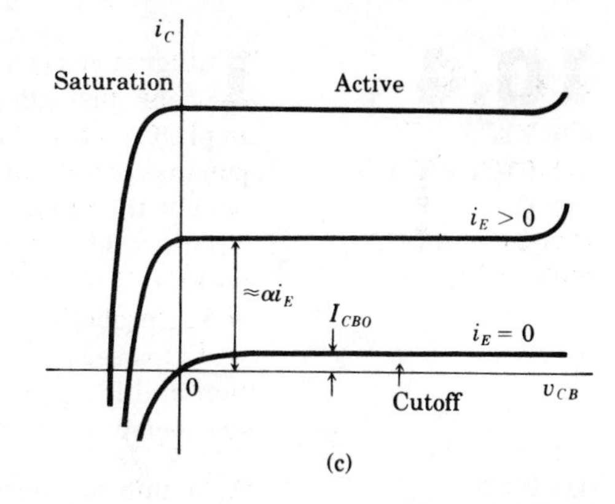

FIGURE 10.3–10
(a) Common-base configuration. (b) Input characteristics.
(c) Output characteristics.

Although the output current is actually somewhat less than the input current, the common-base configuration is still capable of voltage amplification. To demonstrate that capability, consider the *small-signal model* in Fig. 10.3–11 that has input resistance

$$r_e = \alpha \frac{25 \text{ mV}}{I_C} = \frac{25 \text{ mV}}{I_E} \tag{10}$$

whose value will be much smaller than r_π in Eq. (4), Sec. 10.2. (The derivation of this model parallels that of the common-emitter model.) If we connect a large resistance R_C across the output terminals, then $v_{cb} = R_C(-\alpha i_e)$, whereas $-v_{be} = r_e(-i_e)$. Hence, $A_v = v_{cb}/(-v_{be}) = \alpha R_C/r_e$, and we get significant voltage amplification if $R_C > r_e/\alpha$. Physically, the amplification comes about by transferring the current variation i_e from a small resistance r_e to a large resistance R_C. Early transistor circuits depended upon this transfer principle, the name "transistor" itself being a contraction of "transfer resistor."

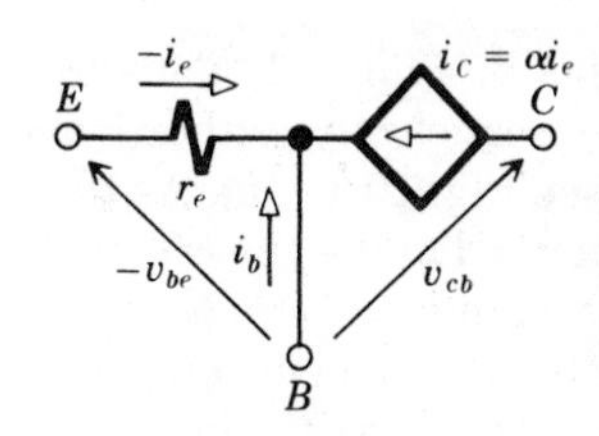

FIGURE 10.3–11
Common-base small-signal model.

Finally, the output characteristics of the common-emitter configuration are obtained from Eq. (8), noting that the horizontal axis in Fig. 10.3–9b must be relabeled with $v_{CB} = v_{CE} - v_{BE}$. When i_C is plotted versus v_{CE}, the knees of the curves become shifted to the right by $v_{BE} \approx V_T$ when $i_B > 0$. Carrying out these steps leads to curves like those previously given in Fig. 10.2–3b.

Exercise 10.3–3 Use Eq. (1), Section 9.1, to justify Eq. (7b).

Exercise 10.3–4 Redraw Fig. 10.2–6 in the common-base configuration with $i_b = i_e/(\beta + 1)$ and derive Eq. (10) by writing $r_e = v_{be}/i_e$.

10.4 DEVICE FABRICATION AND INTEGRATED CIRCUITS †

Integrated circuits and the planar transistor were both invented in 1958, just a decade after the first transistor. Ever since, improvements in planar silicon technology have permitted doubling the number of components per circuit per year, and the cost per component has decreased at nearly the same rate. Today, an advanced, very large-scale integrated circuit might contain over 100,000 components on a chip smaller than a postage stamp and costing only 0.01% of a discrete-device version—which probably could not even be built.

This section surveys some of the major techniques of device fabrication and integrated-circuit construction that have brought about the microelectronics revolution.

DEVICE FABRICATION METHODS

Although germanium was once used extensively in semiconductor devices, and there are some applications for materials such as alloys of gallium, most of the semiconductor industry uses *silicon*—fortunately a very plentiful material. Silicon is usually obtained by the reduction of common sand with carbon in a high-temperature furnace. The resulting intermediate product, a polycrystalline material, is then cast into ingots, which are purified by a process called *zone refining*. A small zone of each ingot is heated by induction to a temperature just above melting. The melted zone is slowly moved along the ingot and any impurities tend to move along with the melted part, leaving the recrystallized areas more nearly pure silicon but still an intermediate polycrystalline material rather than a single crystal.

The next step is to grow a single crystal. This is commonly done by the *Czochralski* method. A small seed crystal is very slowly withdrawn from molten silicon while being slowly rotated (see Fig. 10.4–1a). As the melt condenses, new crystal growth occurs on the seed along the crystalline axes of the original seed crystal. The result is a single crystal *boule* of relatively pure silicon, typically 8 cm in diameter and 25 cm long. By purposely adding appropriate donor or acceptor impurities to the melt, the boule is made *p*- or *n*-type silicon.

One of the methods once used to fabricate semiconductor devices was based on crystal growth from a doped melt. While the crystal was being

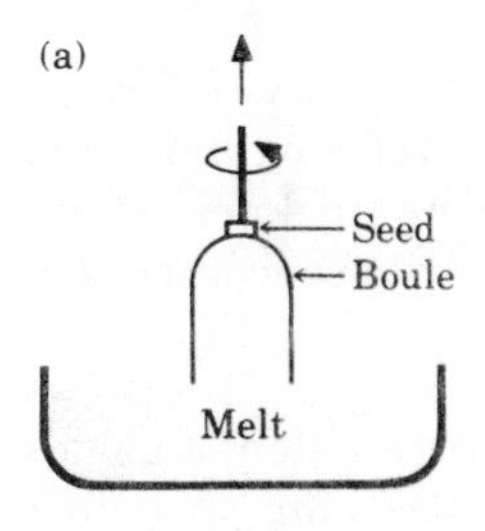

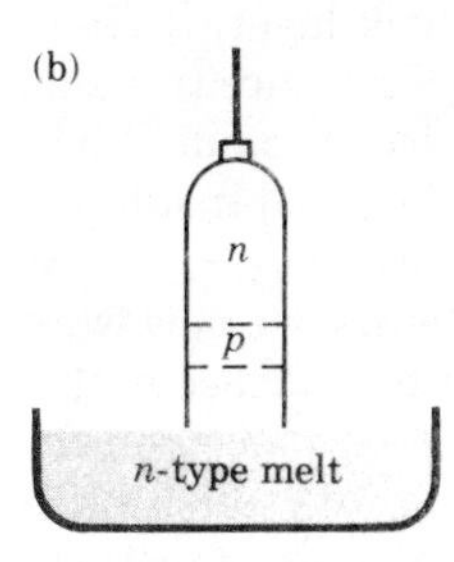

FIGURE 10.4–1
(a) Growing a single crystal. (b) Adding impurities to make a grown-junction device.

slowly pulled, additional impurities were added to the melt, changing it from n-type to p-type, for instance. After a little further growth, the melt could be changed back again, producing a boule with an *npn* sandwich as in Fig. 10.4–1b (a continued alternation of impurity types was commonly employed to utilize the entire boule). Devices made this way are called *grown-junction* transistors or diodes.

But most devices and ICs are now made from a reasonably homogeneous boule of silicon that has been sliced into *wafers* before processing. These wafers are smoothed and polished, to be used as the "raw material" for a variety of additional procedures.

One such procedure is to dot the surface of an n-type wafer with tiny pieces of an acceptor material. When the wafer is heated, each dot partially melts into the surface to produce an alloy whose silicon content increases with depth. There is mostly acceptor metal on the surface, and mostly p-type silicon farther inside, with a curved *pn* junction where the two meet. This process produces an *alloy junction.* A batch of alloy-junction transistors is obtained by putting an opposing set of donor metal pieces on the other face of the wafer, and heating the wafer just long enough for a thin region of n-type silicon to remain between the two junctions, as illustrated in Fig. 10.4–2a. When scribed into tiny squares, broken at the scribe marks, and with proper leads attached, several hundred transistors are produced from one wafer as shown in Fig. 10.4–2b.

Another method for creating *pn* junctions is *impurity diffusion.* A wafer is heated in an inert gas atmosphere containing impurity atoms such as boron (for acceptors) or phosphorus (for donors). The impurity atoms diffuse into the silicon, going from a region of high to low concentration just as electrons and holes diffuse across a junction. For example, the *npn* transistor structure of Fig. 10.3–1a might have been obtained by a two-step process, first diffusing boron into an n-type wafer to get the p-region base and then diffusing phosphorus for the n^+-region emitter.

It is also possible to insert impurities at specific locations inside the crystal via *ion implantation.* This technique, carried out at room temperature, employs a special type of mass spectrometer to bombard the wafer with impurity ions. The energy of the ions determines the depth of penetration.

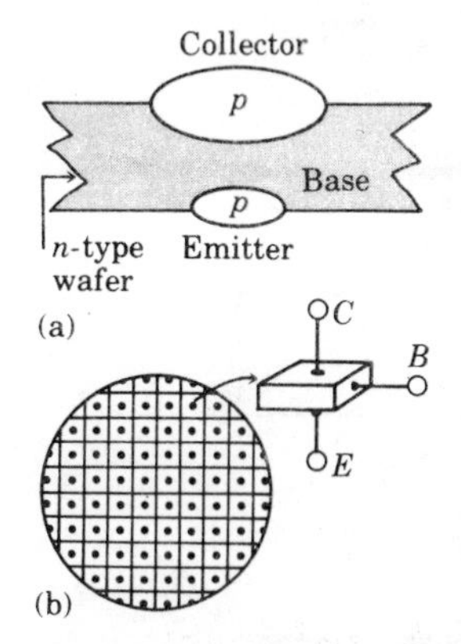

FIGURE 10.4–2
Alloy junction transistors: (a) cross section; (b) scribed wafer producing many transistors.

Finally, a silicon wafer can be used as a *substrate* upon which additional layers of single-crystal silicon are deposited by *epitaxial growth.* In this process the surface of the wafer is exposed to an appropriate gaseous atmosphere at controlled temperatures. The gas is normally hydrogen and silicon tetrachloride, which yields silicon at low concentrations of the tetrachloride. At high concentrations, the reaction reverses and the treatment etches away rather than builds upon the substrate.

Epitaxial processing has several important advantages. When small quantities of selected impurities are added to the gas, a newly formed layer can be made to have any desired concentration of *p* or *n* doping. Since the growth rate is readily predicted, the resulting thickness of the new growth, along with any desired variations of impurity concentration, can be accurately controlled.

PLANAR PROCESSING

Let's now walk through the steps required to fabricate a batch of diodes on a wafer using the technology known as *planar processing*. We start with a heavily doped *n*-type substrate. A relatively thick layer, perhaps 25 μm thick, is epitaxially grown on its upper surface. Exposure to steam at high temperature then causes a surface layer of silicon dioxide (SiO_2) to form. (Silicon dioxide happens to be one of the best electrical insulators and also provides excellent protection for the top surface.)

However, to gain access to the silicon at selected points, we now want to remove some of the oxide. For that purpose we coat the surface with a photosensitive material called *photoresist*. A *mask* is next placed on the photoresist, resulting in the layered arrangement of Fig. 10.4–3a. In this hypothetical example, the mask could be a transparent sheet marked with a regular arrangement of tiny dots at the points where we wish to create diodes. (In practice, the mask would be a photographic negative, produced from large-scale artwork in a special reducing camera.)

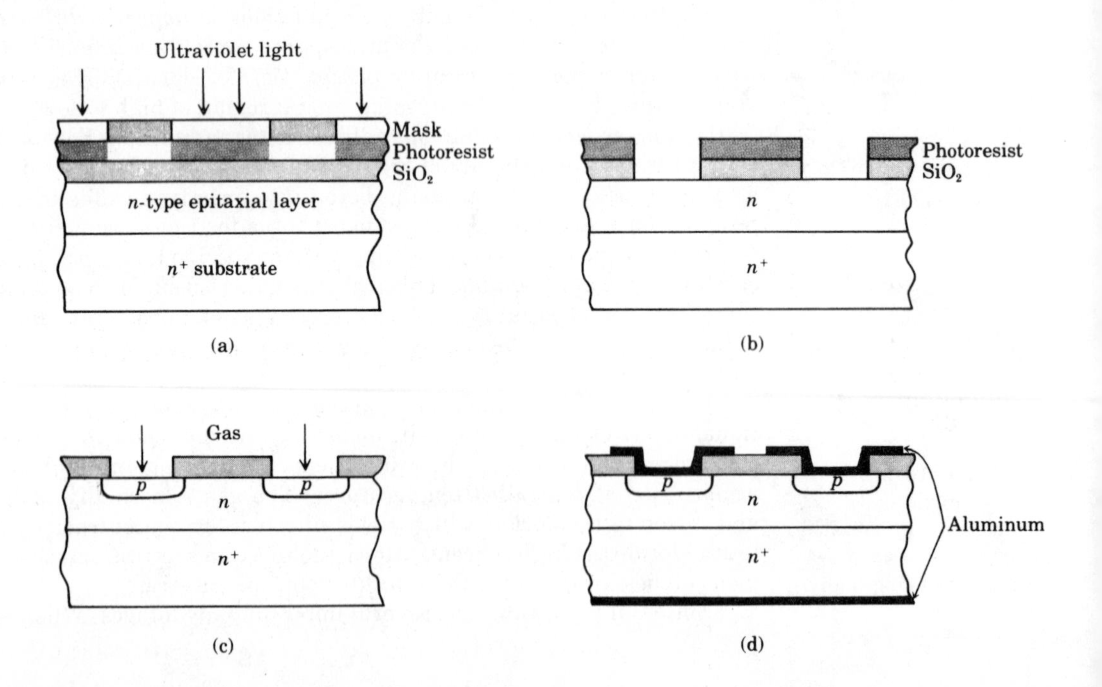

FIGURE 10.4–3
Planar processing steps: (a) masking photoresist on the oxide layer; (b) etching removes masked oxide; (c) diffusion of acceptor impurities produces *p*-type regions; (d) metalization for wire bonding.

We now expose our multilayered sandwich to ultraviolet light, which polymerizes the photoresist in the areas where the mask is transparent. The mask is then removed and the photoresist is dissolved, except where it was exposed. The surface is now ready for an etching solution that removes the oxide from the diode locations (see Fig. 10.4–3b). The remaining photoresist may then be washed away chemically.

The next step is diffusion. Since the oxide is highly resistant to penetration by foreign atoms, high-temperature exposure to an acceptor-impurity gas produces *p*-type regions in the epitaxial layer below the oxide-layer windows. We now have formed *pn* junctions shown in the cross-section view of Fig. 10.4–3c.

The remaining sequence fabricates the individual terminals for wire bonding by a procedure called *metallization*. It starts by exposing both faces of the wafer to evaporating aluminum atoms in vacuum, resulting in a metal film over the entire surface. Photoresist is applied again, followed with a new mask on the upper surface. This mask is opaque except for an array of transparent circles aligned to cover the positions of the diodes. After exposure and etching, metal coats the entire bottom of the wafer, but it covers only areas slightly larger than the apertures in the oxide layer on the upper side (see Fig. 10.4–3d). Note that the circumferential edge of the *pn* junction is still protected from air and moisture by the oxide film.

It is now possible to scribe and break the wafer into hundreds of planar diodes ready for packaging. All of the foregoing processing steps would, generally, be done to many wafers simultaneously, resulting in nearly identical characteristics for an entire batch of, perhaps, ten thousand devices.

INTEGRATED CIRCUITS

Although diodes are relatively simple devices, the very same processes we've just discussed—epitaxial growth, oxidation, impurity diffusion, photo masking, etching, and metallization—are used to make *integrated circuits* (ICs) of varying degrees of complexity. An IC is a silicon *chip* containing numerous transistors and, perhaps, resistors and capacitors—all components having been formed by successive processing steps on a wafer about 5 cm in diameter that yields more than 1000 chips of an area of 2 mm^2. (IC dimensions are usually stated in *mils,* where 1 mil = 0.001 in. = 0.0254 mm.)

Small-scale integration (SSI) would be used for a 20-component op-amp, say, whereas *large-scale integration* (LSI) puts an entire microprocessor with some 10,000 components on a single chip! The benefits derived from integrating many components on an IC chip include low cost, small size, high reliability, and matched characteristics.

To illustrate some of the possibilities, let's consider one *npn* transistor—a small part of an IC on a *p*-type substrate about 0.2 mm thick. The resulting configuration, somewhat idealized, has the cross section shown in Fig. 10.4–4. Fabrication requires six masking steps. All the terminals are on the top, and the substrate is not ordinarily used as an electrode. Instead, it should be connected to the negative terminal of the supply source to prevent conduction through the substrate-to-collector junction. (This junction does, of course, add an effective capacitance to the collector.) The p^+ regions around the outside serve to *isolate* this transistor from its neighbors. These isolation regions are produced by continuing to diffuse acceptor impurities until they penetrate the entire epitaxial layer and join the substrate. The n^+ *buried layer* serves to reduce the effective series resistance of the collector circuit by increasing the conductivity of much of the collector volume.

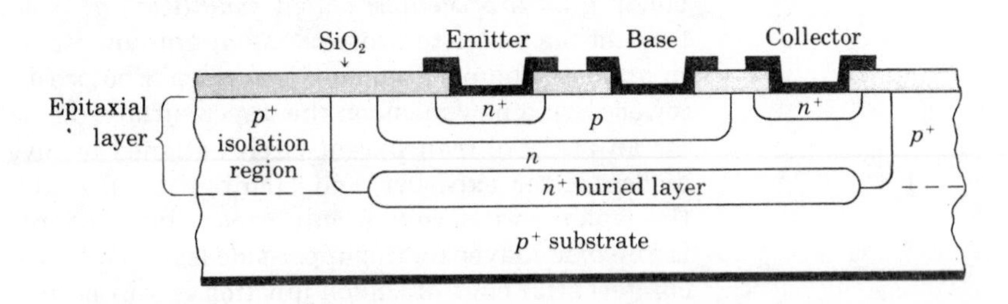

FIGURE 10.4–4
Cross section of an IC *npn* transistor.

The physical size of an integrated bipolar junction transistor would typically be 0.2 mm square. While that may *seem* quite small, the MOSFET device illustrated in Fig. 10.4–5 takes up only 0.04 mm × 0.14 mm of chip area, meaning that about 20 MOSFETs could be put in the place of one BJT. A major space-saving factor here is the *self-isolating* property of MOSFETs. Figure 10.4–5 also shows a low-resistance n-type strip inside the oxide layer that serves as a *buried crossover* for connecting other components.

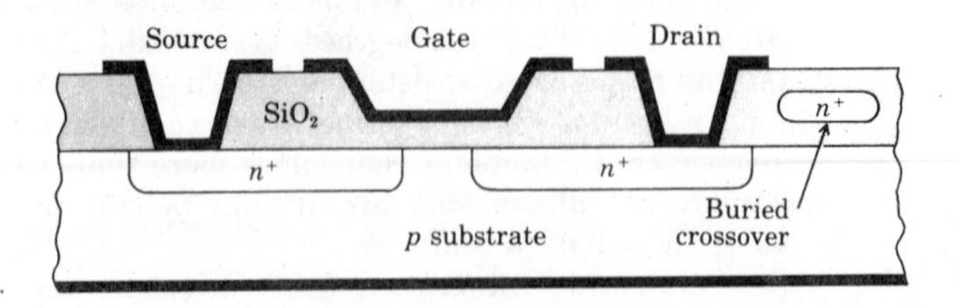

FIGURE 10.4–5
Cross section of an IC enhancement MOSFET.

When IC *resistors* are needed, they may be made along with other devices using the same processing steps. An example of such a resistor is shown in cross section in Fig. 10.4–6a. The value of the resulting resistance will depend on the impurity concentration used, as well as on the

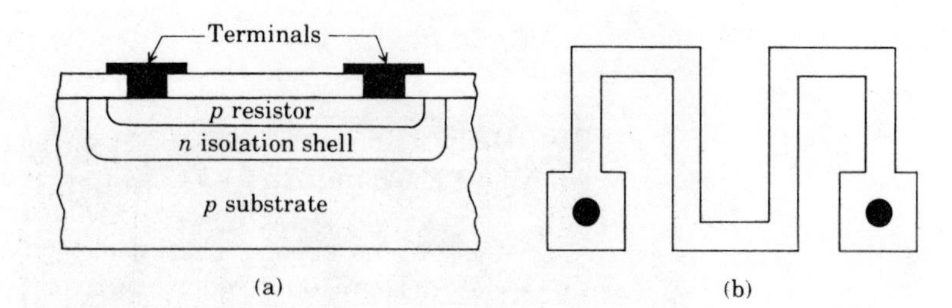

FIGURE 10.4–6
IC resistor: (a) cross section; (b) top view.

shape of its region. To achieve the length necessary for larger resistance values, resistive regions are sometimes folded back and forth several times, as in the top view of Fig. 10.4–6b. The fabrication technique used to make such resistors is relatively inaccurate, but the ratio of values of any two resistances in one IC is likely to match quite well the ratio of resistance values in another similar circuit. Precision resistances are usually created by evaporating a thin alloy film with a low temperature coefficient onto an oxide layer, and then masking and etching the film to the desired configuration. If necessary, the resistance value can be modified in a controlled manner after processing by nibbling away at it with a laser beam during final testing.

Integrated *capacitors* with relatively small capacitance values are most easily made by using the capacitance of a reverse biased junction. However, these components suffer from variations in their capacitance value with applied voltage (the varactor effect), and from the presence of diode behavior if the applied voltage forward-biases the junction. Much larger and more stable capacitors are formed of thin metallic films over silicon dioxide as a dielectric, as in Fig. 10.4–7. Such capacitors do require substantial areas on the substrate, and have values in the neighborhood of 500 pF per square millimeter. (One square millimeter is quite a bit of "real estate" to devote to one circuit element.)

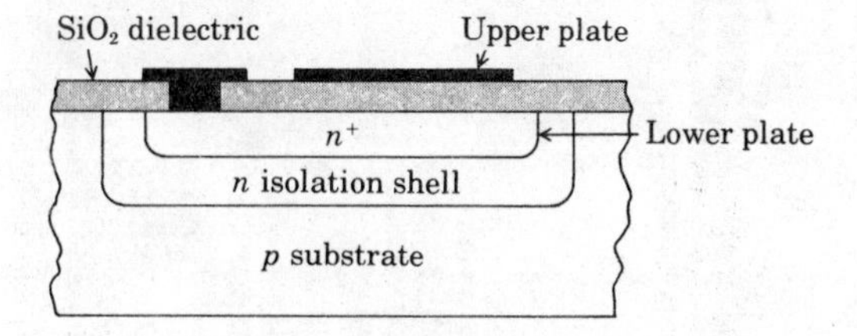

FIGURE 10.4–7
Cross section of IC capacitor.

But integrated resistors may require even larger areas than capacitors, so special switching techniques have been developed recently to make capacitors behave like RC circuits. Other kinds of active circuits may be used to convert IC capacitance into equivalent inductance, since inductors cannot be implemented effectively in IC form.

(a)

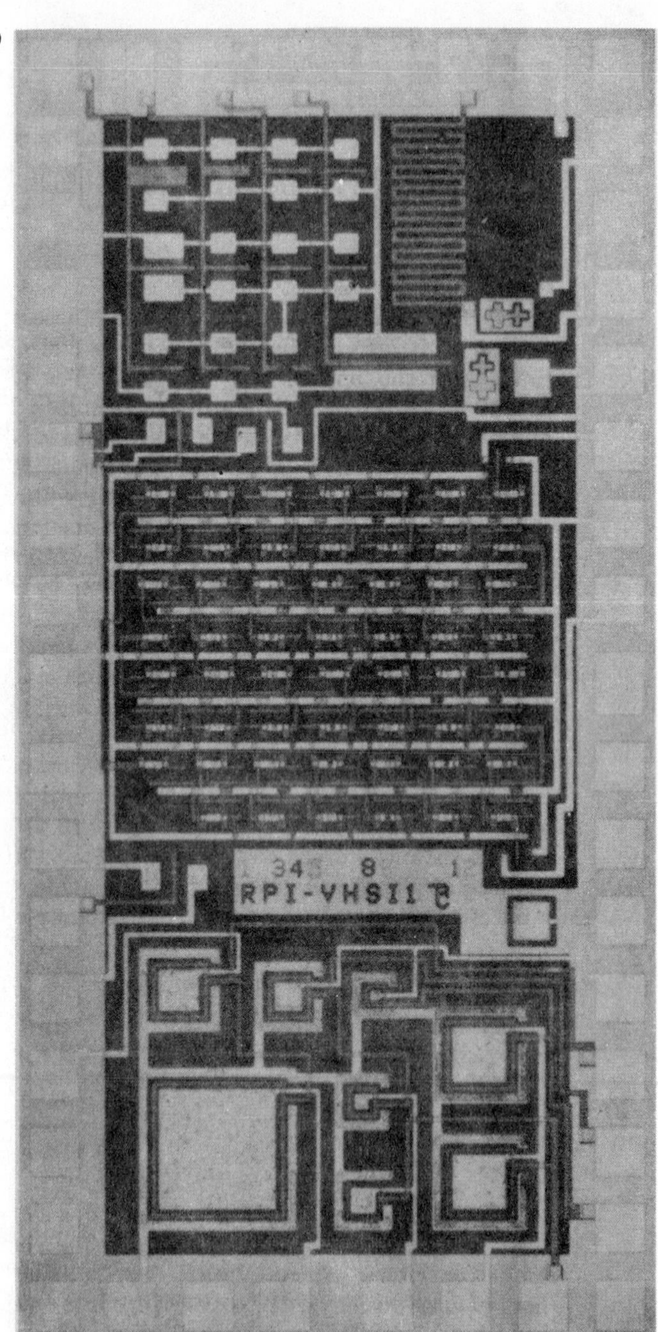

FIGURE 10.4–8
(a) Microphotograph of an integrated circuit containing 133 devices on a 1 mm × 2 mm chip with 46 bonding pads around the periphery for external connections. All devices are enhancement or depletion MOSFETs, except resistor in upper right corner.(b) Detail of lower portion showing eight MOSFETs of various sizes.

Medium and large-scale ICs generally include a network of metallic connecting bridges that join their devices. These bridges form an intricate, often artistic pattern as seen in the magnified top view of a typical chip in Fig. 10.4–8. It's rather sad that those views must be concealed by protective packaging.

Of the many IC packaging techniques, perhaps the most popular is the *dual-in-line package* (DIP) illustrated in Fig. 10.4–9. It consists of a

(b)

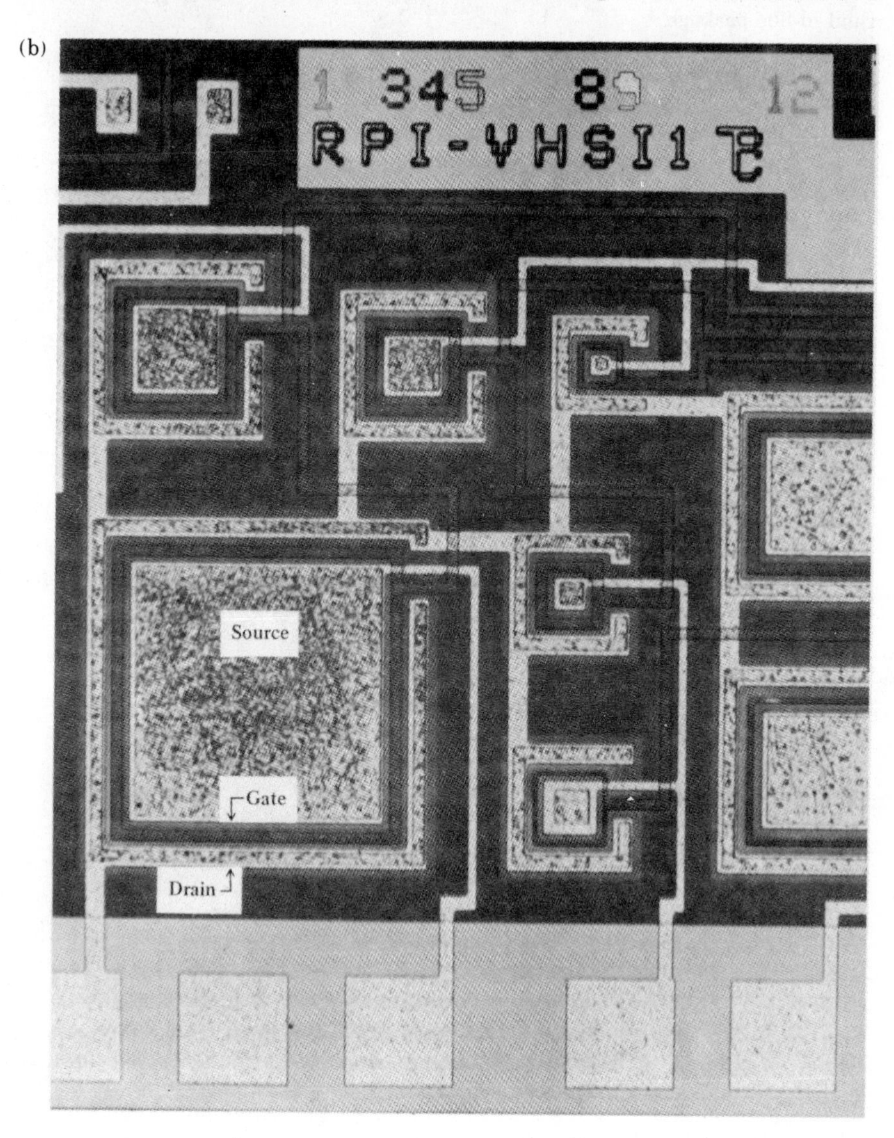

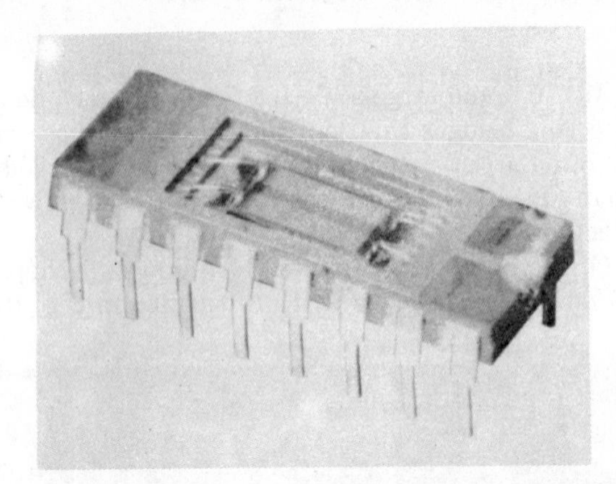

FIGURE 10.4–9
Dual in-line package with top removed. (Courtesy of Texas Instruments.)

rectangular plastic or ceramic case enclosing the IC, with protruding pin terminals that make it look something like a centipede. These pins can be soldered directly to the external circuitry, or inserted into standard sockets that, in turn, are either soldered or wirewrapped for connection to other sockets. This particular unit has 16 pins, and is about 8 mm wide and 18 mm long. Larger DIPs with 24 to 64 pins accommodate the many external connections needed for an LSI chip such as a microprocessor (see, for example, the photograph on page 268).

PROBLEMS

10.1–1 Sketch i_D and v_{DS} versus t for the circuit in Fig. 10.1–9a when $v_{GS} = 2.5 + \sin \omega t$ V.

10.1–2 Repeat the previous problem with $v_{GS} = 4.5 + \sin \omega t$ V.

10.1–3 The characteristic curves for the MOSFET in Fig. P10.1–3 are as plotted in Fig. 10.1–7a. What value of v_{in} yields $v_{out} = 5$ V if $R_D = 0$ and $R_K = 2$ kΩ? Hint: Figure 10.1–7b still applies, but $v_{in} \neq v_{GS}$ and $v_{out} \neq v_{DS}$.

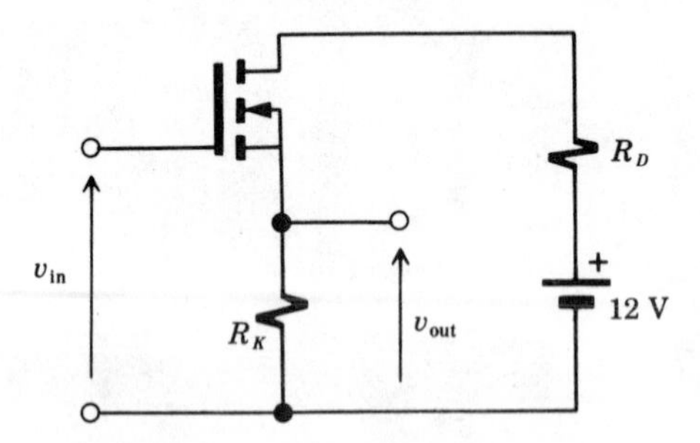

FIGURE P10.1–3

10.1–4 Repeat the previous problem with $R_D = R_K = 1$ kΩ.

10.1–5 Let $K = 10$ mA, $V_T = 2.5$ V, and $v_{GS}(t) = 4 + 0.25 \sin \omega t$ V in Fig. 10.1–9a. Use Eq. (1) to obtain an expression for $i_D(t)$. Sketch your result and compare it with Fig. 10.1–9b.

10.1–6 Repeat the previous problem with $v_{GS}(t) = 3.75 + 0.25 \sin \omega t$ V.

10.1–7 Suppose a MOSFET is made into a two-terminal element by connecting the gate to the drain terminal. Confirm that the MOSFET will operate in the active region, and obtain an expression for v_{DS} in terms of i_D.

10.1–8 Use Eq. (3b) to plot R_{DS} versus v_{GS}/V_T. Label the vertical axis in terms of $r_{mo} = 1/g_{mo}$. Also include a second horizontal axis indicating the value of v_{max}/V_T, where v_{max} is the largest value of v_{DS} that satisfies Eq. (3a).

10.1–9 Suppose the MOSFET in Example 10.1–3 is replaced by one having $K = 4.5$ mA and $V_T = 2$ V. Find the new values of i_D, v_{DS}, g_m, and A_v, keeping $V_{GS} = 4$ V.

10.1–10 Repeat the previous problem with $K = 36$ mA and $V_T = 3$ V.

10.1–11 Let $v_{GS} = V_{GS} + v_{gs}$ in Eq. (1), and show that $i_D = I_D + i_d$, where $i_d \approx g_m v_{gs}$, provided that $|v_{gs}| \ll 2(V_{GS} - V_T)$.

‡ **10.1–12** The circuit in Fig. P10.1–12 uses an *active load* (T_2) in place of a resistor to obtain *linear* amplification. Obtain an expression for v_{out} in terms of v_{in} and the restrictions on v_{in} to ensure both MOSFETs are in the active region. Assume T_1 and T_2 have the same value of K but different threshold voltages, say V_1 and V_2.

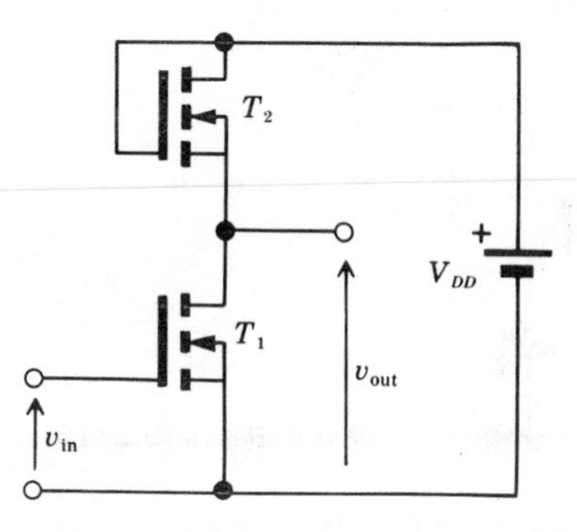

FIGURE P10.1–12

‡ **10.1–13** Repeat the previous problem, assuming T_1 and T_2 have the same threshold voltage but different values of K, say K_1 and K_2.

10.2–1 Use the large-signal model for the active region to evaluate i_B, i_C, and v_{CE} in Fig. P10.2–1 when $R_B = 965$ kΩ, $R_C = 10$ kΩ, and $R_E = R_L = 0$.

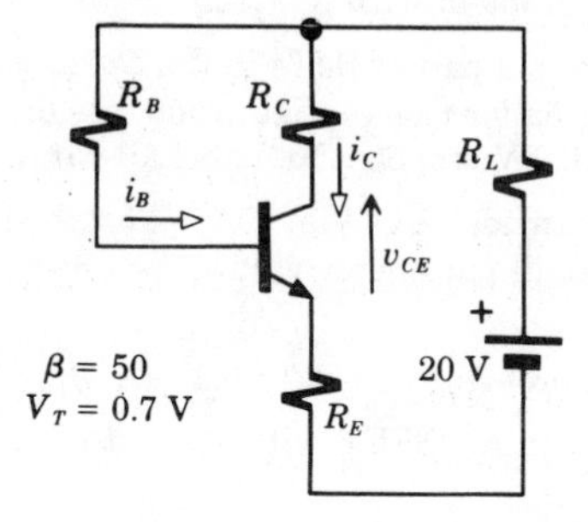

FIGURE P10.2–1

10.2–2 Repeat Problem 10.2–1 with R_B = 455 kΩ, R_E = 10 kΩ, and $R_C = R_L = 0$.

10.2–3 Repeat Problem 10.2–1 with R_B = 455 kΩ, R_L = 10 kΩ, and $R_E = R_C = 0$.

10.2–4 Repeat Problem 10.2–1 with R_B = 710 kΩ, $R_E = R_C$ = 5 kΩ, and $R_L = 0$.

10.2–5 Let R_C = 10 kΩ and $R_E = R_L = 0$ in Fig. P10.2–1. Obtain a condition on R_B such that the transistor is saturated.

‡ **10.2–6** Repeat the previous problem with $R_E = R_C$ = 5 kΩ, and $R_L = 0$. Hint: You will need to solve two simultaneous equations for i_B and i_C.

10.2–7 Show that the transistor in Fig. P10.2–1 cannot be saturated or cut off if $R_C = 0$ and $R_B \neq 0$.

10.2–8 The circuit in Fig. P10.2–8 acts like a *current source* that supplies a constant current I to the variable load R_L.

(a) Assuming active operation, show that I is independent of R_L and find the value of R_E such that I = 20 mA.

(b) Find the largest value of R_L for which I remains constant at 20 mA when V_s = 25 V.

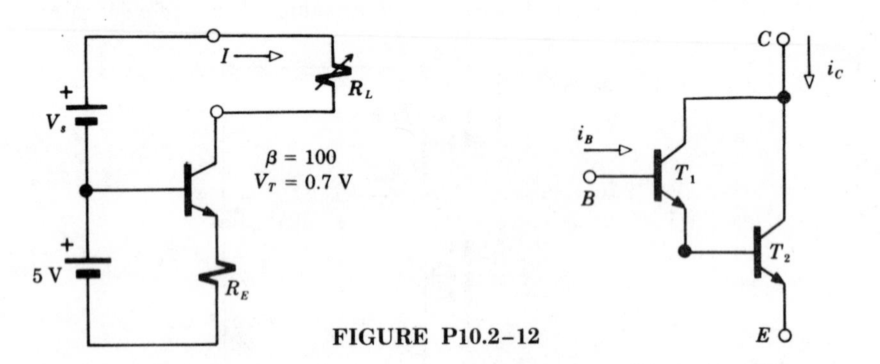

FIGURE P10.2–8

FIGURE P10.2–12

10.2–9 The load resistance R_L in Fig. P10.2–8 varies from 0 to 300 Ω. Find values for R_E and V_s so that I = 100 mA and the transistor remains in its active region, despite the variations of R_L.

10.2–10 Let Fig. 10.2–7a have V_B = 2 V, R_B = 26 kΩ, β = 50, and V_T = 0.7 V. Find values for R_C and V_{CC}, such that V_{CE} = 10 V and $v_{ce} = -10v_b$.

10.2–11 Repeat the previous problem with β = 100.

10.2–12 Figure P10.2–12 shows a pair of BJTs in the *Darlington configuration*. Obtain a model like Fig. 10.2–5a for this configuration, assuming both transistors are active, and have V_T = 0.7 V and β = 50. Label all parameter values on your model.

10.2–13 Obtain a small-signal model, like Fig. 10.2–6, for the Darlington configuration in Fig. P10.2–12. Label all parameter values, assuming I_B = 10 μA and $\beta_1 = \beta_2 =$ 50.

10.2–14 Repeat the previous problem with I_B = 0.1 mA, β_1 = 20, and β_2 = 50.

10.3–1 An *n*-channel depletion MOSFET with v_{DS} = 10 V is found to have i_D = 8 mA when $v_{GS} = 0$ and $i_D = 2$ mA when $v_{GS} = -2$ V. Find i_D and the condition on v_{DS} for active-region operation if $v_{GS} = -1$ V.

10.3–2 Repeat the previous problem with $v_{GS} = +1$ V.

10.3–3 An n-channel JFET with $v_{DS} = 10$ V is found to have $i_D = 0$ when $v_{GS} \leq -6$ V and $i_D = 4$ mA when $v_{GS} = -3$ V. Find i_D and the condition on v_{DS} for active-region operation, if $v_{GS} = -1.5$ V.

10.3–4 What is the maximum value of i_D for the JFET in the previous problem?

‡ **10.3–5** Derive Eq. (3) starting from Eq. (5), Section 10.1.

10.3–6 The transistor in Fig. P10.3–6 is operating in its active region, with $\beta = 40$, $V_T = 0.7$ V, and $I_{CBO} \approx 0$. Find i_E and v_{CB} when $R_E = 5$ kΩ and $R_B = 0$.

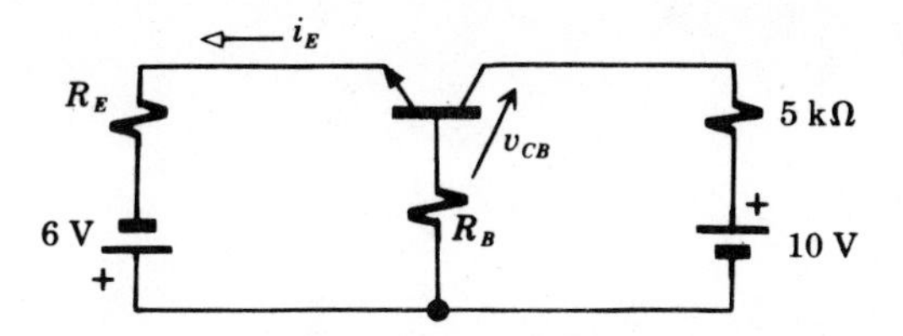

FIGURE P10.3–6

10.3–7 Repeat Problem 10.3–6 with $R_E = 0$ and $R_B = 150$ kΩ.

10.3–8 Repeat Problem 10.3–6 with $R_E = 5$ kΩ and $R_B = 50$ kΩ.

10.3–9 In Fig. 10.3–11, show that $i_c = g_m(-v_{be})$, where $g_m = I_C/(25 \text{ mV})$.

11

NONLINEAR ELECTRONIC CIRCUITS

Broadly speaking, engineers classify electronic circuits as linear or nonlinear. Linear circuits utilize electronic devices that operate in their active regions, usually for the purpose of signal amplification as detailed in the next chapter. The nonlinear circuits considered in this chapter exploit the *switching* action of electronic devices for a variety of purposes other than amplification.

We begin with simple diode and transistor circuits whose switching characteristics are employed in such signal-processing applications as waveshaping and gating. Both analog and digital gates are included. We also examine the effects of energy storage in switching circuits. The resulting switching transients—although a nuisance in some applications—can be put to practical use in timing circuits and waveform generation when combined with comparators. Rectifier circuits for AC-to-DC conversion are discussed in the closing section.

OBJECTIVES

After studying this chapter and working the exercises, you should be able to do each of the following:

- Design simple diode clippers and clamps (Section 11.1).
- Analyze the ON and OFF states of a BJT switch (Section 11.2).

- Show how diodes or FETs can be used to build an analog gate (Sections 11.1 and 11.2).
- Explain the operating principles of simple digital gates and flip-flops (Sections 11.1 and 11.2).
- Analyze an op-amp comparator or trigger, and explain the differences between bistable, monostable, and astable circuits (Sections 11.2 and 11.3). †
- Calculate the time intervals associated with an *RC* timing circuit (Section 11.3). †
- Design a simple rectifier circuit with a smoothing capacitance or inductance (Section 11.4). †

11.1 DIODE SWITCHING CIRCUITS

A number of useful circuits employ diodes as switching devices. Applications include waveshapers, digital and analog gate circuits, and clamps. For simplicity, we assume ideal diodes, except when otherwise stated.

CLIPPERS AND WAVESHAPING

Consider the circuit in Fig. 11.1–1a. Obviously, current flows (diode ON) only when $v > V_0$, and if we view the resistance as a "current to voltage converter" then v_R plotted versus v has the same shape as i versus v. Figure 11.1–1b displays this particular *voltage transfer characteristic*. One could also plot v_D versus v using the fact that $v_D + v_R + V_0 = v$.

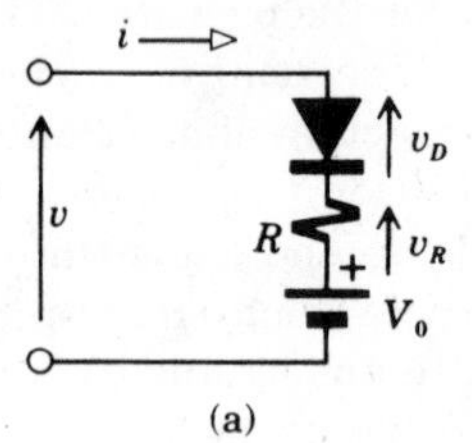

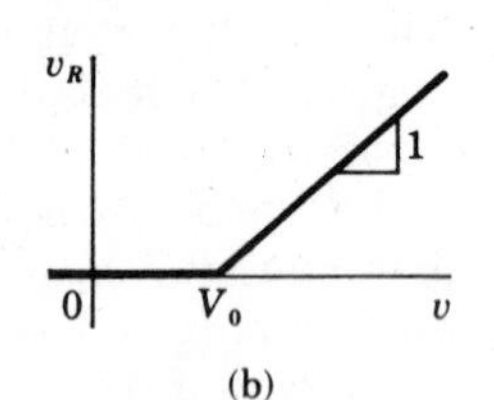

FIGURE 11.1–1
(a) Clipper circuit.
(b) Transfer curve.

Now suppose the input terminals are connected to a sinusoidal voltage source $v = V_m \sin \omega t$ whose peak value V_m exceeds V_0. The diode conducts and v_D equals zero whenever $V_m \sin \omega t > V_0$, so $v_R = v - V_0$. Conversely, the diode is OFF and $v_R = 0$ whenever $V_m \sin \omega t < V_0$, so $v_D = v - V_0$. The resulting waveforms are sketched in Fig. 11.1–2.

These characteristics are typical of a class of circuits called *clippers*. In a sense, the circuit has clipped off the top of the input wave in producing v_D. The waveform v_R can be interpreted as the input wave with all parts below a certain value clipped off. Similar results ensue when the

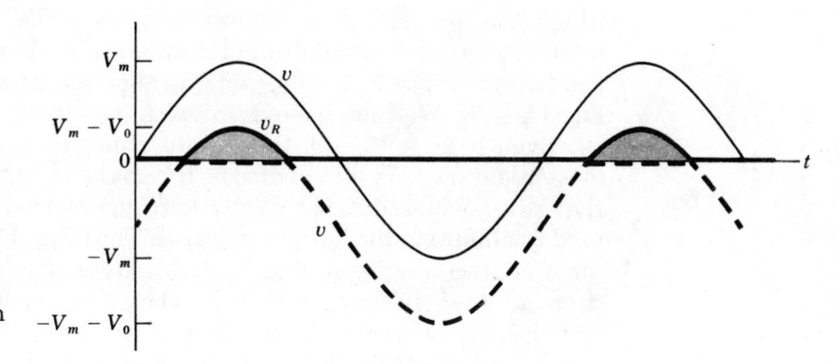

FIGURE 11.1–2
Clipper waveforms with a sinusoidal input.

output voltage is the sum of v_R and v_D, v_R and V_0, or v_D and V_0, the latter obtained by rearranging the elements.

A two-sided clipper that clips both top and bottom of the input requires *two* diodes, as in Fig. 11.1–3a. The transfer curve shows that v_{out} is constrained to lie between $-V_1$ and $+V_2$ regardless of the value of v_{in}. The action of this clipper is described mathematically by

(2)

R
v_{in}
V_2
V_1
v_{out}

(a)

v_{out}
V_2
$-V_1$
1
0
V_2
$-V_1$
v_{in}

(b)

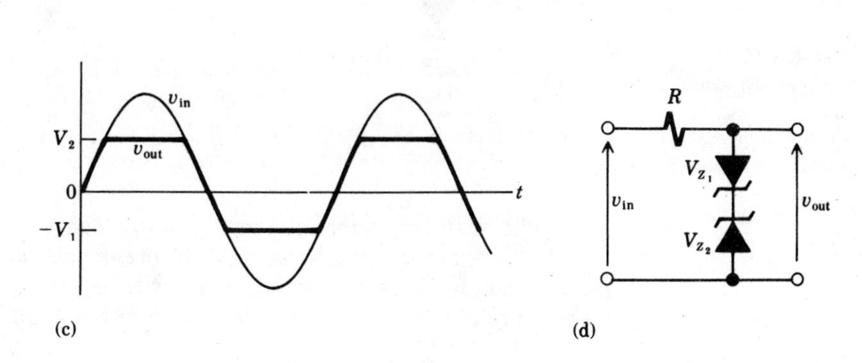

FIGURE 11.1–3
(a) Two-sided clipper.
(b) Transfer curve.
(c) Waveforms.
(d) Implementation with Zener diodes.

which you can confirm by examining each of the three ranges of v_{in}. A relatively large sinusoidal input waveform would be top-clipped at $+V_2$ and bottom-clipped at $-V_1$, yielding the "squarish" output sketched in Fig. 11.1–3c. We then have a *waveshaping* circuit.

Even if $V_1 = V_2 = 0$, this circuit clips at about ± 0.7 V due to the *threshold voltage* V_T of real diodes. By extension of that reasoning, we see that reverse-biased *Zener* diodes with appropriate values of V_Z can be used to eliminate the battery voltages. Thus Fig. 11.1–3d could have the same characteristics as Fig. 11.1–3a. Note that the Zener diodes are "front-to-front" in series, rather than being in parallel. The Zener equivalent circuit in Fig. 9.2–4c should help you understand why.

Symmetrical two-sided clippers with $V_1 = V_2$ are sometimes known as *limiters*. Another application of diode clipping is the generation of *piecewise-linear functions*, as discussed in the following example.

Example 11.1–1
Function generator

Suppose we want to produce an output voltage roughly equal to the square root of a time-varying (but nonnegative) input voltage. The plot made up of straight line segments in Fig. 11.1–4a approximates the desired function $v_{out} \approx \sqrt{v_{in}}$ for $0 \leq v_{in} \leq 9$ V.

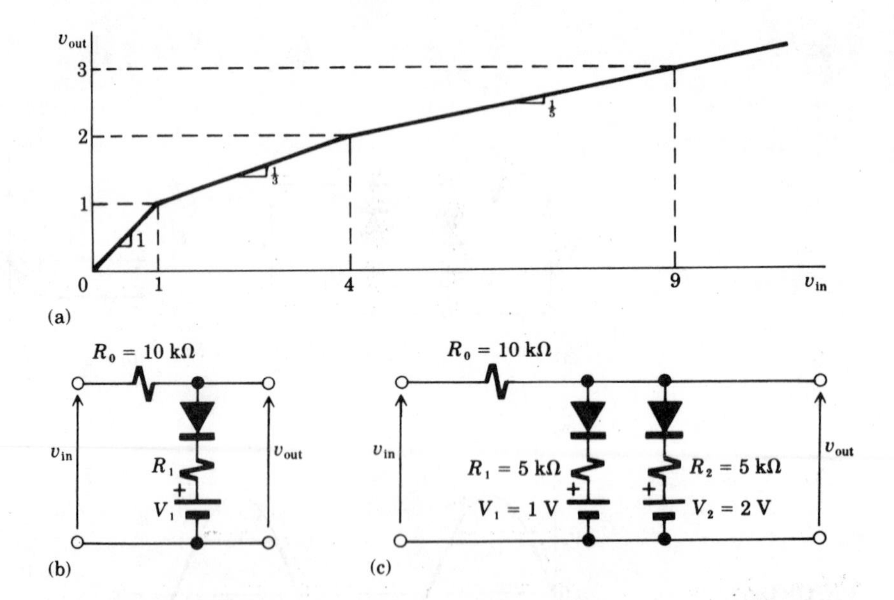

FIGURE 11.1–4
(a) Approximation for $v_{out} = \sqrt{v_{in}}$.
(b) Partial circuit.
(c) Complete circuit.

The first two segments can be implemented by the circuit of Fig. 11.1–4b, which has $v_{out} = v_{in}$ up to the break point at $v_{out} = V_1$. Above the break point, $v_{out} = V_1 + a_1(v_{in} - V_1)$ with $a_1 = R_1/(R_0 + R_1)$. We should therefore take $V_1 = 1$ V and $R_1 = R_0/2 = 5$ kΩ. Since the plot has a second break point at $v_{out} = 2$ V and slope $a_2 = \frac{1}{5}$ for $v_{in} > 4$ V, the complete circuit requires another diode-resistance-battery branch as diagrammed in Fig. 11.1–4c. The 2-V battery establishes the second break point, above which the slope will be $a_2 = R_{eq}/(R_0 + R_{eq})$ with $R_{eq} = R_1 \| R_2$.

Exercise 11.1–1

Add a third segment to Fig. 11.1–4a, extending the approximation to $v_{in} = 16$ V; then devise the corresponding addition to Fig. 11.1–4c.

DIGITAL DIODE GATES

Most of the diode circuits we have considered so far had a single source whose variations constituted the input signal to the circuit. Any other sources we used were assumed to be fixed in value, although in practice, of course, they could also have been functions of time if a design required such special behavior. We are now going to discuss a class of circuits which may have several input signals. If these are *binary digital signals,* they are defined as having only *two* allowed values, say 0 V and 5 V; the output is likewise restricted to the same two values—referred to as LOW and HIGH.

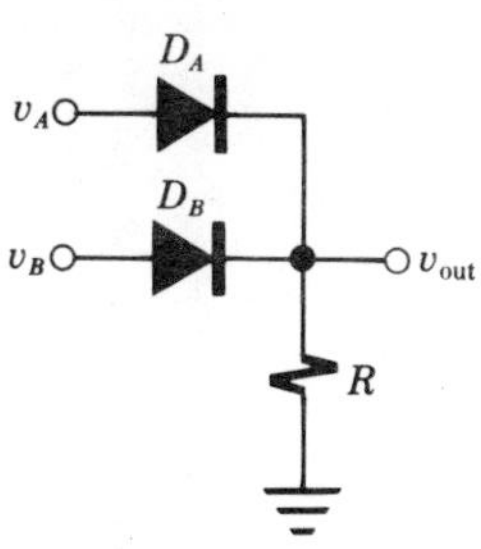

FIGURE 11.1–5
OR gate.

Consider, for example, the circuit of Fig. 11.1–5. If v_A and v_B are both LOW (0 V), then neither diode conducts and $v_{out} = 0$ V. But if v_A is HIGH (+5 V), diode D_A conducts and $v_{out} = 5$ V, regardless of whether v_B is HIGH or LOW. By the same reasoning, v_{out} is HIGH when v_B is HIGH, regardless of v_A. Consequently, v_{out} will be HIGH when either v_A *or* v_B *or* both are HIGH; we therefore call this a diode OR circuit or an OR *gate.*

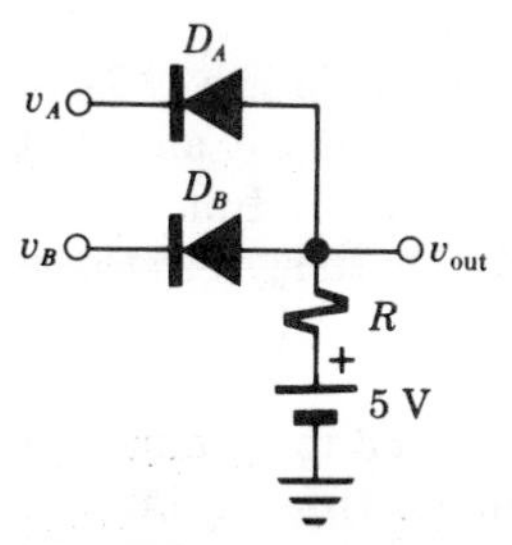

FIGURE 11.1–6
AND gate.

The similar circuit of Fig. 11.1–6 has a slightly different function. Here, whenever v_A or v_B or both are LOW, D_A or D_B or both conduct, and v_{out} is also LOW. The only way v_{out} can be HIGH is for both v_A *and* v_B to be HIGH. Hence, we have a diode AND circuit or AND gate.

With additional diodes, we could build AND or OR gates for more than two inputs. And the output of one gate could be an input to another gate. However, loading effects and the threshold voltage of real diodes generally cause these more complicated gate circuits to fail miserably, a problem overcome with the help of transistors as discussed in the next section. We'll discuss applications of gate circuits for digital logic systems in Chapters 15 and 16.

Example 11.1–2

Let's clarify the difference between OR and AND gates. Suppose the OR-gate circuit (Fig. 11.1–5) has binary input signals v_A and v_B, as shown in Fig. 11.1–7a. If we assume ideal diodes, the output signal equals 5 V whenever either of the two inputs equals 5 V; and $v_{out} = 0$ only when both inputs are zero. This output waveform is shown in Fig. 11.1–7b. If the same two signals were applied to the AND-gate circuit (Fig. 11.1–6), then $v_{out} = 5$ V only when both v_A and v_B are 5 V, as in Fig. 11.1–7c.

The horizontal dashed lines in these figures show the effect of nonideal diodes with $V_T = 0.7$ V in the conducting state. This drop reduces

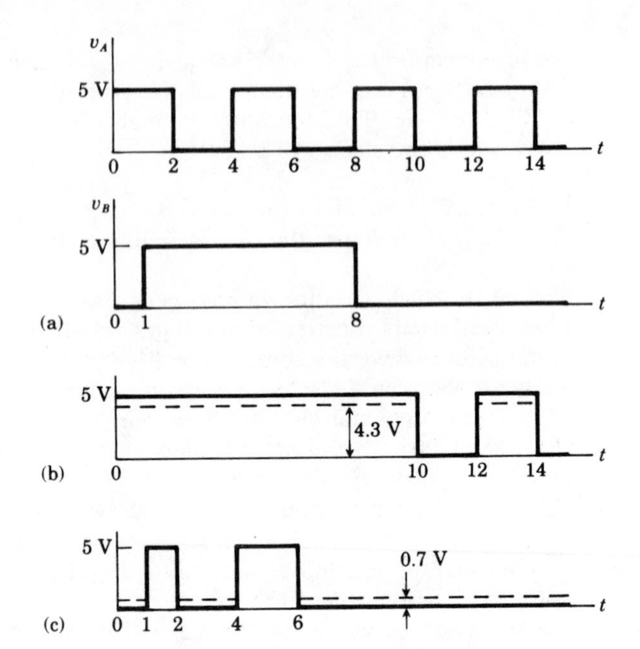

FIGURE 11.1–7 (a) Binary input signals. (b) OR-gate output. (c) AND-gate output.

the high-level output of the OR gate to $5.0 - 0.7 = 4.3$ V and raises the low-level output of the AND gate to 0.7 V.

Exercise 11.1–2

Let the OR gate have $R = 1\ \text{k}\Omega$, and model the diodes as in Fig. 9.2–1c with $R_f = 100\ \Omega$ and $V_T = 0.7$ V. Draw the equivalent circuit and find v_{out} when $v_A = 5$ V and $v_B = 0$ V. Repeat for $v_A = v_B = 5$ V. Hint: In the latter case, use Norton equivalents for each series battery-resistance branch.

ANALOG DIODE GATES

Unlike a digital signal, an *analog signal* is not restricted to specific values and usually has more-or-less smooth and continuous time variations. Figure 11.1–8a represents an *analog gate circuit* with $v_{\text{out}} = v_{\text{in}}$ when the switch is closed (gate ON) and $v_{\text{out}} = 0$ when the switch is open (gate OFF). In other words, the gate extracts *samples* of the analog input signal, as illustrated by Fig. 11.1–8b. Rapid sampling, of course, requires high-speed electronic devices rather than simple mechanical switches. Analog sampling is commonly used in analog-to-digital converters, and has a number of related applications in communications, control, and instrumentation systems.

To build an electronic analog gate, we need an auxiliary digital signal that controls the switching device. On such implementation is the *diode-quad* gate diagrammed in Fig. 11.1–9, where the control voltage v_C alternates between $+10$ V and -10 V, and we also use its *inverse* $\bar{v}_C \triangleq -v_C$. When $v_C = +10$ V (so $\bar{v}_C = -10$ V), we expect all four diodes to con-

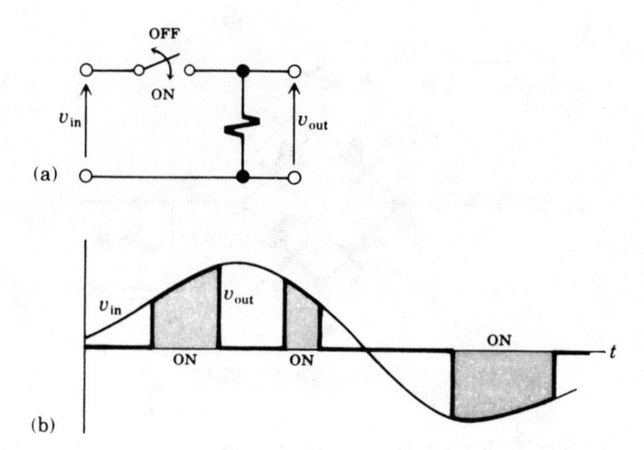

FIGURE 11.1–8
(a) Analog gate.
(b) Waveforms.

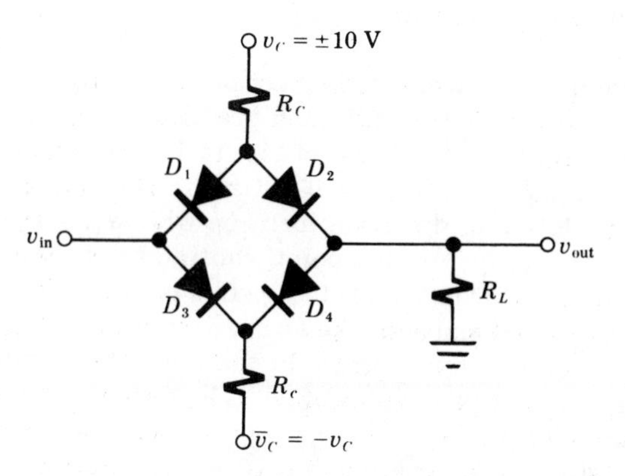

FIGURE 11.1–9
Diode-quad implementation of an analog gate.

duct and become short circuits, which puts v_{out} and v_{in} at the same potential and, therefore, $v_{out} = v_{in}$. (This is true even for nonideal but identical diodes.) On the other hand, we expect the diodes to be reverse biased when $v_C = -10$ V, so their open-circuit condition isolates the output from the input and $v_{out} = 0$.

We have, however, ignored the possibility that the value of v_{in} might alter the diode biases and change the gating action. For instance, suppose that $v_{in} > 10$ V when $v_C = +10$ V and the gate should be ON. Referring to Fig. 11.1–10, we see that D_1 and D_4 are actually reverse biased, so $i_{out} = v_C/(R_C + R_L)$ and $v_{out} = R_L i_{out} < 10 \text{ V} \neq v_{in}$. A similar problem arises when v_{in} has an excessive negative value. Consequently, the input must be limited to

$$|v_{in}| \leq \frac{R_L}{R_C + R_L} |v_C| \tag{2}$$

to ensure that $v_{out} = v_{in}$ when the gate is ON.

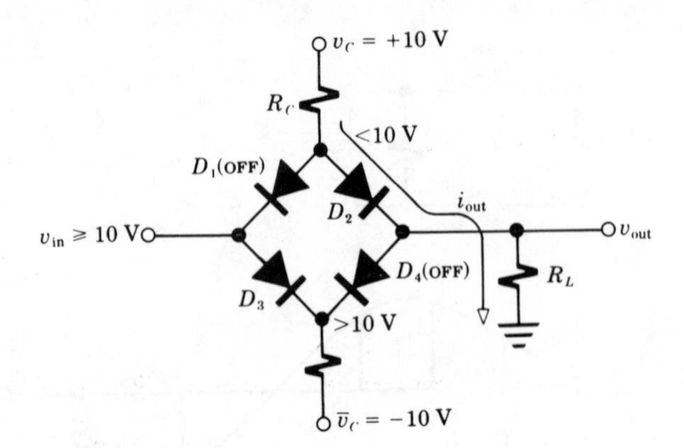

FIGURE 11.1–10

Exercise 11.1–3 Justify Eq. (2) by finding the largest value of v_{in} in Fig. 11.1–9 such that D_2 and D_3 are ON while D_1 and D_4 are at their break points (zero current and zero voltage drop).

PEAK DETECTORS AND CLAMPS

Combining diode switching action with capacitance energy storage leads to numerous interesting and practical circuits. Perhaps the simplest of these is the *peak detector* of Fig. 11.1–11a, whose diode prevents the condition $v_{out} < v_{in}$. If $v_{out} = 0$ initially and v_{in} has the waveform sketched in Fig. 11.1–11b, the diode turns on when $v_{in} = 0$ and v_{out} follows v_{in} up to the first peak. At this point, continuity of voltage across the capacitor tends to hold the value of v_{out} while v_{in} drops. The diode then becomes reverse biased and acts like an open circuit until v_{in} rises above the current value of v_{out}, and so forth. In this way, the circuit detects and stores the largest prior peak value of v_{in}. (The dashed curve represents possible discharge of the capacitor through its leakage resistance or a parallel load resistance.) A variation of this circuit performs *envelope detection* in AM radios.

Another variation is the *clamp circuit* in Fig. 11.1–12a. Its diode prevents any positive excursions of v_{out} and permits C to charge up to the first positive peak value of v_{in}, such as V_p in Fig. 11.1–12b. (Meanwhile, $v_{out} = 0$ since the diode is ON.) When v_{in} drops below V_p, the diode turns OFF and we can analyze the situation by first assuming that $R \to \infty$. We

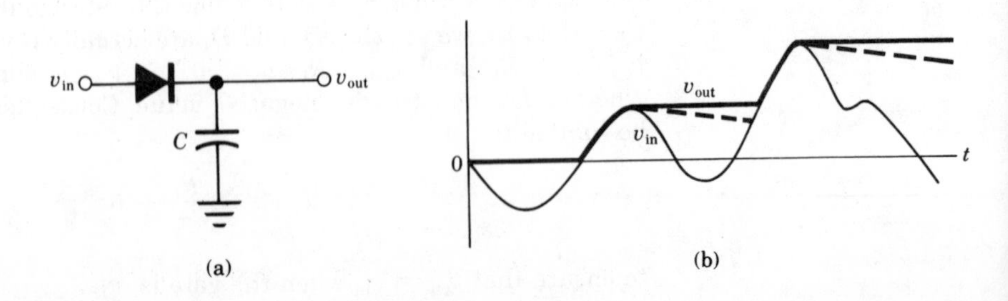

FIGURE 11.1–11
(a) Peak detector.
(b) Waveforms.

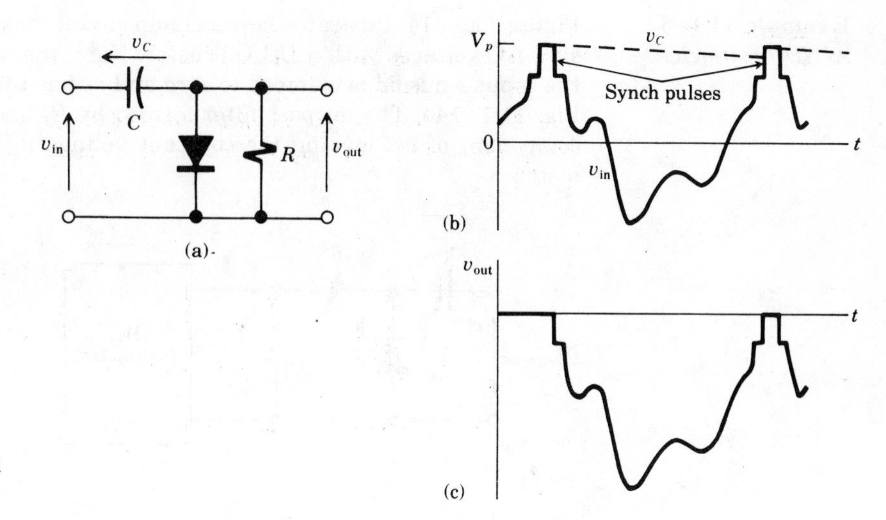

FIGURE 11.1–12
(a) Clamp circuit.
(b) Periodic input waveform. (c) Clamped output waveform.

then have an open circuit across the output and no discharge path for the capacitor, so $v_C = V_p$ and $v_{out} = v_{in} - v_C = v_{in} - V_p$. Essentially the same result applies with a finite resistance, provided that v_{in} is a periodic signal with period

$$T \ll RC \tag{3}$$

Under this condition, the capacitance has negligible impedance at all frequency components of v_{in}, and v_C decays only slightly until the next peak. This next peak turns the diode ON and recharges C; then the cycle repeats itself. Thus, the output waveform sketched in Fig. 11.1–12c has the same shape as v_{in} but is pushed downward so that the positive peaks have been *clamped* to zero.

Incidentally, the input signal shown here represents the *video waveform* after amplification in a television receiver. Video amplifiers generally remove any DC signal component, so v_{in} has a zero average value, and the top of the synchronization pulses are at $V_p > 0$. These "synch" pulses convey the reference zero voltage level for the TV picture, as well as controlling the horizontal sweep circuit. Clamping the top of the pulses to zero, therefore, restores the proper DC level that was lost in amplification.

To clamp positive peaks at some fixed nonzero voltage $V_B < V_p$, we simply put a battery in series with the diode, as in Fig. 11.1–13, creating a *biased clamp* circuit. If Eq. (3) holds, the voltage across the capacitance builds up to $v_C \approx V_p - V_B$ and

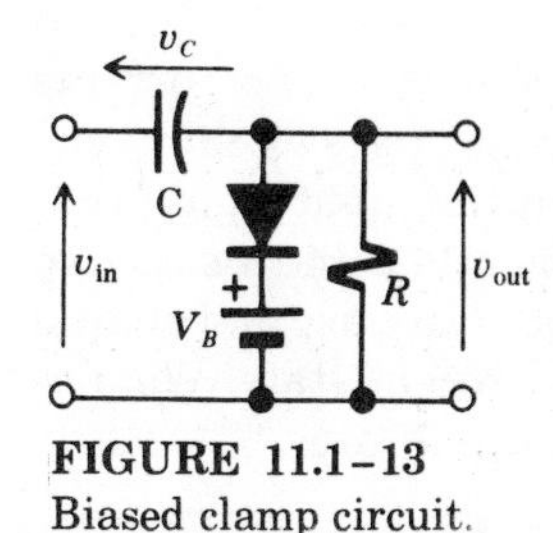

FIGURE 11.1–13
Biased clamp circuit.

$$v_{out} \approx v_{in} - (V_p - V_B) \tag{4}$$

Inverting the diode gives a circuit that clamps negative peaks to V_B.

Example 11.1–3
AC/DC voltmeter

Figure 11.1–14 illustrates how a clamp circuit makes it possible to measure AC voltages with a DC voltmeter. Here, the negative peak $-V_m$ of the input sinusoid is clamped to zero and v_D has the waveform shown in Fig. 11.1–14b. The lowpass filter formed by R_1 and C_1 removes the AC component of v_D, leaving the constant voltage $v_{av} = V_m$ applied to the meter.

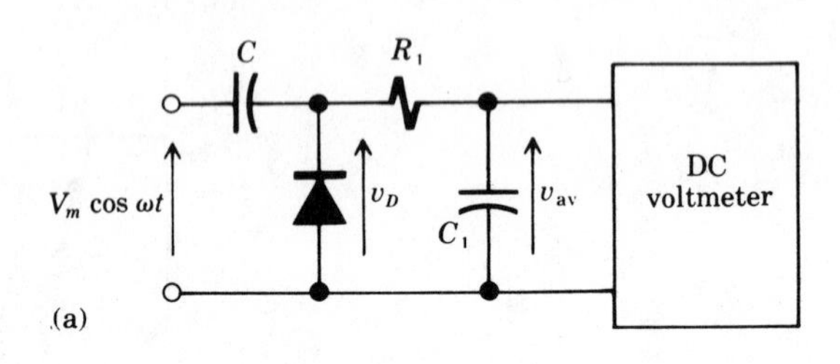

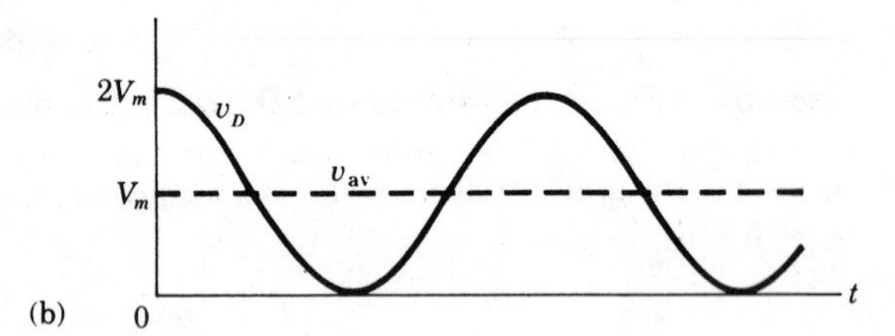

FIGURE 11.1–14
Circuit and waveforms for an AC/DC voltmeter.

Exercise 11.1–4

Sketch the waveforms v_C and v_{out} in Fig. 11.1–12a when

$$v_{in} = \begin{cases} 5 \sin \dfrac{2\pi t}{T} & 0 \leq t \leq T \\ 10 \sin \dfrac{2\pi t}{T} & T \leq t \leq 2T \\ 5 \sin \dfrac{2\pi t}{T} & 2T \leq t \leq 3T \end{cases}$$

Start at $t = 0$ with $v_C = 0$, and assume $RC \gg T$.

11.2 TRANSISTOR SWITCHING CIRCUITS

FETs and BJTs have an OFF state in which essentially no currents flow, and a strongly conducting ON state. Transistor switching circuits are generally operated by sudden changes in the input signal that transfers the device from one of these states to the other. In this section we show how transistor switches can improve both analog and digital gates, and how a circuit can "remember" its most recent state. We also explore some of the consequences of switching at rapid rates.

BJT SWITCHES AND GATES

Consider the BJT circuit of Fig. 11.2–1a, where $v_{out} = v_{CE}$. By convention, we have omitted the source of the supply voltage V_{CC} and show only the connecting terminal, the other terminal being connected to ground. We use the large-signal BJT models to obtain the transfer curves for i_C and v_{out} versus v_{in}.

Any input voltage less than the base-emitter threshold V_T leaves the transistor in *cutoff*—as represented by Fig. 11.2–1b—so $v_{CE} = V_{CC}$ and $i_C = 0$ when $v_{in} < V_T$. As v_{in} increases, the transistor passes through its active region (where $i_C = \beta i_B$) and eventually becomes saturated. Figure 11.2–1c represents the *saturated* state, including the small but nonzero voltage $v_{CE} = V_{sat}$. Here, the collector current attains its maximum value $i_C = (V_{CC} - V_{sat})/R_C$, provided that $i_B > i_C/\beta$. Since $i_B = (v_{in} - V_T)/R_B$, saturation occurs when

$$v_{in} > v_{in-sat} = V_T + \frac{R_B}{\beta R_C}(V_{CC} - V_{sat}) \qquad (1)$$

Typically, $V_{sat} \approx 0.2$ V, and it has negligible effect if $V_{CC} \gg V_{sat}$.

FIGURE 11.2–1
(a) BJT switching circuit. (b) Cutoff state. (c) Saturated state.

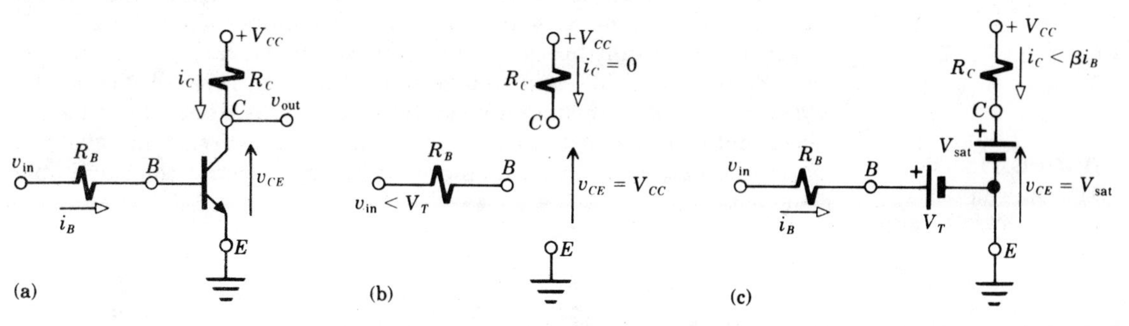

Including the linear behavior of the BJT in its active region, the transfer curves for our circuit are as plotted in Fig. 11.2–2. The active interval between cutoff and saturation is not particularly useful for switching purposes. But the active range may be reduced by appropriate choice of β, R_C, R_B, and V_{CC}. Note that the power $p = v_{CE} i_C$ that is dissipated in the transistor is zero when cutoff and nearly zero when saturated.

With a narrow active region, the BJT circuit becomes a switch with current OFF when $v_{in} < V_T$ and current ON when $v_{in} > v_{in-sat}$. Such a circuit can also function as a clipper or waveshaper, producing relatively square output waveforms from a sinusoidal input voltage (see Fig. 10.1–8). However, the output voltage always changes in a direction opposite of the input voltage changes. This polarity-inversion effect is exploited in digital logic applications where the BJT switch functions as an

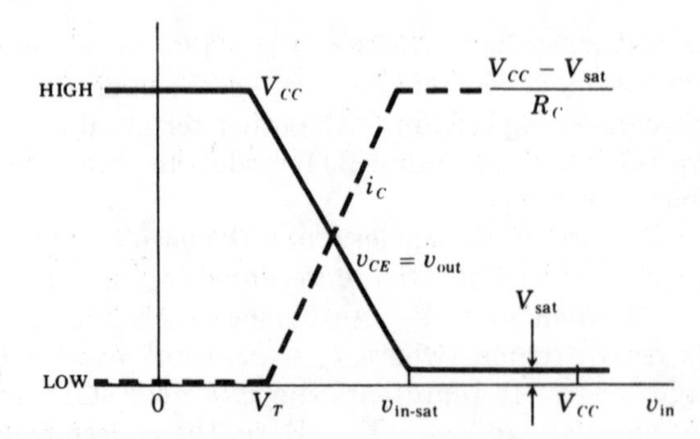

FIGURE 11.2–2
Transfer curves for BJT switch.

inverter or NOT gate, with LOW and HIGH voltage levels of approximately zero and $+V_{CC}$, respectively.

Figure 11.2–2 shows that if v_{in} is LOW, then v_{out} will be HIGH, whereas a HIGH input yields a LOW output. In either case, the output of an inverter equals the logical inverse of the input.

Putting a diode OR gate at the input of an inverter, as in Fig. 11.2–3a, creates a NOT-OR gate, in the sense that v_{out} is NOT HIGH (that is, LOW) when v_A or v_B or both are HIGH. Such circuits are called NOR gates. (In practice, the OR-gate resistance R would be omitted, since the BJT inverter fundamentally operates on input base current rather than on voltage.) Similarly, a diode AND gate combined with an inverter creates the NOT-AND or NAND gate diagrammed in Fig. 11.2–3b. A second inverter at the output of a NOR or NAND gate cancels the first inversion and gives an OR or AND gate with better performance characteristics than a simple diode gate.

FIGURE 11.2–3
DTL gates: (a) NOR (NOT-OR) gate; (b) NAND (NOT-AND) gate.

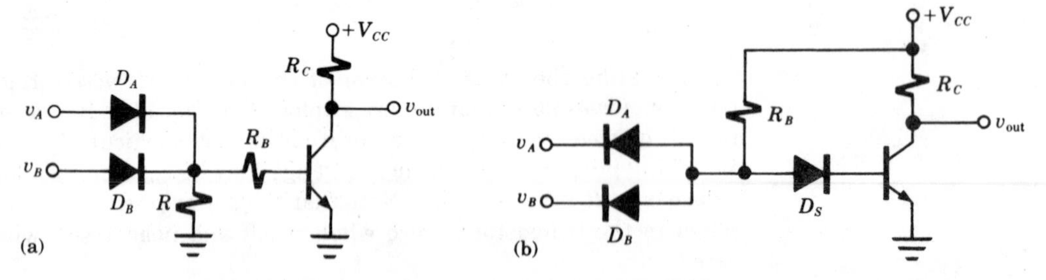

The NOR and NAND gates we have described are members of a family of logic devices called DTL, for *diode-transistor-logic*. The term *family* implies that an output terminal from any member of the family may be connected directly to input terminals of other family members without additional interfacing circuitry. Although integrated-circuit DTL gates are still available, the family is now considered obsolete, having been supplanted by newer technologies with various advantages.

Example 11.2–1
TTL NAND *gate*

Transistor-transistor logic (TTL) has been the most popular logic family for several years. We introduce the major features of this technology by examining the somewhat simplified NAND gate shown in Fig. 11.2–4a.

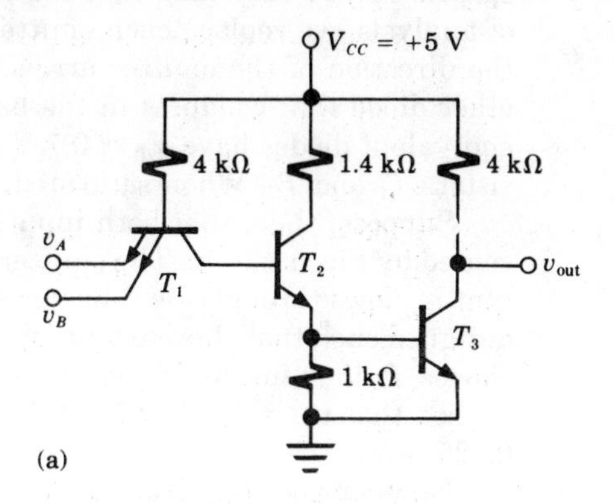

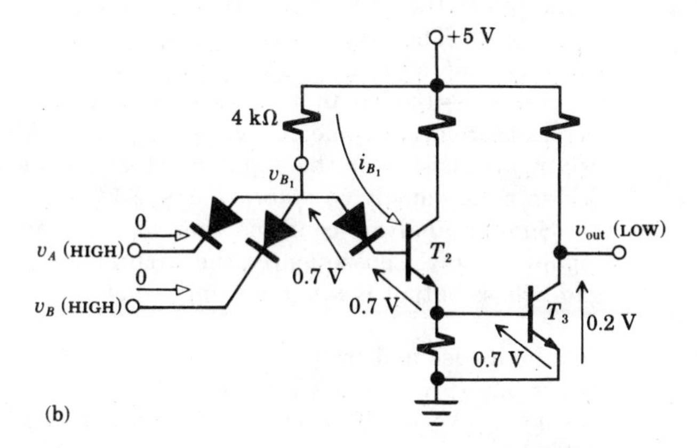

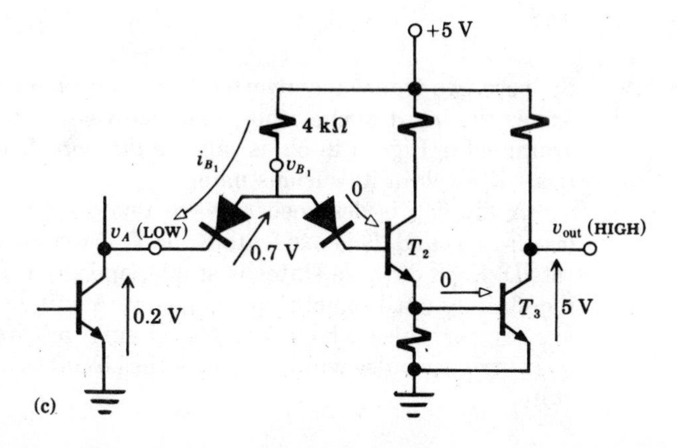

FIGURE 11.2–4
TTL NAND gate:
(a) circuit diagram;
(b) LOW output condition;
(c) HIGH output condition.

The two inputs are applied to a *multi-emitter* transistor T_1, which could have as many as eight emitters. This special transistor acts essentially like the diodes in a DTL gate (Fig. 11.2–3b), but takes up much less space on an IC chip than that required for separate diodes. For purposes of analysis, we replace each emitter of T_1 with a diode that conducts in the direction of the emitter arrow; we also replace the collector by another diode that conducts in the base-to-collector direction. All of these equivalent diodes have $V_T \approx 0.7$ V, as do the base-emitter paths of transistors T_2 and T_3. When saturated, T_2 and T_3 have $v_{CE} = V_{sat} \approx 0.2$ V.

Suppose, then, that both inputs are HIGH compared to v_{B_1}, as represented by Fig. 11.2–4b. The emitter diodes of T_1 will be OFF, but base current i_{B_1} flows through the collector of T_1 into the base of T_2. The circuit is designed such that this current is sufficient to saturate T_2, which in turn causes T_3 to saturate and yields $v_{out} = 0.2$ V (LOW). Working backward, we see that $v_{B_1} = 0.7 + 0.7 + 0.7 = 2.1$ V and $i_{B_1} = (5 - 2.1)V/4\ \text{k}\Omega = 0.725$ mA.

Now consider the situation in Fig. 11.2–4c, where one input has been connected to the saturated output of another TTL gate that establishes $v_A = 0.2$ V. (The other input has no effect, and is omitted here.) The emitter diode of T_1 goes ON and the base voltage drops to $v_{B_1} = 0.2 + 0.7 = 0.9$ V—too low to provide base current for T_2 and T_3. These transistors therefore become cut off, giving $v_{out} = 5$ V (HIGH). Note that the driving transistor at the input must accept or *sink* current from the driven gate, namely $i_{B_1} = (5 - 0.9)V/4\ \text{k}\Omega \approx 1$ mA.

Similar analysis confirms that v_{out} will be HIGH when any or all inputs are LOW. Consequently, the circuit does indeed function as a NAND gate whose output is NOT HIGH only when all inputs are HIGH.

Exercise 11.2–1

Let the diodes and transistor in Fig. 11.2–3a have $V_T = 0.7$ V, and let $V_{sat} = 0.2$ V, $R_B = 5.6\ \text{k}\Omega$, $R_C = 1\ \text{k}\Omega$, and $V_{CC} = 5$ V.
(a) Plot v_{out} versus v_A when $v_B = 0$ V and $\beta = 100$.
(b) Obtain a condition on β such that v_{out} is LOW when v_A is HIGH, and v_B is LOW.

THE ELEMENTARY FLIP-FLOP

Suppose two transistor inverters are connected so that the output of one drives the input of the other, and vice versa. The resulting circuit, diagrammed in Fig. 11.2–5a, is called a *flip-flop*. A description of its properties will explain its curious name.

A flip-flop is designed to have two distinct states: either T_1 is ON (saturated) and T_2 is OFF (cutoff); or T_1 is OFF cutoff), and T_2 is ON (saturated). Each of these states is stable, and the flip-flop stays in one state indefinitely until something changes it. A suitable "something" is a voltage trigger pulse which *flips* the circuit into its other state. A subsequent trigger pulse will then cause the circuit to *flop* back into the former state.

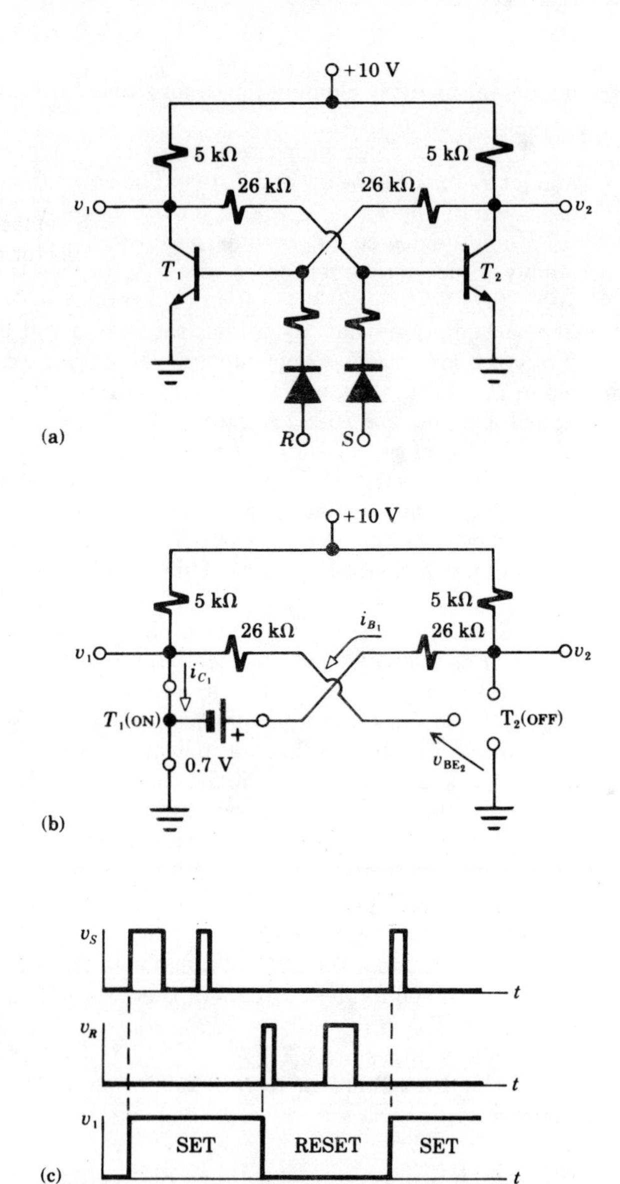

FIGURE 11.2–5
(a) Flip-flop circuit.
(b) One of the two stable states.
(c) Illustrative waveforms.

A flip-flop is said to be *bistable* in view of its two stable states. It is also said to have *memory* because it "remembers" the state into which it has been triggered, just as a light switch on the wall "remembers" whether it is OFF or ON and stays that way until someone flips the switch to the other position. This property makes the flip-flop an important memory element for timers (Section 11.3) and for digital logic circuits (Section 15.3).

To confirm these properties, temporarily ignore terminals R and S, and let T_1 initially be ON and T_2 OFF. The equivalent circuit then becomes as shown in Fig. 11.2–5b where, for the model of T_1, we have taken V_T = 0.7 V and neglected V_{sat} because of its small value compared to the 10-V supply. Since no current goes into T_2, the base current for T_1 is $i_{B_1} = (10 - 0.7)\text{ V}/(5 + 26)\text{ k}\Omega = 0.3\text{ mA}$, while $i_{C_1} \approx 10\text{ V}/5\text{ k}\Omega = 2\text{ mA}$. Our assumption that T_1 is saturated is valid if $i_{C_1} < \beta_1 i_{B_1}$ or $\beta_1 > 2\text{ mA}/0.3\text{ mA} = 6.7$, a very modest value for current gain. Since $v_1 \approx 0$ when T_1 is saturated, no base current can flow to T_2, for if it did the voltage drop across the 26-kΩ resistance would produce $v_{BE_2} = v_1 - 26\text{ k}\Omega \times i_{B_2} < 0$. Therefore, T_2 is in fact cutoff, with $v_{BE_2} = 0$, $i_{B_2} = 0$, $i_{C_2} = 0$, and $v_2 = 10\text{ V} - 5\text{ k}\Omega \times 0.3\text{ mA} = 8.5\text{ V}$. By symmetry, identical arguments hold when we let T_2 be ON initially and T_1 OFF.

One way to flip the circuit out of the state of Fig. 11.2–5b is to turn T_2 ON, which will then turn T_1 OFF. This could be done by applying a positive *trigger* or *set pulse* at terminal S in Fig. 11.2–5a. (The diode here serves to disconnect the flip-flop from the trigger source except during triggering.) A second set pulse would have no effect, but a positive pulse at R would *reset* the circuit to its original condition. The waveforms in Fig. 11.2–5c illustrate a possible sequence of set/reset operations.

Simple circuit modifications permit the use of negative trigger pulses at the S and R terminals or at the collector terminals. In practice, small "speed-up" capacitors would be put in parallel with the 26-kΩ resistors to decrease the time required for a transition between states.

Exercise 11.2–2

Two NOR gates like those in Fig. 11.2–3a (except with R omitted) are made into a flip-flop by connecting each collector terminal to the other A terminal and using the B terminals for triggering.
(a) Draw the equivalent circuit in the form of Fig. 11.2–5b, and find i_{B_1} and v_2 if $V_{CC} = 5$ V, $R_C = 1$ kΩ, $R_B = 35$ kΩ, $V_{sat} = 0.2$ V, and all threshold voltages are 0.7 V.
(b) If the transistors have $\beta = 100$, what is the minimum triggering voltage?

FET SWITCHES AND CMOS GATES

For switching-circuit applications, FETs offer two advantages compared to BJTs. First, being a voltage-controlled device, a FET does not draw current from a control source applied to its gate terminal. Second, a FET has the ability to carry current in either direction through the drain-source channel. (In fact, the functional roles of the drain and source in a symmetrically fabricated FET actually depend on the direction of current flow.) However, the channel resistance R_{DS} of a FET operating in the ohmic region may result in a voltage drop considerably greater than the drop across a saturated BJT. Consequently, FET switches work best when limited to relatively low currents in the ON or ohmic state.

Consider, for example, the analog gate circuit in Fig. 11.2–6 where an n-channel enhancement MOSFET serves as a switch. If the control voltage v_C is less than the threshold voltage V_T, so $v_{GS} < V_T$, the FET will be cutoff and $v_{\text{out}} = 0$. If v_C has a large positive value, the FET will be in its ohmic region and act like a small resistance R_{DS}. Therefore,

$$v_{\text{out}} = \frac{R_L}{R_L + R_{DS}} v_{\text{in}}$$

so $v_{\text{out}} \approx v_{\text{in}}$ and $v_{DS} \approx 0$ if $R_L \gg R_{DS}$, which ensures that i_D will be small. In either state of the switch, no current is required from the control source v_C.

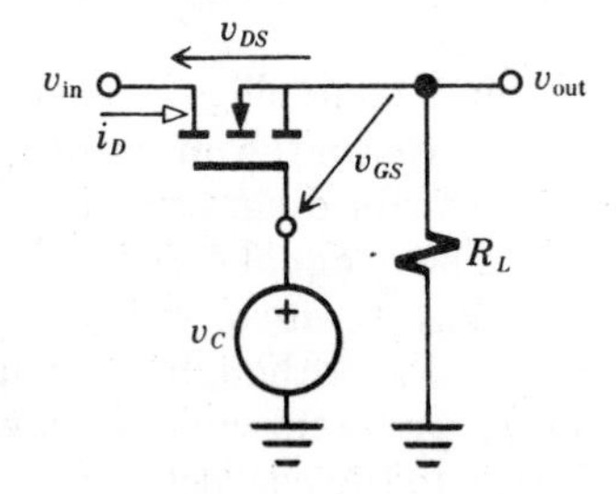

FIGURE 11.2–6
Analog gate with a FET switch.

A disadvantage of the simple FET switch is that R_{DS} varies with $v_{GS} = v_C - v_{\text{out}}$ and produces some nonlinear distortion in the output signal. We solve this problem by using *complementary* n-channel and p-channel MOSFETs, arranged as shown in Fig. 11.2–7a. This arrangement, a CMOS ("see-moss") *switch* or *gate,* is symbolized by Fig. 11.2–7b.

FIGURE 11.2–7
(a) CMOS switch. (b) Symbol. (c) Analog multiplexer with two inputs.

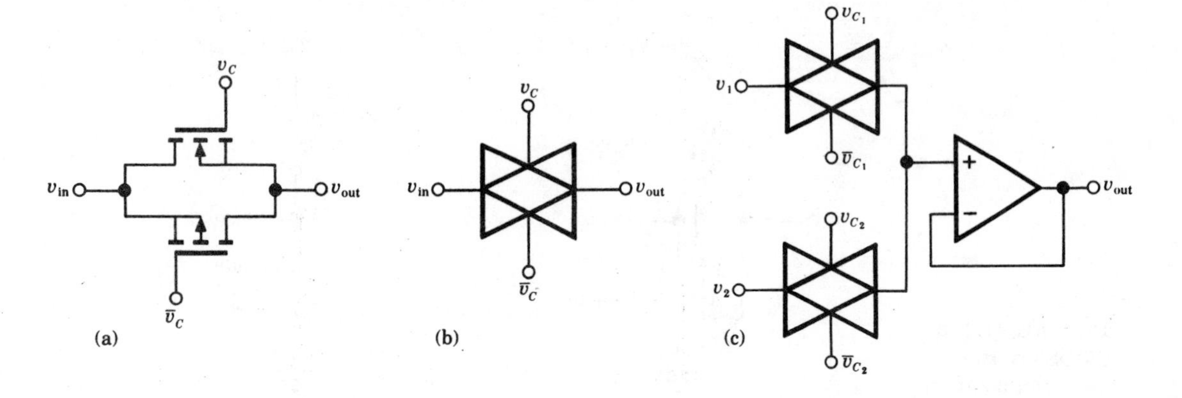

Like a diode quad circuit (Fig. 11.1–9), the CMOS switch requires two control voltages, v_C and $\bar{v}_C = -v_C$. When $v_C > 0$, the switch acts like a constant resistance of a few hundred ohms or less, and transmits any signal voltage in the range

$$-|v_C| \leq v_{\text{in}} \leq |v_C|$$

Conversely, it acts like an open circuit when $v_C < 0$.

Figure 11.2–7c diagrams a simple *analog multiplexer* composed of CMOS gates and an op-amp voltage follower to minimize loading effects. If we make v_{C_2} negative when v_{C_1} is positive, and vice versa, the output will consist of sequential samples of the two input signals. This concept is readily extended to circuits with three or more inputs. However, the relatively slow switching speed of CMOS devices limits the minimum sample duration to about 50 ns, whereas a diode quad can extract samples as brief as 1 ns.

Integrated-circuit CMOS technology is also used for digital gates in the CMOS logic family. To explain the operating principles of this family, let's consider the CMOS inverter diagrammed in Fig. 11.2–8a where the complementary MOSFETs have equal threshold voltages $V_T < V_{DD}$. If we apply a HIGH input, $v_A \approx V_{DD}$, then $v_{SG} = V_{DD} - v_A \approx 0 < V_T$ so T_p must be cut off, while $v_{GS} = v_A > V_T$. Although we would expect T_n to conduct, the cutoff condition of T_p prevents current flow from the supply and, therefore, $i_D = 0$. Referring to the characteristic curves in Fig. 10.1–4 reveals that, with $i_D = 0$ and $v_{GS} > V_T$, T_n must have $v_{DS} = 0$. Thus, a HIGH input produces a LOW output, $v_{\text{out}} = v_{DS} = 0$. In essence, this inverter acts like the two switches in Fig. 11.2–8b, with T_p being an open circuit and T_n a short circuit—albeit one that carries no current. Applying a LOW input, $v_A \approx 0$, simply reverses the open and closed switches and yields $v_{\text{out}} = V_{DD}$, provided that any load connected to the output terminal draws negligible current.

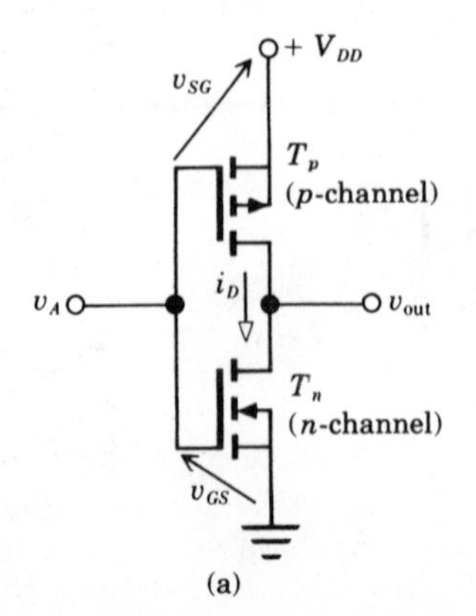

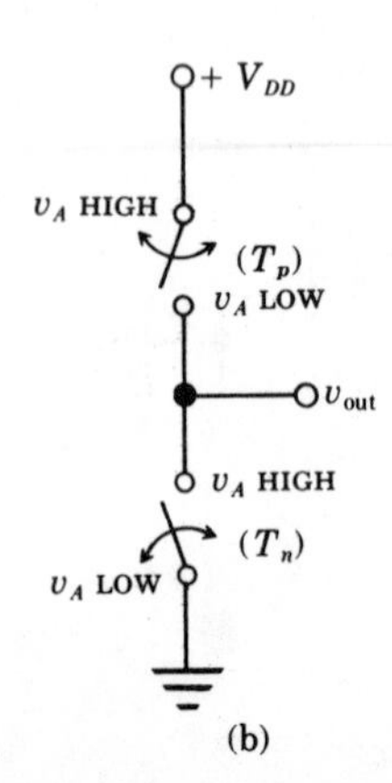

FIGURE 11.2–8 CMOS inverter: (a) circuit diagram; (b) switch model.

CMOS NOR and NAND gates are fabricated by putting additional MOSFETs in series or parallel with the inverter configuration. For instance, Fig. 11.2–9 shows the circuit diagram and switch model of a two-input NOR gate.

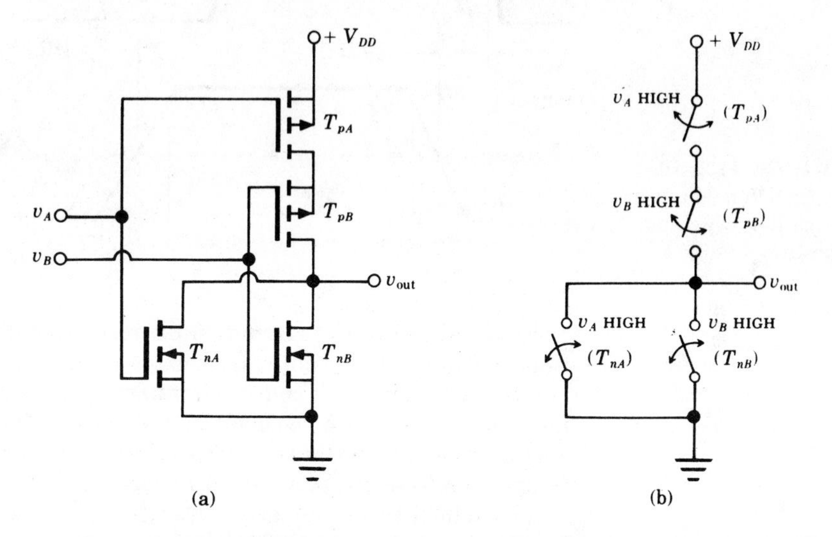

FIGURE 11.2–9 CMOS NOR gate: (a) circuit diagram; (b) switch model.

Because MOSFETs require less space than BJTs or resistors on an integrated-circuit chip, CMOS logic readily lends itself to very large-scale integration (VLSI)—with more than 10,000 components on a single chip! Another CMOS feature is its low power dissipation since, as we've seen, a CMOS gate requires virtually no current from the supply. These two advantages make CMOS logic well suited for applications such as pocket calculators and digital watches.

Exercise 11.2–3

(a) Use the switch model to confirm that the circuit in Fig. 11.2–9 has the required properties of a NOR gate.
(b) Devise a switch model and CMOS circuit for a NAND gate.

SWITCHING TRANSIENTS

High-speed switches and gate circuits are purposely designed without added capacitance or inductance because of the "sluggishness" these energy-storage elements introduce. All electronic devices have some unavoidable energy storage, however, primarily due to stray capacitance and stored charge carriers. Consequently, a finite transition time is required to switch a device from one state to another.

The way in which a typical integrated circuit gate responds to an input signal is illustrated in Fig. 11.2–10. The average time delay between the 50% points of the input and output signals is called the *propagation delay time* t_{pd}. Since propagation delay is often different for transitions in opposite directions, both low-to-high (LH) and high-to-low (HL)

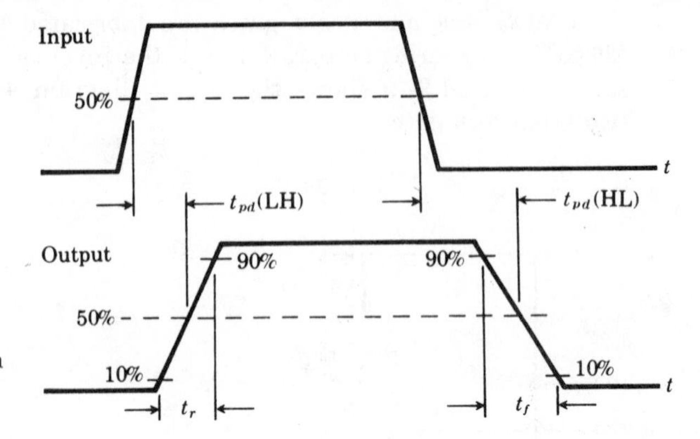

FIGURE 11.2–10 Propagation delays in a transistor switching circuit.

times are defined. The rise and fall times are defined in the usual way, from the 10% to 90% points. In most gates, all these times are dependent on the gate loading, of course. Manufacturers usually specify typical or maximum times for a specific load, which includes the typical input capacitance of a given number of similar gates. Propagation delays may range from 5–50 ns, and rise/fall times run from 1–100 ns.

Switching transients also arise when a transistor drives an external load that includes energy storage. Large capacitive loads seldom occur in practice, but various electromechanical devices such as relays, stepping motors, and type hammers in printers are quite inductive. The following example illustrates resulting transient effects.

Example 11.2–2
BJT relay driver

The *RL* load in Fig. 11.2–11a represents the coil of an electromechanical device being driven by a BJT switch. A small current applied to the base of the transistor produces a large collector current that energizes the coil and actuates the device. The diode is included to prevent a voltage spike and possible damage to the transistor when the coil is de-energized.

Assuming that i_B has been zero for a long time, the operating point for $t < 0$ is at $i_C = 0$, $v_{CE} = 12$ V, as shown on the characteristic curves in Fig. 11.2–11b. When i_B jumps to 1 mA at $t = 0$ the transistor becomes saturated, but the inductance requires that $i_C(0^+) = 0$, so the operating point moves to the origin—the only point where $i_B = 1$ mA and $i_C = 0$. Since v_{CE} equals zero at this point, we can draw the equivalent circuit of Fig. 11.2–11c with the transistor replaced by a short circuit and the reverse-biased diode omitted entirely. Hence, $i_C(t)$ is a step-response transient with time constant $\tau = 1\ \text{H}/200\ \Omega = 5$ ms, and it reaches a final value of $12\ \text{V}/200\ \Omega = 60$ mA after about 25 ms.

If i_B is then turned off, the transistor becomes cut off and the operating point jumps to $i_C = 0$, $v_{CE} = 12$ V. But current must continue to flow in the inductance, so the diode comes into action at this time, as in

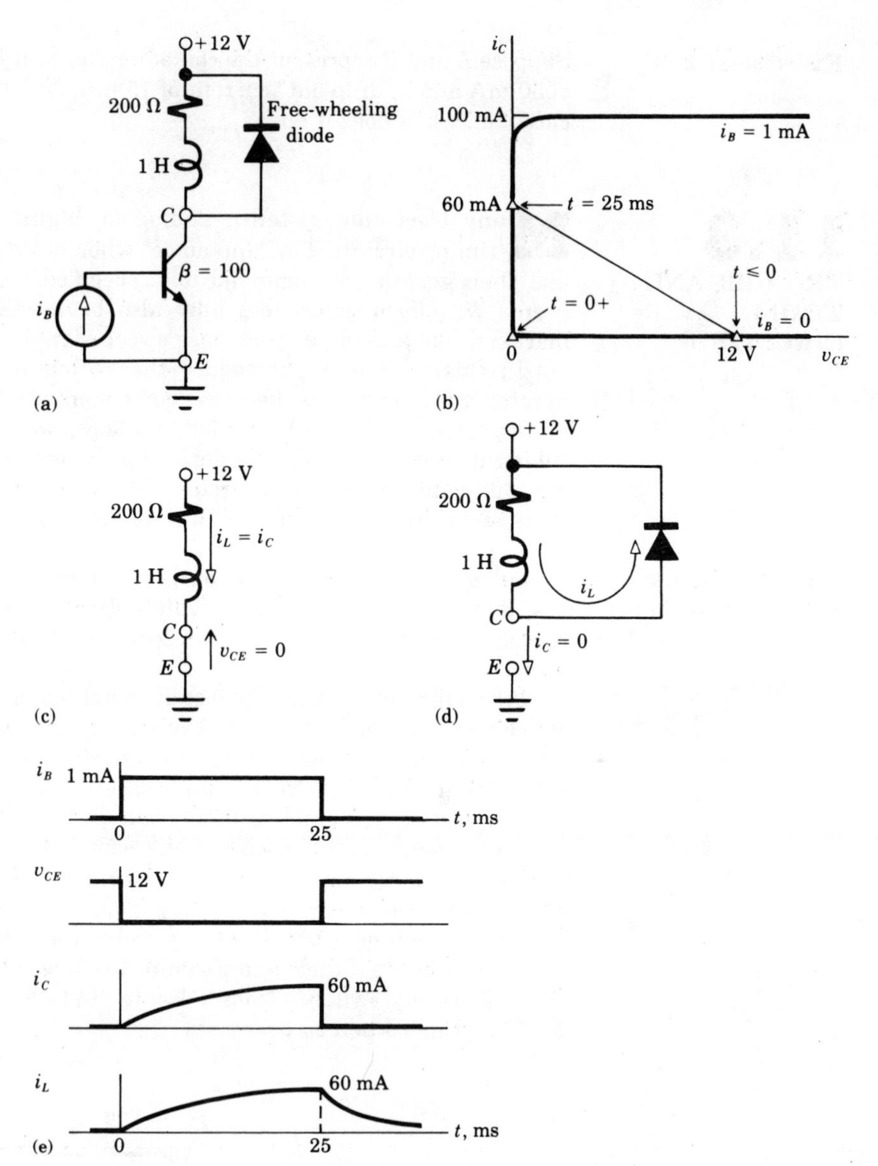

FIGURE 11.2–11
BJT relay driver with free-wheeling diode: (a) circuit; (b) operating points; (c) saturated state; (d) cutoff state; (e) waveforms.

Fig. 11.2–11d, and i_L decays to zero through the diode with time constant τ.

The various waveforms for this circuit are plotted in Fig. 11.2–11e. Note that i_C has a discontinuity at $t = 25$ ms since the diode maintains continuity for i_L. Diodes used for this type of function are said to be *free-wheeling*.

Exercise 11.2–4

Suppose L and R represent the coil of a relay that has a "pull-in" current of 50 mA and a "drop-out" current of 15 mA. At what time does the relay close and when does it open?

11.3 TRIGGER AND TIMING CIRCUITS †

Many electronic systems, analog or digital, require trigger and timing circuits that "announce" when a certain event has occurred and then generate a command at a specified time interval after that event. Waveform generators may also be needed to provide periodic markers, like a clock, or time-base sweep signals.

In this section, we introduce the switching device called a comparator, which serves as the active component of trigger circuits. By combining triggers, RC circuits, and flip-flops, we can build a variety of timing devices and waveform generators. Consistent with modern practice, our emphasis will be on circuits that incorporate readily available IC units rather than assemblies of discrete components.

COMPARATORS AND TRIGGERS

A *comparator* is a device whose output switches between two possible values, depending on the relative value of the input. The sudden output transition caused by switching can serve as a *trigger* signal to some other device.

Although not particularly fast at switching, an op-amp approximates comparator action if we omit feedback to the inverting terminal and apply a fixed reference voltage to its noninverting terminal, as diagrammed in Fig. 11.3–1a. The difference voltage is then $v_d = V_{ref} - v_{in}$, and the plot of v_{out} versus v_d (previously given in Fig. 4.3–1c) can be redrawn to obtain the input-output characteristics of Fig. 11.3–1b. Except for the narrow region of width $2V_{CC}/A$ centered at $v_{in} = V_{ref}$, this op-amp comparator saturates at $v_{out} = \pm V_{CC}$, depending on whether $v_{in} < V_{ref}$ or $v_{in} > V_{ref}$. The linear transition region between the two saturated states will be negligible if the op-amp's gain A is large enough that $2V_{CC}/A \ll |V_{ref}|$. Thus, v_{out} switches from $+V_{CC}$ to $-V_{CC}$ as v_{in} increases past V_{ref}, and vice versa when v_{in} decreases.

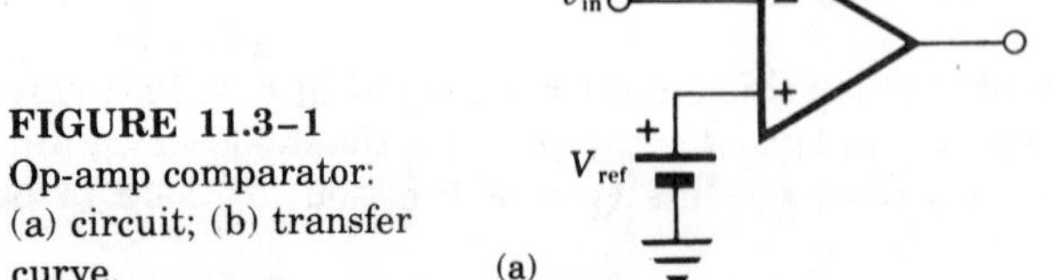

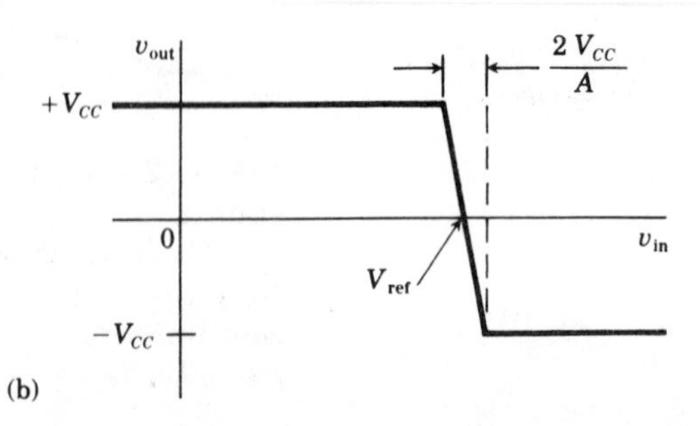

FIGURE 11.3–1 Op-amp comparator: (a) circuit; (b) transfer curve.

To be more realistic, however, we should account for the noise and other small voltage variations that cause v_{out} to switch at random between its two limits whenever $v_{in} \approx V_{ref}$. Such random switching is undesirable for most triggering applications. A better comparator circuit should have a *hysteresis* characteristic like that plotted in Fig. 11.3–2a. Here there are two reference voltages so arranged that v_{out} switches downward only when v_{in}, after having increased past V_1, equals V_2. Conversely, v_{out} switches upward only when v_{in}, after having decreased past V_2, equals V_1. For $V_1 < v_{in} < V_2$, the device has two stable states just like a flip-flop, and the value of v_{out} depends on the past behavior of v_{in}. Figure 11.3–2b illustrates how this hysteresis effect eliminates unwanted triggering by noise variations superimposed on the input signal. (For comparison purposes, try sketching the corresponding output from Fig. 11.3–1a with V_{ref} midway between V_1 and V_2.)

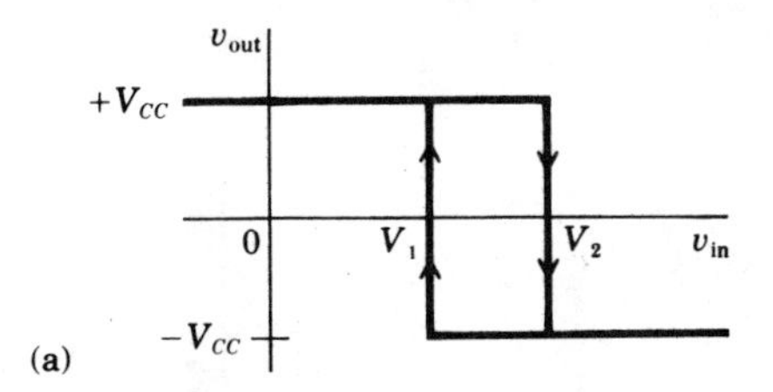

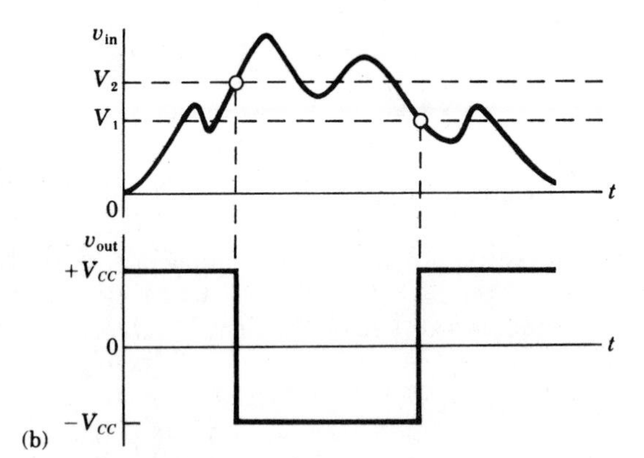

FIGURE 11.3–2
Comparator with hysteresis: (a) transfer curve; (b) typical waveforms.

Two possible ways of adding hysteresis to an op-amp comparator are shown in Fig. 11.3–3. Both circuits incorporate a feedback connection to the *noninverting* terminal so that v_p depends on v_{out}. We find the triggering points V_1 and V_2 by solving $v_d = v_p - v_n = 0$ with $v_{out} = \pm V_{CC}$. In Fig. 11.3–3a, for instance, $v_p = V_0 + R_2 v_{out}/(R_1 + R_2)$ and

$v_n = v_{in}$. Thus, the condition

$$v_d = V_0 \pm \frac{R_2 V_{CC}}{R_1 + R_2} - v_{in} = 0$$

holds when $v_{in} = V_1$ or $v_{in} = V_2$, where

$$V_1 = V_0 - \frac{R_2 V_{CC}}{R_1 + R_2} \qquad V_2 = V_0 + \frac{R_2 V_{CC}}{R_1 + R_2} \tag{1}$$

The width of the hysteresis band $V_2 - V_1$ is easily controlled by adjusting R_2, but using a battery to establish the midpoint V_0 is not very convenient—unless, of course, we want $V_0 = 0$. The other circuit in Fig. 11.3–3 is a more practical implementation; it can be analyzed following the method just outlined.

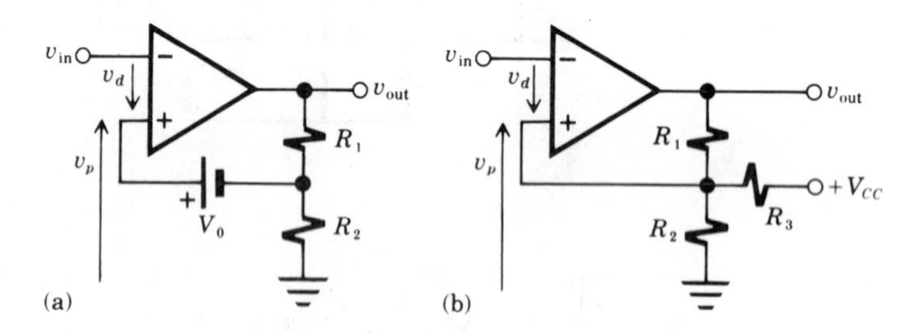

FIGURE 11.3–3 Implementations of a comparator with hysteresis.

IC comparators are units similar to op-amps, but are designed for much faster switching. Since they normally drive digital circuits, their output voltage values are compatible with the corresponding logic family—that is, v_{out} goes HIGH when $v_p > v_n$ and goes LOW when $v_p < v_n$. A comparator with hysteresis is called a *regenerative comparator* or a *Schmitt trigger*. Most logic families include a device consisting of an inverter and flip-flop that produces hysteresis. These devices, however, have only one input terminal and fixed trigger levels V_1 and V_2. Figure 11.3–4 shows the symbol and typical transfer characteristics of a CMOS-family Schmitt-trigger inverter.

Exercise 11.3–1 Show that Fig. 11.3–3b has $V_1 = (R_1 - R_3)\, V_{CC}/R_0$ and $V_2 = (R_1 + R_3)\, V_{CC}/R_0$, where $R_0 = (R_1 R_2 + R_1 R_3 + R_2 R_3)/R_2$. Hint: Use superposition to find v_p with $v_{out} = \pm V_{CC}$.

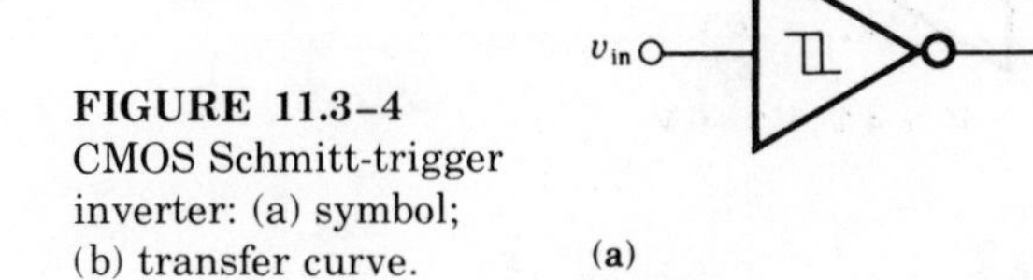

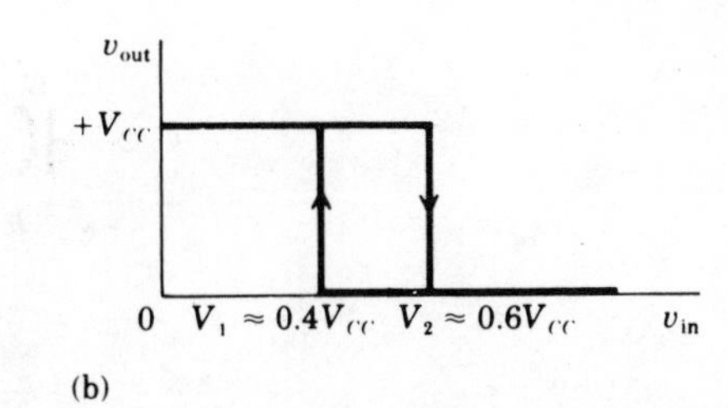

FIGURE 11.3–4 CMOS Schmitt-trigger inverter: (a) symbol; (b) transfer curve.

TIMERS AND MONOSTABLES

Digital systems and various electronic instruments often require controllable *time delays,* or pulses with controllable *duration.* As a simple example of how this can be accomplished, suppose we connect the input to a CMOS Schmitt-trigger inverter through an *RC* circuit as in Fig. 11.3–5a. The waveforms sketched in Fig. 11.3–5b show that an input pulse whose duration is large compared to $\tau = RC$ produces an output pulse inverted and delayed by $t_d \approx \tau$. Hence, we can control the time delay by adjusting either *R* or *C*. Interchanging *R* and *C* yields the circuit and waveforms of Fig. 11.3–6. Here, the leading edge of the input pulse immediately produces an inverted output pulse of controllable duration $D \approx \tau$, while the trailing edge has no effect. (However, the negative spike in v_R may dam-

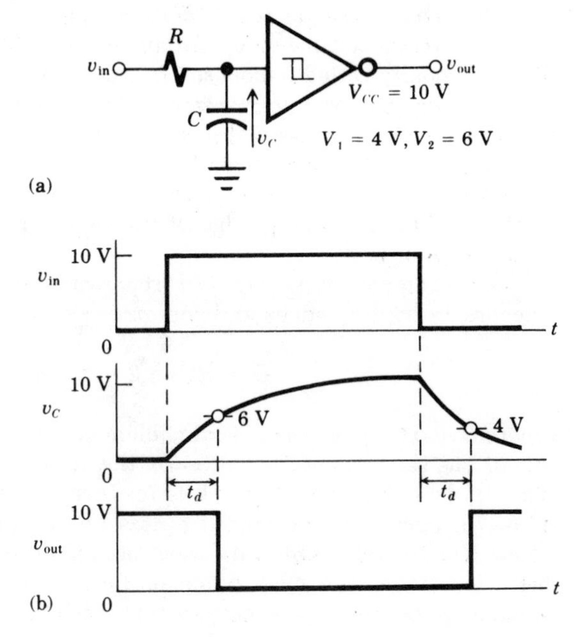

FIGURE 11.3–5 (a) Time-delay circuit. (b) Waveforms.

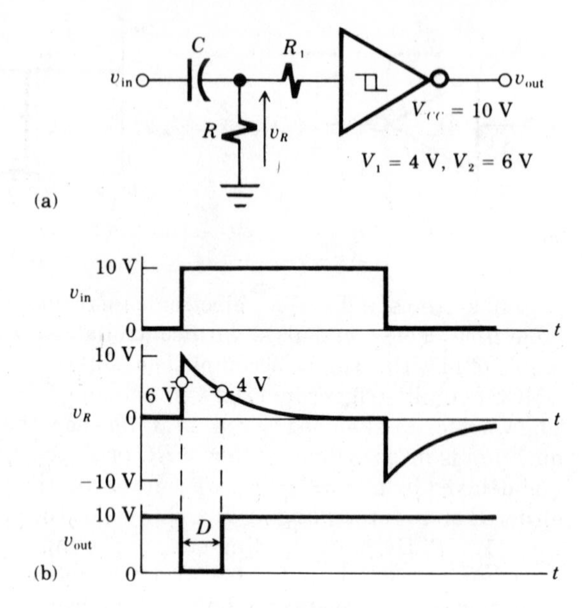

FIGURE 11.3–6
(a) Pulse-forming circuit. (b) Waveforms.

age the Schmitt trigger so we include the large resistance R_1 for current-limiting purposes.) If what we want instead are noninverted pulses at the output of either of these two circuits, all we have to do is connect the Schmitt trigger to a CMOS NOT gate.

More versatile time-delay circuits are possible with a special type of flip-flop usually called a *monostable,* but also known as a *delay multivibrator,* a *one-shot,* or a *univibrator.* This device is arranged so that it can be triggered from its normal or RESET state into a temporary SET state, from which it will trigger itself back to the RESET state after a prescribed time delay. Depending on the particular unit, triggering may take place at the leading or trailing edge of the input, and the SET state may be either HIGH or LOW.

For example, Fig. 11.3–7a represents a TTL monostable that triggers on trailing edges and generates SET pulses of duration

$$D = RC \ln 2 \approx 0.7RC$$

where R and C are external timing elements. If $D = 0.5$ ms and the input signal has the form of Fig. 11.3–7b, the output waveform is the train of short pulses shown in Fig. 11.3–7c. Increasing D to 4 ms, as in Fig. 11.3–7d, lengthens the output pulses and eliminates one of them, because the second input pulse now occurs during a SET interval. This behavior makes the monostable useful as a *contact-bounce eliminator* when v_{in} comes through a mechanical switch that closes and then momentarily bounces open again.

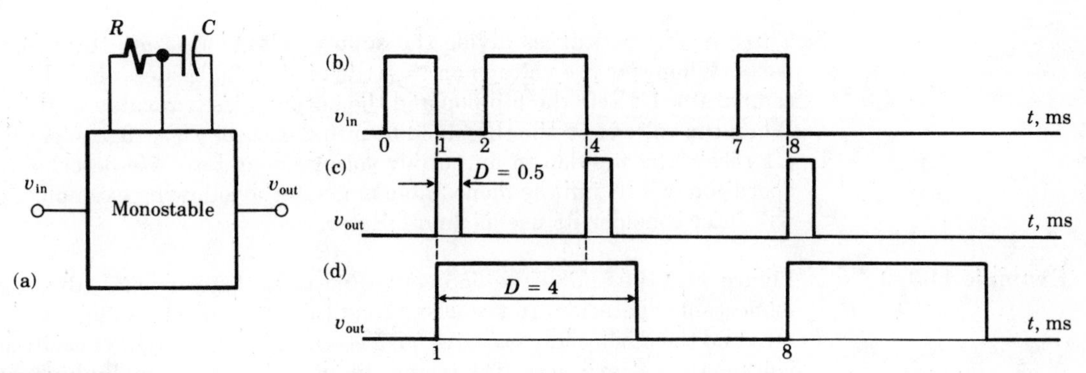

FIGURE 11.3–7 A monostable and typical waveforms.

Some IC monostables also have a RESET input that, when triggered, returns the circuit to its normal state even if the SET time has not elapsed. Other units known as *retriggerable monostables* will start a new SET-state interval every time they are triggered.

The circuits just discussed either are digital ICs or use IC gates as prime components. There is also available a specialized integrated circuit, usually classified as an "interface" or "miscellaneous" IC, rather than a digital IC. This type of IC is simply called a *timer,* and the best known one is designated by the number 555. (The 556 timer consists of two 555 timers in one package.) The 555 is designed to be flexible enough for a great variety of applications. As a monostable, its useful timing period ranges from about one microsecond to one minute. It is easily connected so that the duration or position of its output pulses can be controlled by an input signal, and it has sufficient power-handling capability to operate an electromechanical relay or drive a power transistor. The supply voltage V_{CC} can be anywhere in the 5–15 V range.

Figure 11.3–8 is a somewhat simplified block diagram of the 555.

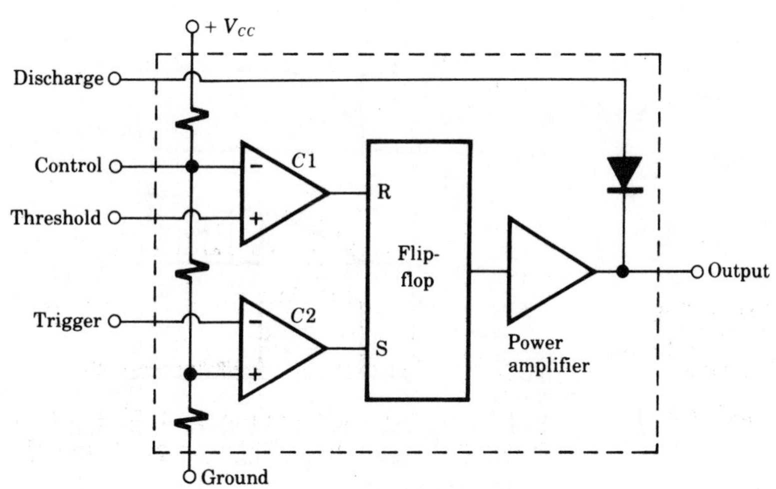

FIGURE 11.3–8 The 555 timer.

Three equal resistances divide the supply voltage V_{CC} into three equal parts. Whenever the voltage on the trigger terminal is less than $V_{CC}/3$, comparator $C2$ sets the flip-flop and the output goes to a value near V_{CC}. When the voltage on the threshold terminal exceeds $2V_{CC}/3$, comparator $C1$ resets the flip-flop to an output voltage near zero. We describe the operation of a 555 in the monostable mode in the following example, and will later consider its use as an oscillator.

Example 11.3–1

Figure 11.3–9a shows the 555 connected to an external RC circuit for monostable operation. In the RESET condition, the output voltage is near zero and the diode causes the discharge terminal and also the capacitance voltage to be nearly zero. The trigger input voltage is normally high, but must drop to a value below $V_{CC}/3$ in order to start the process, as indicated at $t = 0$ in Fig. 11.3–9b. The flip-flop becomes SET and the output voltage rises, disconnecting the diode so that C now charges through R toward V_{CC}. When v_C reaches $2V_{CC}/3$, the flip-flop RESETS and awaits another trigger, while v_C is quickly pulled down to nearly zero as the diode conducts.

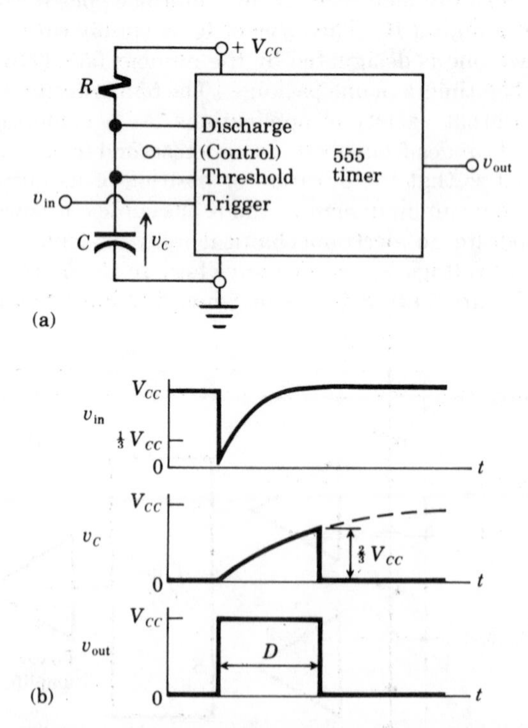

FIGURE 11.3–9
(a) Monostable circuit.
(b) Waveforms.

Exercise 11.3–2

Redraw the waveforms in Fig. 11.3–5b and 11.3–6b, taking the duration of the input pulse to be small compared to RC.

Exercise 11.3–3

Show that the duration of the output pulse in Fig. 11.3–9b is given by $D = RC \ln 3 \approx 1.1\, RC$.

SWITCHING OSCILLATORS AND WAVEFORM GENERATION

The term *oscillator* usually refers to linear amplifiers with feedback arranged to generate sinusoidal signals (as we will detail in Section 13.3). For our purposes here, we will examine *switching* or *relaxation oscillators* that produce periodic but nonsinusoidal signals such as square or triangular waves for use in analog or digital instrumentation. Such circuits are said to be *astable* in that they continually switch between two temporary states.

The circuit in Fig. 11.3–10a consists of a *regenerative comparator* driving an op-amp *integrator* (from Example 5.1–2) whose output becomes part of the comparator's difference voltage $v_d = (R_2 v_a + R_1 v_b)/(R_1 + R_2)$. Setting $v_d \leq 0$ and $v_a = -V_{CC}$ gives the condition $v_b \leq R_2 V_{CC}/R_1$, while $v_d \geq 0$ and $v_a = +V_{CC}$ gives $v_b \geq -R_2 V_{CC}/R_1$. (You may find it interesting to compare a plot of v_a versus v_b with Fig. 11.3–2a.) Since $v_a = \pm V_{CC}$, v_b will be a linear ramp with slope $dv_b/dt = -v_a/RC = \mp V_{CC}/RC$.

The resulting waveforms drawn in Fig. 11.3–10b are easily found by assuming $v_a = -V_{CC}$ and $v_b = -R_2 V_{CC}/R_1$ at $t = 0$. The negative value of v_a causes v_b to rise linearly until it hits $+R_2 V_{CC}/R_1$ at $T_1 = \Delta v_b/(dv_b/dt) = 2R_2 RC/R_1$. At this point v_a switches to $+V_{CC}$ and v_b de-

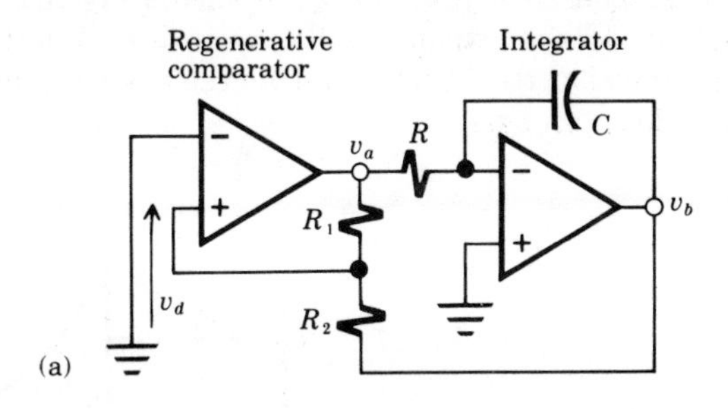

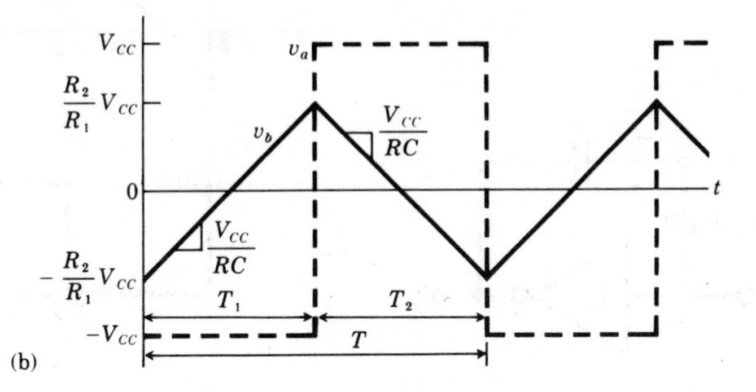

FIGURE 11.3–10 Square/triangular wave generator: (a) circuit; (b) waveforms.

creases linearly until it again reaches $-R_2 V_{CC}/R_1$, completing one full cycle of operation. It's clear that T_2 equals T_1 and that the repetition frequency is

$$f = \frac{1}{T_1 + T_2} = \frac{R_1}{4R_2 RC} \tag{2}$$

We therefore have a combination square/triangular wave generator whose frequency is readily adjustable up to the switching-speed limitation of the comparator.

To obtain the high-frequency *clock* signal needed for a digital system, one must use inverter gates rather than op-amps. Figure 11.3–11 shows a typical CMOS clock circuit with $V_{CC} = 10$ V, so that the Schmitt-trigger inverters switch at $V_1 = 4$ V and $V_2 = 6$ V. (As before, R_1 serves as a protective element and draws negligible current, providing that $R_1 \gg R$.) The waveforms are best explained by assuming $v_a = 0$ and $v_{out} = 10$ V at $t = 0^-$ when v_b decreases to 4 V. Both inverters then trigger at $t = 0$, causing v_b to drop to $4 - 10 = -6$ V, since the voltage across C cannot change instantaneously. Now we have $v_{out} = 0$ and $v_a = 10$ V, so v_b rises toward $+10$ V with time constant $\tau = RC$ until it reaches the 6-V trigger level and both inverters again change states at $t = T/2$. The 10-V jump of v_{out} immediately increases v_b to 16 V, from which it decays exponentially toward zero until it reaches the 4-V trigger level at $t = T$. (Calculating the clock frequency $f = 1/T$ is left for you to do in an exercise.)

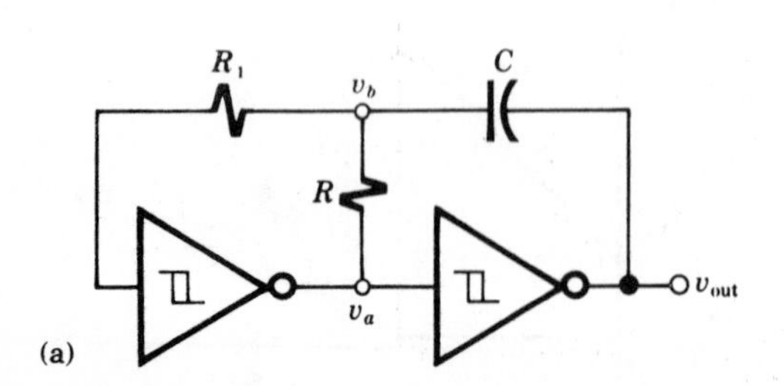

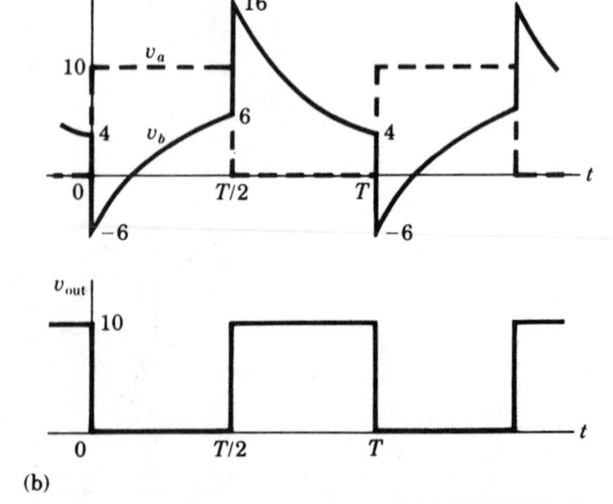

FIGURE 11.3–11
(a) CMOS clock circuit.
(b) Waveforms.

Finally, we consider the 555 timer so connected that it's a self-triggering or *astable* circuit (Fig. 11.3–12a). The capacitance charges through R_A and R_B until $v_C = 2V_{CC}/3$, which resets the flip-flop via the threshold connection. The output and discharge terminals are then at zero and the capacitance discharges directly through R_B. When v_C has decayed to $V_{CC}/3$, the trigger terminal sets the flip-flop, and the cycle starts anew.

Figure 11.3–12b shows the steady-state waveforms. Their asymmetry reflects the two different time constants $\tau_1 = (R_A + R_B)C$ and $\tau_2 = R_B C$, meaning that $T_1 > T_2$. We can easily compute T_2 from the fact that $\frac{2}{3}V_{CC}e^{-T_2/\tau_2} = \frac{1}{3}V_{CC}$, giving $T_2 = \tau_2 \ln 2$. We similarly find that $T_1 = \tau_1 \ln 2$ and therefore that

$$f = \frac{1}{T_1 + T_2} = \frac{1}{(\tau_1 + \tau_2)\ln 2} \approx \frac{1.44}{(R_A + 2R_B)C} \tag{3}$$

whose practical value ranges from about 0.1 Hz to 100 kHz. If $R_A \ll R_B$, then $T_1 \approx T_2$ and the output waveform will be nearly square. A time-varying voltage applied to the control terminal changes f by altering the reset value, and the circuit becomes a *voltage-controlled oscillator* (VCO) or *voltage-to-frequency converter*.

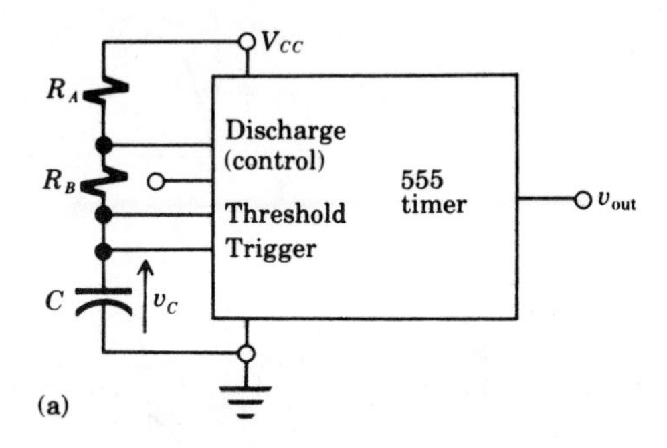

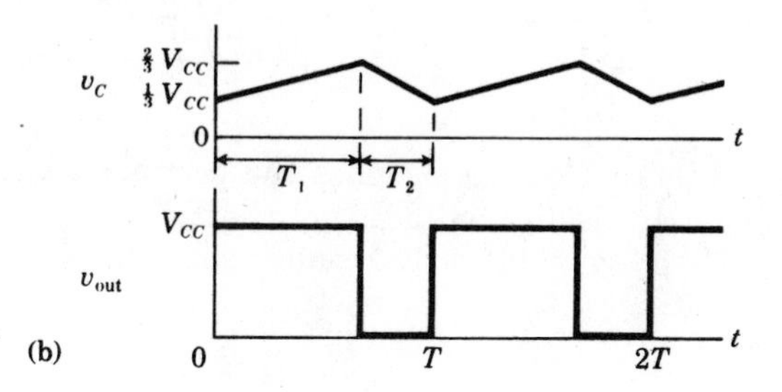

FIGURE 11.3–12
(a) Astable circuit.
(b) Waveforms.

Exercise 11.3–4 Show that both half cycles in Fig. 11.3–11b have the same duration, and calculate $f = 1/T$. Hint: Use Eq. (2), Section 8.2, to find T_1.

11.4 RECTIFIER CIRCUITS

Although AC electric power is more readily available than DC, numerous circuits and devices require constant voltage or at least unidirectional current for proper operation. In this section we discuss the principles of diode rectifier circuits that carry out AC-to-DC conversion for the purpose of supplying electronic apparatus, DC motors, battery chargers, and so forth. We also introduce controlled rectifiers that can control power, as well as perform AC-to-DC conversion.

HALF-WAVE AND FULL-WAVE RECTIFIERS

The basic *half-wave rectifier* was considered previously in Example 9.1–1. Its circuit diagram is repeated as Fig. 11.4–1a, with the addition of an input transformer whose turns ratio would be chosen to provide an appropriate value for the AC secondary voltage. The transformer also isolates the rectifier from the incoming AC line. Let's examine the current through the load resistance R, taking $v_s = V_m \sin \omega t$ with $\omega = 2\pi/T$ and assuming an ideal diode.

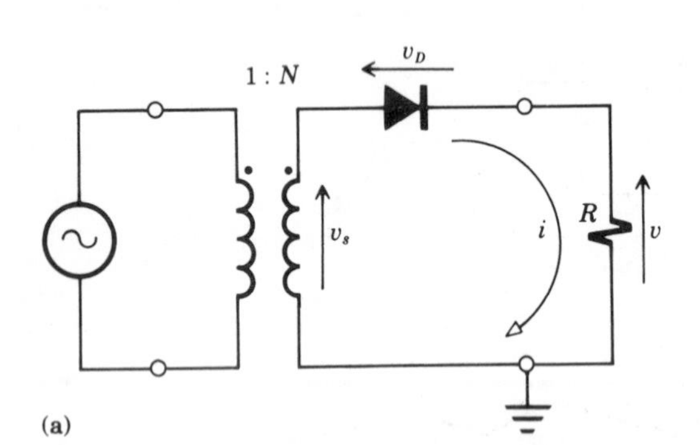

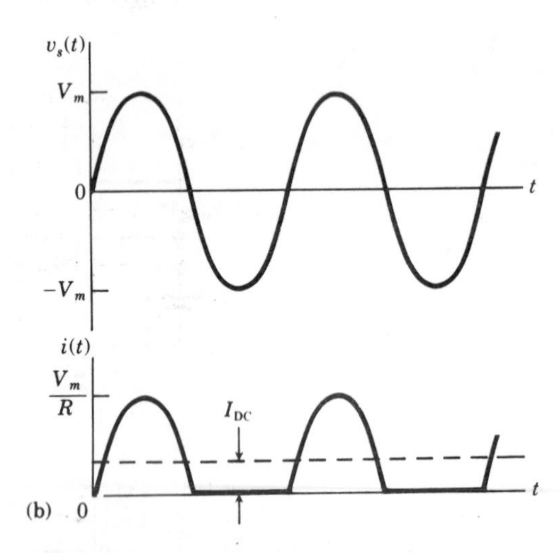

FIGURE 11.4–1
(a) Half-wave rectifier.
(b) Waveforms.

Figure 11.4–1b sketches the pertinent waveforms and emphasizes that $i(t)$ pulsates, but does not change direction. It therefore has a nonzero average value or DC component defined in general by

$$I_{DC} \triangleq \frac{1}{T} \int_0^T i(t)\, dt \tag{1}$$

Since $i(t) = (V_m/R) \sin \omega t$ for $0 \le t \le T/2$, but is zero for the remaining half cycle, the integration yields

$$I_{DC} = \frac{1}{T}\int_0^{T/2} \frac{V_m}{R} \sin \omega t \, dt = \frac{\omega}{2\pi}\int_0^{\pi/\omega} \frac{V_m}{R} \sin \omega t \, dt$$
$$= \frac{V_m}{\pi R} \tag{2}$$

Therefore, I_{DC} is about one-third of the peak current V_m/R and the DC component of the load voltage equals $RI_{DC} = V_m/\pi$.

This simple rectifier could be used for battery charging, electroplating, and similar applications requiring unidirectional but not necessarily constant current. However, the nonsinusoidal alternating current component $i(t) - I_{DC}$ represents unwanted and wasteful power dissipation in the load. For a quantitative measure of the rectification efficiency, we note that the DC load power is $P_{DC} = RI_{DC}^2$, whereas the total average power delivered to the load is

$$P = \frac{1}{T}\int_0^T Ri^2(t) \, dt$$

This quantity also equals the average power from the AC source if we neglect any transformer and diode losses. The ratio of useful DC power to total average power can then be computed, with the result

$$\frac{P_{DC}}{P} = \frac{4}{\pi^2} \tag{3}$$

giving an efficiency of approximately 40%. To improve efficiency, we turn to full-wave rectification.

Two implementations of a *full-wave rectifier* are shown in Fig. 11.4–2. Both circuits are arranged such that $i(t) = i_1(t) + i_2(t)$, where $i_1(t)$ flows during the positive half cycles of v_s, and $i_2(t)$ flows during the negative half cycles. (You might like to check this assertion by tracing the current paths for the two conditions $v_s > 0$ and $v_s < 0$.) It should be clear that the DC value of $i(t)$ is now precisely twice that of a half-wave rectifier:

$$I_{DC} = \frac{2V_m}{\pi R} \tag{4}$$

It's less obvious that power efficiency also has been doubled, so that $P_{DC}/P = 8/\pi^2 \approx 80\%$.

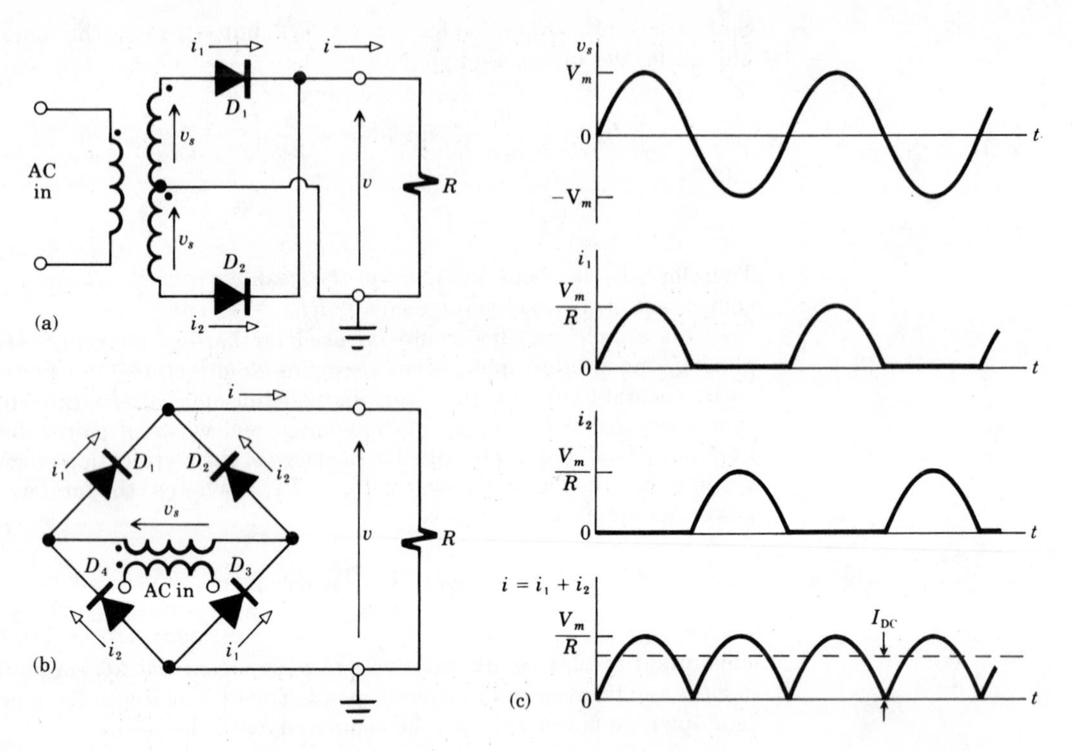

FIGURE 11.4–2
Full-wave rectifiers: (a) Center-tapped circuit; (b) bridge circuit; (c) waveforms.

Despite identical *external* performance, the two full-wave circuits have some practical *internal* differences. Figure 11.4–2a requires a special center-tapped transformer, but gets by with only two diodes. However, the *peak inverse voltage* (PIV) across a nonconducting diode equals $2V_m$ rather than V_m, as in the half-wave rectifier (see Fig. 9.1–2c). The diodes, therefore, must have a higher reverse-breakdown rating. The bridge configuration of Fig. 11.4–2b requires four diodes rated for PIV $\geq V_m$.

Exercise 11.4–1

Suppose all three rectifier circuits just discussed have $R = 100\ \Omega$, diodes rated at 400 PIV, 10 turns on the transformer primary, and a line voltage of 120 V(rms). Taking $V_m = 400$ V in the half-wave rectifier gives $I_{DC} = 400/\pi 100 \approx 1.27$ A and requires $(400/\sqrt{2}\ 120) \times 10 \approx 24$ turns on the secondary. The DC power is $RI_{DC}^2 \approx 162$ W, and requires an average AC input power $P = \pi^2 P_{DC}/4 = 400$ W in absence of transformer and diode losses. The bridge rectifier (Fig. 11.4–2b) could also have $V_m = 400$ V and 24 secondary turns, producing $I_{DC} = 2 \times 400/\pi 100 \approx 2.55$ A and $P_{DC} \approx 684$ W with $P = 800$ W.

For the center-tapped rectifier of Fig. 11.4–2a, we must use a transformer with 12 turns on each side of the center tap so that $V_m = \frac{1}{2}\,\text{PIV} =$

200 V and, consequently, $I_{DC} = 2 \times 200/\pi 100 \approx 1.27$ A. We now have $P_{DC} \approx 162$ W, just as we had from the half-wave rectifier, but here it is derived from $P = 200$ W thanks to improved efficiency.

Example 11.4–1 Repeat the previous example, but specify the PIV rating, transformer secondary turns, and AC input power, so that $I_{DC} = 4$ A for all three circuits.

Exercise 11.4–2 Sketch $i^2(t)$ from Fig. 11.4–1b; determine its average value by inspection; and show that $P = V_m^2/4R$. Then derive Eq. (3).

DC POWER SUPPLIES

Half-wave or full-wave rectification is the first step in AC-to-DC conversion. The next step involves an energy-storage element to smooth out the time variations of the rectified waveform. Consider, for example, a half-wave rectifier with capacitance in parallel with the load as in Fig. 11.4–3a. If $RC \gg T$, we have what is essentially a *peak detector* (like Fig. 11.1–11a), and the load voltage v stays nearly constant at the peak value V_m, as contrasted with the DC voltage component V_m/π without the capacitor.

To make a more detailed analysis, we draw upon the steady-state waveforms in Fig. 11.4–3b, taking $v_s = V_m \cos 2\pi ft$ with $f = 1/T$. Our assumption that $RC \gg T$ means that the load voltage decays a small amount V_r, called the peak-to-peak *ripple voltage,* before the diode conducts again and recharges the capacitance to V_m. Since the recharging duration Δt will be small compared to T, $V_m - V_r \approx V_m e^{-T/RC} \approx V_m(1 - T/RC)$ and $V_r \approx TV_m/RC = V_m/fRC \ll V_m$. Alternatively, writing $V_m/R \approx V_{DC}/R = I_{DC}$, we obtain

$$C = \frac{I_{DC}}{fV_r} \tag{5}$$

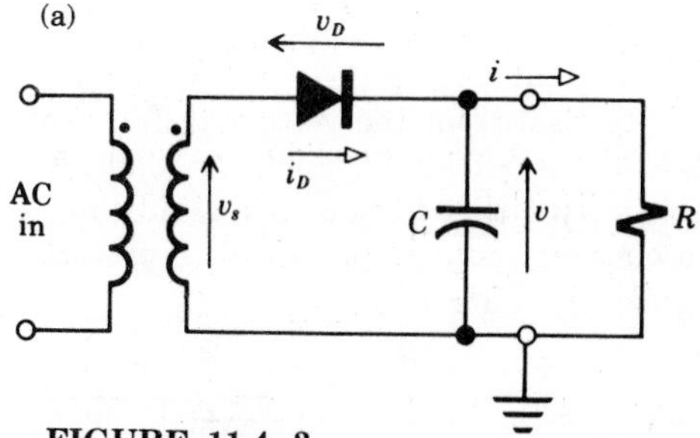

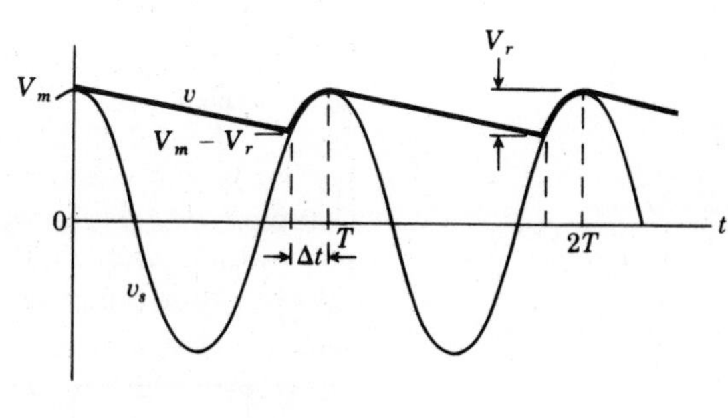

(b)

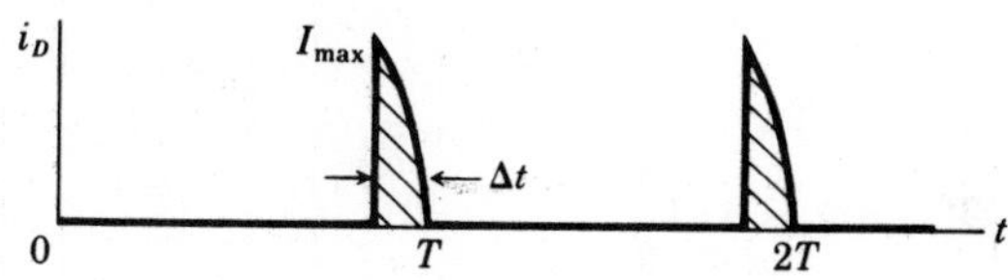

(c)

FIGURE 11.4–3
Half-wave rectifier with smoothing capacitance: (a) circuit; (b) voltage waveforms; (c) diode current.

which gives the value of C for a specified DC current, line frequency, and ripple voltage. The peak value of v_s should then be

$$V_m = V_{DC} + \tfrac{1}{2}V_r \tag{6}$$

to account for the fact that $V_{DC} \approx V_m - \frac{1}{2}V_r$.

Now consider the diode: It must withstand a PIV of about $2V_m$ since $v_D = v - v_s \approx -2V_m$ when $v_s = -V_m$. The diode must also handle a *peak forward current*

$$I_{max} \approx \frac{720°}{\theta} I_{DC} \tag{7a}$$

where

$$\theta = \cos^{-1}\left(\frac{V_m - V_r}{V_m}\right) \tag{7b}$$

which is the *conduction angle* in the sense that $\Delta t/T = \theta/360°$. These relations are derived from Fig. 11.4–3b and the sketch of $i_D(t)$ in Fig. 11.4–3c, together with the observation that the nearly triangular area $\frac{1}{2}I_{max}\,\Delta t$ represents the charge Δq required to raise v from $V_m - V_r$ back to V_m.

It should be obvious that full-wave rectification will further reduce the ripple voltage, other factors being equal. And more sophisticated filter circuits do better jobs of smoothing than single capacitors. But high-quality electronic power supplies also incorporate a *regulator circuit* for additional smoothing and to maintain the DC voltage level despite variations of current demanded by the load. By way of illustration, Fig. 11.4–4 shows a series regulator comprised of a BJT and a reverse-biased Zener diode. If the filtered voltage v always exceeds V_{DC} by a sufficient amount, current through R_B will divide between the Zener and the transistor's base. This keeps the Zener voltage at V_Z and $v_{BE} \approx 0.7$ V, so that $V_{DC} = V_Z - 0.7$ V. The BJT then compensates for the voltage difference $v_{CE} = v - V_{DC}$ when v changes due to the effects of ripple or loading. Complete IC regulators are available with either fixed or adjustable output voltage, and usually include a current-limiting provision to protect the regulator (and the load) from excessive current.

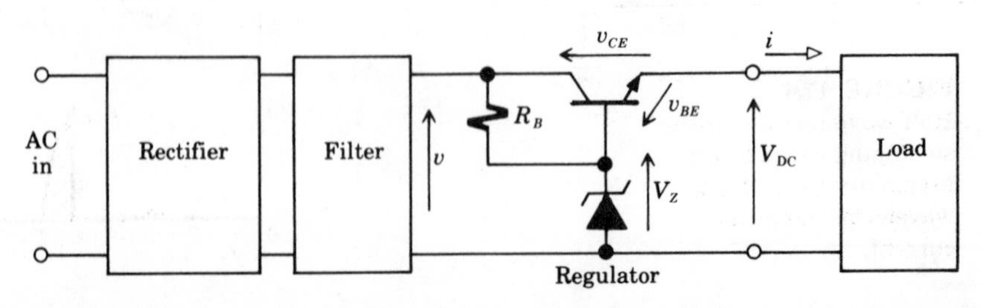

FIGURE 11.4–4
DC power supply with series regulator.

DC motors and related devices generally differ from electronic loads in three respects: They have appreciable inductance; they require more power; and they need constant current rather than constant voltage. When powering such devices, it is often possible to rely on the inherent inductance for current smoothing, without any external capacitance. The power-supply circuit is then easier to build, but harder to analyze.

Figure 11.4–5a diagrams a half-wave rectifier modified for an LR load. Diode D_1 conducts throughout the positive half cycle of v_s and then turns OFF. At this point, the *free-wheeling* diode D_2 turns ON to permit continuity of current through the inductance, and i goes around the path formed by L, R, and D_2 during the negative half cycle of v_s. The full cycle repeats periodically, with D_1 and D_2 switching, or *commutating,* each time v_s crosses zero. The waveforms in Fig. 11.4–5b show that the load voltage v has a half-rectified shape and, therefore, average value V_m/π. Since there can be no DC voltage component across the inductance, the average voltage across R must equal V_m/π and so $I_{DC} = V_m/\pi R$—just like a simple half-wave rectifier. Nonetheless, the load-current variations can be much smaller.

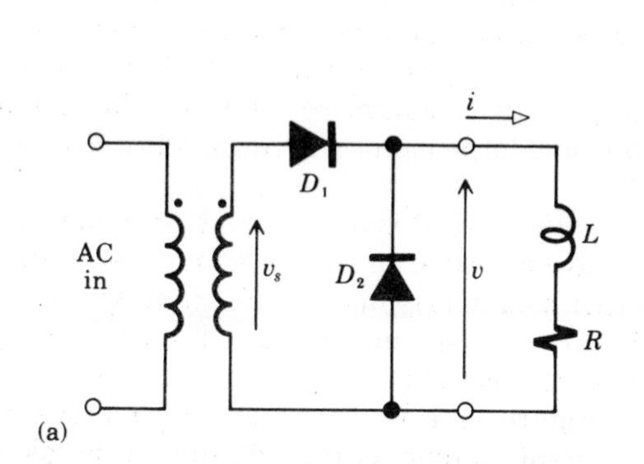

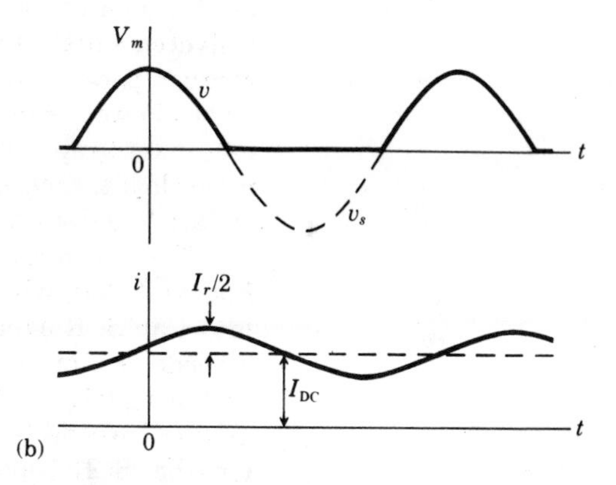

FIGURE 11.4–5 Half-wave rectifier with inductive load and free-wheeling diode: (a) circuit; (b) waveforms.

A rough estimate of the peak-to-peak *ripple current* I_r is obtained by approximating the load voltage with the first two terms of its Fourier series, namely

$$v(t) \approx \frac{V_m}{\pi} + \frac{V_m}{2} \cos \omega t$$

which follows from Example 7.4–1. We then invoke superposition to write

$$i(t) \approx \frac{V_m}{\pi R} + \frac{I_r}{2} \cos (\omega t + \theta)$$

Standard AC circuit analysis then yields

$$I_r = \frac{V_m}{\sqrt{R^2 + (\omega L)^2}} = \frac{\pi}{\sqrt{1 + (\omega L/R)^2}} I_{DC} \tag{8}$$

Thus $I_r \ll I_{DC}$ if $\omega L/R \gg \pi$—a condition that also makes our results quite accurate, because the higher harmonic components of i will be insignificant.

Exercise 11.4–3

Specify all parameter values and diode ratings such that Fig. 11.4–3a has $v = 10 \pm 0.1$ V when $R = 200\ \Omega$ and $f = 60$ Hz.

CONTROLLED RECTIFIERS

The average power in a resistive load may be controlled by adjusting the value of an additional resistance in series with both the load and the source. This method wastes power, however, and the adjustable resistance may have to be quite large to handle the power it dissipates. A more efficient way of controlling power is to use a switch. For instance, if a 2-kW load is connected to its source only half the time, the average power delivered equals 1 kW. By varying the ratio of ON time to OFF time, the average power can be adjusted from 0 to 2 kW, and with an efficiency of 100%. Power transistors effectively perform this switching function at reasonably rapid rates for low and medium power loads. But for high power loads, such as welders and large motor windings, controlled rectifiers are usually preferable.

The *silicon controlled rectifier* (SCR) is a four layer *pnpn* device (see Fig. 11.4–6a), which is always operated in the switching mode. In that mode, an SCR behaves either as a nearly open circuit from anode (A) to cathode (K), or as a diode carrying current from anode to cathode. The switching trigger that makes it conduct is a short current pulse from the gate to cathode terminals. Figure 11.4–6b shows the schematic symbol for the SCR; and the current-voltage curve (excluding breakdown regions) is illustrated in Fig. 11.4–6c.

The usual operating procedure is to apply a forward voltage, $v_{AK} > 0$. Then, at an appropriate time *fire* the device with a short gate current

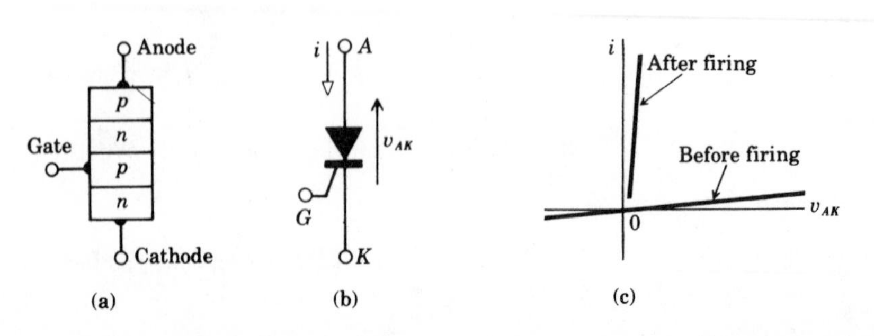

FIGURE 11.4–6 Silicon controlled rectifier: (a) structure; (b) symbol; (c) i-v curve.

pulse. It then behaves as a diode so long as forward current flows. A short period of negative v_{AK} "resets" the SCR, and it regains its insulating ability. These characteristics make the SCR very useful for controlling the average current in rectifier circuit loads by adjusting firing time. The use of an AC source automatically turns off the device or devices used once each cycle.

Consider, for instance, the circuit of Fig. 11.4–7a, which is Fig. 11.4–5, but with D_1 replaced by an SCR. Two operating patterns are shown in the waveforms of Fig. 11.4–7b. One waveform shows the SCR gate fired at 45° of the cycle, while the other (the dashed curves) corresponds to 135° firing. The average value of v is clearly much higher for earlier firings, resulting in load current curves that have substantially

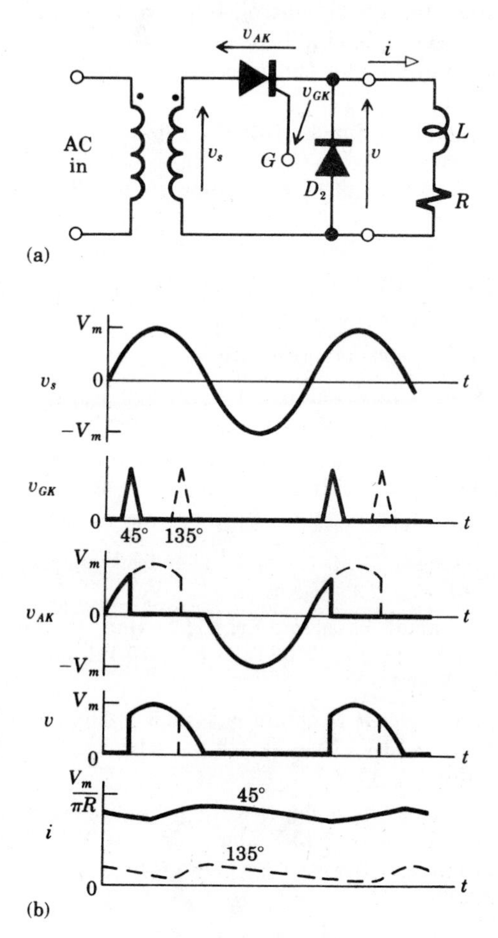

FIGURE 11.4–7 Half-wave rectifier with SCR: (a) circuit; (b) waveforms for two firing angles.

different average values. The average current may be varied from zero to the maximum ($V_m/\pi R$) by phasing back the firing point from 180° to 0°.

The SCR is one member of a class of power switching devices called *thyristors*. Another member of this class is the *triac,* which behaves very much like an SCR except that it can conduct in *either direction* after a gate pulse. This makes it useful for controlling the rms current through AC loads. Small units are employed in home appliances such as incandescent light dimmers and variable-speed kitchen mixers.

Thyristors, when used in circuits such as those just discussed, have one serious disadvantage. The sudden onset of current flow at a point in the cycle when a voltage is already high can cause radiating electromagnetic interference. One solution to this problem is *zero voltage switching:* Instead of supplying power during one fourth of a cycle, for instance, power is supplied for one full cycle out of every four. (This works quite well for motor speed control, but produces annoying flickering in light dimmers.) A single-chip IC is now available that performs most of the circuit requirements for the zero voltage switching mode.

Exercise 11.4–4 Let D_1 and D_2 in Fig. 11.4–2b be replaced by SCRs S_1 and S_2. Sketch the waveforms of $i_1(t)$, $i_2(t)$, and $v(t)$; and find V_{DC}, assuming that S_1 fires when $v_s = +V_m$ and S_2 fires when $v_s = -V_m$.

PROBLEMS

11.1–1 Suppose the diode is reversed in Fig. 11.1–1a and $V_0 = 10$ V. Plot v_R versus v, and sketch v_R and v_D when $v = 15 \sin \omega t$.

11.1–2 Repeat the previous problem with $V_0 = -10$ V.

11.1–3 Let $V_1 = V_2 = 10$ V in Fig. 11.1–3a, and let the diodes have $R_f = R/4$ and $V_T = 0$. Plot v_{out} versus v_{in}, and sketch v_{out} when $v_{in} = 15 \sin \omega t$.

11.1–4 Replace the Zener diodes in Fig. 11.1–3d with the equivalent circuit from Fig. 9.2–4c. Then plot v_{out} versus v_{in} when $V_{Z_1} = 10$ V, $V_{Z_2} = 5$ V, and $R_{Z_1} = R_{Z_2} = 0$.

11.1–5 Repeat the previous problem with $R_{Z_1} = R_{Z_2} = R$.

11.1–6 Rearrange the elements in Fig. 11.1–4b to obtain $v_{out} = v_{in} - 5$ V when $v_{in} < 5$ V and $v_{out} = 0.5v_{in} - 2.5$ V when $v_{in} > 5$ V.

11.1–7 Design a circuit based on Fig. 11.1–4 to approximate the function $v_{out} = \frac{10}{3} \log_{10} (v_{in} + 1)$ for $v_{in} \geq 0$. Take $R_0 = 10$ kΩ, and make the function exact at $v_{in} = 0$, 1, 4, and 9 V.

11.1–8 Modify Fig. 11.1–4c to obtain $v_{out} = 3.5 + 0.5v_{in}$ for $7 < v_{in} < 13$ V, and $v_{out} = 10$ V for $v_{in} > 13$ V. Then sketch v_{out} when v_{in} is the triangular wave in Fig. 7.4–1a with $A = 15$ V.

‡ **11.1–9** Repeat the previous problem with the additional provision that $v_{out} = -3.5 + 0.5v_{in}$ for $-7 > v_{in} > -13$ V and $v_{out} = -10$ V for $v_{in} < -13$ V. (Circuits with similar characteristics are used in function generators to shape triangular waves into sinusoids.)

11.1–10 Figure P11.1–10 represents a cascade of digital gates like those in Figs. 11.1–5 and 6. The diodes have $R_f = 0$ and $V_T = 0.7$ V. The input terminals of the first gate have been tied together, so there are four input-voltage combinations: $v_1 = v_2 = 0$; $v_1 = 0$ and $v_2 = 5$ V; $v_1 = 5$ V and $v_2 = 0$; $v_1 = v_2 = 5$ V. Find the value of v_{out} for each input combination when

(a) Both gates are OR gates.

(b) Gate 2 is changed to an AND gate.

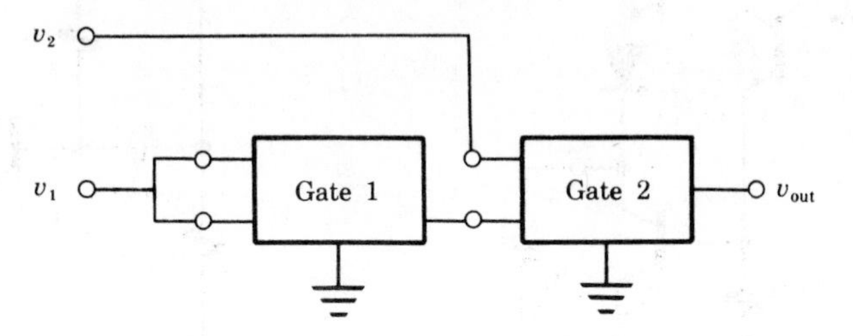

FIGURE P11.1–10

11.1–11 Repeat the previous problem when

(a) Both gates are AND gates.

(b) Gate 2 is changed to an OR gate.

11.1–12 The circuit in Fig. 11.1–9 becomes a *shunt-switch* analog gate when the positions of the input and ground terminals are interchanged. Draw this circuit and qualitatively describe its operation.

11.1–13 In Fig. 11.1–13, let $v_{in} = V_m \cos 2\pi t/T$, $V_B = 3$ V, and $RC \gg T$. Sketch the steady-state waveform of v_{out} when $V_m = 2$ V and $V_m = 4$ V. Repeat with the diode reversed.

11.1–14 Repeat the previous problem with $V_B = -3$ V.

11.1–15 The circuit in Fig. 11.1–14a becomes a *voltage doubler* when R_1 is replaced by a diode that conducts from left to right. Find the steady-state voltage across C_1 and across the added diode.

11.2–1 Let the BJT switch in Fig. 11.2–1 have $V_{CC} = 5$ V, $V_T = 0.7$ V, $V_{sat} = 0.2$ V, and $\beta = 20$. If v_{in} switches between 0 and 5 V, and $i_B \leq 0.1$ mA, what are the minimum values of R_B and R_C for proper operation?

11.2–2 Repeat the previous problem with v_{in} switching between 0 and 2 V.

11.2–3 Start with $i_{B1} = 0.725$ mA to find all the other currents in Fig. 11.2–4b. Then calculate the minimum values of β_2 and β_3 to ensure saturation of T_2 and T_3.

11.2–4 Let the 1.4-kΩ resistor in Fig. 11.2–4a have an arbitrary value R. If $\beta_2 = 50$, what is the condition on R to ensure that T_2 saturates in Fig. 11.2–4b?

11.2–5 Figure P11.2–5 shows a TTL gate with additional elements forming a *totem-pole output*. All transistors have $V_T = 0.7$ V, $V_{sat} = 0.2$ V, and $\beta = 50$. Diode D has $V_T = 0.7$ V. Show that the HIGH output state is $v_{out} < 3.6$ V when T_2 is cut off. Then explain the purpose of D when T_2 is saturated so v_{out} should be LOW.

11.2–6 Consider the TTL gate described in Problem 11.2–5. Let $R_L = 200$ Ω, and let T_2

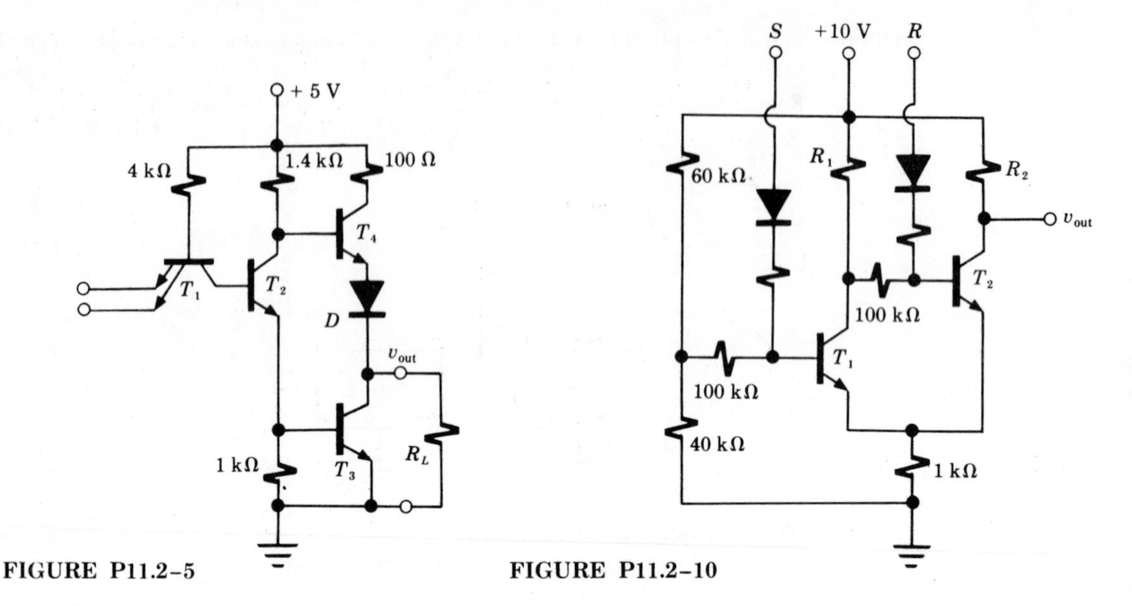

FIGURE P11.2–5

FIGURE P11.2–10

be in cutoff. Determine whether T_4 is active or saturated, and find the value of v_{out}.

11.2–7 Repeat the previous problem with $R_L = 1\ \text{k}\Omega$.

11.2–8 Suppose the BJTs in Fig. 11.2–5a are replaced with *n*-channel enhancement MOSFETs with $K = 10$ mA and $V_T = 5$ V. Confirm bistable operation, and calculate the HIGH and LOW output voltages. Hint: Use Eq. (3b), Section 10.1.

11.2–9 Repeat the previous problem with $K = 8$ mA and $V_T = 4$ V.

11.2–10 Figure P11.2–10 is an *emitter-coupled flip-flop*. The transistors have $V_T = 0.5$ V, $V_{sat} = 0$, and $\beta \gg 1$. Confirm bistable operation, and find the two values of v_{out} if $R_1 = 3\ \text{k}\Omega$ and $R_2 = 1\ \text{k}\Omega$.

‡ **11.2–11** Consider the flip-flop described in the previous problem. Find values for R_1 and R_2 for bistable operation with $v_{out} = 4$ V when T_2 is saturated.

‡ **11.2–12** The diagram of an *emitter-coupled logic gate* (ECL) is given in Fig. 15.4–5a. The circuit is designed so that T_{1A}, T_{1B}, and T_2 never saturate, and T_3 and T_4 are always active. All transistors have $V_T = 0.8$ V and $V_{sat} = 0.3$ V. The base currents of T_2, T_3, and T_4 are very small. Verify that T_{1A} and T_{1B} are cut off and T_2 is active when both inputs are LOW (-1.8 V); find the resulting value of v_X. Then find v_X when either input is HIGH (-0.8 V), and identify the logic function. Hint: You may omit T_3 entirely and temporarily add 5.2 V to all voltage levels to eliminate negative values. Also note that i_1 or i_2 can be found from v_E.

‡ **11.2–13** Repeat the previous problem for $\bar{v}_X$ instead of v_X. You may omit T_4.

11.3–1 Show, by finding V_1 and V_2, that Fig. 4.3–6a becomes a regenerative comparator if the input terminals are interchanged.

11.3–2 Use the results of Exercise 11.3–1 to find element values so that Figure 11.3–3b triggers at $V_1 = 0$ and $V_2 = 4$ V.

11.3–3 Repeat the previous problem for triggering at -3 V and -4 V.

11.3–4 Let $R = 10$ kΩ and $C = 0.2$ μF in Fig. 11.3–5. Calculate t_d and the minimum input-pulse duration for normal operation.

11.3–5 Show that $D \approx 0.92RC$ in Fig. 11.3–6.

11.3–6 Design a system that produces a 1-ms pulse that starts 2 ms after the trailing edge of an input pulse. Use monostables like Fig. 11.3–7a with $R = 10$ kΩ.

11.3–7 Repeat Problem 11.3–6 with the addition of another 1-ms output pulse that starts right after the trailing edge of the input pulse.

11.3–8 Sketch v_a and v_b in Fig. 11.3–10 when $R = R_1 = 10$ kΩ, $R_2 = 5$ kΩ. $C = 1$ μF, and $V_{CC} = 10$ V.

11.3–9 Let $C = 0.1$ μF, $T_1 = 2T_2$, and $f = 1$ kHz in Fig. 11.3–12. What are the values of R_A and R_B?

11.3–10 The 555 data sheet lists the requirement $R_A \geq V_{CC}/(200 \text{ mA})$ for astable operation. Find appropriate values of R_A, R_B, and C so that $T_1 \leq 1.01T_2$ and $f = 1$ Hz if $V_{CC} = 10$ V.

11.3–11 Let $R_A = R_B = 10$ kΩ, $C = 0.05$ μF, and $V_{CC} = 12$ V in Fig. 11.3–12a. Also let a 9-V source be connected to the control terminal. Sketch the waveform v_C and calculate the frequency $f = 1/(T_1 + T_2)$. Compare this frequency with the value of f when the control terminal "floats" at $\frac{2}{3}V_{CC}$.

11.3–12 Repeat the previous problem with a 7-V source connected to the control terminal.

‡ **11.3–13** Let the noninverting terminal of the integrator in Fig. 11.3–10a be connected to a voltage source V_s.

(a) Show that the slopes of v_b become $+(V_{CC} + V_s)/RC$ and $-(V_{CC} - V_s)/RC$.

(b) Sketch v_a and v_b when $V_s = 3$ V and the other element values are as in Problem 11.3–8.

11.4–1 Suppose $R = 4$ Ω in Fig. 11.4–2a and we want $I_{DC} = 10$ A. The high-current diodes have $V_T = 1$ V and $R_f = 0.2$ Ω. Estimate the required peak value of v_s.

11.4–2 Repeat the previous problem for Fig. 11.4–2b.

11.4–3 Find V_r, V_{DC}, and I_{max} in Fig. 11.4–3 when $V_m = 170$ V, $f = 60$ Hz, $C = 1000$ μF, and $R = 500$ Ω.

11.4–4 Repeat Exercise 11.4–3 with $v = 500 \pm 10$ V and $R = 1$ kΩ.

11.4–5 Modify Eqs. (5)–(7) to apply to a *full-wave* rectifier.

11.4–6 Let diodes D_2 and D_3 in Fig. 11.4–2b be replaced with capacitors so that $RC \gg 4T$. Find the maximum values of v, v_C, and v_D. Then sketch v and compare with Fig. 11.4–3b.

11.4–7 Suppose the regulator in Fig. 11.4–4 has $R_B = 100$ Ω, $V_Z = 5.7$ V, $V_T = 0.7$ V, and $\beta = 50$.

(a) Find the minimum required value of v when $i = 1$ A.

(b) Find the power dissipated by the Zener diode when $i = 1$ A and $i = 0$. Take $v = 10$ V.

11.4–8 Field windings of large DC motors are often rated for 100 V. Show that this is consistent with the average value of v in Fig. 11.4–5 when the rms value of v_s is 220 V.

11.4–9 Let Fig. 11.4–5 have $V_m = 170$ V, $f = 60$ Hz, $L = 0.5$ H, and $R = 4\ \Omega$. Find I_{DC} and I_r.

11.4–10 Consider Fig. 11.4–7 with $V_m = 170$ V, $R = 10\ \Omega$, and $\omega L \gg R$. Find the average value of i if the SCR fires at 90°.

‡ **11.4–11** Show that if $\omega L \gg R$ in Fig. 11.4–7, then $I_{DC} \approx (1 - \cos\theta)V_m/2\pi R$ where θ is the firing angle.

12

ELECTRONIC AMPLIFIER CIRCUITS

A complete amplifier circuit generally consists of several transistors serving different purposes. Figure 12.0 illustrates this point with the schematic diagram of an amplifier designed to take a weak audio signal, say from a phonograph pickup, and deliver sufficient power to drive the output loudspeaker. The overall voltage amplification might be around 40, with a reasonably uniform frequency response over the 50-Hz–10-kHz range. To achieve those goals, the amplifier is divided into three *stages*.

At the input end, transistor T_1 and its associated circuitry functions as a *preamplifier* (or "preamp") that provides an appropriate load to the source. We classify this stage as a *small-signal amplifier* because its signal variations will be only a small fraction of the supply voltage. At the output end, transistors T_3 and T_4 work together to form a *power amplifier* with large signal variations that, in fact, are limited primarily by the value of the supply voltage. The *driver* stage (transistor T_2) provides the necessary interface between the preamplifier and the power amplifier stages. Other features of this amplifier include tone and volume controls (C_1 and R_1) and a feedback connection (R_{13} and C_6) from output to preamplifier.

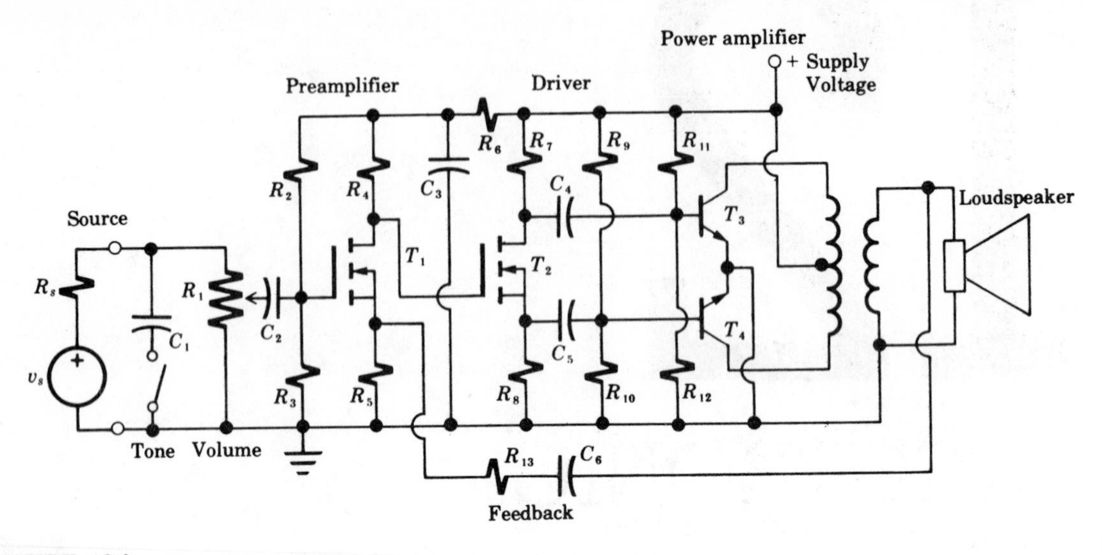

FIGURE 12.0
Three-stage audio amplifier.

This chapter introduces the operating principles and design techniques of electronic amplifier circuits. We start with small-signal amplifiers and proceed to power amplifiers. Along the way, we'll at least touch on all the features of our illustrative audio amplifier, plus some additional special-purpose circuits. Primary emphasis, however, is given to voltage amplification with *n*-channel FETs or *npn* BJTs.

OBJECTIVES

After studying this chapter and working the exercises, you should be able to do each of the following:

- Use appropriate models to find the operating point and small-signal characteristics of a FET or BJT amplifier (Section 12.1).
- Design a simple small-signal amplifier to meet stated requirements (Section 12.1).
- Describe the characteristics of feedback amplifiers, followers, and differential amplifiers (Section 12.2).
- State the factors that determine the low-frequency and high-frequency response of an *RC*-coupled amplifier, and calculate the cutoff frequencies (Section 12.3).†
- Given the ratings of a power transistor, design a transformer-coupled amplifier to deliver maximum output power (Section 12.4).†

12.1 SMALL-SIGNAL AMPLIFIERS

We pointed out in Chapter 10 that transistors produce reasonably linear amplification if restricted to small voltage and current variations. Here we investigate simple but practical FET and BJT circuits that function as linear small-signal amplifiers. Attention is given to the small-signal amplification per se, and to the necessary bias and coupling arrangements. Beside being building blocks in multistage amplifiers, these circuits are important in their own right for such applications as interfacing between a signal source and a measuring or display instrument.

FIXED-BIAS ENHANCEMENT MOSFET AMPLIFIERS

To begin with, every small-signal amplifier must be biased to operate at a suitable point in the active region of the transistor. It must also have provision for coupling the signal at input and output. Brute-force biasing and direct coupling using auxiliary batteries was illustrated in Chapter 10 (see Figs. 10.1–9 and 10.2–7). A more practical scheme involves *biasing resistors* and *coupling capacitors,* as shown in Fig. 12.1–1a with

FIGURE 12.1–1 *RC*-coupled MOSFET amplifier: (a) circuit; (b) waveforms.

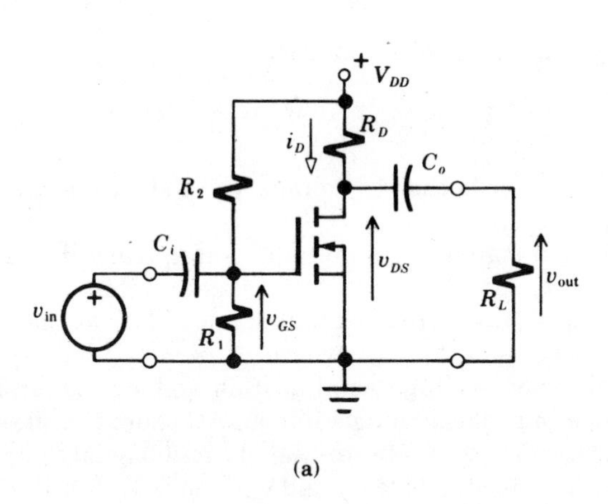

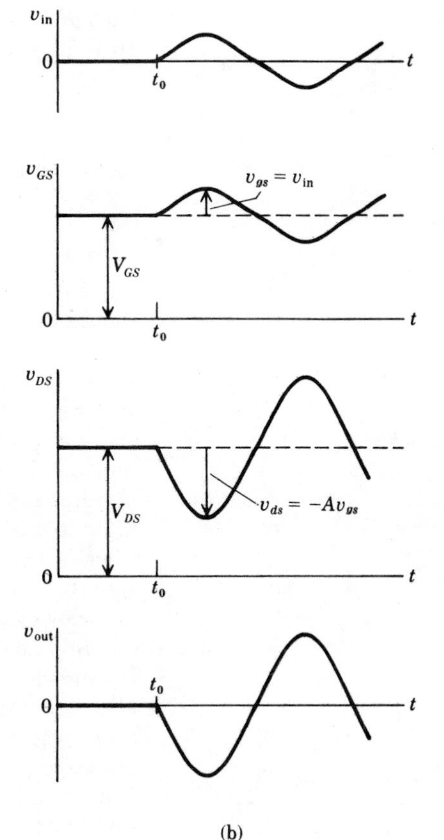

an n-channel enhancement MOSFET. Such configurations are known as *RC-coupled* amplifiers, to distinguish them from those with direct coupling or transformer coupling.

The operation of this circuit can be understood by first letting $v_{in} = 0$ and recalling that capacitance C_i acts as a *DC block* and that the MOSFET draws virtually no gate current. Thus, R_1 and R_2 form a voltage divider across the supply voltage V_{DD} and establish a *fixed* bias voltage

$$V_{GS} = \frac{R_1}{R_1 + R_2} V_{DD} \tag{1}$$

The value of V_{GS} yields the desired DC or Q-point current I_D via

$$V_{GS} = V_T \left(1 + \sqrt{\frac{I_D}{K}}\right) \tag{2}$$

with K and V_T being the MOSFET's parameters from Eq. (1), Section 10.1. The resulting Q-point voltage is

$$V_{DS} = V_{DD} - R_D I_D \tag{3a}$$

and must satisfy the active-state condition

$$V_{DS} > V_{DS_{\min}} = V_{GS} - V_T \tag{3b}$$

Capacitance C_o then blocks the DC voltage V_{DS} and leaves $v_{out} = 0$, as it should be when $v_{in} = 0$.

Now let the input signal be a sinusoidal voltage turned on at time t_0, as in Fig. 12.1–1b. By superposition, the total gate voltage becomes $v_{GS} = V_{GS} + v_{gs}$ where $v_{gs} = v_{in}$ if C_i has negligible impedance at the signal frequency. Likewise, $v_{DS} = V_{DS} + v_{ds}$, where $v_{ds} \approx -Av_{gs}$ with $-A$ representing the small-signal voltage amplification and sign inversion. If C_o also has negligible impedance, it acts like an AC short-circuit, such that $v_{out} = v_{ds} = -AV_{in}$. We thus obtain the desired signal amplification without the auxiliary biasing battery and with no DC offset in the output signal.

However, note carefully the two critical assumptions: linear operation and negligible capacitive impedance. The latter is valid for all signal frequencies of interest, providing that $C \to \infty$ so $|\underline{Z}_C| = 1/\omega C \to 0$ except at $\omega = 0$. Consequently, for the time being, we will consider all coupling capacitors to be perfect AC short-circuits and DC open-circuits. Later on, the effects of finite capacitance are examined in the context of amplifier *frequency response*. The assumption of linear operation holds to the extent that signal variations are limited to a nearly linear portion of the

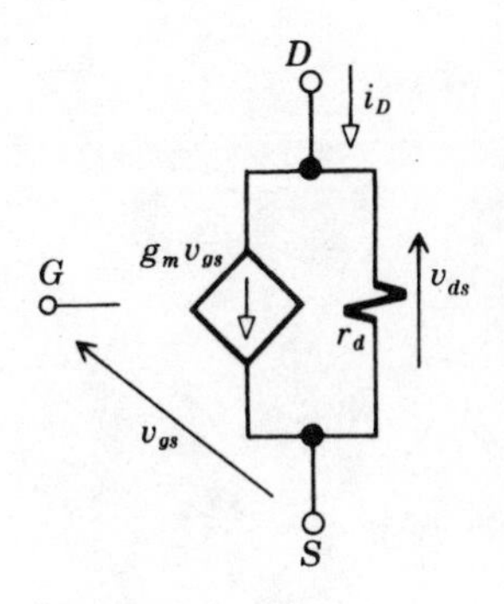

FIGURE 12.1–2
Small-signal FET model including dynamic resistance r_d.

transfer curve. Obtaining larger signal variations may require the use of *feedback* to minimize the resulting nonlinear distortion—another topic for future consideration.

To analyze the signal variations in our fixed-bias amplifier, we first invoke the *small-signal model* of a MOSFET diagrammed in Fig. 12.1–2. This differs from the theoretical model (Fig. 10.1–11b) by the inclusion of a *dynamic resistance* r_d that accounts for the slight upward slope of a real MOSFET's active-region characteristic curves. The dynamic resistance has a value in the range of 1–50 kΩ and varies somewhat with the Q point. The transconductance g_m depends strongly on the Q-point, since from Eqs. (1) and (6), Section 10.1,

$$g_m = \frac{2K}{V_T}\left(\frac{V_{GS}}{V_T} - 1\right) = \frac{2}{V_T}\sqrt{KI_D} \tag{4}$$

which typically equals a few millimhos. (For later use, the same model can represent a depletion MOSFET or JFET, except that a JFET has much larger dynamic resistance $r_d \geq 100$ kΩ.)

Next we insert the model into the circuit diagram, temporarily omitting the load R_L, and replace the capacitances and DC supply with short circuits, Fig. 12.1–3a. Combining the parallel resistances then gives the *small-signal equivalent circuit* of Fig. 12.1–3b. Clearly, the *input* and *output resistances* are

$$R_i = R_1 \| R_2 \qquad R_o = R_D \| r_d \tag{5a}$$

Then, since $v_{\text{out-oc}} = (R_D \| r_d)(-g_m v_{gs}) = -g_m R_o v_{\text{in}}$, we define

$$\mu \triangleq \left|\frac{v_{\text{out-oc}}}{v_{\text{in}}}\right| = g_m R_o \tag{5b}$$

which represents the open-circuit (unloaded) amplification factor. Finally, to include possible loading effects at input and output, we perform a Norton-to-Thévenin conversion of the controlled source and draw the complete circuit, Fig. 12.1–3c, with source and load resistances.

In practice, the bias resistances R_1 and R_2 can and should be quite large—say 100 kΩ to 10 MΩ—to ensure a *large input resistance* and very little input loading. Therefore, if $R_i \gg R_s$ so that $v_{gs} \approx v_s$, the overall small-signal voltage gain will be

$$A_v = \frac{v_{\text{out}}}{v_s} \approx -\frac{R_L}{R_L + R_o}\,\mu \tag{6}$$

which follows from a simple chain expansion. Unfortunately, the *output*

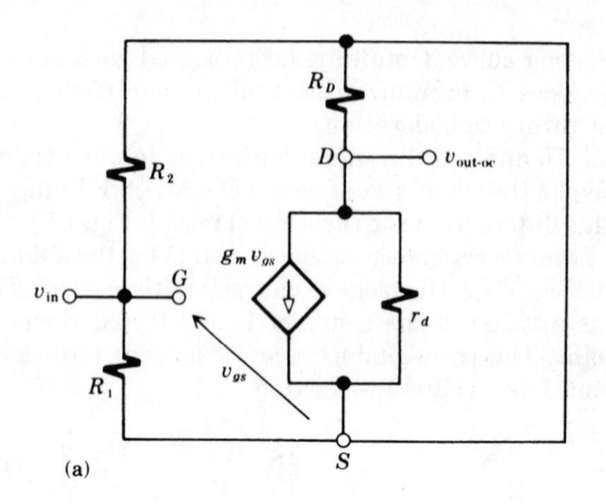

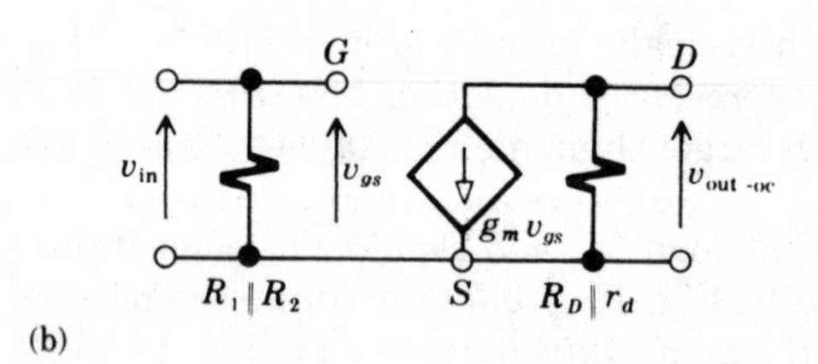

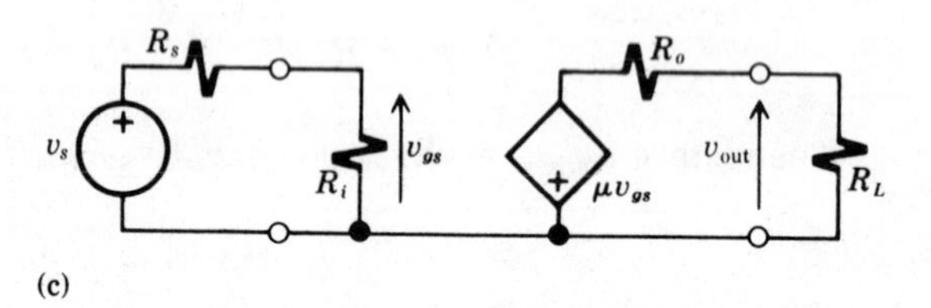

FIGURE 12.1–3
(a) Small-signal model of FET amplifier.
(b) Equivalent circuit.
(c) Two-port model.

loading may be appreciable because $|A_v| < \mu$ and $\mu = g_m R_o$; hence, the output resistance must be sufficiently large that $R_o > |A_v|/g_m$.

If one MOSFET amplifier does not produce enough gain, we might connect two identical stages in the *cascade* arrangement of Fig. 12.1–4a. The small-signal equivalent circuit is shown in Fig. 12.1–4b. Assuming $R_i \gg R_s$ and $R_i \gg R_o$, we see that $v_{gs_1} \approx v_s$ and $v_{gs_2} \approx -\mu v_{gs_1} \approx -\mu v_s$; thus,

$$A_v \approx -\frac{R_L}{R_L + R_o}\mu(-\mu) = \frac{R_L}{R_L + R_o}\mu^2 \tag{7}$$

which, of course, does not have a polarity inversion. Similar expressions are easily derived for nonidentical stages or more than two cascaded

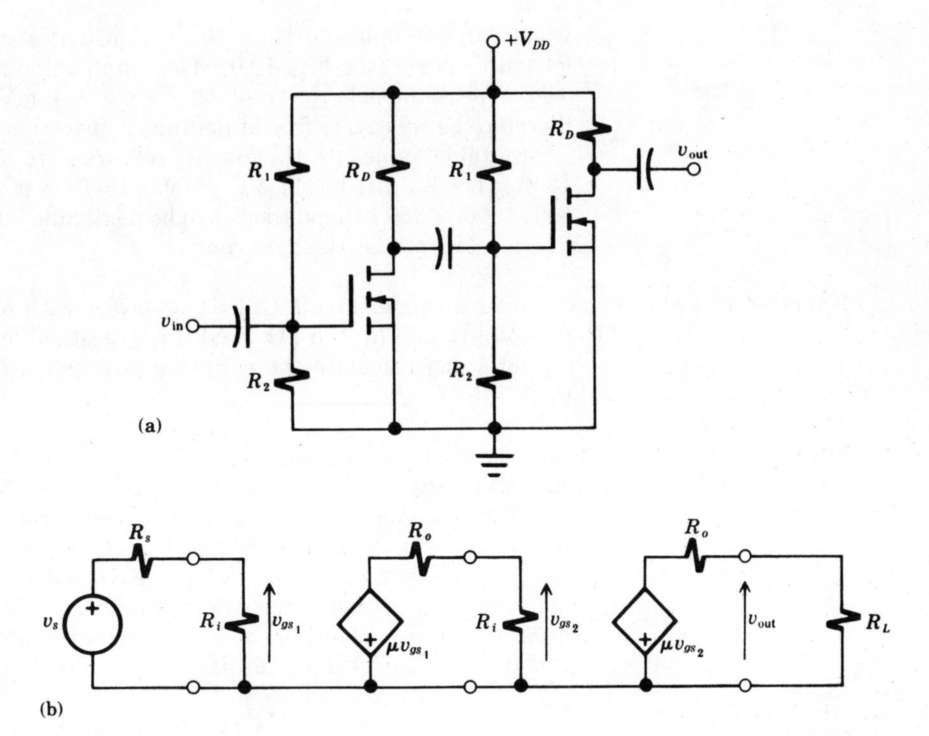

FIGURE 12.1–4 (a) Cascade of two MOSFET stages. (b) Small-signal equivalent circuit.

stages. But we must be sure that the amplified voltage variations do not exceed the *small-signal* limit.

Example 12.1–1

Suppose you need an amplifier with $A_v = 50$ for the interface between a 4-mV rms signal source having $R_s = 10\ \text{k}\Omega$, and an instrument whose input impedance is 5 kΩ. You have available a 9-V battery and two MOSFETs with $K = 16$ mA, $V_T = 2$ V, and $r_d = 40\ \text{k}\Omega$. You select $I_D = 1$ mA (to conserve battery life) and calculate $V_{GS} = 2.5$ V, $V_{DS_{\min}} = 0.5$ V, $R_D < 8.5\ \text{k}\Omega$, and $g_m = 4$ m℧ from Eqs. (2)–(4). You then observe that a single-stage amplifier will not suffice, because it would require $R_o = R_D \| r_d > |A_v|/g_m = 12.5\ \text{k}\Omega$, and the resulting DC voltage drop across R_D would far exceed $V_{DD} = 9$ V.

On the other hand, a cascade of two identical stages provides more than ample gain if, for instance, $R_D = 3\ \text{k}\Omega$ so $V_{DS} = 9 - 3 = 6\ \text{V} > 0.5$ V. Then $R_o = 3\|40 \approx 2.79\ \text{k}\Omega$, $\mu \approx 11.2$, and putting $R_L = 5\ \text{k}\Omega$ in Eq. (7) shows that

$$A_v \approx \frac{5}{7.79}(11.2)^2 \approx 80$$

which can be trimmed to $A_v = 50$ with the help of an input potentiometer for gain control (see Fig. 12.0). The amplified output signal will swing over a peak-to-peak range of $50 \times 2\sqrt{2} \times 4$ mV ≈ 0.5 V and should therefore be relatively free of nonlinear distortion.

Suitable values for the biasing resistors are $R_1 = 500$ kΩ and $R_2 = 1300$ kΩ $= 1.3$ MΩ to get $V_{GS} = (500/1800) \times 9$ V $= 2.5$ V. Then $R_i = 500 \| 1300 \approx 360$ kΩ and there will be negligible input loading assuming, say, a 1-MΩ pot for the gain control.

Exercise 12.1–1

Consider a single-stage MOSFET amplifier with $K = 16$ mA, $V_T = 2$ V, $r_d = 40$ kΩ, and $R_L = 5$ kΩ. Find the largest value of R_D consistent with Eq. (3b), and calculate the resulting gain when $I_D = 1$ mA and $V_{DD} = 18$ V.

SELF-BIASED FET AMPLIFIERS

Turning to amplifiers with depletion MOSFETs or JFETs, we find that the voltage-divider scheme of Fig. 12.1–1a cannot provide the required *negative* bias voltage $V_{GS} < 0$. Instead, one might use the circuit diagrammed in Fig. 12.1–5, where resistor R_K raises the DC potential of the FET's source terminal to $V_K = R_K I_D$. (We use the symbol R_K to avoid possible confusion with R_s, the internal resistance of a signal source.) Resistor R_1 carries essentially zero DC current and merely serves to keep $v_G \approx 0$ when $v_{\text{in}} = 0$. As a result,

$$V_{GS} = -V_K = -R_K I_D$$

and we choose R_K, such that

$$V_{GS} = V_P \left(1 - \sqrt{\frac{I_D}{I_{DSS}}}\right) \tag{8a}$$

where I_D is the desired Q-point current and V_P and I_{DSS} are the device

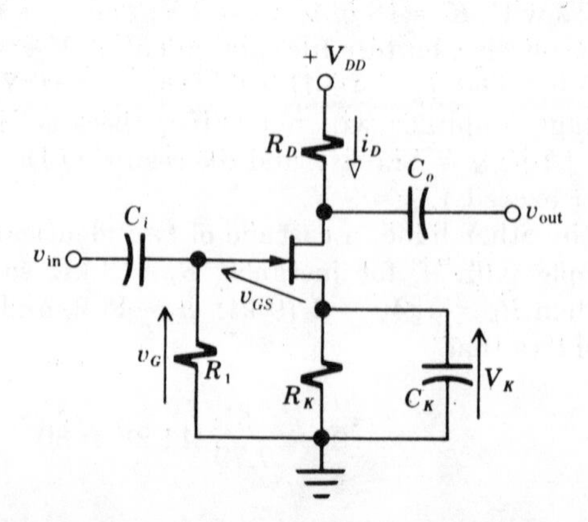

FIGURE 12.1–5
JFET amplifier.

parameters. The corresponding transconductance is given by

$$g_m = -\frac{2I_{DSS}}{V_P}\left(1 - \frac{V_{GS}}{V_P}\right) = -\frac{2}{V_P}\sqrt{I_{DSS}I_D} \tag{8b}$$

and active-state operation requires

$$V_{DS} > V_{DS_{\min}} = V_{GS} - V_P \tag{8c}$$

See Eqs. (1)–(3), Section 10.3, for the source of these expressions. Also remember that $V_P < 0$ and $V_P < V_{GS} < 0$.

With $v_{\text{in}} \neq 0$, the bypass capacitor C_K shorts out any signal variations across R_K, thereby removing it from the equivalent circuit. Thus, since the device model is the same as that of Fig. 12.1–2, our previous small-signal analysis still holds, and we can use Eqs. (5)–(7) for Fig. 12.1–5 with the minor change that $R_i = R_1$.

Resistance R_K creates a *self-biasing* effect in the sense that V_{GS} depends on I_D, in contrast to the *fixed* bias voltage in Fig. 12.1–1a. An even better self-bias configuration is the circuit shown in Fig. 12.1–6. This circuit has less sensitivity to the device parameters, whose values are often only approximately known, and can be used with enhancement or depletion MOSFETs or JFETs. Accordingly, we have introduced a new symbol here to represent an arbitary n-channel FET. As before, the bypass capacitor C_K keeps R_K out of the small-signal circuit, permitting us to concentrate entirely on the biasing.

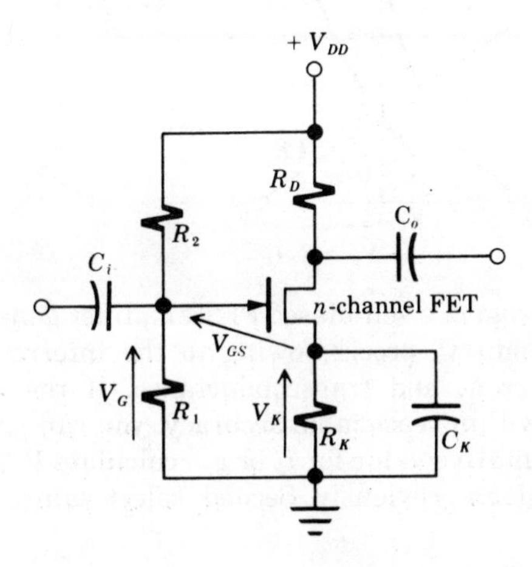

FIGURE 12.1–6
Self-biased amplifier with n-channel FET.

The generality of Fig. 12.1–6 comes from the fact that, when $v_{\text{in}} = 0$,

$$V_{GS} = V_G - V_K = \frac{R_1}{R_1 + R_2}V_{DD} - R_K I_D \tag{9}$$

which can be positive or negative, depending on the element values. The self-biasing feature comes about because V_{GS} decreases when I_D increases, but we know that decreasing the gate voltage of a FET in the active region causes the drain current to decrease. These two effects, then, tend to cancel out each other and stabilize the operating point despite possible parameter changes. Stabilizing the operating point is important for keeping the FET in the active region and for stabilizing the small-signal amplification.

For a graphical demonstration of self-biasing action, take an enhancement MOSFET whose active transfer characteristic $i_D = K[(v_{GS}/V_T) - 1]$ is sketched in Fig. 12.1–7. The operating point Q will be the intersection of this curve with the *bias line,* constructed from Eq. (9) by computing the end points $v_{GS} = V_G = R_2 V_{DD}/(R_1 + R_2)$ when $i_D = 0$ and $i_D = V_G/R_K$ when $V_{GS} = 0$, just like that of a load line. The self-bias line therefore has slope $-1/R_K$, in contrast to the vertical fixed-bias line at $v_{GS} = V_{GS}$. Now suppose the FET actually turns out to have different parameter values, corresponding to the dashed curve. The operating point would then shift markedly to Q'_{fb} with fixed bias but only slightly to Q'_{sb} with self bias. The larger the value of R_K, the smaller the Q-point shift.

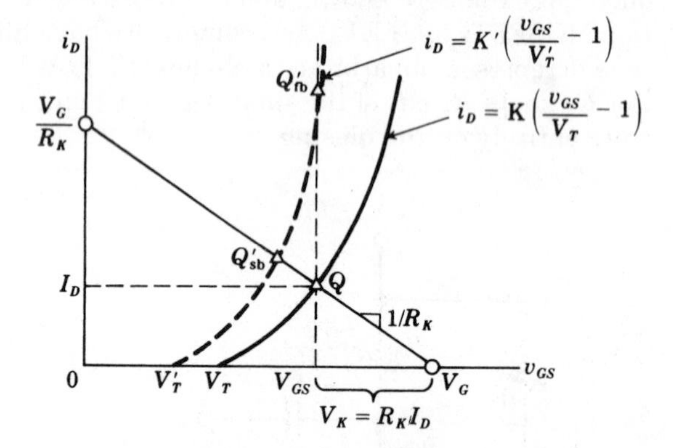

FIGURE 12.1–7 Q-point shift for a self-biased and fixed-bias MOSFET amplifier.

The design of a self-biased FET amplifier generally involves an iterative "cut-and-try" process owing to the interrelationship between the Q-point current and transconductance. If the device parameters are known to within reasonable accuracy, you can proceed as follows. First, choose a tentative value for I_D or g_m, calculate V_{GS} and $V_{DS_{\min}}$ from the expressions given previously. Second, select values for V_{DD}, R_D, and R_K to satisfy

$$V_{DD} = (R_D + R_K)\, I_D + V_{DS} > V_{GS} + R_K I_D \tag{10a}$$

with V_{DS} being at least one volt greater than $V_{DS_{\min}}$. Taking $R_K \geq 0.2\ R_D$

provides some protection against parameter changes. For a depletion MOSFET or JFET, you have the additional constraint $R_K > -V_{GS}/I_D$ (why?). You may then need to go back and modify your choice of I_D to get the desired values of $\mu = g_m R_o$ and $R_o = R_D \| r_d$. Third, select R_1 and R_2 such that

$$\frac{R_1}{R_1 + R_2} = \frac{V_{GS} + R_K I_D}{V_{DD}} \tag{10b}$$

and such that $R_i = R_1 \| R_2$ has an appropriate value.

Example 12.1–2

A self-biased amplifier is to be designed for $\mu \approx 10$, $R_o \leq 5$ kΩ, and $R_i \geq$ 100 kΩ, using a JFET with $I_{DSS} = 18$ mA, $V_P = -6$ V, and $r_d = 200$ kΩ. Let's start by assuming that $R_D = 5$ kΩ, so that $R_o = 5 \| 200 \approx 4.9$ kΩ. We will need $g_m \approx 2$ m℧ to obtain $\mu \approx 10$. Equations (8a)–(8c) then yield $I_D = 2$ mA, $V_{GS} = -4$ V, and $V_{DS_{\min}} = 2$ V. If we take $R_K = 3$ kΩ (so $R_K > -V_{GS}/I_D$), Eq. (10a) requires that $V_{DD} = 16 + V_{DS} > 2$ V, and we might choose $V_{DD} = 20$ V and $V_{DS} = 4$ V. Equation (10b) then becomes $R_1/(R_1 + R_2) = 2/20$, suggesting $R_1 = 200$ kΩ and $R_2 = 1800$ kΩ. Hence, $R_i = 200 \| 1800 = 180$ kΩ > 100 kΩ. Figure 12.1–8 shows the circuit diagram with the DC voltage values.

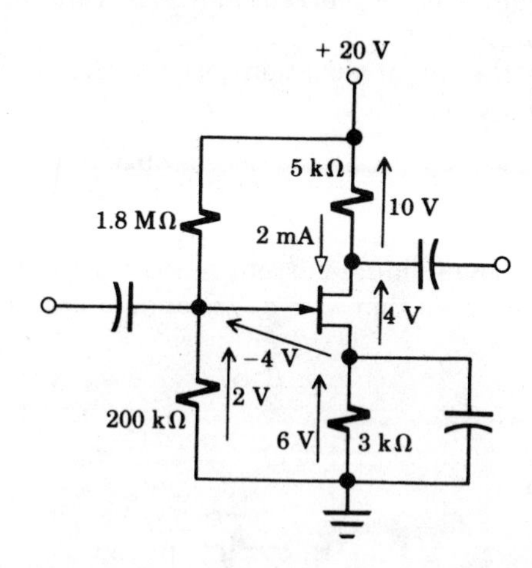

FIGURE 12.1–8 Circuit for Example 12.1–2.

Exercise 12.1–2

Repeat the previous example, using an enhancement MOSFET with $K =$ 16 mA, $V_T = 4$ V, and $r_d = 40$ kΩ. Take $R_D = 5$ kΩ.

BJT AMPLIFIERS

Figure 12.1–9 shows a simple fixed-bias BJT amplifier and the standard notation for such circuits. Functionally, resistor R_B provides a DC *bias current* I_B and thereby establishes the Q point at $i_C = I_C$ and $v_{CE} = V_{CE}$.

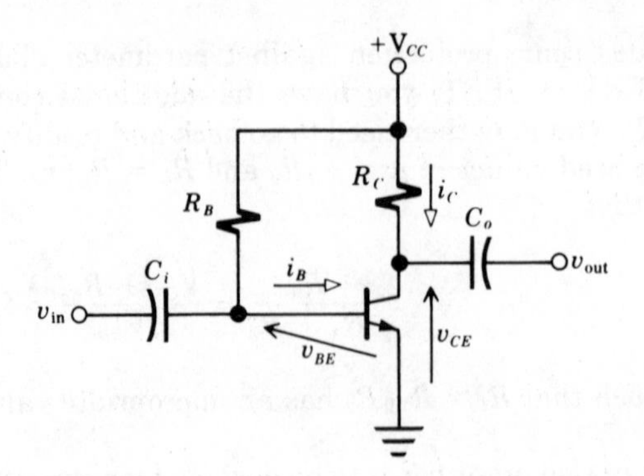

FIGURE 12.1–9
Fixed-bias BJT amplifier.

Since $v_{BE} \approx V_T$ and $v_{CE} > 0$ for active-state operation, R_B and R_C must be such that

$$V_{CC} - R_B I_B = V_T \qquad V_{CE} = V_{CC} - R_C I_C > 0$$

The base and collector currents are related via

$$I_C = \beta_0 I_B + I_{CEO}$$

where β_0 denotes the DC current gain. The collector cutoff current I_{CEO} is usually negligible.

To study the signal variations, we use the small-signal BJT model of Fig. 12.1–10a where

$$r_\pi = \beta \frac{25 \text{ mV}}{I_C} \tag{11a}$$

and β is the small-signal current gain. For direct comparison with the FET model, observe that $i_b = v_{be}/r_\pi$ and, hence, that

$$i_c = \beta(v_{be}/r_\pi) = g_m v_{be}$$

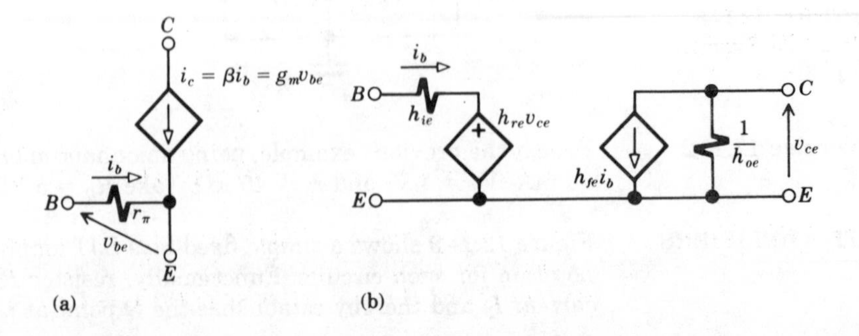

FIGURE 12.1–10
(a) Small-signal BJT model. (b) Common-emitter hybrid parameter model.

with the transconductance given by

$$g_m = \frac{\beta}{r_\pi} = \frac{I_C}{25 \text{ mV}} \tag{11b}$$

A BJT operated at $I_C = 1$ mA will then have $g_m = 1/25 = 0.04$ ℧ $=$ 40 m℧, which suggests that BJT amplifiers produce significantly *more voltage gain* than FET amplifiers. Unfortunately, they also have much *lower input resistance* due to the small value of r_π; for example, $r_\pi =$ 2.5 kΩ if $I_C = 1$ mA and $\beta = 100$.

A more precise BJT model is shown in Fig. 12.1–10b, labeled with the *common-emitter hybrid parameters.* The parameters h_{ie} and h_{fe} correspond to r_π and β, respectively. The dynamic resistance $1/h_{oe}$ plays the same role as r_d in our FET model, but is generally large enough to have little effect in BJT circuits. The remaining parameter h_{re} represents a small *reverse* coupling that, again, can be ignored in most cases. These parameters are the ones listed by device manufacturers on specification sheets. We will, however, stick with our simplified model (Fig. 12.1–8a) and its parameters $r_\pi = h_{ie}$, $\beta = h_{fe}$, and $g_m = h_{fe}/h_{ie}$.

Proceeding as we did with a FET amplifier, we put the small-signal model in the bias circuit and replace the capacitances and DC supply with short circuits to obtain Fig. 12.1–11a. Thus, if $R_B \gg r_\pi$,

$$R_i \approx r_\pi \qquad R_o = R_C \tag{12a}$$

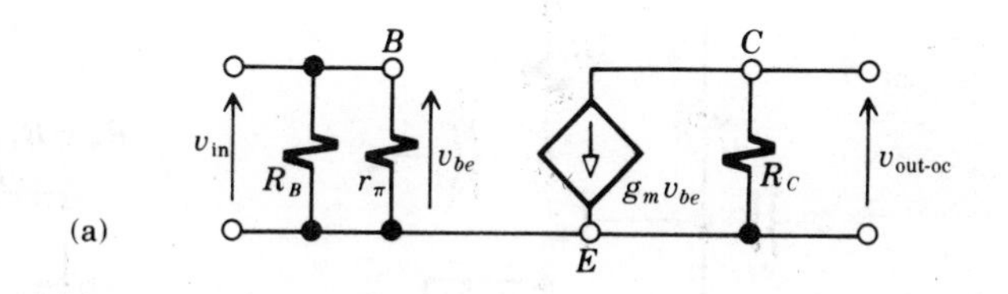

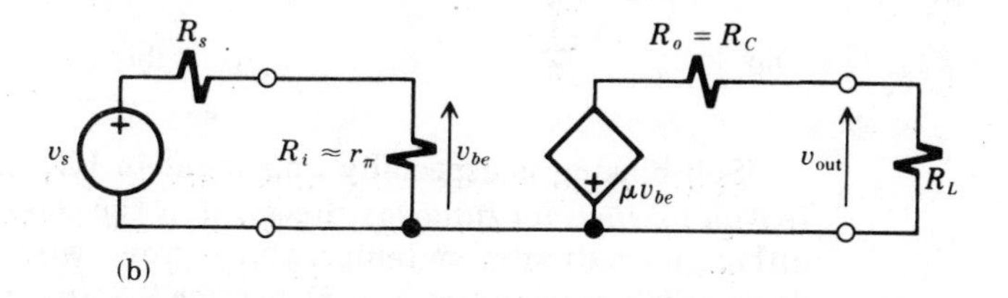

FIGURE 12.1–11
(a) Small-signal equivalent circuit of BJT amplifier.
(b) Two-port model.

and $v_{\text{out-oc}} = R_C(-g_m v_{be})$ so

$$\mu = g_m R_C = \beta \frac{R_C}{r_\pi} \tag{12b}$$

Then, including input and output loading in Fig. 12.1–11b,

$$A_v = -\left(\frac{r_\pi}{r_\pi + R_S}\right)\left(\frac{R_L}{R_L + R_C}\right)\mu \tag{13}$$

The extension of this analysis to the case of cascaded stages should be relatively obvious.

The self-biasing configuration can be also used for BJT amplifiers, as shown in Fig. 12.1–12a. The relationship between the biasing elements and I_B is brought out by Fig. 12.1–12b. Here we have substituted the large-signal model for the active BJT and formed the DC Thévenin equivalent circuit looking back from the base to obtain the open-circuit voltage $V_B = R_1 V_{CC}/(R_1 + R_2)$ and equivalent resistance $R_B = R_1 \| R_2$. Straightforward analysis then yields

$$I_B = \frac{V_B - V_T - R_E I_C}{R_B + R_E} \tag{14}$$

revealing that I_B decreases (or increases) and tends to stabilize I_C if it turns out to be higher (or lower) than expected.

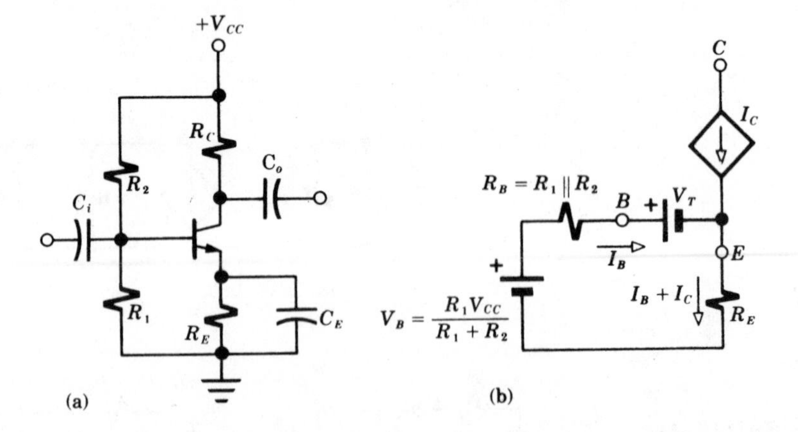

FIGURE 12.1–12 (a) Self-biased BJT amplifier; (b) Large-signal model for bias analysis.

Self-biasing is especially important in BJT circuits due to their potential for *thermal runaway*. Recall that the collector cutoff current I_{CEO}, although small at room temperature, grows with temperature just like a diode's drift current; see Eq. (5), Section 9.3. In a fixed-bias circuit, the increased value of I_C can increase the power dissipated by the transistor, which further increases the temperature and causes I_{CEO} and I_C to increase even more—and so forth until the transistor could be destroyed by overheating. The self-bias circuit largely prevents this effect, as well as compensating for different values of β_0.

Given β_0, V_T, and the desired value of I_C and assuming $I_B \approx I_C/\beta_0 \ll I_C$, a handy method for designing a self-biased BJT amplifier goes as follows. First, select R_C to yield the desired amplification. Second, choose V_{CC}, V_{CE}, and R_E such that

$$V_{CC} - (R_C + R_E)I_C = V_{CE} \tag{15a}$$

Lacking information to the contrary, take V_{CE} to be a few volts and $R_E I_C \approx 0.2\ V_{CC}$. Third, let $R_B \approx 10\ R_E$ to compute

$$V_B = (R_E + R_B/\beta_0)I_C + V_T \tag{15b}$$

from which

$$R_2 = R_B \frac{V_{CC}}{V_B} \qquad R_1 = \frac{R_2 R_B}{R_2 - R_B} \tag{15c}$$

A more elaborate process would be necessary if power efficiency is also a major concern.

Example 12.1–3

Suppose the source in Example 12.1–1 had $R_s = 1$ kΩ and could work into $R_i \approx r_\pi = 2.5$ kΩ. We then might use a BJT with $\beta = 100$ and $I_C = 1$ mA, thereby allowing us to get by with a single-stage amplifier thanks to the higher BJT voltage gain. In particular, inserting $R_s = 1$ kΩ, $R_L = 5$ kΩ; and the values for r_π and β into Eq. (13) show that $|A_v| \approx 50$ when $R_C \approx 2.2$ kΩ. If $V_{CC} = 9$ V, $\beta_0 = 100$, and $V_T = 0.7$ V, the preceding design procedure yields $R_E \approx 2$ kΩ, $R_2 \approx 62$ kΩ, and $R_1 \approx 30$ kΩ.

Exercise 12.1–3

Show that a cascade of two identical BJT stages has

$$A_v \approx \frac{R_L(\beta R_C)^2}{(r_\pi + R_s)(r_\pi + R_C)(R_L + R_C)}$$

Exercise 12.1–4

Confirm the bias design results in Example 12.1–3. Then let $I_C = \beta_0 I_B + I_{CEO}$ in Eq. (14) to find I_B and I_C when $I_{CEO} = 0.2$ mA.

12.2 FEEDBACK AND DIFFERENTIAL AMPLIFIERS

Feedback is probably the single most powerful tool in the hands of an experienced electronics engineer. It is possible to design a negative-feedback amplifier that has less distortion, higher input resistance, lower output resistance, wider frequency response, and less parameter sensitivity than a similar circuit without feedback. But these benefits carry with them two penalties: reduced gain, and the possibility of unstable behavior if the negative feedback should turn into positive feedback. Here, we will look at a few *RC*-coupled amplifier circuits that illustrate how negative feedback brings about improved performance. We'll also examine a simple differential amplifier, the keystone of the modern op-amp. (Chapter 13 presents a general treatment of feedback theory and applications, including positive feedback.)

FEEDBACK AMPLIFIERS

Consider a self-biased BJT amplifier without a bypass capacitor paralleling R_E (see Fig. 12.2–1a). As R_E is no longer bypassed, it must be included in the small-signal equivalent circuit of Fig. 12.2–1b, where $R_B = R_1 \| R_2$. We now have a signal voltage v_f across R_E, and

$$v_{be} = v_{\text{in}} - v_f \tag{1}$$

Note that v_f will be proportional to $v_{\text{out-oc}}$; Equation (1), then, says that a fraction of the output has been *fed back* and *subtracted* from the input to form v_{be}. Omitting the bypass capacitor therefore yields an amplifier with *negative feedback.*

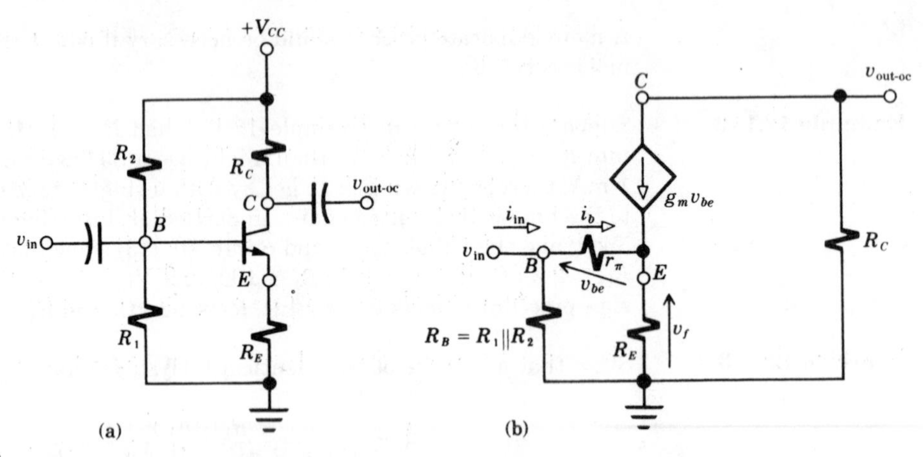

FIGURE 12.2–1
BJT feedback amplifier: (a) circuit diagram; (b) small-signal equivalent circuit.

This seemingly trivial circuit modification has profound effects on performance. In particular, under the reasonable conditions

$$\beta = g_m r_\pi \gg 1 \qquad R_E \gg \frac{1}{g_m}$$

the input resistance *increases* from $R_i \approx r_\pi$ to

$$R_i \approx R_B \| \beta R_E \tag{2a}$$

while the open-circuit amplification *decreases* from $\mu = g_m R_C$ to

$$\mu \approx \frac{R_C}{R_E} \tag{2b}$$

The output resistance stays the same, $R_0 = R_C$.

Obviously, increased input resistance is desirable to minimize input loading, a common problem in BJT amplifiers. (However, since $R_B = R_1 \| R_2$ now plays an important role in R_i, we may wish to alter our previous bias-design procedure and take $R_B \gg R_E$.) Less obvious is the practical significance of reduced amplification—until one recognizes that Eq. (2b) does *not* involve the transconductance g_m. Consequently, the feedback amplifier has a stable and easily adjusted gain, equal to a resistance ratio and independent of the active device. (Recall the similar results for op-amp circuits with feedback that we discussed in Section 4.3.) Since BJT circuits often provide more than enough gain, the reduced amplification is a bargain price to pay for the other benefits.

The preceding expressions for R_i and μ follow from Eq. (1) combined with $v_f = R_E(g_m v_{be} + i_b) = R_E(g_m + 1/r_\pi)v_{be}$, to obtain

$$v_{be} = \frac{v_{\text{in}}}{g_m R_E + (R_E/r_\pi) + 1} \tag{3}$$

Then, using $i_{\text{in}} = (v_{\text{in}}/R_B) + (v_{be}/r_\pi)$ and $g_m r_\pi = \beta$, we get

$$R_i = \frac{v_{\text{in}}}{i_{\text{in}}} = \left(\frac{1}{R_B} + \frac{1}{\beta R_E + R_E + r_\pi}\right)^{-1} \tag{4a}$$

and

$$\mu = \left|\frac{v_{\text{out-oc}}}{v_{\text{in}}}\right| = \left|\frac{R_C(-g_m v_{be})}{[g_m R_E + (R_E/r_\pi) + 1]v_{be}}\right|$$
$$= \frac{R_C}{R_E + (R_E/\beta) + (1/g_m)} \tag{4b}$$

which reduce to Eqs. (2a) and (2b) when $\beta \gg 1$ and $R_E \gg 1/g_m$. For future use, also note that

$$\frac{v_f}{v_{\text{in}}} = \frac{1 + (1/\beta)}{1 + (1/\beta) + (1/g_m R_E)} \tag{5}$$

as found from Eqs. (1) and (3).

Next, consider a FET amplifier with negative feedback introduced by omitting the bypass capacitor. Figure 12.2–2 shows this amplifier's schematic diagram and small-signal equivalent circuit. To save labor, let's assume that $r_d \gg R_D$ so we can ignore the FET's dynamic resistance. Then

Fig. 12.2–2b becomes identical to the BJT circuit (Fig. 12.2–1b) with $R_C = R_D$, $v_{be} = v_{gs}$, etc., and $r_\pi \to \infty$, the latter corresponding to $\beta = g_m r_\pi \to \infty$. Making these replacements in Eq. (4b) immediately yields

$$\mu = \frac{R_D}{R_K + (1/g_m)} \tag{6}$$

and we see that feedback reduces and stabilizes the gain of a FET amplifier, especially if $R_K \gg 1/g_m$ so $\mu \approx R_D/R_K$. The input resistance remains unchanged at $R_i = R_1 \| R_2$, which was large anyhow.

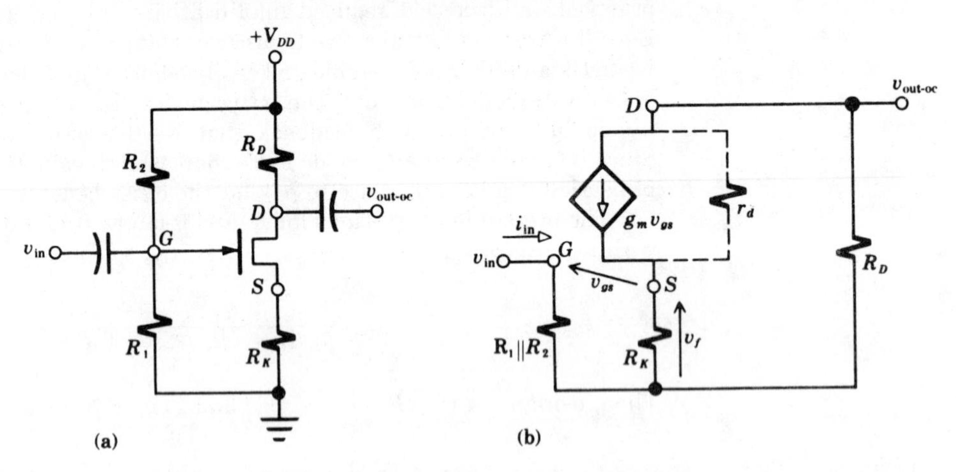

FIGURE 12.2–2
FET feedback amplifier:
(a) circuit diagram;
(b) small-signal equivalent circuit.

The major benefit of feedback here is the *reduction of nonlinear distortion,* a matter of concern in both BJT and FET amplifiers, but a more serious concern with FETs due to their inherently nonlinear transfer curves. (Compare, for instance, Figs. 10.1–7b and 10.2–4b.) For a qualitative understanding of how feedback reduces distortion, let the output without feedback be $v_{\text{out–oc}} = -\mu_o(v_{\text{in}} + v_d)$, where v_d represents a distortion component not present at the input. If we add feedback that reduces the gain to $\mu_f < \mu_o$ and simultaneously increases the input signal by a factor of μ_o/μ_f, the output becomes

$$v_{\text{out–oc}} = -\mu_f\left(\frac{\mu_o}{\mu_f}v_{\text{in}} + v_d\right) = -\mu_o\left(v_{\text{in}} + \frac{\mu_f}{\mu_o}v_d\right)$$

and the distortion has been reduced by $\mu_f/\mu_o < 1$. Of course, this assumes a distortionless preamplifier to increase the input to the feedback amplifier. But the output signal of the preamp is only $(\mu_f/\mu_o)v_{\text{in}}$ and could easily fall within the linear small-signal range, as contrasted with $\mu_o v_{\text{in}}$ at the feedback amplifier's output.

If you return to Fig. 12.0 you'll see a similar strategy implemented by feedback resistor R_{13}, which couples a portion of the final output voltage back to resistor R_5 in the preamplifier. A related feature of Fig. 12.0 is the *decoupling* network consisting of R_6 and C_3, which prevents unwanted feedback from the power amplifier to the preamp.

Example 12.2–1 Suppose you need to develop a 2.5-V rms signal across a high-impedance load, given a source with $v_s = 50$ mV rms. You try a FET amplifier having $g_m = 5$ mV, $R_D = 10$ kΩ, and $r_d \gg R_D$, so that $|A_v| \approx \mu_o = g_m R_D = 50 = 2.5$ V/50 mV, as required. Unfortunately, due to the rather large voltage swing, the output also contains a distortion component equal to 20% of the desired signal. To minimize the distortion, you settle on the arrangement represented by Fig. 12.2–3 where feedback via $R_K = 2$ kΩ reduces the amplification to

$$\mu_f = \frac{10 \text{ k}\Omega}{2 \text{ k}\Omega + 1/(5 \text{ m℧})} = \frac{10}{2.2}$$

which is compensated by a preamplifier with gain $\mu_o/\mu_f = 50 \times 2.2/10 = 11$. The preamp's output voltage is $11\, v_s = 0.55$ V rms, well within the linear range.

The distortion in the feedback amplifier is modeled by adding a distortion current $g_m v_d$ to the controlled drain current $g_m v_{gs}$. Thus

$$v_{gs} = 11\, v_s - v_f \qquad v_f = 2 \text{ k}\Omega \times 5 \text{ m℧}(v_{gs} + v_d)$$

which combine to give $v_{gs} = v_{\text{in}} - (10/11)v_d$. Consistent with the given information and the properties of FET amplifiers, all loading effects are negligible and $v_{\text{out}} \approx v_{\text{out-oc}}$. The output voltage, then, is

$$v_{\text{out}} = 10 \text{ k}\Omega\,[-5 \text{ m℧}\,(v_{gs} + v_d)] = -50\,(v_s + \tfrac{1}{11}\,v_d)$$

FIGURE 12.2–3 Circuit for Example 12.2–1.

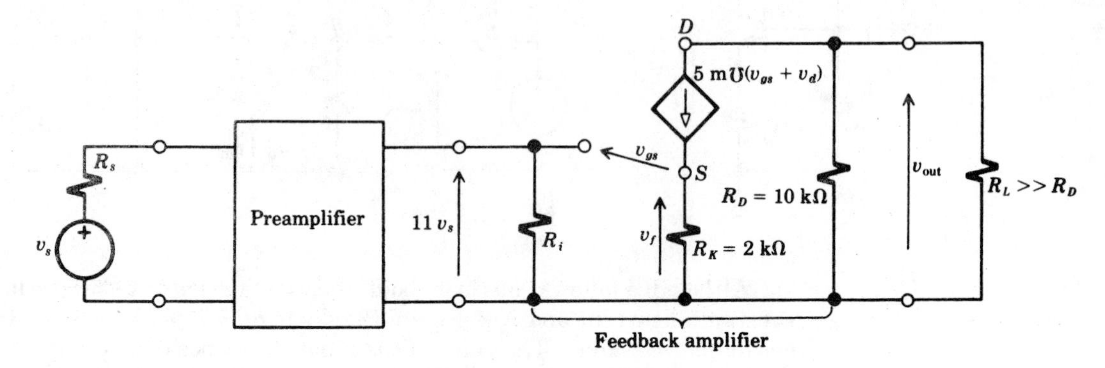

As predicted, the distortion has been reduced by the factor $\mu_f/\mu_o = \frac{1}{11}$ and now equals just 1.8% of the amplified source signal.

Exercise 12.2-1 Use Eqs. (4a) and (4b) to calculate R_i and μ when $R_C = 4\ \text{k}\Omega$, $R_E = 1\ \text{k}\Omega$, $R_1 = 50\ \text{k}\Omega$, $R_2 = 12.5\ \text{k}\Omega$; and the BJT has $I_C = 2.5$ mA and $\beta = 100$. Compare your results to those for the same circuit with a bypass capacitor.

FOLLOWERS AND PHASE SPLITTERS

BJT and FET amplifiers, with or without feedback, have relatively high output resistance and may not be suitable for driving a low-impedance load. When such output loading is a problem, we can use a BJT circuit like that in Fig. 12.2-4a (or the corresponding FET circuit). The BJT circuit differs from that of Fig. 12.2-1 in that R_C has been omitted and the output is taken across R_E. Thus, $v_{\text{out-oc}} = v_f$ and Eq. (5) gives

$$\mu = \frac{1 + (1/\beta)}{1 + (1/\beta) + (1/g_m R_E)} \approx 1 \tag{7}$$

so $v_{\text{out-oc}} \approx v_{\text{in}}$ with no polarity inversion.

We call this circuit an *emitter follower* because the signal voltage at the emitter terminal essentially "follows" the input signal variations. The transistor itself is in a *common-collector* configuration, the collector terminal in Fig. 12.2-4b being common to both the small-signal input and output. Likewise, the FET version is called a *source follower,* which refers to the voltage at the source terminal of the FET, with the FET in a *common-drain* configuration.

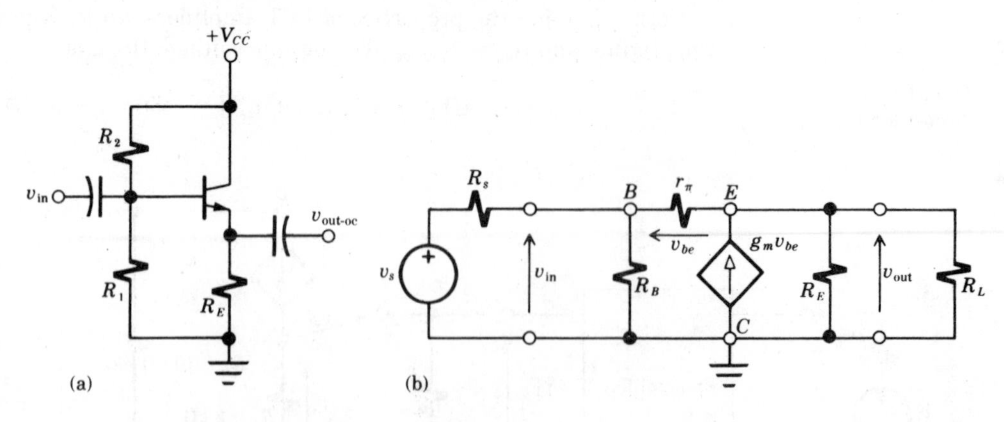

FIGURE 12.2-4
(a) Emittter follower.
(b) Small-signal equivalent circuit.

Although a follower has just about unity voltage gain, it does produce substantial current and power gain, thanks to *high input resistance* and *low output resistance.* The value of the input resistance of an emitter follower is given by Eq. (4a) with R_E replaced by $R_E \| R_L$, since the load resistance is now in parallel with R_E (see Fig. 12.2–4b). Thus

$$R_i \approx R_B \| \beta(R_E \| R_L) \tag{8}$$

which assumes $\beta \gg 1$ and $R_E \| R_L \gg 1/g_m$.

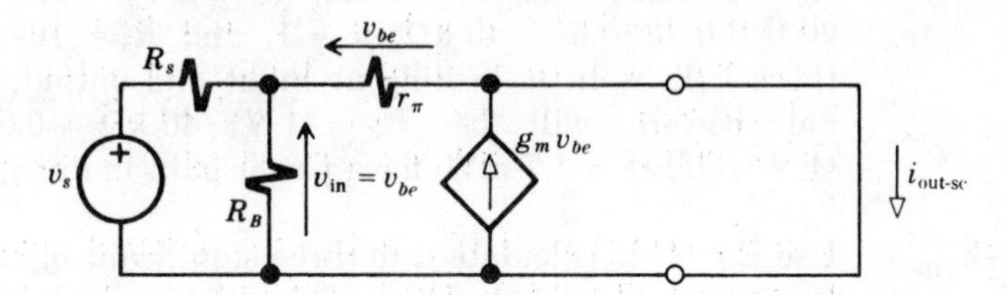

FIGURE 12.2–5

To determine the Thévenin equivalent output resistance, we use the general definition $R_o = v_{out-oc}/i_{out-sc}$ with v_{out-oc} and i_{out-sc} expressed in terms of the source voltage v_s. A well-designed emitter follower would have $R_i \gg R_s$ and $\mu \approx 1$ so $v_{in} \approx v_s$ and $v_{out-oc} = \mu v_{in} \approx v_s$. As for the short-circuit output current, we redraw Fig. 12.2–4b with $R_L = 0$ and observe that the shorted output bypasses R_E and results in $v_{in} = v_{be}$, as shown in Fig. 12.2–5. Clearly, $i_{out-sc} = g_m v_{be} + v_{be}/r_\pi$ and $v_{be} = (R_B \| r_\pi)v_s/[(R_B \| r_\pi) + R_s]$ so, after a little algebra,

$$\frac{i_{out-sc}}{v_s} = \frac{g_m + \dfrac{1}{r_\pi}}{1 + \dfrac{R_s}{R_B} + \dfrac{R_s}{r_\pi}}$$

Therefore,

$$R_o = \frac{v_{out-oc}}{i_{out-sc}} \approx \frac{v_s}{i_{out-sc}} \approx \frac{1}{g_m}\left(1 + \frac{R_s}{r_\pi}\right) \tag{9}$$

under the usual condition $R_B \gg R_s$. Typically, R_o will be less than 100 Ω, in contrast with the several kilohms output resistance of an ordinary amplifier.

A follower circuit by itself could be used to interface a high-impedance source with a low-impedance load when the unloaded source voltage is large enough. Otherwise, to get both voltage gain and power gain, the follower could be used as the output stage of a cascade.

A special-purpose circuit related to the follower is the *phase splitter* diagrammed in Fig. 12.2–6. Here, $R_E = R_C = R$ and there are two outputs

$$v_1 \approx v_{in} \qquad v_2 \approx -v_{in}$$

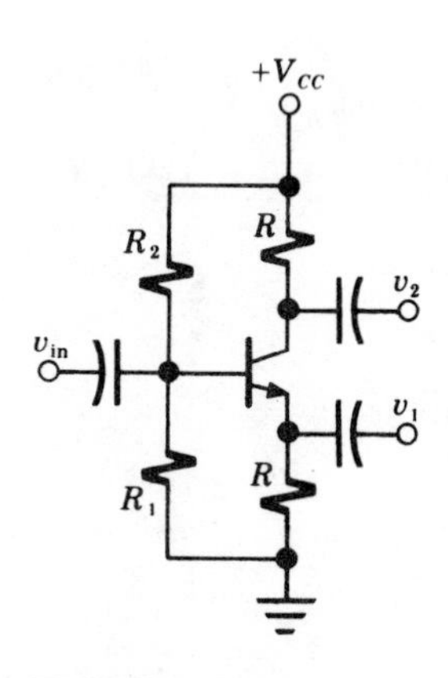

FIGURE 12.2–6
BJT phase splitter.

as obtained from Eqs. (7) and (2b). The audio amplifier back in Fig. 12.0 has a MOSFET phase splitter (T_2) that drives the power-amplifier stage.

(Note, by the way, the direct coupling from T_1 to T_2, made possible by designing R_4, R_5, and R_8 to produce an appropriate gate bias for T_2.)

Example 12.2-2

A 1-V rms source having $R_s = 2\ \text{k}\Omega$ is to be interfaced with a 600-Ω load. If we use an emitter follower with $I_C = 2.5$ mA, $\beta = 100$, $R_E = 3\ \text{k}\Omega$, and $R_B = 200\ \text{k}\Omega$, then $r_\pi = 1\ \text{k}\Omega$, $g_m = \beta/r_\pi = 100\ \text{m℧} = 1/(10\ \Omega)$, $R_i = 200 \| 100\ (3 \| 0.6) = 40\ \text{k}\Omega$, $\mu \approx 1$, and $R_o \approx 10\ (1 + \frac{2}{1}) = 30\ \Omega$. Thus, there will be little loading at input and output, and the average signal powers will be $P_{\text{in}} \approx (1\ \text{V})^2/40\ \text{k}\Omega = 0.025$ mW and $P_{\text{out}} \approx (1\ \text{V})^2/600\ \Omega \approx 1.7$ mW, for a power gain in excess of 60.

Exercise 12.2-2

Use Eq. (7) to calculate μ to three significant figures for the emitter follower just discussed. Then calculate A_v, including input and output loading.

Exercise 12.2-3

Redraw Figs. 12.2-4 and 12.2-5 for FET follower, including the dynamic resistance r_d. Show directly that $\mu = 1/(1 + 1/g_m R_K')$, where $R_K' = R_K \| r_d$, and that $R_o = \mu/g_m$.

DIFFERENTIAL AMPLIFIERS †

Figure 12.2-7 diagrams one implementation of a differential amplifier. This rather formidable-looking circuit consists of two FETs connected

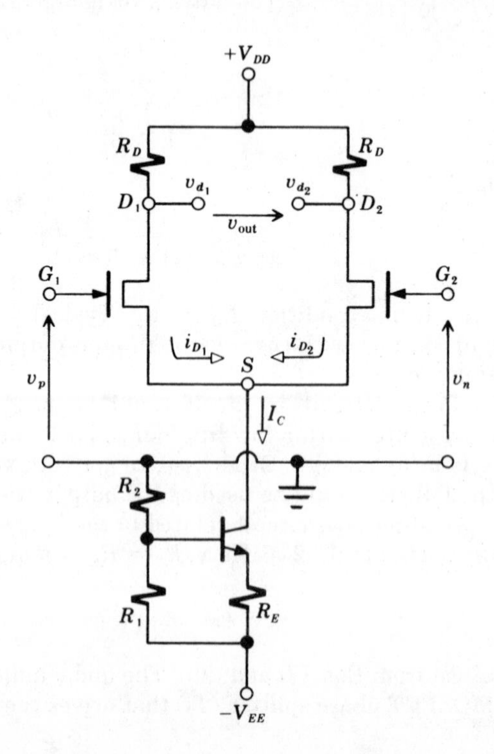

FIGURE 12.2-7
Differential amplifier circuit.

together at their source terminals to the collector of a BJT. The BJT serves as a *constant-current source,* keeping the value of $I_C = i_{D_1} + i_{D_2}$ fixed, while the FETs (which could also have been BJTs) do the actual amplifying. The two input voltages at G_1 and G_2 have been labeled v_p and v_n, respectively, underscoring the relationship between this circuit and an operational amplifier. The output may be taken to be v_{d_1} or v_{d_2}, or $v_{\text{out}} = v_{d_2} - v_{d_1}$ as shown. In any case, the circuit's symmetry results in an output voltage proportional to the *difference* $v_p - v_n$.

For the small-signal analysis, we assume the FETs are identical and operating in the active region. (Integrated-circuit fabrication justifies the former, and we will address the matter of biasing shortly.) The equivalent circuit then takes the form of Fig. 12.2–8a, where V_{DD} has been replaced by a short circuit to ground and the constant-current source becomes an *open circuit,* since it will not permit any signal variations in I_C. (A more precise analysis would include a large resistance from terminal S to ground, which alters the results only slightly.) Converting the con-

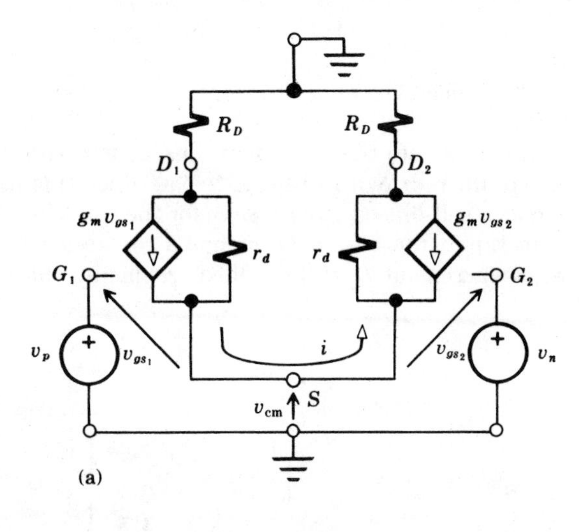

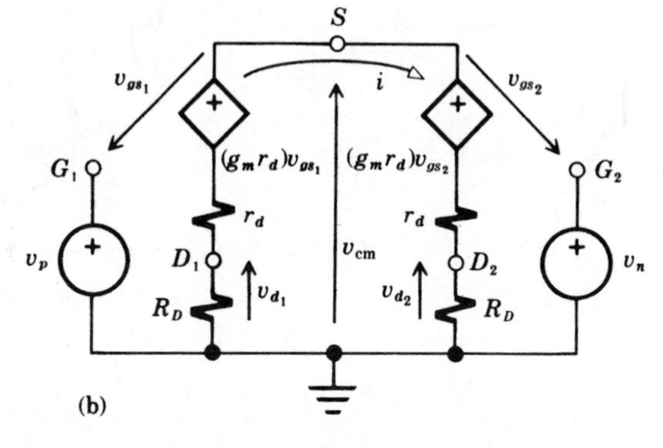

FIGURE 12.2–8
Small-signal equivalent circuits of a differential amplifier.

trolled current sources to Thévenin-equivalent voltage sources and turning the FET loop upside down yields Fig. 12.2–8b. We now easily see that

$$v_{gs_1} = v_p - v_{\rm cm} \qquad v_{gs_2} = v_n - v_{\rm cm}$$

and the current around the loop is

$$i = \frac{(g_m r_d)v_{gs_1} - (g_m r_d)v_{gs_2}}{2R_D + 2r_d} = \frac{g_m r_d}{2(R_D + r_d)}(v_p - v_n)$$

independent of the *common-mode voltage* $v_{\rm cm}$. Therefore letting $R'_D = R_D r_d/(R_D + r_d) = R_D \| r_d$,

$$v_{d_2} = R_D i = \tfrac{1}{2} g_m R'_D (v_p - v_n) \qquad v_{d_1} = R_D(-i) = -v_{d_2}$$

and

$$v_{\rm out} = v_{d_2} - v_{d_1} = g_m R'_D (v_p - v_n) \tag{10}$$

so we, indeed, have a differential amplifier with unloaded gain $\mu = g_m R'_D$.

Let's turn to the bias conditions, considering first the BJT portion of the circuit redrawn in Fig. 12.2–9a. Since this has the same configuration as a self-biased circuit, save for the collector connection and the lack of an input, the base current has a constant value I_B and fixes the collector current at $I_C = \beta_0 I_B$. Next, replacing the BJT by its large-signal

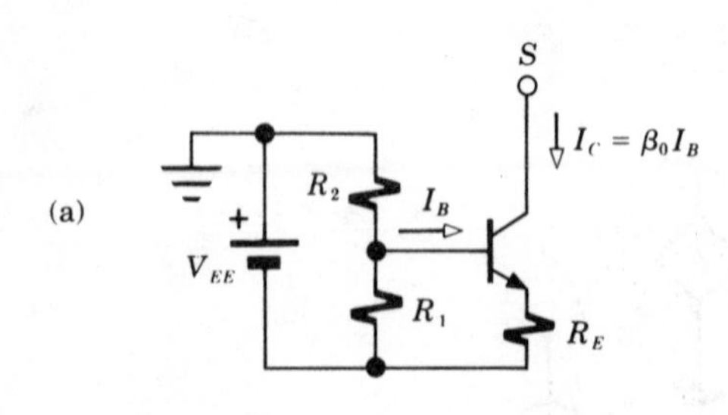

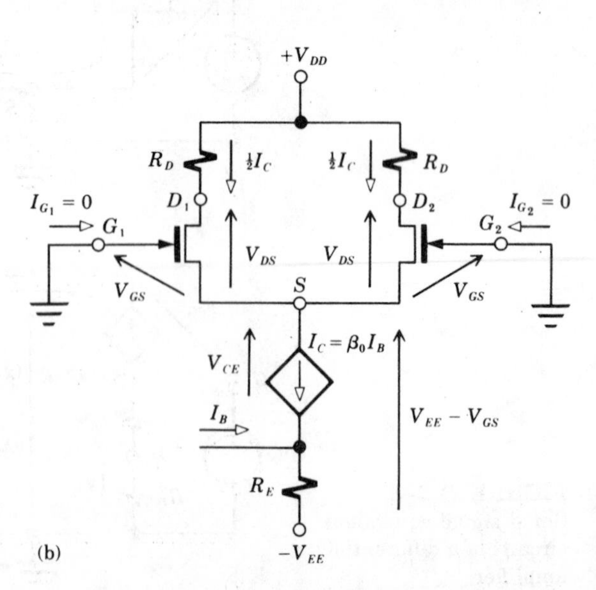

FIGURE 12.2–9
(a) BJT constant-current source.
(b) DC conditions in a differential amplifier.

model, we have the complete DC circuit of Fig. 12.2–9b where terminals G_1 and G_2 are grounded to represent $v_p = v_n = 0$. From symmetry, I_C divides equally between the two FETs, each carrying $I_D = \frac{1}{2}I_C$. Therefore, providing that V_{DS} and V_{CE} satisfy the conditions for active-state operation of the FETs and BJT, the potential at terminal S will automatically be such that V_{GS} is consistent with the value of I_D and the FET parameters.

A special advantage of the differential amplifier is that it can be *direct coupled,* thus eliminating coupling capacitors and permitting the amplifier to handle very low signal frequencies—possibly even down to DC. Any unwanted DC offset voltage at the output is removed by a simple level-shifting circuit. Two DC supply voltages are needed to allow for positive and negative signal swings at input and output.

A complete *operational amplifier* is block-diagrammed in Fig. 12.2–10. The differential amplifier at the input is followed by a high-gain amplifier, which is often another differential amplifier. A level shifter resets the DC voltage at the input of the output driver, which in turn develops the large-signal output. With additional refinements, this op-amp circuit may have a dozen or more transistors.

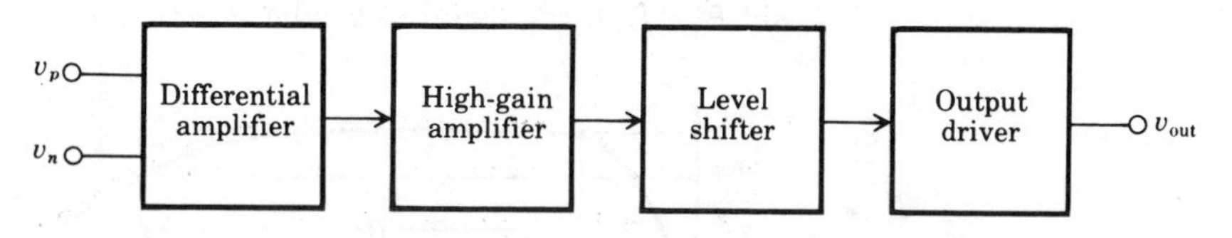

FIGURE 12.2–10 Block diagram of an operational amplifier.

Exercise 12.2–4

Let the FETs in Fig. 12.2–9b be enhancement MOSFETs with $K = 18$ mA, $V_T = 3$ V, and $r_d = 40$ kΩ, and let the BJT have $\beta = 50$ and $I_C = 4$ mA. Find all the DC voltages if $R_D = 10$ kΩ, $R_E = 3$ kΩ, and $V_{DD} = V_{EE} = 20$ V. Use your results to justify the assertion that $v_{D_1} = v_{D_2} = v_{out} = 0$ when $v_p = v_n = 0$. Then calculate the value of μ.

12.3 AMPLIFIER FREQUENCY RESPONSE

In this section we investigate factors that cause the gain of an amplifier to vary with signal frequency. Following an overview of frequency-response characteristics, we present circuit models for low-frequency and high-frequency behavior and develop some approximate formulas for computing the cutoff frequencies of simple single-stage and cascade amplifiers.

FREQUENCY-RESPONSE CHARACTERISTICS

The gain $|A_v(f)|$ of an RC-coupled amplifier varies with frequency f due, primarily, to two situations. At low signal frequencies, the coupling and bypass capacitors begin to have significant impedance, which reduces the gain and, eventually, causes $|A_v(f)| \to 0$ as $f \to 0$. At high frequencies, internal capacitances representing energy storage within the transistors become important, for they tend to short out the signal and cause $|A_v(f)| \to 0$ as $f \to \infty$. (The "high/low" tone control in Fig. 12.0 utilizes the same principle, in that C_1 shorts out the high-signal frequencies when the switch is closed.) With both effects taken into account, the typical gain curve $|A_v(f)|$ versus f looks like Fig. 12.3–1a.

In this figure we see a *midfrequency* or *midband* region with maximum gain $|A_v(f)| \approx |A_0|$, this being the gain calculated in previous sections where we ignored all capacitances. We also see *low-frequency* and *high-frequency* regions below and above the lower and upper *cutoff frequencies* f_ℓ and f_u. By convention, the cutoff frequencies are taken to be where

$$|A_v(f)| = \frac{1}{\sqrt{2}}|A_0| \approx 0.707\,|A_0|$$

so the voltage amplification is down by about 30% or, equivalently, 20 log $|A_v(f)/A_0| \approx -3$ dB. Such a frequency response would be satisfactory providing all the significant frequency components of the signal fall in the midband range $f_\ell < f < f_u$. The amplifier's *bandwidth* is $B = f_u - f_\ell$, and $B \approx f_u$ in the usual case where $f_\ell \ll f_u$.

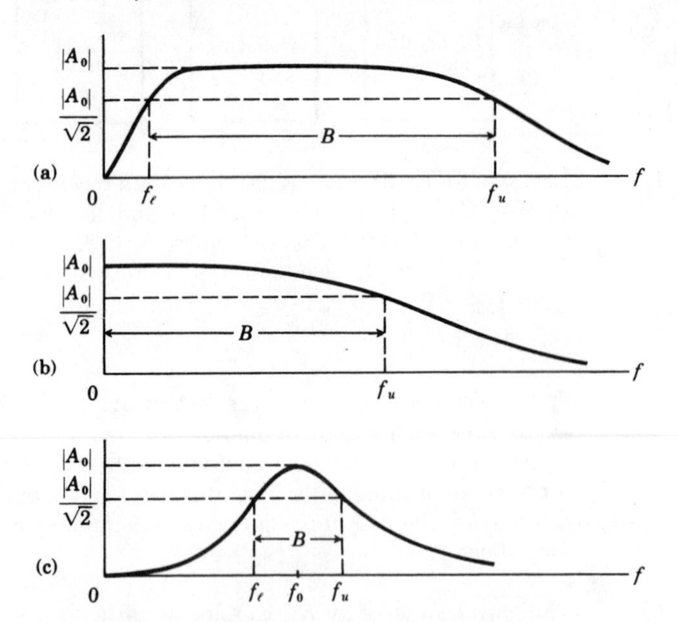

FIGURE 12.3–1
Amplifier gain curves:
(a) *RC*-coupled;
(b) direct-coupled;
(c) tuned.

An audio amplifier has relatively modest frequency-response requirements, with $f_\ell \approx 20$–200 Hz and $f_u \approx 4$–20 kHz. In contrast, the video amplifier in a TV set must handle frequencies over the range from

30 Hz to 4.2 MHz, justifying its designation as a *wideband* amplifier. Besides the "flat" amplitude response, wideband amplifiers need a carefully controlled phase shift to preserve any sharp corners in the signal waveform.

If the signal contains very low frequencies, the amplifier should be *direct-coupled* to obtain a frequency characteristic like Fig. 12.3–1b where $B = f_u$ since $f_\ell = 0$. In the absence of any external circuitry, an operational amplifier has this type of response with $|A_0| \approx 10^5$ and $B \approx 10$ Hz. Feedback makes it possible to exchange gain for bandwidth, but the *gain-bandwidth product* remains constant.

Another type of frequency-response characteristic is found in a *tuned* amplifier, as sketched in Fig. 12.3–1c. This device amplifies a narrow band of frequencies centered around a resonant frequency f_0. Radio and TV receivers employ both fixed-tuned amplifiers and tunable amplifiers, the latter having an adjustable value of f_0.

The essential concepts of tuned amplifiers and direct-coupled amplifiers were discussed in Sections 7.2 and 12.2, respectively. The remainder of this section, therefore, deals specifically with low-frequency and high-frequency response.

LOW-FREQUENCY RESPONSE

Amplifier coupling capacitors generally find themselves in an equivalent circuit like that in Fig. 12.3–2a, which is a *highpass filter*. The amplitude ratio from Eq. (9), Section 7.1, becomes

$$|\underline{H}(f)| = \frac{|\underline{V}_2|}{|\underline{V}_1|} = \frac{R_2}{R_1 + R_2} \frac{(f/f_\ell)}{\sqrt{1 + (f/f_\ell)^2}}$$

$$= \frac{|A_0|}{\sqrt{1 + (f_\ell/f)^2}} \tag{1a}$$

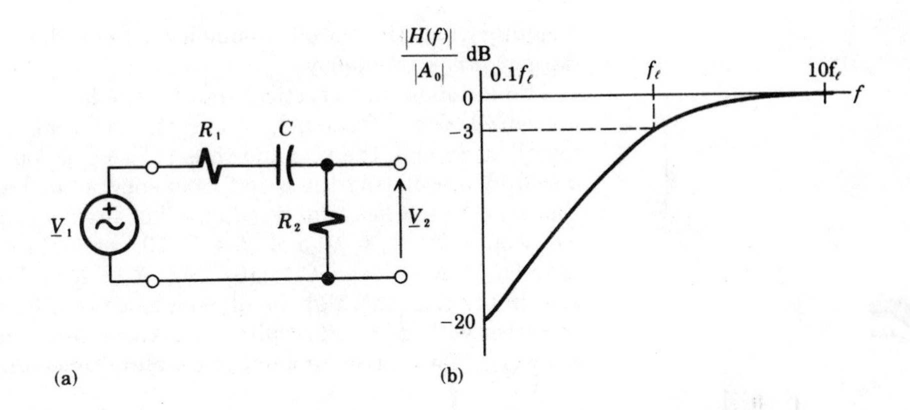

FIGURE 12.3–2

where $|A_0| = R_2/(R_1 + R_2)$ is the "gain" and the lower cutoff frequency is

$$f_\ell = \frac{1}{2\pi(R_1 + R_2)C} \tag{1b}$$

We could also write $f_\ell = 1/2\pi\tau$, with $\tau = (R_1 + R_2)C$ being the circuit's time constant. Figure 12.3–2b sketches the normalized gain $|H(f)|/|A_0|$ in dB versus f on a logarithmic axis, that is, a Bode plot. (See Fig. 7.1–6 for a sketch of the phase shift.)

Now consider the small-signal equivalent circuit of Fig. 12.3–3a, an amplifier including input and output coupling capacitors. With appropriate values of R_i, μ, and R_o, this circuit could represent any one of several possible FET or BJT amplifiers. Since both the input and output sides act like highpass filters, direct extension of Eq. (1) gives the amplitude response as the product

$$|A_v(f)| = \frac{|\underline{V}_{out}|}{|\underline{V}_s|} = \frac{|A_0|}{\sqrt{1 + (f_i/f)^2}\sqrt{1 + (f_o/f)^2}} \tag{2a}$$

where

$$f_i = \frac{1}{2\pi(R_s + R_i)C_i} \qquad f_o = \frac{1}{2\pi(R_L + R_o)C_o} \tag{2b}$$

and the mid-frequency gain is $|A_0| = R_i R_L \mu/[(R_i + R_s)(R_L + R_o)]$. Figure 12.3–3b gives the corresponding Bode plot assuming $f_o \gg f_i$, in which case $|A_v(f)|/|A_0| = 1/\sqrt{2}$ (-3 dB) at $f \approx f_o$, but the curve has another "break" at $f = f_i$. In general, if the two *break frequencies* f_i and f_o differ by at least an order of magnitude, then

$$f_\ell \approx \begin{cases} f_o & f_o \geq 10 f_i \\ f_i & f_i \geq 10 f_o \end{cases} \tag{3}$$

meaning that the cutoff frequency essentially equals the dominant (larger) break frequency.

Equation (3) has practical significance for two reasons. First, certain undesired side effects may occur if the break frequencies are close together. Second, the size and thus the cost of coupling capacitors is reduced by associating the cutoff frequency with the side of Fig. 12.3–3a that has the *smaller total resistance*. For instance, if $R_L + R_o < R_s + R_i$, we would let $f_o \approx f_\ell$ and $f_i < f_\ell/10$, as obtained by taking $C_o \approx [2\pi f_\ell(R_L + R_o)]^{-1}$ and $C_i > 10/[2\pi f_\ell(R_s + R_i)]$. Equations (2b) and (3) also imply that the high input resistance of a FET amplifier or a BJT amplifier with feedback results in a better low-frequency response, that is a lower f_ℓ, for a given amount of coupling capacitance.

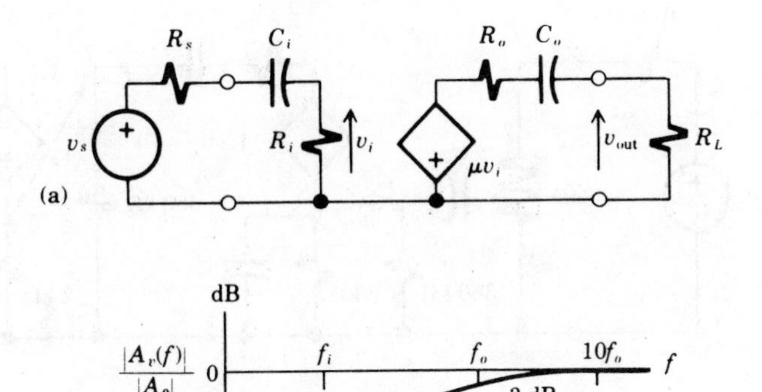

FIGURE 12.3-3
(a) RC-coupled amplifier. (b) Low-frequency response.

We have not yet taken account of the bypass capacitor, which—if present—introduces negative feedback at low frequencies. Advanced analysis reveals that this may create another break frequency

$$f_{\text{bypass}} \approx \frac{1}{2\pi[R\|(1/g_m)]C} \tag{4}$$

where C stands for the bypass capacitor (C_K or C_E) and R the resistor being bypassed (R_K or R_E). Theoretically, the bypass capacitor has the greatest potential influence on low-frequency response, owing to the small value of the equivalent resistance $R\|(1/g_m)$. Fortunately, thanks to the small DC voltage across C, we can get by with a very large but inexpensive electrolytic capacitor—say 100–1000 μF—so that $f_{\text{bypass}} \ll f_\ell$. (However, if designing for mass production, one would want to explore the economics of other alternatives.)

A similar strategy applies to the design of a cascade amplifier having n capacitors and break frequencies $f_1, f_2, \ldots, f_n$. We conservatively assume that

$$f_\ell \approx f_1 + f_2 + \cdots + f_n \tag{5}$$

and choose the capacitors to yield a dominant break frequency, with all other break frequencies being smaller by an order of magnitude or more.

Example 12.3–1 Figure 12.3–4 represents the cascade from Example 12.1–1 modified for self biasing with $R_K = 2$ kΩ. The FETs have $g_m = 4$ m℧, and we need to choose the capacitors to get $f_\ell \leq 30$ Hz. Since C_5 is in series with the smallest resistance, we associate it with the dominant break frequency by taking $C_5 > (2\pi \times 30 \text{ Hz} \times 7.8 \text{ k}\Omega)^{-1} \approx 0.7\ \mu$F. We then take $C_3 >$

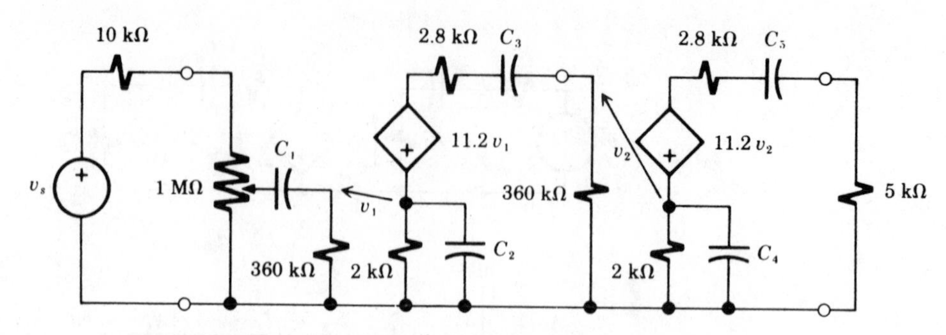

FIGURE 12.3–4 Circuit for Example 12.3–1.

$10/(2\pi \times 30 \times 363 \times 10^3) \approx 0.15\ \mu\text{F}$, and similarly for C_1. Finally, using Eq. (4) with $R_K \| (1/g_m) \approx 222\ \Omega$, the choice of $C_2 = C_4 = 250\ \mu\text{F}$ gives $f_{\text{bypass}} < 3$ Hz.

Exercise 12.3–1 Calculate values for C_i and C_o so the emitter follower circuit in Example 12.2–2 has $f_\ell \approx 50$ Hz.

Exercise 12.3–2 Consider an amplifier with n identical break frequencies f_n. Generalize Eq. (2a) to show that $f_\ell = f_n/\sqrt{2^{1/n} - 1}$.

HIGH-FREQUENCY RESPONSE

The high-frequency response of an *RC*-coupled or direct-coupled amplifier depends primarily upon the internal transistor capacitances indicated in the *high-frequency hybrid-π* models of Fig. 12.3–5 (these models have been somewhat simplified for our purposes). At very high frequencies, say above 10 MHz, one should also include stray capacitances between wires and other effects that greatly complicate amplifier analy-

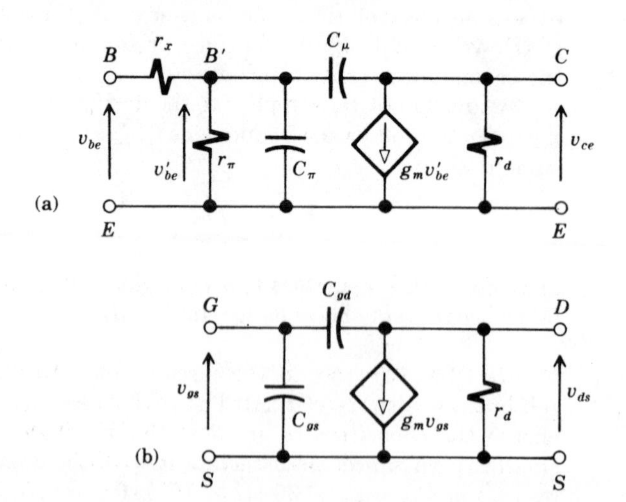

FIGURE 12.3–5 High-frequency hybrid-π models: (a) BJT; (b) FET.

sis and design. We will see that these capacitances result in a potential trade-off between gain and bandwidth as measured by the upper cutoff. Consequently, lower gain amplifiers (for example, FET and feedback circuits) tend to have better high-frequency response.

The BJT model includes two capacitances, $C_\pi \approx 30-500$ pF and $C_\mu \approx 1-20$ pF, and an additional resistance $r_x \approx 20-100\ \Omega$, the lower element values being typical of a transistor fabricated specifically for high-frequency applications. Known as the *base-spreading resistance*, r_x accounts for the small ohmic resistance of the base region; it must be included at high frequencies since the control voltage v'_{be} may differ appreciably from the input voltage v_{be}. Two other related BJT parameters are the *beta-cutoff frequency* f_β and the *transition frequency* f_T, given by

$$f_\beta = \frac{1}{2\pi r_\pi (C_\pi + C_\mu)} \qquad f_T = \beta f_\beta = \frac{g_m}{2\pi (C_\pi + C_\mu)} \tag{6}$$

which are the frequencies where the short-circuit current gain drops to $i_c/i_b = \beta/\sqrt{2}$ and $i_c/i_b = 1$, respectively. For example, if $C_\pi = 100$ pF, $C_\mu = 4$ pF, $r_\pi = 1$ kΩ, and $g_m = 100$ m℧, then $f_\beta \approx 1.5$ MHz and $f_T \approx 150$ MHz. These numbers have significance because the hybrid-π model becomes invalid at frequencies greater than about $f_T/3$ and the BJT is no longer a useful amplification device.

The hybrid-π FET model (Fig. 12.3–5b) is less complex. It consists of the low-frequency elements plus capacitances $C_{gs} \approx 5-50$ pF and $C_{gd} \approx 0.5-5$ pF. However, we will focus our attention on BJT amplifiers, observing that the corresponding FET results can be obtained by letting $C_\pi = C_{gs}$, $C_\mu = C_{gd}$, $r_x = 0$, and $r_\pi \to \infty$.

Figure 12.3–6a shows the high-frequency equivalent circuit of a BJT

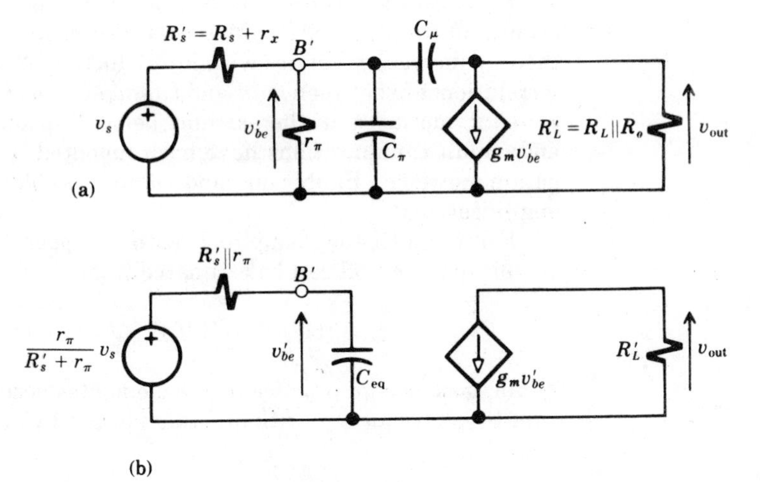

FIGURE 12.3–6 (a) High-frequency equivalent circuit of BJT amplifier. (b) Simplified circuit with Miller-effect capacitance C_{eq}.

amplifier (with $R_B \gg r_\pi$) after combining the series resistances $R'_s = R_s + r_x$ at the input and the parallel resistances $R'_L = R_L \| R_o$ at the output. The feedback via C_μ produces the *Miller effect* discussed in Example 5.3–3 and increases the equivalent input capacitance to

$$C_{eq} \approx C_\pi + (1 + g_m R'_L) C_\mu \tag{7}$$

This is based on the midband approximation $v_{out}/v'_{be} \approx -g_m R'_L$ for the Miller-effect multiplication of C_μ. If $g_m R'_L > 1$, the limitation $f < f_T/3$ virtually ensures that very little current reaches the output through C_μ, and our circuit simplifies to that of Fig. 12.3–6b where we've performed a Thévenin conversion on the elements to the left of C_{eq}. In view of the *lowpass* filter configuration on the input side, it immediately follows from Eq. (7), Section 7.1, that

$$|A_v(f)| = \frac{|A_0|}{\sqrt{1 + (f/f_u)^2}} \tag{8a}$$

with

$$A_0 = -\frac{r_\pi}{R'_s + r_\pi} g_m R'_L \tag{8b}$$

$$f_u = \frac{1}{2\pi (R'_s \| r_\pi) C_{eq}} \tag{8c}$$

which are the amplifier's midband gain and upper cutoff frequency. The frequency-response curve will have the same shape as $|\underline{H}(f)|$ in Fig. 7.1–3 or 7.1–7.

Careful examination of Eqs. (7) and (8c) reveals that the circuit designer has little control over f_u, save through the choice of the transistor and the value of R'_L. Moreover, decreasing R'_L to decrease C_{eq} and thereby increase f_u carries with it a reduction of A_0 per Eq. (8b). This interrelationship between *gain* and *bandwidth* (as measured by f_u) is illustrated numerically in the example below. Various sophisticated devices and circuit configurations have been invented to overcome, in part, the gain-bandwidth limitation and other problems of high-frequency amplification.

For a multistage amplifier with n upper break frequencies, the overall upper cutoff can be estimated from

$$f_u \approx [(1/f_1)^2 + (1/f_2)^2 + \cdots + (1/f_n)^2]^{-1/2} \tag{9}$$

Again, good design practice favors elements chosen to produce one dominant break frequency, with all others larger by about an order of magnitude.

Example 12.3–2 Suppose a BJT amplifier has $R_s' = r_\pi = 1\text{ k}\Omega$, $C_\pi = 100$ pF, $C_\mu = 4$ pF, and $g_m = 100$ m℧. If $R_L' = 2\text{ k}\Omega$, then $C_{eq} = 904$ pF, $f_u \approx 350$ kHz, and $|A_0| = 100$. Decreasing R_L' to 200 Ω drops the gain to $|A_0| = 10$, but gives $C_{eq} = 184$ pF and $f_u \approx 1.7$ MHz.

Exercise 12.3–3 Calculate $|A_0|$ and f_u for a FET amplifier with $R_s = 1\text{ k}\Omega$, $R_L' = 4\text{ k}\Omega$, $g_m = 5$ m℧, $C_{gs} = 20$ pF and $C_{gd} = 2$ pF.

Exercise 12.3–4 Combine Eqs. (6)–(8) to obtain the *gain-bandwidth product* of a BJT amplifier in the form

$$|A_0|f_u = \frac{(R_L'/R_s')f_T}{1 + 2\pi R_L' C_\mu f_T}$$

12.4 POWER AMPLIFIERS

We conclude our tour of electronic amplifier circuits with a look at the concepts, problems, and techniques associated with power amplifiers—amplifiers capable of producing more than about one-half watt of output signal power. A preliminary discussion of an elementary amplifier sets the stage for examining the popular transformer-coupled and push-pull circuits used in power amplification.

LIMITATIONS AND EFFICIENCY

A power amplifier is a *large-signal* amplifier in the sense that the output voltage swing will be a sizable fraction of the supply voltage. We therefore immediately anticipate a problem with *nonlinear distortion*. To minimize distortion, one should select an amplifying device with reasonably linear characteristics over the required signal swing. The power BJT is an obvious choice, but a specially constructed power MOSFET called a VMOS or VFET also has good large-signal linearity. In any case, the complete amplifier should incorporate *negative feedback* to an earlier stage for purposes of "mopping up" residual distortion. With that understanding, we can make the simplifying assumption of *linear* large-signal behavior. We will also be dealing exclusively with BJT circuits; power FET circuits differ in biasing, but not in operating principles.

To introduce some of the important concepts, let's consider the elementary BJT power amplifier of Fig. 12.4–1a where the input is an AC signal and the output is the AC component at the collector. The idealized characteristic curves are sketched in Fig. 12.4–1b, along with the load line. Because we expect that the input signal has equal positive and negative excursions and we want the largest possible output excursions, the Q point is taken at the *center* of the load line, namely $V_{CE} = V_{CC}/2$ and $I_C = V_{CC}/2R_C$, obtained by appropriate setting of the bias I_B. Figure 12.4–1c shows the corresponding waveforms $i_C(t) = I_C + I_m \cos \omega t$ and $v_{CE}(t) =$

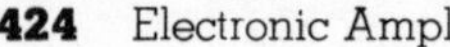

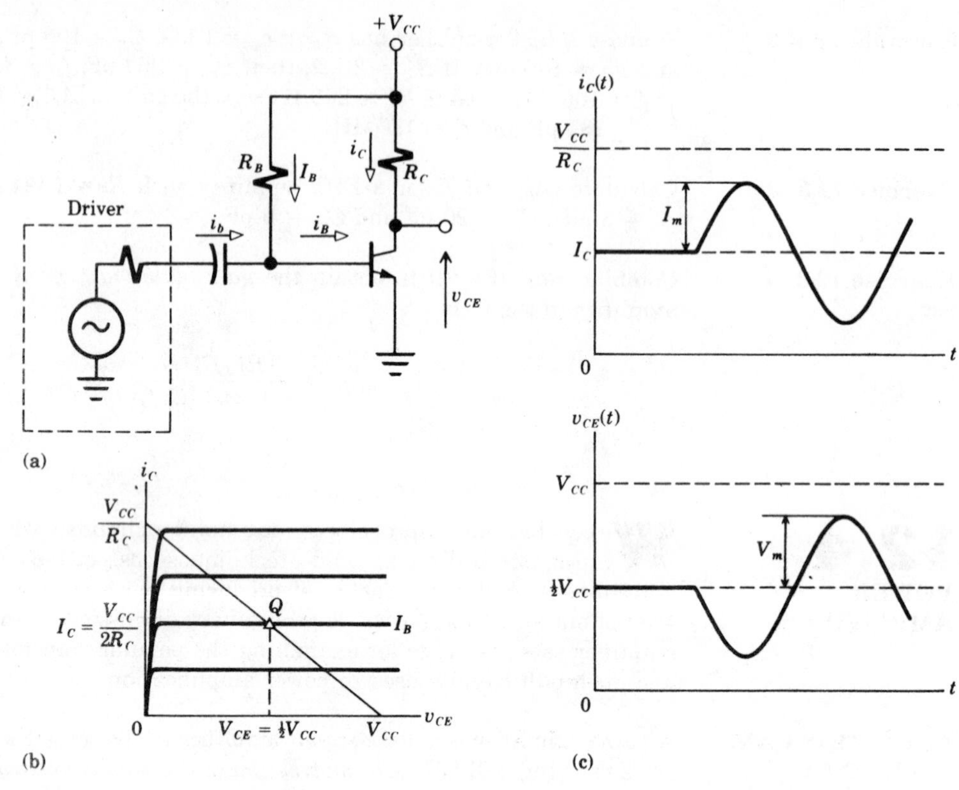

FIGURE 12.4–1
BJT power amplifier: (a) circuit; (b) Q point for maximum symmetric signal swing; (c) waveforms.

$V_{CE} - V_m \cos \omega t$, whose AC amplitudes are limited by

$$I_m \leq I_C = \frac{V_{CC}}{2R_C} \qquad V_m \leq V_{CE} = \frac{V_{CC}}{2} \tag{1}$$

The maximum amplitudes $I_m = I_C$ and $V_m = V_{CE}$ neglect saturation effects, a valid simplification if V_{CC} is much larger than the value of v_{CE} at saturation.

The centered Q point thereby permits *maximum symmetric signal swing*, with $i_C(t) \approx I_C + I_C \cos \omega t$, and so forth. Any other Q point would result in lower limits on I_m and V_m. And maximum symmetric swing is a generally desirable condition, since most signals of interest, although not sinusoidal, do have equal peak excursions. Of course, the preceding stage must develop sufficient output to drive the power amplifier. For the circuit at hand, this requires $i_b(t) = (I_m/\beta) \cos \omega t$, from which one can determine the necessary small-signal gain of the driver.

Let's now calculate the amplifier's *efficiency,* defined as the ratio of average output signal power to the total DC power from the supply. The AC current through R_C is $I_m \cos \omega t$ so, from Eq. (1),

$$P_{\text{out}} = \tfrac{1}{2} I_m^2 R_C \leq \left(\frac{V_{CC}}{2R_C}\right)^2 R_C = \frac{V_{CC}^2}{8R_C} \tag{2}$$

If we neglect the small amount of power dissipated in the bias resistor R_B, the supply provides the DC power

$$P_{\text{DC}} = V_{CC} I_C = \frac{V_{CC}^2}{2R_C} \tag{3}$$

Therefore,

$$\text{Eff} = \frac{P_{\text{out}}}{P_{\text{DC}}} \leq \frac{1}{4}$$

meaning that the amplifier has, at best, an efficiency of only 25%. We call this the maximum *theoretical* efficiency, since a typical input signal generally produces less output power and the actual efficiency will be considerably below 25%.

Another important consideration is the average power dissipated in the transistor itself, denoted P_C and called the *collector dissipation.* Taking account of the DC power $R_C I_C^2$ dissipated by R_C, it follows that

$$P_C = (v_{CE} i_C)_{\text{av}} = P_{\text{DC}} - (P_{\text{out}} + R_C I_C)^2$$

whose maximum value P_{CO} occurs when $P_{\text{out}} = 0$. Thus, with no input signal, $v_{CE} = V_{CE}$, $i_C = I_C$, and

$$P_{CO} = V_{CE} I_C = \frac{V_{CC}^2}{4R_C} \tag{4}$$

so the ratio of maximum collector dissipation to maximum output signal power is

$$\frac{P_{CO}}{P_{\text{out-max}}} = 2$$

a quantity known as the amplifier's *figure of merit.* A better design would have a *lower* figure of merit, corresponding to less collector dissipation for a given value of $P_{\text{out-max}}$.

Consider the significance of these expressions. Suppose we want $P_{\text{out}} = 5$ W when $R_C = 40\ \Omega$. This will require $V_{CC} \geq \sqrt{1600} = 40$ V and $P_{\text{DC}} \geq 20$ W, from Eqs. (2) and (3). The transistor must then be rated for

$v_{CE_{max}} \geq V_{CC} \geq 40$ V, $i_{C_{max}} \geq V_{CC}/R_C \geq 1.0$ A, and $P_{C_{max}} \geq 10$ W. It's clear that this simple circuit does not make very effective use of either the transistor or the power supply. Moreover, adding a bypassed emitter resistor for stability, or connecting the load through a DC blocking capacitor would only further degrade the performance. (Shortly, we'll look at two improved designs for power amplification.)

Our final consideration here relates *transistor ratings* to *Q*-point selection. Every transistor has maximum allowable values for i_C, v_{CE}, and P_C, which altogether define the *safe operating region* plotted on the i_C-v_{CE} plane in Fig. 12.4–2. Note that the collector rating plots as a hyperbola, $v_{CE}\, i_C = P_{C_{max}}$. The *Q* point of an amplifier must fall within this region; otherwise the transistor could be damaged or destroyed. The value of $P_{C_{max}}$ actually is controlled by the junction temperature, so the transistor should be mounted on a *heat sink* to keep the temperature down. Even with a heat sink, the transistor's temperature will rise, and it may be necessary to *derate* (lower) the specified collector power in accordance with data supplied by the manufacturer. For instance, if an amplifier requires $P_{CO} = 10$ W and the anticipated operating temperature corresponds to a derating factor of 40%, then the transistor must have $P_{C_{max}} \geq 10 \text{ W}/0.4 = 25$ W.

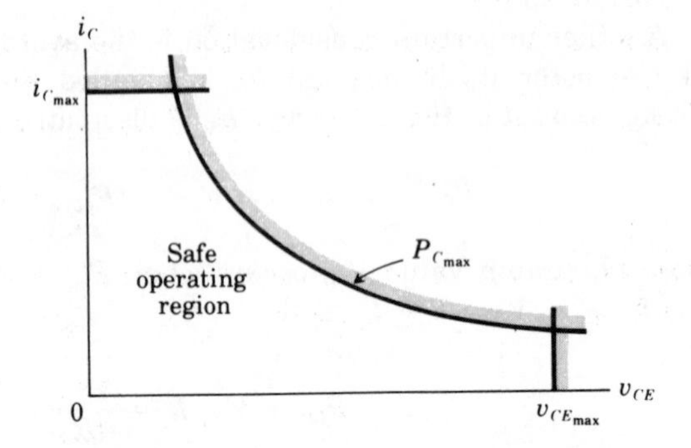

FIGURE 12.4–2
Safe operating region of a BJT.

Example 12.4–1
Power amplifier with RC-coupled load

Let's look at an example of the degradation brought on by capacitive load coupling. Take the circuit in Fig. 12.4–3a with a centered *Q* point at $V_{CE} = 20$ V and $I_C = 0.5$ A. Since all of the DC current flows through R_C, $P_{DC} = 20$ W and $P_{CO} = 10$ W as computed before. A sinusoidal input signal with maximum amplitude produces $i_C(t) \approx 0.5 + 0.5 \cos \omega t$ and would yield $P_{out} = 5$ W if all of the AC component went through R_L. In actuality, the AC current divides equally between the equal resistances R_C and R_L, so $i_{out} = -0.25 \cos \omega t$ and $P_{out} = \frac{1}{2}(0.25 \text{ A})^2 \times 40\ \Omega = 1.25$ W. Consequently, the amplifier's efficiency drops to $P_{out}/P_{DC} = 0.0625$ (6.25%), and its figure of merit is $P_{CO}/P_{out\text{-}max} = 8$.

The division of the AC current also reduces the output voltage amplitude to 40 Ω × 0.25 A = 10 V rather than $V_m \approx V_{CE} = 20$ V. This effect is interpreted in Fig. 12.4–3b, which shows an *AC load line* with slope $-1/(R_C \| R_L)$ intersecting the Q point on the DC load line. The AC load line represents the possible values of $i_C(t)$ and $v_{CE}(t)$ when the coupling capacitor acts like a short circuit. With maximum symmetric signal swing, $v_{CE}(t) = 20 - 10 \cos \omega t$ and $v_{\text{out}} = v_{CE}(t) - V_{CE} = -10 \cos \omega t$.

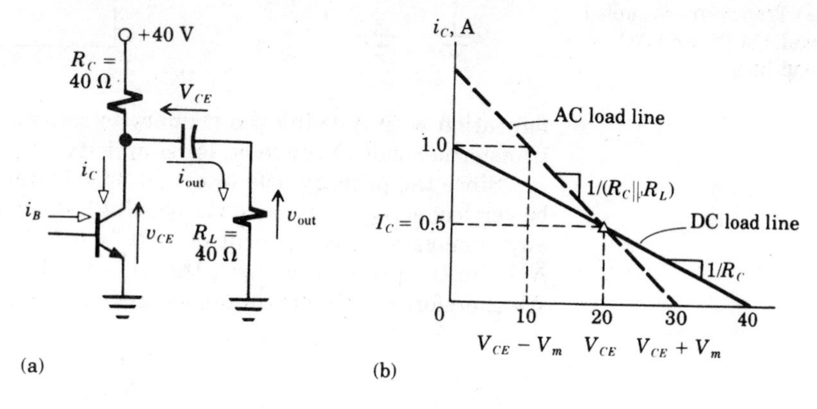

FIGURE 12.4–3
(a) Power amplifier with RC-coupled load.
(b) DC and AC load lines.

Finally, for the sake of completeness, we determine the bias and input specifications, assuming that $i_C(t) = \beta i_B(t)$. If $\beta = 20$—a typical value for a power BJT—the bias circuit must provide $I_B = I_C/\beta = (0.5\text{ A})/20 = 25$ mA; and, likewise, the input signal current must have an amplitude of 25 mA.

Exercise 12.4–1

The capacitive-coupled power amplifier is improved somewhat by moving the Q point to the center of the AC load line. This is accomplished if

$$I_C = \frac{V_{CC}}{R_{\text{AC}} + R_{\text{DC}}} \qquad V_{CE} = \frac{R_{\text{AC}} V_{CC}}{R_{\text{AC}} + R_{\text{DC}}}$$

where R_{AC} and R_{DC} stand for the equivalent AC and DC resistances seen from the collector. Redraw Fig. 12.4–3b with this change, using the element values from Example 12.1–4; confirm that the Q point falls at the center of the AC load line. Then calculate P_{out}, Eff, and P_{CO}/P_{out}, taking $i_C(t) \approx I_C + I_C \cos \omega t$.

TRANSFORMER-COUPLED AMPLIFIERS

A more efficient power amplifier is the transformer-coupled circuit of Fig. 12.4–4a. To reflect the fact that most practical loads have relatively small values of R_L, the transformer has been drawn in the *step-down* con-

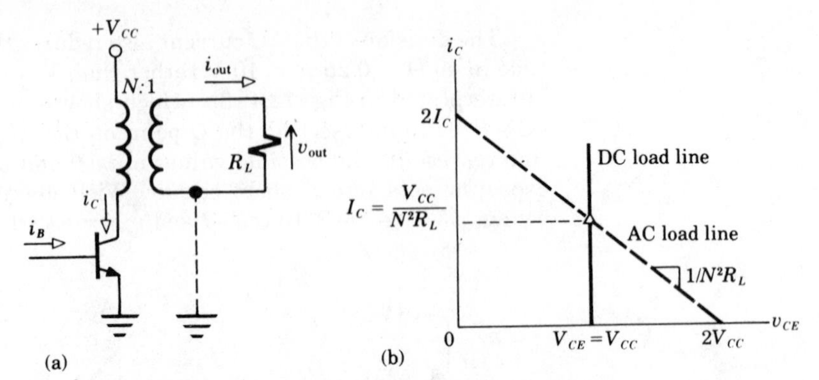

FIGURE 12.4–4
(a) Transformer-coupled load. (b) DC and AC load lines.

figuration with N being the primary-to-secondary turns ratio. An ideal transformer will be assumed for simplicity.

Since the primary acts like a DC short circuit, the DC load line will be vertical at $V_{CE} = V_{CC}$ and the Q point is determined entirely by I_B. But any time-varying component of $i_C(t)$ "sees" the *referred* load resistance N^2R_L in the primary, causing the AC load line to have slope $-1/N^2R_L$. We therefore obtain maximum symmetric signal swing by taking

$$I_C = \frac{V_{CC}}{N^2R_L} \tag{5}$$

so the Q point bisects the AC load line, as shown in Fig. 12.4–4b. Note that it is now possible for $v_{CE}(t)$ to exceed the supply voltage V_{CC}; indeed, $v_{CE}(t)$ equals $2V_{CC}$ at those instants when $i_C(t) = 0$. You can easily derive equation (5) from the slope and desired intercepts of the AC load line, observing that $2I_C = (2V_{CC})/(N^2R_L)$.

If $i_C(t) = I_C + I_m \cos \omega t$, the AC component transformed to the secondary yields

$$i_{out} = \pm NI_m \cos \omega t \qquad v_{out} = \pm NI_m R_L \cos \omega t$$

where $I_m \leq I_C = V_{CC}/N^2R_L$. The sign depends on which terminal of the secondary is grounded or taken as the reference, thanks to the DC-isolation property of transformers. The average output signal power then is

$$P_{out} = \tfrac{1}{2}(NI_m)^2 R_L \leq \frac{V_{CC}^2}{2N^2R_L} \tag{6}$$

while the DC power from the supply is

$$P_{DC} = V_{CC}I_C = \frac{V_{CC}^2}{N^2R_L} \tag{7}$$

Therefore,

$$\text{Eff} = \frac{P_{\text{out}}}{P_{\text{DC}}} \le \frac{1}{2}$$

so transformer coupling gives a maximum theoretical efficiency of 50%, twice that of our elementary amplifier. The figure of merit, however, remains the same at

$$\frac{P_{CO}}{P_{\text{out-max}}} = 2$$

since $P_{CO} = V_{CE} I_C = P_{\text{DC}}$.

The analysis of a power amplifier with a *real* transformer becomes considerably more complicated, since one must include resistance and inductive reactance in both the primary and secondary. For a well-designed transformer at midband frequencies, the major effect will be the primary's winding resistance, which causes the DC load line to slope slightly to the left. At lower and higher frequencies, the inductive reactances and stray capacitances become important, giving a frequency-response curve something like that of Fig. 12.4–5. Consequently, although transformer coupling may be acceptable for audio amplifiers, it has serious drawbacks for wideband or high-frequency power amplification.

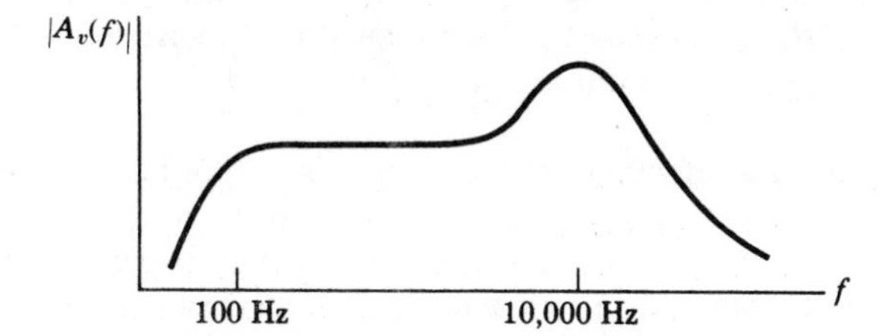

FIGURE 12.4–5 Typical frequency response with transformer coupling.

Example 12.4–2 *Power-amplifier design*

Suppose we want to deliver as much power as possible to an 8-Ω load using transformer coupling and a power BJT having $i_{C_{\max}} = 2.0$ A, $v_{CE_{\max}} = 50$ V, and $P_{C_{\max}} = 10$ W after derating. Figure 12.4–6 shows the safe operating region and an AC load line that touches the 10-W hyperbola at the Q point, so $P_{CO} = P_{C_{\max}}$ and the figure of merit tells us that $P_{\text{out}} = \frac{1}{2}P_{CO} = 5$ W if $I_m = I_C$. Equation (6) then gives $V_{CC}/N = \sqrt{2R_L P_{\text{out}}} = \sqrt{80}$, which could be satisfied by various combinations of V_{CC} and N.

But the maximum voltage and current ratings impose two additional constraints: $2V_{CC} \le 50$ V and $2I_C = 2(V_{CC}/N^2 R_L) \le 2$ A, or $V_{CC} \le 25$ V and $V_{CC}/N^2 \le 8$ V. Substituting $V_{CC} = N\sqrt{80}$ leads to the bounds

$$\frac{\sqrt{80}}{8} = 1.12 \le N \le \frac{25}{\sqrt{80}} = 2.80$$

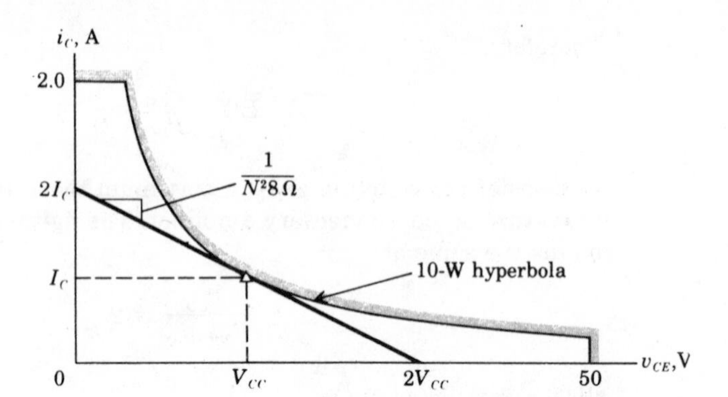

FIGURE 12.4–6 Safe operating region and AC load line for Example 12.4–2.

so, for instance, a transformer with $N = 2$ will do quite nicely. The power supply must then have $V_{CC} = 2\sqrt{80} \approx 18$ V and produce $I_C \approx 18\text{ V}/(2^2 \times 8\ \Omega) = 0.56$ A.

Exercise 12.4–2 Suppose we also want to minimize the DC supply current in the preceding example. What should be the values of N, V_{CC}, and I_C?

Exercise 12.4–3 Let the circuit in Fig. 12.4–4a include an emitter resistor R_E with a bypass capacitor, so that the DC voltage drop across R_E results in $V_{CE} < V_{CC}$. Redraw Fig. 12.4–4b for this case, assuming $\beta \gg 1$ and $R_E < N^2R_L$. Then use the formulas stated in Exercise 12.4–1 to locate the Q-point for maximum symmetric signal swing.

PUSH-PULL AMPLIFIERS

The transformer-coupled amplifier, like our elementary amplifier, must be biased such that $I_C \geq I_m$, and we have maximum collector dissipation when the circuit is idling with no signal present. That situation is especially wasteful in view of the fact that typical nonsinusoidal signals have frequent gaps and low-level intervals and, in effect, are "off" as much as "on." For the modest price of one more transistor, we can greatly improve the actual efficiency of a power amplifier by using a *class B push-pull circuit* whose operating principles are illustrated by the circuit diagram and waveforms in Fig. 12.4–7.

The basic idea behind this circuit is to divide the work between T_1 and T_2, with T_1 amplifying the positive signal excursions and T_2 amplifying the negative excursions. Consequently, both transistors can be biased such that $i_C = 0$ and $P_{DC} = P_C = 0$ in the idle state ($v_{in} = 0$). With $v_{in} \neq 0$, the driver stage develops two voltages, $v_1 = V_T + v_{in}$ and $v_2 = V_T - v_{in}$, where V_T is the base-emitter threshold voltage of the transistors, assumed to be identical. It then follows that i_{B_1} and i_{C_1} are proportional to the positive variations of v_{in} but equal zero when $v_{in} < 0$; conversely, i_{B_2}

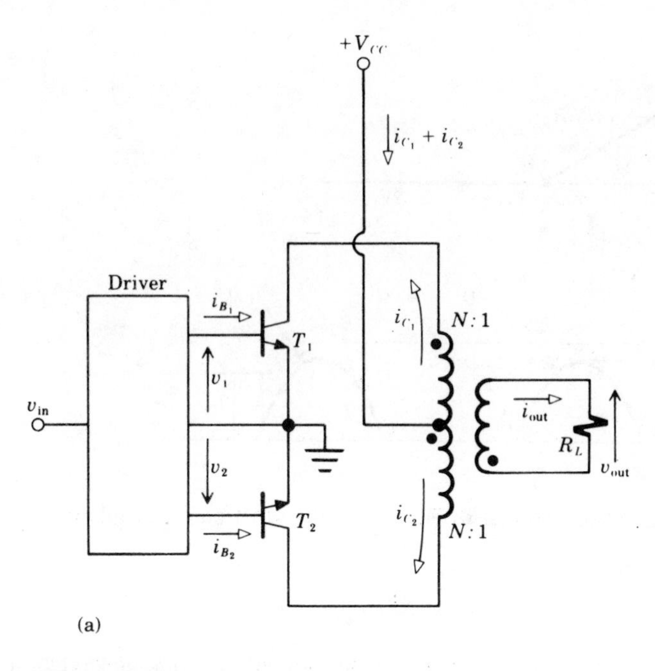

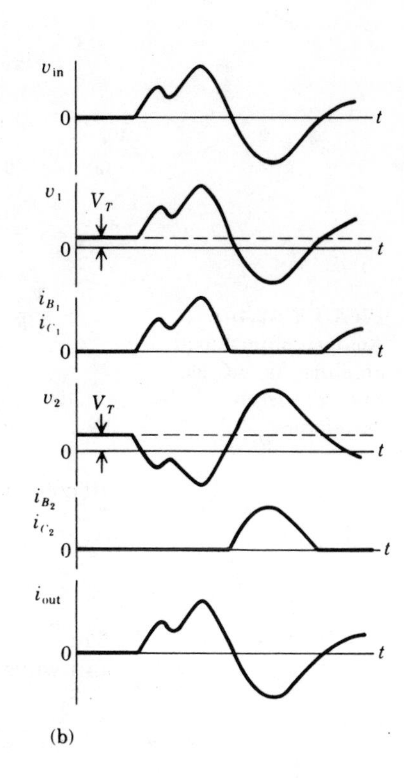

FIGURE 12.4–7
Push-pull amplifier:
(a) circuit;
(b) waveforms.

and i_{C_2} are proportional to the negative signal variations and equal zero when $v_{in} > 0$. Since i_{C_1} and i_{C_2} flow in opposite directions through their respective halves of the center-tapped output transformer, $i_{out} = Ni_{C_1} - Ni_{C_2} = N(i_{C_1} - i_{C_2})$ and v_{out} will have the same waveshape as v_{in}. On the other hand, the total current from the supply is $i_{C_1} + i_{C_2}$ and will be proportional to $|v_{in}|$, and similar to a full-wave rectifier.

We call this a *push-pull* circuit because T_1 "pushes" the positive excursions of the output current, while T_2 "pulls" the negative excursions. The designation *class B* means that each transistor conducts for only half a cycle when the input is sinusoidal. All the previous amplifiers we have examined are *class A,* which means that their transistors conduct throughout the full cycle of the signal.

To analyze the performance of the push-pull amplifier, we must first observe that each transistor has its Q point at $I_C = 0$, and $V_{CE} = V_{CC}$ and sees the referred AC resistance N^2R_L. The AC load line therefore takes the form of Fig. 12.4–8a. A sinusoidal input produces the half-wave rec-

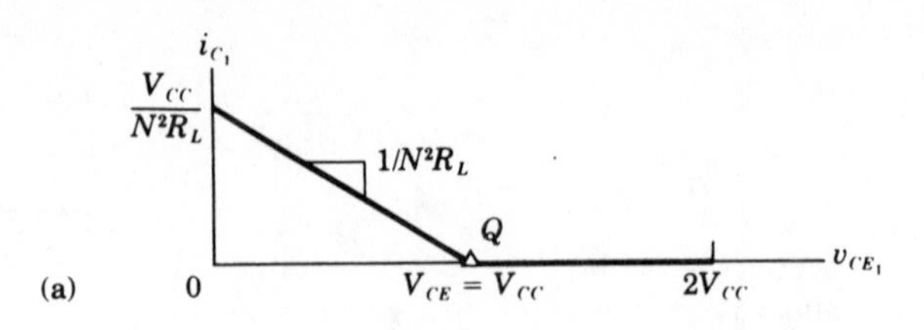

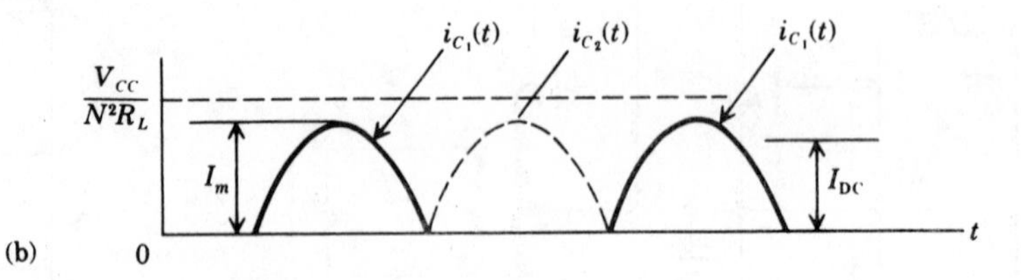

FIGURE 12.4–8 Analysis of push-pull amplifier: (a) AC load line; (b) current waveforms.

tified collector currents shown in Fig. 12.4–8b, with equal amplitudes

$$I_m \leq \frac{V_{CC}}{N^2 R_L} \tag{8}$$

Thus, the average (DC) current from the supply is the same as that in a full-wave rectifier:

$$I_{DC} = (i_{C_1} + i_{C_2})_{av} = \frac{2}{\pi} I_m$$

The average output signal power and DC supply power are then

$$P_{out} = \tfrac{1}{2}(NI_m)^2 R_L \leq \frac{V_{CC}^2}{2N^2 R_L} \tag{9}$$

$$P_{DC} = V_{CC} I_{DC} = \frac{2}{\pi} V_{CC} I_m \leq \frac{2V_{CC}^2}{\pi N^2 R_L} \tag{10}$$

giving an efficiency of

$$\text{Eff} = \frac{\pi N^2 R_L I_m}{4V_{CC}} \leq \frac{\pi}{4}$$

whose theoretical maximum is about 78.5%.

But the real pay-off comes in the form of greatly reduced collector dissipation. As noted above, $P_C = 0$ with no signal present. With a signal applied, we compute P_C from $2P_C = P_{DC} - P_{out}$, which takes account of both transistors. Simple calculus then shows that P_C has the maximum value

$$P_{CO} = \frac{V_{CC}^2}{\pi^2 N^2 R_L} \tag{11}$$

occurring when $I_m = 2V_{CC}/\pi N^2 R_L$. Therefore, the figure of merit for a class-B, push-pull amplifier is

$$\frac{P_{CO}}{P_{\text{out-max}}} = \frac{2}{\pi^2}$$

an improvement by roughly a factor of 10 compared to class A amplifiers. The practical impact of this result is that a push-pull amplifier using two of the 10-W transistors from Example 12.4–2 could deliver much more power than the 5-W output of a transformer-coupled amplifier.

The required driver stage can be implemented with the addition of another center-tapped transformer or a *phase splitter*—the latter being the choice in our illustrative audio amplifier of Fig. 12.0. Alternatively, the need for transformers is avoided altogether if we use a pair of *complementary* transistors, as in Fig. 12.4–9a. In this circuit T_1 is an *npn* transistor, whereas T_2 is a *pnp* transistor with characteristics identical to those of T_1, except that i_{B_2} and i_{C_2} are negative quantities that naturally follow the negative signal excursions. The two diodes provide the base-emitter offset voltages equivalent to V_T in Fig. 12.4–7b. The functions of C_i, R_B, and C_o should be apparent.

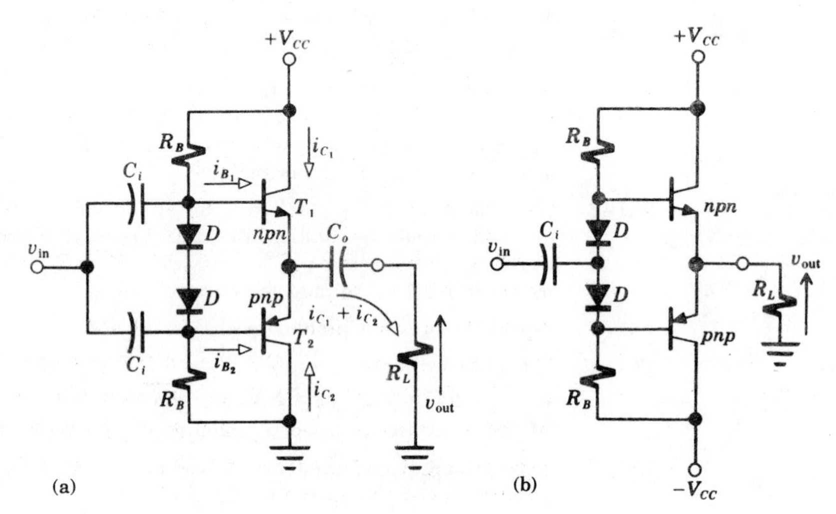

FIGURE 12.4–9 Complementary-symmetry push-pull amplifiers: (a) *RC*-coupled input; (b) direct-coupled input.

Eliminating the output transformer reduces cost and improves the frequency response. Going a step further, Fig. 12.4–9b shows a *direct-coupled* complementary-symmetry push-pull amplifier whose low-frequency response will be superior to the *RC*-coupled circuit (why?). Like all direct-coupled amplifiers, it requires two supply voltages to produce positive and negative output variations.

Exercise 12.4–4

Consider a push-pull amplifier using two transistors whose specifications are stated in Example 12.4–2.

(a) Show that the voltage and current limitations require that $P_{\text{out}} \leq 25$ W.

(b) If $R_L = 8\ \Omega$ and $P_{\text{out}} = 25$ W, what are the corresponding values of V_{CC}, N, P_{DC}, and P_{CO}?

PROBLEMS

12.1–1 In Fig. 12.1–1a, let $R_1 = 600$ kΩ, $R_2 = 300$ kΩ, $R_D = 1$ kΩ, and $V_{DD} = 9$ V. Find the operating point, g_m, and μ, assuming that the transistor has $K = 15$ mA, $V_T = 4$ V, and $r_d = 50$ kΩ.

12.1–2 Repeat the previous problem with $K = 12$ mA and $V_T = 3.5$ V.

12.1–3 Consider a three-stage version of Fig. 12.1–4 with $R_1 = 200$ kΩ, $R_2 = 120$ kΩ, $R_D = 27$ kΩ, $R_s = 1$ kΩ, $R_L = 10$ kΩ, $V_{DD} = 6$ V, $K = 5$ mA, $V_T = 2$ V, and $r_d = 100$ kΩ. Find the overall gain.

12.1–4 In Fig. 12.1–6, let $V_{DD} = 12$ V, $R_1 = 300$ kΩ, $R_2 = 900$ kΩ, and $R_K = R_D = 2$ kΩ. The transistor is a JFET with $V_P = -2$ V.

(a) Find the value of I_{DSS} so that $V_K = 4$ V.

(b) Find the operating point if I_{DSS} is twice the value found in (a).

(c) Make a sketch like Fig. 12.1–7 for the conditions in (a) and (b).

12.1–5 Redraw Fig. 12.1–7 for a JFET whose transfer curve is given in Fig. 10.3–5b. Extend the bias line to get Q at $V_{GS} < 0$. Then draw another bias line through Q and the origin, corresponding to the circuit in Fig. 12.1–5. Which bias circuit has less sensitivity to the JFET parameters?

12.1–6 The transistor in Fig. 12.1–6 is an enhancement MOSFET with $K = 20$ mA, $V_T = 2.5$ V, and $r_d = 40$ kΩ. Given that $V_{DD} = 30$ V and $I_D = 4$ mA, select values for the resistors to obtain $A_v \approx -10$ when $R_s = R_L = 5$ kΩ. Draw the circuit diagram and label all voltage and current values.

12.1–7 Repeat the previous problem with $K = 10$ mA.

12.1–8 In Fig. 12.1–9, let $V_{CC} = 10$ V, $V_T = 0.7$ V, $\beta = 100$, and $R_C = 1$ kΩ.

(a) Find R_B so that $V_{CE} = 5$ V, and calculate the corresponding value of g_m.

(b) What are the values of V_{CE} and g_m if β increases by 50%?

12.1–9 Repeat the previous problem with the upper end of R_B connected to the collector terminal rather than going to V_{CC}.

12.1–10 Two BJT amplifiers like Fig. 12.1–12a are connected in cascade with $R_C = R_E =$

8 kΩ, R_1 = 47 kΩ, R_2 = 70 kΩ, V_{CC} = 12 V, V_T = 0.7 V, and β = 100. The DC voltage at the collector is 8 V. Find the overall gain with R_L = 8 kΩ and R_s = 0.

12.1–11 In Fig. 12.1–12, let V_{CC} = 12 V, V_T = 0.7 V, β = 100, R_C = 1 kΩ, R_E = 400 Ω, and R_1 = 20 kΩ. Also let V_C be the DC voltage at the collector.

(a) Find R_2 so that V_C = 7 V.

(b) Find the new value of V_C if β doubles.

12.1–12 Repeat the previous problem with β = 50.

12.1–13 Derive Eqs. (15b) and (15c) from Fig. 12.1–12b.

12.1–14 The transistor in Fig. 12.1–12 has β = 100, and V_T = 0.7 V. Given that V_{CC} = 15 V and I_C = 2.5 mA, select values for the resistors to obtain $A_v \approx -30$ when $R_s = R_L$ = 1 kΩ. Draw the circuit diagram and label all DC voltage and current values.

12.1–15 Repeat the previous problem with β = 80 and I_C = 2 mA.

12.2–1 Suppose $R_B = 10R_E$, β = 100, and $R_E \gg 1/g_m$ in Fig. 12.2–1. Find values for R_E and R_C so that $R_i \approx$ 10 kΩ and $\mu \approx 5$. Also obtain the condition on I_C so that $R_E \gg 1/g_m$.

12.2–2 Repeat the previous problem with $R_B = 20R_E$.

12.2–3 The FET in Fig. 12.2–2 has g_m = 5 m℧ and I_D = 2 mA. Select values for R_D and R_K so that $\mu \approx 4$ and $V_{DS} \geq 6$ V when V_{DD} = 20 V. Then calculate μ with g_m reduced by 50%.

12.2–4 Repeat the previous problem with g_m = 6 m℧ and I_D = 2.5 mA.

12.2–5 Obtain an expression for v_f/v_{in} in Fig. 12.2–2b, including the effect of r_d. Then show that

$$\mu = \frac{R_D}{R_K + (R_K + R_D + r_d)/g_m r_d}$$

12.2–6 Both transistors in Fig. P12.2–6 have β = 100 and V_T = 0.7 V.

(a) Find R_2 so that I_1 = 1 mA, and show that $I_2 \approx$ 22 mA.

(b) Evaluate v_{out}/v_s.

12.2–7 Repeat Problem 12.2–6(b) with C_1 removed. Use Eq. (4b) for T_1.

12.2–8 Let the BJTs in Fig. P12.2–6 be replaced with enhancement MOSFETS with K = 50 mA, V_T = 2.5 V, and r_d = 40 kΩ.

(a) Find R_2 so that I_1 = 1 mA, and show that $I_2 \approx$ 20 mA.

(b) Evaluate v_{out}/v_s. Hint: See Exercise 12.2–3.

12.2–9 Repeat Problem 12.2–8(b) with C_1 removed. Use the expression in Problem 12.2–5 for T_1.

12.2–10 Suppose the FETs in Fig. 12.2–8 have g_m = 4 m℧ but unequal dynamic resistances r_{d_1} = 50 kΩ and r_{d_2} = 30 kΩ.

(a) Find v_{d_2} in terms of v_p and v_n when R_D = 10 kΩ.

(b) Let $v_p = v_2 + \frac{1}{2}v_1$ and $v_n = v_2 - \frac{1}{2}v_1$ to put your result in the form $v_{d_2} = A_1 v_1 + A_2 v_2$, where A_1 is the *differential gain* and A_2 is the *common-mode gain*. Evaluate A_1/A_2.

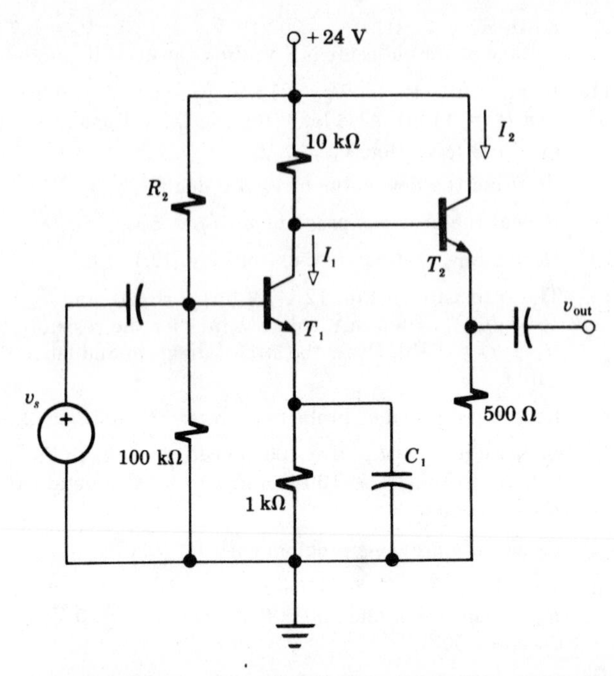

FIGURE P12.2–6

12.2–11 Suppose the current source in Fig. 12.2–7 is replaced by a 10-kΩ resistance, which then appears between S and ground in Fig. 12.2–8. Repeat the previous problem with identical FETs that have $g_m = 5$ m℧ and $r_d = 40$ kΩ.

‡ **12.2–12** Let the FETs in Fig. 12.2–7 be replaced with BJTs. The element values are $V_{DD} = -V_{EE} = 15$ V, $R_D = 20$ kΩ, $R_E = 5$ kΩ, $R_1 = 30$ kΩ, and $R_2 = 60$ kΩ. All transistors have $\beta = 100$ and $V_T = 0.7$ V.

(a) Show that $I_C \approx 0.83$ mA.

(b) Find $(v_{d_2} - v_{d_1})/v_{in}$ and $R_i = v_{in}/i_{in}$ when $v_{in} = v_p - v_n$.

12.3–1 Select values for the capacitors so that the JFET amplifier in Example 12.1–2 has $f_\ell \approx 40$ Hz when $R_s = R_L = 5$ kΩ.

12.3–2 Repeat the previous problem for the BJT amplifier in Example 12.1–3.

12.3–3 The amplifier in Fig. 12.0 has $R_s = R_1 = R_2 = R_3 = 100$ kΩ, $C_1 = 500$ pF, and $C_2 = 0.05$ μF. The tone switch is closed and the volume control is set at its maximum.

(a) Estimate f_ℓ if C_1 acts like an open circuit.

(b) Estimate f_u if C_2 acts like a short circuit.

12.3–4 Repeat the previous problem with the volume control set at the midpoint of the potentiometer.

12.3–5 The amplifier in Example 12.1–2 is driven from a source with $R_s = 20$ kΩ and has no external load. A 500-pF capacitor is to be added to reduce f_u. The FET capacitances are negligible. Find f_u when the capacitor is connected from D to ground, from G to ground, and from G to D.

12.3–6 Five identical amplifiers like the circuit in Example 12.3–2 are connected in cascade. Find the overall gain and f_u when each stage has $R'_L = 200\ \Omega$.

12.3–7 Repeat the previous problem with $R'_L = 400\ \Omega$.

‡ **12.3–8** Estimate f_u for an emitter follower with $R_E = R'_s = r_\pi = 1\ \text{k}\Omega$, $C_\pi = 100$ pF, $C_\mu = 4$ pF, $\beta = 100$, and $R_L \gg R_E$. Hint: Draw the equivalent circuit.

12.4–1 Derive the expressions for V_{CE} and I_C given in Exercise 12.4–1.

12.4–2 Suppose $I_C = 1$ A in Fig. 12.4–3a, and R_C is replaced with a large inductance. Find the maximum AC output power and the efficiency.

12.4–3 In Fig. 12.4–4a, let $V_{CC} = 24$ V, $I_C = 1$ A, $N = 3$, and $R_L = 4\ \Omega$. Draw the AC and DC load lines. Then find the maximum AC output power and the corresponding efficiency. What value of N would be optimum if nothing else is changed?

12.4–4 Repeat Example 12.4–2 with a 100-Ω load.

12.4–5 Repeat Example 12.4–2 with a 2-Ω load.

12.4–6 Suppose the circuit in Fig. 12.4–4a includes $R_E = 10\ \Omega$ with a bypass capacitor. The maximum AC output power is 3 W, $R_L = 6\ \Omega$, and $N = 2.5$. Find V_{CC} if the operating point is at the midpoint of the AC load line.

12.4–7 Find the maximum efficiency of a class B push-pull amplifier when the input is a square wave like Fig. 7.4–1b.

12.4–8 The phase splitter in Fig. 12.2–6 is to be used in the driver of a push-pull amplifier. Find the value of R for maximum symmetric swing of v_1 and v_2 when $V_{CC} =$ 24 V and $I_C = 2$ mA. You may ignore V_{sat} and the offset V_T.

‡ **12.4–9** Find the efficiency of a class B push-pull amplifier operated at 50% of its maximum output power. The input is sinusoidal.

PART III

Feedback and digital control ensure that this industrial robot makes rapid and precise spot welds. (Courtesy of Cincinnati Milacron)

ANALOG AND DIGITAL SYSTEMS

13

FEEDBACK AND CONTROL SYSTEMS

Having considered various devices and circuits that form the building blocks of an electrical system, we now shift our emphasis to the broader perspective of *systems engineering*. This means that we will be concerned with the interconnection of functional blocks to achieve a specified overall system performance, rather than with the internal details of the individual blocks. And, of course, we'll work primarily with block diagrams rather than circuit diagrams.

Most systems of any complexity involve the use of *feedback*. In these systems, a measure of the output is fed back and combined with the input to yield an overall performance that would be difficult or costly to achieve without the help of feedback. We have already seen some of the benefits of feedback in electronic circuits. Here we discuss feedback systems in general. Our topics include the distinction between positive and negative feedback, the advantages and disadvantages of feedback, and the role of feedback in *automatic control systems*. An optional section looks at the problem of stability and oscillations in feedback systems.

OBJECTIVES

After studying this chapter and working the exercises, you should be able to do each of the following:

- Describe the general structure of a feedback system and the functions of its building blocks, distinguishing between positive and negative feedback (Section 13.1).
- List the specific advantages and disadvantages of feedback (Sections 13.1 and 13.2).
- Find the overall gain for a system having one or more feedback loops (Sections 13.1 and 13.2).
- Design a simple feedback system to achieve specified closed-loop performance relative to parameter variation, disturbances, or dynamic response (Section 13.2).
- State the conditions for stability in a negative feedback system and for oscillation in a positive feedback system (Section 13.3). †

13.1 FEEDBACK CONCEPTS

Suppose we want to design a system whose output $y(t)$ has a prescribed linear relationship to the input $x(t)$. If we connect the system's building blocks in cascade, as in Fig. 13.1–1a, the signal flows directly through them, and the output of any one unit depends entirely on its input and not on the effects of subsequent blocks. Consequently, we can do little or nothing to compensate for any unforeseen imperfections or disturbances in the system. We can minimize such problems, however, by adding a feedback path from output to input, as in Fig. 13.1–1b, so that unwanted output perturbations are corrected after they are compared with the input.

Figure 13.1–1b shows the general structure and essential features of any feedback system. There is a *forward path* that approximates the in-

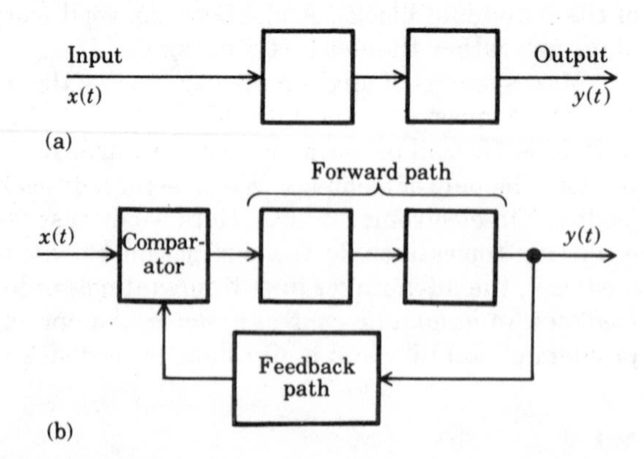

FIGURE 13.1–1
(a) Cascade system.
(b) Feedback system.

tended input-output relationship and, perhaps, consists of two or more cascaded units. There is a *feedback path* that conveys some measure of the actual output back to the input. And there is a *comparator* whose output drives the forward path by comparing the input with the information from the feedback path. The nature of the comparison determines whether the system has positive or negative feedback.

NEGATIVE FEEDBACK

Comparison takes the form of *subtraction* in a negative feedback system, the comparator's output being proportional to the *difference* between the input and feedback signals. If the output is too small, subtraction yields a large positive difference or error signal which, applied to the forward path, tends to increase the output—and vice versa when the output is too large. Negative-feedback systems are *self-correcting* or *goal-seeking* in that the error signal automatically compensates for various forward-path imperfections.

Consider the simple but highly informative example of the feedback amplifier diagrammed in Fig. 13.1–2. This system is intended to produce the amplified output $y = Kx$. It consists of a forward-path amplifier having gain $y/x_e = A \gg K$, a voltage or current divider in the feedback path giving the *feedback signal* $y_f = y/K$, and a subtractor forming the *error signal* $x_e = x - y_f$. By convention the subtractor is shown as a summing junction (a summation sign inside a circle) with a minus sign to indicate subtraction.

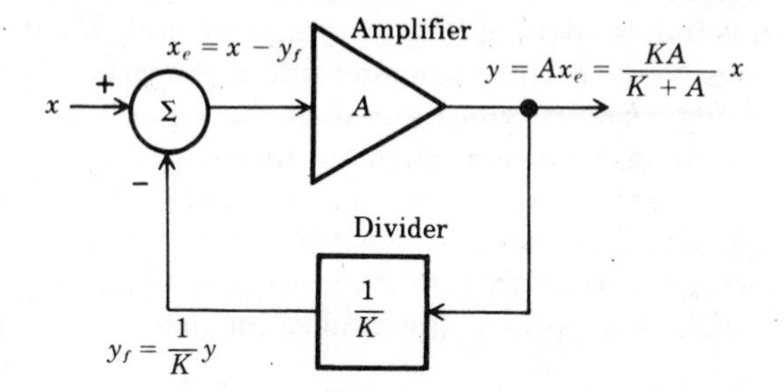

FIGURE 13.1–2
Amplifier with negative feedback.

Since $y_f = y/K$, the summing-junction equation is

$$x_e = x - y_f = x - \frac{1}{K}y$$

and, at the output,

$$y = Ax_e = A\left(x - \frac{1}{K}y\right) = Ax - \frac{A}{K}y$$

Regrouping terms then yields

$$y = \frac{A}{1 + (A/K)} x = \frac{KA}{K + A} x \tag{1}$$

But $A \gg K$ implies that $K + A \approx A$ and, to a good approximation,

$$y = Kx \tag{2}$$

Similar manipulations were used in Section 4.3 to analyze the noninverting op-amp circuit, which was our first encounter with negative feedback. Here we have broadened the scope to include *any* amplifier that may be modeled by Fig. 13.1–2.

The deceptively simple result of Eq. (2) has tremendous significance, not so much in what it contains but in what it does *not* contain—the amplifier gain A. Most op-amps and other types of electronic amplifiers without feedback are somewhat like dragons: They are *powerful* ($A \gg 1$), but *unreliable* (the gain may change unpredictably with time, temperature, frequency, and so forth). Voltage or current dividers, on the other hand, are *meek* ($1/K < 1$) but *reliable*. Combining these two units in a negative-feedback configuration "tames the dragon" and yields a compromise system having *moderate* but *reliable* gain K.

We say that negative feedback has reduced the *sensitivity* to forward-path parameter variations in that changing values of A do not affect the output, providing that $A \gg K$. By the same token, the system's *bandwidth* may be greater than that of the amplifier alone since K is independent of frequency and y equals Kx at any frequency f, as long as $A \gg K$. A third dragon-taming property of feedback is the *reduction of disturbances,* since any disturbance voltage occurring in the forward path tends to be cancelled out by comparison to the input.

Generalizing from this example, we conclude that a reliable negative-feedback path with gain $1/K < 1$ wrapped around an unreliable forward path with gain $A \gg 1$ comes close to giving us the best of both worlds. The specific advantages include:

- Reduced sensitivity to parameter variations in the forward path.
- Improved frequency response and increased bandwidth.
- Reduction of disturbances in the forward path.

In addition, feedback can be arranged so that it modifies the input and output resistances of an amplifier circuit as we noted in Section 12.2.

The price we pay for these benefits—and a minor disadvantage of negative feedback—is lower *overall amplification.* A more serious disadvantage is that negative feedback has the potential to turn into positive feedback.

Example 13.1–1 The subtraction required for negative feedback may be achieved through differential amplification (as in op-amp circuits), or it may be achieved by taking advantage of the inherent polarity inversion of transistors and other amplifying devices. As an example of the latter, consider the FET feedback amplifier in Fig. 12.2–2 whose small-signal circuit model is repeated as Fig. 13.1–3a.

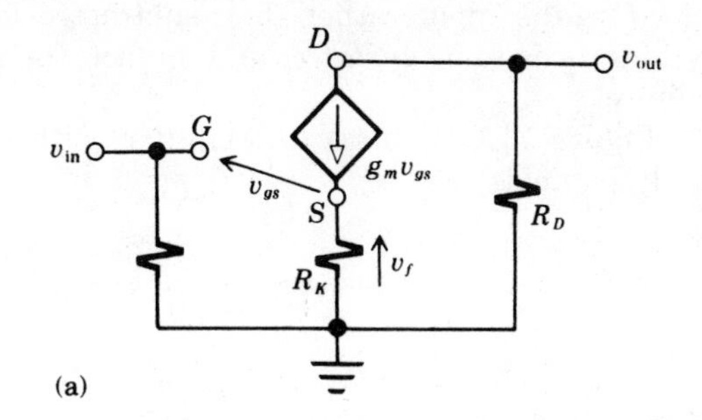

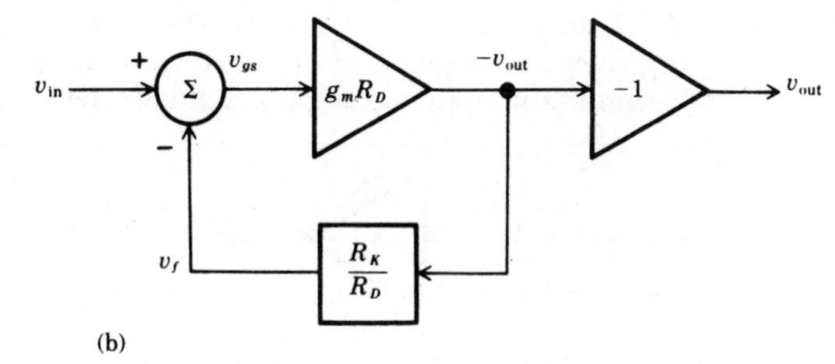

FIGURE 13.1–3 FET feedback amplifier: (a) circuit model; (b) block-diagram model.

Analyzing this circuit from our current perspective, we note that

$$v_{gs} = v_{in} - v_f \qquad v_{out} = -g_m R_D v_{gs} \qquad v_f = g_m R_K v_{gs} = \frac{R_K}{R_D}(-v_{out})$$

If we interpret v_f and v_{gs} as feedback and error signals, respectively, these equations logically lead us to the block-diagram model of Fig. 13.1–3b where the polarity inversion has been represented by an inverting amplifier at the output. Comparing this diagram with that of Fig. 13.1–2, we see that the forward and feedback gains are $A = g_m R_D$ and $1/K = R_K/R_D$. Thus, from Eq. (1),

$$v_{out} = -\frac{g_m R_D}{1 + g_m R_K} v_{in}$$

and $v_{out} \approx -(R_D/R_K)v_{in}$ if $g_m R_K \gg 1$ (so $A \gg K$) Under the latter condition, the amplifier's performance is independent of any variations of the FET's transconductance g_m.

The final expression for v_{out} could also have been obtained by manipulating the circuit equations from Fig. 13.1–3a. But the block-diagram model of Fig. 13.1–3b more clearly brings out the presence of feedback and the influence of the circuit parameters.

POSITIVE FEEDBACK

Positive feedback refers to the situation in which a feedback signal is *added* to the input rather than subtracted from it. Consequently, the system tends to be *unstable;* and, in fact, the gain may actually become infinite!

Figure 13.1–4 shows an amplifier with positive feedback. Routine analysis yields

$$y = \frac{A}{1 - (A/K)} x = \frac{KA}{K - A} x \tag{3}$$

which is identical to Eq. (1) except for the sign in the denominator. Rather confusingly, the *negative sign* means *positive* feedback, and vice versa. That negative sign turns out to be the crux of the problem, for if $A = K$ then $KA/(K - A) \to \infty$ and the system has no value as an amplifier, since a vanishingly small input would drive the forward-path amplifier into saturation and produce a grossly distorted output.

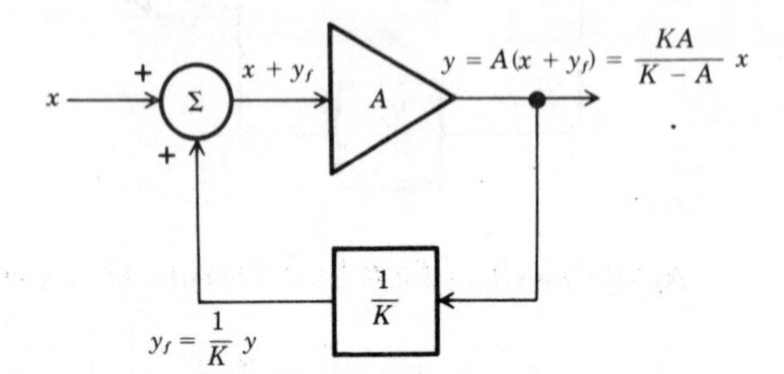

FIGURE 13.1–4 Amplifier with positive feedback.

To illustrate this effect, let's suppose that $A = K$ in Fig. 13.1–4. Any applied input value x initially produces $y = Ax$ and $y_f = (1/K)\,Ax = x$, so $x_e = x + x = 2x$. But then $y = A(2x)$, which is fed back to produce $y_f = 2x$, and $x_e = x + 2x = 3x$—and so forth around and around the loop with x_e and y continually increasing. All this takes place virtually instantaneously and, in theory, the output would become *infinitely large.* Growth of the output in an actual system is limited by nonlinearities or system self-destruction whichever happens first.

Given the inherent unreliability of forward-path amplifiers, we can seldom be sure of avoiding the condition $A = K$. Therefore, positive feedback is intentionally used only in very special applications. One such application is the *positive-feedback oscillator,* where infinite gain produces self-sustaining output oscillation without any input at all. The annoying whistle heard from a public-address system when the microphone is too close to the loudspeaker demonstrates unwanted positive-feedback oscillation in action.

Exercise 13.1–1

Calculate y/x for a negative-feedback amplifier with $K = 5$ when $A = 400$, 200, and 10. Repeat for positive feedback when $A = 400$, 5.2, and 4.8.

AUTOMATIC CONTROL AND OTHER FEEDBACK SYSTEMS

The advantages of negative feedback we have seen in amplifier design play an even greater role in *automatic control systems* where the forward path includes a physical process often more dragon-like than those of electronic amplifiers. In fact, feedback's self-correcting nature puts the *automatic* into automatic control. Applications range from consumer appliances, to manufacturing and chemical processing, to space vehicle guidance.

Consider, for example, a home heating system represented schematically by Fig. 13.1–5. You set the desired temperature on the thermostat, where it is compared with the actual temperature. If the actual temperature is too low, the thermostat turns on the furnace, which supplies heat until the temperature has risen appropriately. Subsequent changes—say, due to a drop in outside temperature—are sensed by the thermostat and automatically corrected. In a similar way, the exposure-control system built into a camera adjusts the iris until the light level, as sensed by a photocell, equals the value required for the film speed and shutter setting.

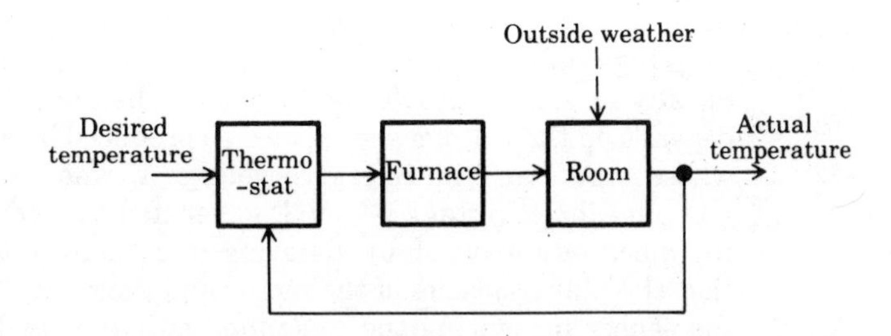

FIGURE 13.1–5 Temperature control system.

These two examples indicate certain differences between feedback control systems and feedback amplifiers. The general block diagram of a control system in Fig. 13.1–6 makes those differences more apparent. Here, the forward path consists of a process or *plant* performing the useful work under the governance of a *controller.* The *controlled output* from the plant is monitored by a transducer or *sensor* that generates a

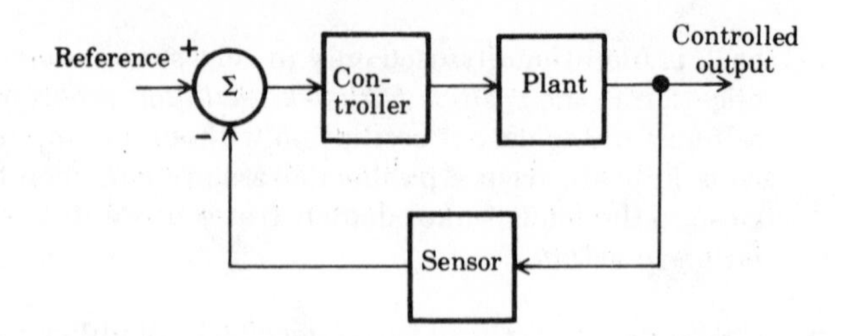

FIGURE 13.1–6 Automatic control system.

feedback signal for comparison with the desired or *reference* value. The various "signals" can take many different physical forms—temperature, light level, even the position of a knob. The system could even be totally devoid of electronic components. Nonetheless, the design strategies for electrical feedback systems are mathematical tools that apply equally well to nonelectrical systems.

Besides the categories just mentioned, feedback occurs naturally or can be incorporated into such diverse settings as physiology, economics, ecology, and politics. The human body contains literally hundreds of complex feedback systems controlling eye focus, body temperature, etc. (Certain physical disorders have been traced to the interruption of the feedback path or to positive-feedback instabilities.) The economic system, too, has numerous feedback loops interrelating money supply, interest rates, prices and wages. Thus it is that electrical engineers have found new roles in the emerging fields of biomedical engineering, urban studies, and public systems.

Example 13.1–2
Speed control system

Figure 13.1–7a represents an electromechanical system with feedback controlling the rotational speed y of a certain machine driven by a DC motor. Under steady-state conditions, the motor speed is given by $y = (\beta i + x_d)/D$, where i is the motor current, β the electromechanical coupling coefficient, and D the friction drag constant. The term x_d stands for any external disturbance torque on the machine. The motor current depends on the voltage $v = A(x - y_f)$ produced by a differential amplifier whose inputs are the reference voltage x, which sets the desired speed, and a feedback voltage $y_f = y/K$ generated by a tachometer connected to the machine's drive shaft. Relating this diagram to Fig. 13.1–6, we see that the plant consists of the motor and machine, that the tachometer is the sensor measuring the controlled output y, and that the differential amplifier serves as the controller, as well as performing the comparison between the reference input x and the feedback voltage y_f.

If the motor operates in the armature-controlled mode, its steady-state equivalent circuit takes the form of Fig. 13.1–7b. Here, the controlled voltage source $v_a = \beta y$ represents the so-called "back emf" of the armature, with β being the same coupling coefficient as in the speed

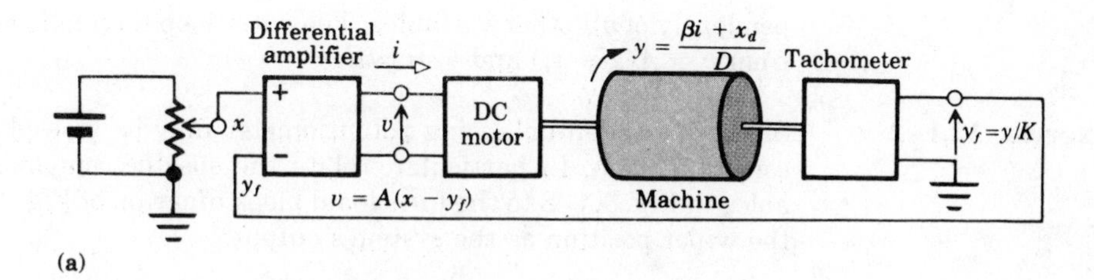

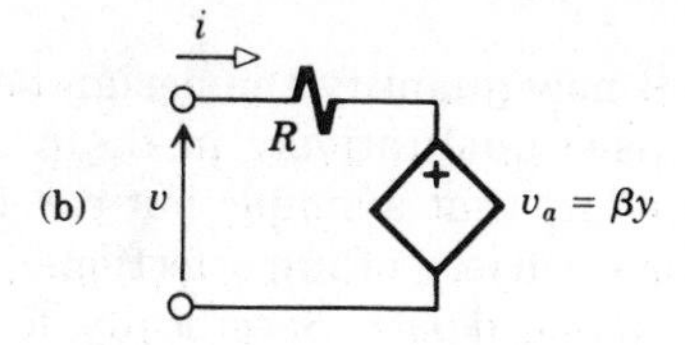

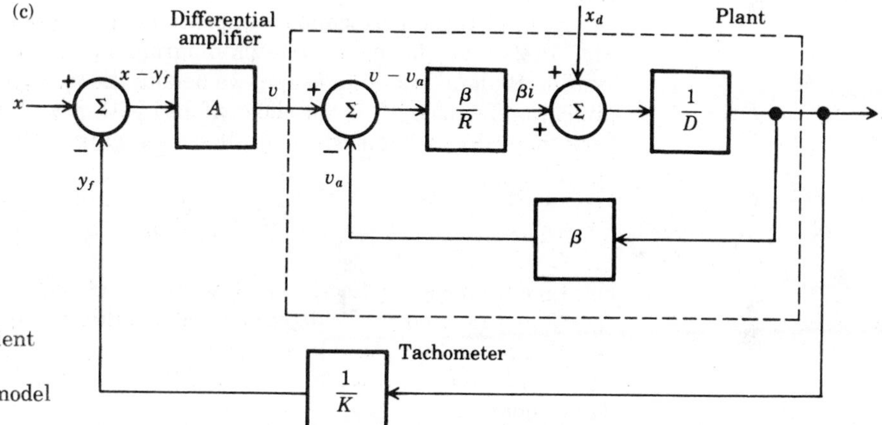

FIGURE 13.1–7
(a) Speed control system. (b) Equivalent circuit of motor. (c) Block-diagram model of system.

equation. This back emf will be discussed later in conjunction with electromechanical energy conversion. For the present, we point out that it results in an additional feedback path because the speed y depends on i; but, from the equivalent circuit, $i = (v - v_a)/R$, which depends on y.

To devise a block-diagram model of the entire system, bringing out both feedback paths and the effect of the disturbance torque x_d, we start with the plant (motor plus machine) and write

$$y = \frac{1}{D}(\beta i + x_d) \qquad \beta i = \frac{\beta}{R}(v - v_a) \qquad v_a = \beta y$$

which yields the inner loop in Fig. 13.1–7c. Note that the plant has one output y but *two* inputs, v and x_d, because the disturbance x_d acts independently of all other variables. The outer loop then follows immediately from $v = A(x - y_f)$ and $y_f = y/K$.

Exercise 13.1–2

Show that a self-balancing potentiometer may be viewed as a feedback control system. In particular, relate the specific components and variables in Fig. 3.4–8 to the functional block diagram of Fig. 13.1–6, taking the wiper position as the system's output.

13.2 FEEDBACK SYSTEM ANALYSIS

We'll now quantify the design benefits of negative feedback we discussed qualitatively in the previous section. Those benefits stem from the fact that a simple but carefully chosen feedback path compensates for a variety of imperfections in the more complicated and unpredictable forward path. Specifically, feedback reduces sensitivity to parameter variations and disturbance inputs, while it improves the dynamic behavior.

Each of these performance improvements is investigated here with Fig. 13.2–1 as the general block-diagram model of a linear feedback system. By long-standing usage, we denote the forward-path gain or amplification factor by G (not to be confused with power gain) and the feedback factor by H. Thus, the overall system gain with feedback is

$$G_F = \frac{G}{1 + GH} \tag{1}$$

obtained from Eq. (1), Section 13.1, with $A = G$ and $1/K = H$. In general, G_F, G, and H are *transfer functions* and should be written as $G_F = G_F(s)$, etc., to account for possible energy storage. This transfer-function notation will become important when we get to frequency response and transient behavior.

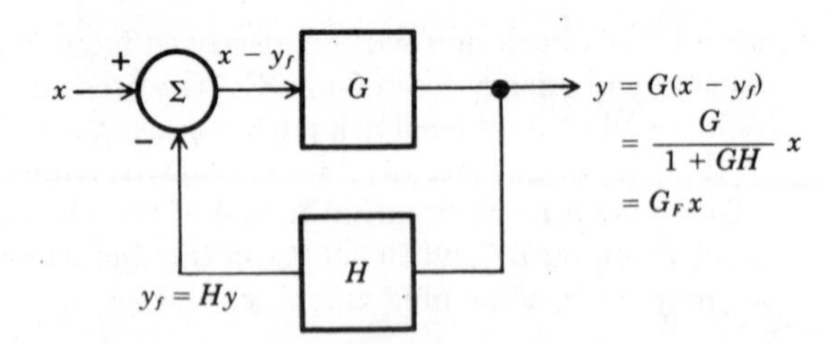

FIGURE 13.2–1 General feedback system.

PARAMETER VARIATIONS AND SENSITIVITY

Suppose we need several voltage amplifiers with gain $K(1 \pm \epsilon)$, where $\pm K\epsilon$ is the permissible small gain difference among them. Further, suppose we have available a batch of simple transistor amplifiers with gain G anywhere in a wide range $G_{\min} \leq G \leq G_{\max}$ caused by the variation of transistor parameters from unit to unit. Clearly, this task is made-to-order for the use of feedback, providing certain conditions are met.

To carry out the design we start with the feedback-amplifier gain $G_F = G/(1 + GH)$ and note that the feedback gain H must be such that

$$K(1 - \epsilon) \leq \frac{G}{1 + GH} \leq K(1 + \epsilon) \tag{2}$$

If we assume that $G_{\min} > K > 1$, Eq. (2) leads to *upper* and *lower bounds* on H, namely

$$\frac{1}{K(1 + \epsilon)} - \frac{1}{G_{\max}} \leq H \leq \frac{1}{K(1 - \epsilon)} - \frac{1}{G_{\min}} \tag{3}$$

If the upper bound exceeds the lower bound, then any value of H in between meets the requirements and all the feedback amplifiers will satisfy Eq. (2) despite the variation of G.

For a numerical illustration, take $K = 20$, $\epsilon = 0.05$ ($\pm 5\%$), $G_{\min} = 300$ and $G_{\max} = 600$. Inserting these into Eq. (3) yields

$$\frac{1}{21} - \frac{1}{600} = 0.0459 \leq H \leq \frac{1}{19} - \frac{1}{300} = 0.0493$$

and if we use $H = 0.048$, we get $G_{\max}/(1 + G_{\max}H) = 20.13$ and $G_{\min}/(1 + G_{\min}H) = 19.48$. Therefore, $19.48 \leq G_F \leq 20.13$, even though $300 \leq G \leq 600$. Since $H < 1$, the feedback network can be our good old reliable voltage divider.

This strategy has obvious merit for mass-production manufacturing. But even in one-of-a-kind applications, feedback helps reduce the effect of slowly changing parameter values in the forward path that are caused by aging, temperature changes, and so forth. Specifically, a small *fractional change* dG/G of the forward-path gain produces a fractional change for the overall system given by

$$\frac{dG_F}{G_F} = S\,\frac{dG}{G} \tag{4a}$$

where

$$S = \frac{1}{1 + GH} \tag{4b}$$

which is called the *sensitivity* of G_F with respect to G. Since G is a changing parameter, we use its nominal or average value to calculate the sensitivity S. The feedback gain H, of course, is assumed to be constant.

Negative-feedback systems usually have $1 + GH \gg 1$, so $S \ll 1$ and the overall gain G_F hardly changes at all, despite the variations of G in the forward path. For instance, suppose a chemical process suffers from random temperature changes that cause G to drift over the range from 45 to 55. Taking $G_{av} = 50$, the fractional change is $dG/G \approx (55 - 45)/50 = 0.2 = 20\%$—this being such a large variation that we can only approximate the differential change. If the system has a constant feedback gain of $H = 0.18$, then $1 + GH \approx 1 + 50 \times 0.18 = 10$, $G_F \approx \frac{50}{10} = 5$, and Eq. (4) gives $S \approx \frac{1}{10}$ and $dG_F/G_F \approx \frac{1}{10} \times 0.2 = 2\%$. Therefore, G_F will stay within the range of 4.95 to 5.05. Figure 13.2–2 shows the variations of G and G_F versus temperature.

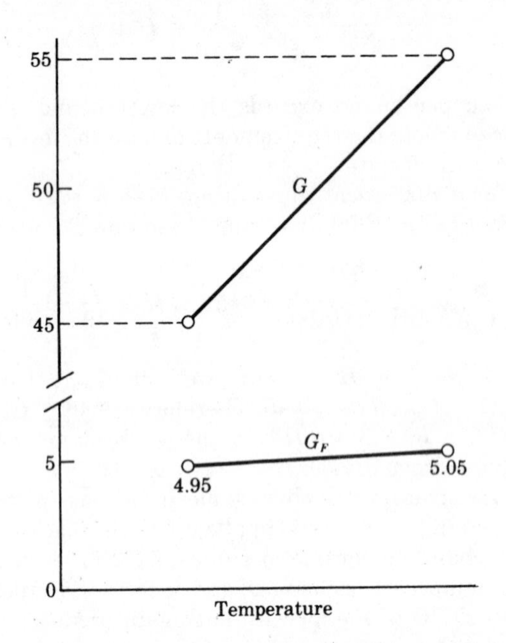

FIGURE 13.2–2

To derive the sensitivity equation, we use Eq. (4a) in the form

$$S = \frac{dG_F/G_F}{dG/G} = \frac{G}{G_F}\frac{dG_F}{dG} \tag{5}$$

which is the basic definition of sensitivity. Differentiating G_F with respect to G yields

$$\frac{dG_F}{dG} = \frac{d}{dG}\left(\frac{G}{1 + GH}\right) = \frac{(1 + GH) - GH}{(1 + GH)^2} = \frac{1}{(1 + GH)^2}$$

and therefore

$$S = G\frac{1 + GH}{G}\frac{1}{(1 + GH)^2} = \frac{1}{1 + GH}$$

in agreement with Eq. (4b).

REDUCTION OF DISTURBANCES

Often a *disturbance signal* x_d enters somewhere in the forward path of a feedback system—produced, perhaps, by internal noise, external interference, or distortion. We model this as diagrammed in Fig. 13.2–3a, where the forward-path gain G has been broken into two blocks with $G_1G_2 = G$ and the disturbance injected between them.

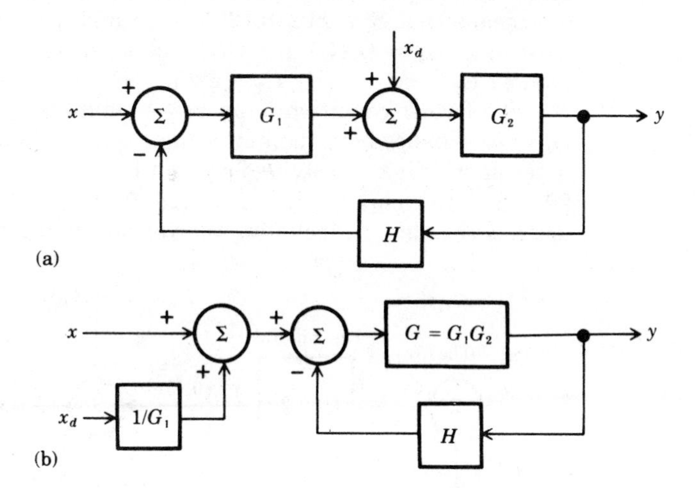

FIGURE 13.2–3
(a) Feedback system with disturbance x_d in forward path;
(b) Equivalent diagram.

Simple manipulation of equations readily leads to the equivalent configuration shown in Fig. 13.2–3b, where x_d enters at the input through a gain $1/G_1$. The total output is thus

$$y = \frac{G}{1 + GH}\left(x + \frac{1}{G_1}x_d\right) = G_F x + \frac{G_F}{G_1}x_d \tag{6}$$

This result, or Fig. 13.2–3b, shows that feedback has reduced the disturbance by the factor $1/G_1$ relative to the input. Effective disturbance suppression requires $G_1 \gg 1$; in other words, amplify the input as much as possible before the disturbance adds to it—a very logical thing to do, but sometimes a hard thing to achieve.

Example 13.2–1

A certain power amplifier is supposed to produce an 8-V output from a 1-V input. However, its voltage amplification drifts randomly from 7.5 to 8.5, and there is also an unwanted 2-V power-supply hum at the output. A preamplifier and a feedback network arranged like Fig. 13.2–3a have been proposed to reduce these effects. The problem is to determine values for G_1 and H such that the output hum does not exceed 0.1 V and $G_F = 8$ with, at most, $\pm 1\%$ variation.

First tackling the hum suppression, we note that $x_d = 2\text{ V}/G_2 = 0.25$ V, since the power amplifier has a nominal gain $G_2 = 8$. Then, from Eq. (6), we find the output hum with feedback is $(G_F/G_1)x_d \leq 0.1$ V, so we want $G_1 \geq G_F x_d/0.1 = 8 \times 0.25/0.1 = 20$.

Next we look at parameter variation, assuming G_1 to be essentially constant; therefore $dG/G = d(G_1 G_2)/G_1 G_2 = dG_2/G_2 = (8.5 - 7.5)/8 = 0.125$. The requirements call for $dG_F/G_F \leq 2\% = 0.02$, which, from Eq. (4), means that $S \leq 0.02/0.125 = 0.16$ and that $1 + GH = 1/S \geq 6.25$. Now, since $G_F = G_1 G_2/(1 + GH) = 8$, we have a second condition on G_1, namely $G_1 = 8(1 + GH)/G_2 \geq 6.25$.

The former condition happens to dominate in this case, and taking $G_1 = 20$ satisfies both the hum-suppression and parameter-variation requirements. Finally, solving for H via $1 + GH = G/G_F$ with $G = G_1 G_2 = 20 \times 8$, we obtain $H = \frac{1}{8} - \frac{1}{160} = 0.11875$. Figure 13.2–4 summarizes this design, including the various voltages.

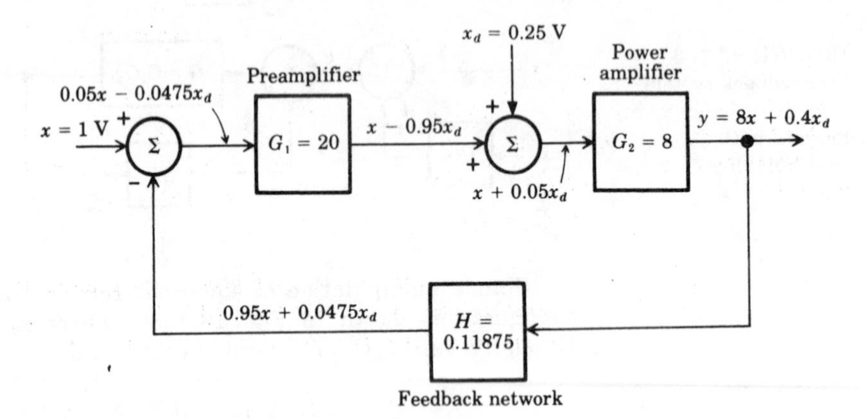

FIGURE 13.2–4
Amplifier system for Example 13.2–1.

Exercise 13.2–1

Given a process with $x_d = 0.02$ whose gain varies from 1.9 to 2.1, design a process control system, similar to Fig. 13.2–4, that has $0.995 \leq G_F \leq 1.005$ and whose output disturbance does not exceed 0.005.

BANDWIDTH AND TRANSIENT RESPONSE

Negative feedback tends to smooth out any frequency variations of the forward-path gain, just as it smooths out other types of variations. We can therefore use feedback to increase *bandwidth* by reducing the frequency-dependent gain variations—and, of necessity, by reducing the maximum gain. Moreover, increased bandwidth results in decreased time constants and a more rapid *transient response.* In general, then, feedback makes it possible to *swap excess gain for improved dynamic behavior.*

To put these assertions on a quantitative footing, we let the feedback system in Fig. 13.2–5 have the forward-path transfer function

$$G(s) = \frac{A_0}{1 + \tau_0 s} \tag{7a}$$

which represents a first-order system with maximum gain $G(0) = A_0$ and time constant τ_0. The corresponding frequency response is that of a lowpass filter,

$$\underline{G}(f) = \frac{A_0}{1 + j(f/B_0)} \tag{7b}$$

whose bandwidth (or cutoff frequency) is

$$B_0 = \frac{1}{2\pi\tau_0} \tag{7c}$$

a relationship we previously encountered in Chapter 8. Equation (7b) follows from (7a) by setting $s = j2\pi f$.

Now, assuming the feedback path has constant gain H, the transfer function of the feedback system will be

$$G_F(s) = \frac{G(s)}{1 + G(s)H} = \frac{A_0}{1 + \tau_0 s + A_0 H} = \frac{A_0/(1 + A_0 H)}{1 + \tau_0 s/(1 + A_0 H)}$$

so that

$$G_F(s) = \frac{A_F}{1 + \tau_F s} \tag{8a}$$

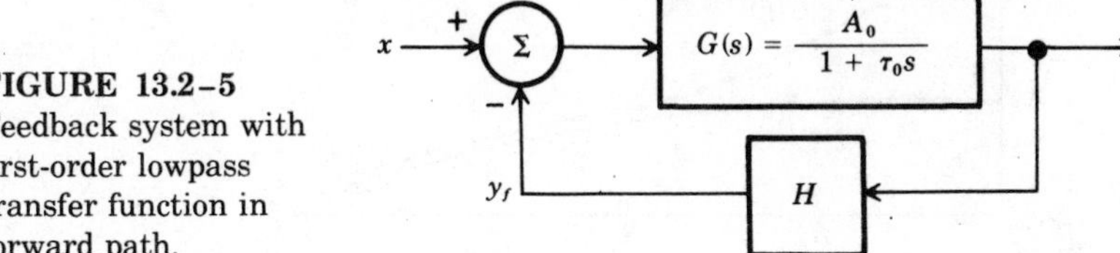

FIGURE 13.2–5
Feedback system with first-order lowpass transfer function in forward path.

and

$$G_F(f) = \frac{A_F}{1 + j(f/B_F)} \tag{8b}$$

where we have let

$$A_F = \frac{A_0}{1 + A_0 H} \qquad \tau_F = \frac{\tau_0}{1 + A_0 H} \qquad B_F = (1 + A_0 H)B_0 \tag{8c}$$

Comparing these expressions with Eq. (7) reveals that the feedback system has reduced gain $A_F < A_0$, reduced time constant $\tau_F < \tau_0$, and increased bandwidth $B_F > B_0$.

Figure 13.2–6a illustrates the step-response transient with and without feedback, showing that feedback has speeded up the response in the sense that it gets to a given percentage of its final value in less time than it does without feedback. Figure 13.2–6b plots the amplitude ratios $|\underline{G}(f)|$ and $|\underline{G}_F(f)|$, and vividly brings out the increased bandwidth *and* the reduced gain of the feedback system. Similar results apply for systems with a highpass or bandpass characteristic in the forward path.

It is important to observe from Eq. (8c) that the *gain-bandwidth product* with feedback equals the gain-bandwidth product without feedback, that is,

$$A_F B_F = A_0 B_0 \tag{9}$$

This equation directly states the exchange of gain for bandwidth and serves as a valuable design guideline. For instance, if a certain op-amp has $A_0 = 10^4$ and $B_0 = 20$ Hz, its gain-bandwidth product is $A_0 B_0 = 2 \times$

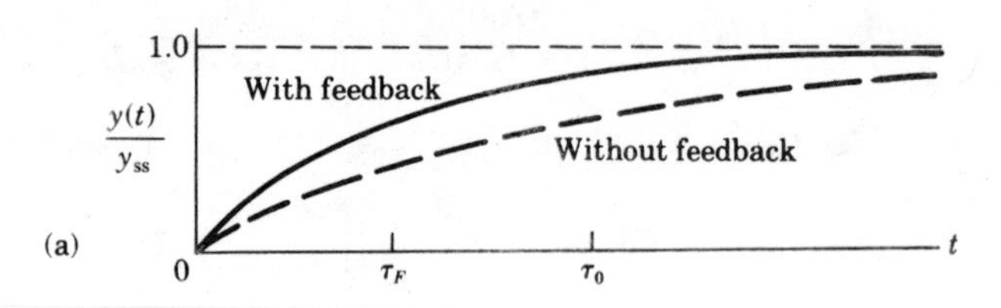

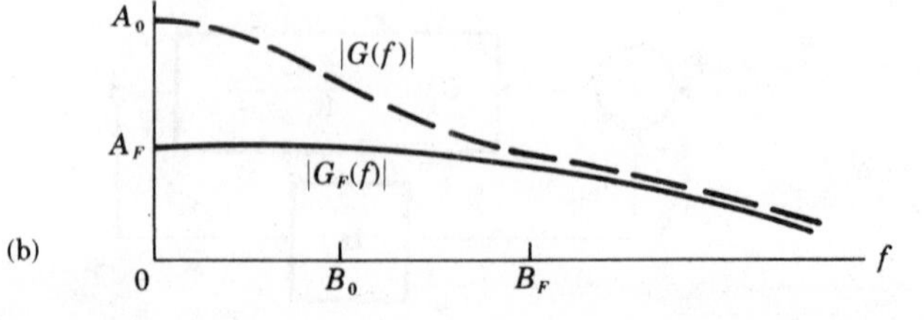

FIGURE 13.2–6 Improved dynamic performance with feedback: (a) step response; (b) frequency response.

10^5. We can then increase the bandwidth to $B_F = 5$ kHz using feedback, but the gain must drop to $A_F = A_0 B_0 / B_F = 40$. The maximum practical bandwidth for this particular amplifier is $B_F = 200$ kHz, corresponding to $A_F = 1$, for if $B_F > 200$ kHz, then $A_F < 1$.

Exercise 13.2–2

Suppose the power amplifier in Fig. 13.2–4 has $G_2(s) = 8/(1 + 10^{-3}s)$. Find B_F, assuming that G_1 and H have the constant values indicated.

BLOCK-DIAGRAM REDUCTION

Some systems have multiple forward and/or feedback paths that create loops within loops. For the purpose of applying our previous analysis and design results, we must reduce a system's block diagram to an equivalent single loop. Take, as a simple example, the diagram in Fig. 13.2–7a, where the letters in each block represent constant gains or transfer functions. Invoking Eq. (1), we find the inner loop can be replaced by one block with equivalent gain $G_{2F} = G_2/(1 + G_2 H_2)$. We then have a single-loop system, Fig. 13.2–7b, with $H = H_1$ in the feedback path and three blocks in cascade in the forward path, so that $G = G_1 G_{2F} G_3$.

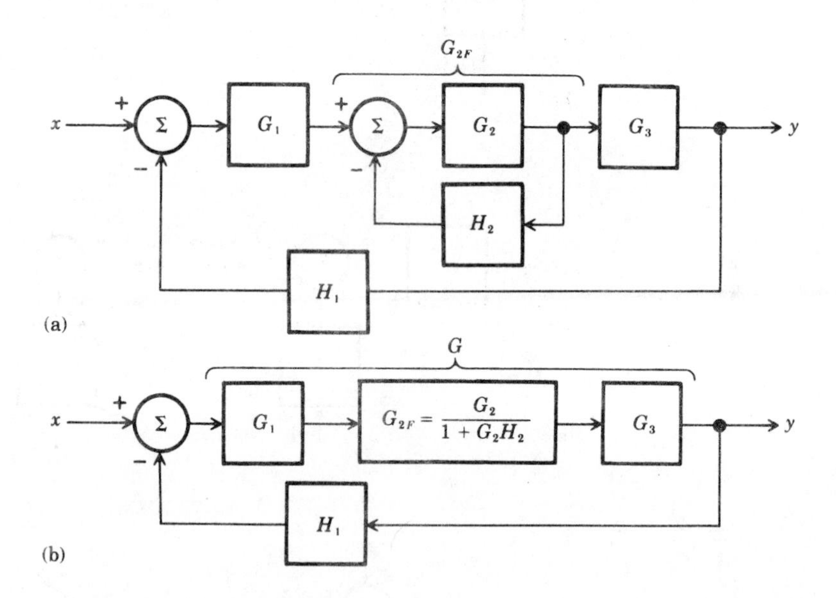

FIGURE 13.2–7
(a) System with two feedback loops;
(b) Equivalent single-loop system.

The reduction of more complicated diagrams may involve additional manipulations. Specifically, Fig. 13.2–8 gives the techniques for
(a) Combining parallel blocks.
(b) Moving a pick-off point.
(c) Moving a summing junction.
(d) Combining or decomposing a summing junction.

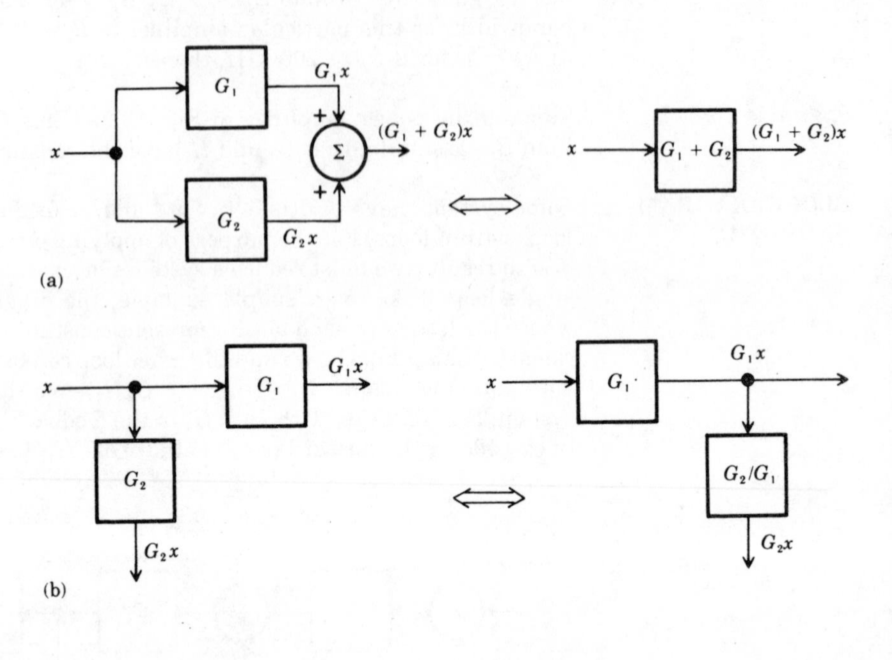

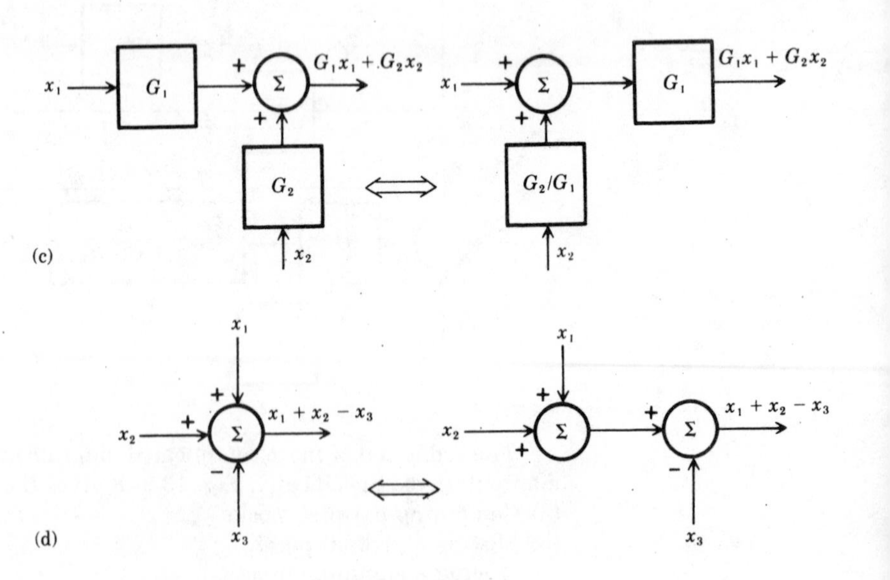

FIGURE 13.2–8 Block-diagram reduction techniques.

These manipulations merely reflect different but equivalent ways of writing the same input-output relationship. In Fig. 13.2–8c, for instance, the output of the left-hand diagram is $G_1x_1 + G_2x_2$ and the output of the right-hand diagram is $G_1[x_1 + (G_2/G_1)x_2] = G_1x_1 + G_2x_2$ as required. Similarly, Fig. 13.2–8d reflects the fact that the order of summation does not matter as long as we keep track of which inputs, if any, carry a negative sign.

The following example illustrates block-diagram reduction techniques.

Example 13.2–2

The block diagram of the machine speed-control system described in Example 13.1–2 is repeated as Fig. 13.2–9a with two changes: numerical values have been provided for all parameters, except the amplifier gain A; and the transient behavior of the machine has been included in the form of the transfer function $0.01/(1 + 3s)$, implying a time constant $\tau_0 =$ 3 sec. We seek the value of A that will speed up the feedback system's transient response so that $\tau_F = 0.5$ sec. To that end, the diagram must be reduced to the form of Fig. 13.2–5.

Although we could reduce the inner loop as in Fig. 13.2–7, a better approach here is to combine both of the feedback paths. We do this by moving the middle summing junction to the left of block A, changing the feedback gain from 10 to $10/A$ according to the manipulation in Fig. 13.2–8c. This puts two adjacent summing junctions at the input that can be manipulated to yield Fig. 13.2–9b (since the order of summation does not matter). Now the two feedback gains are in parallel and can be added (as in Fig. 13.2–8a), while the gains of the two cascade blocks in the forward path can be multiplied.

Finally, we arrive at Fig. 13.2–9c by moving the disturbance-summing junction to the input. This manipulation is also based on Fig. 13.2–8c, but note that the direct entry of x_d in Fig. 13.2–9b is equivalent to a block with unity gain, which becomes $1/2A$ after relocation. (The same manipulation was implicit in Fig. 13.2–3 when we first dealt with disturbance inputs.)

From our reduced diagram in Fig. 13.2–9c we see that

$$G(s) = 2A \times \frac{0.01}{1 + 3s} = \frac{0.02A}{1 + 3s}$$

$$H = \frac{1}{4} + \frac{10}{A}$$

so that $G(s)$ has the form of Eq. (7a) with $A_0 = 0.02A$. It therefore follows from Eq. (8c) that obtaining $\tau_F = 0.5 \text{ sec} = \tau_0/6$ requires $1 + A_0H = 6$ or

$$0.02A\left(\frac{1}{4} + \frac{10}{A}\right) = 5$$

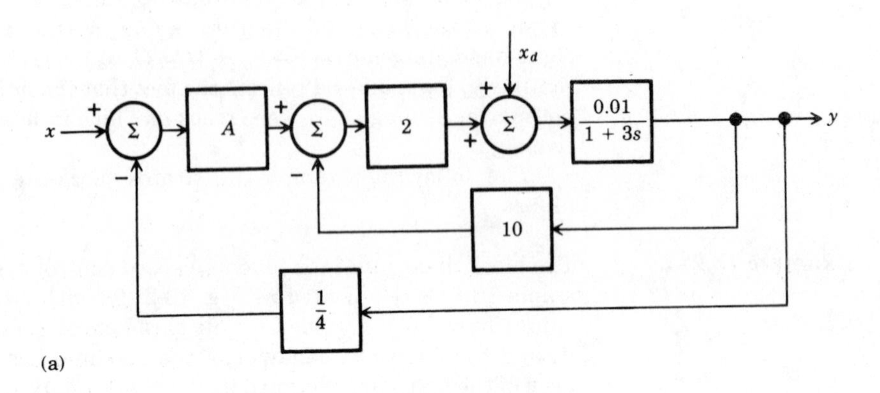

(a)

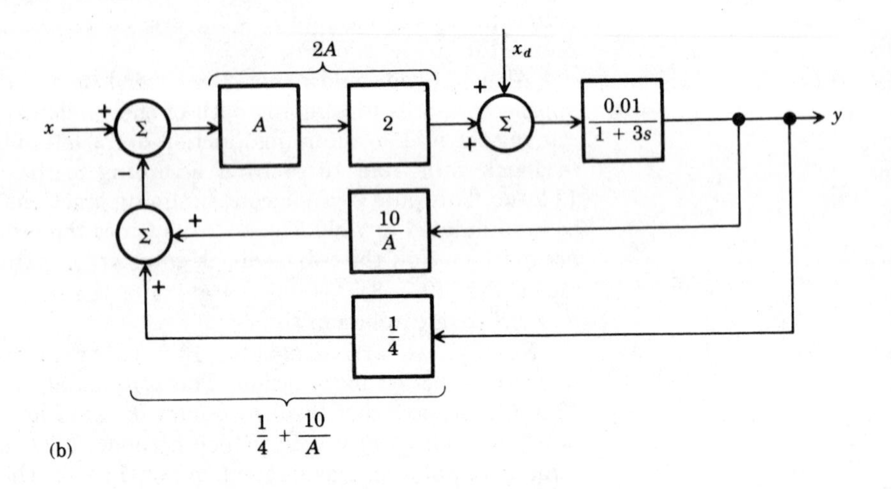

(b)

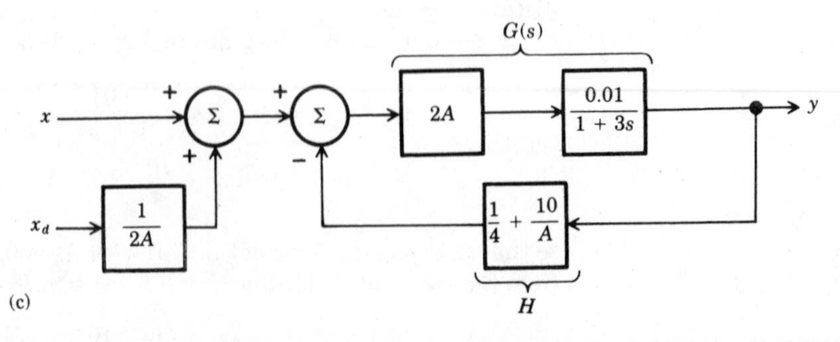

(c)

FIGURE 13.2–9 Block diagrams for Example 13.2–2: (a) initial diagram; (b) combining feedback paths; (c) reduced diagram.

whose solution yields $A = 960$. This large gain will also significantly reduce the effects of the disturbance x_d.

Exercise 13.2–3

As an alternative to Fig. 13.2–7b, show that Fig. 13.2–7a can be reduced to a single loop with $G = G_1 G_2 G_3$ and $H = H_1 + H_2/G_1 G_3$. Then prove that both equivalent diagrams yield the same expression for G_F.

13.3 STABILITY AND OSCILLATIONS †

We observed in Section 13.1 that an amplifier with positive feedback may become *unstable,* in the sense of having infinite closed-loop gain. Here we look more closely at the question of feedback-system stability and the related topic of oscillations in feedback systems. Our starting point is the relationship between stability and natural response, based on the properties of transfer functions.

TRANSFER FUNCTION POLES AND ZEROS

Figure 13.3–1 represents an arbitrary linear system with transfer function $H(s)$ such that the input $x = X_{\text{in}} e^{st}$ produces the forced response $y = H(s) X_{\text{in}} e^{st}$ at the output. Whether the system in question is a simple two-port network or a multiloop feedback configuration, its overall transfer function usually may be expressed as

$$H(s) = \frac{b_m s^m + b_{m-1} s^{m-1} + \cdots + b_1 s + b_0}{a_n s^n + a_{n-1} s^{n-1} + \cdots + a_1 s + a_0}$$

where the a and b are constants involving the element values, n is the order of the system, and the denominator is the *characteristic polynomial.*

In principle, we can compute the n roots of the characteristic polynomial and the m roots of the numerator polynomial and rewrite $H(s)$ in the factored form

$$H(s) = K \frac{(s - z_1)(s - z_2) \cdots (s - z_m)}{(s - p_1)(s - p_2) \cdots (s - p_n)} \tag{1}$$

The roots p_i and z_i are known as the *poles* and *zeros* of $H(s)$, respectively, and they may be complex quantities, whereas the constant K is always a real quantity. The computation of poles and zeros is a routine task when $n \leq 2$ and $m \leq 2$. For higher-order systems, however, one must usually resort to numerical analysis carried out by computer or programmed calculator.

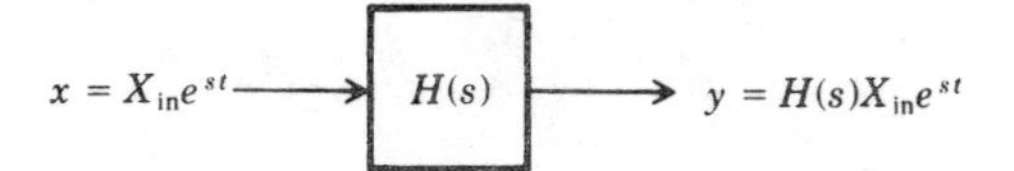

FIGURE 13.3–1

For a simple example, take the two-port network of Fig. 13.3–2a with input current x and output voltage y. Then

$$H(s) = Z(s) = \frac{(sL + R)(1/sC)}{sL + R + 1/sC} = \frac{1}{C}\frac{s + R/L}{s^2 + (R/L)s + 1/LC}$$

$$= 4000\frac{s + 160}{s^2 + 160s + 10^4}$$

$$= 4000\frac{s - (-160)}{[s - (-80 + j60)][s - (-80 - j60)]}$$

so $K = 4000$, $z_1 = -160$, $p_1 = -80 + j60$, and $p_2 = -80 - j60$. We summarize this information graphically in the *pole-zero diagram* of $H(s)$, Fig. 13.3–2b, a complex *s-plane* with axes Re[s] and Im[s] on which we mark an ✕ at each pole location and a o at each zero location. Such diagrams convey all the data regarding the transfer function, save for the constant K.

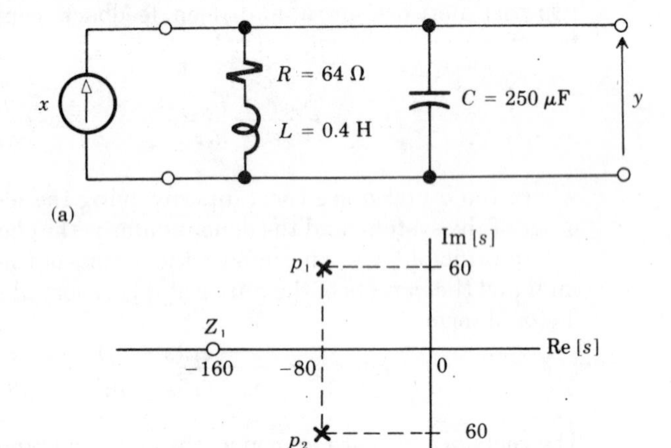

FIGURE 13.3–2
(a) Two-port network.
(b) Pole-zero diagram.

The names "poles" and "zeros" come from the nature of the forced response when s equals the value of a pole or zero. Suppose, for instance, that the input has $x = X_{\text{in}}e^{z_1 t}$; by definition of $H(s)$, the forced output will be

$$y = H(z_1)X_{\text{in}}e^{z_1 t} = 0$$

because $H(z_1) = 0$ when we set $s = z_1$ in Eq. (1b). Therefore, we have zero output when s equals z_1 or any other zero of $H(s)$. On the other hand, $H(s)$

"blows up" when s equals any pole p_i of $H(s)$—since $H(p_i) \to \infty$—and the forced output theoretically becomes infinitely large.

Another and more significant interpretation of the poles relates to the *natural response,* for the poles are roots of the characteristic polynomial and therefore describe the natural behavior in absence of any forcing input. Specifically, if $x = 0$ and there is initial stored energy at $t = 0$, then

$$y = y_N = A_1 e^{p_1 t} + A_2 e^{p_2 t} + \cdots + A_n e^{p_n t} \tag{2}$$

assuming no repeated roots. (Repeated roots would be treated as in Eq. (12), Section 8.3.) The pole locations in the s-plane tell us immediately the shape of the natural-response waveforms.

If the denominator of $H(s)$ contains a quadratic factor of the form

$$(s - p_1)(s - p_2) = s^2 + 2\alpha s + \omega_0^2 \tag{3}$$

and if $0 < \alpha < \omega_0$, the poles p_1 and p_2 fall on a semicircle of radius ω_0, as shown in Fig. 13.3–3a. This corresponds to an underdamped system whose natural response contains a decaying oscillation at the natural frequency $\omega_N = \sqrt{\omega_0^2 - \alpha^2}$—see Eq. (10), Section 8.3. Figure 13.3–3b and c give the pole locations for critical damping ($\alpha = \omega_0$) and overdamping ($\alpha > \omega_0$), respectively. The *damping ratio*

$$\zeta \triangleq \alpha/\omega_0 \tag{4}$$

is sometimes used to identify these three cases: $\zeta < 1$ means underdamping; $\zeta = 1$ means critical damping; and $\zeta > 1$ means overdamping. In any case, as long as $\zeta > 0$, both poles will be to the left of the imaginary axis, and the natural behavior decays exponentially. This observation provides an important clue about s-plane pole locations and the *stability* of a system.

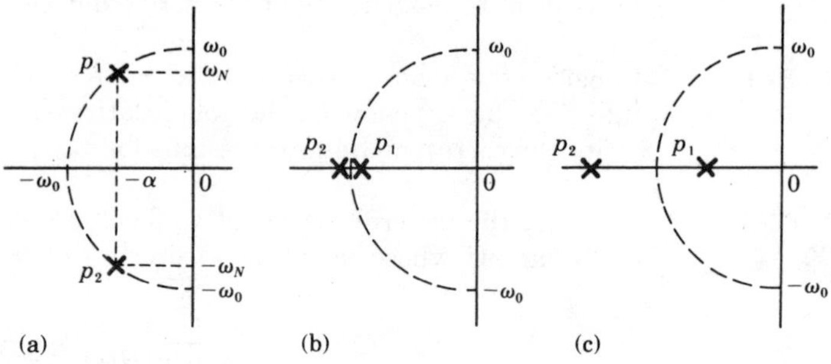

FIGURE 13.3–3 Pole locations for roots of $s^2 + 2\alpha s + \omega_0^2$: (a) underdamped, $\alpha < \omega_0$; (b) critically damped, $\alpha = \omega_0$; (c) overdamped $\alpha > \omega_0$.

A system is said to be *stable* when all the poles lie entirely within the *left half plane*. Under that condition, all the components of the natural response die away as time increases, so $y_N \to 0$ as $t \to \infty$. But any pole within the *right half plane* represents a growing time function such that $y_N \to \infty$ as $t \to \infty$. We then say the system is *unstable*.

For instance, if $H(s) = K/(s^2 + 7s - 30)$, then $s^2 + 7s - 30 = (s + 10)(s - 3)$ so $p_1 = -10$ while $p_2 = +3$ and the natural response includes the *growing exponential* $A_2 e^{+3t}$. Similarly, if the parameter α in Eq. (3) turns out to be negative and $|\alpha| < \omega_0$, then $p_1, p_2 = +|\alpha| \pm j\omega_N$, which corresponds to the *growing oscillation* $Ae^{|\alpha|t} \cos(\omega_N t + \theta)$ shown in Fig. 13.3–4. In the case of zero damping ($\alpha = 0$), the poles fall exactly on the imaginary axis and we have a natural-behavior oscillation with a constant amplitude that neither grows nor decays. Stability in the sense that $y_N \to 0$ as $t \to \infty$ therefore requires all poles to be within the left half of the s plane.

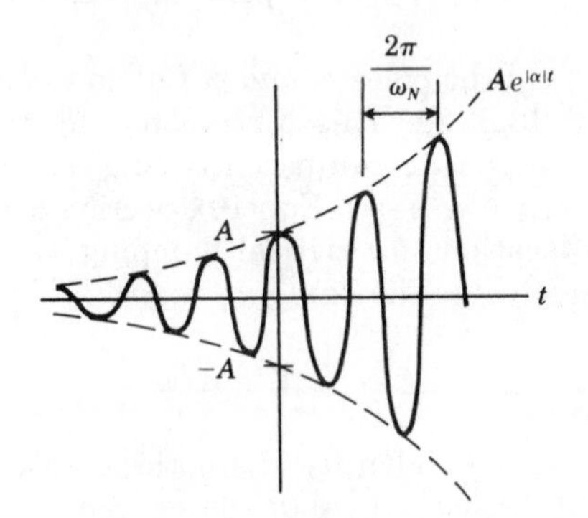

FIGURE 13.3–4
Growing oscillation $Ae^{|\alpha|t} \cos(\omega_N t + \theta)$.

Passive systems are never unstable, for they do not contain the inexhaustible energy source needed to support an endlessly increasing natural response. However, amplifier circuits and control systems are *active* systems that require a power supply for normal operation. Such systems may become unstable in that the natural response grows with time until limited by nonlinearities or other effects that void the system model—or perhaps until destruction of the active device.

Exercise 13.3–1

Consider the function $H(s) = 1/(s^2 - as + 100)$. Draw s-plane diagrams like Fig. 13.3–3, showing the pole locations for $a = 12$, 20, and 25. Then write the corresponding expressions for $y_N(t)$.

STABILITY OF FEEDBACK SYSTEMS

Consider the generalized negative feedback system diagrammed in Fig. 13.3–5a, and whose *closed-loop transfer function* is

$$G_F(s) = \frac{G(s)}{1 + G(s)H(s)} = \frac{G(s)}{1 + F(s)} \tag{5}$$

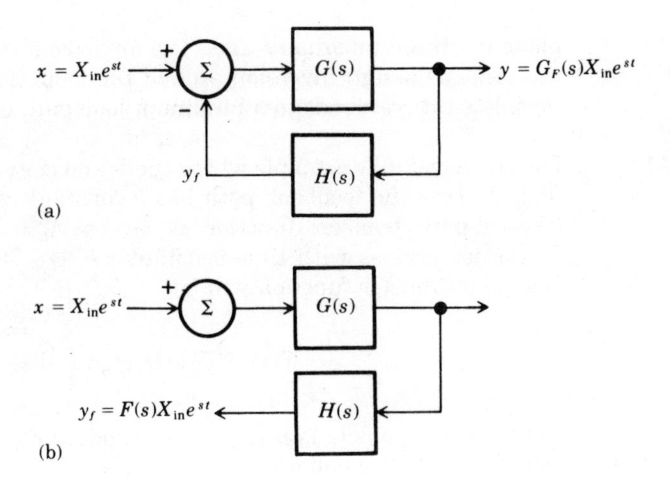

FIGURE 13.3–5

where $F(s) \triangleq G(s)H(s)$. With the help of Fig. 13.3–5b, we interpret $F(s)$ as the *open-loop transfer function* from the input around to the disconnected summing junction. Setting the denominator of $G_F(s)$ equal to zero then gives the characteristic equation in terms of $F(s)$ as

$$1 + F(s) = 0 \tag{6}$$

The roots of Eq. (6) are the *closed-loop poles* needed to determine the stability of the feedback system.

The apparent simplicity of Eq. (6) tends to conceal the serious difficulties often encountered when analyzing or designing feedback systems. To cope with these problems, several high-powered techniques have been devised that allow one to predict the closed-loop stability from the open-loop transfer function $F(s)$. Two of those techniques—the root-locus and frequency-response methods—will be illustrated shortly.

Systems with *positive* feedback clearly have the potential for unstable behavior. But even a "safe" *negative* feedback system may become unstable when the *phase shift* around the loop turns negative feedback into positive feedback at a particular frequency. In that case, the natural response will include a sinusoidal oscillation with constant or growing amplitude.

We can test for this *oscillatory instability* using the AC transfer function $\underline{F}(f)$ obtained by setting $s = j2\pi f$ in $F(s)$. Specifically, a negative feedback system will be unstable if there exists a frequency f_0 such that

$$|\underline{F}(f_0)| \geq 1 \qquad \sphericalangle \underline{F}(f_0) = \pm 180° \tag{7}$$

which is equivalent to a complex pair of closed-loop poles in the right half plane or on the imaginary axis. The angle condition $\sphericalangle \underline{F}(f_0) = \pm 180°$ is the same as a sign inversion around the loop; the magnitude condition $|\underline{F}(f_0)| \geq 1$ corresponds to a minimum loop gain of unity.

Example 13.3–1

For an elementary example whose open- and closed-loop poles are easy to find, suppose the feedback path has a constant gain $H(s) = A_H$ and the forward-path transfer function is $G(s) = A_G/(s + \alpha)$, representing a first-order process with time constant $\tau = 1/\alpha$. Taking $K = A_G A_H$, the open-loop transfer function will be

$$F(s) = G(s)H(s) = \frac{K}{s + \alpha}$$

so the open-loop pole is $p = -\alpha$. Next, using Eq. (6), we get the closed-loop characteristic equation

$$1 + \frac{K}{s + \alpha} = 0$$

which yields the closed-loop pole location at

$$s = -(K + \alpha)$$

Note that the closed-loop pole equals the open-loop pole if $K = 0$.

Figure 13.3–6 shows the open-loop pole and the s-plane path of the closed-loop pole for $K > 0$ and $K < 0$. This *root-locus diagram* helps us study the stability of the feedback system. We keep α fixed, but allow $K = A_G A_H$ to vary in recognition of the fact that forward-path gain variation is a common occurrence. We see from the root locus that the closed-loop pole moves to the left along the real axis for any positive value of K; thus, the system will always be stable if $K > 0$. But if K happens to be negative, the closed-loop pole moves to the right and enters the right half plane when $K < -\alpha$; the system is then unstable. This potential instability could have been anticipated, since negative K actually corresponds to *positive* feedback.

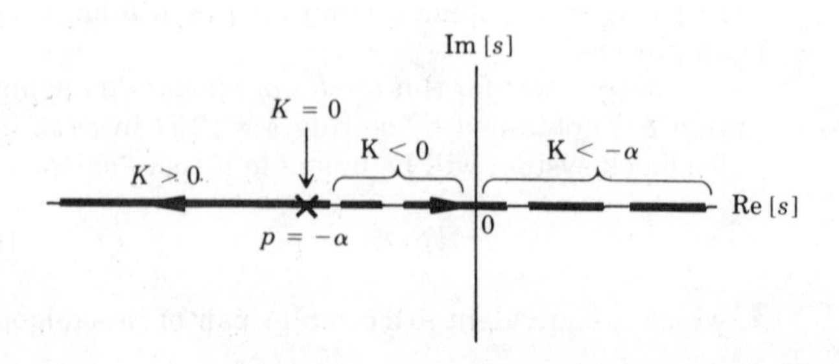

FIGURE 13.3–6
Root locus for Example 13.3–1.

Example 13.3–2 With higher-order systems, potential instability becomes more subtle and difficult to detect. As a case in point, consider the third-order open-loop transfer function

$$F(s) = \frac{K}{s^3 + 38s^2 + 400s + 1056} = \frac{K}{(s + 4)(s + 12)(s + 22)}$$

whose poles are at -4, -12, and -22. Applying Eq. (6), the closed-loop poles are the roots of the cubic equation

$$s^3 + 38s^2 + 400s + 1056 + K = 0$$

Solving this equation for different values of K would obviously be a tiresome chore.

Fortunately, however, with the help of special rules or computer programs, it is possible to sketch the root locus directly from the *open-loop poles,* the result being given in Fig. 13.3–7, taking $K \geq 0$. This diagram reveals that two of the roots leave the real axis and become complex conjugates as K increases, crossing into the right half plane at $s = \pm j20$ when $K = 15{,}200$. The natural response must then include a growing oscillation at the natural frequency $\omega_N \geq 20$ rad/s.

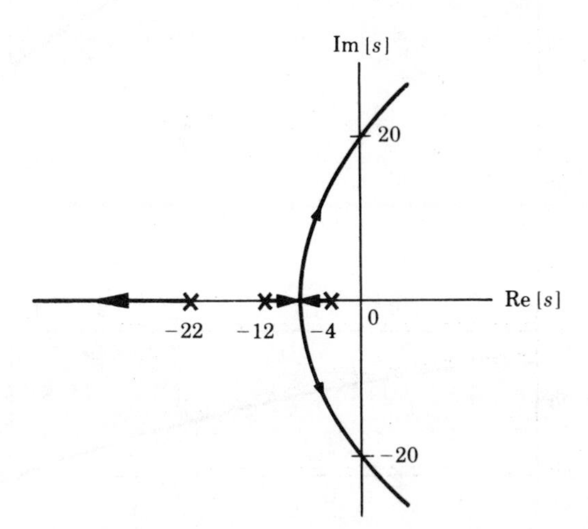

FIGURE 13.3–7
Root locus for Example 13.3–2.

In view of this oscillatory instability, we could use a *frequency-response* analysis based on Eq. (7) as an alternative to root-locus analysis. The frequency-response method has the added advantage of not even needing the open-loop pole locations; it only requires the AC transfer function $\underline{F}(f)$. A handy and informative way of carrying out the method is to draw a *Bode diagram* of $\underline{F}(f)$, plotting $|\underline{F}(f)|_{\text{dB}} = 10 \log |\underline{F}(f)|^2$ and

$\measuredangle \underline{F}(f)$ versus frequency on a logarithmic axis. Then, recalling that unity gain becomes 0 dB, we look for any frequency $f = f_0$ that simultaneously satisfies the conditions $|\underline{F}(f_0)| \geq 0$ dB and $\measuredangle \underline{F}(f_0) = \pm 180°$.

Figure 13.3–8 shows the Bode diagram for the system in question, with $K = 30{,}000$ and $K = 3{,}000$. (The angle plot happens to be independent of K.) Since $\measuredangle \underline{F} = -180°$ at $f_0 = 20/2\pi \approx 3$ Hz, the system is unstable when $K = 30{,}000$ because $|\underline{F}(f_0)| \approx 6$ dB > 0 dB, but stable when $K = 3{,}000$ because $|\underline{F}(f_0)| \approx -14$ dB < 0 dB. In the latter case we say that the system has a *gain margin* of 14 dB, meaning that the system remains stable for any gain increase less than 14 dB.

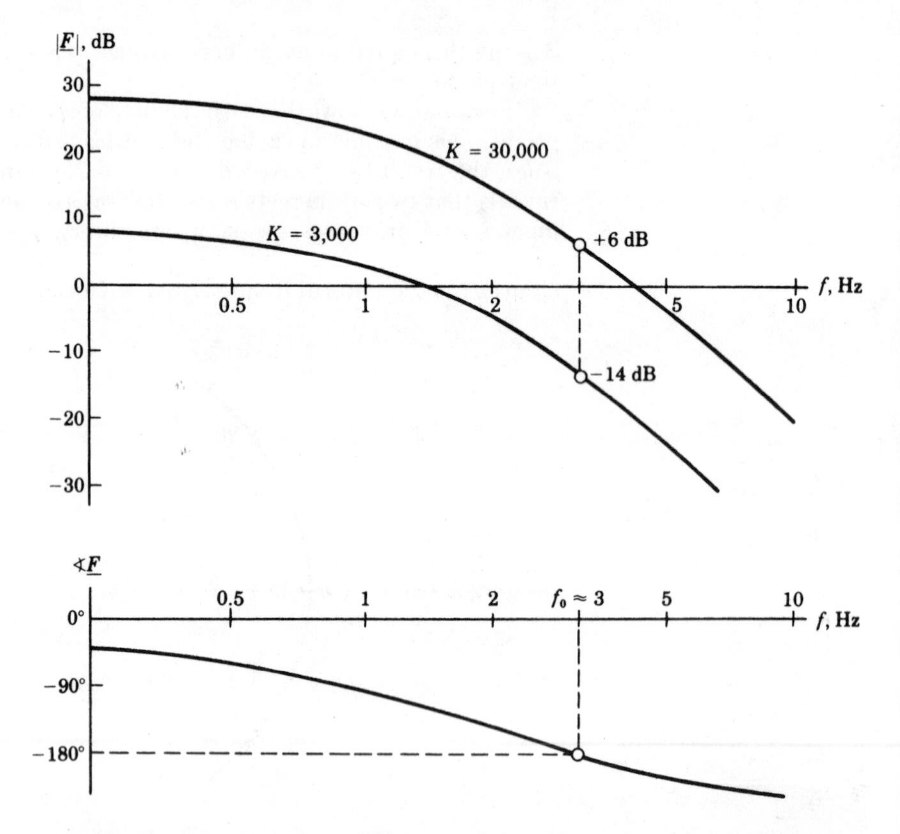

FIGURE 13.3–8 Bode diagram for Example 13.3–2.

Exercise 13.3–2 Obtain an expression for the closed-loop poles when $F(s) = Ks/(s^2 + 2\alpha s + \omega_0^2)$ with $\alpha > 0$. Sketch the root locus for $K \geq 0$ and show that the system is stable unless $K < -2\alpha$.

Exercise 13.3–3 If a complex function has the form $\underline{F} = K/(a + jb)$, where a, b, and K are real quantities and $K > 0$, then $\measuredangle \underline{F} = \pm 180°$ only when $a < 0$ and $b = 0$. Use this observation to find f_0, such that $\measuredangle \underline{F}(f_0) = \pm 180°$ when $F(s) =$

$K/(s^3 + 130s^2 + 3600s)$ with $K > 0$. Then obtain the condition on K for stability.

FEEDBACK OSCILLATORS

On the other side of the stability coin we have the *feedback oscillator,* a simple method for generating sinusoidal waveforms with the help of positive feedback. As diagrammed in Fig. 13.3–9a, the closed loop contains an amplifier, limiter, and feedback network—but no external input. This system will produce a sinusoidal output at frequency f_0 if

$$\underline{F}(f_0) = \underline{G}(f_0)\underline{H}(f_0) = 1 \angle 0^\circ \tag{8}$$

and if this condition holds only at one frequency. We justify Eq. (8) with the help of Fig. 13.3–9b where the loop has been broken and an external sinusoid $x = A \cos 2\pi f_0 t$ is applied to the amplifier's input. The signal returning through the feedback path will then be exactly the same as the input when $\underline{F}(f_0) = 1$, so we can close the loop and let the system carry on without continuing the external input. Positive feedback thereby creates a self-sustaining oscillation.

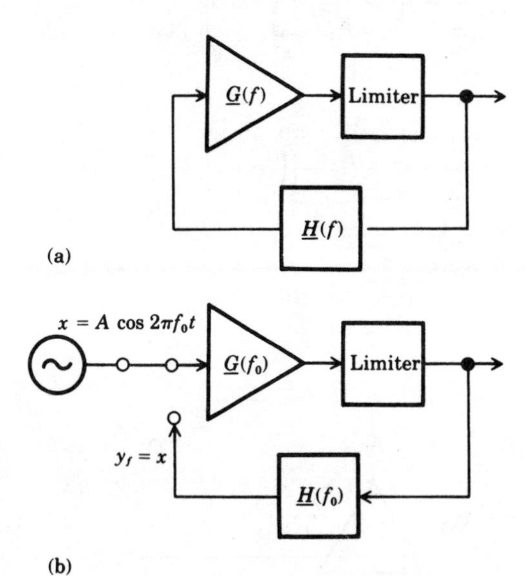

FIGURE 13.3–9 (a) Feedback oscillator. (b) Equivalent open-loop system.

In practice, the starting signal comes from the internal noise present in any amplifier, and we take $\underline{F}(f_0) > 1$ to ensure that the oscillation builds up in amplitude. The limiter then stabilizes the steady-state amplitude at some fixed level. (This stabilization is sometimes carried out by saturation within the amplifier itself.) Usually, the amplifier has a relatively constant gain and no phase shift other than a possible sign inversion, so $\underline{G}(f) = \pm K$ with $K > 1$. The feedback network must then be frequency selective with the property that $\underline{H}(f_0) = \pm 1/K$ at the desired oscillation frequency. There are numerous circuit realizations of the positive-feedback principle. We'll examine two of them that use an ideal op-amp as the active device.

The oscillator circuit in Fig. 13.3–10a has a noninverting op-amp with $\underline{G}(f) = (R_2 + R_1)/R_1$ in the forward path. The feedback transfer function is

$$\underline{H}(f) = \frac{\underline{V}_f}{\underline{V}_{\text{out}}} = \frac{\underline{Z}_3}{\underline{Z}_3 + \underline{Z}_4} = \frac{1}{3 + j(2\pi fRC - 1/2\pi fRC)}$$

The phase shift goes from $+90°$ to $-90°$ as f goes from zero to infinity. Since $\sphericalangle \underline{G}(f) = 0°$, $\sphericalangle \underline{F}(f) = 0°$ only when $\sphericalangle \underline{H}(f) = 0°$, which occurs at

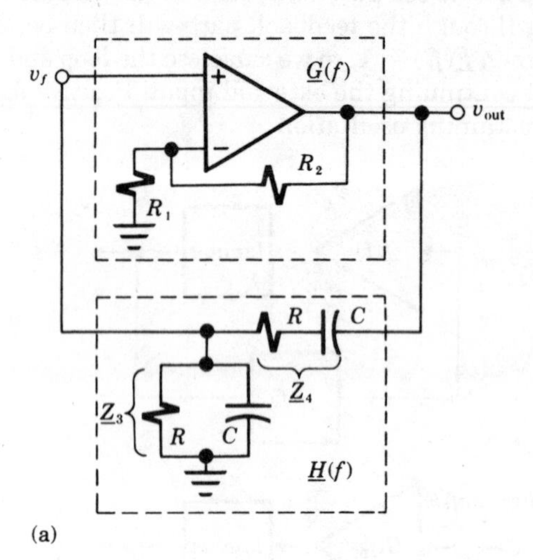

(a)

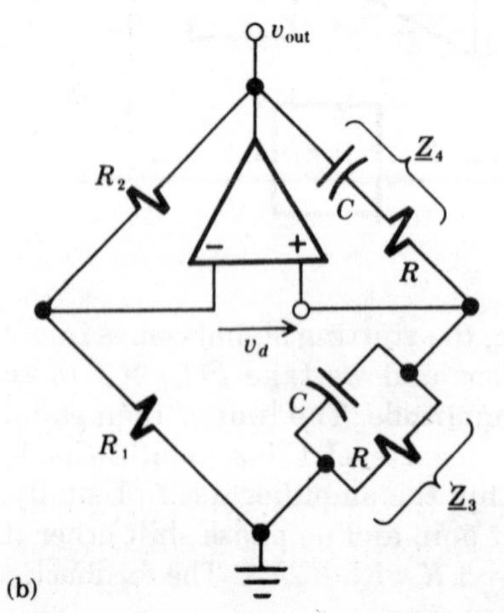

(b)

FIGURE 13.3–10
Wien-bridge oscillator.

the frequency

$$f_0 = \frac{1}{2\pi RC} \tag{9a}$$

where $\underline{H}(f_0) = \frac{1}{3}$ and $\underline{F}(f_0) = \underline{G}(f_0)\underline{H}(f_0) = (R_2 + R_1)/3R_1$. Equation (8), then, is satisfied if

$$R_2 = 2R_1 \tag{9b}$$

and the oscillation frequency can be varied over a fairly wide range using mechanically coupled adjustable resistors or capacitors in the feedback network.

This circuit is called a *Wien bridge oscillator* in view of its bridge configuration, as brought out by redrawing the circuit in the form of Fig. 13.3–10b. The oscillation conditions correspond to a *balanced* bridge with $v_d = 0$, which is consistent with the virtual short across the input of an ideal op-amp. Amplitude limiting can be accomplished if we use a small incandescent lamp or other temperature-sensitive element for R_1, and make R_2 slightly greater than the room-temperature value of R_1. Then, as the oscillation builds up, current through R_1 causes its resistance to increase until the circuit reaches steady-state conditions with $2R_1 = R_2$.

Another bridge circuit is the simplified *crystal oscillator* shown in Fig. 13.3–11a. Here, the feedback network incorporates a special quartz crystal (XTAL), in which the piezoelectric effect and mechanical vibrations combine to produce resonance phenomena represented by the equivalent electrical circuit of Fig. 13.3–11b. Figure 13.3–11c sketches the crystal's impedance variation with frequency, showing a series resonance at f_s and a parallel resonance at f_p. The crystal acts like a small resistance R_0 at a frequency f_0, slightly above f_s, so $\underline{H}(f) = R_3/(R_3 + R_0)$ and the circuit oscillates if $(R_2 + R_1)/R_1 \geq (R_3 + R_0)/R_3$. No other oscillation frequency is possible due to the increased value of $|\underline{Z}|$ and/or $\measuredangle \underline{Z} \neq 0°$ for $f \neq f_0$.

Crystal oscillators are used primarily in communication systems and similar applications where we need a precisely controlled high-frequency sinusoid. (With ordinary circuit elements, any temperature variation changes the values of inductance and capacitance and results in excessive frequency *drift*.) To obtain a different oscillation frequency, we insert another crystal. Some citizen's band (CB) radios, for instance, have a separate crystal for each channel frequency. Alternatively, a single crystal oscillator may provide a stable *reference frequency* from which other frequencies are derived.

Exercise 13.3–4 The bridge in Fig. 13.3–10b is balanced when $\underline{Z}_3/\underline{Z}_4 = R_1/R_2$. Use this condition to derive the oscillation conditions in Eq. (9).

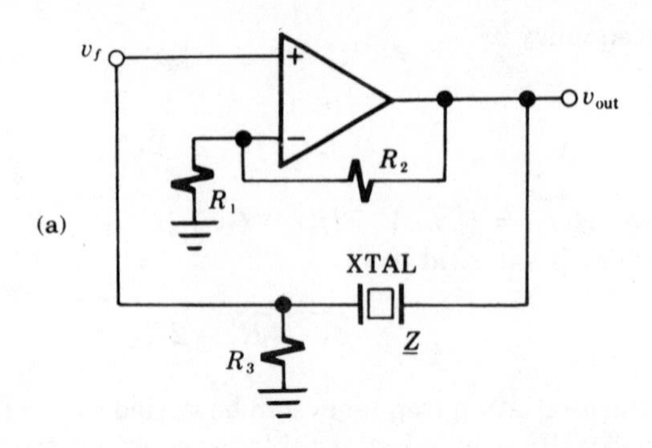

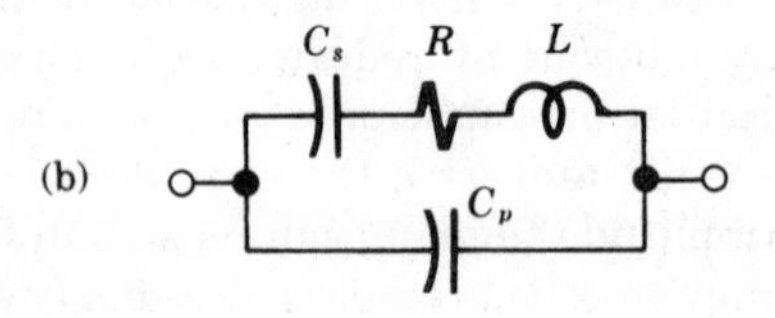

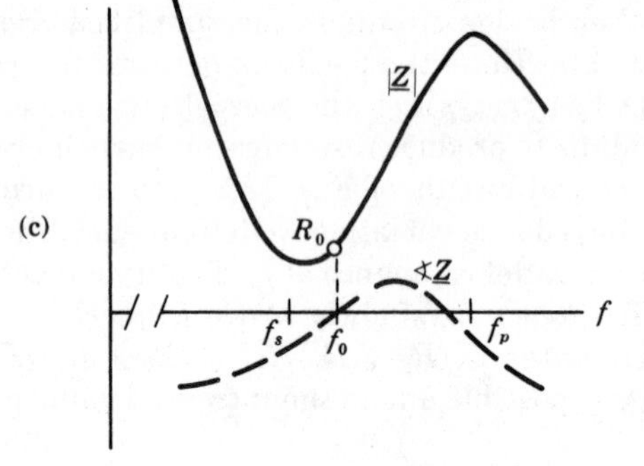

FIGURE 13.3–11
(a) Crystal oscillator. (b) Equivalent circuit of crystal. (c) Crystal impedance.

PROBLEMS

13.1–1 Obtain a condition on A in terms of K so that the approximation in Eq. (2) does not exceed the exact value in Eq. (1) by more than 1%.

13.1–2 Use an appropriate series expansion to approximate Eq. (1) in the form $y \approx K[1 - (K/A)]x$ when $|K/A| < 1$.

13.1–3 Some feedback amplifiers have the form of Fig. P13.1–3. Write x_e in terms of x and y, and show that $y/x = K_{\text{in}}KA/(K + A)$.

13.1–4 Let the op-amp in Fig. 4.3–1a have $i_p = i_n = 0$, like an ideal op-amp, but let the gain A be finite so that $v_{\text{out}} = A(v_p - v_n)$. A block-diagram representation might then have $v_p - v_n$ at the output of a summing junction.

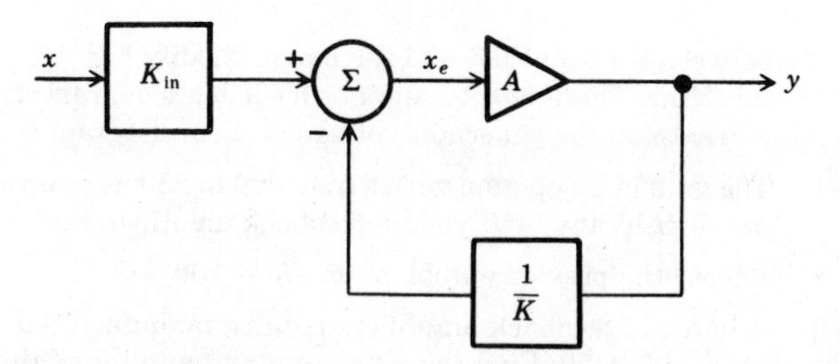

FIGURE P13.1–3

(a) Use this approach to construct a feedback model of the noninverting amplifier in Fig. 4.3–4a.

(b) Calculate v_{out}/v_{in}, taking $R_F = 9R_1$ and $A = 1000$.

13.1–5 Repeat the previous problem for the inverting amplifier in Fig. 4.3–6a, where $v_n = (R_F v_{in} + R_1 v_{out})/(R_F + R_1)$. Your diagram should have the form of Fig. P13.1–3.

13.1–6 Repeat Example 13.1–1 for the BJT amplifier in Fig. 12.2–1b. Hint: $v_f = R_E(g_m + 1/r_\pi)v_{be}$.

13.1–7 Suppose the op-amp described in Problem 13.1–4 is connected as a voltage follower (Fig. 4.3–5), but the input terminals are accidentally interchanged. Obtain a feedback diagram similar to Fig. P13.1–3 and show that the circuit has positive feedback.

13.1–8 Repeat the previous problem for an inverting amplifier (Fig. 4.3–6a) with interchanged input connections.

13.1–9 Figure P13.1–9 is the equivalent circuit of a generator whose source resistance R_s causes the output voltage to vary with the current i. Draw the block diagram of a feedback system that includes a differential amplifier and a potentiometer arranged to control v. Let v_{ref} be the input and v the output, and treat $R_s i$ as a disturbance input. (Your diagram is used in Problem 13.2–6.)

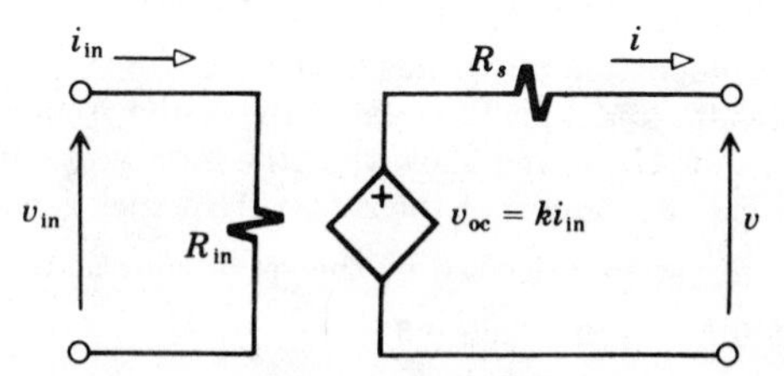

FIGURE P13.1–9

13.1–10 The self-biased BJT amplifier in Fig. 12.1–12 incorporates negative feedback to stabilize the DC collector current $I_C = \beta_0 I_B + I_{CEO}$. Use Eq. (4), Sect. 12.1, to obtain a feedback model similar to Fig. P13.1–3. Take $V_B - V_T$ as the input and I_C as the output, and treat I_{CEO} as a disturbance input. (Your diagram is used in Problem 13.2–14.)

13.1–11 Suppose the motor in Example 13.1–2 requires more current than the amplifier can supply. A generator like the one in Problem 13.1–9 might then be inserted between the amplifier and the motor. Modify Fig. 13.1–7c to include the generator. Hint: There will be another feedback loop, reflecting the effect of the motor current i on the generator voltage v. (Your diagram is used in Problem 13.2–15.)

13.2–1 The gain of an op-amp varies from 200 to 250 as temperature changes. What values of H, if any, will yield a feedback amplifier with $G_F = 100 \pm 10$?

13.2–2 Repeat the previous problem for $G_F = 100 \pm 2$.

13.2–3 A batch of feedback amplifiers is to be manufactured with $H = 0.004$ such that $G_F = 200 \pm 20$. Find the maximum and minimum allowable values of G.

‡ **13.2–4** Parameter variations in the feedback path are generally more serious than forward-path variations. Obtain an expression for the sensitivity $S_H = (H/G_F)dG_F/dH$ and show that $S_H \approx -1$ if $GH \gg 1$.

13.2–5 Let the system in Fig. 13.2–3 have $G_2 = 5.0$. Find values for G_1 and H such that $y = 20x + 0.1x_d$.

13.2–6 Let the system in Problem 13.1–9 have $v_{\text{ref}} = 5$ V, $R_{\text{in}} = 60\ \Omega$, $k = 240$, and $R_s = 2\ \Omega$. Find A and $H = 1/K$ such that $v = 100$ V when $i = 0$ and $v = 99$ V when $i = 12.5$ A.

13.2–7 When two lowpass amplifiers are in cascade, like Fig. 13.1–1a, and the bandwidth of the first amplifier is much greater than that of the second, the overall gain equals the product of the individual gains and the overall bandwidth equals the bandwidth of the second amplifier. Use these facts to design a cascade of two feedback amplifiers that achieves an overall gain of 4000 and a 5-kHz bandwidth when the individual forward-path amplifiers have $A_0 = 10^4$ and $B_0 = 100$ Hz. Specify the value of H for each unit.

13.2–8 Repeat the previous problem for an overall gain of 200 and a 25-kHz bandwidth.

13.2–9 A certain lowpass amplifier has $B_0 = 2$ kHz and its low-frequency gain A_0 varies from 40 to 60. Find the value of H so that $B_F \geq 10$ kHz, $dG_F/G_F \leq 0.1$, and A_F is as large as possible. Calculate the resulting values of A_F, B_F, and dG_F/G_F.

13.2–10 Repeat the previous problem with $B_F \geq 6$ kHz.

13.2–11 A tuned bandpass amplifier has $G(s) = A_0 2\pi B_0 s/(s^2 + 2\pi B_0 s + \omega_0^2)$, where B_0 is the bandwidth and A_0 is the gain at the center frequency $f_0 = \omega_0/2\pi$. Obtain the expression for $G_F(s)$ and show that the gain and bandwidth with feedback are given by Eq. (8c), and that the center frequency remains the same.

13.2–12 Consider a system described by the set of equations

$$y = w + 6u \qquad u = x - 0.5y$$
$$w = 20v \qquad v = 3u - 0.2w$$

Draw the block-diagram representation with x and y as the input and output, respectively. Then use block-diagram reduction to show that $y/x = 1.8$.

13.2–13 Repeat the previous problem for

$$y = 9v \qquad v = 5u - 3x \qquad u = x - 0.2y$$

Hint: Any of the summing junctions in Fig. 13.2–8 could have a minus sign at one of the inputs.

‡ **13.2–14** Let the circuit in Problem 13.1–10 have $\beta_0 = 50$. Reduce the block diagram to the form of Fig. 13.2–3b with $x_d = I_{CEO}$. Then find values for R_B and R_E, such that $I_C = 2$ mA $+ 0.1I_{CEO}$ when $V_B - V_T = 4$ V.

‡ **13.2–15** Let the system in Problem 13.1–11 have $k = 300$, $R_{\text{in}} = 60\ \Omega$, $R_s = 2.5\ \Omega$, $R = 10\ \Omega$, $\beta = 20$, and $D = 8$. Reduce the block diagram to the form of Fig. 13.2–3b. Then find values for the amplifier gain A and the tachometer gain $H = 1/K$, such that $y = 4x + 0.005x_d$.

13.3–1 Consider $H(s) = (s + d)/[s^2 + 2\alpha s + (\alpha^2 - d^2)]$ with $\alpha > 0$. Plot the pole-zero diagram when $d = \alpha/3$ and when $d = 3\alpha$.

13.3–2 Repeat the previous problem for $H(s) = (s + d)/[s^2 - 2\alpha s - (d^2 - \alpha^2)]$.

13.3–3 Use Eq. (12), Section 8.3, to show that a system with $p_1 = p_2 = 0$ is unstable.

13.3–4 The factors $(s - z_1)$ and $(s - p_1)$ will cancel in Eq. (1) if $z_1 = p_1$, so that we have *pole-zero cancellation*. Show that this property is exploited by the compensated probe in Example 5.3–5. Specifically, obtain an expression for $H(s) = V_{\text{out}}/V_{\text{in}}$ in the form of Eq. (1) when $R_p = 9R$ but C_p has an arbitrary value. Then let $C_p = C/9$.

13.3–5 A *zero* in the right half plane does not imply an unstable system. As an example, consider the passive circuit in Fig. P7.1–8 with $Z_1 = 1/sC$ and $Z_2 = R$. Find $H(s) = V_{\text{out}}/V_{\text{in}}$ and show that $p = -1/RC$ but that $z = +1/RC$.

13.3–6 Consider a feedback system with $H(s) = K$ and an *unstable* forward-path function $G(s) = 100/(s - 20)$. Use Eq. (6) to show that the system will be stable if $K > 0.2$.

13.3–7 Repeat the previous problem with $G(s) = -5s/(s - 20)$.

13.3–8 Consider the open-loop function $F(s) = K(s - 2)$, which has a zero in the right half plane. Find the closed-loop pole and the condition on K for stability.

13.3–9 Repeat the previous problem with $F(s) = K(s - 2)/(s + 6)$.

13.3–10 A feedback system with $F(s) = K/s^2$ will oscillate at frequency $f_0 = \sqrt{K}/2\pi$.
(a) Confirm this assertion using Eq. (6). (b) Show that $\underline{F}(f_0) = 1\ \underline{/180°}$.

‡ **13.3–11** A feedback system has $\underline{F}(f) = K/(1 + jf)^3$. Draw the Bode plot and find the upper limit of K_{dB} for stability.

‡ **13.3–12** Repeat the previous problem for $\underline{F}(f) = K/[(1 + jf)(1 + jf/10)^2]$.

13.3–13 Let the microphone of a PA system be $\ell = 2$ m away from the loudspeaker. A sinusoidal signal $y = A \cos \omega t$ at the loudspeaker's output will arrive at the microphone delayed in time by $t_0 = \ell/u$, where $u \approx 344$ m/sec is the velocity of sound in air. Assuming sufficient gain and no phase shift in the PA system, use Eq. (8) and Eq. (5), Section 6.1, to determine the possible oscillation frequencies.

‡ **13.3–14** An oscillator can be constructed by connecting in a loop three identical lowpass op-amp filters like Fig. 7.1–11b. Then $\underline{F}(f) = [\underline{H}(f)]^3$, where $\underline{H}(f)$ is given by the expression following Eq. (10), Section 7.1. Show that this system will oscillate at $f_0 = \sqrt{3}/2\pi R_F C_F$, providing that $R_F = 2R_1$.

‡ **13.3–15** Figure P13.3–15 is an *RC phase-shift oscillator*. The op-amp section has

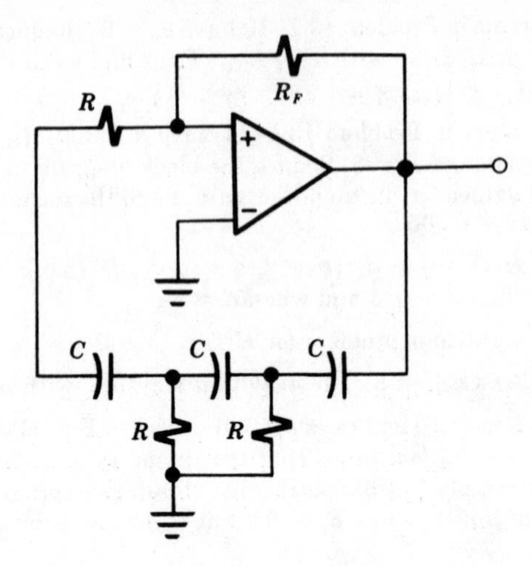

FIGURE P13.3-15

$\underline{G}(f) = -R_F/R$ and the phase-shift network can be shown to have

$$H(f) = \frac{(\alpha f)^3}{(\alpha f)^3 - 5\alpha f - j(6\,\alpha^2 f^2 - 1)}$$

where $\alpha = 2\pi RC$. Show that $f_0 = 1/\sqrt{6}\,\alpha$, providing that $R_F = 29R$. Hint: See Exercise 13.3-3.

14

SIGNAL PROCESSING AND COMMUNICATION SYSTEMS

An electrical *signal* is a voltage or current waveform whose time variations correspond to some desired *information*. The information in question, usually from a nonelectrical source, has been converted into electrical form to take advantage of the relative ease and flexibility of processing and transmitting electrical quantities. This chapter deals with systems designed to handle information-bearing signals for purposes of measurement and processing, as in an instrumentation system, or for long distance transmission, as in a communication system. Regardless of their particular details, all of these systems involve certain basic concepts and share certain problems.

Foremost among signal concepts is *spectral analysis:* the representation of signals in terms of their frequency components. That concept serves as a unifying thread throughout this chapter. Specific topics include the problems of attenuation, distortion, interference, and noise, as well as processing techniques such as equalization, filtering, sampling, modulation, and multiplexing. We'll concentrate on analog signals, and some of the discussions will be descriptive rather than analytical to keep the mathematics at an appropriate level. Even so, the ideas developed here have value in their own right and also establish a framework for understanding the increasingly important digital signal techniques presented in Chapter 16.

OBJECTIVES

After studying this chapter and working the exercises, you should be able to do each of the following:

- Interpret the spectrum of a signal, and relate it to the concept of bandwidth (Section 14.1).
- Given the model of a linear transmission system, calculate the attenuation and state the conditions for distortionless signal recovery (Section 14.2).
- Sketch typical spectra at various points in a system employing product modulation and filtering (Section 14.3).
- State the conditions such that a signal can be sampled and then reconstructed from its sample values (Section 14.3).
- Describe the major causes of interference and noise, and techniques for minimizing their adverse effects (Section 14.4). †
- Identify the advantages and disadvantages of various types of modulation methods, including pulse modulation, and estimate the required transmission bandwidths (Sections 14.3 and 14.5). †

14.1 SIGNALS, SYSTEMS, AND SPECTRAL ANALYSIS

To put the various topics of this chapter in perspective, we begin with an overview of analog signal systems. Then we introduce spectral analysis and the related concept of signal bandwidth.

SIGNAL TRANSMISSION AND PROCESSING

Figure 14.1–1a diagrams a system for measuring and recording the temperature of an industrial oven. The thermistor inside the oven is a sensing device whose electrical resistance R depends upon the temperature. This thermistor forms part of a simple bridge circuit designed to be balanced when $R = R_Q$ at the normal operating temperature T_Q. Temperature changes above or below T_Q cause the bridge to become unbalanced and produce, after differential amplification, a voltage $x(t)$ as sketched in Fig. 14.1–1b. We call $x(t)$ an *analog* signal because the voltage variations are similar or analogous to the temperature variations. A cable connects the amplifier to the recording location where a lowpass filter removes any AC "hum" that has been picked up. The filtered output is applied to a chart recorder, and it might also be used for automatic temperature control.

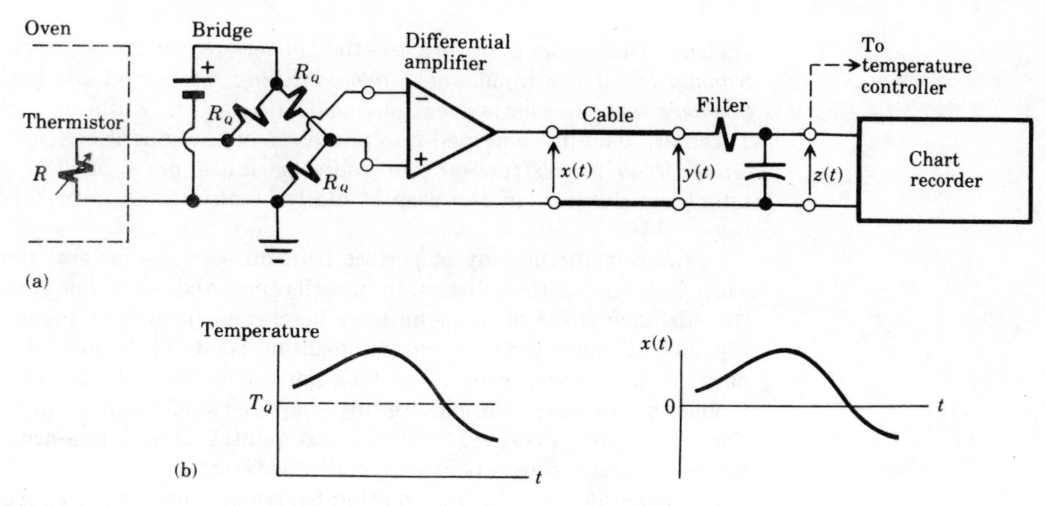

FIGURE 14.1–1
Temperature recording system.

Figure 14.1–2, a generalization from this example, shows the *functional elements* found in most signal-handling systems. The information sought resides in a physical quantity (temperature, acoustical pressure, light intensity, etc.) at a source some distance away from its destination—across the room, say, or perhaps even across the solar system. Initially, a sensor or *transducer* converts the information to electrical form. Auxiliary *conditioning equipment* may work together with the input transducer in order to generate an appropriate signal. Conversely, another transducer at the destination produces the desired output presentation—a digital display, sound wave, video image, or what have you. The thermistor bridge and chart recorder carry out these conversion functions in Fig. 14.1–1a; in a voice communication system, they would be carried out by a microphone at the input and a loudspeaker at the output.

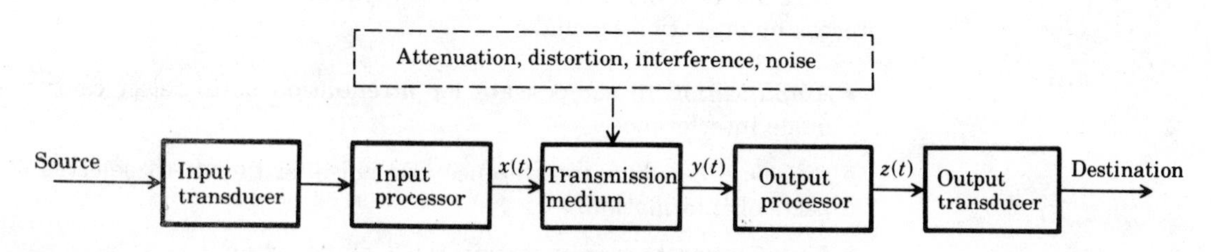

FIGURE 14.1–2
Functional elements of a signal-handling system.

The *transmission medium* serves as the electrical connection spanning the gap between source and destination. This medium might be a short pair of wires or, for longer distances, a coaxial cable or a propagating electromagnetic wave—the latter requiring a *modulating transmitter* at the input and a *demodulating receiver* at the output. A *two-way* transmission system obviously needs both units at each end. These are usually combined into one piece of equipment called a *transceiver* (*trans*mitter/*receiver*) in radio parlance or a *modem* (*modu*lator/*dem*odulator) in the case of digital transmission over telephone lines.

Standing in the way of perfect transmission are several problems caused by attenuation, distortion, interference, and noise. For simplicity, the effects of these phenomena have been concentrated at the center of Fig. 14.1–2, since the transmission medium is often the most vulnerable part of a system, especially long-haul transmission systems. But problems may enter the picture anywhere between source and destination, and can be expected even in instrumentation and data-acquisition systems when distance is not a significant factor.

In general, then, our information-bearing signal arrives at the destination in the form

$$y(t) = K\,\tilde{x}(t - t_d) + n(t) \tag{1}$$

where K represents the attenuation, $\tilde{x}(t - t_d)$ is a distorted and time-delayed version of $x(t)$, and $n(t)$ represents added interference and noise. *Attenuation,* caused by losses within the system, reduces the size or "strength" of the signal, whereas *distortion* is any alteration of the waveshape itself due to energy storage and/or nonlinearities. *Interference* is contamination by extraneous signals, usually man-made, while *noise* comes from natural sources usually internal to the system. Not all of these effects would be serious in every system, but any one of them could pose a major challenge to the design engineer.

To handle these problems, successful information recovery almost always calls for *signal processors* at the input and output. Common signal processing operations include:

- *Amplification* to compensate for attenuation or to reject common-mode interference.
- *Filtering* to reduce interference and noise, or to extract selected aspects of information.
- *Equalization* to correct certain types of distortion.
- *Frequency translation* or *sampling* to obtain a signal that better suits the characteristics of the system.
- *Multiplexing* to accommodate two or more signals in one system.

To this list can be added a host of other operations that enhance the quality of information recovery and presentation—linearizing, averaging, compressing, peak detecting, thresholding, counting, and timing, to name a few.

In the rest of this chapter, we deal with analog signal transmission and illustrative processing techniques. Keep in mind, however, that the versatile and inexpensive microprocessor now permits simple or sophisticated digital processing operations of all sorts.

SPECTRAL ANALYSIS

Throughout previous discussions we have spoken of signals as voltages and currents that are functions of *time,* which is their natural "domain." However, we may also represent a signal in the *frequency domain,* wherein we view it as being made up of a number of sinusoidal components at various frequencies. By doing so, we obtain valuable insight for system analysis and design.

Any *sinusoidal* waveform such as $x_1(t) = A_1 \cos(2\pi f_1 t + \theta_1)$ is completely characterized by three parameters: its *amplitude* A_1, *frequency* f_1, and *phase angle* θ_1. Figure 14.1–3 conveys the same information in two parts: a plot of amplitude versus frequency, called the *amplitude spectrum;* and a plot of phase angle versus frequency, called the *phase spectrum.* Since the signal in question has just one frequency f_1, the amplitude spectrum has a line of height A_1 at $f = f_1$, while the phase spectrum has a line of height θ_1 at $f = f_1$. Taken together, the two plots constitute the spectrum of the signal, and we say that Fig. 14.1–3 is the *frequency-domain representation* of $x_1(t)$.

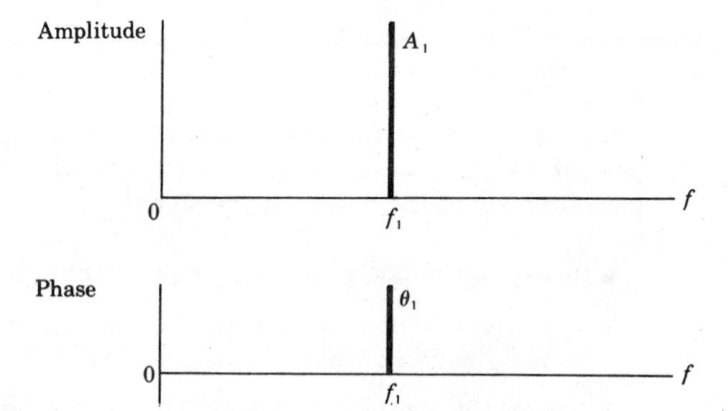

FIGURE 14.1–3 Amplitude and phase spectrum of a sinusoid.

Now consider a more complicated signal consisting of a DC term, plus several sinusoids at various frequencies, say

$$x(t) = 3 - 2 \cos 2\pi 10t + \sin 2\pi 20t$$

which is sketched in Fig. 14.1–4a and looks rather like a sawtooth wave. This signal, too, can be represented in the frequency domain if we first convert each term to the form $A \cos(2\pi ft + \theta)$ where, by convention, the amplitude must be a positive quantity. The DC term, being a constant, becomes a zero-frequency sinusoid with $A = 3$ and $\theta = 0$; the other two components are converted by adding appropriate phase angles, namely

$$-2 \cos 2\pi 10t = 2 \cos(2\pi 10t - 180°)$$

$$\sin 2\pi 20t = \cos(2\pi 20t - 90°)$$

Figure 14.1–4b is the resulting spectrum of $x(t)$. Such diagrams are called *line spectra* because they consist of lines at discrete frequencies.

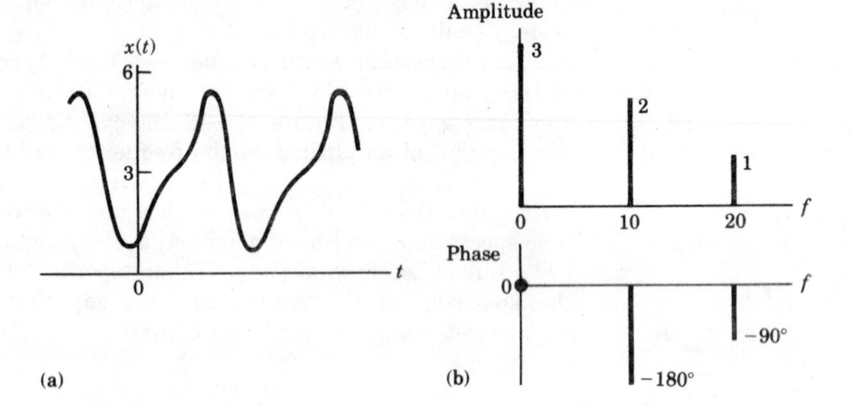

FIGURE 14.1–4 (a) Signal composed of sinusoids. (b) Line spectrum.

Note that the amplitude spectrum provides the bulk of the frequency-domain representation, since it shows what frequencies are contained in the signal and what their relative proportions are. Moreover, the amplitude spectrum provides us with three useful pieces of information about the nature of the signal:

- Its average value, which equals the amplitude of the DC component.
- The extent of slow or "smooth" time variations, which correspond to its low-frequency components.
- The extent of rapid or "sudden" time variations, which correspond to its high-frequency components.

Thus, from the amplitude spectrum in Fig. 14.1–4b, we can predict that the signal has an average value of 3 (from the amplitude at $f = 0$) and relatively smooth time variations (from the absence of components above

$f = 20$ Hz). The phase spectrum becomes significant primarily when we need to know the actual shape of the waveform.

Clearly, it is possible to construct the spectrum of any signal made up entirely from sinusoids, a conclusion that takes on greater impact in view of the *Fourier-series theorem.* This theorem states that almost any *periodic signal* that repeats itself every T seconds can be expanded as an infinite sum of sinusoids whose frequencies are *harmonics* (integer multiples) of the *fundamental frequency*

$$f_0 \triangleq \frac{1}{T} \tag{2}$$

We will not dwell on the theory here, but you might like to consult Section 7.4 for the Fourier-expansion formulas.

Example 14.1–1 The waveform in Fig. 14.1–5a is a rectangular pulse train with pulse duration $D = \frac{1}{3}$ ms and period $T = 1$ ms $= 10^{-3}$ sec. It will therefore contain only those frequencies that are integer multiples of the fundamental $f_0 = 1/10^{-3} = 1$ kHz. The results of Example 7.4–2 then lead to the spectrum plotted in Fig. 14.1–5b.

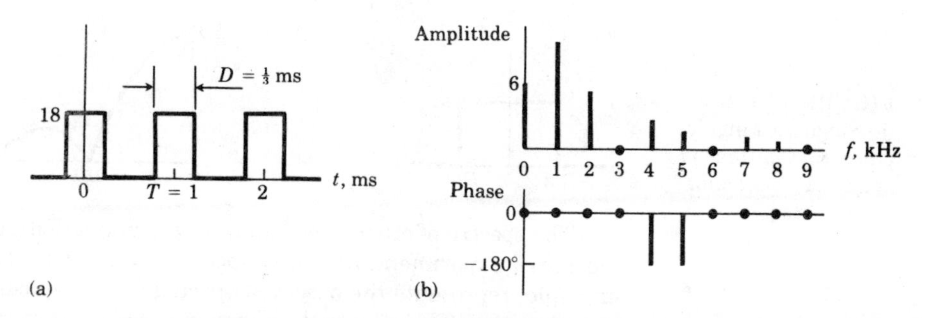

FIGURE 14.1–5 (a) Rectangular pulse train. (b) Line spectrum.

We see that the second harmonic (at $2f_0 = 2$ kHz) has about half the amplitude of the fundamental, while every third harmonic ($3f_0$, $6f_0$, . . .) happens to be completely absent. The amplitude of the "zeroth" harmonic (at $f = 0$) equals the average value of the signal, as expected, and the presence of many high-frequency harmonics corresponds to the sudden "jumps" in the waveform. Going back to Fig. 7.4–3 shows how this pulse train can be built up from its frequency components.

Exercise 14.1–1 Construct the line spectra for the following signals, taking $f_0 = \omega_0/2\pi =$ 100 Hz:

$$x_1(t) = \tfrac{3}{4} + \cos \omega_0 t - \tfrac{1}{3} \cos 3\omega_0 t + \tfrac{1}{5} \cos 5\omega_0 t$$

$$x_2(t) = \sin \omega_0 t - \tfrac{1}{9} \sin 3\omega_0 t$$

Now sketch the waveforms by graphically summing their frequency components. You will find that $x_1(t)$ approximates a rectangular pulse train with $D = T/2$, and that $x_2(t)$ approximates a triangular wave.

SIGNAL BANDWIDTHS

Although we have dealt only with periodic signals, the concept of spectral analysis also applies to *nonperiodic* signals. The major difference is that the spectrum of a nonperiodic signal turns out to be a smooth curve as a function of frequency, rather than lines at specific frequencies; this means that the signal energy is spread over a continuous frequency range instead of being concentrated in discrete sinusoidal components.

Consider, for instance, the *single* rectangular pulse of duration D in Fig. 14.1–6a. Intuitively, we know we could obtain this nonrepeating waveform from a rectangular pulse train by letting $T \to \infty$ so that all the pulses vanish except the one at $t = 0$. But if $T \to \infty$, then $f_0 \to 0$ and the spectral lines must merge into a continuum. Theoretical analysis using the *Fourier integral* shows that the resulting amplitude spectrum will be as plotted in Fig. 14.1–6b, where the value at $f = 0$ now equals the pulse's *area* hD. Note that there is relatively little spectral content for $f > 1/D$.

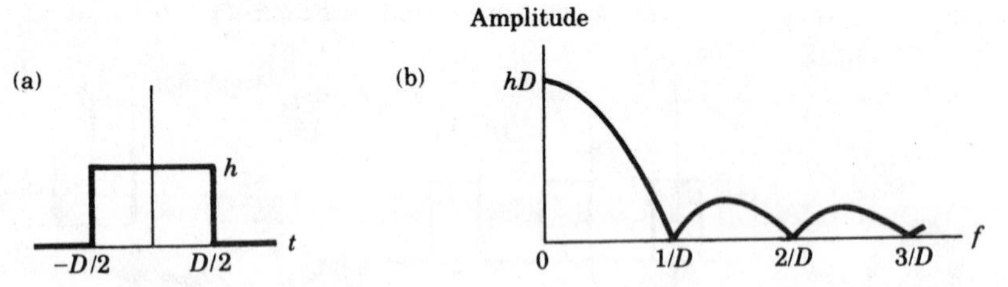

FIGURE 14.1–6 (a) Single rectangular pulse. (b) Continuous amplitude spectrum.

The spectra of other, more complicated nonperiodic signals have been measured experimentally using spectrum analyzers. Figure 14.1–7, for example, represents the observed spectrum of a typical voice signal.

Despite different signal shapes, the pulse train, single pulse, and voice signal all have one feature in common: Their amplitude spectra progressively decrease as $f \to \infty$. Consequently, we can identify some point $f = W$, below which all significant frequency content falls—any higher-frequency components being relatively unimportant to waveform reproduction. We therefore might take $W \approx 3.2$ kHz for a typical voice signal, judging from Fig. 14.1–7, while a high-fidelity audio signal has $W \approx 10$–20 kHz. We call these *lowpass signals* and refer to W as the *signal bandwidth,* by analogy to the frequency response of a lowpass filter (see Fig. 7.1–3). By the same reasoning, the *system bandwidth* B required for transmission or other processing of such a signal must satisfy

$$B \geq W \tag{3}$$

in order to preserve the significant high frequencies.

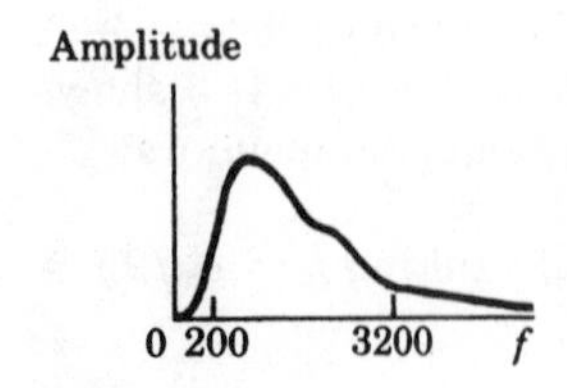

FIGURE 14.1–7 Amplitude spectrum of a typical voice signal.

A case of special interest is the bandwidth of a *single pulse* of duration D that does not necessarily have a rectangular shape. As a rough but handy rule of thumb, we usually take

$$W \approx 1/D \tag{4}$$

so that if $D = 5\ \mu s = 5 \times 10^{-6}$ sec, then $W \approx 0.2 \times 10^{6}$ Hz $= 200$ kHz. This reciprocal relationship means that a short pulse has a large bandwidth, and a long pulse has a small bandwidth. Equation (4) agrees qualitatively with the rectangular pulse spectrum (Fig. 14.1–6) but ignores components above $f = 1/D$. Therefore, preserving the higher frequencies needed for the square corners requires $W > 1/D$.

Although the value of W for a particular signal is somewhat arbitrary, the concept of signal bandwidth plays a major role in signal processing and communication. On the one hand, it allows us to group together all possible waveforms that have the same bandwidth, without needing to study each one individually. And on the other hand, W serves as a measure of *information rate* in the sense that a large-bandwidth signal conveys more information per unit time than a small-bandwidth signal does. Consider, for instance, the difference between the potential information in a video signal with $W \approx 4$ MHz and a voice signal with $W \approx 3.2$ kHz. (A quantitative expression for digital information rate is formulated in Chapter 16.)

Example 14.1–2
TV signal

The waveform in Fig. 14.1–8a represents one horizontal sweep of a monochromatic (black-and-white) television signal. The 5-μs pulse synchronizes the sweep circuit at the receiver, and the $\frac{1}{4}$-μs "blip" near the

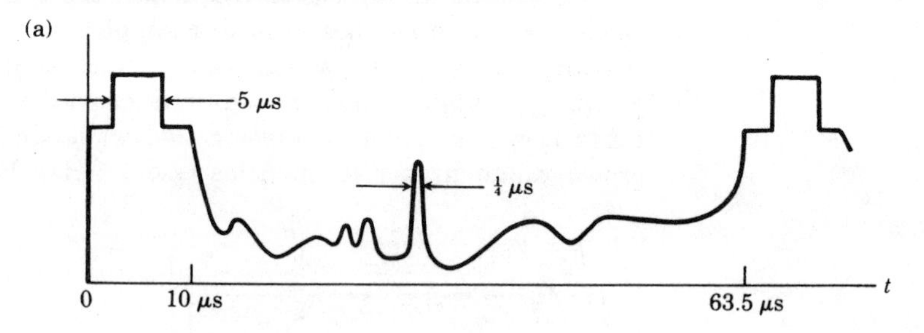

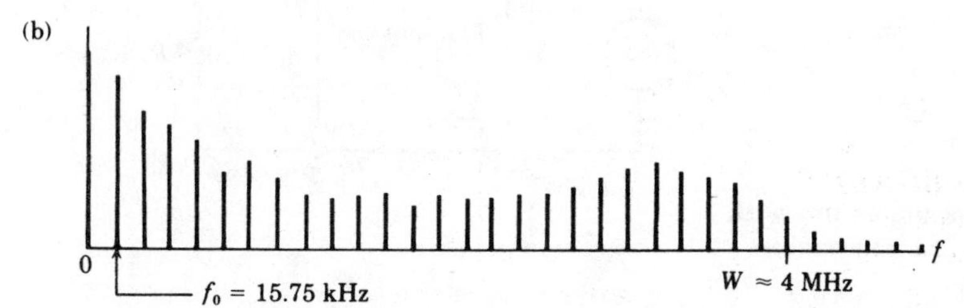

FIGURE 14.1–8
(a) TV signal.
(b) Amplitude spectrum.

center corresponds to the smallest image detail that can be reproduced. We will assume that the waveform repeats periodically with $T \approx 63.5\ \mu s$. The amplitude spectrum sketched in Fig. 14.1–8a consists of harmonics of $f_0 = 1/T \approx 15.75$ kHz and extends up to $W \approx 4$ MHz, where we have used Eq. (4) with $D = \frac{1}{4}\ \mu s$ for the smallest pulse duration. (Through an ingenious arrangement, the additional signals required for a color image are inserted in the gaps between the spectral lines of the black-and-white signal, so that the bandwidth doesn't need to be increased.)

Exercise 14.1–2

With the help of Fig. 7.4–3, estimate the value of W for a rectangular pulse train with $D = T/3$ and $T = 30\ \mu s$ when (a) we are concerned only with counting the number of pulses in a given time, and when (b) we must distinguish between essentially rectangular and nonrectangular pulses.

14.2 ATTENUATION AND DISTORTION

We have qualitatively described the problems of signal attenuation and distortion. In this section, we address these topics quantitatively, starting with the properties of transmission lines, then going on to the corrective processing that can be achieved by amplifiers and equalizers.

TRANSMISSION LINES, ATTENUATION, AND AMPLIFICATION

Figure 14.2–1 represents a transmission line of length ℓ connecting a sinusoidal source to a load resistance at the destination. We let $x(t) = A \cos 2\pi ft$ be the voltage across the line's input terminals and let P_{in} be the average input power. We want to find the resulting output voltage $y(t)$ and power P_{out} delivered to the load.

Depending on the signal frequency, the transmission line itself has several possible configurations: a simple pair of wires—either "open" (parallel) or twisted—works well at frequencies below a few hundred kilohertz; coaxial cables are required for good transmission in the megahertz range; and hollow pipes called waveguides must be used at microwave and higher frequencies ($f \geq 1$ GHz). Fortunately, all of these

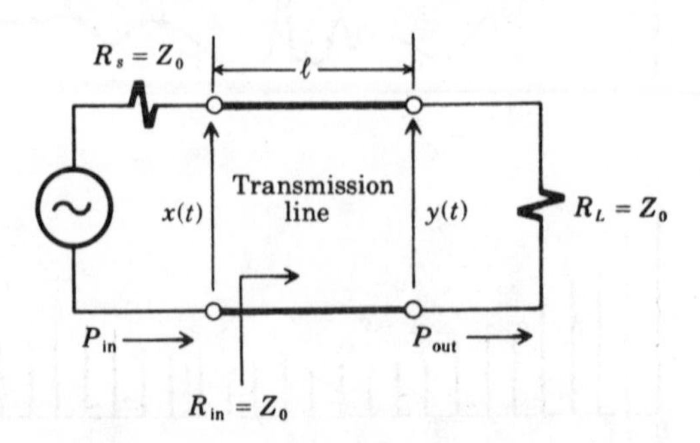

FIGURE 14.2–1 Transmission line with matched impedances.

transmission devices have the same, relatively simple model when used under their normal operating conditions. A similar model applies to radio-wave transmission, as discussed in Section 14.5.

An important transmission-line parameter is its *characteristic impedance* Z_0, which relates the voltage and current at any point along the line. For proper operation, Z_0 should be a real (noncomplex) quantity at the signal frequency; and the source and load resistances should be *matched* to it, meaning that $R_s = Z_0$ and $R_L = Z_0$. Otherwise, an impedance *mismatch* at the load ($R_L \neq Z_0$) would cause a portion of the signal energy to be reflected and travel back toward the source where, if $R_s \neq Z_0$, another portion would be reflected in the forward direction—and so on, back and forth along the line. Not only do the reflected components detract from effective signal transmission, but they may actually damage the line or the source. *Impedance matching* at both ends *eliminates reflections* and also *maximizes power transfer* from source to line and from line to load, a desirable state of affairs for information transfer.

Under matched conditions, the output signal has the form

$$y(t) = |H(f)|\, A \cos\,[2\pi ft + \theta(f)] \tag{1}$$

where $|H(f)|$ is the amplitude ratio of the transmission line and $\theta(f)$ its phase shift. Both of these quantities generally depend on the line length ℓ, as well as on the signal frequency f. The phase shift $\theta(f)$ has a negative value, representing the small amount of time t_d required for the signal to travel from source to destination. Specifically, from Eq. (5), Section 6.1,

$$t_d = -\frac{\theta(f)}{360° f} \tag{2}$$

whose significance will be brought out when we consider signal distortion. The amplitude ratio $|H(f)|$ has a value less than unity, representing energy lost as ohmic heating in the line. This energy loss accounts for the transmission attenuation.

To express attenuation in terms of signal power, we use the fact that $R_{\text{in}} = Z_0$ under matched conditions, and note that the rms input voltage $A/\sqrt{2}$ corresponds to $P_{\text{in}} = A^2/2Z_0$. Then, with $R_L = Z_0$, $P_{\text{out}} = |H(f)|^2 A^2/2Z_0 = |H(f)|^2 P_{\text{in}}$, so that

$$\frac{P_{\text{out}}}{P_{\text{in}}} = |H(f)|^2 = \frac{1}{L} \tag{3}$$

By introducing the *transmission power loss* $L = 1/|H(f)|^2 > 1$, we emphasize that $P_{\text{out}} = P_{\text{in}}/L < P_{\text{in}}$. Equation (3) is independent of the characteristic impedance Z_0 because of impedance matching. It is also valid for nonsinusoidal waveforms when $|H(f)|$ is essentially constant.

The loss of a transmission line increases *exponentially* with distance ℓ and can be written as

$$L = 10^{(\alpha\ell/10)} \tag{4a}$$

in which the parameter α is the line's *attenuation coefficient* in decibels per unit length. Expressing L in dB, similar to power gain, then yields

$$L_{\mathrm{dB}} = 10 \log \frac{P_{\mathrm{in}}}{P_{\mathrm{out}}} = \alpha\ell \tag{4b}$$

This equation provides a very simple way of calculating attenuation and explains, in part, why communication engineers usually work with decibels. Typical values of α range from 0.05 to 100 dB/km, depending on the type of cable and the frequency.

But dB values tend to obscure how rapidly output power falls off as distance increases. We underscore this effect by restating Eq. (4) in the form

$$\frac{P_{\mathrm{out}}}{P_{\mathrm{in}}} = 10^{-(L_{\mathrm{dB}}/10)} = 10^{-(\alpha\ell/10)}$$

which is plotted versus ℓ in Fig. 14.2–2.

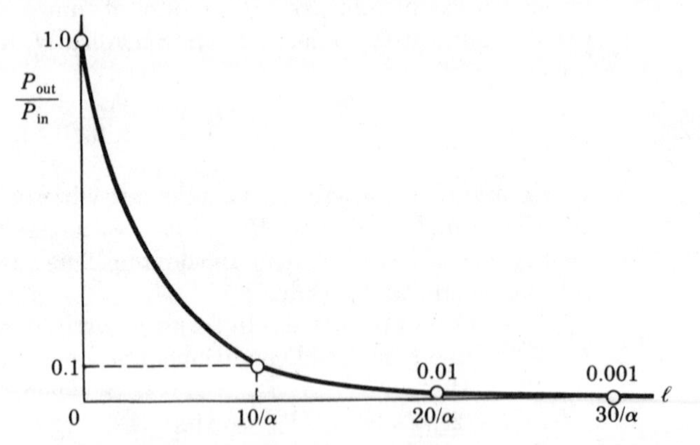

FIGURE 14.2–2
Power ratio $P_{\mathrm{out}}/P_{\mathrm{in}} = 1/L$ versus distance.

Transmission loss can be overcome with the help of one or more amplifiers connected in cascade with the line. Figure 14.2–3 gives the block diagram of a system with a transmitting amplifier, or *preamplifier*, at the source, a receiving amplifier at the destination, and an additional amplifier called a *repeater* inserted at some intermediate point. All amplifiers, of course, must be impedance matched for maximum power transfer.

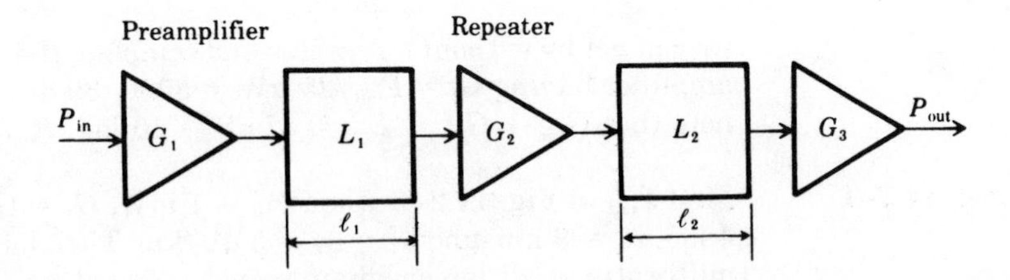

FIGURE 14.2–3 Transmission system with preamplifier and repeater.

We find the final output power by a simple chain calculation. If the preamplifier's power gain is G_1, then its output power will be $G_1 P_{in}$, which results in $G_1 P_{in}/L_1$ at the input to the repeater, and so forth. Thus,

$$\frac{P_{out}}{P_{in}} = \frac{G_1 G_2 G_3}{L_1 L_2} \tag{5}$$

which shows that we can compensate for the line loss and get $P_{out} \geq P_{in}$ if $G_1 G_2 G_3 \geq L_1 L_2$.

A low-loss system might need only one amplifier, usually at the destination. However, noise considerations often call for a preamplifier to boost the signal level before the noise becomes significant. Long-distance transmission generally requires several repeaters, as in the case of transcontinental telephone links that have literally hundreds of repeaters between source and destination.

All such cascade systems are readily analyzed or designed using equations similar to Eq. (5). But one must take care not to exceed the maximum power rating of any amplifier, for an overdriven amplifier would produce nonlinear distortion. One must also observe the maximum ratings of the transmission line, for it could be damaged by excessive heat dissipation or high-voltage rupture. (The latter actually happened to one of the first transatlantic cables.)

Example 14.2–1

Suppose we want $P_{out} = 200$ mW at a point $\ell = 23$ km from a source producing $P_{in} = 4$ mW. The transmission line has $\alpha = 2$ dB/km, so $L_{dB} = 2 \times 23 = 46$ dB and $L = 10^{46/10} \approx 40{,}000$. We therefore need a total amplifier power gain $G = P_{out}/(P_{in}/L) = 2 \times 10^6$, which is to be obtained using adjustable-gain amplifiers that have a minimum input sensitivity of 1 μW (due to internal noise) and a maximum output power rating of 1 W.

If we put all the amplification at the output, the signal power $P_{in}/L = 0.1$ μW at the end of the line would be below the amplifier noise level. If we put all the amplification at the input, the signal power $GP_{in} = 8$ kW would exceed the amplifier power rating and would also probably destroy the transmission line. However, using a preamplifier with $G_1 = 1\text{ W}/P_{in} = 250$ (or about 24 dB) gives $G_1 P_{in}/L = 25$ μW $\gg 0.1$ μW, so we can get by without a repeater and complete the system with an output amplifier having $G_2 = P_{out}/25\ \mu W = 8000$ (39 dB). To check this design, note that $G_{1_{dB}} + G_{2_{dB}} - L_{dB} = 17\ dB = 10 \log (P_{out}/P_{in})$.

Exercise 14.2–1

Find P_{out} in Fig. 14.2–3 when $P_{in} = 1$ mW, $G_1 = G_2 = G_3 = 26$ dB, $\ell_1 = 4$ km, $\ell_2 = 9$ km, and $\alpha_1 = \alpha_2 = 5$ dB/km. Then label the signal levels in milliwatts at all intermediate points.

LINEAR DISTORTION

Besides attenuation, we must also deal with the signal distortion that occurs in transmission. In a *linear* system, distortion is caused entirely by energy-storage effects and can be described in terms of the system's *transfer function* $\underline{H}(f) = |H(f)| \underline{/\theta(f)}$, as diagrammed in Fig. 14.2–4 where $x(t)$ is now an arbitrary information-bearing signal. For further generality, $\underline{H}(f)$ could also include any relevant energy storage in the input transducer and conditioning equipment. We investigate distortion by combining the concepts of signal spectrum and transfer function.

Suppose $x(t)$ has a sinusoidal component of amplitude A_1 and phase θ_1 at frequency f_1. Drawing upon Eqs. (1) and (2), we find the resulting output component will have amplitude $|H(f_1)|A_1$ and phase $\theta_1 + \theta(f_1)$, where $\theta(f_1)$ reflects a time delay $t_d = -\theta(f_1)/360°f_1$. If the spectrum of $x(t)$ contains more than one sinusoidal component, superposition tells us that the output spectrum will consist of the same frequencies, but with each amplitude multiplied by the corresponding value of $|H(f)|$ and each phase angle shifted by $\theta(f)$. (The assumption of a linear system guarantees that no other frequency components will appear in the output signal.)

Distortionless transmission requires that the relative proportions of the amplitude spectrum be preserved at the output and that all components be delayed by the same amount of time. Consequently, the conditions on $\underline{H}(f)$ are

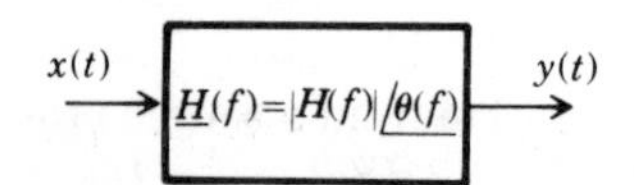

FIGURE 14.2–4

$$|H(f)| = K \tag{6a}$$

$$\theta(f) = -360° t_d f \tag{6b}$$

where K and t_d are constants whose values are more or less arbitrary, within reason. When both conditions hold, any input $x(t)$ produces the output

$$y(t) = Kx(t - t_d) \tag{7}$$

This is an *undistorted* version of $x(t)$, differing only by the attenuation factor K and time delay t_d as illustrated in Fig. 14.2–5.

Since the spectrum of an arbitrary signal might extend to arbitrarily high frequencies, distortionless transmission would require Eq. (4) to hold over all frequency, $0 \leq f \leq \infty$. But energy storage in a real transmis-

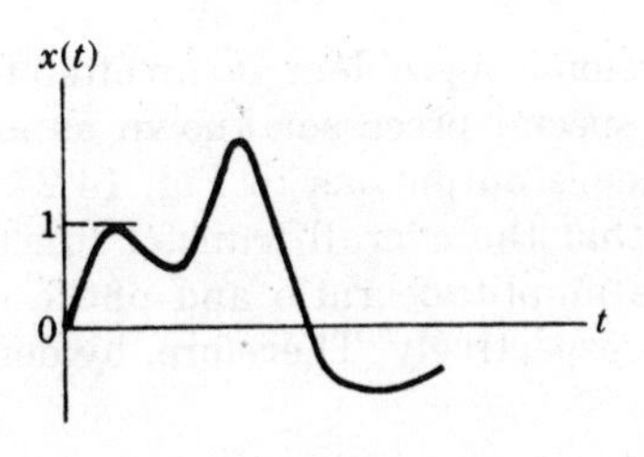

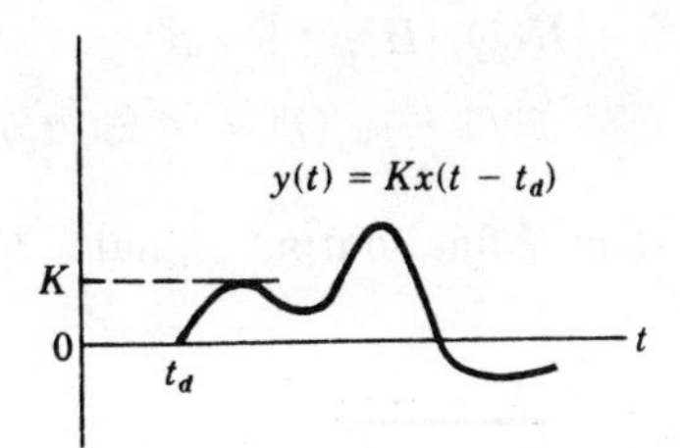

FIGURE 14.2–5 Distortionless transmission.

sion system always results in a high-frequency "rolloff" characteristic, such that $|H(f)| \to 0$ as $f \to \infty$. Consequently, *perfect* waveform reproduction is physically impossible! However, a system can be designed with $\underline{H}(f)$ approximating Eq. (4) over a finite bandwidth B. Such a system would satisfactorily reproduce any input signal with bandwidth $W \le B$, and B can be called the *transmission bandwidth.*

If $\underline{H}(f)$ does not satisfy Eq. (6a) over the frequency range of interest, the resulting output suffers from *amplitude* or *frequency distortion,* in the sense that different frequency components will have different amounts of attenuation. Similarly, if $\underline{H}(f)$ does not satisfy Eq. (6b), the output will have *phase* or *delay distortion,* with different frequencies being delayed by different amounts of time. Both of these effects usually occur together and are put under the general heading of *linear distortion,* as contrasted to the distortion caused by a nonlinear system.

Exercise 14.2–2

Suppose a system has the lowpass frequency response $\underline{H}(f) = 1/[1 + j(f/f_{co})]$ with $f_{co} = 20$ kHz.

(a) Calculate K and t_d at $f = 10$ kHz, 20 kHz, and 30 kHz to support the assertion that $B \approx f_{co}$.

(b) If the input is a single pulse, what is the condition on the pulse duration D for a reasonably undistorted output? Compare your result with Eq. (9), Section 8.2.

Exercise 14.2–3

Demonstrate delay distortion produced by a *constant* phase shift by sketching the output signal when $x_2(t)$, from Exercise 14.1–1, is transmitted by a system with $\underline{H}(f) = 1\,\underline{/90^\circ}$ for all f.

EQUALIZATION †

When linear distortion is a problem, it can often be cured or at least minimized by using a special processor known as an *equalizer* that is connected at the system's output, as in Fig. 14.2–6. The strategy here is based on the fact that the overall transfer function will be the product $\underline{H}(f)\underline{H}_{eq}(f)$, whose amplitude ratio and phase shift are $|H(f)|\,|H_{eq}(f)|$ and $\theta(f) + \theta_{eq}(f)$, respectively. Therefore, by designing the equalizer so that

$$
\begin{aligned}
|H(f)|\,|H_{eq}(f)| &= K' \\
\theta(f) + \theta_{eq}(f) &= -360° t'_d f
\end{aligned}
\tag{8}
$$

we obtain the *undistorted* final output signal $z(t) = K'x(t - t'_d)$.

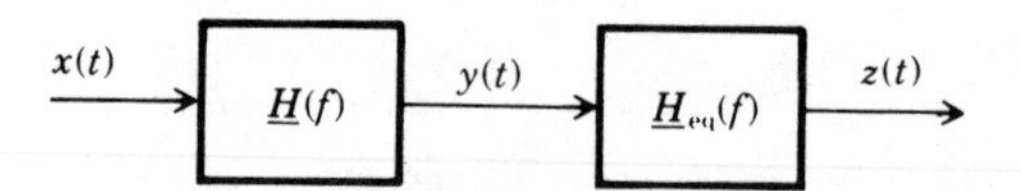

FIGURE 14.2–6
System with equalizer.

Besides its applications in signal transmission, equalization proves helpful whenever energy storage in a transducer or some other system component causes linear distortion. Audio equalizers, for instance, allow you to adjust the amplitude ratio over several frequency bands to compensate for both electrical and acoustical frequency distortion. Incidentally, phase equalization is not critical in audio systems because the human ear has relatively little sensitivity to delay distortion.

Nonlinear distortion, on the other hand, presents a serious problem that often defies correction. For example, if a system has the nonlinear transfer characteristic $y(t) = Kx^2(t)$, then $y(t) \geq 0$ even when $x(t) < 0$, and nothing can be done at the output to distinguish between the positive and negative values of $x(t)$. As a general rule, then, significant nonlinear distortion must be prevented rather than corrected. Less extreme cases can be treated with a nonlinear processor called a linearizer.

Example 14.2–2

Figure 14.2–7 plots $\underline{H}(f)$ from Exercise 14.2–2, along with the transfer function of an equalizer that increases the transmission bandwidth to about 40 kHz. The equalized system has $K' \approx 0.4$ and $t'_d \approx 3\ \mu s$, since $\theta(f) + \theta_{eq}(f) = -45°$ at $f = 40$ kHz.

Exercise 14.2–4

Consider a system having $\underline{H}(f)$ given by Eq. (9), Section 7.1, with $f_{co} =$ 100 Hz. Sketch $|H_{eq}(f)|$ and $\theta_{eq}(f)$ to get $K' = 0.2$ and $t_d = 0$ for 50 Hz $\leq f \leq$ 200 Hz.

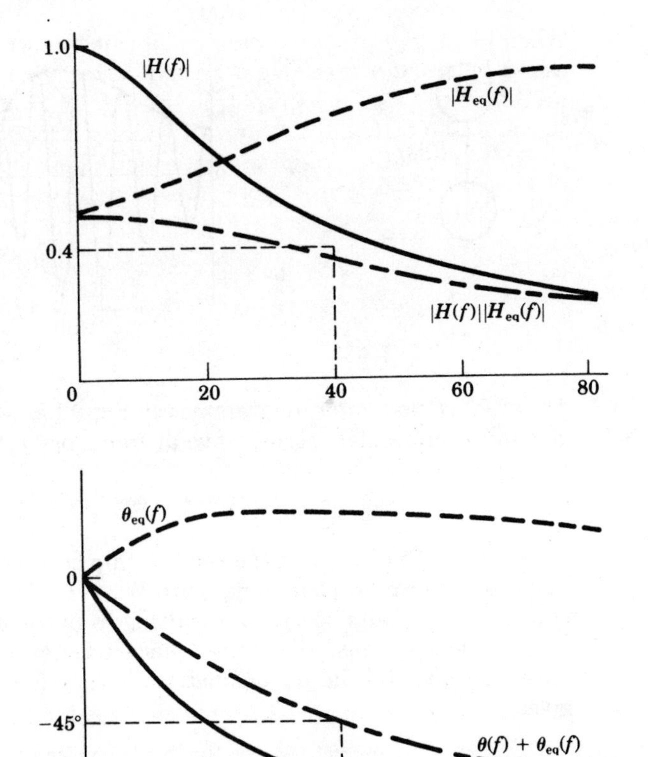

FIGURE 14.2–7 Transfer functions for Example 14.2–2.

14.3 MODULATION, SAMPLING, AND MULTIPLEXING

Modulation is a processing operation that impresses the information from a signal $x(t)$ on a *carrier* waveform whose characteristics suit the particular application. The carrier may be a sinusoid, in which case we get the phenomenon of *frequency translation*. Or the carrier may be a pulse train, which requires that the signal be *sampled* as part of the modulation process. Frequency translation and sampling have many important uses, and both lend themselves to *multiplexing* that permits one transmission system to handle two or more information-bearing signals simultaneously.

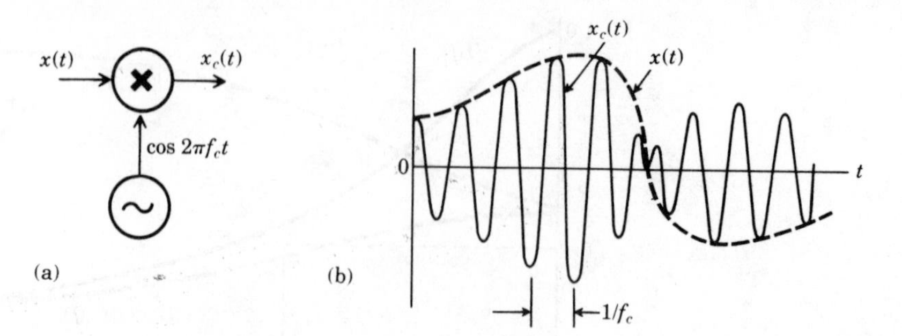

FIGURE 14.3–1
(a) Product modulator.
(b) Waveforms.

FREQUENCY TRANSLATION AND PRODUCT MODULATION

The *product modulator* diagrammed in Fig. 14.3–1a multiplies the signal $x(t)$ and a sinusoidal carrier wave at frequency f_c to produce

$$x_c(t) = x(t)\cos 2\pi f_c t \tag{1}$$

Figure 14.3–1b illustrates the relationship between $x_c(t)$ and $x(t)$, taking the latter to be a lowpass signal with $W \ll f_c$. The modulated wave $x_c(t)$ now has a *bandpass* spectrum resulting from frequency translation.

Frequency translation takes place whenever sinusoids are multiplied. Specifically, the trigonometric identity for the product of cosines gives

$$\cos 2\pi f_1 t \times \cos 2\pi f_2 t = \tfrac{1}{2}\cos 2\pi(f_2 - f_1)t + \tfrac{1}{2}\cos 2\pi(f_2 + f_1)t \tag{2}$$

so the product consists of the *sum and difference frequencies* $f_2 + f_1$ and $f_2 - f_1$. Applying this result to the process at hand, suppose $x(t)$ contains a component $A_m \cos 2\pi f_m t$, which we multiply by a carrier wave with $f_c \gg f_m$ to get

$$A_m \cos 2\pi f_m t \times \cos 2\pi f_c t = \frac{A_m}{2}\cos 2\pi(f_c - f_m)t + \frac{A_m}{2}\cos 2\pi(f_c + f_m)t$$

the various waveforms and line spectra being shown in Fig. 14.3–2. Note that the product contains neither f_m nor f_c, but consists instead of a *pair* of lines offset from f_c by $\pm f_m$. The low frequency f_m has, as a result, been translated to the higher frequencies $f_c \pm f_m$.

Now let $x(t)$ be an arbitrary lowpass signal with the typical amplitude spectrum of Fig. 14.3–3a. The amplitude spectrum of the modulated wave $x_c(t)$ will now contain *two sidebands* each of width W on either side of f_c, as shown in Fig. 14.3–3b. We therefore have a signal that could be

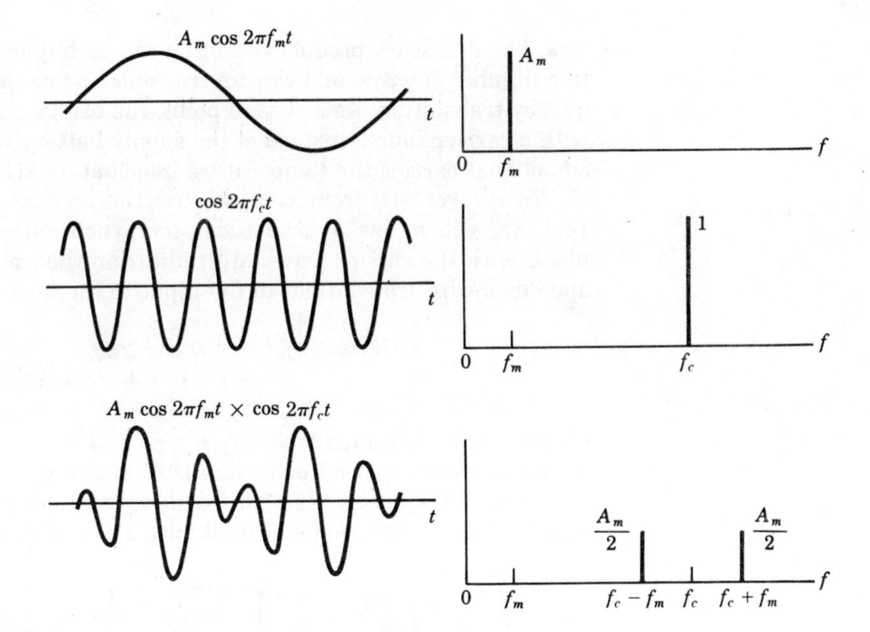

FIGURE 14.3–2 Waveforms and line spectra in frequency translation.

transmitted over a bandpass system with a minimum bandwidth of

$$B = 2W \tag{3}$$

which is precisely *twice* the bandwidth of the modulating signal. For obvious reasons, this process bears the name *double-sideband modulation* (DSB). If bandwidth must be conserved, either the lower or upper sideband may be removed by filtering to get *single-sideband modulation* (SSB) with $B = W$.

The frequency-translation aspect of product modulation, together with the relative lack of restrictions on f_c, allows an engineer to minimize distortion and other problems by putting the carrier frequency at a point where the system has favorable characteristics. This is especially helpful when $x(t)$ contains important DC and low-frequency components that would be lost in a system with transformer coupling or coupling capaci-

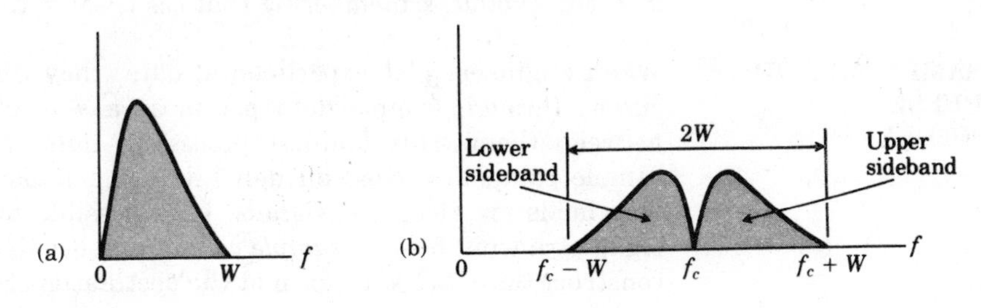

FIGURE 14.3–3 Amplitude spectra in double-sideband modulation: (a) lowpass modulating signal; (b) bandpass modulated signal.

tors. The necessary product operation can be implemented electronically in a number of ways, and certain transducers are easily modified for frequency translation. As a case in point, the bridge circuit in Fig. 14.1–1a with a carrier source instead of the supply battery would produce a DSB signal that carries the temperature information $x(t)$.

To recover $x(t)$ from $x_c(t)$, we use the *product demodulator* of Fig. 14.3–4a, which has a local oscillator synchronized in frequency and phase with the carrier wave. Multiplication then produces both upward and downward translation, so the input to the lowpass filter will be

$$\begin{aligned} x_c(t) \cos 2\pi f_c t &= x(t) \cos^2 2\pi f_c t \\ &= \tfrac{1}{2}\, x(t) + \tfrac{1}{2}\, x(t) \cos 2\pi(2f_c)t \end{aligned}$$

which follows from Eq. (2) with $f_1 = f_2 = f_c$. The first term is proportional to $x(t)$, while the second looks like DSB at carrier frequency $2f_c$. Therefore, if the filter rejects the high-frequency components and passes $f \leq W$, the filtered output has the desired form $z(t) = Kx(t)$.

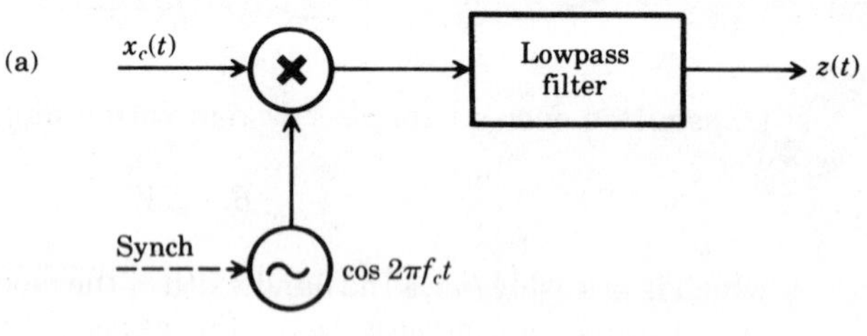

FIGURE 14.3–4 (a) Product demodulation. (b) Spectrum before lowpass filtering.

Exercise 14.3–1 Sketch the amplitude spectrum of $x_c(t)$ when $x(t) = 12 \cos 2\pi 100t + 8 \cos 2\pi 150t$ and $f_c = 600$ Hz. Now list all the frequencies in the product $x_c(t) \cos 2\pi 500t$, remembering that $\cos(-\phi) = \cos \phi$.

SAMPLING AND PULSE MODULATION

When engineers plot experimental data, they frequently draw smooth curves through sample data points as a way of interpolating values between them. This familiar process is quite accurate, provided the sample points are "close enough." Rather astonishingly, the same property holds for electrical signals. It is possible to sample an electrical signal, transmit just the *sample values,* and use them to interpolate or reconstruct the *entire* waveform at the destination. Sampling also makes it possible to convert an analog signal to *digital* form, permitting the use of digital processing methods.

Figure 14.3–5a shows a simple switching sampler and waveforms. The switch alternates between the two contacts at the *sampling frequency* $f_s = 1/T_s$ Hz. It dwells at the upper contact for a short interval $D \ll T_s$ and extracts a sample piece of the input signal $x(t)$ every T_s seconds. Since the lower contact is grounded, the output sampled waveform $x_s(t)$ looks like a train of pulses whose tops carry the sample values of $x(t)$. We can analyze this process and prove that $x(t)$ can be recovered from $x_s(t)$ by modeling sampling as *multiplication* in the form

$$x_s(t) = x(t)s(t) \tag{4}$$

where the *switching function* $s(t)$ is nothing more than a unit-height rectangular pulse train, as in Fig. 14.3–5b. Fourier expansion of the periodic switching function yields

$$s(t) = a_0 + a_1 \cos 2\pi f_s t + a_2 \cos 2\pi(2f_s)t + \cdots$$

with $a_0 = D/T_s$ and $a_n = (2/\pi n) \sin (\pi\, Dn/T_s)$ for $n = 1, 2, \ldots$ (see Table 7.4–1). Upon inserting this expansion, Eq. (4) becomes

$$\begin{aligned} x_s(t) = a_0 x(t) &+ a_1 x(t) \cos 2\pi f_s t \\ &+ a_2 x(t) \cos 2\pi(2f_s)t + \cdots \end{aligned} \tag{5}$$

an awesome-looking result but easily interpreted if we go to the frequency domain.

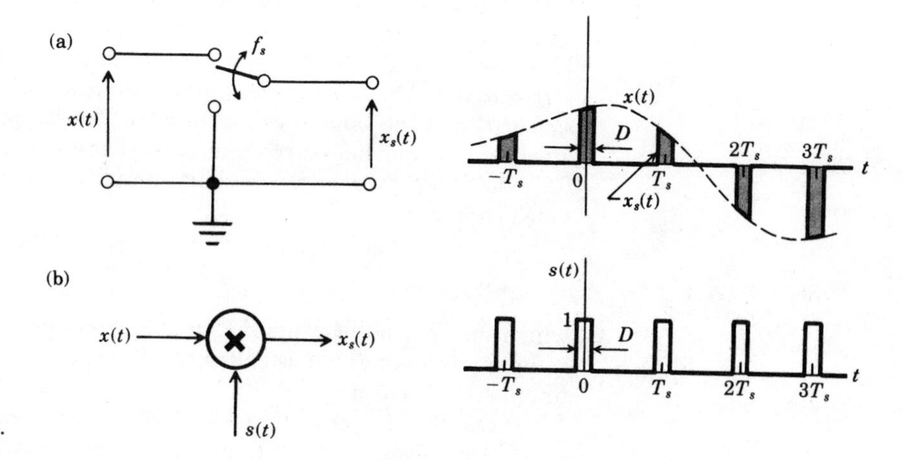

FIGURE 14.3–5
(a) Switching sampler.
(b) Model using switching function $s(t)$.

Suppose that $x(t)$ has the lowpass amplitude spectrum of Fig. 14.3–6a. The corresponding spectrum of the sampled signal $x_s(t)$ sketched in Fig. 14.3–6b is based on a term-by-term examination of Eq. (5) with the assumption $f_s \geq 2W$. The first term has the same spectrum as $x(t)$ scaled by the factor a_0. The second term is identical to product modulation with a scale factor a_1 and carrier frequency f_s, so it has a double-sideband spectrum over the range $f_s - W \leq f \leq f_s + W$. The third and all other terms have the same DSB interpretation with progressively higher carrier frequencies, $2f_s$, $3f_s$, etc.

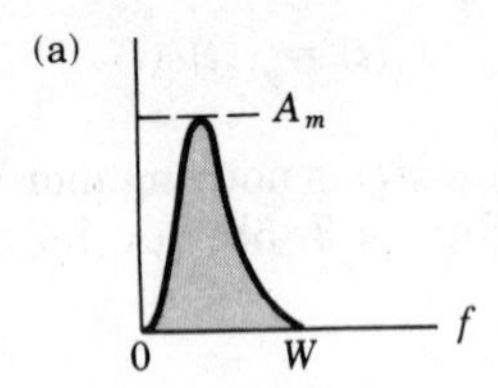

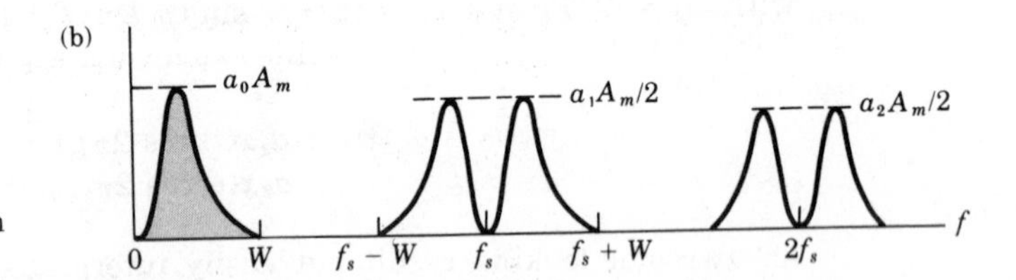

FIGURE 14.3–6
(a) Lowpass amplitude spectrum. (b) Spectrum of sampled signal.

Now observe that not one of the spectral components in Fig. 14.3–6b overlaps the others and that the spectrum in the range $0 \leq f \leq W$ has exactly the same shape as the spectrum of $x(t)$. Therefore, if the sampled signal $x_s(t)$ is applied to a lowpass filter whose bandwidth equals W, the filter removes all the high-frequency stuff and produces at its output $z(t) = a_0 x(t)$. This means that the *entire waveform* $x(t)$ *has been reconstructed* from the sample values in $x_s(t)$. The sample duration D does not affect our conclusions (except as it appears in a_0), and therefore $x(t)$ is fully described by arbitrarily short samples with spacing $T_s = 1/f_s$, provided that

$$f_s \geq 2W \tag{6}$$

which ensures no overlapping in the spectrum of $x_s(t)$. The minimum sampling frequency $2W$ is called the *Nyquist rate,* and $1/2W$ is the maximum sample spacing.

If Eq. (6) does not hold, the resulting spectral overlap causes any high signal frequency $f' > f_s/2$ to appear in the output at a lower frequency

$|f_s - f'| < W$. This phenomenon is known as *aliasing*. To prevent aliasing, the signal $x(t)$ should be processed by a lowpass filter with bandwidth $B_p \leq f_s/2$ *before* sampling.

Figure 14.3–7a shows the elements of a typical *pulse modulation* system. The pulse generator produces a pulse train with the sample values carried by the pulse *amplitude* (PAM), *duration* (PDM), or relative *position* (PPM), as illustrated in Fig. 14.3–7b. At the destination, the modulated pulses are converted back to sample values for reconstruction by lowpass filtering.

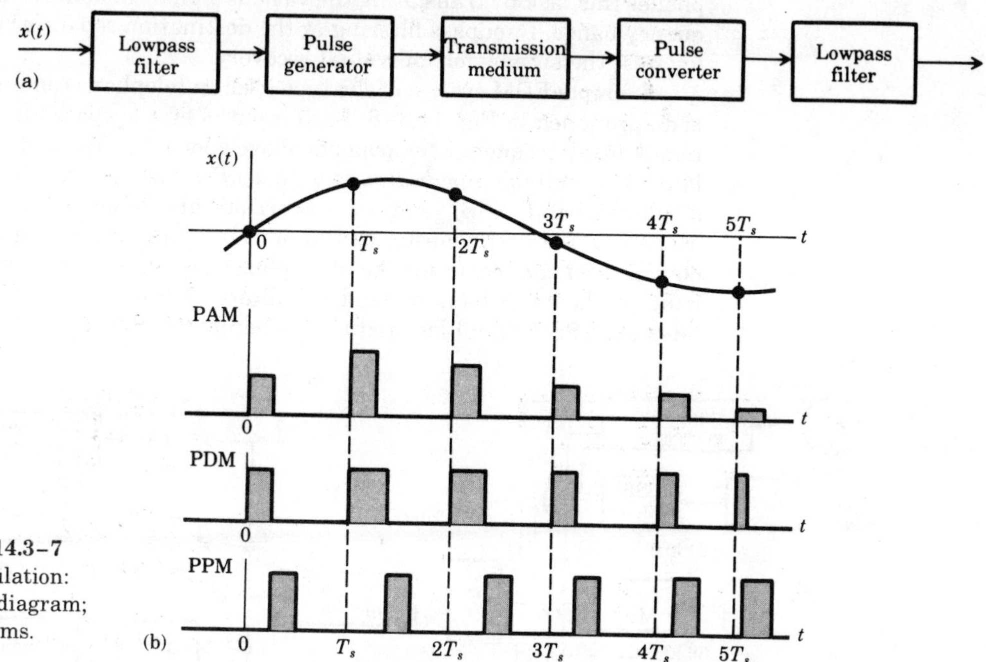

FIGURE 14.3–7 Pulse modulation: (a) system diagram; (b) waveforms.

Pulse-duration and pulse-position modulation have the advantage of being immune to *nonlinear distortion*, since the pulse is either ON or OFF. (This property is also exploited in conjunction with the optoisolator mentioned in Section 9.4.) In exchange, however, the transmission bandwidth must be

$$B \geq 1/D \gg W \tag{7}$$

to accommodate pulses with duration $D \ll T_s \leq 1/2W$. The pulse-modulated wave may be frequency translated for transmission over a bandpass system, which again doubles the bandwidth requirement.

Exercise 14.3–2 Make a careful plot of the signal $x(t) = 3\cos 2\pi 10t - \cos 2\pi 30t$, which approximates a square wave with $W = 30$ Hz. Mark the sample points at $t = 0, \frac{1}{60}, \frac{2}{60}, \ldots, \frac{6}{60}$, corresponding to $T_s = 1/2W$, and convince yourself that $x(t)$ could be recovered from these samples. Then mark the sample points at $t = 0, \frac{1}{40}, \frac{2}{40}, \ldots$, corresponding to $T_s > 1/2W$ and $f_s < 2W$; a smooth curve drawn through these points will demonstrate aliasing.

MULTIPLEXING SYSTEMS

The term *multiplexing* refers to sending two or more signals together on one transmission facility. *Frequency-division multiplexing* (FDM) accomplishes this task by translating the various signals to nonoverlapping frequency bands. Bandpass filtering at the destination separates (or *demultiplexes*) the signals for individual recovery.

A simple FDM system of the type used in telephone communications is diagrammed in Fig. 14.3–8. Each input is first lowpass filtered (LPF), removing all frequency components above about 3.2 kHz, and then modulated onto individual *subcarriers* with 4-kHz spacing. The modulation is single-sideband (SSB), and all subcarriers are generated by harmonic operation on a 4-kHz master oscillator. Summing the SSB signals, plus a 60-kHz *pilot carrier*, forms the multiplexed signal, with the typical spectrum as shown. A bank of bandpass filters (BPF) at the destination isolates each SSB signal for product demodulation. Synchronization is pro-

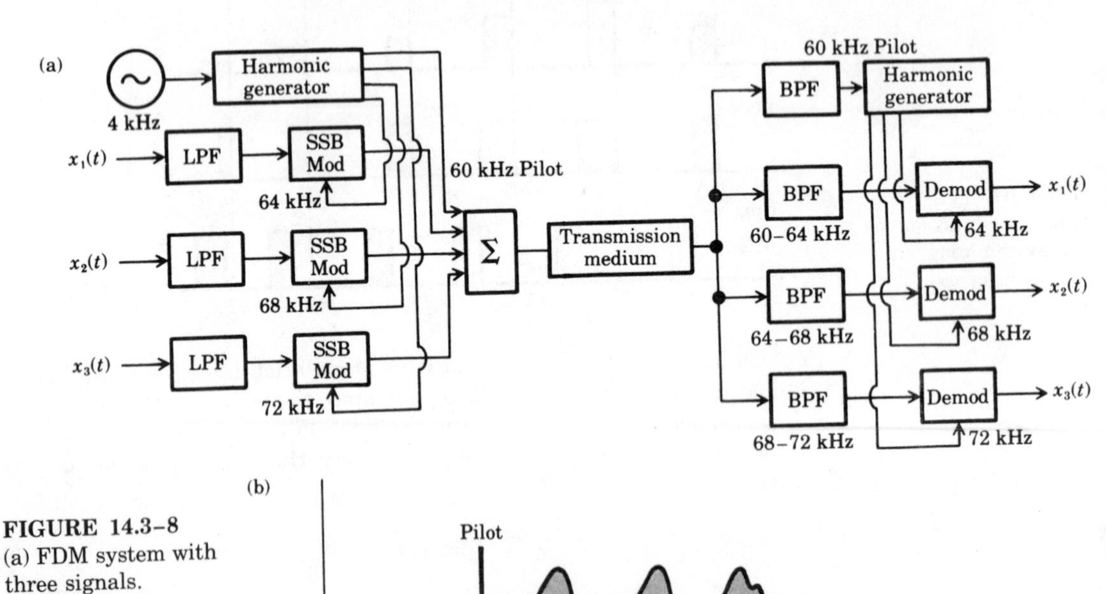

FIGURE 14.3–8
(a) FDM system with three signals.
(b) Spectrum of multiplex signal with pilot.

vided by deriving the local-oscillator waveforms from the pilot carrier. As many as 3600 telephone signals have been multiplexed in this manner. (The FDM system for FM stereo broadcasting will be outlined in Section 14.5.)

Time-division multiplexing (TDM) takes advantage of the fact that a sampled signal is OFF most of the time. The intervals between samples are therefore available for the insertion of samples from other signals.

A rudimentary TDM system has the structure diagrammed in Fig. 14.3–9a, where the three input signals are assumed to have equal bandwidths W. An electronic switch or *commutator* sequentially extracts one sample from each input every T_s seconds, producing a multiplexed waveform with interleaved samples, as in Fig. 14.3–9b. A similar switch at the receiver separates and distributes the samples to a bank of lowpass filters for individual signal reconstruction. More sophisticated systems convert the sample values to pulse modulation before multiplexing and might include carrier modulation after multiplexing.

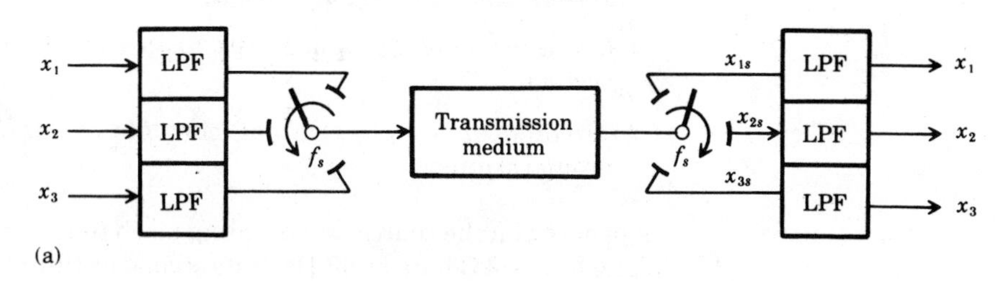

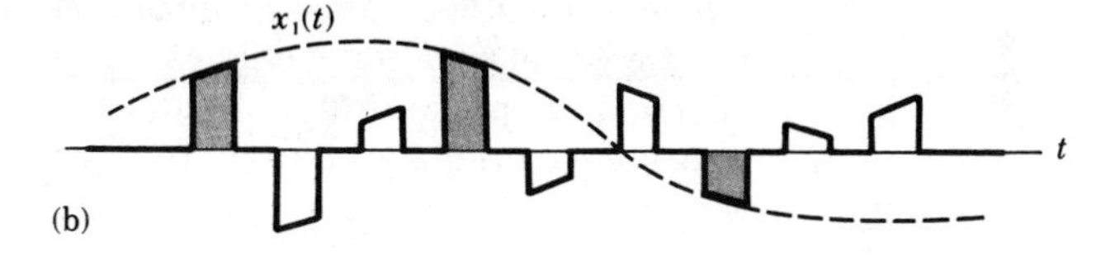

FIGURE 14.3–9
(a) TDM system with three signals.
(b) Multiplexed waveform.

TDM has the advantage of simpler implementation than FDM, especially in view of the wide availability of integrated switching circuits. But, of course, the two commutating switches must be synchronized precisely for successful operation.

Exercise 14.3–3

A data telemetry system is to be designed to handle three different signals with bandwidths $W_1 = 1$ kHz, $W_2 = 2$ kHz, and $W_3 = 3$ kHz. Find the transmission bandwidth required using

(a) FDM with DSB subcarrier modulation.

(b) TDM with pulse duration $D = T_s/6$.

14.4 INTERFERENCE AND NOISE †

An information-bearing signal often becomes contaminated by externally generated interference or by internally generated noise. This section describes some of the major causes of interference and noise and some methods for dealing with their effects.

INTERFERENCE, SHIELDING, AND GROUNDING

Figure 14.4–1 depicts a simple instrumentation system intended to amplify and display the signal $x(t)$ generated by a transducer. The system's electrical environment includes several sources of potential interference such as AC power lines, rotating machinery, thunder storms, radio transmitters, and other electrical and electronic equipment. Interfering signals from these sources enter the system primarily through four mechanisms:

- *Capacitive coupling,* through stray capacitance C between the system and an external voltage.
- *Magnetic coupling,* through mutual inductance M between the system and an external current.
- *Radiative coupling,* from electromagnetic radiation impinging on the system.
- *Ground-loop coupling,* from currents flowing between different ground points.

Depending on the source and coupling mechanism, interference may take many forms: AC hum at 60 Hz or its second or third harmonic; higher frequency "whistles," pulses, and other intelligent-appearing signals; or erratic waveforms commonly called "static." You can hear some of these effects if you put an AM/FM radio near a fluorescent light, an automobile ignition, or even a seemingly innocent calculator.

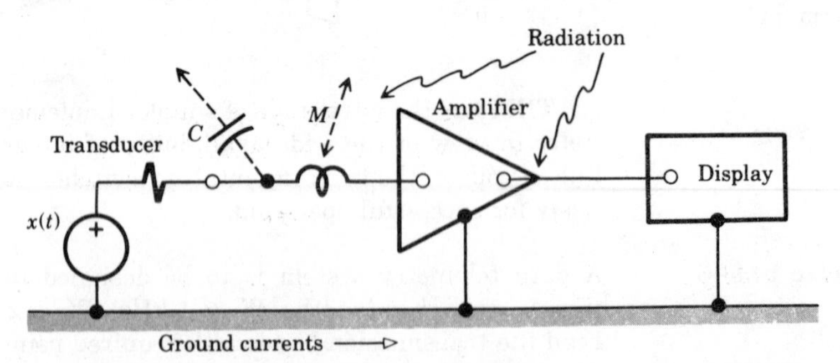

FIGURE 14.4–1 Instrumentation system with interference.

Obviously, if possible, we should turn off any offending sources in the vicinity of the system. Then, to minimize coupling from the inevitable remaining sources, all exposed elements should be enclosed within conducting *shields,* as illustrated by Fig. 14.4–2. (This drawing assumes that the amplifier, its power supply, and all other instruments are shielded by their own metal cases.) When held at a common potential, these shields greatly reduce most types of interference. However, low-frequency magnetic coupling can induce undesired current flow through the shields themselves, so we have to interrupt the shield connection, say at the amplifier's output, to avoid a closed-loop current path. Extreme instances of interference by magnetic coupling may necessitate a layer of special magnetic shielding material in addition to the conducting material.

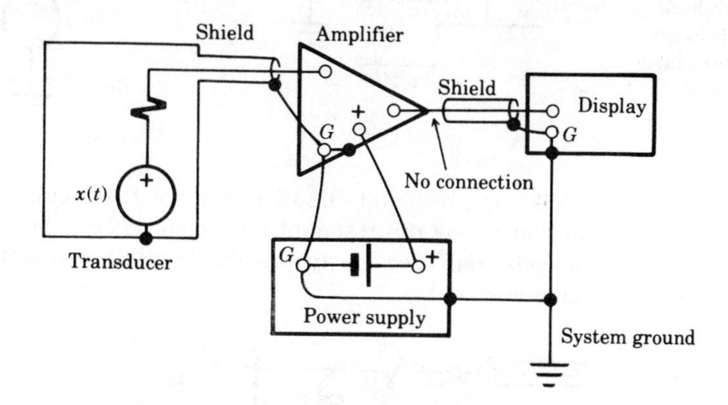

FIGURE 14.4–2 Shielding and grounding arrangement.

Note that grounding terminals (labeled G), the equipment cases, and the shields have all been tied together at a *single ground point* in Fig. 14.3–4 to prevent interference caused by ground-loop currents. In contrast, the three ground points in Fig. 14.4–1 would be at different potentials when the ground currents (usually at 60 Hz) flow through the small resistances between them, producing an apparent interference voltage source in series with $x(t)$. Of course, any instrument connected to an AC power line should be grounded for safety reasons.

If the transducer has a local ground that cannot be disconnected, we must resort to an arrangement like Fig. 14.4–3a, where a separate *ground strap* connects the local ground G' to the system ground point. (Braided-wire straps are preferred for their small mutual inductance.) There will still be a residual voltage v_{cm} that could produce a ground-loop current through the shield, so we disconnect it from the amplifier. The resulting equivalent source circuit is as diagrammed in Fig. 14.4–3b, and includes wiring and shield resistances. We see that v_{cm} becomes a *common-mode voltage* and, if bothersome, can be rejected by differential amplification (see Figs. 4.3–10 and 4.3–11).

Interference with frequency components outside the range of the desired signal can also be removed by a *notch* or *band-stop* filter. For in-

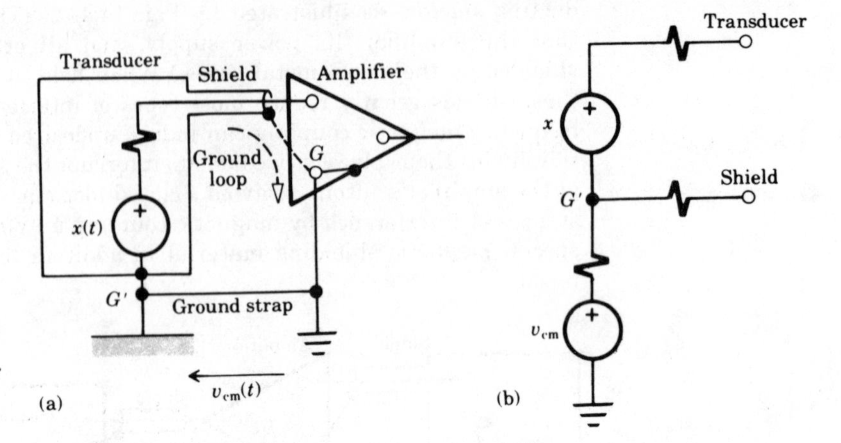

FIGURE 14.4–3
(a) System with local ground at transducer.
(b) Model of transducer with common-mode voltage.

stance, the resonant circuit in Fig. 14.4–4 would short out interference voltages near the resonant frequency $1/2\pi\sqrt{LC}$. Such filtering should precede amplification to prevent possible saturation by a large interference signal.

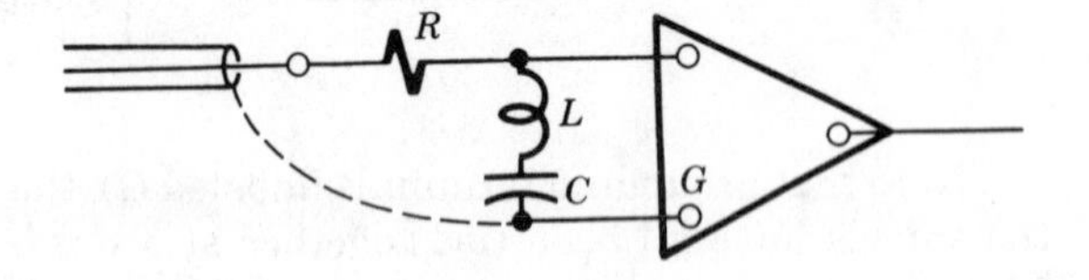

FIGURE 14.4–4
Notch filter to reduce interference.

A fifth interference coupling mechanism is illustrated in Fig. 14.4–5 where two instruments are powered from the same DC supply. Current peaks drawn by one instrument will appear via the source resistance R_s as variations of the terminal voltage v and thus be coupled to the other instrument. This form of interference can be suppressed by adding a *decoupling capacitor* which, together with R_s, acts as a lowpass filter and tends to maintain constant terminal voltage. (Similarly, to prevent un-

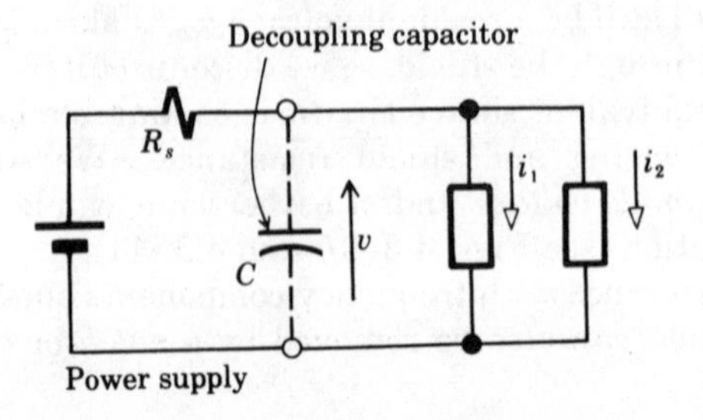

FIGURE 14.4–5
Power supply with decoupling capacitor.

wanted feedback, amplifier circuits like Fig. 12.0 often include RC decoupling in the power-supply connection between the preamplifier and the remaining stages.)

Exercise 14.4–1

Suppose the transducer in Fig. 14.4–1 has 1 kΩ resistance and produces a 0.5-mV (rms) signal. Assume infinite input impedance for the amplifier, and calculate the rms interference voltage caused by
(a) Capacitive coupling via $C = 5$ pF to a 120-V (rms), 60-Hz source.
(b) Magnetic coupling via $M = 10$ nH to a 50-A (rms), 60-Hz source. See Eq. (6), Section 5.2, regarding mutual inductance.

ELECTRICAL NOISE

When a system requires a great deal of amplification for the processing of a very weak signal, the resulting output often includes internally generated noise along with the amplified signal. Every electrical system, as a matter of fact, produces noise although usually so small that it goes unnoticed. To explain the omnipresent nature of noise, we return to our picture of a metallic lattice with electrons dancing about in random thermal motion (Fig. 2.3–4). Each vibrating electron constitutes a tiny current, and the net effect of billions of random electron currents in a resistive material produces the phenomenon called *thermal noise*. Since we cannot build an electrical system without electrons and resistance, thermal noise is inevitable—like death and taxes.

In view of the unpredictable behavior of a thermal noise waveform $n(t)$, typically illustrated by Fig. 14.4–6a, we must resort to *average* properties for its quantitative description. The average value of $n(t)$ equals zero, so a more useful quantity is the *root-mean-square value* n_{rms} that we obtain by averaging $n^2(t)$ over a long time interval and taking the square root of the result. This quantity has the same interpretation relative to average power as the rms value of a sinusoidal waveform. Specifically, if $n(t)$ appears as a voltage across some resistance R, the *average noise power*

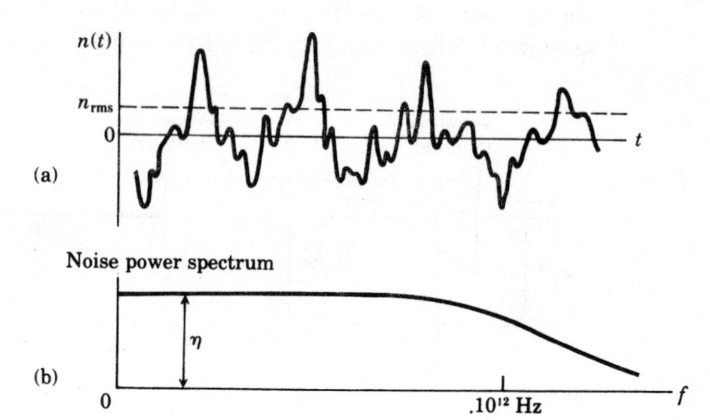

FIGURE 14.4–6
Thermal noise:
(a) typical waveform;
(b) power spectrum.

will be

$$N = \frac{n_{\text{rms}}^2}{R} \tag{1}$$

Similarly, for a noise current, we would write $N = Rn_{\text{rms}}^2$.

Unlike a sinusoid, the spectrum of thermal noise power is *uniformly spread* over frequency up to the infrared region around 10^{12} Hz, as plotted in Fig. 14.4–6b. This spectral distribution is attributed to the fact that there will be equal numbers of electrons vibrating at every frequency, on the average, and thus $n(t)$ contains all electrical frequencies in equal proportion. Thermal noise is also called *white noise,* by analogy to white light, which contains all visible frequencies in equal proportion.

The constant η in Fig. 14.4–6b stands for the *noise power spectral density,* expressed in terms of power per unit frequency (W/Hz). We interpret this concept with the help of Fig. 14.4–7a, where resistance R—a thermal noise source—is connected to an amplifier with a matched input resistance. The amplifier is presumed to be noiseless, with power gain G and bandwidth B. Under these conditions, the output noise power is

$$N_{\text{out}} = G\eta B$$

meaning that the amplifier accepts power ηB falling within the passband and amplifies it by the factor G. We then write the source noise power as

$$N = \eta B \tag{2}$$

which still includes B because we always "see" thermal noise through the limiting bandwidth of one instrument or another. Equation (2) states that N equals the *area* under the spectral-density curve between the instrument's lower cutoff f_ℓ and upper cutoff $f_u = f_\ell + B$, Fig. 14.4–7b, assuming a reasonably "square" passband. (With a more rounded frequency response, such as a first-order filter, one must replace B with a somewhat larger quantity called the *noise equivalent bandwidth* to ac-

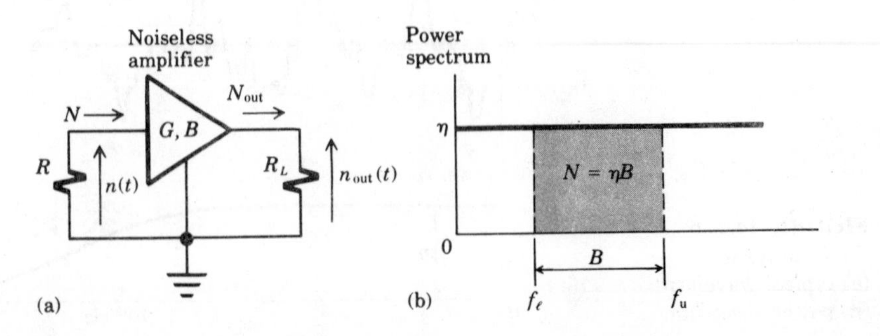

FIGURE 14.4–7 (a) (b)

count for noise components outside the 3-dB bandwidth that get through the filter.)

As to the value of the density constant η, statistical theory shows that

$$\eta = kT \tag{3}$$

where k is Boltzmann's constant and T the source *temperature* in degrees kelvin. A hot resistance is therefore noisier than a cool one, which agrees with our notion of thermally agitated electrons. Note that η and N do not depend on the resistance R, although the rms noise voltage does. Specifically, combining Eqs. (1)–(3) yields

$$n_{\text{rms}} = \sqrt{RkTB} \tag{4}$$

for a thermal source connected to a matched resistance. The open-circuit voltage would be twice this value.

Prior to amplification, thermal noise is exceedingly minute. For instance, take the case of *room temperature* $T_0 \approx 290$ K (17°C) at which

$$\eta_0 = kT_0 \approx 4 \times 10^{-21} \text{ W/Hz} \tag{5}$$

so the noise power in bandwidth $B = 1$ MHz is only $N = \eta_0 B = 4 \times 10^{-15}$ W and Eq. (4) predicts $n_{\text{rms}} = 2\ \mu\text{V}$ if $R = 1\ \text{k}\Omega$. But an amplifier would amplify this noise and contribute additional noise of its own.

Amplifier noise comes from both thermal sources (resistances) and *nonthermal* sources (semiconductor devices). Although nonthermal noise is not related to physical temperature and does not necessarily have a uniform spectrum, it is still convenient to speak of an amplifier's *noise temperature* T_a as a measure of "noisyness" referred to the input. We thus write output noise power caused only by the amplifier in the form

$$N_a = GkT_aB$$

which includes the fact that N_a depends on the gain and bandwidth. For the more general situation of Fig. 14.4–8 with input noise $N = \eta B = kTB$ from a source at temperature T, N_a *adds* to the amplified source noise and

$$N_{\text{out}} = GN + N_a = Gk(T + T_a)B \tag{6}$$

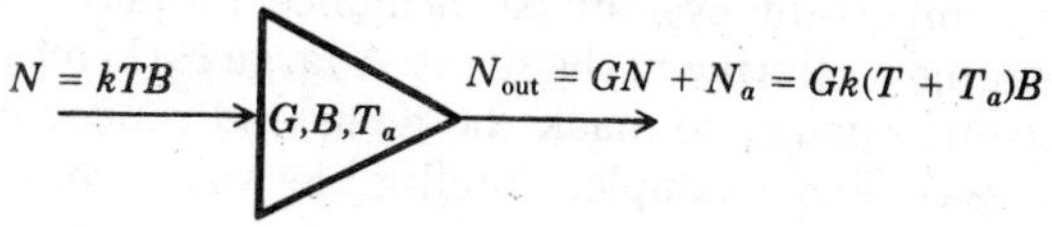

$N = kTB$ → [G, B, T_a] → $N_{\text{out}} = GN + N_a = Gk(T + T_a)B$

FIGURE 14.4–8 Model of a noisy amplifier.

Garden-variety amplifiers have $T_a \gg T_0$, which means that they are very noisy, not physically hot. Therefore, if $T = T_0$, $N_{out} \approx N_a$, and the amplifier noise dominates the source noise—a common occurrence.

Figure 14.4–9 shows the variation of noise temperature with frequency for a typical nonthermal source. The pronounced low-frequency rise reflects several phenomena lumped together under the heading *one-over-f* ($1/f$) *noise.* Transistors generate $1/f$ noise, and so do certain transducers, especially photodiodes and other optical sensors. In addition, slow and random *equipment drifts* have an effect that can be modeled as $1/f$ noise. Whenever the noise temperature of a source or amplifier varies appreciably with frequency, the value to be used in Eq. (3), (4), or (6) is the *average* temperature over the frequency range of interest.

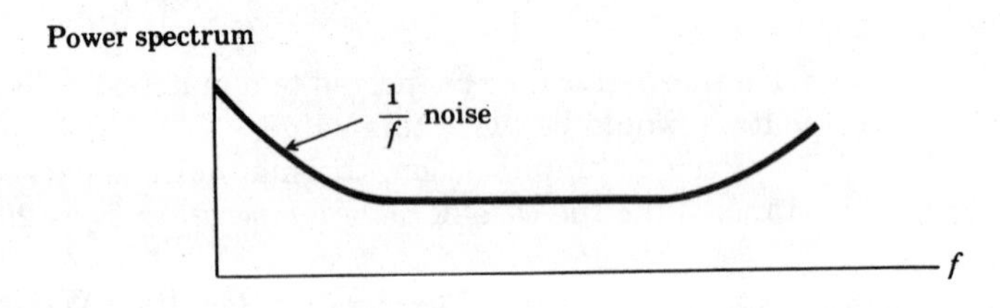

FIGURE 14.4–9 Power spectrum of nonthermal noise.

Exercise 14.4–2 A certain video amplifier has $G = 10^6$ (60 dB), $B = 5$ MHz, and $T_a = 80\ T_0$. Find the rms noise voltage across a 1-kΩ load if $T = T_0$.

Exercise 14.4–3 An amplifier is found to have $N_{out} = 580\ \mu W$ when $T = T_0$. The source is then immersed in liquid nitrogen, so $T \approx 80$ K, and N_{out} drops to 475 μW. Use these results to calculate the amplifier's noise temperature T_a.

SIGNALS IN NOISE

Now let an information signal from a source with $P_{in} = S$ be transmitted via an attenuating medium having loss L, resulting in S/L for the average signal power at the destination. An amplifier with power gain G yields $S_{out} = GS/L$ but also produces output noise power. We therefore speak of the *signal-to-noise ratio* (SNR)

$$\frac{S_{out}}{N_{out}} = \frac{GS/L}{Gk(T + T_a)B} = \frac{S}{Lk(T + T_a)B} \tag{7}$$

in which G has canceled out.

The signal-to-noise ratio—usually expressed in decibels—serves as an important system performance measure, for it tells us the signal strength relative to the noise. A large ratio means that the signal may be strong enough to mask the noise and render it inconsequential (or vice versa). For example, intelligible voice communication is possible if

$S_{out}/N_{out} \geq 20$ dB, but a reasonably noise-free television image must have $S_{out}/N_{out} \geq 50$ dB. Lower values than these would cause noticeable "static" in the voice signal or a "snowy" TV picture.

Examining Eq. (7) reveals that good performance requires a large value for S and/or small values for L, $T + T_a$, and B. Remember, however, that the amplifier's bandwidth B should not be less than the signal bandwidth W. Therefore, we expect noise to be more troublesome when we're dealing with large-bandwidth signals.

If the noise temperature varies with frequency, we can improve system performance by using *frequency translation* to put the signal in a less noisy frequency band. This strategy has particular merit for the case of a lowpass signal whose spectrum falls in the range of an amplifier's $1/f$ noise. Figure 14.4–10a diagrams the implementation with product modulation and demodulation, and Fig. 14.4–10b illustrates the noise reduction in terms of areas under the spectral density curve. (It turns out that synchronized product demodulation doubles the final signal-to-noise ratio, which compensates for the bandwidth doubling of product modulation.) Drawing upon the frequency-translation property of sampling, one normally uses a pair of synchronized switches or *choppers* in place of the multipliers in Fig. 14.4–10a.

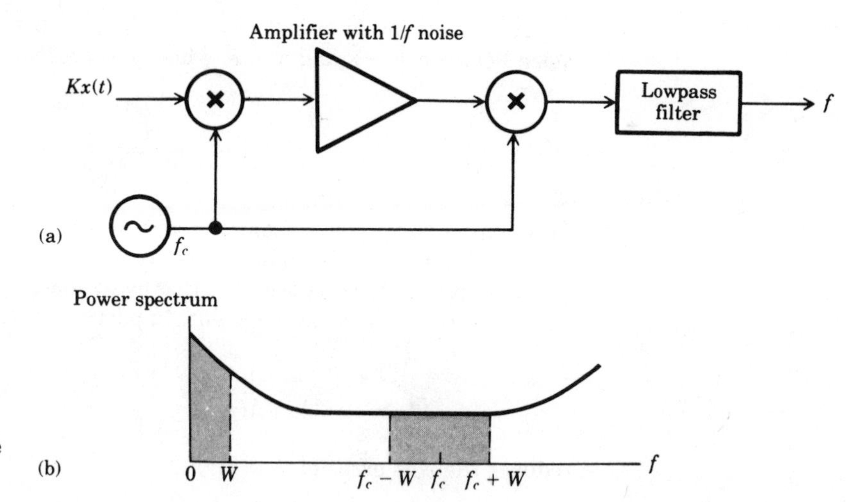

FIGURE 14.4–10 Frequency-translation system (or chopper amplifier) to reduce the effect of $1/f$ noise.

Another way of improving signal-to-noise ratios when the noise density is higher in one region of the spectrum than another is the *preemphasis-deemphasis* technique. If prior to noise contamination, we selectively emphasize (amplify) the portion of the signal spectrum that falls in the high noise region then, after contamination, we can deemphasize the signal-plus-noise, thereby restoring the signal spectrum and reducing the noise. In disk recording, for instance, the higher signal frequencies are emphasized so that a lowpass deemphasis filter will suppress the high-frequency surface noise (or hiss) during playback.

Sometimes the signal in question is a *constant* whose value we seek, as in a simple measurement system. Since a constant corresponds to a DC component (or average value) and since noise usually has zero average value, the measurement accuracy will be enhanced by a lowpass filter with the smallest attainable bandwidth B. (Lowpass filtering, then, partially carries out the operation of *averaging*.) But some noise will get through the filter and cause the processed signal $z(t)$ to fluctuate about the true value x, as in Fig. 14.4–11.

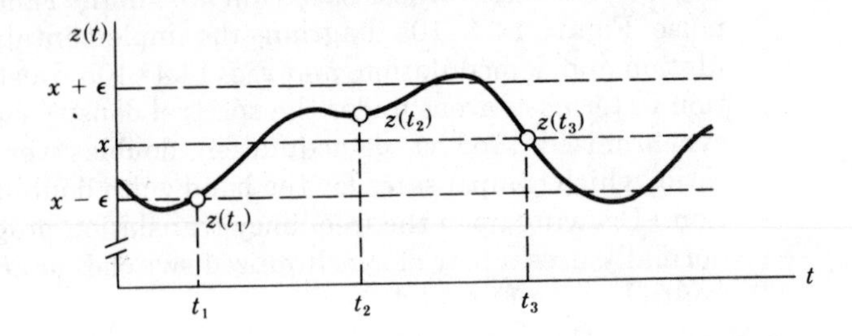

FIGURE 14.4–11
Constant signal with noise fluctuations.

Any one sample of $z(t)$, say $z_1 = z(t_1)$, is then expected to be someplace between $x - \epsilon$ and $x + \epsilon$, where ϵ represents the *rms error* and is given by

$$\epsilon = \frac{x}{\sqrt{S_{\text{out}}/N_{\text{out}}}} \tag{8}$$

This expression follows from the fact that N_{out} is proportional to the square of the rms noise and S_{out} is proportional to x^2. If we take M different samples of $z(t)$ spaced in time by at least $1/B$ seconds, the noise-induced errors tend to average out. In particular, the arithmetic average

$$z_{\text{av}} = \frac{1}{M}(z_1 + z_2 + \cdots + z_m)$$

reduces the rms error to

$$\epsilon_M = \frac{1}{\sqrt{M}}\,\epsilon \tag{9}$$

equivalent to reducing the bandwidth to B/M.

Averaging techniques can also be applied to a sinusoidal signal of known frequency whose amplitude is to be measured. One would then use a narrow *bandpass* filter or a special processor called a *lock-in amplifier*. Other, more sophisticated methods have been devised for extracting information from signals deeply buried in noise, making it possible in essence to "hear a hummingbird in a hurricane." Most of these methods involve digital processing.

Example 14.4–1

A radio antenna at the top of a tower is connected to the receiver by a cable with loss $L = 20$. Given that $S = 1$ nW (10^{-9} W) at the antenna terminals and that $B = 5$ kHz, we seek the condition on the receiving amplifier's noise temperature such that $S_{out}/N_{out} \geq 10^5$. Making the reasonable assumption that the cable is at room temperature, we can write $k(T + T_a) = kT_0(1 + T_a/T_0)$; and substituting numerical values into Eq. (7) leads to $(1 + T_a/T_0) \leq 25$ or $T_a \leq 24\ T_0$, a condition not hard to satisfy.

Exercise 14.4–4

Repeat the previous calculations with $B = 5$ MHz, and show that even a noiseless amplifier would not satisfy the requirement.

Exercise 14.4–5

A measurement system has $S_{out}/N_{out} = 20$ dB and $B = 5$ Hz. How long must the output be observed and averaged to attain an accuracy of $\pm 0.5\%$?

14.5 RADIO COMMUNICATION SYSTEMS †

Having addressed general aspects of signal transmission and processing, we close this chapter with a brief look at the more salient features of radio communication systems. We'll describe the pertinent characteristics of radio propagation, the two most common types of broadcast modulation (AM and FM), and some of the associated hardware. The information-bearing signal represented by $x(t)$ could be a single analog waveform, a pulse-modulated or digital waveform, or a multiplexed signal. Furthermore, many of the techniques developed for radio systems have been adapted to a variety of other uses, such as data telemetry and FM tape recording.

RADIO TRANSMISSION

Radio transmission requires transmitting and receiving *antennas* at the source and destination, as represented in Fig. 14.5–1. When a transmitting antenna is driven by a sinusoidal carrier in the frequency range from about 10 kHz to 100 GHz (10^{11} Hz), an electromagnetic radio wave is radiated and propagates through space without the help of a transmission line. Besides the obvious advantage of not needing a physical connection between source and destination, the total radio transmission loss may be substantially less than that of a transmission-line system.

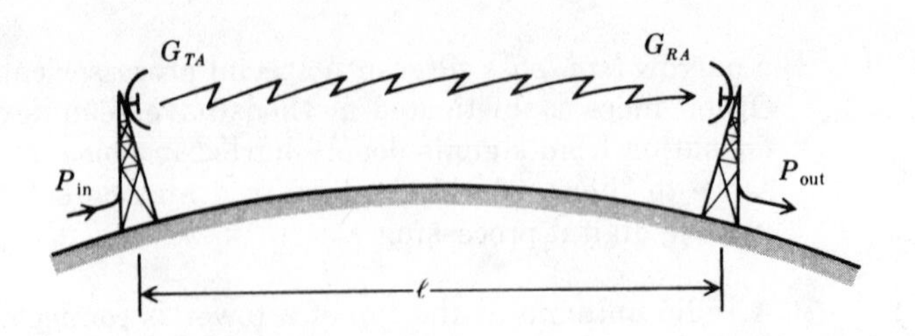

FIGURE 14.5–1 Line-of-sight radio transmission.

Specifically, the power loss on a line-of-sight radio path is

$$L = \frac{P_{in}}{P_{out}} = \frac{1}{G_{TA}G_{RA}}\left(\frac{4\pi\ell}{\lambda}\right)^2 \tag{1}$$

where G_{TA} and G_{RA} are the power gains of the transmitting and receiving antennas. The parameter λ is the *wavelength* of the radio wave, related to the carrier frequency f_c by

$$f_c\lambda = c \approx 3 \times 10^5 \text{ km/s} \tag{2}$$

with c being the velocity of light. We see that radio transmission loss differs from that of a transmission line in two respects: It increases as ℓ^2 instead of exponentially, and it may be compensated, in part, by the antenna gains.

Antenna gain depends on both shape and size. *Dipole antennas,* commonly used at lower radio frequencies, consist of a rod or wire of length $\lambda/10$ to $\lambda/2$ and have $G_A = 1.5\text{–}1.64$ (1.8–2.1 dB). *Horn antennas* and *parabolic dishes* give much more gain at higher frequencies, since they have

$$G_A = \frac{4\pi A_{ap}}{\lambda^2} \tag{3}$$

where A_{ap} is the aperture area. This equation is valid only when $A_{ap} \gg \lambda^2$ and hence $G_A \gg 1$. However, Eq. (1) does not hold if $G_{TA}G_{RA} > (4\pi\ell/\lambda)^2$, implying that $L < 1$, since P_{out} cannot exceed P_{in}. Physically, antenna gain comes from power focusing and collecting, like an optical lens rather than a power amplifier, and Eq. (3) assumes perfect antenna alignment.

Of course, power amplifiers and repeaters may be used in a radio system to increase the final output signal level, and short transmission lines are needed to make connections to the antennas. All of these system components should be impedance matched.

Equations (1) through (3) also apply to optical-frequency transmission where the electromagnetic wave takes the form of a coherent light beam. Figure 14.5–2 shows the radio and optical spectrum, including the special designations of various frequency bands and their representative applications.

Regardless of the particular details, radio transmission is inherently a *bandpass* process in the sense of having a relatively narrow bandwidth B nominally centered at the carrier frequency f_c, similar to the frequency

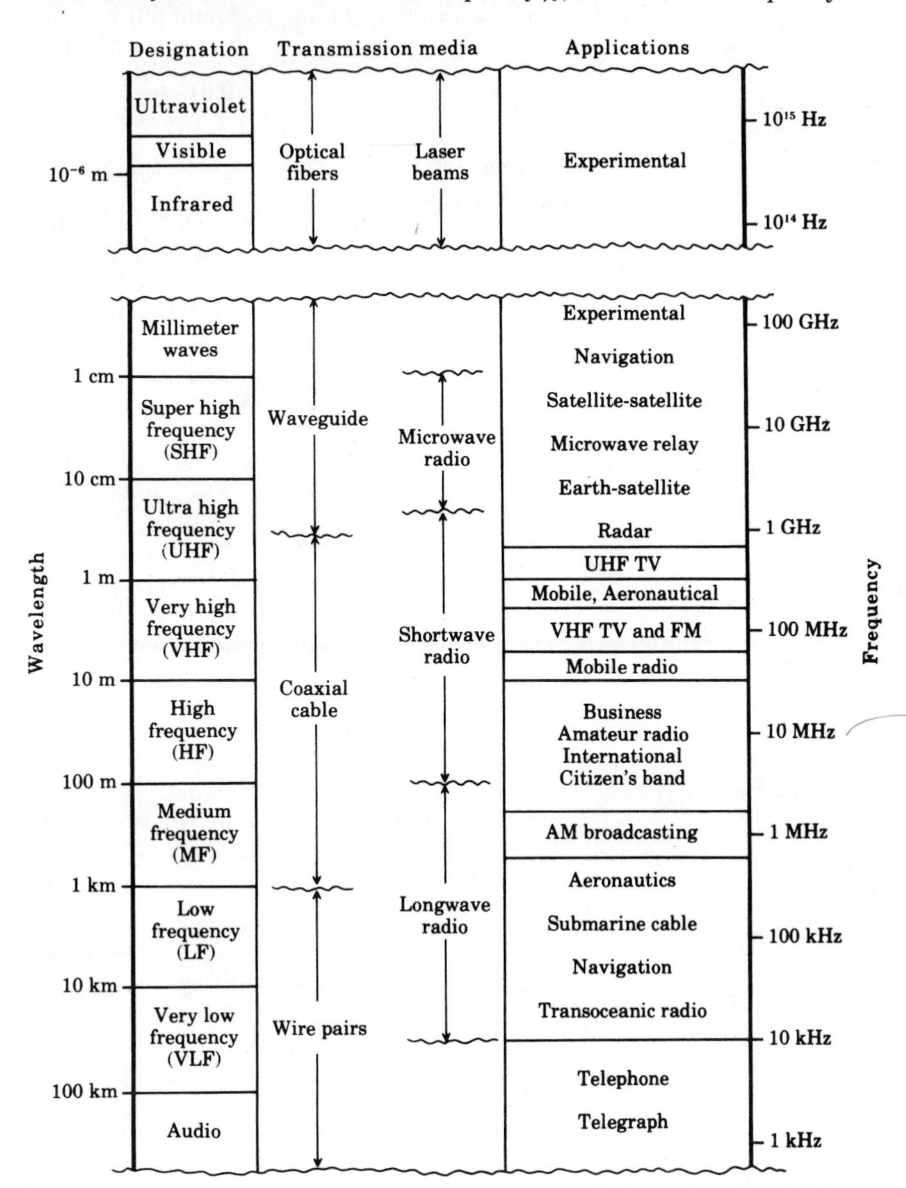

FIGURE 14.5–2
The electromagnetic spectrum. (From *Communication Systems* by A. B. Carlson. Copyright © by McGraw-Hill. Used with permission of McGraw-Hill Book Co.)

response of a tuned circuit (Fig. 7.2–5). Furthermore, practical considerations result in a significant connection between the values of B and f_c. Radio antennas, bandpass amplifiers, and other hardware associated with bandpass transmission require very careful design to avoid serious distortion if the bandwidth B must be either very large or very small compared to f_c. The *fractional bandwidth* B/f_c becomes a key design factor and, as a rule of thumb, it should fall within

$$0.01 \leq \frac{B}{f_c} \leq 0.1 \tag{4}$$

Larger and smaller values *can* be achieved, to be sure, but generally at great expense.

An immediate implication of Eq. (4) is that *large signal bandwidths require high carrier frequencies* for transmission, say $f_c \geq 10B$. This is one reason why TV signals are transmitted at $f_c \approx 100$ MHz, while AM radio signals get by at $f_c \approx 1$ MHz. Similarly, much of the current interest in optical communication systems stems from the tremendous bandwidth potential ($\approx 10^{12}$ Hz) and the correspondingly high information rate.

Example 14.5–1

A signal with a bandwidth of 150 MHz is to be transmitted 40 km. We could use a microwave carrier at $f_c = 20 \times 150$ MHz $= 3$ GHz, for which $\lambda = 10^{-4}$ km $= 0.1$ m, and circular-aperture parabolic antennas with a radius of 0.5 m, so $G_{TA} = G_{RA} = 4\pi(\pi\ 0.5^2)/0.1^2 \approx 1000$ (30 dB). The transmission loss is then $L = 10^{-6}(4\pi \times 40/10^{-4})^2 = 2.5 \times 10^7$ or 74 dB. This may seem to be a considerable loss, but a microwave transmission line (waveguide) would typically have $\alpha = 4$ dB/km and $L_{dB} = 160$ dB when $\ell = 40$ km. By the way, 40 km is about the maximum possible distance for line-of-sight transmission over smooth terrain without using tall antenna towers.

Exercise 14.5–1

Suppose a certain radio system and a transmission-line system both have $L_{dB} = 60$ dB when $\ell = 12$ km. What is the loss of each if the distance is increased to 24 km?

AMPLITUDE MODULATION

Clearly, some sort of modulation is required to put the information from a lowpass signal on a high-frequency radio carrier. Product modulation (DSB or SSB) is one possibility, with SSB allowing more radio channels in a given frequency band. However, successful product demodulation depends critically on synchronizing the local oscillator, a tricky and expensive proposition. In point-to-point communication (one transmitter, one receiver) synchronization may be worth the trouble, but in broadcasting systems (one transmitter, *many* receivers) the added cost per receiver is hard to justify and we look for another form of modulation that simplifies the hardware required for demodulation. One answer is the familiar amplitude modulation.

Amplitude modulation (AM) differs from DSB by the addition of a carrier-frequency component, so that

$$\begin{aligned} x_c(t) &= x(t)mA_c \cos 2\pi f_c t + A_c \cos 2\pi f_c t \\ &= [1 + mx(t)]A_c \cos 2\pi f_c t \end{aligned} \tag{5}$$

where the constant m is called the *modulation index.* The spectrum then looks like Fig. 14.3–3b with a line of height A_c at $f = f_c$. Adding the carrier component does not change the bandwidth, still given by $B = 2W$, but it does change the modulated waveform. It also roughly doubles the total power, which is a price paid at the transmitter for simplified demodulation.

Figure 14.5–3a represents a typical $x(t)$ and Fig. 14.5–3b is the corresponding $x_c(t)$, assuming

$$-1 \le mx(t) \le 1 \tag{6}$$

so that $1 + mx(t)$ is never negative. Under this condition the *envelope* of $x_c(t)$, indicated by the dashed line, has exactly the same shape as $x(t)$. Thus, we can demodulate AM simply by extracting the envelope of $x_c(t)$, a process called *envelope detection,* and involving nothing more than a few circuit elements.

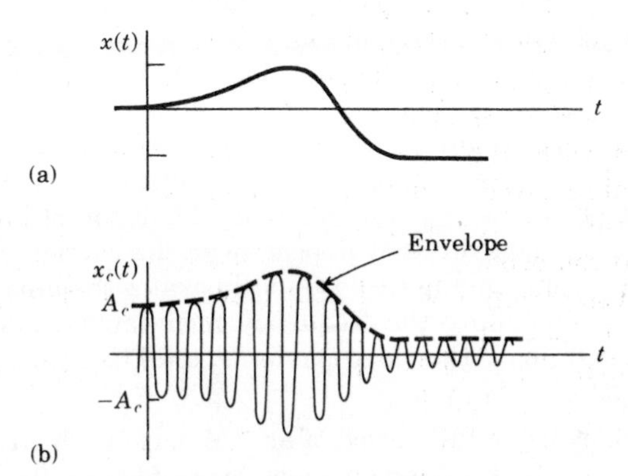

FIGURE 14.5–3
AM waveforms.

Figure 14.5–4a gives the circuit of a simple envelope detector. The diode together with C_1 and R_1 act like a halfwave rectifier so that $x_1(t)$ follows the envelope of $x_c(t)$, as in Fig. 14.5–4b. Then C_2 serves as a DC block so the output voltage has positive and negative variations closely

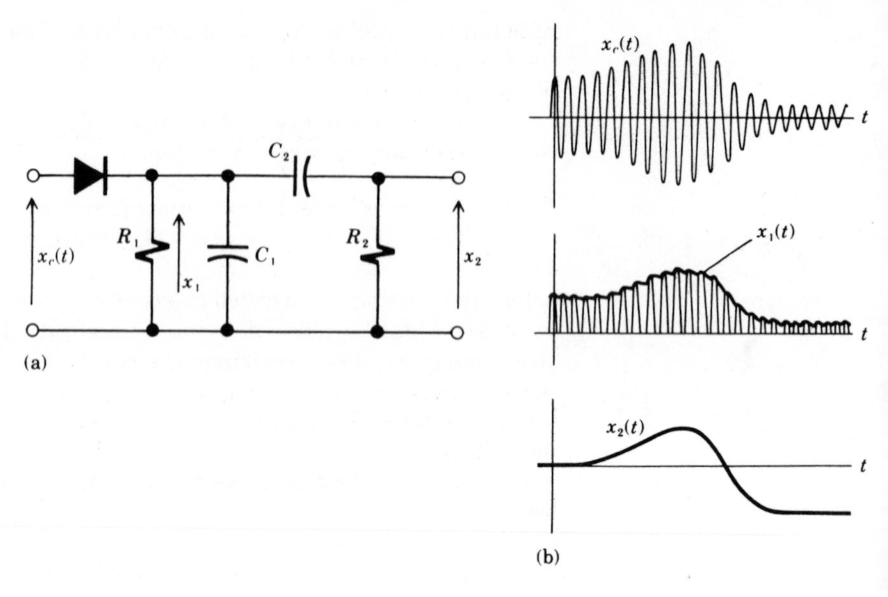

FIGURE 14.5–4
(a) Envelope detector.
(b) Waveforms.

approximating the signal $x(t)$. Because of the necessary DC block, enveloped-detected AM is unsatisfactory when $x(t)$ contains significant DC and low-frequency components.

Often $x_1(t)$ is also lowpass filtered to produce a voltage proportional to the carrier amplitude A_c. This voltage is fed back to earlier stages of the receiver as an *automatic volume control* (AVC) signal that compensates for any fading in the received signal level.

The bandwidth-doubling effect of AM is a serious liability for the transmission of large-bandwidth signals. *Television broadcasting* therefore employs a modified scheme called *vestigial-sideband modulation* (VSB) with all but a vestige of the lower sideband of the translated video signal removed. Complete sideband suppression (or SSB) would be unsatisfactory due to an accompanying distortion of important low-frequency video components that appear near the carrier frequency in the translated spectrum. Figure 14.5–5 illustrates the broadcast spectrum of a TV signal, including the frequency modulated audio on a carrier that is 4.5 MHz above the video carrier. A total bandwidth of 6 MHz is allocated to each TV channel.

Exercise 14.5–2

(a) Sketch one full period of an AM wave with $x(t) = \cos 2\pi f_m t$, $f_m \ll f_c$, and $m = 1$. Draw the envelope by connecting the positive peaks of $x_c(t)$. (b) Repeat (a) with $m = 2$, in which case the carrier is *over-modulated* and the envelope does not have the same shape as $x(t)$.

SUPER-HETERODYNE RECEIVERS

A radio receiver must perform three general functions: tuning (station selection), demodulation, and amplification. However, most receivers—including commercial AM, FM, and TV sets—are somewhat more com-

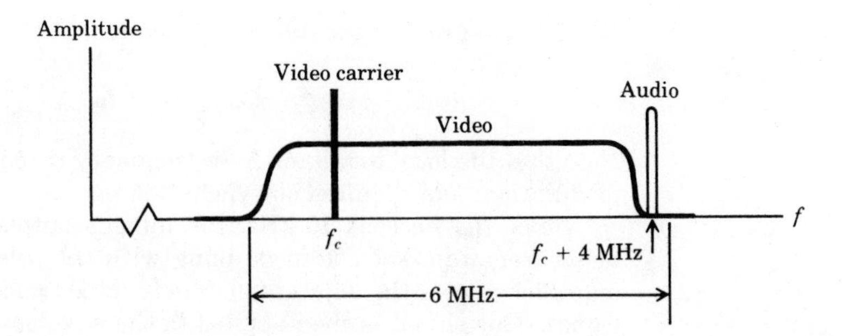

FIGURE 14.5–5 Broadcast spectrum of TV signal.

plicated, largely due to fractional-bandwidth design considerations. To illustrate these concepts, we'll describe the common *superheterodyne* (or "superhet") AM radio shown in Fig. 14.5–6.

AM broadcasting stations are assigned carrier frequencies from 540 kHz to 1600 kHz with 10-kHz spacing. (This allows for $B \approx 10$ kHz or $W = B/2 \approx 5$ kHz, which provides moderate audio fidelity.) Since a radio antenna picks up all of these signals, the first function of the receiver is to select the desired station. This is partly accomplished by a tunable bandpass filter called the *radio-frequency* (RF) *stage*. If we desire the station at $f_c = 1000$ kHz, for example, the RF stage may actually pass 975–1025 kHz if its bandwidth is $B_{RF} \approx 0.05 f_c = 50$ kHz. Considerable amplification and filtering, then, is necessary before the desired signal can be demodulated.

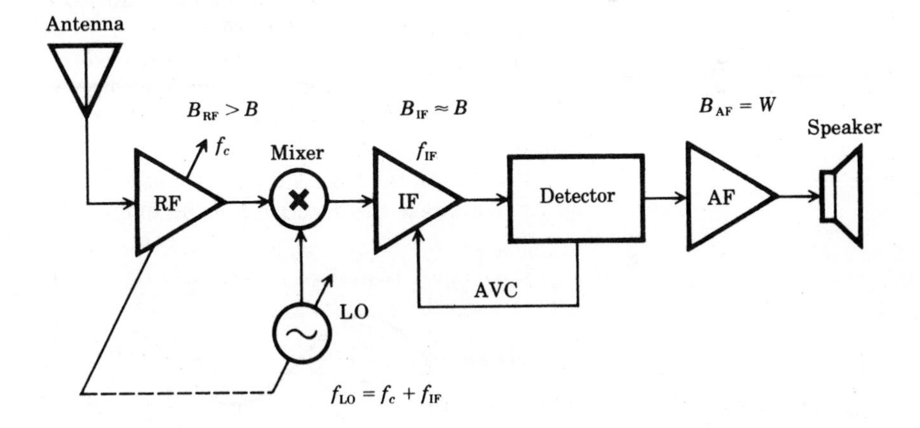

FIGURE 14.5–6 Superheterodyne AM radio.

But high-gain tunable amplifiers with small fractional bandwidths are difficult to build. Instead, the superhet has a fixed-tuned *intermediate-frequency* (IF) *amplifier*, usually with center frequency $f_{IF} = 455$ kHz and bandwidth $B_{IF} = 10$ kHz, so $B_{IF}/f_{IF} \approx 0.02$. A frequency translator consisting of a *mixer* (or multiplier) and *local oscillator* translates the desired f_c down to f_{IF}. Thus, f_{LO} is adjusted simultaneously with the

RF stage to produce the difference frequency

$$|f_{LO} - f_c| = f_{IF} \tag{7}$$

Note that the local oscillator is for frequency translation, not product demodulation, and requires no synchronizing.

Since B_{RF} exceeds 10 kHz, the mixer's output will include components from *adjacent channels* along with the selected channel. The IF amplifier rejects the adjacent-channel signals and amplifies the desired signal. This signal is then applied to the envelope detector and, finally, the demodulated signal is further amplified by an *audio-frequency* (AF) amplifier, which drives the loudspeaker.

Exercise 14.5–3

What *two* values of f_c satisfy Eq. (7) when f_{LO} = 995 kHz and f_{IF} = 455 kHz? The higher one is called the *image frequency* and must be removed by the RF stage before it gets to the mixer.

FREQUENCY MODULATION

An alternative to modulating the amplitude of the carrier wave is *frequency modulation* (FM). FM might be accomplished with a voltage-controlled oscillator (VCO) whose output has a constant amplitude but the variable instantaneous frequency

$$f(t) = f_c + mx(t) \tag{8}$$

with m being the modulation index. Figure 14.5–7 shows a modulating signal $x(t)$ and the resulting FM waveform whose frequency swings above and below f_c in proportion to $x(t)$. Compared to AM, FM has the desirable attribute of suppressing interference and noise—at the price of a much larger bandwidth. *Phase modulation* (PM) is similar to FM, but has less effective noise suppression.

Because of the time-varying frequency in an FM signal, it is very difficult to analyze the spectrum of the modulated waveform, and exact re-

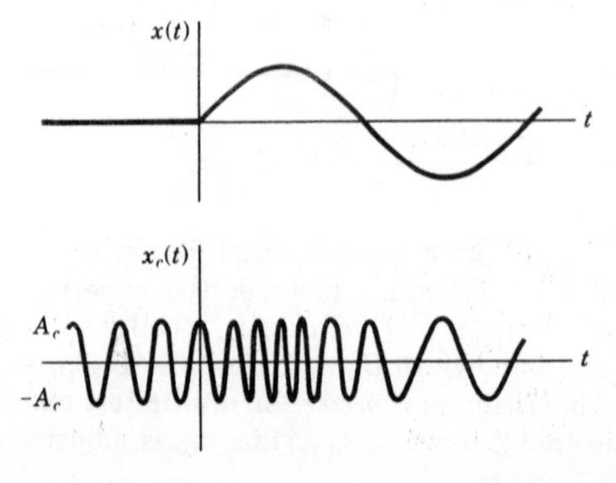

FIGURE 14.5–7
FM waveforms.

sults are known only for certain special cases. These results indicate that the spectrum has sidebands extending much further than $\pm W$ on either side of f_c. Thus, $B \gg 2W$, and we say that FM is *wideband* modulation. For an estimate of the FM transmission bandwidth, we use the *maximum frequency deviation*

$$\Delta f = m\,|x(t)|_{\max}$$

Then if $\Delta f \geq 2W$,

$$B \approx 2\,\Delta f + 4W \tag{9}$$

For example, commercial FM uses $\Delta f = 75$ kHz and $W = 15$ kHz, so $\Delta f = 5W$ and $B \approx 2 \times 75 + 4 \times 15 = 210$ kHz.

To understand how noise is suppressed in FM, first look at the demodulation process, which is usually accomplished by *FM-to-AM conversion*. Suppose $x_c(t)$ is applied to a simple tuned circuit with center frequency $f_0 > f_c$, as in Fig. 14.5–8a. As the instantaneous frequency swings above and below f_c, the frequency variations are converted by the ampli-

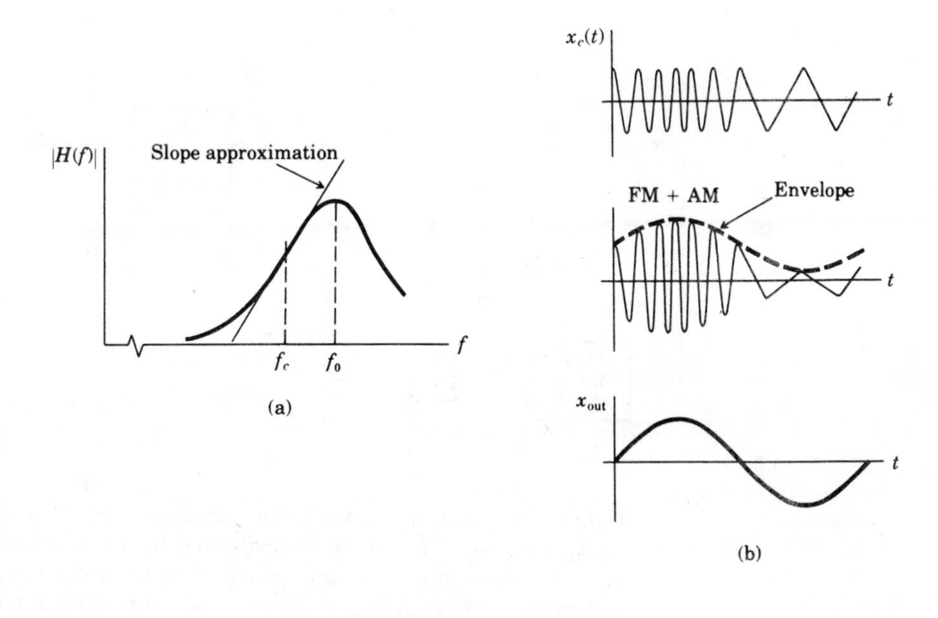

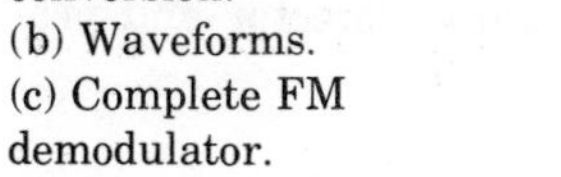

FIGURE 14.5–8
(a) FM-to-AM conversion.
(b) Waveforms.
(c) Complete FM demodulator.

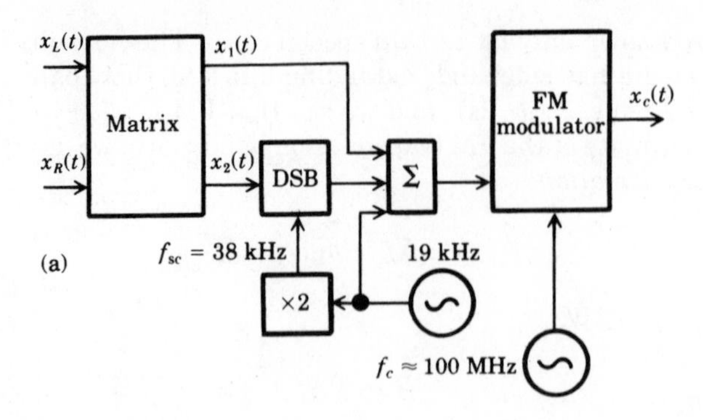

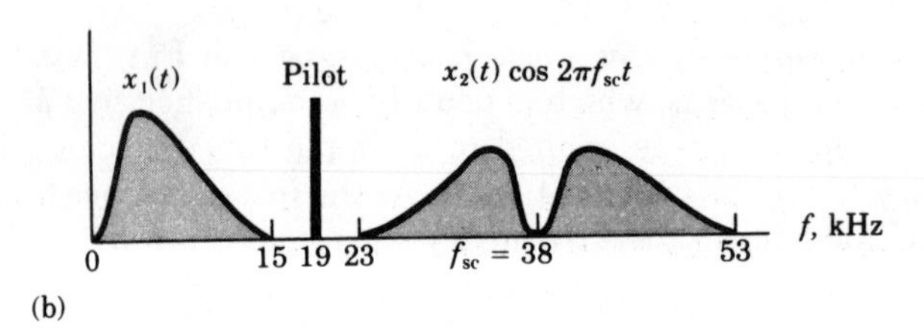

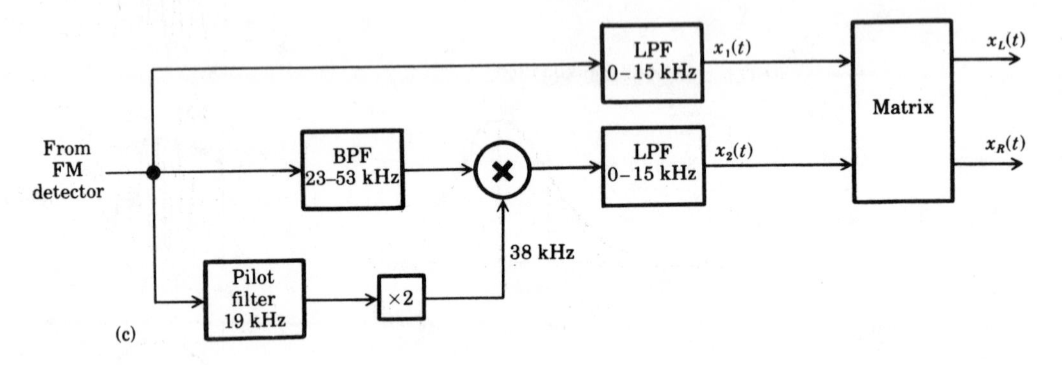

FIGURE 14.5–9
FM stereo multiplexing:
(a) transmitter;
(b) multiplexed baseband spectrum;
(c) receiver.

tude ratio, to amplitude variations shown in Fig. 14.5–8b. The amplitude variations, in turn, can be demodulated by an envelope detector. A complete FM demodulator (often called a *frequency discriminator*) consists of a limiter, FM-to-AM converter, and envelope detector, as in Fig. 14.5–8c. The limiter removes any spurious amplitude variations from $x_c(t)$ caused by interference and noise. The more refined *balanced discriminator* eliminates the need for DC blocking in the envelope detector and provides an excellent low-frequency response.

From the above description it follows that the demodulated signal is proportional to the frequency deviation in $x_c(t)$, the larger the deviation the bigger the output signal. When noise is present, we can increase the output signal simply by increasing the frequency deviation; this requires a larger transmission bandwidth, but does not require more transmitted signal power since $x_c(t)$ has constant amplitude independent of the modulation. Increasing the frequency deviation thus reduces the output noise compared to the signal; therefore, FM has the characteristic of *wideband noise reduction*. Commercial monophonic FM with $B \approx 14\ W$ has a noise-reduction factor of about 640 (28 dB). Unfortunately stereophonic transmission wipes out most of the FM noise reduction, which explains why most stereo FM receivers include a provision for switching to monophonic reception for weak signals.

Finally, we describe the multiplexing system used for FM stereophonic broadcasting, starting with the transmitter diagrammed in Fig. 14.5–9a. The left- and right-channel signals, having $W \approx 15$ kHz, are first *matrixed* to form

$$x_1(t) = x_L(t) + x_R(t) \qquad x_2(t) = x_L(t) - x_R(t)$$

so that the sum signal $x_1(t)$ heard on a monophonic receiver will not have any sound "gaps". The difference signal $x_2(t)$ is DSB modulated onto a 38-kHz subcarrier obtained by doubling the frequency from a 19-kHz oscillator. DSB has been employed here to ensure good fidelity at the lower audio frequencies. Combining $x_1(t)$, $x_2(t) \cos 2\pi f_{sc} t$, and the 19-kHz pilot yields the so-called *baseband signal* of Fig. 14.5–9b, which is applied to the FM modulator for radio transmission with $f_c = 88$–108 MHz.

A stereo FM receiver, Fig. 14.5–9c, recovers the baseband signal by FM demodulation. The pilot synchronizes product demodulation of the DSB component, and another matrix reproduces $x_L(t)$ and $x_R(t)$ from $x_1(t)$ and $x_2(t)$. The portion of the receiver prior to the FM demodulator (not shown) is usually a superheterodyne with $f_{IF} = 10.7$ MHz.

Exercise 14.5–4 The TV audio signal of Fig. 14.5–5 is frequency-modulated with $\Delta f = 25$ kHz and has $W \approx 10$ kHz. What percentage of the channel bandwidth is occupied by the audio signal?

PROBLEMS

14.1–1 Use Eq. (1), Section 7.4, and Table 7.4–1 to draw the spectrum of a triangular wave (Fig. 7.4–1a) with $T = 2$ ms and $A = \pi^2/8$. Then estimate W by neglecting any frequency components whose amplitudes are less than 10% of that of the component having the largest amplitude.

14.1–2 Repeat Problem 14.1–1 for a square wave (Fig. 7.4–1b) with $T = 10$ ms and $A = \pi/4$.

14.1–3 Repeat Problem 14.1–1 for a full-rectified cosine wave (Fig. 7.4–1d) with $T =$ 0.1 ms and $A = \pi/2$.

‡ **14.1–4** Repeat Problem 14.1–1 for a rectangular pulse train (Fig. 7.4–1e) with $T =$ 100 μs, $D = 25$ μs, and $A = 4$.

14.2–1 A transmission system with $P_{in} = 5$ mW consists of a preamplifier, cable, and output amplifier intended to produce $P_{out} = 2$ W. The amplifiers are rated for 4-W maximum output power. The cable has $\alpha = 1.5$ dB/km. If $\ell = 18$ km, what should be the values of G_1 and G_2 (in dB) so that G_1 is as large as possible?

14.2–2 Repeat the previous problem assuming the cable has a 1-W power limitation.

14.2–3 Suppose the distance in Problem 14.2–1 is to be made as large as possible, subject to the condition that the output amplifier requires at least 1 mW at its input. Find G_1, ℓ, and G_2.

14.2–4 Suppose the cable sections in Fig. 14.2–3 were obtained by cutting a single cable with loss L and length $\ell = \ell_1 + \ell_2$. Show that $L_1 L_2 = L$.

14.2–5 Figure P14.2–5 represents a long-haul *repeater system* consisting of M identical cable sections and M identical amplifiers, so $L_1 = L_2 = \cdots = L_M$ and $G_1 = G_2 = \cdots = G_M$. If $P_{out} = P_{in} = 100$ mW, $\alpha = 1.8$ dB/km, and the amplifiers require at least 25 μW at their input terminals, what is the minimum number of amplifiers and the corresponding amplifier gain in dB needed to span a distance of 3000 km?

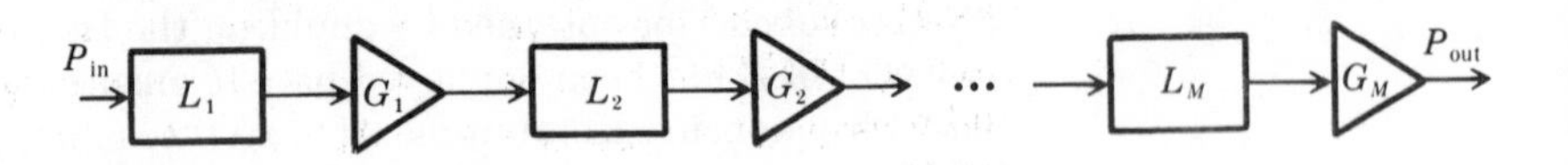

FIGURE P14.2–5

14.2–6 Suppose the output of an otherwise perfect transmission system includes an *echo*, so that $y(t) = x(t) + ax(t - t_\Delta)$, where t_Δ is the echo's time delay. (Echoes commonly occur in radio transmission, but they may also be caused by reflections on a mismatched cable.) Take $x(t) = Xe^{st}$ and show that $H(s) = y(t)/x(t) = 1 + ae^{-st_\Delta}$. Then let $s = j2\pi f$ to obtain

$$\underline{H}(f) = 1 + ae^{-j\phi}$$

where $\phi = 2\pi t_\Delta f = 360° t_\Delta f$.

14.2–7 Let the system in the previous problem have $a = 0.5$ and $t_\Delta = \frac{1}{4}$ ms. Use Eq. (6) to evaluate K and t_d at $f = 0$, 1, 2, and 3 kHz. What is the physical meaning of the negative value of t_d when $f = 3$ kHz?

14.2–8 Apply Euler's theorem to $\underline{H}(f)$ in Problem 14.2–6 to show that an echo produces *ripples* in the amplitude ratio $|H(f)|$. Then obtain expressions for the frequencies at which $|H(f)|$ is maximum and minimum.

‡ **14.2–9** Figure P14.2–9 is the diagram of a simple *tapped-delay-line equalizer* (also called a *transversal filter*), which purposely introduces delayed echoes to achieve a rippled amplitude ratio. The method outlined in Problem 14.2–6 gives the transfer function as $H_{eq}(f) = a + e^{-j\phi} + ae^{-j2\phi}$, where $\phi = 360° t_\Delta f$. Manipulate this expression to obtain $|H_{eq}(f)| = 1 + 2a\cos\phi$ and $\theta_{eq}(f) = -\phi$.

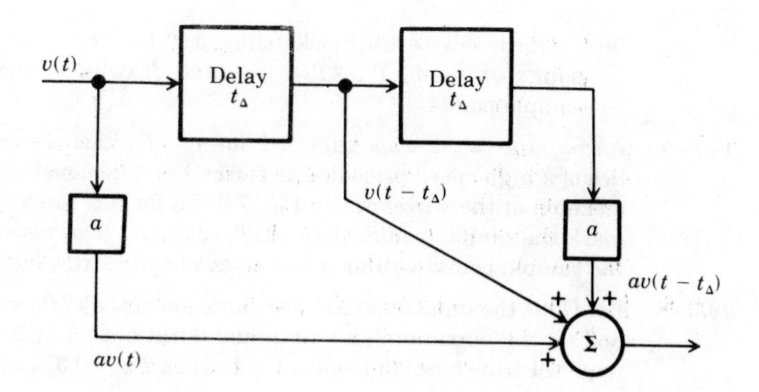

FIGURE P14.2–9

14.3–1 Consider the test signal $x(t) = 6 \cos 2\pi f_m t + 2 \cos 2\pi 3 f_m t$. Taking $f_m = 100$ Hz, draw the spectrum of $x_c(t)$ for DSB modulation with $f_c = 1$ kHz. Then remove all the components above or below f_c to obtain the spectrum for upper-sideband SSB and lower-sideband SSB.

14.3–2 Repeat the previous problem with $f_m = 600$ Hz. (See the reminder in Exercise 14.3–1 about "negative" frequencies.)

14.3–3 Suppose the oscillator in Fig. 14.3–4a is poorly synchronized and produces $\cos\,[2\pi(f_c + \Delta f)t + \Delta\phi]$. Find $z(t)$ when $x_c(t) = x(t) \cos 2\pi f_c t$, corresponding to DSB. Then let $x(t) = 4 \cos 2\pi 1000t$ and consider what happens to $z(t)$ when

(a) $\Delta f = 200$ and $\Delta\phi = 0$;

(b) $\Delta f = 0$ and $\Delta\phi = 90°$.

14.3–4 Repeat the previous problem for SSB, taking $x_c(t) = 2 \cos 2\pi(f_c \pm 1000)t$.

14.3–5 Figure P14.3–5 is a simplified *speech scrambler* that converts a voice signal $x(t)$ into an unintelligible signal $\tilde{x}(t)$. (This system can be incorporated in a telephone to ensure communication privacy.)

(a) Find $\tilde{x}(t)$ when $x(t)$ is the test signal in Problem 14.3–1 with $f_m = 800$ Hz. Compare the input and output spectra.

(b) Using your results, sketch the spectrum of $\tilde{x}(t)$ when $x(t)$ has the spectrum shown in Fig. 14.1–7. Why is $\tilde{x}(t)$ unintelligible?

(c) Explain how $x(t)$ can be recovered from $\tilde{x}(t)$.

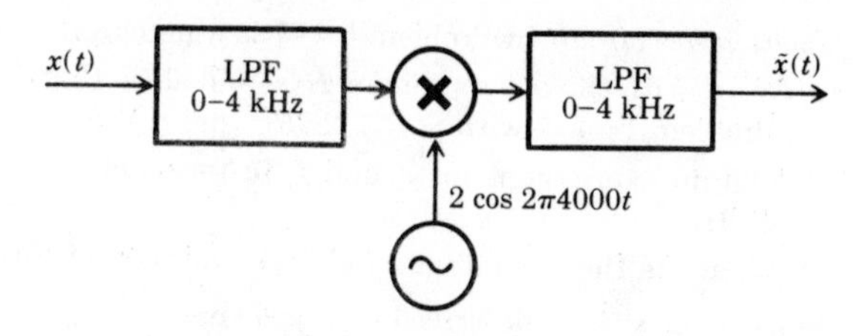

FIGURE P14.3–5

14.3–6 Suppose $x(t) = A_m \cos 2\pi Wt$ has $W = 6$ kHz and is sampled at the rate $f_s = 8$ kHz < 2 W. The sampled waveform is then processed by a 4-kHz lowpass filter.

(a) Make a sketch similar to Fig. 14.3–6 to show that the filtered output frequency will be $f_s - W = 2$ kHz because of aliasing.

(b) Confirm this result by sketching $x(t)$ for $0 \le t \le 3/W$, marking the sample points at $t = 0, T_s, 2T_s, \ldots$, and drawing a smooth curve through the sample points.

14.3–7 A *sampling oscilloscope* takes advantage of aliasing to produce a low-speed replica of a high-speed periodic waveform. For a demonstration of this process, make a sketch of the waveform in Fig. 7.4–2a for $0 \le t \le 12T$. Then let $T_s = \frac{9}{8}T$ and mark the sample points at $t = 0, T_s, 2T_s, \ldots$. A smooth curve drawn through the sample points will have the same shape as $x(t)$, but with period $9T$.

14.3–8 Let D_0 be the unmodulated pulse duration of the PDM waveform in Fig. 14.3–7b, and let the maximum and minimum durations be $D_0 \pm \frac{1}{3}D_0$. Find the minimum required transmission bandwidth B when $D_0 = 0.3T_s$ and $f_s = 8$ kHz.

14.3–9 Let D be the pulse duration of the PPM waveform in Fig. 14.3–7b, and let the maximum position shift be $\pm 2D$. Calculate the minimum transmission bandwidth B by considering the maximum permitted value of D when $f_s = 8$ kHz.

14.3–10 Let the TDM system in Fig. 14.3–9 have $B = 300$ kHz. Find the maximum number of voice signals with $W = 4$ kHz that can be transmitted.

14.3–11 Repeat the previous problem with the additional constraint that the TDM waveform must be OFF at least 50% of the time.

14.3–12 Repeat Problem 14.3–10, assuming the pulses have position modulation as described in Problem 14.3–9.

14.4–1 Many *notch filters* for interference rejection have a transfer function of the form

$$H(f) = \frac{1 - (f/f_0)^2}{1 - (f/f_0)^2 + j2\alpha(f/f_0)}$$

so $\underline{H}(f_0) = 0$, whereas $\underline{H}(0) = \underline{H}(\infty) = 1$. The 3-dB frequencies on either side of the notch are related by

$$f_u f_\ell = f_0^2 \qquad f_u - f_\ell = 2\alpha f_0$$

(a) Show that the filter in Fig. 14.4–4 has this frequency response, with $f_0 = 1/2\pi\sqrt{LC}$ and $\alpha = (R/2)\sqrt{C/L}$.

(b) If $R = 100\ \Omega$, find values for L and C so that $f_\ell \approx 110$ Hz and $f_u \approx 130$ Hz.

14.4–2 Repeat Problem 14.4–1(b) for $f_\ell \approx 58$ Hz and $f_u \approx 62$ Hz. Why would this be difficult to achieve?

14.4–3 The *twin-tee circuit* in Problem 5.4–7 is a notch filter.

(a) Start with the expression for $H(s)$ and show that $\underline{H}(f)$ has the form given in Problem 14.4–1 with $f_0 = 1/2\pi RC$ and $\alpha = 2$.

(b) Obtain expressions for f_ℓ and f_u in terms of f_0. Evaluate them taking $f_0 = 60$ Hz.

(c) What are the advantages and disadvantages of this filter?

14.4–4 Suppose the system described by Eq. (7) has $S_{out}/N_{out} = 40$ dB when $S = 10$ mW and $B = 5$ kHz. If other values are held fixed, what value of S is required to upgrade the system for high-fidelity audio transmission with $S_{out}/N_{out} = 60$ dB and $B = 20$ kHz?

14.4–5 What value of S is sufficient if the system in the previous problem is downgraded to $S_{out}/N_{out} = 30$ dB and $B = 3$ kHz?

14.4–6 A repeater system like Fig. P14.2–5 usually has $G_1/L_1 = G_2/L_2 = \cdots = G_M/L_M = 1$. If the amplifiers have $T_a \gg T_0$, then all other noise sources are negligible, so $S_{out}/N_{out} \approx P_{in}/ML_1kT_aB$. The total path loss is $L = L_1^M$, assuming identical cable sections.

(a) Suppose this system has $L = 10^{24}$ and $S_{out}/N_{out} = 500$ when $M = 6$. Find S_{out}/N_{out} if M is increased to 12, L being held fixed.

(b) Repeat (a) with M reduced to 4.

14.4–7 Let the system in Problem 14.4–6 have $P_{in} = 20$ mW, $T_a = 10T_0$, $B = 5$ MHz, and $L = 10^{30}$. What value of M is needed to get $S_{out}/N_{out} \geq 10^4$?

14.4–8 A method for measuring an amplifier's noise temperature goes as follows: First, connect a thermal source at temperature T_0 and measure N_{out}. Then increase the source temperature to T_R, such that the output power doubles. Obtain an expression for T_a in terms of T_R and T_0.

14.4–9 The *noise figure* F of an amplifier is defined as $F = 1 + T_a/T_0$. Obtain an expression for S_{out}/N_{out} in terms of F, the input signal power S, the input noise power N, and the input noise temperature T. Show that your result simplifies to $S_{out}/N_{out} = S/FN$ when $T = T_0$.

14.4–10 Two noisy amplifiers having the same bandwidth are connected in *cascade* to get the overall gain $G = G_1G_2$. The individual noise temperatures are T_{a_1} and T_{a_2}. Obtain an expression for the total output noise power when the input noise to the first amplifier is $N = kTB$. Compare your result with Eq. (6) to show that the effective noise temperature of the cascade is $T_a = T_{a_1} + T_{a_2}/G_1$.

14.4–11 Derive the expression for S_{out}/N_{out} in Problem 14.4–6.

‡ **14.4–12** The circuit in Problem 7.1–5 can be used to *preemphasize* high frequencies.

(a) Sketch the asymptotic Bode plot of H_{dB}. Then find values for K and f_{co} so that $|H(f)| \approx 4|H(0)|$ for $f \geq 10$ kHz.

(b) Show that an RC lowpass filter with an additional resistance in series with C is an appropriate *deemphasis* filter.

14.5–1 A bandpass signal with $f_c = 500$ MHz is to be transmitted a distance $\ell = 15$ km. Calculate L in dB assuming dipole antennas with $G_{TA} = G_{RA} = 2$ dB.

14.5–2 Suppose identical parabolic antennas with radius r are used in the previous problem. Find r (in meters) such that $L = 70$ dB.

14.5–3 A radar system uses pulses of duration D to modulate a carrier of frequency f_c. The system can distinguish between targets spaced by $d \geq cD$. In view of Eq. (4) and the bandwidth-doubling of modulation, what is the minimum practical value of f_c so that $d_{min} = 30$ m?

14.5–4 A satellite broadcasts a TV signal ($B \approx 5$ MHz) on a 4-GHz carrier to a ground station 37,000 km away. The satellite has $P_{in} = 5$ W and $G_{TA} = 30$ dB. The ground station has a parabolic antenna of radius r connected to an amplifier with noise temperature $T + T_a = 5T_0$. Use Eq. (7), Section 14.4, to find the value of r (in meters) needed to get $S_{out}/N_{out} = 40$ dB.

14.5–5 When a signal consists of sinusoids at different frequencies, its average power is proportional to the sum of the squared amplitudes. Find the ratio of carrier-frequency power to total power in an AM wave with $x(t) = \cos 2\pi f_m t$ and $m = 1$.

14.5–6 Let $x(t)$ be the test signal in Problem 14.3–1.

(a) Show that Eq. (6) requires $m \leq \frac{1}{8}$.

(b) Repeat the power-ratio calculation in the previous problem using this signal with $m = \frac{1}{10}$.

14.5–7 Figure P14.5–7 shows a way to generate AM using a nonlinear device with $x_{\text{out}} = x_{\text{in}} + ax_{\text{in}}^2$. Take $x(t) = A_m \cos 2\pi f_m t$ and show that $x_c(t)$ has the form of Eq. (5), providing that the filter passes $f_c - f_m$ to $f_c + f_m$ and that $f_c > 3f_m$.

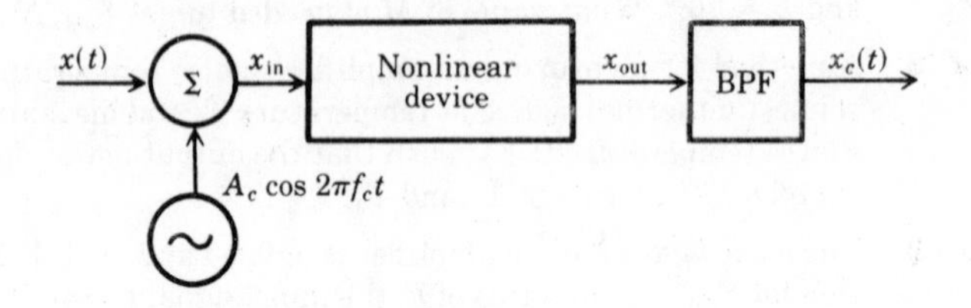

FIGURE P14.5–7

14.5–8 Suppose the nonlinear device in Fig. P14.5–7 has $x_{\text{out}} = ax_{\text{in}}^2$. Modify the diagram to obtain AM at the filter's output.

14.5–9 Repeat Problem 14.5–4 assuming FM carrier modulation with $\Delta f = 25$ MHz, so $B \approx 70$ MHz, but the FM demodulator increases $S_{\text{out}}/N_{\text{out}}$ by a factor of 640.

15

DIGITAL LOGIC

This chapter begins our investigation of logic and digital systems, systems that essentially operate on "numbers" rather than "signals." Digital electronics is perhaps the fastest growing and most exciting field in electrical engineering, for each new day seems to bring another technological advance that lowers cost, reduces size, improves performance, and expands the scope of applications. These applications range from pocket calculators and giant computers to household appliances, scientific instruments, and industrial controls.

Logic variables and functions, and their relationship to binary and decimal numbers are fundamental to digital systems, so we begin our discussion with these concepts. We then devote separate sections to combinational circuits built up from logic gates, and sequential circuits involving flip-flops as memory units. Our work will be illustrated by simple examples, the more elaborate systems deferred to Chapter 16. The closing section describes major IC logic families that implement the building blocks introduced here.

OBJECTIVES

After studying this chapter and working the exercises, you should be able to do each of the following:

- Draw the symbol and truth tables for an AND, OR, NOT, exclusive-OR, NAND, and NOR gate (Sections 15.1 and 15.2).

- Perform binary-to-decimal and decimal-to-binary number conversions (Section 15.1).
- Use the rules of Boolean algebra to simplify a given logic function, and draw the corresponding gate circuit (Section 15.2).
- Construct the truth table for a logic function; derive its sum-of-products or product-of-sums expression; and synthesize the function with NAND or NOR gates (Section 15.2).
- Explain the operation of an *RS* flip-flop and the various types of clocked flip-flops (Section 15.3).
- Analyze the action of a counter or shift register by sketching the relevant waveforms (Section 15.3).

15.1 DIGITAL LOGIC CONCEPTS

Digital systems differ fundamentally from analog systems in that they operate with discrete or *digitized* variables. So, although an analog signal might vary anywhere over a continuous range, a digital signal is restricted to only a few specified and clearly separated values. Accordingly, we say that a digital system has a *finite number of allowed states.*

By restricting a system to distinct states, we obviously lose the ability to represent intermediate values between states. But in exchange, we gain three potential advantages of digital systems over analog systems:

- More economical, flexible, and compact system implementation.
- Improved reliability in the face of hardware imperfections, spurious signals, etc.
- The ability to make logical decisions, carry out digital computations, and store results.

The first two advantages reach their zenith in a *binary* system which has just *two* states—the fewest number of states a digital system can have. (A system with one state never changes, and a system with no states can't exist!) Consequently, almost all modern digital systems are binary.

This section introduces binary systems as switching circuits capable of making logical decisions that have only two alternatives. We define the basic logic units known as AND, OR, and NOT gates, and develop the connection between logic variables, binary digits, and decimal numbers. We also distinguish combinational and sequential operation.

LOGIC VARIABLES AND SWITCHES

The circuit in Fig. 15.1–1a is a simple binary system, for the light is either ON or OFF depending on whether the switch is UP or DOWN. The circuit is drawn to emphasize that we will think of the switch as the cause or

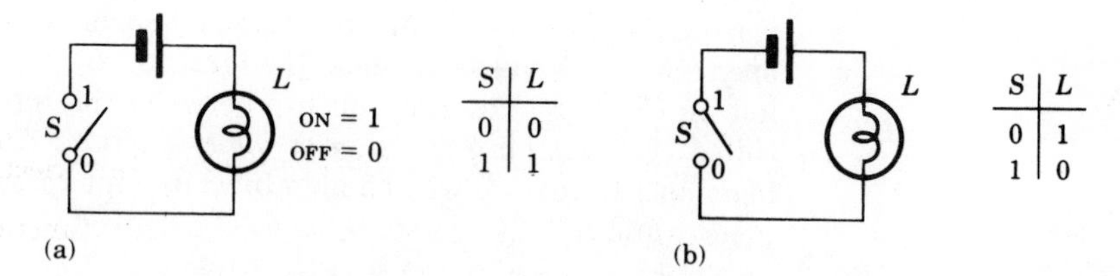

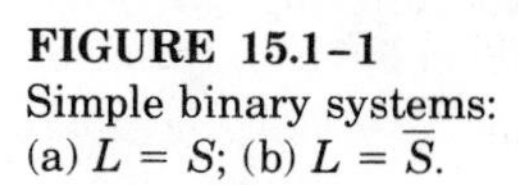

FIGURE 15.1–1
Simple binary systems: (a) $L = S$; (b) $L = \overline{S}$.

input, and the light as the effect or output. We represent the states of the light and switch by *logic variables* L and S, and identify specific states with the *binary digits* 1 and 0 rather than with words like ON/OFF or UP/DOWN. In particular, we write $L = 1$ to indicate that the light is ON, while $L = 0$ means that it's OFF. Likewise, the switch has been labeled 1 for UP and 0 for DOWN. Using these binary digits to represent the values of logic variables has mathematical significance as well as notational convenience.

Clearly, if $L \neq 1$ then it must be true that $L = 0$, and vice versa, since a logic variable has only two possibilities. But what about the cause-and-effect relationship between L and S? That relationship depends on the circuit connection and may be displayed by a simple tabulation of the possible values of S and the corresponding values of L. It is both obvious and trivial that $L = S$ in Fig. 15.1–1a because the light is OFF ($L = 0$) when the switch is DOWN ($S = 0$), and the light is ON ($L = 1$) when the switch is UP ($S = 1$). If the switch happens to be reversed, as in Fig. 15.1–1b, we would write $L = \overline{S}$ to symbolize that $L = 1$ when $S = 0$, and the converse.

Now suppose three switches, S_1, S_2, and S_3, control L as in Fig. 15.1–2a. Since each switch has two states, the connected switches have $2 \times 2 \times 2 = 2^3 = 8$ possible conditions enumerated in the accompanying table. Of those 8 conditions, the light goes ON only if $S_1 = 0$ *and* either $S_2 = 1$ *or* $S_3 = 1$ *or* both, which you can easily check. Note, by the way, that Fig. 15.1–2b lists every possible set of input-output values, something that could not be done for an analog system. Such tabulations are called *truth tables*.

Figure 15.1–2a is a nontrivial but simple example of a switching circuit that makes a logical decision. For instance, it might turn on a

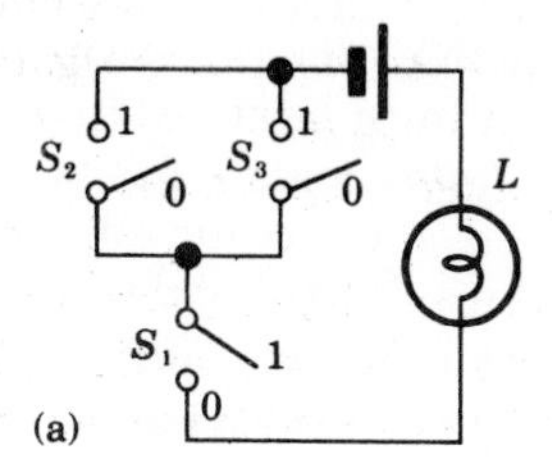

(b)

S_1	S_2	S_3	L
0	0	0	0
0	0	1	1
0	1	0	1
0	1	1	1
1	0	0	0
1	0	1	0
1	1	0	0
1	1	1	0

FIGURE 15.1–2
(a) Logic system with three inputs. (b) Truth table.

warning light ($L = 1$) in an automobile whenever the driver's door is open ($S_1 = 0$) and the key is in the ignition ($S_2 = 1$) or the headlights are left on ($S_3 = 1$). The relationship between the light and the switches is called the *switching function* or *logic function*. After defining the basic logic operations, we will be able to write an expression for L in terms of S_1, S_2, and S_3.

LOGIC OPERATIONS AND GATES

A logic gate is a switching circuit (usually electronic) that operates on one or more input variables to produce an output variable. All variables are binary or logic variables, limited to the values 1 and 0. The three basic operations defined here are those of the AND, OR, and NOT gates.

Figure 15.1–3 gives the schematic symbol and truth table for an AND gate with two inputs, A and B. By definition, the output X equals 1 only when $A = 1$ *and* $B = 1$. We represent this by writing

$$X = A \cdot B \tag{1a}$$

where $A \cdot B$ stands for "A AND B". The truth table lists the four possible input combinations and the resulting output of the two-input AND gate, namely

$$0 \cdot 0 = 0 \qquad 0 \cdot 1 = 0 \qquad 1 \cdot 0 = 0 \qquad 1 \cdot 1 = 1 \tag{1b}$$

which has the appearance of ordinary multiplication. Correspondingly, in the case of three or more inputs, the output $X = A \cdot B \cdot C \cdots$ equals 1 only when *all* inputs equal 1. The AND gate is, then, an all-or-nothing operation.

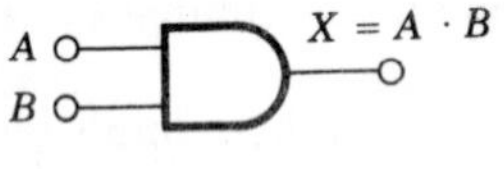

A	B	X
0	0	0
0	1	0
1	0	0
1	1	1

(b)

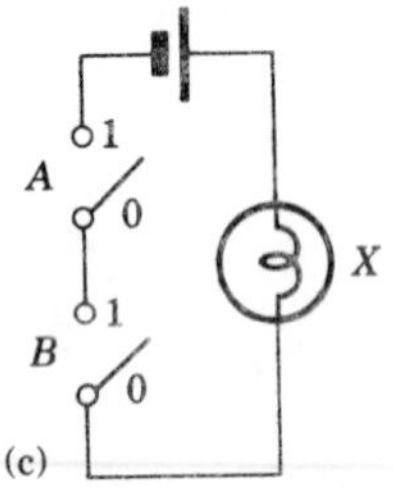

FIGURE 15.1–3 Two-input AND gate: (a) symbol; (b) truth table; (c) equivalent switching circuit.

To build an AND gate with mechanical switches, we just connect them in series so that all switches must be closed to complete the electrical circuit, as in Fig. 15.1–3c. Some logic circuits still include mechanical switches, primarily to handle large currents. However, modern high-speed gates employ electronic switching devices, the inputs and outputs being HIGH/LOW voltage levels. (Simple diode and transistor gate circuits were described in Chap. 11.) From this point on, Fig. 15.1–3a with two or more inputs will be used to represent any AND-gate implementation.

An OR gate, Fig. 15.1–4, is the equivalent of connecting switches in parallel, since $X = 1$ if $A = 1$ *or* $B = 1$ *or* both. Symbolically, we write

$$X = A + B \tag{2}$$

where $A + B$ stands for "A OR B." The truth table reveals that

$$0 + 0 = 0 \qquad 0 + 1 = 1 \qquad 1 + 0 = 1 \qquad 1 + 1 = 1 \tag{2b}$$

which has the appearance of ordinary addition except for $1 + 1 = 1$. If this equation puzzles you, you should keep in mind that we are dealing with logic variables that represent the states of a switching circuit. Thus, $1 + 1 = 1$ merely reflects the fact that two closed switches in parallel are the same electrically as one closed switch. By direct extension of this reasoning, it follows that an OR gate with three or more inputs is an any-or-all operation and that the output $X = A + B + C \ldots$ equals 1 if any one or more inputs equals 1.

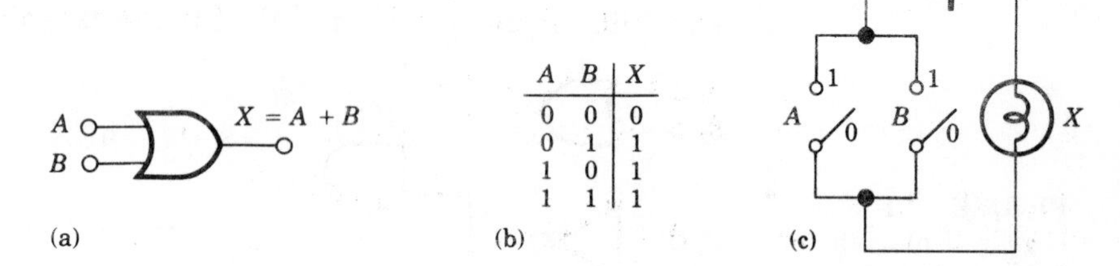

FIGURE 15.1–4 Two-input OR gate: (a) symbol; (b) truth table; (c) equivalent switching circuit.

Finally, the NOT gate or *inverter* in Fig. 15.1–5a is a single-input gate whose output is the opposite state of the input. We express this in the form

$$X = \overline{A} \tag{3a}$$

where $\overline{A}$ stands for "NOT A"—also called the *negation* or *complement* of A. Clearly, the NOT operation changes 1 to 0 and 0 to 1, so

$$\overline{1} = 0 \qquad \overline{0} = 1 \tag{3b}$$

The inherent voltage-inverting action of transistors makes them ideally suited for electronic NOT gates (See Section 11.2). The mechanical version

A — $X = \overline{A}$

(a)

A	X
0	1
1	0

(b)

A — $\overline{A}$ — $\overline{\overline{A}} = A$

(c)

FIGURE 15.1–5 (a) NOT gate or inverter. (b) Truth table. (c) Double negation by two inverters.

could be a "normally-closed" switch like the one in Fig. 15.1–1b whose closed position is labeled 0, so we get $L = 1 = \overline{0}$ when $S = 0$.

If two NOT gates are connected in the cascade arrangement of Fig. 15.1–5c, the final output equals the original input. We represent this property in general by writing

$$\overline{\overline{A}} = \overline{(\overline{A})} = A \tag{4}$$

since the second negation cancels the first.

Example 15.1–1

Let's return to the switching circuit in Fig. 15.1–2 and describe it using the three basic logic operations. An inspection of the circuit diagram or truth table shows that $L = 1$ only when $S_1 = 0 = \overline{1}$ and $S_2 = 1$ or $S_3 = 1$ or $S_2 = S_3 = 1$. For a more compact statement, let $A = \overline{S_1}$ and $B = S_2 + S_3$; then $L = 1$ when $A = 1$ and $B = 1$. We therefore write the logic function

$$L = A \cdot B = \overline{S_1} \cdot (S_2 + S_3)$$

which expresses L in terms of the logic variables S_1, S_2, and S_3. This switching circuit is thus equivalent to the gate diagram in Fig. 15.1–6.

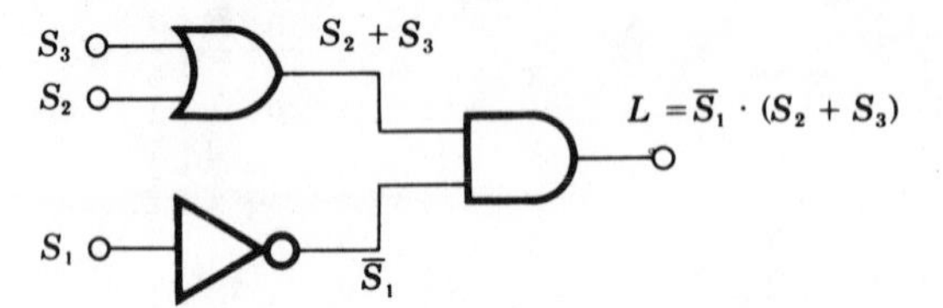

FIGURE 15.1–6
Logic-gate diagram for Example 15.1–1.

To check our logic function, take the fourth line of the truth table in Fig. 15.1–2 where $S_1 = 0$, $S_2 = 1$, and $S_3 = 1$. Inserting these values into the expression gives $L = \overline{0} \cdot (1 + 1) = 1 \cdot (1 + 1) = 1 \cdot 1 = 1$, as required. You may find it interesting and informative to check out other lines of the truth table.

Exercise 15.1–1

Construct a truth table, logic function and gate diagram for a switching circuit that turns off the heating element ($H = 0$) in an industrial oven when the temperature exceeds an upper limit ($U = 1$) or when the conveyor is off ($C = 0$) and, simultaneously, the temperature exceeds a lower limit ($L = 1$). Hint: Take U, C, and L as the inputs, and let H be the output of a NOT gate whose input equals $\overline{H}$.

BINARY AND DECIMAL NUMBERS

Besides making logical decisions, gate circuits are capable of carrying out *arithmetic operations*. Typical circuits that perform addition, subtraction, and multiplication will be described in Section 16.3. Here, we explain the underlying concept of treating logic variables as *binary digits*—called *bits,* for short—and grouping bits together to form *binary numbers.* By drawing upon the positional notation of number systems, we can convert back and forth between the binary numbers inside an arithmetic circuit and the decimal numbers of the outside world.

We construct a number in the decimal system using the ten digits 0, 1, . . . 9, and the position of each digit signifies a power of 10. To clarify, 105.8 actually means

$$105.8_{10} = 1 \times 10^2 + 0 \times 10^1 + 5 \times 10^0 + 8 \times 10^{-1}$$

The subscript 10 has been tagged onto 105.8 to emphasize that we are talking about a *base*-10 number. Binary numbers have the same structure except that we use only two digits, 0 and 1, and digit position signifies a power of 2. Thus,

$$110.1_2 = 1 \times 2^2 + 1 \times 2^1 + 0 \times 2^0 + 1 \times 2^{-1}$$

where the dot must be called the *binary point* (rather than decimal point) because this is a *base*-2 number. It then follows that

$$110.1_2 = (4 + 2 + 0 + 0.5)_{10} = 6.5_{10}$$

and binary-to-decimal conversion involves nothing more than summing (in base 10) the expansion of the binary number.

Now consider an ordered sequence of logic variables such as $B_2B_1B_0B_{-1}$. This sequence constitutes a *binary word,* since each variable (or bit) B_i has the value of 0 or 1. To interpret the word as a *binary number,* we let the subscript i denote the *weight* 2^i assigned to each bit or, equivalently, the bit position relative to the binary point. The rule for converting our binary number $B_2B_1B_0B_{-1}$ to the corresponding decimal quantity x_{10} then becomes

$$x_{10} = B_2 \times 2^2 + B_1 \times 2^1 + B_0 \times 2^0 + B_{-1} \times 2^{-1} \tag{5}$$

This expression is easily generalized for a binary word with an arbitrary number of bits before or after the binary point.

Decimal-to-binary conversion is somewhat more difficult, especially for non-integer numbers. Small decimal integers may be converted by counting in binary—like counting on your elbows instead of your fingers. Large decimal integers may be converted by a procedure sometimes referred to as *double-dabble.* You successively divide the decimal integer by 2, and write down the remainder (0 or 1) after each division; the remainders taken in reverse order, then spell out the binary word. Application of this procedure to 105_{10}, for instance, gives

		Remainder
2	105	1
	52	0
	26	0
	13	1
	6	0 ↑
	3	1
	1	1
	0	

Hence, reading the remainders from bottom to top yields $105_{10} = 1101001_2$.

Decimal fractions are converted using *reverse* double-dabble. You multiply successively by 2 and extract the integer carries (0 or 1) to form the binary fraction in forward order. With the exception of decimal fractions that equal sums of powers of $\frac{1}{2}$, exact conversion requires an infinite string of bits. Take 0.8_{10}, for example:

	Carry
$2 \times 0.8 = 1.6$	1
$\times\ 0.6 = 1.2$	1
$\times\ 0.2 = 0.4$	0 ↓
$\times\ 0.4 = 0.8$	0
$\times\ 0.8 = 1.6$	1
$\times\ 0.6 = 1.2$	1
⋮	⋮

By noting the repeating pattern in this case, we obtain $0.8_{10} = 0.11001100 \ldots_2$. Truncating after six bits yields the approximation

$$0.8_{10} \approx 0.110011_2 = 0.796875_{10}$$

which has an error of about 0.4%. Such *truncation* or *round-off errors*, although much smaller, often show up in the output of digital-computer calculations. Because fractional parts lead to unwieldy binary numbers, we'll limit our further discussion to integers.

Table 15.1–1 lists the binary words for 0 through 15_{10}. We see from the table that 2-bit words are needed to represent the decimal integers 2

TABLE 15.1–1

Decimal	*Binary*	*Hexadecimal*
0	0	0
1	1	1
2	10	2
3	11	3
4	100	4
5	101	5
6	110	6
7	111	7
8	1000	8
9	1001	9
10	1010	A
11	1011	B
12	1100	C
13	1101	D
14	1110	E
15	1111	F

and 3, 3-bit words are needed for 4 through 7, and so forth. In general, an n-bit binary integer can represent any decimal integer up to

$$x_{\max} = 2^n - 1 \tag{6}$$

Consequently, on the bottom line of the table we find $1111_2 = 15_{10} = 2^4 - 1$. Leading zeros are added when a long word represents a small decimal number, as in $0011_2 = 3_{10}$.

The table also brings out the natural pattern of binary integers, as follows: The least significant bit (LSB) alternates between 0 and 1 down its column, the next bit alternates in pairs, the next in groups of four, and the most significant bit (MSB) in an n-bit word would alternate in groups of 2^{n-1}. This same pattern is usually adopted for the input variables in a truth table; see, for example, Figs. 15.1–2b and 15.1–3b.

Digital computers work with binary words as long as 32 or even 64 bits, usually subdivided into 8-bit units called *bytes*. Intermediate printouts for diagnostic purposes may then be presented in *hexadecimal* (base-16) numbers, without going through binary-to-decimal conversion. To form a hexadecimal number, you separate the binary word into 4-bit groups (adding leading zeros if needed), and replace each group with the appropriate hexadecimal digit from Table 15.1–1. Note that the letters A through F are used to represent 10_{10} through 15_{10}. This procedure, and the decimal equivalent, is illustrated by

$$10111010111_2 = 0101 \quad 1101 \quad 0111 = 5D7_{16}$$

$$5D7_{16} = 5 \times 16^2 + 13 \times 16^1 + 7 \times 16^0 = 1495_{10}.$$

Octal (base-8) numbers are handled in a similar fashion, with 3-bit groups and the octal digits 0 through 7. Thus

$$10111010111_2 = 010 \quad 111 \quad 010 \quad 111 = 2727_8$$

$$2727_8 = 2 \times 8^3 + 7 \times 8^2 + 2 \times 8^1 + 7 \times 8^0 = 1495_{10}$$

Octal notation is found in some minicomputers and microprocessor systems.

Other systems—notably calculators and digital instruments—often employ a different process: Single decimal digits are sequentially encoded to or decoded from a group of 4 (or more) binary digits, *one decimal digit at a time*. This makes it possible, for instance, for encoding to start as soon as you press any digit key on your calculator, instead of waiting for the complete decimal number to be keyed in.

Such digit-by-digit coding is called *binary coded decimal* (BCD), and there are several possible versions. Table 15.1–2 displays the common 8421 BCD code, which consists of 4 bits with weights $2^3 = 8$, $2^2 = 4$, $2^1 = 2$, and $2^0 = 1$, reading from left to right. (This is also called "natural" BCD, since the weights are the same as for direct binary/decimal conversion.) An m-digit decimal number is then encoded as a sequence of m BCD code words, one word for each decimal digit. As an example, 1495_{10} becomes 0001 0100 1001 0101 in BCD, for a total of 16 bits—as contrasted with our earlier result $1495_{10} = 10111010111_2$, which has only 11 bits. The extra bits in BCD is a price one pays for digit-by-digit coding and relatively simple coding circuits.

TABLE 15.1–2
BCD code

Decimal Digit	*Code Word* $B_3B_2B_1B_0$
0	0000
1	0001
2	0010
3	0011
4	0100
5	0101
6	0110
7	0111
8	1000
9	1001

Example 15.1–2
BCD encoding and decoding circuits

To illustrate the role of gate circuits in numerical coding, let's consider the case of a BCD *encoder* represented by Fig. 15.1–7a. There are ten input terminals, $D_0, D_1, \ldots, D_9$, corresponding to the decimal digits, but

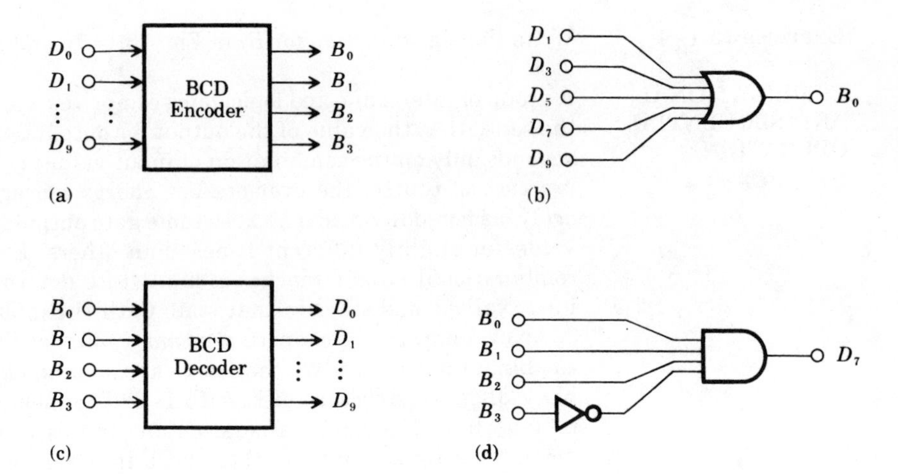

FIGURE 15.1–7 *BCD* encoding and decoding circuits.

only one input is activated at any one time. For each possible input, the encoder produces at the output a 4-bit word $B_3B_2B_1B_0$ drawn from Table 15.1–2. We can easily design the internal logic with the help of the table, by noting that $B_0 = 1$ whenever D_1 or D_3 or D_5 or D_7 or D_9 is "keyed." We therefore write the logic function $B_0 = D_1 + D_3 + D_5 + D_7 + D_9$, which is implemented by the five-input OR gate in Fig. 15.1–7b. The complete encoder contains similar OR gates for the other three outputs.

The BCD *decoder* in Fig. 15.1–7c performs the opposite function of the encoder, producing one of the ten outputs from a given 4-bit input. For instance, Table 15.1–2 shows that the decoder should activate D_7 only when $B_3B_2B_1B_0 = 0111$, so we want $D_7 = 1$ when $B_3 = 0$ and $B_2 = B_1 = B_0 = 1$. The corresponding logic function is $D_7 = \overline{B_3} \cdot B_2 \cdot B_1 \cdot B_0$, implemented in Fig. 15.1–7d with an AND gate plus inverter to obtain $\overline{B}_3$ when $B_3 = 0$. The complete decoder consists of 10 such circuits.

BCD encoders and decoders are available as standard MSI (medium-scale integration) units. Encoders often have a *priority* feature that automatically selects the highest decimal digit if two or more inputs happen to be activated simultaneously.

Exercise 15.1–2 Verify each of the following conversions:
(a) $10101011_2 = 171_{10} = AB_{16} = 253_8$ (b) $0.1011_2 = 0.6875_{10}$;
(c) $200_{10} = 11001000_2$ (d) $0.38_{10} \approx 0.01100001_2$.

Exercise 15.1–3 (a) Calculate or estimate the largest decimal integer that can be represented by a 32-bit word. (You may use the handy approximation $2^{10} = 1024 \approx 10^3$.)
(b) Repeat (a), assuming 8 BCD words.

Exercise 15.1–4 Write the logic function for B_1 in Fig. 15.1–7a and for D_4 in Fig. 15.1–7c.

COMBINATIONAL AND SEQUENTIAL OPERATION

A group of interconnected logic gates comprises a *combinational circuit* in the sense that the value of the output (or outputs) at any particular time depends only on the combination of input values at that same instant. In practice, of course, the ever-present energy storage causes *propagation delay,* as noted in Section 11.2, so some gate outputs will attain their final values at slightly different times than others. Eventually, however, a combinational circuit reaches a final state determined entirely by the input values, and stays in that state until the inputs change.

Now suppose we connect a *memory unit* (such as a flip-flop) and a combinational circuit. We then have a *sequential circuit* with the general block-diagram structure of Fig. 15.1–8. The memory unit accepts some outputs from the combinational circuit, stores their values for various amounts of time, and feeds them back into the combinational circuit as additional inputs. Consequently, the external output values depend not only on the current input but also on *earlier* input values that have been operated on, stored in memory, and reinserted. Putting this another way: a sequential circuit carries out successive combinational operations based on the current input and memory states. The output may then continue to change even though the external inputs remain constant. To ensure successful operation despite varying propagation delays, virtually all sequential circuits require *synchronization* from a timing signal called the *clock*.

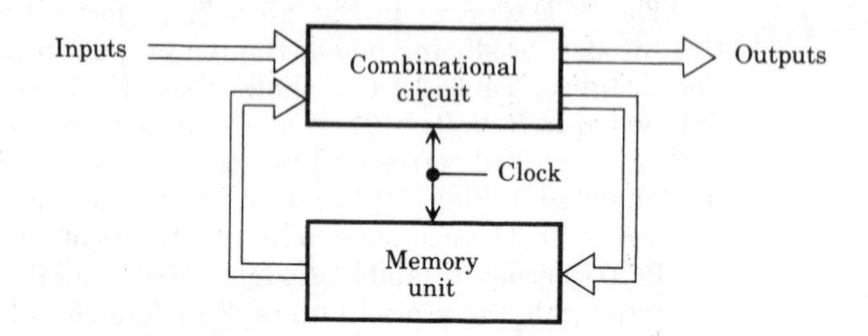

FIGURE 15.1–8
Sequential circuit.

Applications of sequential circuits range from simple counters and registers (discussed in Section 15.3) to extremely complex memory units and microprocessor systems (discussed in Section 16.4).

15.2 COMBINATIONAL CIRCUITS

This section presents the mathematical tools needed to manipulate, simplify, and understand logic expressions that describe combinational circuits. We will summarize the rules of Boolean algebra, and apply them to the design of combinational logic circuits—including the important case of NAND/NOR logic.

BOOLEAN ALGEBRA

Boolean algebra departs from ordinary algebra in two major aspects: Variables are restricted to just two values, 0 and 1; and the basic operations consist of the logical OR, AND, and NOT as previously defined. Some Boolean relationships have a familiar appearance, such as

$$A + 0 = A \qquad B \cdot 1 = B \qquad A \cdot (B + C) = A \cdot B + A \cdot C$$

On the other hand, many valid Boolean equations look rather startling at first, $1 + 1 = 1$ being a case in point.

Rather than presenting a formal treatment of the subject, Table 15.2–1 lists selected "rules" of Boolean algebra that differ from commonplace algebraic expressions. Following the usual convention, we omit the dot symbol for the AND operation except where required for clarity or emphasis. Thus, in Rule 10, $(A + B)(A + C) = A + BC$ means the same thing as $(A + B) \cdot (A + C) = A + B \cdot C$. In absence of parentheses, the AND operation takes precedence over the OR operation—like "multiplication" before "addition."

TABLE 15.2–1 Selected Boolean rules

1	$A + 1 = 1$
2	$A + A = A$
3	$A \cdot A = A$
4	$A + \overline{A} = 1$
5	$A \cdot \overline{A} = 0$
6	$\overline{A + B} = \overline{A} \cdot \overline{B}$
7	$\overline{AB} = \overline{A} + \overline{B}$
8	$A + AB = A$
9	$A + \overline{A}B = A + B$
10	$(A + B)(A + C) = A + BC$
11	$AB + BC + \overline{A}C = AB + \overline{A}C$
12	$(A + B)(B + C)(\overline{A} + C) = (A + B)(\overline{A} + C)$

Any valid Boolean relation may be proved by constructing a truth table. This method is especially easy for the one- and two-variable relations (Rules 1 through 9), as demonstrated in Fig. 15.2–1 for Rule 9. Other relations can then be derived using the one-variable rules and algebraic manipulations based on the usual commutative, associative, and distributive laws. To illustrate, we prove Rule 8 by writing

$$\begin{aligned} A + AB &= A \cdot 1 + A \cdot B \\ &= A \cdot (1 + B) && \text{(Distributive law)} \\ &= A \cdot 1 && \text{(Rule 1)} \\ &= A \end{aligned}$$

Some proofs, however, require considerable inspiration and/or perspiration!

A	$\bar{A}$	B	$\bar{A}B$	$A + \bar{A}B$	$A + B$
0	1	0	0	$0 + 0 = 0$	$0 + 0 = 0$
0	1	1	1	$0 + 1 = 1$	$0 + 1 = 1$
1	0	0	0	$1 + 0 = 1$	$1 + 0 = 1$
1	0	1	0	$1 + 0 = 1$	$1 + 1 = 1$

FIGURE 15.2–1 Truth table proving rule 9.

Rules 8 through 12 exhibit *absorption* effects of Boolean algebra. By this we mean that certain terms disappear from one side of an equation because they are absorbed in other terms, and do not influence the result. For instance, $A + AB$ equals the value of A regardless of the value of B.

Rules 6 and 7 for the expansion of the complement of the OR and AND operations are particular cases of a more general principle named *DeMorgan's theorem:*

> The complement of any Boolean expression may be obtained by complementing each variable and interchanging ANDs and ORs.

Suppose, for example, that $X = C + \bar{C}D$ and we want to find $\bar{X}$. (Temporarily ignore Rules 9 and 6, which show that $C + \bar{C}D$ simplifies to $C + D$ and so $\bar{X} = \overline{C + D} = \bar{C}\bar{D}$.) DeMorgan's theorem tells us that

$$\bar{X} = \overline{C + (\bar{C} \cdot D)} = \bar{C} \cdot (\bar{\bar{C}} + \bar{D}) = \bar{C}(C + \bar{D})$$

since $\bar{\bar{C}} = C$. But $\bar{C}(C + \bar{D}) = \bar{C} \cdot C + \bar{C} \cdot \bar{D} = 0 + \bar{C} \cdot \bar{D}$, per Rule 5, so $\bar{X} = \bar{C}\bar{D}$ in agreement with Rules 9 and 6.

Finally, all the rules still apply when a variable is consistently replaced by its complement or by a group of variables. Thus, if $A = \bar{C}$ and $B = D\bar{E}$ in Rule 9, then $C + CD\bar{E} = \bar{C} + D\bar{E}$, and so forth.

Example 15.2–1

Suppose the specification for a certain logic system leads to the rather awesome expression

$$X = D\,(\overline{DE} + F)\,\bar{F} + \bar{E}F$$

whose direct implementation is diagrammed in Fig. 15.2–2a. This circuit requires eight gates and numerous internal connections, implying significant hardware and assembly costs. Clearly, we should look for possible simplifications from Boolean absorption, for they would reduce the complexity and costs.

To seek out a simpler equivalent function, we expand the first term and proceed as follows:

$$\begin{aligned} D(\overline{DE} + F)\bar{F} &= D \cdot \overline{DE} \cdot \bar{F} + D \cdot F \cdot \bar{F} \\ &= D(\bar{D} + \bar{E})\bar{F} + DF\bar{F} && \text{(Rule 7)} \\ &= (D\bar{D})\bar{F} + D\bar{E}\bar{F} + D(F\bar{F}) \\ &= 0 \cdot \bar{F} + D\bar{E}\bar{F} + D \cdot 0 && \text{(Rule 5)} \\ &= D\bar{E}\bar{F} \end{aligned}$$

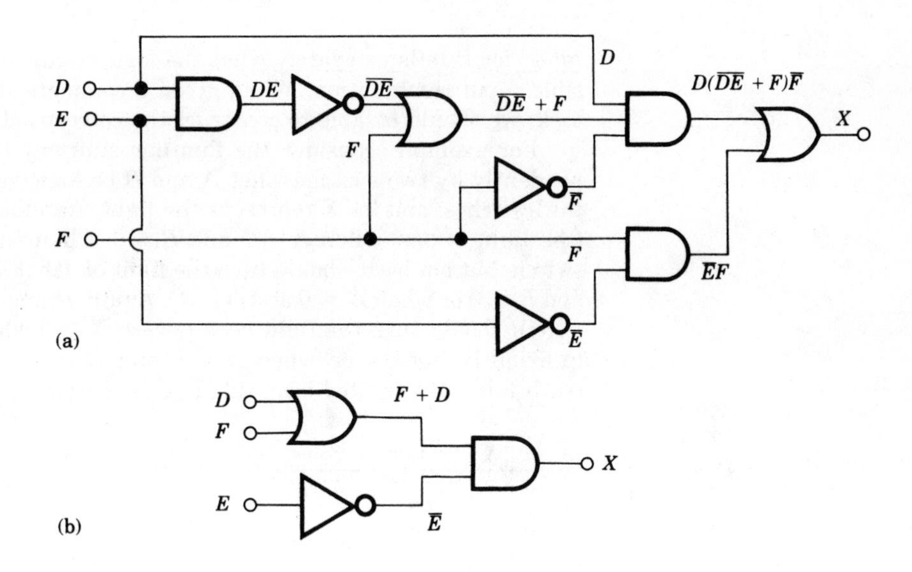

FIGURE 15.2–2 Gate circuits for Example 15.2–1.

Thus

$$X = D\overline{E}\overline{F} + \overline{E}F = \overline{E}(F + \overline{F}D)$$
$$= \overline{E}(F + D) \qquad \text{(Rule 9)}$$

Implementing this reduced expression requires only 3 gates, as shown in Fig. 15.2–2b, a substantial saving compared to the original version.

Another advantage of the simplified circuit is that the "longest path" from input to output goes through just two successive gates, as against the five-gate path length in Fig. 15.2–2a. Minimizing path length reduces potential problems due to propagation delay.

Exercise 15.2–1

(a) Prove rule 6 by constructing its truth table.
(b) Prove rule 10 by algebraic manipulation.
(c) Apply DeMorgan's theorem to find an expression for $\overline{L}$ in Example 15.1–1.

Exercise 15.2–2

Using algebraic manipulation and Boolean rules, show that

$$X = D\overline{(\overline{D} + E)}(\overline{D} + F)$$

reduces to $X = D\overline{E}F$.

SYNTHESIS OF LOGIC FUNCTIONS

So far we have dealt with logic expressions that were given or could be inferred from a simple statement of the task. Such is seldom the actual case, however; and we need a systematic *synthesis* procedure for generating logic functions from scratch. Our starting point will be the *truth table,* for it tells us exactly what the logic circuit must do for every possible input combination. When given a complete statement of the logic task, we should be able to construct the corresponding truth table.

For example, consider the familiar stairway light controlled independently by two switches. Let A and B be logic variables representing the switches, and let X represent the light. Assume initially that $X = 0$ (the light is OFF) when $A = 0$ and $B = 0$. Changing the state of either switch, but not both, should turn the light ON (that is, $X = 1$ when $A = 1$ and $B = 0$ or when $A = 0$ and $B = 1$). Again, changing the state of either switch should turn the light *back* OFF, so $X = 0$ when $A = 0$ and $B = 0$ (previously noted), or when $A = 1$ and $B = 1$. Figure 15.2–3 is the truth table constructed from this line of reasoning.

FIGURE 15.2–3
Truth table for $X = \overline{A}B + A\overline{B}$ (exclusive-OR) showing product and sum terms.

A	B	X	Products	Sums
0	0	0		$A + B$
0	1	1	$\overline{A}B$	
1	0	1	$A\overline{B}$	
1	1	0		$\overline{A} + \overline{B}$

Now, to develop an expression for X from the truth table, we first must see that $X = 1$ when $\overline{A}B = 1$ ($A = 0$ and $B = 1$) or when $A\overline{B} = 1$ ($A = 1$ and $B = 0$), implying that

$$X = \overline{A}B + A\overline{B} \tag{1}$$

The "product" terms $\overline{A}B$ and $A\overline{B}$ are easily found: Simply look at the lines of the truth table for which $X = 1$ and write the logical-AND "product" of all input variables, complementing any variable whose value equals 0 on that line. Equation (1) then reflects the fact that $X = 1$ if either $\overline{A}B = 1$ or $A\overline{B} = 1$. Such *sum-of-products* (SOP) expressions are formed by taking the logical-OR "sum" of the product terms corresponding to $X = 1$.

By the same procedure, we could generate an SOP expression for $\overline{X}$ instead of X. In the case at hand we see that

$$\overline{X} = \overline{A}\,\overline{B} + AB$$

since $\overline{X} = 1$ when $\overline{A}\,\overline{B} = 1$ (top line of the truth table) or when $AB = 1$ (bottom line). Invoking DeMorgan's theorem then yields

$$X = \overline{\overline{X}} = \overline{\overline{A} \cdot \overline{B} + A \cdot B} = (\overline{\overline{A}} + \overline{\overline{B}}) \cdot (\overline{A} + \overline{B}) = (A + B)(\overline{A} + \overline{B}) \tag{2}$$

which is called a *product-of-sums* (POS) expression. The "sum" terms $A + B$ and $\overline{A} + \overline{B}$ correspond to lines of the truth table where $X = 0$; they are formed by writing the "sum" of all input variables, complementing any variable whose value equals 1 on that line.

We therefore have two easy methods for synthesizing a logic function from its truth table. They yield different but equivalent expressions—an SOP or a POS.

The example we have been working with happens to have practical significance beyond switching stairway lights. The truth table in Fig. 15.2–3 defines a new logic operation, known as the *exclusive*-OR (X-OR), represented by

$$X = A \oplus B \tag{3a}$$

where $A \oplus B$ stands for "A OR B but not both." As seen from the truth table, the exclusive-OR operation yields

$$0 \oplus 0 = 0 \qquad 0 \oplus 1 = 1 \qquad 1 \oplus 0 = 1 \qquad 1 \oplus 1 = 0 \tag{3b}$$

which differs from the OR operation in that $1 \oplus 1 = 0$. The exclusive-OR plays a major role in arithmetic logic circuits, and Eqs. (1) and (2) suggest two ways of building an X-OR gate using AND, OR, and NOT gates. (Figure 6.4–6 shows the mechanical-switch version.) For convenience, however, we symbolize an X-OR gate by Fig. 15.2–4, regardless of how the gate is implemented. An X-OR gate usually has only two inputs.

FIGURE 15.2–4
Exclusive-OR gate.

Returning to our synthesis procedures, let's suppose we have a function of three variables related by the truth table in Fig. 15.2–5a, which also lists the product and sum terms obtained immediately by inspection. Since there are three combinations for which $X = 1$, the SOP has three terms, namely

$$X = \overline{A}B\overline{C} + \overline{A}BC + ABC \tag{4a}$$

whereas the POS would have five terms. Furthermore, Boolean absorption leads to additional simplifications of the SOP by grouping the first and second terms as $\overline{A}B\overline{C} + \overline{A}BC = \overline{A}B(\overline{C} + C) = \overline{A}B$ or grouping the second and third terms as $\overline{A}BC + ABC = (\overline{A} + A)BC = BC$. Both simplifications are actually possible if we use the fact that $\overline{A}BC = \overline{A}BC +$

A	B	C	X	Products	Sums
0	0	0	0		$A + B + C$
0	0	1	0		$A + B + \overline{C}$
0	1	0	1	$\overline{A}\,B\,\overline{C}$	
0	1	1	1	$\overline{A}\,B\,C$	
1	0	0	0		$\overline{A} + B + C$
1	0	1	0		$\overline{A} + B + \overline{C}$
1	1	0	0		$A + B + \overline{C}$
1	1	1	1	$A\,B\,C$	

(a)

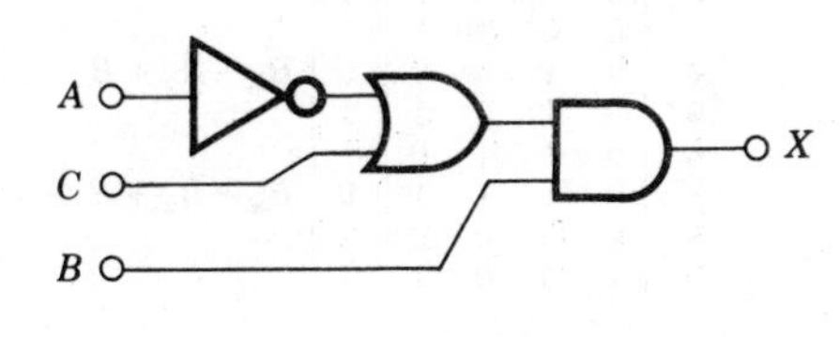

(b)

FIGURE 15.2–5
(a) Truth table for three-variable function. (b) Minimized circuit from sum of products.

$\overline{A}BC$ (from Rule 2) to write

$$X = \overline{A}B\overline{C} + \overline{A}BC + \overline{A}BC + ABC$$
$$= \overline{A}B(\overline{C} + C)\overline{C} + (\overline{A} + A)BC$$
$$= \overline{A}B + BC = (\overline{A} + C)B \qquad (4b)$$

obtaining the final circuit of Fig. 15.2–5b. This is a *minimized* circuit in that it requires the fewest number of gates to produce X.

Example 15.2–2
BCD to seven-segment display

Consider a calculator that does computations in 8421 BCD and presents result as decimal digits in the familiar seven-segment LED display of Fig. 15.2–6. The four-input, seven-output display decoder takes a BCD word ($B_3B_2B_1B_0$) and lights the appropriate segments of the display. We'll synthesize the function for one of the outputs, say segment d, to further illustrate the procedures involved.

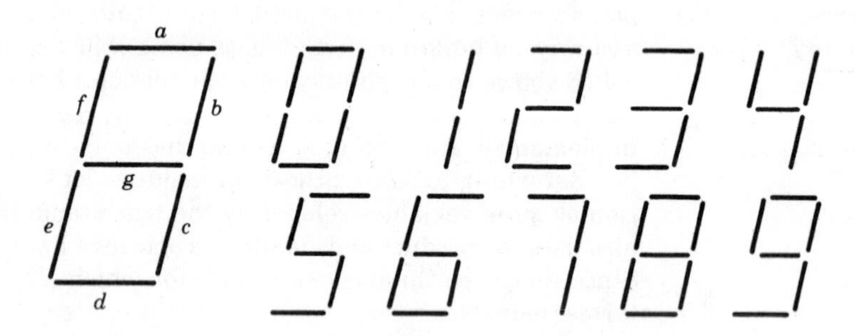

FIGURE 15.2–6 Seven-segment LED display.

FIGURE 15.2–7 Truth table and gate circuit for Example 15.2–2.

Inspecting Fig. 15.2–6 reveals that segment d should be lit for the decimal digits 0, 2, 3, 5, 6, 8, and 9. We therefore let $d = 1$ on the corresponding lines of the truth table, Fig. 15.2–7a. Since there are seven

Digit	B_3	B_2	B_1	B_0	d	Sums
0	0	0	0	0	1	
1	0	0	0	1	0	$B_3 + B_2 + B_1 + \overline{B}_0$
2	0	0	1	0	1	
3	0	0	1	1	1	
4	0	1	0	0	0	$B_3 + \overline{B}_2 + B_1 + B_0$
5	0	1	0	1	1	
6	0	1	1	0	1	
7	0	1	1	1	0	$B_3 + \overline{B}_2 + \overline{B}_1 + \overline{B}_0$
8	1	0	0	0	1	
9	1	0	0	1	1	

(a)

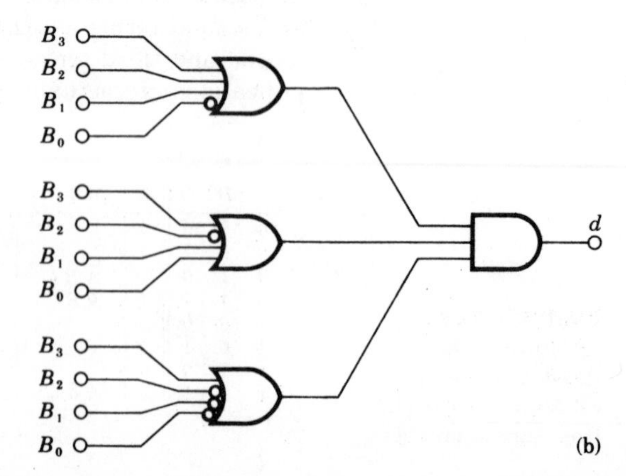

(b)

product terms but only $10 - 7 = 3$ sum terms, the latter should be easier to implement. We thus write the POS as

$$d = (B_3 + B_2 + B_1 + \overline{B}_0)(B_3 + \overline{B}_2 + B_1 + B_0)(B_3 + \overline{B}_2 + \overline{B}_1 + \overline{B}_0)$$

and obtain the gate diagram of Fig. 15.2–7b. To avoid a confusing clutter of crossing leads, this diagram omits those wires that would tie together gates having the same input variables. We have also used "bubbles" at the gate inputs as a shorthand convention for input inverters.

Example 15.2–3
Data selector/multiplexer

Figure 15.2–8a represents a 4-to-1 *data selector* or *digital multiplexer* with four inputs, one output, and a 2-bit control word $S_1 S_0$. The purpose of $S_1 S_0$ is to select which of the inputs appears at the output, as tabulated in Fig. 15.2–8b. We will not attempt to construct the complete truth table because, with $4 + 2 = 6$ input variables, it would have $2^6 = 64$ lines! Instead, we synthesize the gate circuitry by noting that $X = A_0$ only when $S_1 = S_0 = 0$ or $\overline{S}_1 \cdot \overline{S}_0 = 1$, while $X = A_1$ when $\overline{S}_1 \cdot S_0 = 1$, etc. This line of reasoning leads to the simplified SOP

$$X = \overline{S}_1\overline{S}_0A_0 + \overline{S}_1S_0A_1 + S_1\overline{S}_0A_2 + S_1S_0A_3$$

implemented in Fig. 15.2–8c.

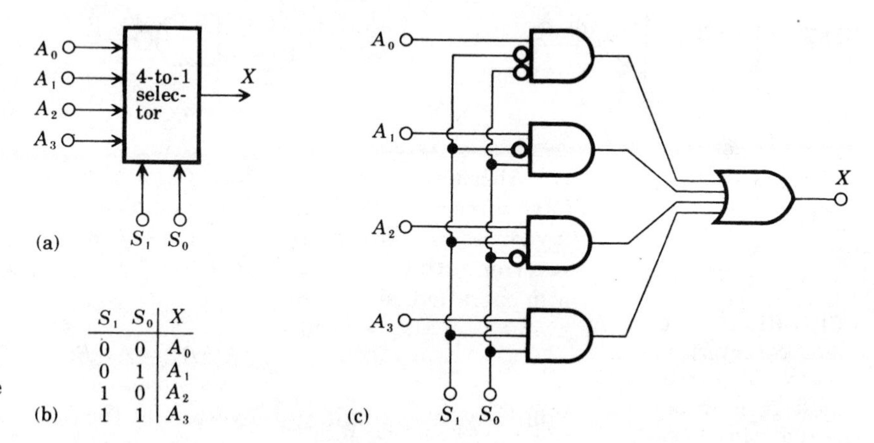

S_1	S_0	X
0	0	A_0
0	1	A_1
1	0	A_2
1	1	A_3

FIGURE 15.2–8
(a) Four-to-one selector. (b) Truth table. (c) Gate implementation.

Commercial data selectors are available as integrated circuits having 2^n inputs and n-bit control words, where $n = 2$, 3, or 4. Some of the applications are explored in problems at the end of the chapter.

Exercise 15.2–3

A logic circuit with inputs A, B, and C is to have $X = 1$ at the output only when $A = B = C$ or when $A = B = \overline{C}$. Construct the truth table, write the SOP, and simplify your results.

Exercise 15.2–4

A digital *demultiplexer* performs the reverse function of Fig. 15.2–8 in that $S_1 S_0$ steers a single input X to one of four output terminals, A_0, A_1, A_2, or A_3. The logic values equal 0 at the nonselected terminals. Devise the corresponding gate diagram.

NAND AND NOR CIRCUITS

Logic gates built with transistors inherently negate the output caused by the natural voltage-inverting action of transistor circuits. Thus, a transistor AND gate is actually a NOT-AND gate like Fig. 15.2–9a, with output

$$X = \overline{A \cdot B} \tag{5a}$$

The truth table in Fig. 15.2–9b follows directly by negating the output of an AND gate. We call this circuit a NAND gate, and represent it by the special symbol in Fig. 15.2–9c where the output bubble stands for negation. Another representation comes about by applying DeMorgan's theorem to Eq. (5a), so

$$X = \overline{A \cdot B} = \overline{A} + \overline{B} \tag{5b}$$

which leads to Fig. 15.2–9d.

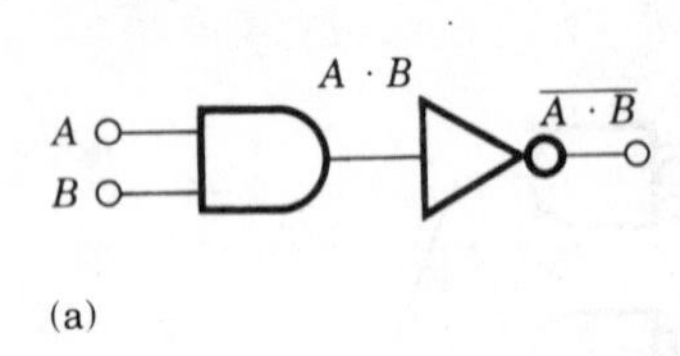

A	B	$\overline{A \cdot B}$
0	0	1
0	1	1
1	0	1
1	1	0

(b)

(c)

(d)

FIGURE 15.2–9
Two-input NAND gate: (a) NOT-AND circuit; (b) truth table; (c) symbol; (d) equivalent symbol.

Although we have considered only two inputs, a NAND gate could have three or more inputs. Or it could have just one input—in which case it reduces to a NOT gate or inverter (Fig. 15.2–10a). Connecting such an inverter at the *output* of another NAND gate (Fig. 15.2–10b) produces the AND operation, since

$$\overline{\overline{(A \cdot B)}} = A \cdot B$$

Similarly, inserting NAND inverters at the *inputs* of another NAND gate

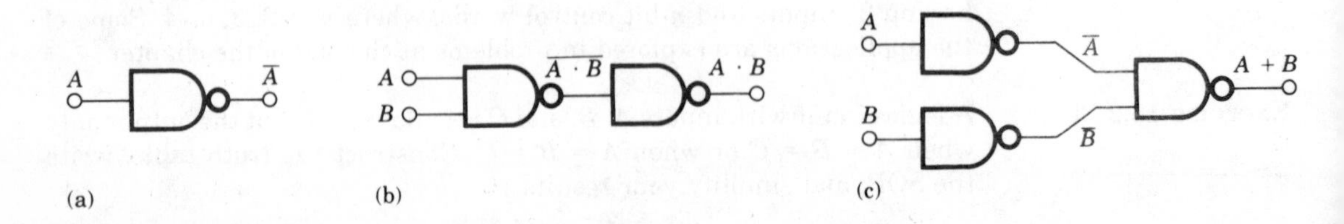

FIGURE 15.2–10
(a) NAND inverter. (b) AND circuit using NAND gates. (c) OR circuit using NAND gates.

(Fig. 15.2–10c) produces the OR operation, since

$$\overline{\overline{A} \cdot \overline{B}} = \overline{\overline{A}} + \overline{\overline{B}} = A + B$$

The NAND gate therefore serves as a *universal* gate in the sense that any logic operation can be achieved using only NAND gates. Consequently, rather than working with a mixture of AND, OR, and NOT gates, you might buy several IC units typically containing four NAND gates each, and interconnect them to suit the task at hand. (An inverter like Fig. 15.2–10a is obtained by tying together all the inputs of a multi-input NAND gate.)

Direct synthesis of NAND-gate circuits turns out to be quite straightforward when you start with an SOP expression for the desired function. To illustrate the procedure, take the exclusive-OR function $X = \overline{A}B + A\overline{B}$ as implemented in Fig. 15.2–11a. (The inverters for $\overline{A}$ and $\overline{B}$ have been omitted for clarity.) The circuit in Fig. 15.2–11b produces the same output, because the negations caused by changing the AND gates to NAND gates have been canceled by negations at the OR-gate inputs. But, from Fig. 15.2–9d, an OR gate with input negations is equivalent to a NAND gate. The final all NAND-gate circuit then becomes as diagrammed in Fig. 15.2–11c, and requires no more gates than the Fig. 15.2–11a circuit. Comparing these two circuits brings out a handy shortcut for NAND-gate synthesis: Simply implement the SOP with NAND gates replacing AND and OR gates.

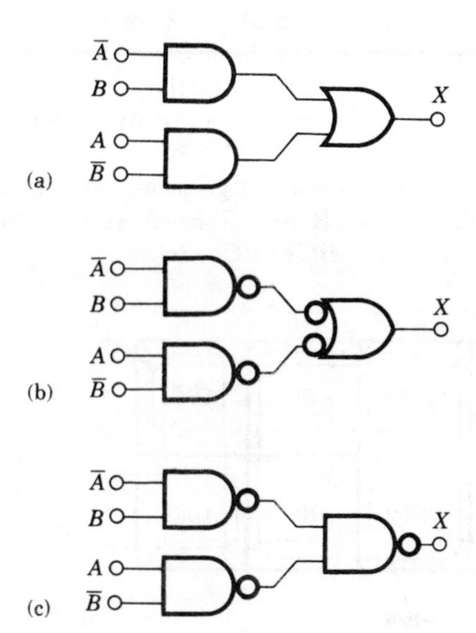

FIGURE 15.2–11 Synthesizing a sum-of-products function using NAND gates.

Another universal gate is the NOR gate, whose symbol and truth table are given in Fig. 15.2–12. This gate performs a NOT-OR operation on its inputs, so

$$X = \overline{A + B} = \overline{A} \cdot \overline{B} \tag{6}$$

A single-input NOR gate acts as an inverter, and the OR and AND operations are obtained by combining NOR gates and NOR inverters. For direct NOR-gate synthesis of a logic function, you start with a POS expression (instead of an SOP) and replace the OR and AND gates with NOR gates.

FIGURE 15.2–12 Two-input NOR gate: (a) symbol; (b) truth table.

(a) A, B → $\overline{A + B}$

(b)

A	B	$\overline{A + B}$
0	0	1
0	1	0
1	0	0
1	1	0

Exercise 15.2–5

(a) Use Eq. (6) to construct NOR-gate diagrams that produce $X = A + B$ and $X = A \cdot B$.
(b) Obtain a NOR-gate circuit for the exclusive-OR by starting with Eq. (2).

MINIMIZATION AND KARNAUGH MAPS †

In view of the rather hit-or-miss nature of simplification by algebraic manipulation, we sometimes need a systematic method for *minimizing* switching functions. We will briefly describe a popular minimization process based on the *Karnaugh map* (pronounced Car-no). This method works well for functions with three or four variables. Minimizing expressions with more than four variables is rather difficult by hand but, fortunately, that task seldom arises. In fact, minimization has become a less important consideration thanks to the low cost of mass-produced IC NAND and NOR gates.

Figure 15.2–13 shows the general map structure for a function of three variables, say A, B, and C. Each square represents one of the possible product terms, ABC, $\overline{A}BC$, etc., and the peculiar arrangement is

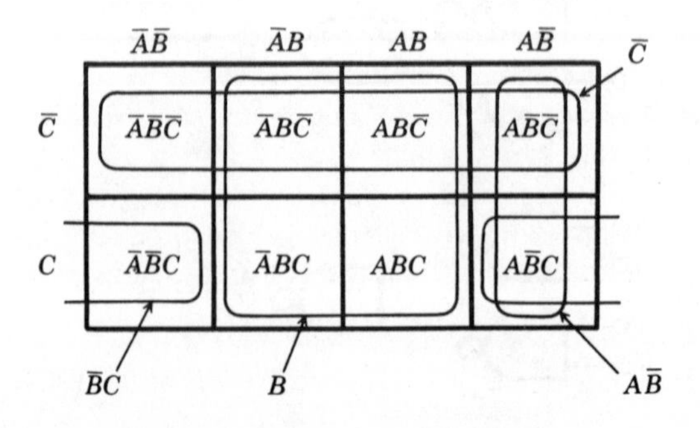

FIGURE 15.2–13 Karnaugh map for three variables.

purposely chosen so that the sum of product terms from any *two adjacent* squares is *independent of one of the variables*. Thus, for instance, the vertical adjacent pair on the far right of the map represents $A\overline{B}\,\overline{C} + A\overline{B}C = A\overline{B}(\overline{C} + C) = A\overline{B}$, independent of the absorbed variable C. We refer to a pair of adjacent squares as a 2-square *subcube* and identify it by drawing a rounded rectangle that includes the two terms. Adjacency also extends beyond the map's edges if we think of opposite edges as being adjacent. To illustrate, the subcube comprising the bottom corners represents $\overline{A}\,\overline{B}C + A\overline{B}C = (\overline{A} + A)\overline{B}C = \overline{B}C$.

Going even further, we find that any subcube of *four* adjacent squares represents a sum of products that is *independent of two variables*. For example, the entire top row of Fig. 15.2–13 represents $\overline{A}\,\overline{B}\,\overline{C} + \overline{A}B\overline{C} + AB\overline{C} + A\overline{B}\,\overline{C} = \overline{A}(\overline{B} + B)\overline{C} + A(B + \overline{B})\overline{C} = (\overline{A} + A)\overline{C} = \overline{C}$, in which A and B have absorbed. You can similarly show that the 4-square subcube in the middle represents $\overline{A}B\overline{C} + AB\overline{C} + \overline{A}BC + ABC = B$, independent of A and C. With a little practice, such reductions will become apparent from inspection of the map—just eliminate those variables that appear with negation in half the squares of the subcube and without negation in the other half.

The minimization process requires an unsimplified or *expanded* sum of products (ESOP) for the desired function, as would be obtained from its truth table. Equations (2) and (4a) are examples of ESOPs, for every sum term contains *all* input variables. You then take a blank Karnaugh map (without lettering inside the squares) and proceed as follows:

1. Map the function by putting 1s in only those squares that correspond to terms in the ESOP.
2. Locate any isolated 1s and draw 1-square subcubes around them.
3. Draw the fewest number of largest possible subcubes around the remaining adjacent 1s until every 1 is "covered" by at least one subcube. Overlapping subcubes are permitted.
4. Identify the reduced terms represented by the subcubes, and form their logical-OR sum to obtain the *minimum* sum of products (MSOP).

If possible and desired, the result may be further simplified by factoring. (Of course, you would not factor if you were seeking an MSOP for NAND-gate synthesis.)

Let's apply our process to the function $X = \overline{A}B\overline{C} + \overline{A}BC + ABC$ from Eq. (4a). The three terms are mapped in Fig. 15.2–14, and the 1s can be covered by two 2-square subcubes. These subcubes represent $\overline{A}B$ and BC, so we write the MSOP as

$$X = \overline{A}B + BC = (\overline{A} + C)B$$

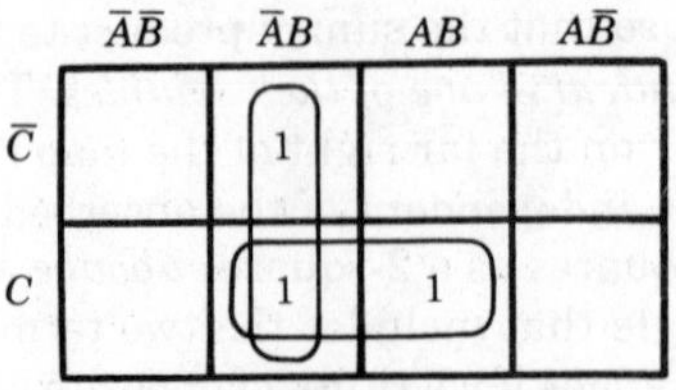

	$\bar{A}\bar{B}$	$\bar{A}B$	AB	$A\bar{B}$
$\bar{C}$		1		
C		1	1	

FIGURE 15.2–14
Map of $X = \bar{A}B\bar{C} + \bar{A}BC + ABC$.

where the final factoring brings the result in agreement with Eq. (4b). Note that the term $\bar{A}BC$ was covered by both subcubes and appears to have been "counted" twice. Such double counting is equivalent to writing $\bar{A}BC = \bar{A}BC + \bar{A}BC$, the trick used before to obtain Eq. (4b).

Minimizing a function of four variables involves the same steps previously taken. Figure 15.2–15 illustrates the map of a four-variable function whose ESOP would contain eleven terms. (We won't even bother writing it out.) The 8-square subcube formed by the top and bottom rows corresponds to $\bar{C}\bar{D} + C\bar{D} = \bar{D}$, the 4-square vertical subcube corresponds to AB, and the 2-square subcube corresponds to $\bar{A}\bar{B}C$. Thus,

$$X = \bar{A}\bar{B}C + AB + \bar{D}$$

with no factoring possible.

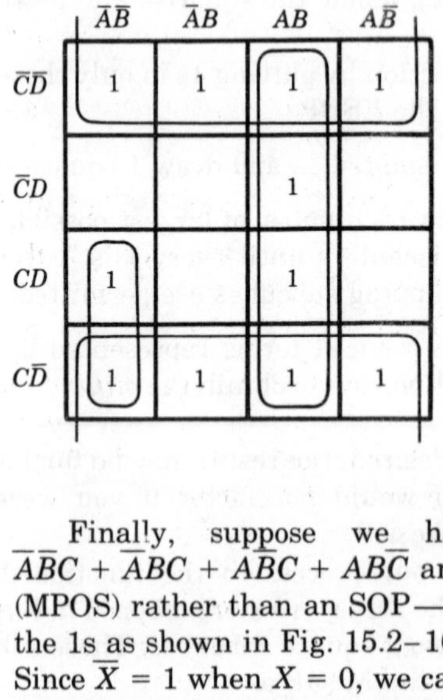

	$\bar{A}\bar{B}$	$\bar{A}B$	AB	$A\bar{B}$
$\bar{C}\bar{D}$	1	1	1	1
$\bar{C}D$			1	
CD	1		1	
$C\bar{D}$	1	1	1	1

FIGURE 15.2–15
Map of a four-variable function.

Finally, suppose we have the three-variable ESOP $X = \bar{A}\bar{B}C + \bar{A}BC + A\bar{B}C + AB\bar{C}$ and we want a minimum *product of sums* (MPOS) rather than an SOP—perhaps for NOR-gate synthesis. We map the 1s as shown in Fig. 15.2–16 and fill the remaining squares with 0s. Since $\bar{X} = 1$ when $X = 0$, we can obtain an MSOP for $\bar{X}$ by ignoring the 1s and covering the 0s instead. This leads to

$$\bar{X} = ABC + \bar{A}\bar{C} + \bar{B}\bar{C}$$

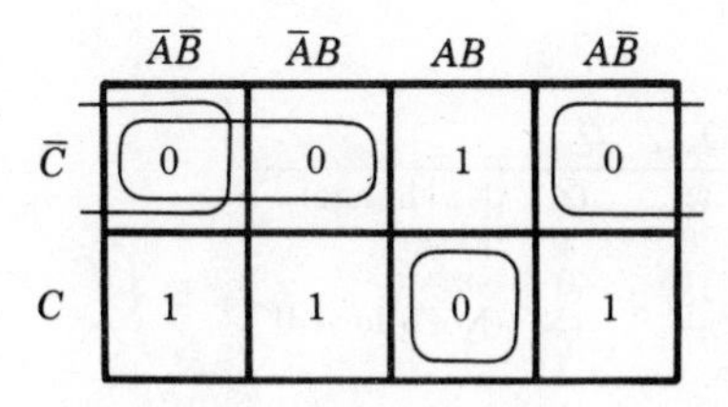

FIGURE 15.2–16 Map coverage for an MPOS.

and application of DeMorgan's theorem yields

$$X = \overline{ABC + \bar{A}\bar{C} + \bar{B}\bar{C}} = (\bar{A} + \bar{B} + \bar{C})(A + C)(B + C)$$

which is the desired MPOS.

Exercise 15.2–6 Map $X = \bar{A}\bar{B}\bar{C} + \bar{A}\bar{B}C + \bar{A}BC + A\bar{B}\bar{C} + A\bar{B}C$ and obtain the MSOP and MPOS.

15.3 SEQUENTIAL CIRCUITS

Sequential digital circuits have *memory:* A current output state depends on previous states of one or more inputs, as well as on current input states. The simplest circuit with memory is a flip-flop, whose basic form and variations will be discussed first. We then describe how groups of flip-flops can be arranged to function as counters, shift registers, and binary memory units.

FLIP-FLOPS

The block symbol in Fig. 15.3–1a represents an ideal *set-reset* (*SR*) *flip-flop,* also called an *SR latch.* It has two outputs Q and $\bar{Q}$, one being the *complement* of the other. The relationships between the inputs and outputs are as follows:

- The input combination $S = 1$ and $R = 0$ produces the SET state: $Q = 1$ and $\bar{Q} = 0$.
- The input combination $S = 0$ and $R = 1$ produces the RESET state: $Q = 0$ and $\bar{Q} = 1$.
- The input combination $S = 0$ and $R = 0$ produces *no change* of state.
- The input combination $S = 1$ and $R = 1$ is *not allowed.*

These relationships are summarized in the accompanying truth table, Fig. 15.3–1b.

Because $S = R = 0$ produces no change at the output, we must use the notation Q_n and Q_{n+1} to stand for the output values *before* and *after* the instant when the inputs change states. (The subscript n will shortly take on added meaning.) The expanded truth table in Fig. 15.3–1c emphasizes the distinction between Q_n and Q_{n+1} by treating Q_n as if it were

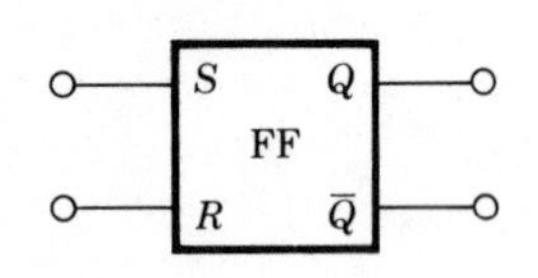

(a)

S	R	Q_{n+1}	$\overline{Q}_{n+1}$	
0	0	Q_n	$\overline{Q}_n$	(No change)
0	1	0	1	(RESET)
1	0	1	0	(SET)
1	1	—	—	(Not allowed)

(b)

S	R	Q_n	Q_{n+1}	
0	0	0	0	(stays RESET)
0	0	1	1	(stays SET)
0	1	0	0	(stays RESET)
0	1	1	0	(RESET)
1	0	0	1	(SET)
1	0	1	1	(stays SET)
1	1	0	—	(Not allowed)
1	1	1	—	

(c)

FIGURE 15.3–1
SR flip-flop: (a) symbol; (b) truth table; (c) expanded truth table.

another input. Here, and in all subsequent flip-flop tables, we omit the complemented output $\overline{Q}_{n+1}$.

The memory capability of an *SR* flip-flop comes from the fact that it retains the previous output state when the inputs change to $S = R = 0$. For example, suppose that $Q = 0$ initially and that $S = R = 0$; the flip-flop will stay in the RESET state until a SET pulse ($S = 1$) comes along. It then "latches" into the SET state and stays there after the SET pulse ends (so $S = R = 0$). Now the flip-flop remembers that it has been set, and ignores any subsequent SET pulses until a RESET pulse ($R = 1$) comes along. (See Fig. 11.2–5c for the voltage waveforms corresponding to this typical sequence of events.) You should carefully examine the properties of an *SR* flip-flop; they are basic to an understanding of sequential circuits.

Figure 15.3–2 illustrates one way of building an *SR* flip-flop using *cross-coupled* NAND gates. (This diagram essentially corresponds to the BJT circuit back in Fig. 11.2–5a.) The lower output terminal has been labeled *P*, rather than $\overline{Q}$, for purposes of analysis. Drawing upon the NAND definitions in Eq. (5), Section 15.2, we obtain logic expressions for the outputs in the form

$$Q = \overline{\overline{S} \cdot P} = S + \overline{P} \qquad P = \overline{\overline{R} \cdot Q} = R + \overline{Q} \tag{1}$$

Therefore, if $S = 1$ and $R = 0$, then $Q = 1 + \overline{P} = 1$ and $P = 0 + \overline{Q} = 0$, which constitute the SET state. It follows from symmetry that we get the RESET state, $P = 1$ and $Q = \overline{1} = 0$, when $S = 0$ and $R = 1$. Finally, with

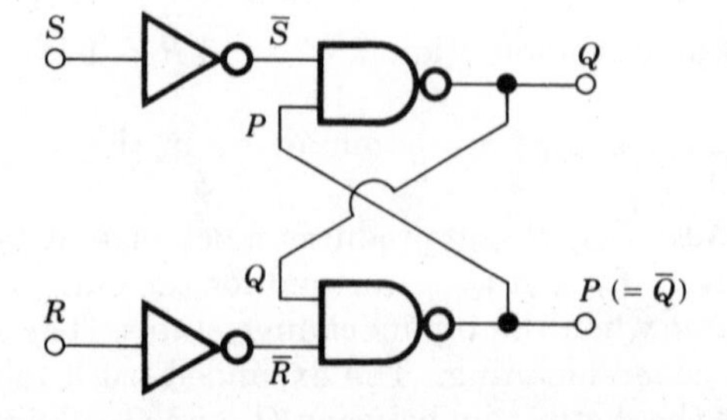

FIGURE 15.3–2
NAND-gate implementation of an *SR* flip-flop.

$S = R = 0$, Eq. (1) predicts $Q = 0 + \overline{P} = \overline{P}$ and $P = 0 + \overline{Q} = \overline{Q}$, meaning that the outputs are again complementary and the state remains unchanged from its previous condition. We have thus confirmed that this circuit satisfies the properties of an *SR* flip-flop. Furthermore, we see that the forbidden input combination $S = R = 1$ would produce *non*complementary outputs, namely $Q = 1 + \overline{P} = 1$ and $P = 1 + \overline{Q} = 1$.

There are several other ways of implementing an *SR* flip-flop, but they all involve a cross-coupled configuration to provide the *feedback* that creates the latching effect. All practical flip-flop circuits also depart from ideal behavior in certain respects. In particular, because of propagation delays, a sudden input change does not produce an instantaneous output response. Consequently, the duration of a SET or RESET pulse must be long enough to achieve the latched condition. Typical latching times are less than a microsecond and pose no problem except at high switching rates.

Sometimes we desire a third input that *gates* the other two flip-flop inputs. Gating might be needed, for instance, to prevent unwanted output state transitions during intervals when the inputs are supposed to be inactive. The *gated SR flip-flop* in Fig. 15.3–3a provides this capability. Setting $G = 1$ *enables* S and R since $Y = \overline{1 \cdot S} = \overline{S}$ and $X = \overline{1 \cdot R} = \overline{R}$, just as in Fig. 15.3–2. However, setting $G = 0$ *disables* S and R, since $Y = \overline{0 \cdot S} = \overline{0}$ and $X = \overline{0 \cdot R} = \overline{0}$ regardless of the values of S and R. The output state therefore remains constant. We represent the disabled case by putting × under S and R on the first line of the truth table, Fig. 15.3–3b, which has $G = 0$ in the first column. These ×s stand for *don't care conditions* in the sense that they could be either 0s or 1s. The rest of the truth table displays the usual *SR* pattern, with the addition of $G = 1$ in the first column.

As a further refinement, a gated flip-flop may be equipped with a *direct preset* (*Pr*) and/or a *direct clear* (*Cr*), shown as dashed lines in Fig. 15.3–3a. These inputs override the gated inputs and allow one to force the flip-flop into an appropriate initial state when $G = 0$. Figure 15.3–3c gives the truth table for the direct inputs. Note that they are deactivated by setting $Pr = Cr = 1$, which returns control to the gated inputs.

FIGURE 15.3–3 Gated *SR* flip-flop with preset and clear: (a) circuit; (b) truth table for gated inputs; (c) truth table for direct inputs.

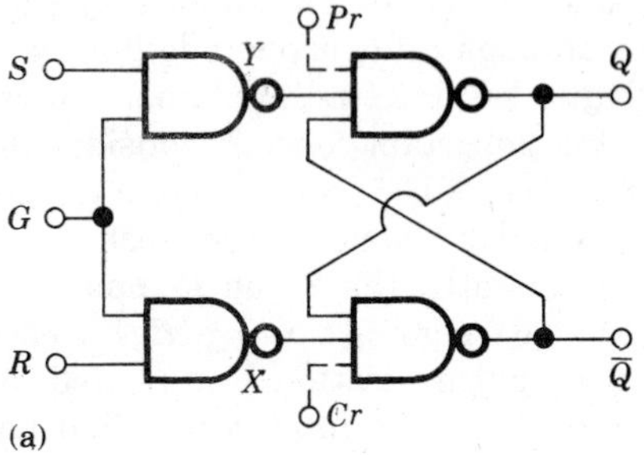

(a)

G	S	R	Q_{n+1}
0	×	×	Q_n
1	0	0	Q_n
1	0	1	0
1	1	0	1
1	1	1	—

(b)

G	Pr	Cr	Q	
×	1	1	Q_{n+1}	(Gated mode)
0	1	0	0	(RESET)
0	0	1	1	(SET)
0	0	0	—	(Not allowed)

(c)

Now let the signal applied to the gate terminal consist of a train of pulses like Fig. 15.3–4, called a *clock waveform*. In accordance with the positive-logic convention, the HIGH signal level during a pulse represents the logic-1 state and the LOW level between pulses represents the logic-0 state. Output state transitions then will be possible only when a clock pulse is present, so we say the flip-flop is *clocked*.

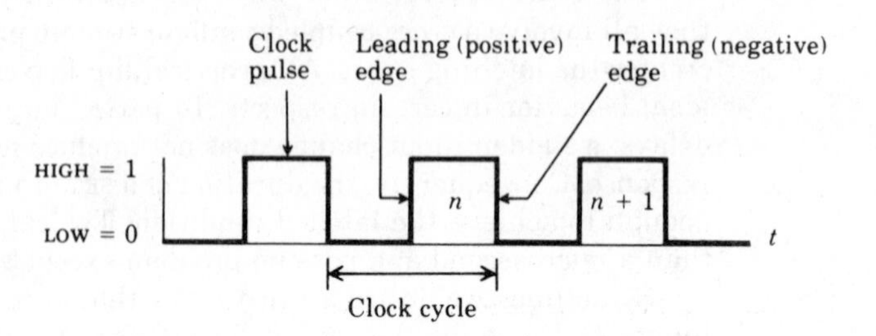

FIGURE 15.3–4
Clock waveform.

Most sequential circuits contain many flip-flops that are *synchronized* by clocking all of them from a master clock. Synchronous clocking is usually necessary for two major reasons:

- It provides periodic disabled intervals between clock pulses, during which the combinational circuits settle down to their final states.
- It provides system timing to accommodate both short operations, completed within one clock cycle, and longer operations that require several clock cycles.

When dealing with a synchronous system, we write S_n, R_n, and Q_n to denote a flip-flop's inputs and output *before the* n*th clock pulse*. Likewise, Q_{n+1} will be the next output state, as long as S_n and R_n remain constant throughout the clock pulse.

However, clocking alone does not eliminate the threat of *glitches*—stray spikes that occasionally appear on input lines. If a glitch occurs *during* a clock pulse, it may cause an erroneous flip-flop transition. The resulting error generally has devastating effects in a sequential system, whose structure and memory tend to propagate the error indefinitely.

A popular method for coping with glitches and similar phenomena involves *edge triggering*. To introduce this important concept, consider the block diagram in Fig. 15.3–5a. Here, the clock (Ck) connects to a trigger device that generates a very short gate pulse whenever the clock waveform switches from HIGH to LOW. Consequently, the S and R inputs are enabled only for a brief instant at the negative-going *trailing edge* of each clock pulse, instead of for the entire clock-pulse duration. Figure 15.3–5b illustrates edge-triggered waveforms, including an input glitch that has been rendered harmless.

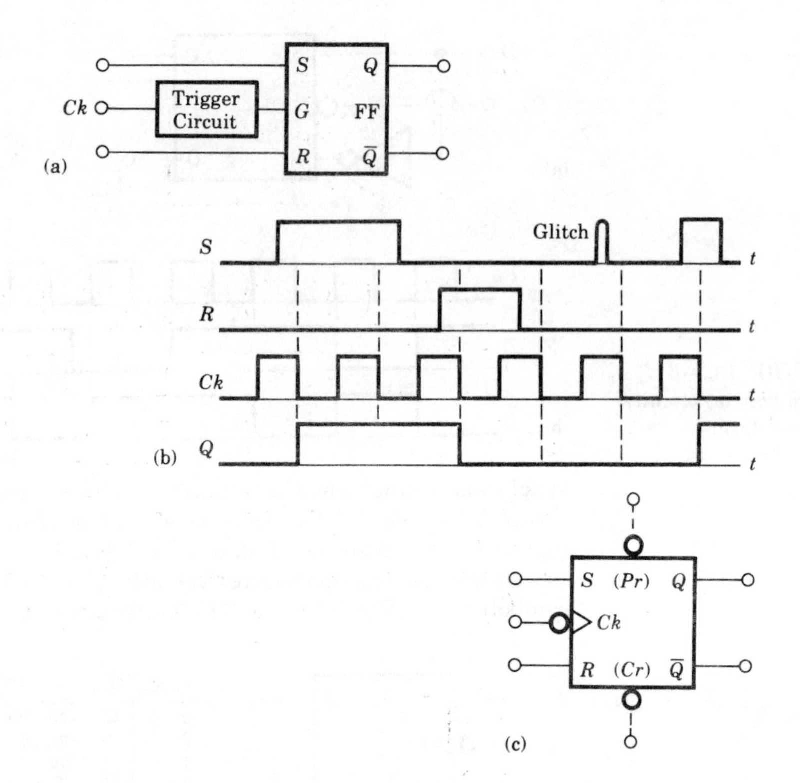

FIGURE 15.3–5 Edge-triggered SR flip-flop: (a) circuit; (b) waveforms; (c) symbol including direct inputs.

There are many different ways to build an edge-triggered flip-flop, most of them being more complicated than our block diagram implies. Nonetheless, the compact symbol in Fig. 15.3–5c has become fairly standard. The negation bubble and the triangle at the clock terminal indicate trailing-edge triggering. *Leading*-edge triggering is indicated by omitting the bubble. The bubbles at the preset and clear terminals, if present, reflect the fact that they act like $\overline{S}$ and $\overline{R}$. (Compare Figs. 15.3–3c and 15.3–1b.) The same symbol is used for the *master-slave* flip-flop, a configuration that also limits output transition times to the leading or trailing edge. For simplicity, we assume throughout the rest of this section that any clocked flip-flop triggers on the trailing edge.

Connecting the S terminal through an inverter to the R terminal of a clocked flip-flop creates a *delay* or *D flip-flop,* shown in Fig. 15.3–6 along with illustrative waveforms. Drawing upon the SR truth table with $S = \overline{R} = D$, we find that $Q_{n+1} = 1$ when $D_n = 1$ and $Q_{n+1} = 0$ when $D_n = 0$. Hence,

$$Q_{n+1} = D_n \tag{2}$$

which says that the output always takes the *previous state* of the input. A

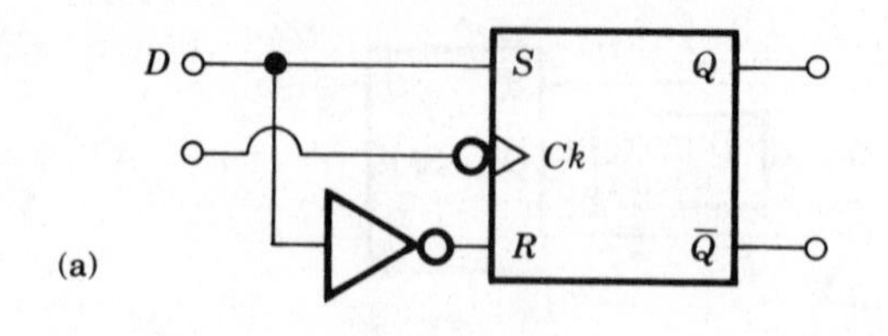

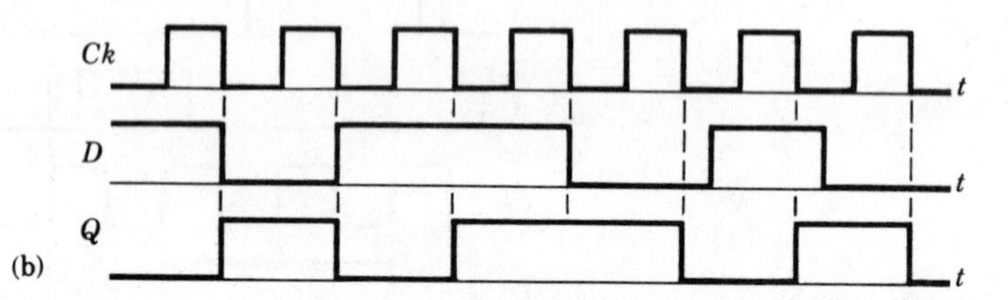

FIGURE 15.3–6
D flip-flop: (a) circuit; (b) waveforms.

synchronous input waveform therefore appears unchanged at the output, except for a *delay* of one clock cycle. An *asynchronous* (unsynchronized) input will be synchronized as well as delayed.

At last we come to the most versatile clocked flip-flop, the *JK flip-flop* symbolized by Fig. 15.3–7a. The truth table in Fig. 15.3–7b indicates

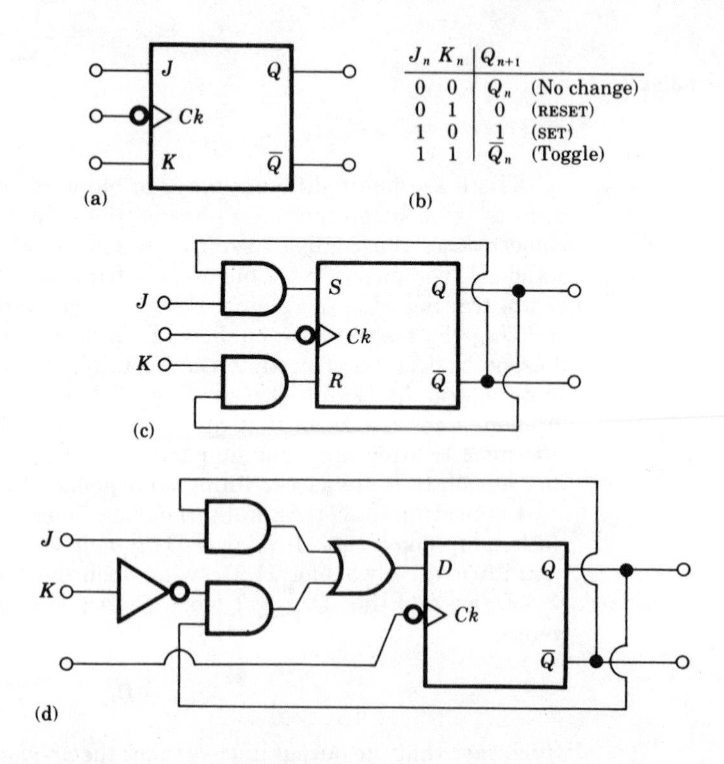

J_n	K_n	Q_{n+1}	
0	0	Q_n	(No change)
0	1	0	(RESET)
1	0	1	(SET)
1	1	$\overline{Q}_n$	(Toggle)

FIGURE 15.3–7
JK flip-flop: (a) symbol; (b) truth table; (c) implementation using an *SR* flip-flop; (d) implementation using a *D* flip-flop.

that this device differs from an *SR* flip-flop in one key respect:

> The input combination $J_n = K_n = 1$ is allowed, and produces the *complement* of the previous state, so $Q_{n+1} = \overline{Q}_n$.

For all other input combinations, a *JK* flip-flop acts like an *SR* flip-flop with $S = J$ and $R = K$. Implementing *JK* behavior requires a second set of feedback connections from the outputs, such as those diagrammed in Figs. 15.3–7c and 15.3–7d.

The distinctive feature of the *JK* flip-flop is its *toggle* mode, obtained by letting $T = J = K = 1$, as in Fig. 15.3–8. The outputs then toggle (switch) back and forth between 0 and 1 after each clock pulse. This toggle property makes sense only because output transitions are limited to one per clock cycle; otherwise, the outputs would continually switch when we held the inputs at $J = K = 1$ throughout a clock pulse. If T is an arbitrary input signal, then

$$Q_{n+1} = \begin{cases} Q_n & T_n = 0 \\ \overline{Q}_n & T_n = 1 \end{cases} \tag{3}$$

which defines a *T flip-flop.*

Commercially available IC flip-flops include the *JK* and *D* types in various packaging arrangements. (See Fig. 15.4–1 for two examples.) The *T* type and clocked *SR* flip-flops are seldom manufactured as separate products, since a *JK* flip-flop can replace either of them. Unclocked flip-flops, usually in the form of an $\overline{SR}$ latch, are available for such asynchronous applications as eliminating contact bounce from mechanical switches.

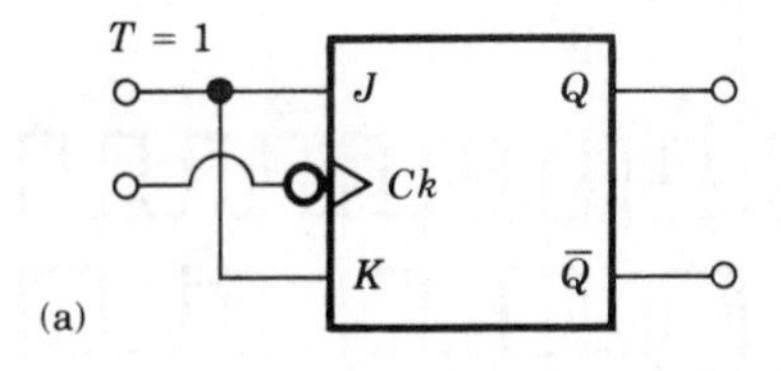

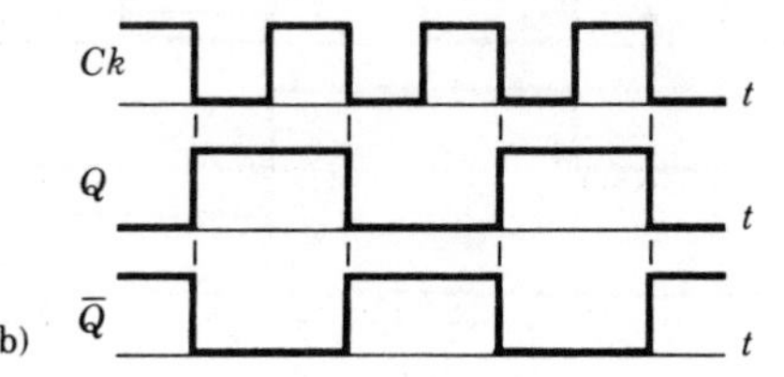

FIGURE 15.3–8
JK flip-flop in the toggle mode: (a) circuit; (b) waveforms.

Exercise 15.3–1 Label the lower output terminal P in Fig. 15.3–3a and write logic expressions for Q and P in terms of X, Y, Pr, and Cr. Then confirm the truth table in Fig. 15.3–3c.

Exercise 15.3–2 Just before an output transition, the circuit in Fig. 15.3–7c has $S_n = J_n \cdot \overline{Q}_n$ and $R_n = K_n \cdot Q_n$. Evaluate S_n and R_n for the eight possible combinations of Q_n, J_n, and K_n, and use Fig. 15.3–1b to show that this circuit has the properties of a JK flip-flop.

COUNTERS

A *counter* is a sequential circuit that has N *possible states,* each state being associated with a number. Counting takes place by causing the circuit to step through its states in numerical order in response to successive input pulses. After N pulses, the counter reverts to its initial state and the counting cycle may repeat. We refer to N as the counter's *modulus.*

To illustrate these concepts, let Fig. 15.3–9a represent a sequential circuit containing four flip-flops. Each flip-flop output ($Q_0, Q_1, \ldots$) has two possible values, and the four outputs taken together form a *4-bit binary word* $Q_3Q_2Q_1Q_0$ corresponding to the state of the circuit. (Don't confuse the positional notation $Q_0, Q_1, \ldots$ with our sequential notation Q_n and Q_{n+1}.) The internal connections, yet to be described, are arranged

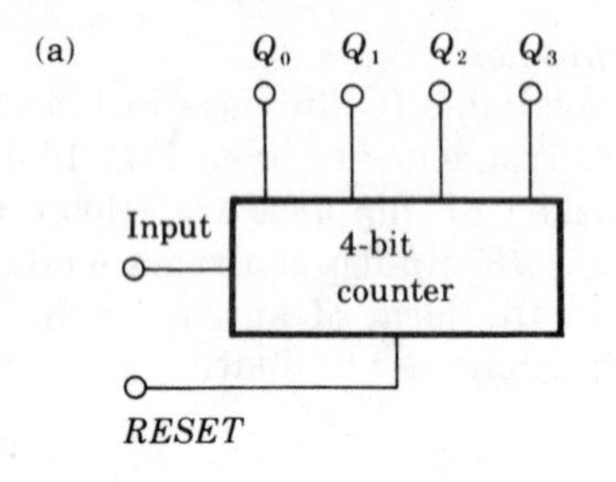

(b)

Number of Input pulses	State Q_3	Q_2	Q_1	Q_0
0	0	0	0	0
1	0	0	0	1
2	0	0	1	0
⋮	⋮	⋮	⋮	⋮
14	1	1	1	0
15	1	1	1	1
16	0	0	0	0
⋮	⋮	⋮	⋮	⋮

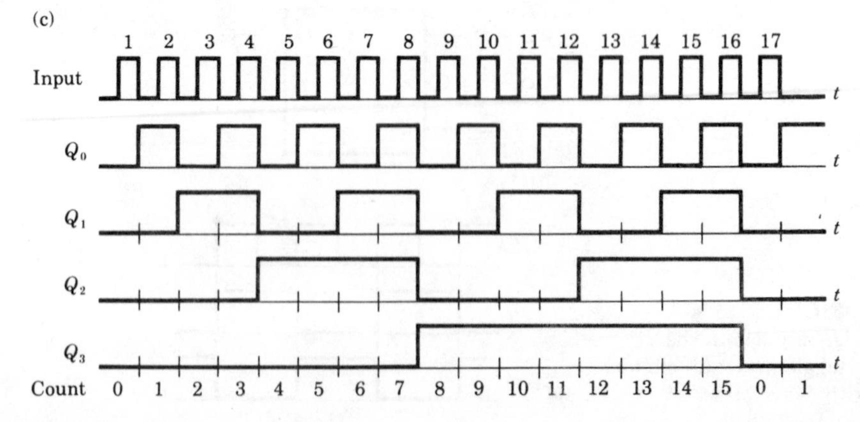

FIGURE 15.3–9 Four-bit binary counter: (a) symbol; (b) counting sequence; (c) timing diagram.

to achieve the following action: A pulse applied at the RESET terminal resets all flip-flops and establishes the initial state $Q_3Q_2Q_1Q_0 = 0000$. The first input pulse produces $Q_3Q_2Q_1Q_0 = 0001$, and each subsequent pulse increases the count in standard binary order, as tabulated in Fig. 15.3–9b. The initial state then reappears after the 16th input pulse, so we have a counter with modulus $N = 16 = 2^4$.

Figure 15.3–9c illustrates the waveforms or *timing diagram* for this 4-bit binary counter. A periodic train of input pulses has been assumed for convenience, but the results would be the same in the case of non-periodic pulses. We have also continued our previous assumption that all flip-flop transitions occur at trailing edges.

If desired, we can detect any particular state of the counter using a simple *decoding circuit*. For instance, an AND gate with inputs Q_3, Q_2, $\overline{Q}_1$, and $\overline{Q}_0$ would produce $X = Q_3 \cdot Q_2 \cdot \overline{Q}_1 \cdot \overline{Q}_0 = 1$ when $Q_3Q_2Q_1Q_0 = 1100 = 12_{10}$, which occurs between the 12th and 13th input pulses. If counting continues beyond 16 pulses, we would again get $X = 1$ after the 28th pulse, the 44th pulse, etc. Similar circuits could decode any other state.

But suppose we need a *decade* counter—a counter that recycles after 9 counts and, including the initial state, has modulus $N = 10$. Our 4-bit counter is easily adapted to this task by decoding the state $Q_3Q_2Q_1Q_0 = 1010 = 10_2$ and generating a *CARRY* pulse which resets the counter. Figure 15.3–10 shows the modified circuit and the relevant portion of the timing diagram. Because of propagation delays, the 1010 state actually appears for a brief time after the 10th pulse, while $CARRY = 1$, until the flip-flops become reset.

The figure also suggests how to connect two or more 4-bit decade counters in *cascade* to obtain a multiple-decade system. A system with M

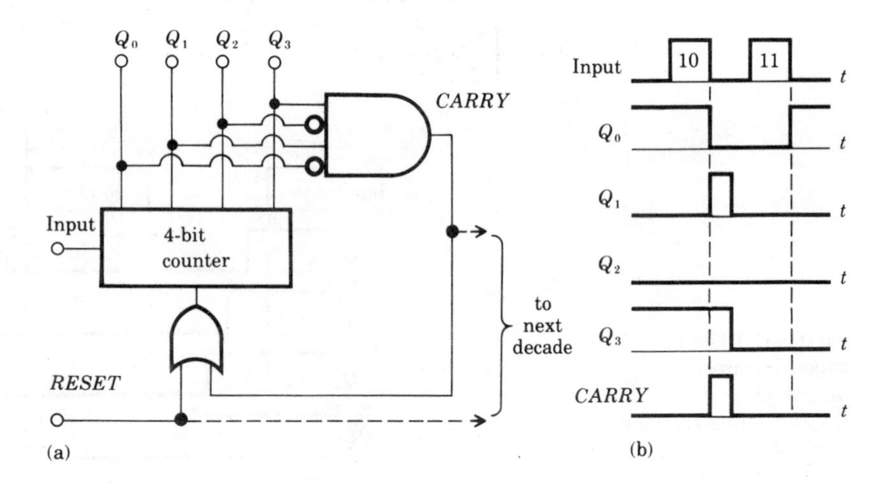

FIGURE 15.3–10
Decade counter:
(a) circuit;
(b) waveforms and *CARRY* pulse.

decades will have $N = 10 \times 10 \times \cdots = 10^M$. The M 4-bit output words then constitute the *BCD code* for the value of the count at any particular time.

Let's generalize our results so far by considering a counter composed of n flip-flops. If sequential input pulses advance the n-bit output word in standard binary order, the counter could have any modulus satisfying the condition

$$N \leq 2^n \tag{4}$$

Values of N less than 2^n require auxiliary decoding gates. To build a system with

$$N = N_1 \times N_2 \times \cdots \tag{5}$$

we merely cascade two or more counters.

As for the internal circuitry of an n-bit counter, Fig. 15.3–11a diagrams a simple implementation using a chain of T (or JK) flip-flops with $T = 1$. The pulses being counted are applied to the clock terminal of FF_0 whose output Q_0 toggles after each input pulse and completes a full HIGH/LOW cycle for every *two* input pulses (see Fig. 15.3–8b). Since Q_0

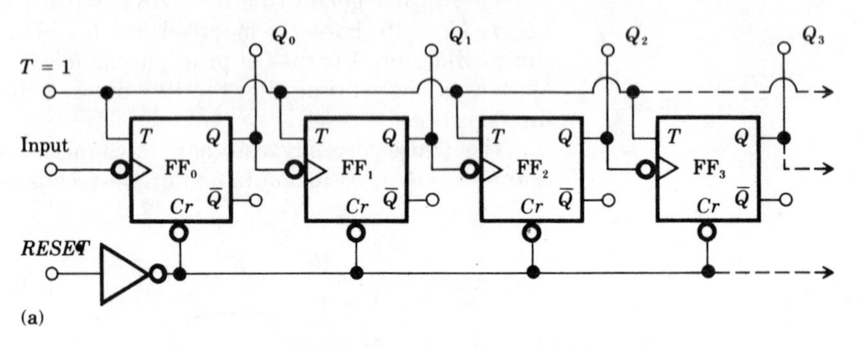

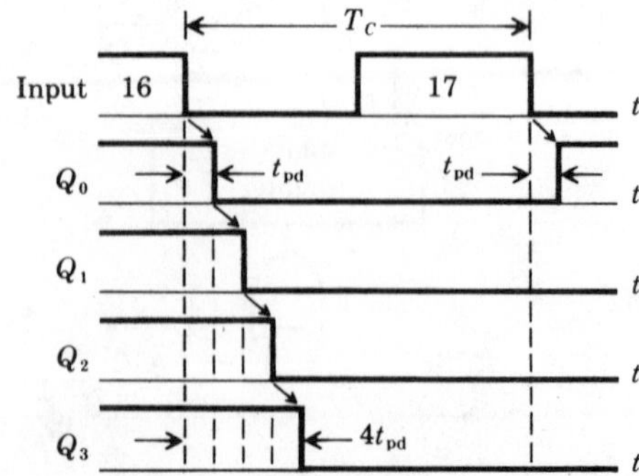

FIGURE 15.3–11
Ripple counter:
(a) circuit;
(b) waveforms showing ripple delay.

goes to the input of FF_1, Q_1 completes a full cycle for every *four* input pulses and so on along the chain. Therefore, a chain of $n = 4$ flip-flops can, indeed, produce the counting sequence and timing diagram in Fig. 15.3–9. The extension to other values of n is straightforward.

This circuit is called a *ripple counter* because the 16th (Nth) pulse produces a downward transition at Q_0, which triggers Q_1, which triggers Q_2, which triggers Q_3—like falling dominos rippling from left to right. But the initial state does not reappear instantaneously, for each flip-flop introduces a *delay* t_{pd} whenever it changes state. Figure 15.3–11b gives a time expansion of the events between pulses 16 and 17 including propagation delay, and reveals that the initial state has been delayed by $4t_{pd}$ after the trailing edge of the 16th pulse. Also note that these waveforms pass through the state $Q_3Q_2Q_1Q_0 = 1100$, so a decoding circuit with $X = Q_3 \cdot Q_2 \cdot \overline{Q}_1 \cdot \overline{Q}_0$ might produce a brief spurious output.

The potential decoding errors of a ripple counter are alleviated by the *synchronous counter* of Fig. 15.3–12. Here, the flip-flops operate in parallel, the input pulses being applied simultaneously to all clock terminals; thus, all output transitions take place at essentially the same time. The other connections and the AND gates are arranged such that Q_1 toggles after $Q_0 = 1$, Q_2 toggles only after $Q_1 \cdot Q_0 = 1$, etc., which yields the desired waveforms and counting sequence. Despite its name, a synchronous counter could be used to count nonperiodic pulses. On the other hand, a ripple counter could not be used if the application requires synchronous flip-flop transitions.

With the help of some additional gates, a ripple or synchronous counter becomes a reversible *up/down* counter. Up/down counters feature a

FIGURE 15.3–12
Synchronous counter.

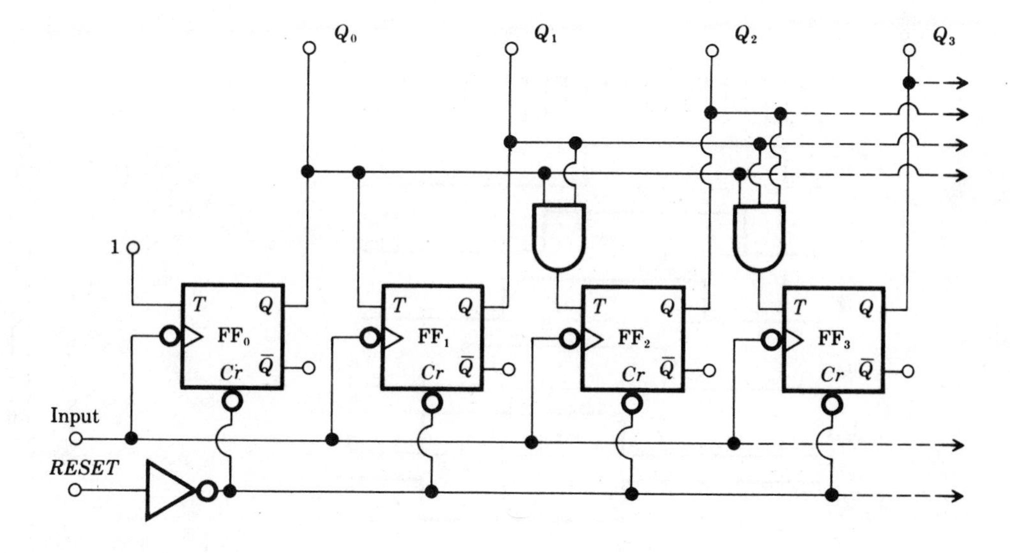

control terminal that allows us to *reverse the counting sequence* when desired. Timing considerations generally favor synchronous counters for up/down counting.

An entirely different principle characterizes the *Johnson counter*, Fig. 15.3–13, which consists of D flip-flops arranged in a ring. The input pulses are applied to all clock terminals, like a synchronous counter, but the complemented output $\overline{Q}_4$ from the last stage goes back to the D terminal of the first stage. (Hence, this circuit is also known as a *twisted ring* or *switched-tail ring counter*.) To understand the waveforms and counting sequence, you must recall that the output of a D flip-flop equals the previous state at the D terminal. If all flip-flops are initially reset, so $\overline{Q}_4 = 1$, the first input pulse produces $Q_1 = 1$, which in turn produces $Q_2 = 1$ after the second pulse, and so forth until $Q_0 = Q_1 = Q_2 = Q_3 = Q_4 = 1$ after

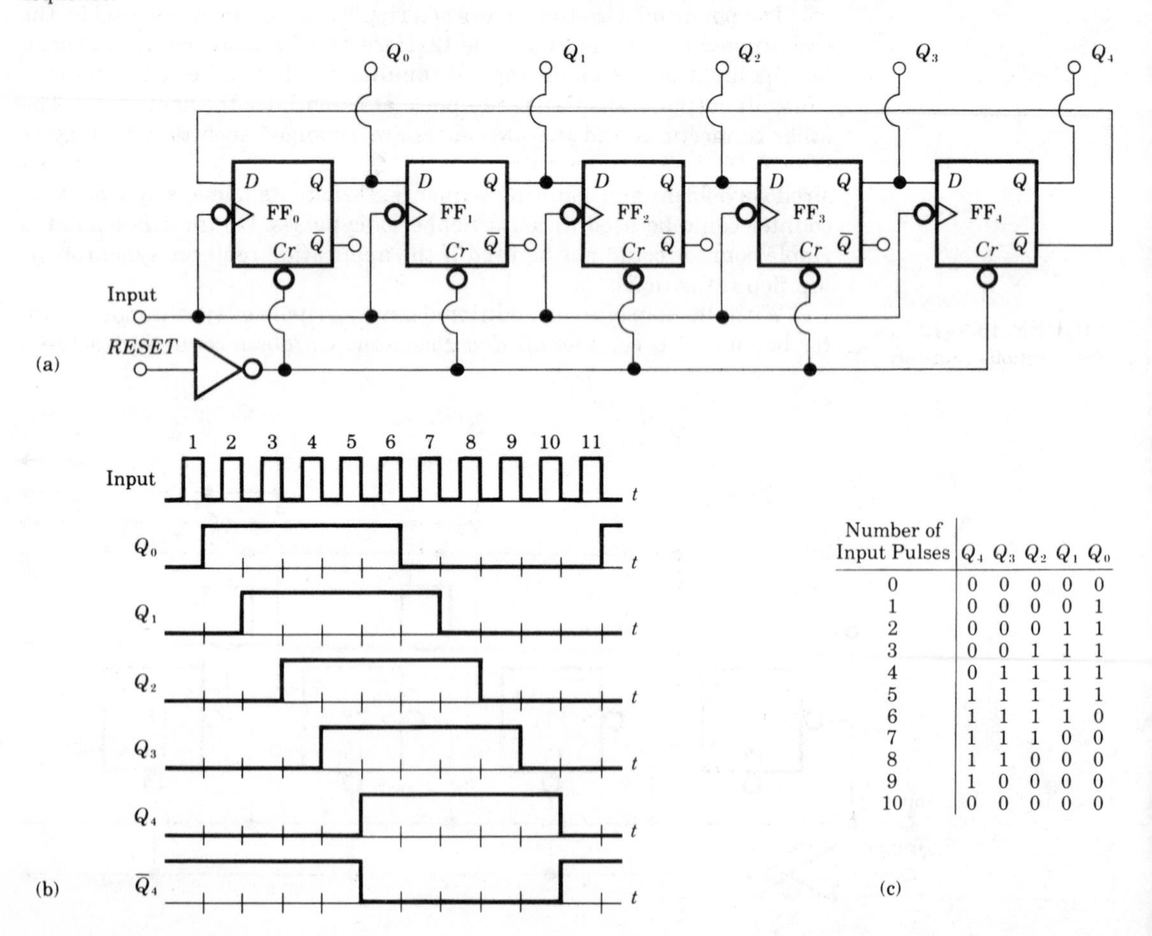

Number of Input Pulses	Q_4	Q_3	Q_2	Q_1	Q_0
0	0	0	0	0	0
1	0	0	0	0	1
2	0	0	0	1	1
3	0	0	1	1	1
4	0	1	1	1	1
5	1	1	1	1	1
6	1	1	1	1	0
7	1	1	1	0	0
8	1	1	0	0	0
9	1	0	0	0	0
10	0	0	0	0	0

FIGURE 15.3–13 Johnson counter: (a) circuit; (b) timing diagram; (c) counting sequence.

the 5th pulse. Now $\overline{Q}_4 = 0$, so $Q_0 = 0$ after the 6th pulse, and the counter arrives back at the initial state after the 10th pulse.

The five-stage Johnson counter is called a *divide-by-10 counter* because the waveforms complete one full cycle for every 10 input pulses. However, this counter requires one more flip-flop than a 4-bit decade counter, and it counts in a nonstandard sequence. In exchange for these disadvantages, it has the advantage of synchronous flip-flop transitions implemented without additional AND gates. The five-stage ring configuration also forms the basis of a handy *divide-by-N counter* that allows you to select any modulus in the range $2 \leq N \leq 10$ using, at most, one auxiliary gate.

A representative listing of standard IC counters consist of items such as

- 4-bit binary and decade counters (ripple or synchronous)
- 4-bit synchronous up/down counters (binary or decade)
- 12-bit ripple counters
- Johnson counters with decoded decimal outputs
- Divide-by-N ring counters

Special features may include: *CARRY* outputs for cascading, *PRESET* inputs for establishing an arbitrary initial state, and built-in decoders for seven-segment displays.

Exercise 15.3–3

A counter with a 60-Hz input produces output pulses once per minute, once per hour, once per day, and once per week. (a) What is the minimum number of flip-flops needed? (b) How many divide-by-N counters are needed, if $N \leq 10$?

Exercise 15.3–4

Verify that the ripple counter in Fig. 15.3–11a counts *down* from 1111 to 0000 if the $\overline{Q}$ terminals are connected to the succeeding clock terminals, still taking $Q_3 Q_2 Q_1 Q_0$ as the output.

REGISTERS

A *register* is a group of flip-flops designed for *temporary storage* of binary data, one bit per flip-flop. D-type flip-flops are particularly well suited to this purpose—so much so that they are also known as *data* flip-flops.

Consider, to begin with, the 4-bit register in Fig. 15.3–14, where the binary data word $B_3 B_2 B_1 B_0$ might come from a combinational circuit. We store this data word by applying a *LOAD* pulse to the clock terminals so that $Q_0 = B_0$, $Q_1 = B_1$, $Q_2 = B_2$, and $Q_3 = B_3$ after the pulse ends. The data word now remains at the Q terminals for as long as desired, barring a power-supply failure, and the complemented word appears at the $\overline{Q}$ terminals (not shown). Later on, a new input word can be "written over" the

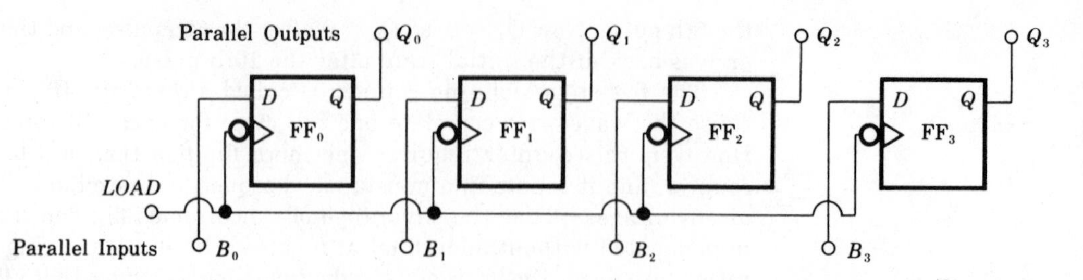

FIGURE 15.3–14
Parallel-input/parallel-output register.

current one by applying another *LOAD* pulse. We describe this register's operation as *parallel-input/parallel-output* since all input bits are stored simultaneously and, having been stored, are available simultaneously at the output terminals.

But suppose a data word arrives in *serial* form, one bit at a time. The storage of serial data requires a *shift register* like Fig. 15.3–15a, where bits shift from one flip-flop to the next after each clock pulse. The incoming bits are fed into the register synchronously with the clock pulses by setting $SHIFT = 1$. We then set $SHIFT = 0$ to disable the clock input,

FIGURE 15.3–15
Shift register:
(a) circuit;
(b) waveforms.

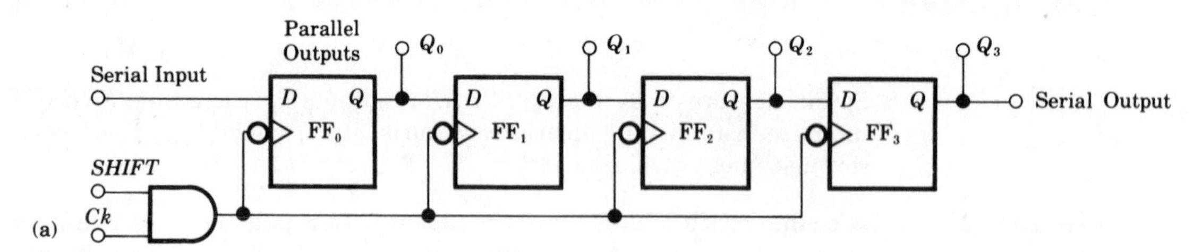

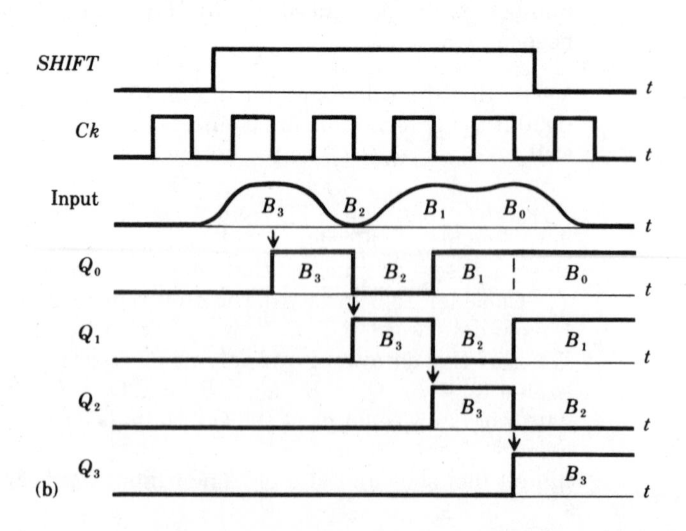

and the entire 4-bit word is available at the parallel output terminals. The waveforms in Fig. 15.3–15b illustrate this *serial-to-parallel* sequence. Note that the first flip-flop cleans up any *distortion* present in the incoming waveform—due, perhaps, to the imperfections of a data transmission system. If, at a later time, we wish to convert the stored word back to serial form, we simply let *SHIFT* = 1 again so the bits appear sequentially at the serial output terminal Q_3.

With the addition of parallel inputs, a shift register becomes capable of *parallel-to-serial conversion* as well as serial-to-parallel conversion. Figure 15.3–16 diagrams an implementation of this feature using 2-to-1 selectors similar to the 4-to-1 selector discussed in Example 15.2–3. Setting *LOAD* = 1 connects the parallel inputs to the *D* terminals, and storage is triggered by the next clock pulse. Setting *LOAD* = 0 and *SHIFT* = 0 then keeps the data word available at the parallel outputs. Serial operation begins when *LOAD* = 0 and *SHIFT* = 1. A *universal register* would have the further capability of *bidirectional* shifting, left to right or right to left. Usually, a master *CLEAR* terminal is also provided to reset all flip-flops.

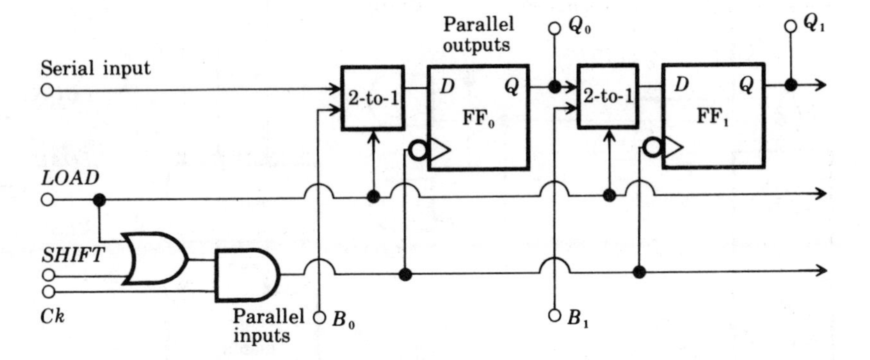

FIGURE 15.3–16 Shift register with serial and parallel inputs.

While we're on the subject of shift registers, we should mention the related concept of *recirculating memory.* Recirculation is obtained by loading a shift register and connecting the serial output back to the serial input, so the data bits continually cycle through the register while the clock ticks. This scheme is essential for a *dynamic* register consisting of *charge-coupled devices* (CCDs) or other memory cells that cannot hold data bits indefinitely. Dynamic memory cells require much less space than flip-flops on an IC chip, which facilitates construction of very long registers. For instance, you can buy an 18-pin unit that contains nine CCD registers, each 1024 bits long, and provides a total storage capacity of 9216 data bits. (Other memory units will be considered in Section 16.4.)

Exercise 15.3–5

Verify that the register in Fig. 15.3–15a would shift bits to the left if we applied the serial input to D_3 and connected Q_3 to D_2, Q_2 to D_1, etc. Then draw the block diagram of a 4-bit bidirectional register that accepts serial data at either end and shifts right or left, depending on the value of a control signal named $RIGHT/\overline{LEFT}$. Hint: Use 2-to-1 selectors.

Example 15.3–1
Frequency counter

The illustrative system in Fig. 15.3–17 includes a flip-flop, counter, divider, and register to carry out digital measurement of the frequency of a sinusoidal input. The sinusoid is applied to a trigger circuit that generates one pulse per cycle, say at each positive-going zero-crossing of the waveform. The *COUNT* signal gates the trigger pulses to a counter for exactly one second, so the decoded displayed count equals the input frequency measured in hertz (cycles per second).

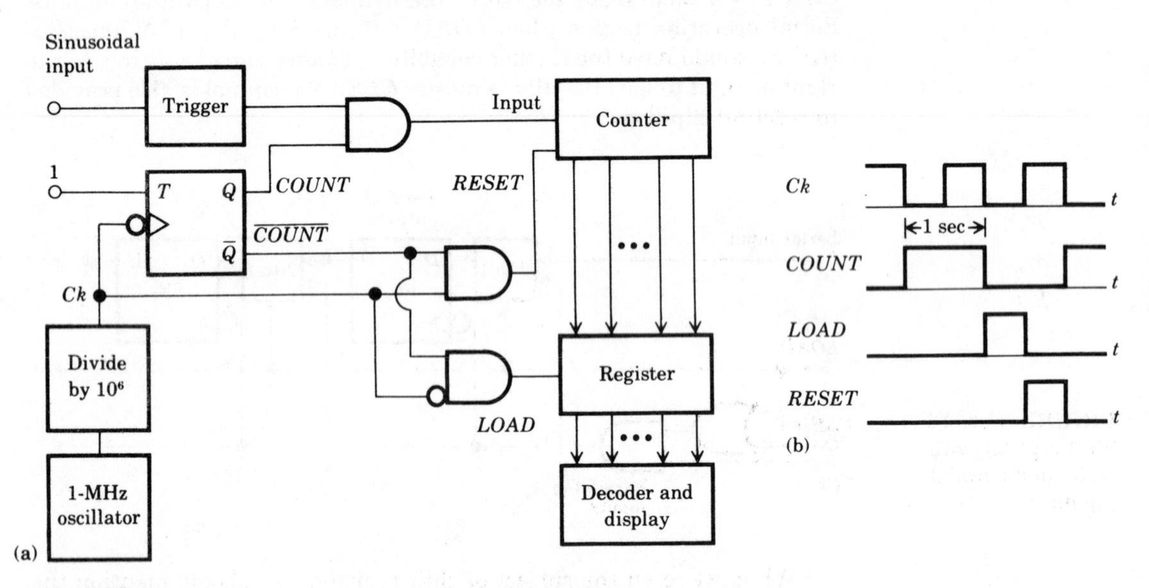

FIGURE 15.3–17
Frequency counter: (a) circuit; (b) timing diagram.

The *COUNT* signal is derived from a highly stable 1-MHz oscillator (usually quartz-crystal controlled) whose output passes through a six-decade divider to become a one-cycle-per-second clock waveform *Ck*. Connecting *Ck* to a flip-flop with $T = 1$ produces a *COUNT* waveform that is HIGH for one second and LOW for the next second, as shown in Fig. 15.3–17b. (If we bypass three decades of the divider, so it divides by 10^3, *COUNT* will be HIGH for one millisecond and the displayed value will equal the input frequency measured in kilohertz.)

Counting ceases when *COUNT* goes LOW, and a *LOAD* pulse transfers the count bits in parallel to the storage register. The register's parallel outputs, in turn, go to the decoder/display unit. Another pulse then resets the counter, and a new counting cycle begins when *COUNT* goes HIGH. Meanwhile, the register continues to hold the previous result. The *LOAD* and *RESET* waveforms are generated from $\overline{COUNT}$, *Ck*, and $\overline{Ck}$ using AND gates.

15.4 IC LOGIC FAMILIES †

The building blocks of digital systems are commercially fabricated in various integrated-circuit forms known as *logic families* or *technologies*. Some illustrative circuits from the DTL, TTL, and CMOS families were presented in Section 11.2, where we described how those circuits operate internally. Now we shift our attention to the *external* characteristics of logic families. In particular, we'll discuss the range of logic functions and special features available within a typical family, and we'll compare relative advantages and disadvantages of the major families.

IC LOGIC FUNCTIONS AND FEATURES

A logic family comprises an assortment of gates, flip-flops, registers, and other devices that are *mutually compatible*. This means that an output from any one family member provides appropriate HIGH and LOW electrical states for the input to any other member. Therefore, subject to a few restrictions, you can build almost any logic network you've designed on paper simply by connecting the corresponding functional devices from one family. In contrast, any connection between members of different families usually requires a special *interface* unit to achieve proper operation. Connections to outside-world devices—keyboards, displays, motors, and so forth—almost always require interfacing.

Compatibility within a logic family is achieved through the use of a basic gate circuit that is replicated in numerous combinations with various modifications to implement different functions. *Small-scale integration* (SSI) produces functions that involve the equivalent of less than a dozen gates. SSI circuits are usually mounted in packages having 14 or 16 external pins, two of which are reserved for the power supply (V_{CC}) and ground (*GND*) connections. To illustrate, Fig. 15.4–1 shows the logic diagrams and pin assignments for the following TTL functional packages:

(a) Triple 3-input NAND gates
(b) 4-wide 3-2-2-3-input AND-OR-INVERT gate
(c) Dual *JK* flip-flops with preset, common clear, and common clock
(d) Quadruple $\overline{S}\overline{R}$ latches

All told, the TTL menu lists some 75 SSI functions.

Medium-scale integration (MSI) produces functions that involve the equivalent of 12 to 100 gates. Typical MSI functions include quadruple 2-input X-OR gates, 16-to-1 data selectors, decade counters, code converters, small registers and memory units, and various arithmetic cir-

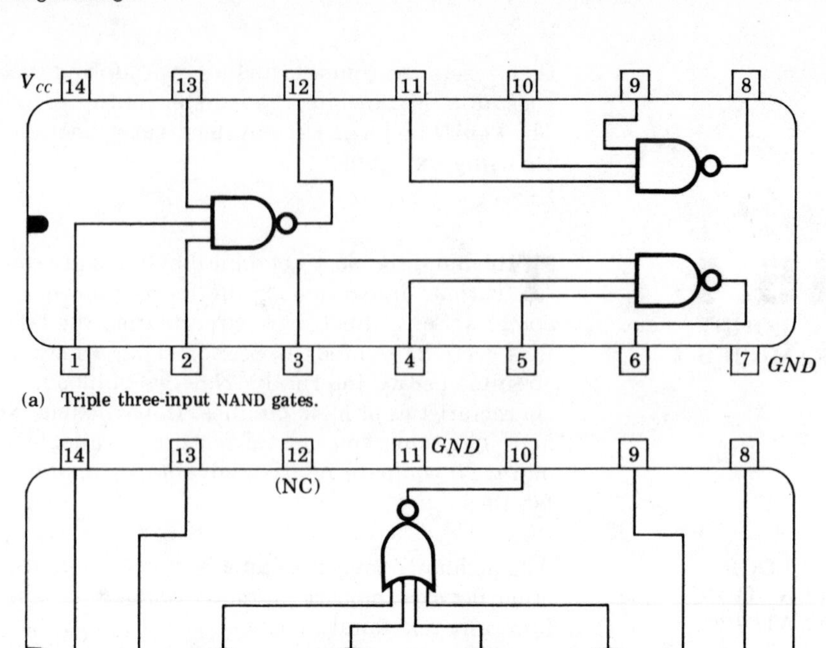

(a) Triple three-input NAND gates.

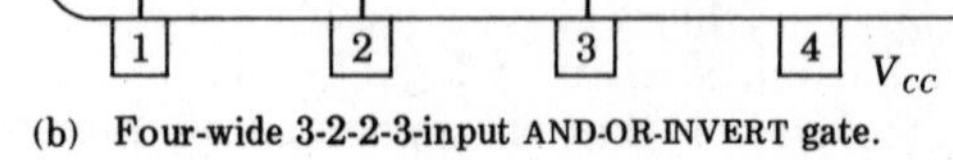

(b) Four-wide 3-2-2-3-input AND-OR-INVERT gate.

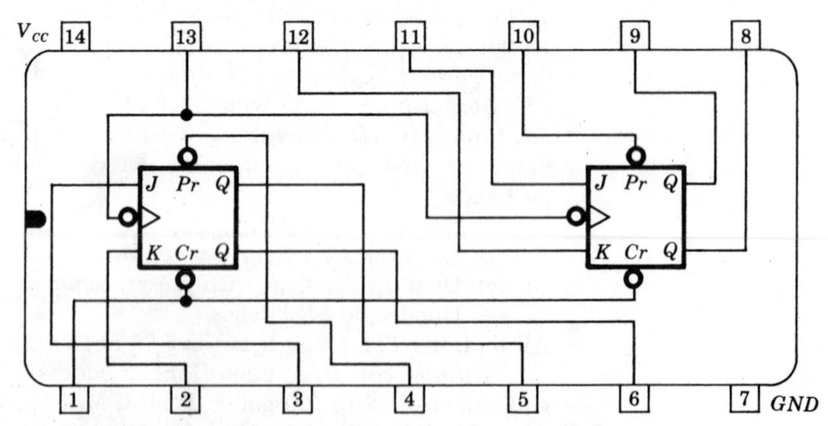

FIGURE 15.4–1
Logic diagrams and pin assignments for typical SSI functions (TTL).

(c) Dual *JK* flip-flops with preset, common clear, and common clock.

FIGURE 15.4–1 (Continued)

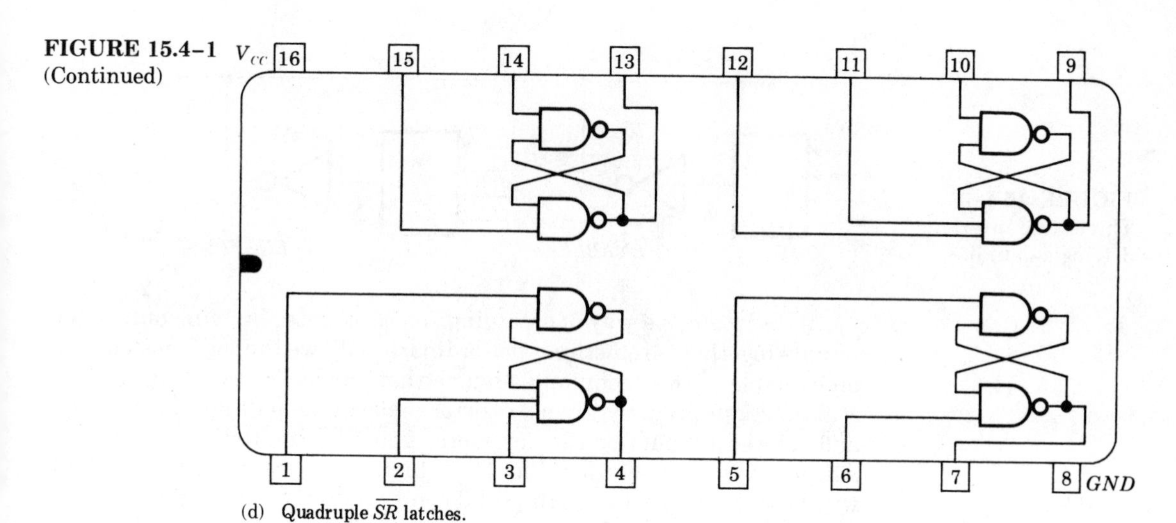

(d) Quadruple $\overline{SR}$ latches.

cuits. Functions that require more than 100 equivalent gates are produced by *large-scale integration* (LSI). Some LSI circuits are intended for specific applications—a digital wrist watch, for example—and do not interface with other logic devices, so their internal circuitry often differs from the basic family design.

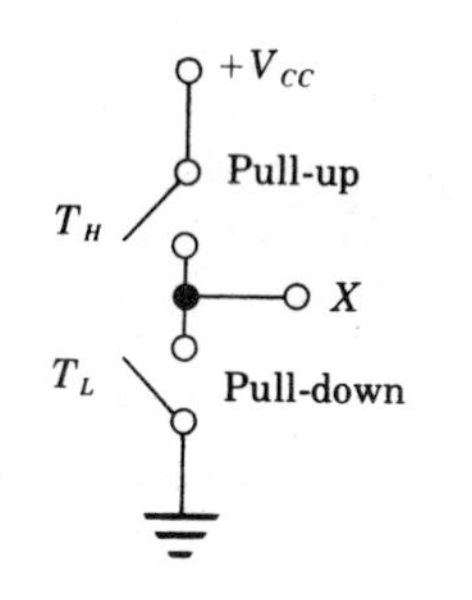

FIGURE 15.4–2 Model of gate output with active pull-up and pull-down.

Two special features permitted by certain technologies are *three-state outputs* and *wired logic*. To explain these features, we will use the switch circuit in Fig. 15.4–2 as a simplified model of the output portion of a gate. The switches, representing transistors, are opened and closed by input circuitry not shown. Such a gate is said to have *active pull-up* and *pull-down* in the following sense: Opening T_L and closing T_H "pulls up" the output X to the HIGH state, whereas opening T_H and closing T_L "pulls down" X to the LOW state. Consequently, under normal conditions, one switch should be open and the other closed.

But suppose both switches happen to be open. We then have a *third output state,* neither HIGH nor LOW, in which the gate is *disabled* and a *high impedance* exists between the output terminal and the rest of the circuit. Three-state gates, such as the inverters in Fig. 15.4–3, are equipped with control terminals that allow us to enable normal operation ($ENABLE = 1$) or to disable the gate and activate the high-impedance state ($ENABLE = 0$). If we enable just one inverter at a time in Fig. 15.4–3, then only one output will be connected to the data line. This arrangement, frequently found in computer systems, acts as a data selector with the additional provision of *electrical isolation* between the data line and the disabled gates.

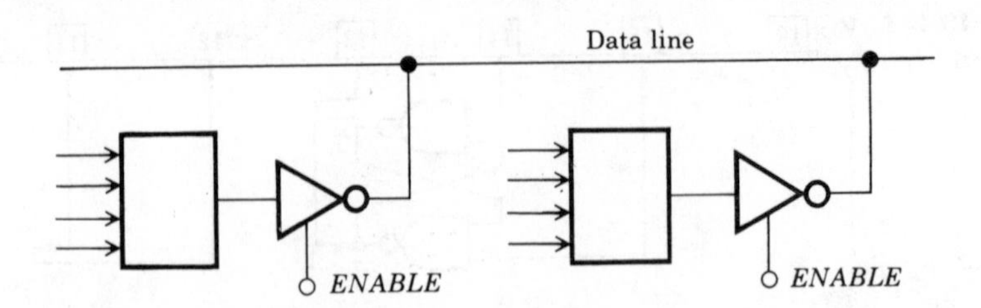

FIGURE 15.4–3
Three-state inverters driving a data line.

Wired logic is a way of obtaining "no-cost" gates by tying output terminals together—something not ordinarily allowed in logic systems. As an example of this technique, assume that our model gate (Fig. 15.4–2) still functions properly when a resistor replaces the pull-up transistor T_H, and let the outputs of two (or more) gates be connected to the same *pull-up resistor* as shown in Fig. 15.4–4a. Clearly, the combined output Z will be HIGH only when both pull-down switches are open, which corresponds to $X = 1$ *and* $Y = 1$. Thus, Z equals $X \cdot Y$, and we have built the *wired*-AND gate symbolized by Fig. 15.4–4b. Similarly, a *wired*-OR gate results if we keep the active pull-up transistors and substitute a *pull-down resistor* for T_L. In either case, however, we lose the separate identities of the individual outputs (Why?).

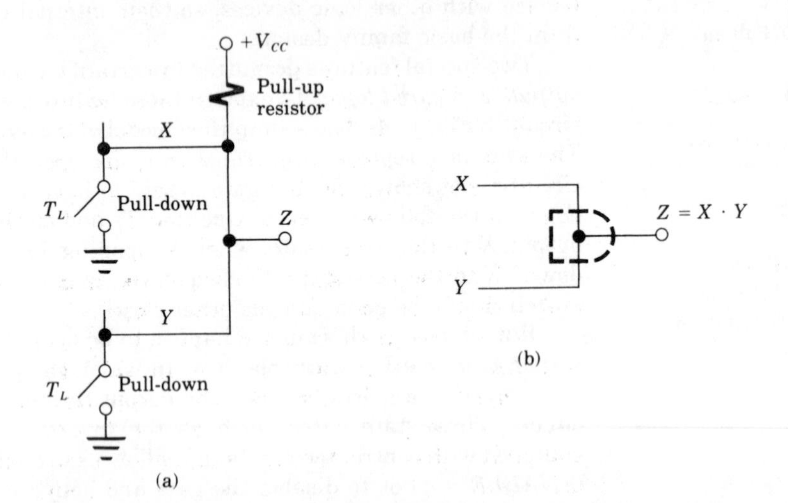

FIGURE 15.4–4
Wired AND: (a) circuit; (b) symbol.

COMPARISON OF LOGIC FAMILIES

Table 15.4–1 lists most of the technologies that have been developed for digital ICs. Some of these technologies are now obsolete or have evolved into improved versions; other technologies are intended for specialized applications; still others have flourished and spawned distinct *subfamilies*. Since we cannot hope to cover all technologies, we'll compare rep-

representative operating properties of the three most widely used: CMOS, TTL, and ECL. We'll also take a brief look at the promising new I^2L technology.

TABLE 15.4–1 Digital IC technologies

Technology	*Description*	*Comment*
RTL	Resistor-transistor logic	Obsolete.
DCTL	Direct-coupled transistor logic	Evolved into I^2L.
DTL	Diode-transistor logic	Replaced by TTL.
HTL	High-threshold logic	DTL modified for noisy environments.
TTL	Transistor-transistor logic	Moderate speed and power consumption; five subfamilies.
ECL	Emitter-coupled logic	High switching speed; three subfamilies.
PMOS NMOS	*p*-channel/*n*-channel MOSFET logic	Evolved into CMOS, but still used for special purposes.
CMOS	Complementary-symmetry MOSFET logic	Low power consumption; two subfamilies.
I^2L	Integrated-injection logic	Adjustable speed/power performance.

It should be emphasized that we are dealing here with *representative* properties rather than specific details. For more complete information about any particular line of IC logic, you should consult the *data book* published by its manufacturer. These books usually include schematic circuits and logic diagrams, along with electrical and mechanical data and descriptions of available functions. (Data books make for very interesting reading, especially with regard to the fine points of practical logic circuits.)

For comparison purposes, Table 15.4–2 summarizes the salient features and typical performance values of the three major families. Entries above the dotted line in each column characterize the entire family, whereas the power/speed data below the dotted line pertain to the indicated subfamily. If you scan the power/speed data from left to right, you'll find that the families are arranged in order of *increasing power consumption* and *decreasing propagation delay* or, equivalently, *increasing switching speed.* We'll return to this important observation shortly.

TABLE 15.4–2 Comparison of Major Logic Families

	CMOS	*TTL*	*ECL*
Basic gate circuit	NOR/NAND	NAND	OR/NOR
Wired logic	Not permitted	AND	OR
Supply voltage	3–15 V	5.0 V	−5.2 V
Nominal LOW state	0 V	0.4 V	−1.8 V
Nominal HIGH state	3–15 V	3.4 V	−0.9 V
Noise margin	1.4–6.8 V	1.5 V	0.2 V
Maximum fan-out (static)	>50	10	25–90
Subfamily	Series 4000B, 10-V supply	Series LS	Series III
Power consumption per gate (static)	0.01 μW	2 mW	60 mW
Propagation delay per gate	25 ns	8 ns	0.7 ns
Power consumption per flip-flop (dynamic)	10 mW	50 mW	220 mW
Maximum flip-flop toggle frequency	10 MHz	45 MHz	500 MHz

Most of the terms in the table should be self-explanatory, but two of them require clarification:

- *Noise margin* is the maximum voltage perturbation that can be tolerated at an input terminal; larger values of input noise may produce ambiguous or erroneous output voltage levels.
- *Fan-out* is the number of input terminals being driven from one output terminal of a device; loading effects caused by excessive fan-out may result in unsatisfactory performance.

Some values have been specified under *static* conditions, meaning that the signals are constant or relatively slow in changing. Energy-storage effects that occur during high-speed *dynamic* operation generally alter the static values in the direction of deteriorating performance.

We begin our comparison by considering the properties of CMOS logic, whose basic circuit consists of MOSFETs arranged as a NOR gate (like Fig. 11.2–9) or a NAND gate. The circuit configuration does not permit wired logic, but three-state outputs can be obtained by incorporating CMOS switches (see Fig. 11.2–7). The table shows that CMOS offers the advantages of good noise margin, large fan-out, and very low power consumption. Additionally, unlike the fixed supply voltages of the other two families, the CMOS supply voltage may be any desired value within the 3–15 V range. These properties readily lend themselves to battery-powered applications such as portable digital instruments. The primary disadvantage of CMOS is its relatively long propagation delay, which limits the maximum switching speed to about 10 MHz.

Another CMOS feature (not listed) is its *high packing density,* measured in terms of equivalent gates per unit area on an IC chip. The high packing density and low power dissipation make CMOS especially well suited to *very large-scale integration* (VLSI) for the fabrication of complete logic systems on a single chip.

When systems must have switching speeds above 10 MHz, TTL may be the appropriate logic family. A simplified version of the basic NAND gate circuit was diagrammed in Fig. 11.2–4. Some TTL units are supplied with *open collectors* on the pull-down transistors, so you can add external resistors to obtain the wired AND. Most units, however, have a *totem-pole output* with active pull-up and pull-down. This configuration delivers more signal power to drive data lines, and it permits the addition of three-state outputs. The voltage drop across the active pull-up transistor accounts for the difference between the 5-V supply and the 3.4-V HIGH state.

Although the five TTL subfamilies have different power/speed values, the family as a whole is characterized by moderate power consumption, moderate speed, moderate noise margin, and limited fan-out. Before the perfection of CMOS logic, TTL was the family of choice for general-purpose applications. Subsequently, CMOS units operated at 5 V have replaced many TTL units, except in situations that require the higher speed or line-driving capability of TTL.

Some systems exceed the speed of TTL and must be implemented with ECL, whose basic gate circuit is shown in Fig. 15.4–5a. The name

FIGURE 15.4–5
ECL OR/NOR gate:
(a) circuit; (b) symbol.

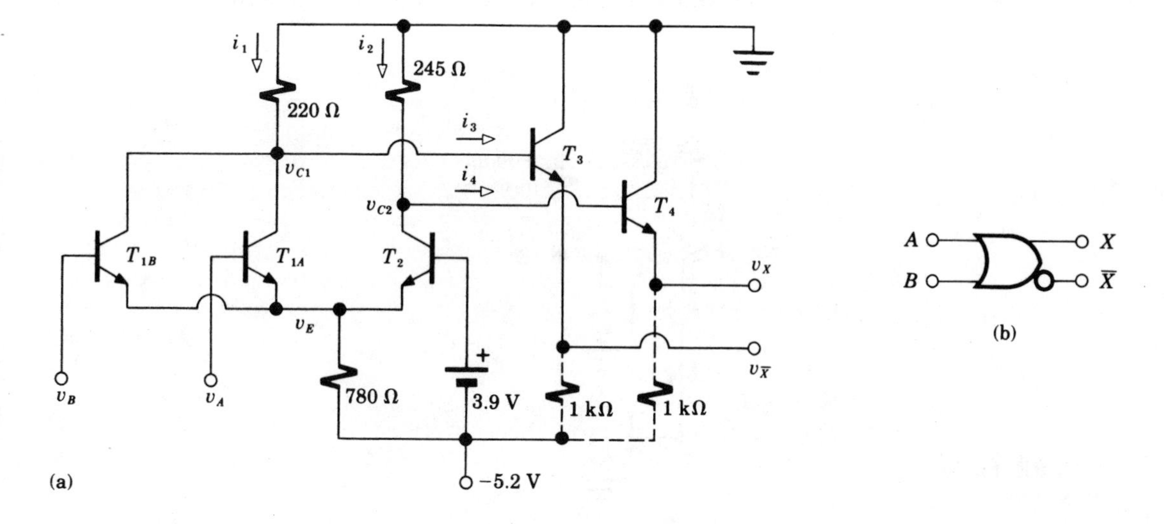

"emitter-coupled logic" reflects the coupled emitters of the input transistors, T_{1A} and T_{1B}, and transistor T_2. The two output transistors T_3 and T_4 provide complementary voltage levels such that $X = A + B$ and $\overline{X} = \overline{A + B}$, as symbolized by Fig. 15.4–5b. (See Problems 11.2–12 and 11.2–13 regarding the circuit analysis.) The *open emitters* at the output connect to external pull-down resistors, permitting you to construct a wired OR. Unfortunately, the high-speed ECL design results in considerable power consumption and small noise margin—disadvantages that limit ECL to rather specialized applications, notably in large computers and high-frequency counters.

The differing power/speed characteristics of CMOS, TTL, and ECL imply a fundamental relationship between *power consumption* and *switching speed*. This relationship stems from the inevitable stray capacitance whose energy storage tends to increase the propagation delay. Rapid switching therefore requires large currents to charge and discharge the capacitance, and large currents result in large power dissipation. The high speed of ECL is achieved by allowing large currents and by avoiding transistor saturation, since a saturated BJT stores charge like a capacitor. Conversely, the low-power CMOS circuits cannot provide the currents needed for rapid switching.

The product of the propagation delay times the static power consumption—called the *delay-power product*—serves as a figure of merit for comparing gate-circuit designs. For example, each of the TTL subfamilies has roughly the same delay-power product; the circuit modifications of one subfamily decrease the propagation delay but increase the power consumption, and vice versa for another subfamily.

Integrated injection logic goes even further in this direction. Its circuit design allows the user to trade off switching speed versus power con-

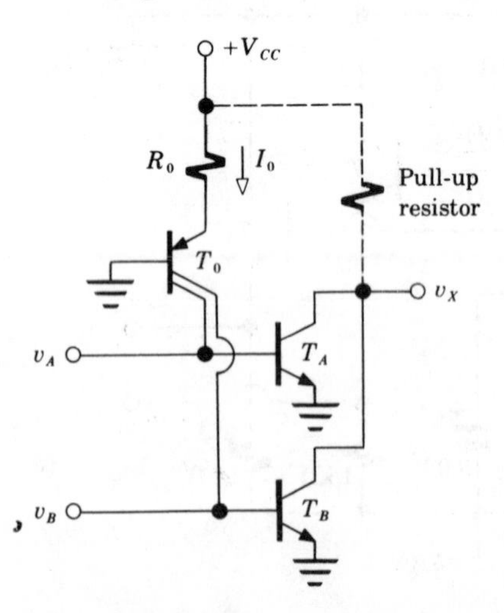

FIGURE 15.4–6
I²L NOR gate.

sumption. The key component in the illustrative I^2L NOR gate of Fig. 15.4–6 is the *multicollector pnp* transistor T_0. This transistor, together with resistance R_0 and the power supply, establishes an *injection current* I_0 that supplies base current to T_A or T_B when v_A or v_B is HIGH. If the delay-power product equals 0.5×10^{-12} W · sec, a typical value, you can adjust I_0 for a power consumption of 5 μW and you'll get $t_{pd} = 0.5 \times 10^{-12}/5 \times 10^{-6} = 100$ ns; or you can adjust I_0 such that $t_{pd} = 1$ ns, and the power consumption becomes $0.5 \times 10^{-12}/10^{-9} = 500$ μW. I^2L also has a high packing density, comparable to CMOS, making it suitable for LSI and VLSI applications ranging from digital watches to calculators and microprocessors.

PROBLEMS

15.1–1 Devise a gate circuit to sound a warning buzzer W in an automobile if the ignition key K is ON and either the driver's seat belt D is not buckled or the passenger's seat belt P is not buckled and a pressure switch S indicates that that seat is occupied.

15.1–2 Devise a gate circuit to produce $M = 1$ when a traffic signal malfunctions, such that all three lights (red, yellow, and green) are ON or none is ON.

15.1–3 Devise a gate circuit that produces $X = 1$ when $A = B = 1$ and $C = 0$, or when $C = 1$ and $D = 0$ or $E = 0$.

15.1–4 Convert the decimal numbers 54, 5.4, and 0.54 to binary. Fractional parts may be truncated after six bits. Hint: $5.4 = 5 + 0.4$.

15.1–5 Repeat the previous problem for 121, 12.1, and 1.21.

15.1–6 Apply the grouping method in reverse to express $E3A_{16}$ and 7072_8 in binary. What are the decimal values?

15.1–7 Show that 0.1011_2, 0.54_8, and $0.B_{16}$ all equal 0.6875_{10}. Then confirm that the grouping method for binary to octal and binary to hexadecimal also works for fractional parts if trailing zeros are added when necessary.

15.1–8 Use the grouping method with leading and trailing zeros to express 10011.11001_2 in octal and hexadecimal.

15.1–9 Suppose a binary number has m bits after the binary point. Write an expression for the smallest nonzero decimal quantity that can be represented. What value of m is required so that the decimal-to-binary truncation error does not exceed $\pm 0.00005_{10}$?

15.1–10 *Floating-point numbers* are stored in a large computer in a form equivalent to $x_{10} = y \times 16^{(z-64)}$, where y is represented by a 24-bit fraction $(B_{-1}B_{-2} \cdots B_{-24})$ and z is represented by a 7-bit binary integer. Estimate the largest value of x_{10}.

15.1–11 Estimate the smallest nonzero value of x_{10} in the previous problem.

15.1–12 The 2421 BCD code differs from the 8421 BCD in that the leading bit has a weight of 2 rather than 8. Make a list like Table 15.1–2 for this code, and show that some digits have two possible code words.

15.1–13 The $84\bar{2}\bar{1}$ BCD code differs from 8421 BCD in that the last two bits have negative weights, -2 and -1, respectively. Make a list like Table 15.1–2 for this code.

15.2–1 Use algebraic manipulation and Boolean rules to prove that $D(E + \overline{F}) + D\overline{E} = D$ and

$$\overline{D\overline{E} + F} = \overline{D}\,\overline{F} + E\overline{F}$$

15.2–2 Repeat Problem 15.2–1 for $D\overline{E}F + DE = DE + DF$ and $\overline{D}\,\overline{DE} = \overline{D}$.

15.2–3 Repeat Problem 15.2–1 for $(D + E + F)(D + E + \overline{F}) = D + E$ and $\overline{\overline{D}(D + E)} = D + \overline{E}$.

15.2–4 Use the truth-table synthesis method to show that $A + B = \overline{A}B + A\overline{B} + AB$ and $AB = (A + B)(A + \overline{B})(\overline{A} + B)$.

15.2–5 A light X is to be controlled independently by *three* switches, A, B, and C. Construct a truth table starting with $X = 0$ when $A = B = C = 0$. Then obtain an SOP expression for X.

15.2–6 A logic system with inputs A, B, and C is to produce $X = B \oplus C$ when $A = 1$ and $X = \overline{BC}$ when $A = 0$. Construct the truth table, obtain the POS expression, and draw a gate diagram similar to Fig. 15.2–7b. (Your truth table will be used in Problem 15.2–19.)

15.2–7 Repeat the previous problem with $X = B + C$ when $A = 0$. (Your truth table will be used in Problem 15.2–20.)

15.2–8 Four persons—A, B, C, and D—are business partners. Use the truth-table synthesis method to design a *voting machine* that produces $X = 1$ only when a majority of the partners vote *YES*. Simplify your expression before drawing the gate circuit.

15.2–9 Farmer A owns a bag of grain B, a chicken C, and a dog D. He has chores to do in rooms 0 and 1 of the barn, and does not want to move B, C, and D each time he goes from one room to the other. However, C will eat B if left alone together, and D will eat C if left alone. Use truth-table synthesis to design a warning machine that produces $E = 1$ when an intended move by the farmer would result in something being eaten. Simplify your expression before drawing the gate circuit.

15.2–10 The 4-to-1 data selector in Example 15.2–3 becomes a *logic function synthesizer* when we let S_1 and S_0 be input variables and let $A_3A_2A_1A_0$ be the control word. If $A_3A_2A_1A_0 = 1110$, for instance, then $X = \overline{S}_1S_0 + S_1\overline{S}_0 + S_1S_0 = S_1 + S_0$ (see Problem 15.2–4). Find the control words that yield $X = S_1\overline{S}_0$, $X = S_1 \oplus S_0$, $X = \overline{S}_0$, and $X = 0$.

15.2–11 Repeat the previous problem for $X = S_1 + \overline{S}_0$, $X = \overline{S_1 \oplus S_0}$, $X = S_1$, and $X = 1$.

15.2–12 Obtain a NAND-gate circuit and a NOR-gate circuit for $X = A(\overline{B} + C)$. Use input inversion bubbles where needed.

15.2–13 Repeat Problem 15.2–12 for $X = A + \overline{B}C$. Hint: See rule 10.

15.2–14 Show that $\overline{A(\overline{BC})} \neq \overline{(\overline{AB})C}$, so the NAND operation is not associative.

15.2–15 Show that $\overline{A + (\overline{B + C})} \neq \overline{(\overline{A + B}) + C}$, so the NOR operation is not associative.

15.2–16 Use rule 7 to verify that Fig. 15.2–11c is equivalent to Fig. 15.2–11a.

15.2–17 Suppose the NAND gates in Fig. 15.2–11c are replaced by NOR gates. Find the resulting expression for X.

‡ **15.2–18** Map the SOP expression from Problem 15.2–5 and show that it cannot be simplified.

‡ **15.2–19** Obtain the MPOS for the function in Problem 15.2–6. Check your result against the truth table.

‡ **15.2–20** Obtain the MPOS for the function in Problem 15.2–7. Check your result against the truth table.

‡ **15.2–21** Modify the truth table for Problem 15.2–8 so that $X = A$ in case of a tie vote. (Person A is the senior partner.) Then obtain the MSOP.

‡ **15.2–22** The 4-bit words 1010 through 1111 are not included in the 8421 BCD code. Obtain an MSOP function that detects invalid BCD words. Let $B_3B_2B_1B_0 = ABCD$ for convenience.

15.3–1 Suppose Fig. 15.3–2 is modified as follows: The NAND gates are replaced by NOR gates; the inverters are omitted; and the S and R terminals are interchanged. Verify that this circuit has the properties of an SR flip-flop. Also find Q and P when $S = R = 1$.

15.3–2 Figure P15.3–2 is a *master-slave SR flip-flop* consisting of two *gated SR* flip-flops. Draw the waveforms Q' and Q for S, R, and Ck as in Fig. 15.3–5b. Let $Q' = Q = 0$ initially. Comment on your results.

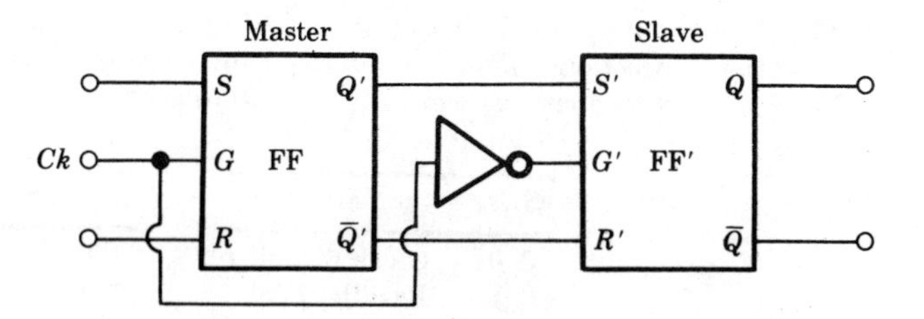

FIGURE P15.3–2

15.3–3 Repeat the previous problem with the inverter at the G terminal rather than the G' terminal.

15.3–4 Construct the expanded truth table, with inputs J_n, K_n, and Q_n, to verify that Fig. 15.3–7d functions as a JK flip-flop.

15.3–5 Draw the complete gate diagram of a JK flip-flop in the form of Fig. 15.3–7c with a master-slave SR flip-flop (Fig. P15.3–2). Use only NAND gates, and eliminate any unnecessary gates.

15.3–6 Construct the expanded truth table, with inputs T_n and Q_n, for a T flip-flop. Then draw the Q waveform that results if a T flip-flop replaces the D flip-flop in Fig. 15.3–6. Take $Q = 0$ initially.

15.3–7 Repeat the previous problem with $Q = 1$ initially.

15.3–8 Suppose a JK flip-flop is modified such that $J = A \oplus B$ and $K = B$, where A and B are the new inputs. Construct a truth table giving Q_{n+1} as a function of A_n and B_n.

15.3–9 Justify the assertion that the negated CARRY-gate inputs (Q_0 and Q_2) can be omitted in Fig. 15.3–10. Hint: Fill in the missing lines in Fig. 15.3–9b.

15.3–10 Draw the diagram of a system that counts to 1800 and resets. This system is to consist of three 4-bit binary counters and gates.

15.3–11 Draw the diagram of a system that carries out the task in Exercise 15.3–3. The system is to consist of 6-bit binary counters and gates.

15.3–12 A synchronous counter with $N = 3$ can be built using two JK flip-flops connected as follows: $K_0 = K_1 = 1, J_0 = \overline{Q}_1$, and $J_1 = Q_0$. The input is applied to both Ck terminals, and Q_1Q_0 is the output state. Construct a table like Fig. 15.3–9b, including J_0 and J_1. Draw the timing diagram with $Q_1Q_0 = 00$ as the initial state.

15.3–13 Suppose that Q_4, rather than $\overline{Q}_4$, is fed back to FF_0 in Fig. 15.3–13a. (We then have an *untwisted ring counter*.) Draw the timing diagram with $Q_4Q_3Q_2Q_1Q_0 = 10000$ as the initial state, and show that $N = 5$.

15.3–14 *A digital waveform generator* can be built using a clock, a counter, and auxiliary gates. Draw the waveform X produced when the outputs in Fig. 15.3–13 are connected such that $X = \overline{Q}_4Q_1 + Q_2\overline{Q}_0 + Q_4\overline{Q}_3$.

‡ **15.3–15** Repeat the previous problem for Fig. 15.3–9 with $X = Q_3\overline{Q}_2 + \overline{Q}_3Q_2\overline{Q}_1 + Q_3Q_2Q_1Q_0$. Then devise a function for X that yields the upper waveform in the photograph on page 8.

15.3–16 Draw the diagram of a 4-bit *recirculating memory* (shift register) that has serial input/output and a control signal named $CIRC/\overline{IN}$ to select between recirculation and serial input.

‡ **15.3–17** Devise a diagram for a 4-bit *universal register* that has parallel inputs, as well as the bidirectional features described in Exercise 15.3–5. The operating mode is selected by a control word S_1S_0 as follows:

S_1S_0	Mode	S_1S_0	Mode
0 0	Clock disabled	1 0	Shift right
0 1	Parallel load	1 1	Shift left

Hint: A 4-to-1 selector can be used for a 3-to-1 selector.

‡ **15.3–18** Design a system similar to Fig. 15.3–17 that measures and displays the approximate time in milliseconds between two pulses. The time between measurements is controlled by the user, so a manual RESET is allowed. The start and stop pulses are on separate lines.

‡ **15.3–19** Repeat the previous problem for start and stop pulses on the same line.

16 DIGITAL SYSTEMS

Culminating our study of electronics, we come at last to systems comprising digital logic circuits. We'll consider systems that are entirely digital, and *hybrid* systems that involve both analog and digital components. The latter have become increasingly important as advances in digital technology—especially the microprocessor—have made it possible to replace analog instrumentation with more compact, more reliable, more flexible, and less expensive digital ICs.

Figure 16.0 represents a general hybrid system with an analog-to-digital converter (ADC) and a digital-to-analog converter (DAC) interfacing the analog and digital sections. Such a system, for instance, might monitor and control the temperature and ventilation throughout a large office building or factory. The analog inputs would include inside temperature and air-quality measurements and outside weather conditions, while the digital inputs would provide the desired temperature levels, time-of-day information, and the control strategies.

The analog processor, in general, carries out any operations that must be done in analog form, but the sophisticated computations and decision-making are left to the digital processor using digitized measurement information coming from the ADC. Some digital outputs then go directly to display and control devices, while other outputs are converted by the DAC to obtain any necessary analog outputs.

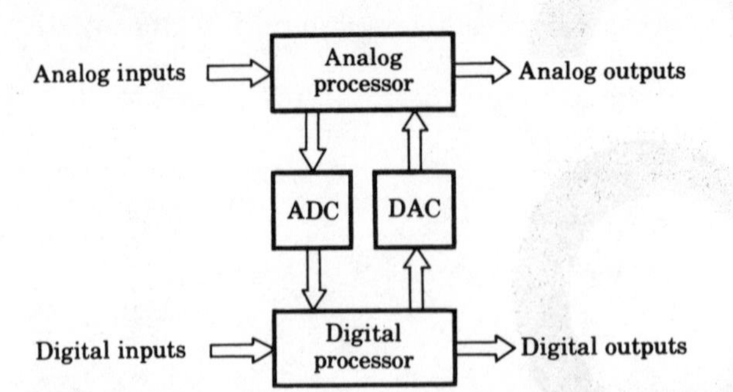

FIGURE 16.0
A hybrid system.

The first two sections of this chapter deal with digital instrumentation and communication, emphasizing the interface between analog and digital signals. The last half is concerned with digital operations per se, from arithmetic circuits to complete microprocessors and computers.

OBJECTIVES

After studying this chapter and working the exercises, you should be able to do each of the following:

- Calculate the resolution of an ADC, and explain the relationship between speed and accuracy in a counter-controlled converter (Section 16.1).
- Write an expression for the analog output in terms of the digital input to a weighted-resistance DAC (Section 16.1).
- State the advantages and problems of digital communication, and determine the bandwidth required for digital signaling at a specified rate (Section 16.2). †
- Draw the block diagram for a PCM system, and analyze its performance (Section 16.2). †
- Draw the gate circuit for a half adder and full adder (Section 16.3). Describe how binary addition circuits can be modified to handle signed numbers, BCD numbers, and multiplication (Section 16.3).
- List the functional sections of a digital computer and a microprocessor, and explain the significance of word length and memory capacity (Section 16.4). †

16.1 DIGITAL INSTRUMENTATION

The purpose of an instrumentation system is to extract desired information regarding some physical quantity. Although the quantity of interest is almost always *analog* in nature, *digital* instrumentation often provides the most convenient means of processing and display. There are two major advantages of digital instrumentation over and above those benefits listed in Section 15.1. First, digital circuits have a *memory* capability that would be difficult or impossible to implement in analog form. Second, the operating characteristics of a *programmable* digital processor are readily altered by changing the program or *software,* whereas one must rebuild the hardware of an analog processor to implement changes.

Putting these advantages to work on analog data obviously requires analog-to-digital and digital-to-analog conversion. So these two topics take up most of this section. The section concludes with the description of a typical data acquisition system with digital processing.

ANALOG-TO-DIGITAL CONVERSION

An analog-to-digital converter (ADC) takes an analog signal, usually a voltage, and produces a digital code word, usually in binary form. If the code word consists of n bits there can be no more than 2^n possible output values, whereas the analog input has an infinite number of possible values within its range. The conversion process, then, inherently involves *round-off* or *quantization* of the analog signal.

To illustrate these concepts, let's suppose we have an analog voltage v_a known always to be within the range 0–8 V, and we want to represent an arbitrary value of v_a by a 2-bit word. Lacking other information to the contrary, we might use the scheme in Fig. 16.1–1a where the 8-V range has been divided into four equal intervals, a quantized value v_q is defined at the midpoint of each interval, and a code word B_1B_0 is assigned to each value of v_q. For example, any value of v_a falling between 2.0 and 4.0 V would be rounded off to $v_q = 3$ V and encoded as $B_1B_0 = 01$. Such digitization always introduces a *quantizing error* $v_a - v_q$, which, in this case, has a maximum value of ± 1.0 V.

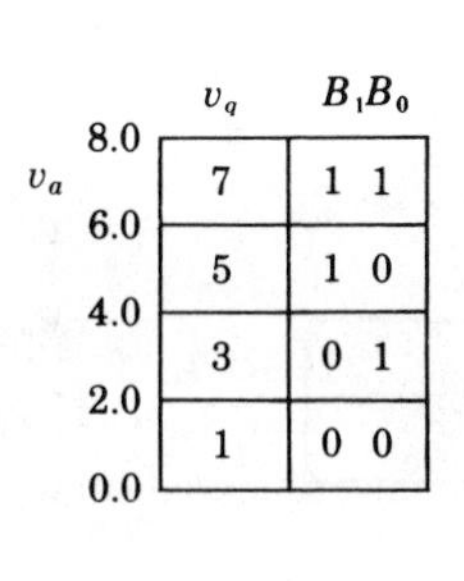

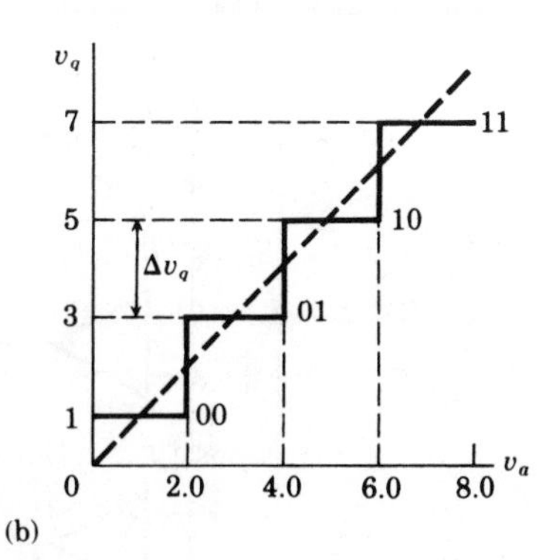

FIGURE 16.1–1 Quantizing and encoding an analog voltage.

Viewed from another angle, the plot of v_q versus v_a in Fig. 16.1–1b shows that we are making a "staircase" approximation of the analog voltage. The steps are $\Delta v_q = 2$ V high, starting from $v_q = 1$ V, and occur at the threshold voltages $v_a = 2.0, 4.0$, and 6.0 V. The quantizing error for a given value of v_a equals the difference between the staircase and the dashed diagonal line.

To improve the approximation and reduce the quantization error, we would need more steps of less height and, consequently, a longer code word. In general, for an n-bit ADC with an analog range of V_R volts, the step height or *resolution* is

$$\Delta v_q = \frac{V_R}{2^n} \tag{1a}$$

giving a quantizing error

$$|v_a - v_q| \leq \frac{1}{2}\,\Delta v_q = \frac{V_R}{2^{n+1}} \tag{1b}$$

(These expressions also hold for a bipolar ADC with $-V_R/2 \leq v_a \leq V_R/2$.) The resolution corresponds to the analog weight of the least significant bit and is often expressed in percent of the full-scale value. Thus, an 8-bit ADC has a resolution of $100/2^8 \approx 0.4\%$ and a maximum quantizing error of $\pm 0.2\%$.

Operationally, the most straightforward of ADC implementations is the *parallel-comparator* ADC diagrammed in Fig. 16.1–2a for $n = 2$. There are $2^n = 4$ equal resistances forming a voltage divider that establishes the reference thresholds for $2^n - 1 = 3$ comparators. The output of each comparator is either LOW or HIGH, depending on whether v_a is below

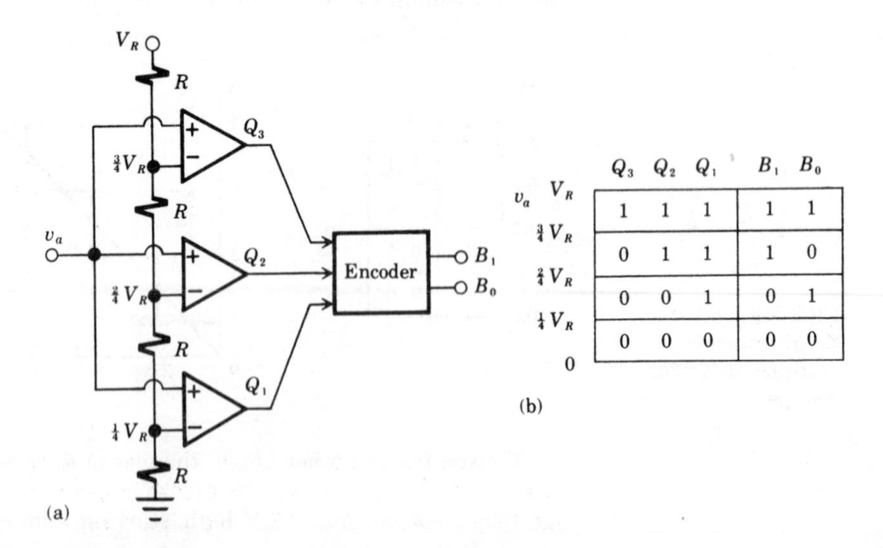

v_a	Q_3	Q_2	Q_1	B_1	B_0
V_R to $\frac{3}{4}V_R$	1	1	1	1	1
$\frac{3}{4}V_R$ to $\frac{2}{4}V_R$	0	1	1	1	0
$\frac{2}{4}V_R$ to $\frac{1}{4}V_R$	0	0	1	0	1
$\frac{1}{4}V_R$ to 0	0	0	0	0	0

FIGURE 16.1–2 Parallel-comparator ADC: (a) circuit; (b) truth table.

or above the comparator's threshold. The encoder then translates the characteristic 3-bit pattern $Q_3Q_2Q_1$ into B_1B_0 according to the truth table in Fig. 16.1–2b. The conversion process takes place almost instantaneously—so rapidly, in fact, that these are known as *flash* converters. Their resolution, however, depends directly upon the number of comparators, and a resolution of 0.4% would require $2^8 - 1 = 255$ comparators! More practically, available IC units typically have 4-bit outputs (6.25% resolution), with conversion rates up to 25 MHz. They would thus be suited to applications calling for very high speed rather than accuracy.

The speed of a flash converter raises another consideration: Variations of v_a may cause the output to change before we have time to process the current value of v_q. Controlling the conversion rate then requires a *sample-and-hold* circuit (S/H) like that in Fig. 16.1–3, where the FET-switch symbol represents any suitable n-channel FET. A positive pulse in the switching wave $s(t)$ turns on the FET switch and allows the capacitance to charge up to the sample value v_{as}. The FET switch then becomes an open circuit and C holds v_{as}, thanks to the high input impedance of the op-amp voltage follower that buffers the circuit from the ADC.

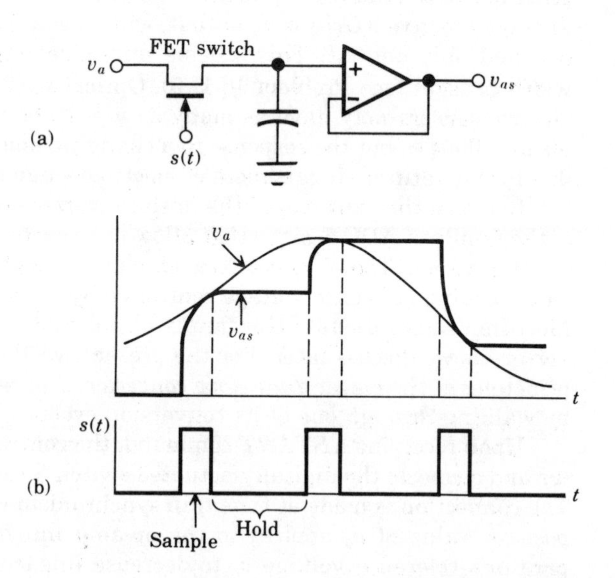

FIGURE 16.1–3
(a) Sample-and-hold (S/H) circuit.
(b) Waveforms.

At the other end of the speed/accuracy spectrum is the *rotational* ADC or *shaft encoder*. This electromechanical device consists of a disk attached to a rotating shaft whose angular position θ is proportional to the analog quantity of interest. A set of fixed-position sensors detect and digitize the angle. The disk of the 3-bit version drawn in Fig. 16.1–4 has

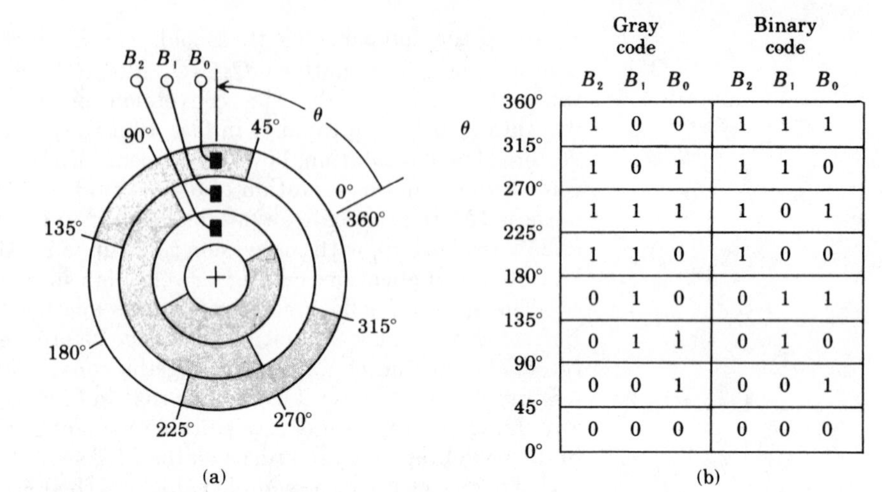

θ	Gray code B_2	Gray code B_1	Gray code B_0	Binary code B_2	Binary code B_1	Binary code B_0
315°–360°	1	0	0	1	1	1
270°–315°	1	0	1	1	1	0
225°–270°	1	1	1	1	0	1
180°–225°	1	1	0	1	0	0
135°–180°	0	1	0	0	1	1
90°–135°	0	1	1	0	1	0
45°–90°	0	0	1	0	0	1
0°–45°	0	0	0	0	0	0

FIGURE 16.1–4
(a) Shaft encoder disk.
(b) Angular codes.

three concentric bands divided into conducting and nonconducting segments. Carbon brushes press on the bands and pick up a DC voltage applied to the conducting segments. The segments in the figure have been arranged to give a *Gray code* output, with successive words that differ by one and only one bit. This scheme minimizes errors caused by finite-width brushes (see Problem 16.1–5). Optical shaft encoders with photoelectric sensors may have as many as $n = 18$ bands, for a resolution of about 0.0004%, but the response to a changing analog input is quite slow due to the natural sluggishness of electromechanical elements.

Between the extremes of the flash converter and shaft encoder, there is the family of ADC designs that allow us to exchange speed for accuracy or vice versa. These converters employ a single comparator with a time-varying reference voltage controlled by a digital clock and counter. Most implementations of this family also include a digital-to-analog converter, as we discuss later. For the present, we'll describe the operating principles of the clever *dual-slope converter,* diagrammed in Fig. 16.1–5, by walking through one of its conversion cycles.

Upon receiving a *START* command, the control unit resets the counter and connects the digitally-actuated switch S to the analog voltage v_a. The connection is made at time t_1 in synchronism with a clock pulse. The positive value of v_a applied to the op-amp integrator causes the comparator's reference voltage v_r to decrease linearly with slope $-v_a/RC$, and Q immediately goes HIGH since $v_r < 0$. The AND gate now passes clock pulses to the counter, which eventually overflows at time $t_2 = t_1 + 2^nT$, T being the clock period. The control unit then switches S to the fixed voltage $-V_R$, so v_r starts increasing linearly with slope $+V_R/RC$ until time t_3 when $v_r = 0^+$ and Q goes LOW. This stops the counting, and the control unit moves S to the grounded position, completing the cycle.

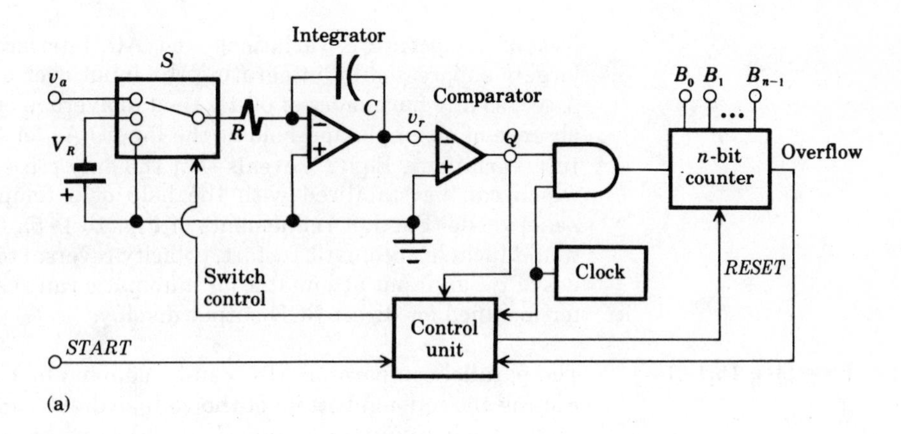

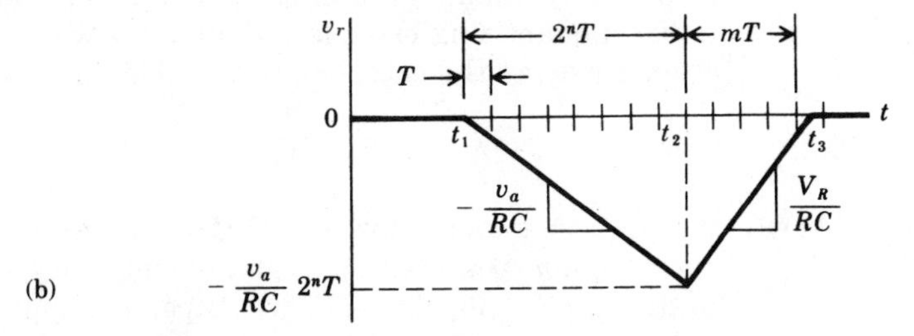

FIGURE 16.1–5 Dual-slope ADC: (a) block diagram; (b) waveform.

Since the counter reset itself at t_2 (Why?), the number of counts m recorded between t_2 and t_3 yields an output code word $B_{n-1} \ldots B_1 B_0$ equivalent to the quantized voltage

$$v_q = \frac{m}{2^n} V_R \tag{2a}$$

where

$$(2^n - 1)\frac{v_a}{V_R} < m \le 2^n \frac{v_a}{V_R} \tag{2b}$$

The quantizing error here stems from the fact that $t_3 - t_2$ does not, in general, equal an integer multiple of T.

Note that the resolution depends on the size of the counter, rather than the number of comparators, but the conversion cycle requires up to 2×2^n clock periods. If $T = 10\ \mu\text{s}$, for example, then taking $n = 4$ yields 6.25% resolution at a conversion rate of $(2^{n+1}T)^{-1} \approx 3$ kHz, whereas $n = 8$ yields 0.4% resolution at a rate of about 0.2 kHz. We have therefore reduced speed in favor of accuracy.

The dual-slope ADC has special advantages for *digital voltmeters* (DVMs) and related instrumentation that must perform reliably in the face of temperature variations and AC interference. Interference is largely suppressed by integrating the input over an integer multiple of $\frac{1}{60}$ sec, so any harmonics of 60 Hz tend to average out. (This explains the absence of a sample-and-hold at the input.) As for the effect of temperature variations, Eq. (2) reveals that the only critical parameter is V_R—which can be stabilized with the help of a temperature-compensated Zener diode. Further refinements of Fig. 16.1–5a for DVM applications would include automatic restart, polarity reversal to permit negative values of v_a, an input attenuator for automatic range selection, and a counter modified for direct BCD output display.

Exercise 16.1–1

The parallel-comparator ADC can be adapted to a bipolar input by connecting the top and bottom of the voltage divider to $+V_R/2$ and $-V_R/2$, respectively. Using this approach, find the number of comparators needed to get a quantizing error less than ± 1.0 V when $-6.0 \text{ V} \le v_a \le 6.0$ V. Then construct the encoder's truth table and plot v_q versus v_a.

Exercise 16.1–2

Show that $t_3 - t_2 = 2^n T(v_a/V_R)$ in Fig. 16.1–5b. Then derive Eq. (2b).

DIGITAL-TO-ANALOG CONVERSION

A digital-to-analog converter (DAC) takes an n-bit digital word and produces a *quantized* analog output. For simplicity, we'll limit our discussion to the case of unipolar voltages. Bipolar outputs involve offset voltages and the representation of signed numbers.

Figure 16.1–6a gives the schematic diagram of a *weighted-resistance* DAC for $n = 4$ bits. The input word is applied to a bank of digitally-actuated switches, such that the output of the ith switch will be a voltage

$$v_i = B_i V_R = \begin{cases} V_R & B_i = 1 \\ 0 & B_i = 0 \end{cases} \tag{3}$$

These voltages constitute the input to a *summing* op-amp with weighted input resistances $R_i = R/2^i$ and feedback resistance R_F. Combining Eq.

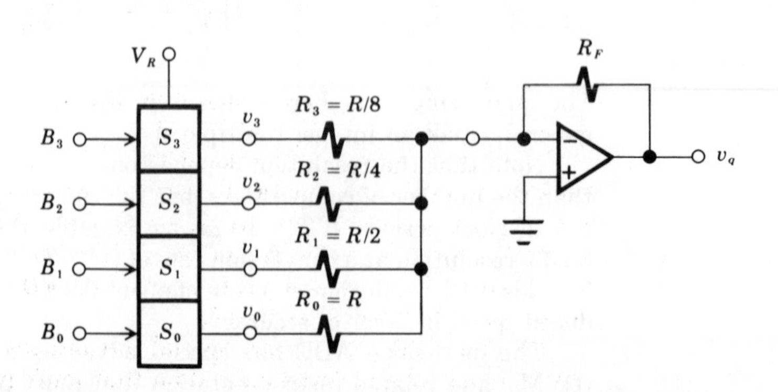

FIGURE 16.1–6 Weighted-resistance DAC.

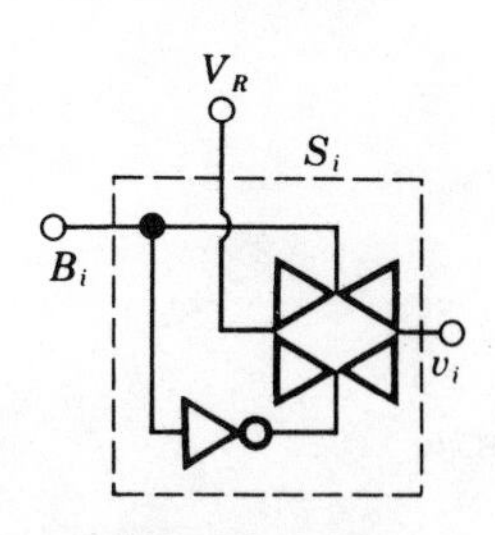

FIGURE 16.1–7
Digital switch using CMOS transmission gate.

(3) with Eq. (6a), Section 4.3, yields the output voltage

$$\begin{aligned} v_q &= -\left(\frac{R_F}{R/8}v_3 + \frac{R_F}{R/4}v_2 + \frac{R_F}{R/2}v_1 + \frac{R_F}{R}v_0\right) \\ &= (8B_3 + 4B_2 + 2B_1 + B_0)\left(-\frac{R_F}{R}V_R\right) \end{aligned} \tag{4}$$

and each bit has been multiplied by the appropriate factor for binary to decimal conversion. Thus, taking $R_F = R/2^n$, we obtain $0 \le |v_q| < |V_R|$ with step size or resolution $\Delta v_q = V_R/2^n$, the same as an n-bit ADC. If desired, we can make v_q exactly equal to the decimal value of the binary word by letting $V_R = -R/R_F$—so $v_q = 5$ when $B_3B_2B_1B_0 = 0101$, etc.

Figure 16.1–7 shows a simple implementation of the necessary digital switches. The CMOS gate acts as an open circuit when B_i is LOW, or as a small resistance r when B_i is HIGH. To prevent unwanted loading effects, v_i must be applied to $R_i \gg r$. In practice, the inverter would be omitted if the input word comes from a register with both B_i and $\overline{B}_i$ available.

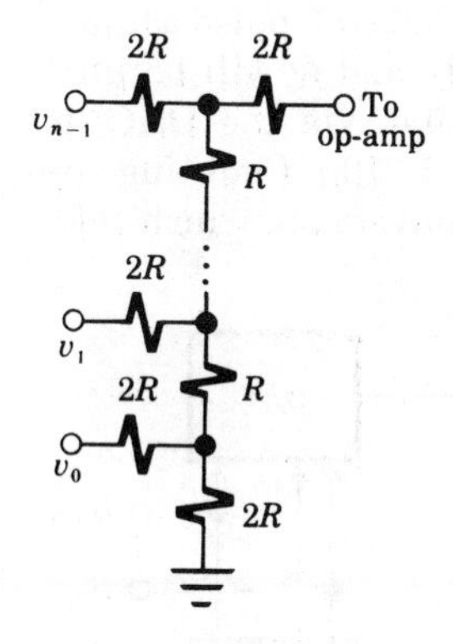

FIGURE 16.1–8
R-2R ladder for DAC.

The weighted-resistance DAC has problems of accuracy and loading when n is large, say $n \ge 8$, which requires a wide range of precision resistors for the R_i's. Those problems are circumvented by the *R-2R ladder* network of Fig. 16.1–8. This network is easily fabricated in IC form, since it involves just two resistance values, their 2:1 ratio being the only critical factor. When an R-$2R$ ladder replaces the weighted resistances in Fig. 16.1–6, the output voltage becomes

$$v_q = (2^{n-1}B_{n-1} + \cdots + 2B_1 + B_0)\left(\frac{-R_FV_R}{2^n \times 3R}\right) \tag{5}$$

The derivation provides a nice exercise in Thévenin/Norton manipulations (see Problem 3.3–11).

Exercise 16.1–3

Consider Fig. 16.1–6 expanded for $n = 16$ input bits. If $R_{15} = 10\ \text{k}\Omega$, what should be the values of R_0 and R_F such that $\Delta v_q = V_R/2^n$?

Exercise 16.1–4

The system in Fig. 16.1–9 is intended to convert a three-digit *decimal* number, such as 8.05, to a quantized analog voltage v_q that goes from 0.00 to 9.99 V in steps of 0.01 V. The decimal digits are represented by BCD words, and each BCD word is applied to a separate DAC whose output has the possible values 0, −1, −2, . . . , −9 V. Write an expression for v_q in terms of v_1, $v_{0.1}$, and $v_{0.01}$; then determine the required values of R_1, $R_{0.1}$, and $R_{0.01}$ in terms of R_F.

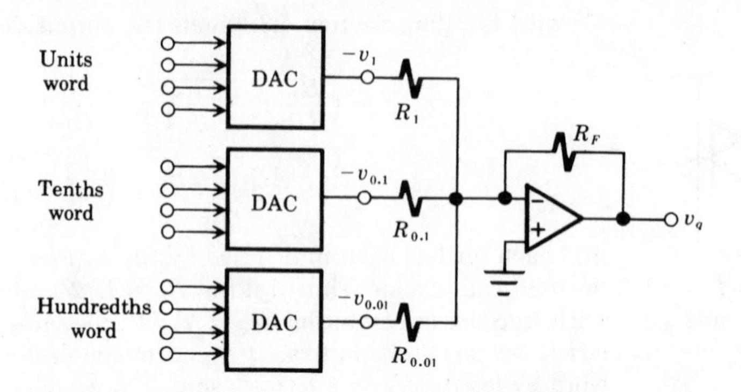

FIGURE 16.1–9
DAC for BCD input.

Example 16.1–1
Counter-controlled ADC

Illustrating how a DAC can be employed in analog-to-digital conversion, Fig. 16.1–10a diagrams a *counter-controlled* ADC whose reference voltage v_r is a staircase or pseudoramp waveform generated by digital-to-analog conversion of the counter's output word. A *START* pulse actuates the S/H and clears the counter so $v_r = 0$ initially and Q will be HIGH. Each clock pulse then increments the counter and, via the DAC, increases v_r in steps of $V_R/2^n$ as plotted in Fig. 16.1–10b. Counting continues until $v_r > v_{as}$ and Q goes LOW to end the conversion, which takes up to 2^n clock periods.

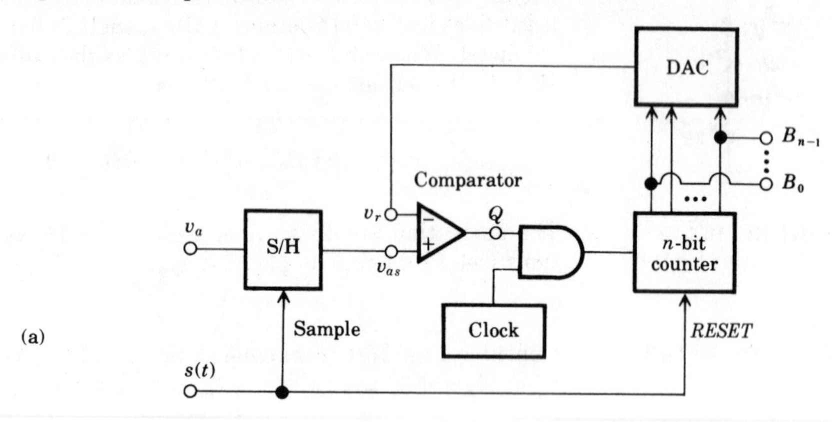

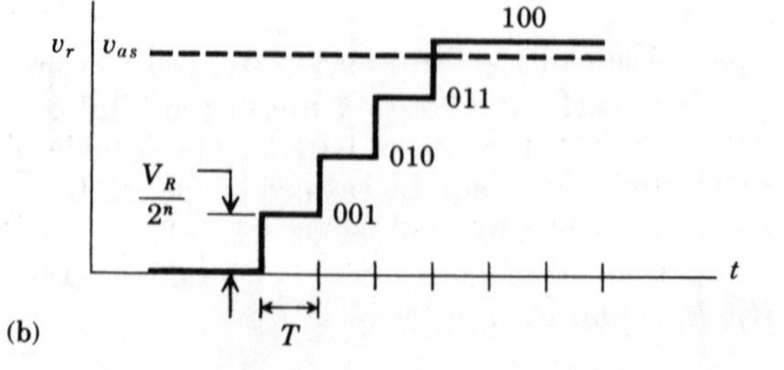

FIGURE 16.1–10
Counter-controlled ADC: (a) block diagram; (b) waveform.

Maximum conversion rates for this ADC are higher than the dual-slope converter, but less than 100 kHz. A *tracking* ADC uses an *up/down* counter to cut conversion time in half, while the more complex *successive-approximation* ADC requires just n clock periods and operates at speeds up to 1 MHz. These ADCs have the speed and resolution appropriate for large-scale digital systems, particularly those with on-line computer data processing.

DATA ACQUISITION AND DIGITAL PROCESSING

Collecting and processing experimental data is a tiresome and error-prone chore when done by hand. (Remember your physics labs?) Fortunately, when suitable analog-signal transducers exist, the rest of the task can be turned over to a modern multichannel *data acquisition system* or *data logger* that has the functional elements shown in Fig. 16.1–11. Such a system might handle 40 or more input quantities, both analog and digital, with a scan rate of perhaps 20 channels per second and 0.01% ADC resolution. Besides collecting, displaying, and printing the data, this system might also carry out various numerical operations and deliver processed data to a computer terminal or peripheral recording device for further analysis.

Several of the blocks in Fig. 16.1–11 have been previously examined, and the internal workings of others will be discussed subsequently. Our purpose here is to present an overview of digital data processing by

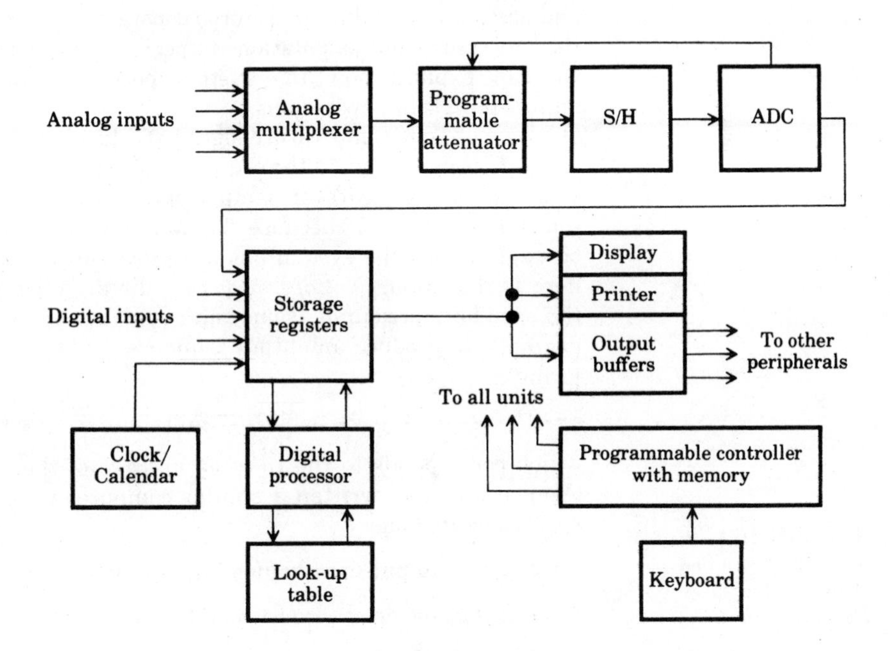

FIGURE 16.1–11 Data acquisition system.

describing typical operations on one of the analog inputs, say the temperature signal $x(t)$, generated by a thermal sensor.

The analog multiplexer periodically samples $x(t)$ and other analog inputs at a rate governed by the controller. The interleaved analog samples then pass through the *programmable attenuator* (see Problem 16.1–8) for digitizing by the ADC. Feedback from the ADC to the attenuator provides automatic ranging, and the digitized data and range information are loaded into storage registers. Other registers hold data obtained from direct digital inputs and time/date information provided by the clock/calendar. The remainder of the system consists essentially of a small digital computer that can be programmed from the keyboard.

The digital processor, in conjunction with the controller, might average the temperature data over a specified number of samples. At a predetermined time (or upon command by the operator), the processed data are displayed, printed, and transferred to peripheral devices. The output includes identifying labels for each data value, along with logging information from the clock/calendar and the keyboard entries. The entire cycle then automatically repeats every few seconds, or perhaps just once a week, according to the instructions stored in the controller.

Clearly the controller and digital processor constitute the "brains" of such a system, and we'll have more to say about the arithmetic and control circuits in Sections 16.3 and 16.4. Right now it's enough to state that the processor is capable of basic arithmetic calculations involving past and present data values and stored constants, and that the controller tells the processor what calculations to perform. Armed with this knowledge, we can explain how the system performs simple but illustrative digital-processing operations.

Suppose our temperature signal $x(t)$ has been sampled and digitized every T seconds to yield the sequence of data points $x_0, x_1, x_2, \ldots$, where x_k stands for the current sample value at $t = kT$. These values differ somewhat from $x(kT)$ because of quantizing, but the quantizing error will be negligible if the system has a high-resolution ADC. The main point here is that *numerical operations* by the digital processor can supplant the need for more cumbersome *analog processing*. It's a trivial matter, for example, to produce an output sequence $y_0, y_1, \ldots$ calculated from the formula

$$y_k = \tfrac{1}{10}\,(x_k + x_{k-1} + \cdots + x_{k-9})$$

which corresponds to the *running average* of the past 10 input values. (You've probably written a similar computer program.) Other common operations include:

- *Scaling,* to put data values into appropriate units.
- *Editing,* to remove questionable values.
- *Linearizing,* to correct the effects of a nonlinear transducer.

Some operations might draw upon conversion factors stored in the look-up table.

In addition—and more significantly—the processor can simulate the action of a *filter*, in the sense that the output sequence corresponds to the sample values that would have been obtained had $x(t)$ been passed through an analog filter before sampling. The system then functions as a *digital filter*. For instance, suppose the controller is programmed to calculate

$$y_k = \frac{1}{\tau + T}(Tx_k + \tau y_{k-1}) \tag{6a}$$

where the constant τ is large compared to T. Rewriting Eq. (6a) in the form

$$\tau\left(\frac{y_k - y_{k-1}}{T}\right) + y_k = x_k \tag{6b}$$

brings out the fact that this expression is a *difference equation*, involving the difference between y_k and the previously computed value y_{k-1}. Further inspection reveals that we have approximated the *differential equation*

$$\tau\frac{dy}{dt} + y = x$$

since $(y_k - y_{k-1})/T = \Delta y/\Delta t \approx dy/dt$. But this differential equation has the same form as that for an RC circuit with input voltage x, output voltage y across the capacitor, and $\tau = RC$—see Eq. (3), Section 5.3. The digital processor, therefore, acts essentially like an RC lowpass filter.

Although you can't be expected to get excited about this crude filter, you should now have some appreciation for the fact that relatively simple digital processors are capable of highly sophisticated filtering characteristics—often well beyond the scope of practical analog implementation. Moreover, the controller can be programmed to change the filter characteristics when the nature of the data changes, thereby creating an automatic *adaptive* filter.

16.2 DIGITAL COMMUNICATION†

Digital systems often involve the transmission of binary words from one point to another. For short distances, *parallel* transmission with a separate line for each bit is preferred. But longer distances involve such costs that bit-by-bit *serial* transmission over a single pair of wires or coaxial cable is usually favored. This section explores the concepts and techniques of digital communication, in contrast to those of analog communication. We also describe how analog waveforms can be transmitted digitally to exploit the advantages of digital signaling.

DIGITAL SIGNALS AND BANDWIDTH

A binary digital signal is a waveform that represents a string of bits, say 10110100 . . . , and might look like Fig. 16.2–1a when it emerges from a logic circuit. For transmission purposes, however, we usually need to make various modifications. In particular, the *bipolar* waveform in Fig. 16.2–1b has the advantage that no power is wasted in a DC component if 1s and 0s occur, on the average, in equal proportion. We call this waveform (and the one in Fig. 16.2–1a) a *synchronous* or *non-return-to-zero* (NRZ) signal because there are no "spaces" between adjacent pulses. As a consequence, a long succession of 1s or 0s causes synchronization problems at the destination unless accompanied by an auxiliary timing signal. Alternatively, inserting half-bit spaces gives the *return-to-bias* (RB) waveform of Fig. 16.2–1c, which is self-synchronizing but at the cost of an increased transmission bandwidth.

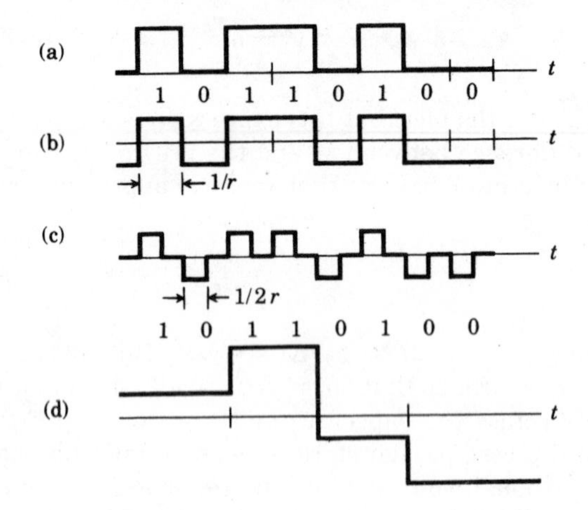

FIGURE 16.2–1 Digital signals: (a) unipolar NRZ; (b) bipolar NRZ; (c) bipolar RB; (d) four-level bipolar NRZ.

Suppose we want an *information rate* of r bits per second (b/sec), so the duration of a 1-bit pulse in an NRZ waveform will be $1/r$ sec. We recall from Section 14.1 that the bandwidth required for reasonable reproduction of a single pulse of duration D is $B \geq 1/D$. Actually, we can get by with about half that value in digital signaling because the pulse shape doesn't matter very much, providing we can still recognize the correct bits. Therefore, the transmission bandwidth for an NRZ signal is

$$B \geq \frac{r}{2} \tag{1}$$

meaning that we can signal at up to 2 b/sec per hertz of bandwidth. A wire-pair with $B = 4$ kHz could then be used for binary signaling at $r \leq$ 8 kb/sec. An RB signal requires twice as much bandwidth for the same information rate since the pulses have duration $\frac{1}{2}r$. By the way, you will often find signaling rates expressed in *bauds,* a term carried over from telegraphy that refers to the number of *level changes per second.* Hence, 1 baud = 1 b/sec in the case of binary signals.

Occasionally, successive bits are grouped in twos or threes and represented by a *multilevel* waveform such as that of Fig. 16.2–1d. For groups of M bits, there must be 2^M different pulse levels, but the pulse duration becomes M/r and the bandwidth requirement is reduced to $B \geq r/2M$. The bit rate then equals M times the baud rate.

As they stand, none of the waveforms in Fig. 16.2–1 would be suitable for an AC-coupled transmission medium. Instead, they would be used to *modulate* the amplitude, frequency, or phase of an appropriate sinusoidal carrier wave; and a corresponding demodulator would recover the signal at the destination. A two-way link, then, needs a modulator/demodulator or *modem* at each end. Modems designed for voice-grade telephone lines operate in a synchronous mode at rates from 600 to 9600 b/sec. Wideband high-speed systems achieve rates from 19.2 kb/sec to more than 200 Mb/sec. These higher rates are appropriate for computer-to-computer links, but far exceed those needed for data originating at a keyboard terminal, DVM, or similar low-speed source.

Low-speed digital transmission usually takes place via a two-wire DC current loop using a format called the American Standard Code for Information Exchange, abbreviated ASCII and pronounced "ask-ee." The ASCII code consists of 10- or 11-bit words. The waveform, shown in Fig. 16.2–2, is unipolar NRZ with current flowing in the HIGH state and no current in the LOW state. The line stays in the HIGH state during the idle time between words to keep the connection "alive." A LOW *start bit* announces the beginning of a word and 8 data bits follow thereafter, terminated by one or two *stop bits* in the HIGH state.

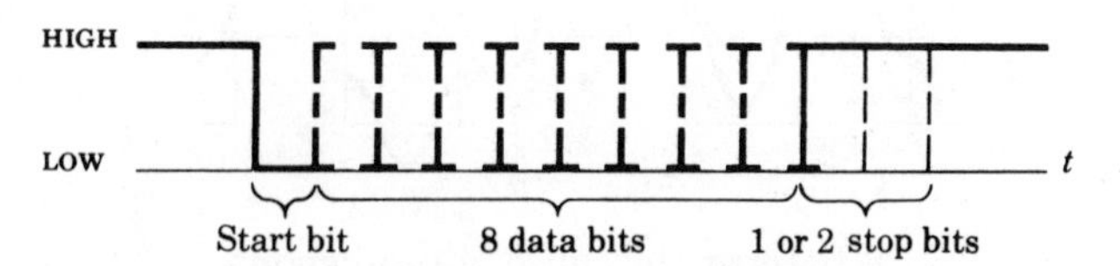

FIGURE 16.2–2
ASCII waveform.

This scheme is said to be *start/stop synchronous* because the bits are uniformly spaced within each word but the time between words is arbitrary. Standard bit rates within ASCII words are 110, 300, 600, and 1200 b/sec, of which only the lowest rate requires the second stop bit. When used for alphanumeric character transmission from a keyboard terminal, the eighth ASCII data bit is actually a *parity* bit that helps detect transmission errors—our next topic.

Exercise 16.2–1 Calculate the minimum time required to transmit 8000 data bits using
(a) A binary RB waveform on a line having $B = 4$ kHz.
(b) An 8-level NRZ waveform on the same line.
(c) A 45-Mb/sec wideband system.
(d) A 110-b/sec ASCII code.

TRANSMISSION ERRORS AND PARITY CODES

As we observed in Chapter 14, any long-distance communication system suffers from the effects of attenuation, distortion, interference, and noise. Consequently, after amplification at the receiving end, an NRZ bipolar signal representing 101101 might look like Fig. 16.2–3. To recover the digital information, we could sample this waveform at the nominal midpoint of each bit interval and call it a 1 if the sample value is positive or a 0 if the sample value is negative. This decision rule results in correct digit identification regardless of signal contaminations, providing they don't alter the polarity at the sample times. We therefore see that binary signals are immune to many of the problems that plague analog communication. (A multilevel digital signal is less "rugged" because the receiver must decide between more than two possible values.)

Occasionally, of course, a large contaminating spike will alter the polarity and cause an *error,* changing a 1 to a 0 or vice versa. Errors resulting from thermal noise turn out to be quite rare at any reasonable value of signal-to-noise ratio. In particular, random-signal theory predicts that a binary communication system with $S/N = 10$ dB has an *error probability* around 10^{-3}, meaning an average of one error per thousand bits. And increasing S/N to 16 dB drops the probability to about 10^{-9}! Therefore, most errors are caused by electrical interference and hardware glitches rather than thermal noise.

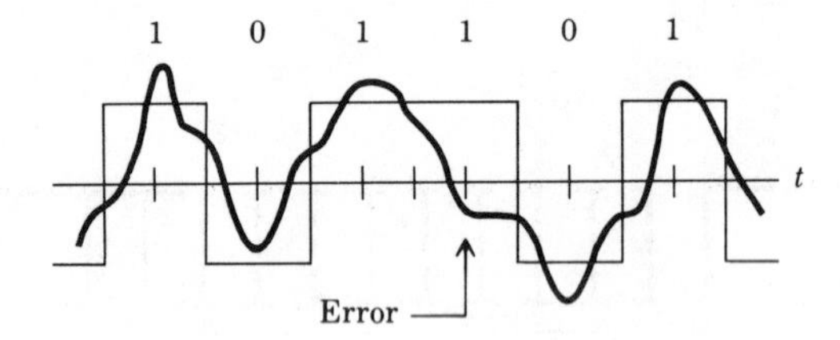

FIGURE 16.2–3 Digital signal with error due to transmission effects.

Whatever the cause, digital errors could produce grave consequences in, say, an aircraft guidance system. Less disastrous but still annoying are errors in data-acquisition or computerized-banking systems—especially if they reduce the balance in *your* account. Accordingly, various *error-detecting codes* have been developed. These codes rely on the fact that if the error probability per bit is relatively small, two or more errors seldom appear in one word. For example, when a system has an easily achieved error probability of 10^{-4} per bit, the probability of more than one erroneous bit in a 4-bit word is less than 10^{-7}.

TABLE 16.2–1 Error-Detecting Codes

Decimal Digit	8421P	2-*out*-*of*-5
0	00000	00011
1	00011	00101
2	00101	00110
3	00110	01001
4	01001	01010
5	01010	01100
6	01100	10001
7	01111	10010
8	10001	10100
9	10010	11000

Table 16.2–1 lists two error-detecting codes for decimal digits. The code labeled 8421P is just like ordinary BCD, except that an extra bit—called a *parity bit* or *check bit*—has been added. This bit is chosen so that the entire word contains an even number of 1s, and we say that the word has *even parity.* If a single error occurs in one of these words, changing a 1 to 0 or 0 to 1, the parity becomes odd. Then we know immediately that the word has an error in it, although we don't know where. The code thus permits single-error detection by *parity checking.* Double errors in one word go undetected because parity remains even in that rare event.

The 2-out-of-5 code also has even parity and, in addition, has exactly two 1s in every word, making it possible to detect multiple errors that change the number of 1s but not the parity. However, encoding and decoding is more difficult with this code than with the 8421P code, which can be handled by using ordinary BCD circuitry plus the simple arrangement of X-OR gates in Fig. 16.2–4.

Once an error has been detected, the entire word must be either discarded or retransmitted. If the data words are essential and retransmission is impractical, perhaps for lack of a two-way communication system, one of the more elaborate *error-correcting codes* is necessary. Such codes generally have more check bits for a given number of data bits, resulting in a larger bandwidth requirement or a smaller equivalent information rate.

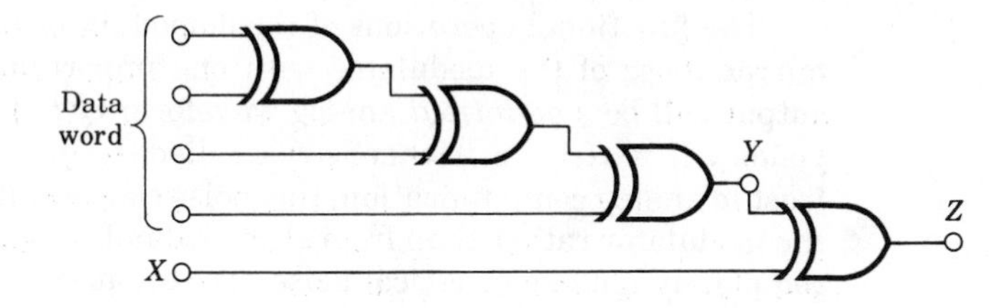

FIGURE 16.2–4 Parity checker/generator.

Exercise 16.2–2

Suppose the following words have been received from a source that employs the 8421P code:

00101 01110 01111 11000

Identify the valid and invalid words. For each invalid word, determine the minimum number of erroneous bits, and list the corresponding possible valid words.

Exercise 16.2–3

The circuit in Fig. 16.2–4 can serve as either a *parity checker,* if X equals the parity bit of the received word, or as a *parity generator,* if $X = 0$.
(a) Use the input words 00101 and 01101 to show that $Z = 0$ indicates even parity and that $Z = 1$ indicates odd parity.
(b) Use the data words 0010 and 0110 to show that Z equals the appropriate parity bit when $X = 0$.

PULSE-CODE AND DELTA MODULATION

Having seen some of the transmission advantages of digital signals, we now consider two modulation methods that make possible *analog* communication in digital form. Both methods trade off increased transmission bandwidth for reduced noise and interference, similar to frequency modulation. Unlike FM, however, the digitized waveforms are readily time-division multiplexed with other digital signals that represent analog or digital information, thus promoting flexibility and efficient use of communication hardware. A case in point—the Bell System's T4 cable carries a diverse mix of digital data and digitized voice and TV signals that adds up to its maximum rate of 274 Mb/sec.

Pulse-code modulation (PCM) draws on the sampling theorem (from Section 14.3), plus analog-to-digital and digital-to-analog conversion (from Section 16.1). The modulator's block diagram in Fig. 16.2–5a has an input lowpass filter to prevent aliasing effects, followed by a sample-and-hold circuit, n-bit ADC, and parallel-to-serial converter. All of the latter are synchronized by the switching wave $s(t)$ at frequency $f_s = 1/T_s \geq 2W$, where W is the nominal bandwidth of the analog input signal $x(t)$. Each sample value of $x(t)$ is quantized and encoded into an n-bit word by the ADC, whose serially-converted output becomes the PCM bit stream. Figure 16.2–5b shows typical waveforms taking $n = 3$ and a bipolar RB output.

The functional operations of the demodulator in Fig. 16.2–5c simply mirror those of the modulator with one important exception: The S/H output will be a *quantized* analog waveform $x_q(t)$. Lowpass filtering then yields $y(t) = x(t) + \epsilon(t)$ where $\epsilon(t)$ is called the *quantization noise*. In contrast to analog communication, this noise comes entirely from the ADC at the modulator rather than from electrical noise sources. (Indeed, we have completely ignored electrical noise here because its only effect will be oc-

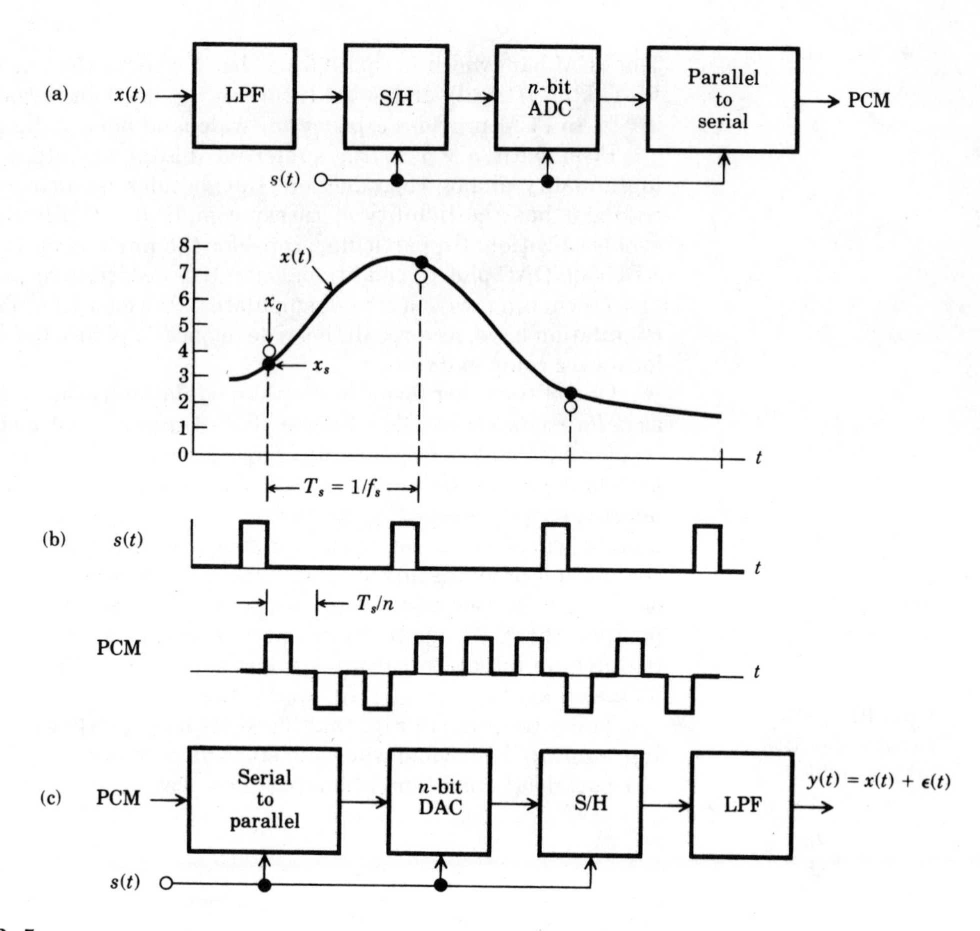

FIGURE 16.2–5
Pulse-code modulation:
(a) modulator;
(b) waveforms;
(c) demodulator.

casional bit errors.) The ratio of output signal power to quantization noise power N_q is given by

$$\frac{S}{N_q} \approx 2^{2n} \tag{2}$$

and is consistent with the fact that the ADC quantizing error is proportional to $1/2^n$.

Equation (2) implies that PCM needs a relatively large value of n to minimize the quantization noise which, in turn, leads to a large *transmission bandwidth* requirement. Specifically, we have $r = nf_s$ b/sec since there are n bits every T_s seconds; so combining $B \geq r/2$ with $f_s \geq 2W$ yields

$$B \geq \tfrac{1}{2}nf_s \geq nW \tag{3}$$

The PCM bandwidth is therefore at least n times the analog bandwidth W. Taking $B = nW$ and substituting in Eq. (2) then reveal that $S/N_q \approx 2^{2(B/W)}$, so PCM provides *exponential* wideband noise reduction.

PCM with $n \geq 7$ is the preferred digital modulation method for high-quality analog reproduction. But, besides requiring a large bandwidth, it has the liability of rather complicated hardware and critical synchronization. (In particular, consider the implementation of the n-bit ADC and DAC plus the additional circuitry needed to recreate a *synchronized* switching wave at the demodulator.) Various other forms of digital modulation have, as a result, been developed to reduce bandwidth and/or hardware complexity.

One of the other forms of digital modulation is the technique known as *delta modulation* (DM). Figure 16.2–6 shows a DM modulator whose bipolar NRZ output becomes, after integration, a *piecewise ramp approximation* $x_r(t)$ for the analog signal $x(t)$. (Some versions use a *stepwise* approximation, accounting for the name *delta* modulation.) The approximation waveform is fed back and compared with $x(t)$, and the bipolar comparator produces an output voltage $\pm V_{CC}$ according to whether $x_r > x$ or $x_r < x$. This voltage is sampled and held every $T_s = 1/f_s$ seconds to produce the DM signal. Since $x_r(t)$ is generated from the output by the op-amp integrator, it always has slope $\mp V_{CC}/RC$ and decreases or increases by $\Delta x_r = V_{CC}T_s/RC$ every T_s seconds.

The waveforms in Fig. 16.2–6b show how $x_r(t)$ tracks $x(t)$, except for the "hunting" condition when $x(t)$ stays more or less constant, and for the *slope-overload* condition when $x(t)$ increases (or decreases) too rapidly.

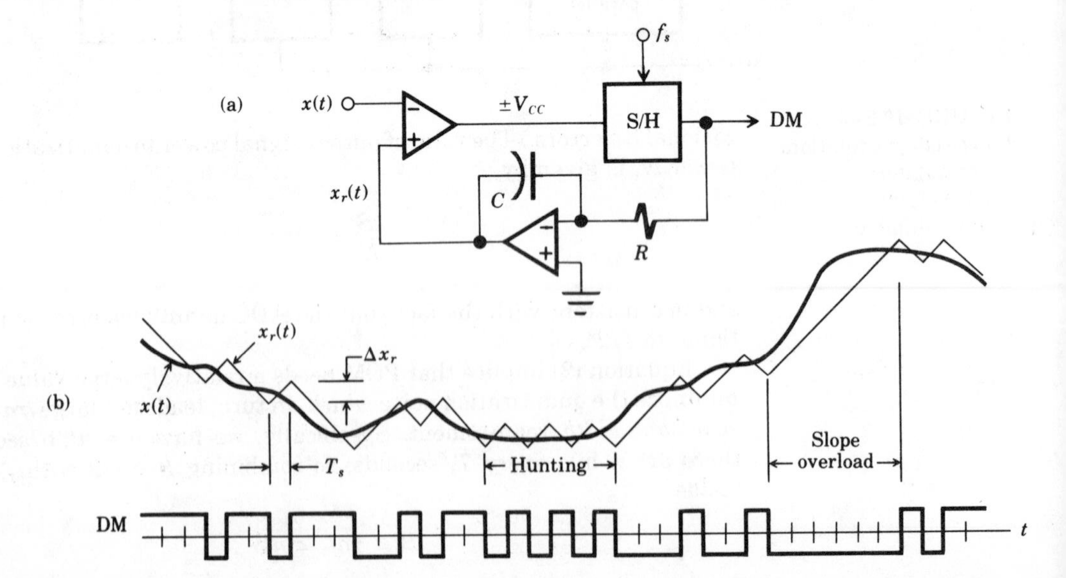

FIGURE 16.2–6 Delta modulation: (a) modulator; (b) waveforms.

The latter effect, rather than the sampling theorem, dictates the minimum sampling frequency f_s. The corresponding bandwidth requirement is then $B \geq f_s/2$ since the DM signal has just one bit per sample.

Clearly, the modulator in Fig. 16.2–6a involves much less equipment than PCM. Furthermore, demodulation requires only an op-amp integrator, recreating $x_r(t)$, followed by a lowpass filter to smooth the output waveform. The "kinks" in $x_r(t)$ produce something like PCM quantization noise. This noise decreases as f_s increases, again implying the trade-off of bandwidth for noise reduction.

Example 16.2–1 Consider a voice signal with $W = 4$ kHz. If the signal is transmitted as a PCM wave with $n = 8$ and $f_s = 2W$, the required bandwidth is $B \geq 8 \times 4$ kHz $= 32$ kHz and $S/N_q \approx 2^{16} \approx 48$ dB. A DM system with same bandwidth and $f_s = 2B = 64$ kHz would yield a lower but still adequate signal-to-noise ratio.

Exercise 16.2–4 An analog data signal with $W = 12$ kHz is to be transmitted by PCM. Find the bandwidth required such that $S/N_q \geq 50$ dB. Hint: $\log_{10} 2^x \approx 0.3x$.

16.3 ARITHMETIC CIRCUITS

We now shift our emphasis to systems that are entirely digital, designed to carry out such *arithmetic computations* as addition, subtraction, and multiplication. Our discussion must necessarily be limited to a few examples that illustrate the basic concepts, for there are literally hundreds of special-purpose arithmetic circuits. On the other hand, we will see that a binary adder can be augmented to perform a variety of operations and, as such, becomes one of the major building blocks of a microprocessor or digital computer.

ADDITION CIRCUITS

Suppose two decimal numbers have been converted to binary words and stored in registers. The binary words can be added by a logic circuit derived from the following rules for the addition of two single bits:

$$\begin{array}{r}0\\+\ 0\\\hline 00\end{array}\qquad\begin{array}{r}0\\+\ 1\\\hline 01\end{array}\qquad\begin{array}{r}1\\+\ 0\\\hline 01\end{array}\qquad\begin{array}{r}1\\+\ 1\\\hline 10\end{array}$$

This *arithmetic addition* (as distinguished from the logical-OR operation) yields two distinct bits, a *sum* bit S and a *carry* bit C. If we view the bits to be added as input variables, say A and B, the output variables S and C are related to them per the truth table in Fig. 16.3–1a. Clearly, $C = AB$ (a logical AND) while $S = A \oplus B$ (the exclusive-OR from Fig. 15.2–4). Hence, the gate circuit diagrammed in Fig. 16.3–1b produces the desired

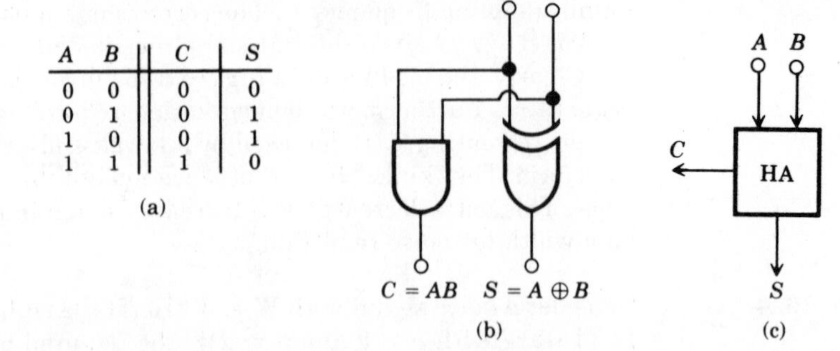

A	B	C	S
0	0	0	0
0	1	0	1
1	0	0	1
1	1	1	0

FIGURE 16.3–1
Half adder: (a) truth table; (b) gate circuit; (c) symbol.

sum and carry for any input combination. We call this important circuit a *half adder* (HA) and represent it symbolically by Fig. 16.3–1c.

Now to see that happens in a complete binary addition, consider the following example, where the addition has been done in two steps, a partial bit-by-bit sum plus the carry bits:

$$\begin{array}{r} 13 \\ +\ 7 \\ \hline 20 \end{array} \quad \Rightarrow \quad \begin{array}{rl} 1101 & \\ +\ 0111 & \\ \hline 1010 & \text{Partial sum} \\ 0101 & \text{Carry bits} \\ \hline 10100 & \text{Result} \end{array}$$

Although a half adder suffices for the least significant bits, we obviously need a more complicated circuit to include possible carries in any subsequent position. The circuit in question is a *full adder* (FA), consisting of two half adders and an OR gate, Fig. 16.3–2. This circuit adds A and B plus an input carry C_{in}, producing a sum bit S and output carry bit C_{out}.

By connecting together a half adder and several full adders, we obtain a circuit that will add two binary words bit by bit. Figure 16.3–3 shows a complete 4-bit adder. The input words to be added are denoted by $A_3A_2A_1A_0$ and $B_3B_2B_1B_0$; $S_3S_2S_1S_0$ is the sum word obtained at the output. Note that the output carry C_0 from the half adder becomes

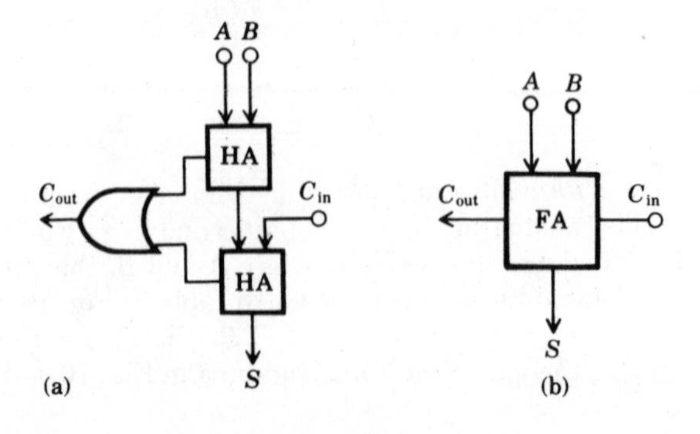

FIGURE 16.3–2
(a) Full adder.
(b) Symbol.

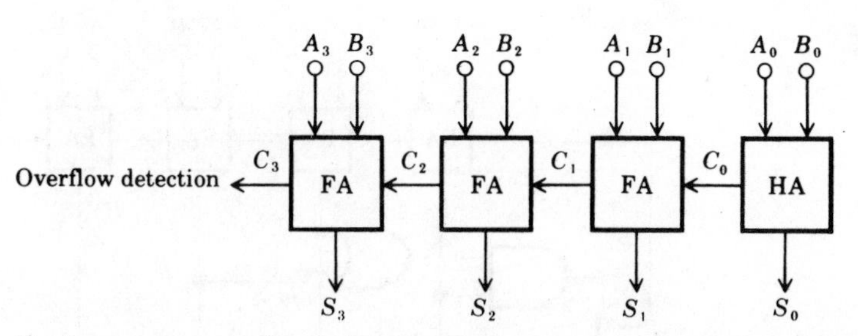

FIGURE 16.3–3
Four-bit parallel adder.

the input carry to the first full adder, and so on, moving from right to left. When the leftmost carry bit C_3 equals 1, the sum of the two 4-bit words is actually a 5-bit word (illustrated in the previous example). This produces an *overflow* condition if the particular system is limited to 4-bit words, in which case C_3 should be applied to an overflow-detection unit.

This circuit is called a *parallel* binary adder because all bits are added almost simultaneously. However, it does take a finite amount of time for the carries to propagate from right to left, a phenomenon known as the *carry-ripple* problem. The more complicated *look-ahead-carry adders* overcome this problem and attain greater speed, typically taking less than 20 ns to add two 16-bit words. Alternatively, to minimize hardware rather than computation time, one could use the *serial adder* of Fig. 16.3–4, which involves just one FA and a D flip-flop for one-bit time delay. Here, the input bits A_k and B_k are entered serially from shift registers, the successive sum bits S_k being delivered to an output shift register. Note the importance of synchronous timing provided by the clock.

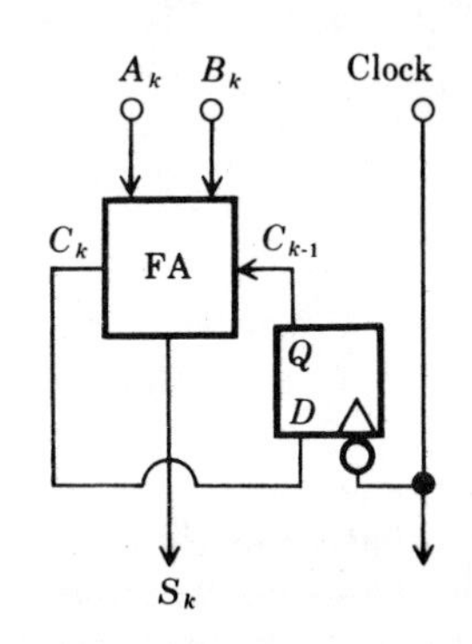

FIGURE 16.3–4
Serial adder.

Addition circuits that operate on BCD words, as in a calculator, must have a slightly different structure. To demonstrate why, consider the following additions in BCD:

$$\begin{array}{r} 6 \\ +\ 3 \\ \hline 9 \end{array} \quad \Rightarrow \quad \begin{array}{r} 0110 \\ +\ 0011 \\ \hline 1001 \end{array} \qquad\qquad \begin{array}{r} 6 \\ +\ 7 \\ \hline 13 \end{array} \quad \Rightarrow \quad \begin{array}{r} 0110 \\ +\ 0111 \\ \hline 1101\ ? \end{array}$$

The first result is correct but the second is wrong because 13_{10} is represented in BCD by *two* words, namely 0001 0011, not 1101. Moreover, 1101 is not even a proper code word (see Table 15.1–2).

These quirks stem from the fact that BCD uses only 10 of the 16 possible 4-bit words, and the remaining 6 words are forbidden or meaningless. To compensate, the adding circuit must detect whenever a sum exceeds 9_{10}; when that happens, the circuit must add $6_{10} = 0110$ to the sum to skip over the 6 forbidden words and to produce a carry. Figure 16.3–5 diagrams a parallel adder with these provisions.

Exercise 16.3–1

Construct a truth table for the full adder. Then show that $S = \overline{A}\overline{B}C_{\text{in}} + \overline{A}B\overline{C}_{\text{in}} + A\overline{B}\overline{C}_{\text{in}} + ABC_{\text{in}}$ and that $C_{\text{out}} = AB + AC_{\text{in}} + BC_{\text{in}}$.

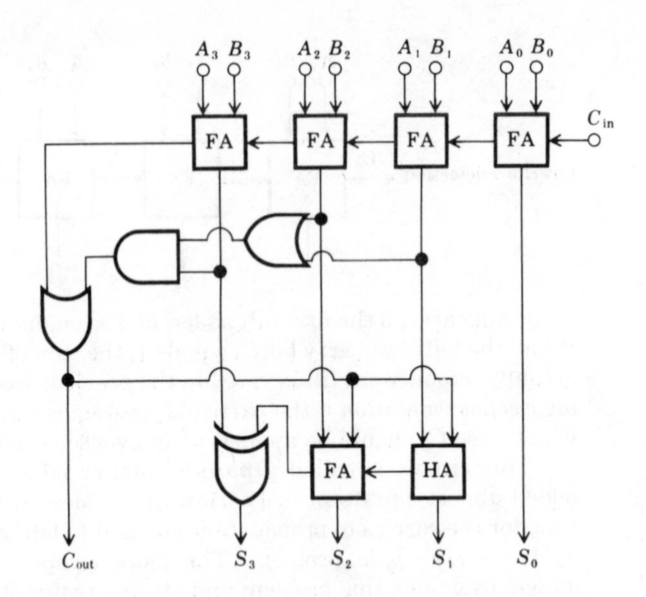

FIGURE 16.3–5
BCD adder.

Exercise 16.3–2 Follow through all logic operations in Fig. 16.3–5 to confirm that the BCD adder produces correct results when $C_{in} = 0$, $A_3A_2A_1A_0 = 1000 = 8_{10}$, and $B_3B_2B_1B_0 = 0111 = 7_{10}$.

SIGNED NUMBERS AND SUBTRACTION

Subtraction and signed numbers go hand in hand in base-2 arithmetic, just as they do in base-10. Thus, although binary subtraction circuits may be used for special applications, general-purpose arithmetic circuits carry out *subtraction* by *adding negative numbers*. This is the method employed by most digital computers, where negative binary numbers are represented in *2s-complement* form. Similar methods are used in calculators and other digital systems that operate on BCD numbers. However, since BCD subtraction is somewhat more complicated, we'll limit our consideration to standard binary numbers.

We form the 2s complement of a binary word simply by complementing each bit (change 0 to 1 and 1 to 0) and adding 1 to the least significant bit. Applying this recipe to 1001 gives 0110 + 1 = 0111 for its 2s complement. But note that $0111_2 = 7_{10}$, whereas $1001_2 = 9_{10}$. We therefore need some way of distinguishing the 2s complement of a negative binary number from the positive binary number with the same bit string. Lacking plus and minus signs in binary notation, we must convey this information with another bit called the *sign bit*.

The usual convention is to put the sign bit first, before the most significant bit of the word, letting 0 stand for "positive" and 1 stand for "negative." A negative binary number then will be represented by its 2s complement preceded by a 1, whereas a positive binary number is preceded by a 0. To illustrate,

$$9_{10} = 0 * 1001_2 \qquad 4_{10} = 0 * 0100_2$$

$$-9_{10} = 1 * 0111_2 \qquad -4_{10} = 1 * 1100_2$$

where, in the interest of clarity, we have separated the sign bit from the rest of each word by an asterisk. (A logic circuit designed to handle such words would, of course, automatically recognize the leading bit as the sign bit.)

An immediate consequence of our representation is increased word length to accommodate the sign bit. In exchange for the extra bit, we can now carry out binary subtraction by adding. To demonstrate this, let's consider the decimal subtraction $9 - 4 = 5$ or, equivalently, $9 + (-4) = 5$. Written as a binary addition we have

$$\begin{array}{r} 9 \\ -\ 4 \\ \hline 5 \end{array} \quad \Rightarrow \quad \begin{array}{r} 0{*}1001 \\ +\ 1{*}1100 \\ \hline 10{*}0101 \end{array}$$

After dropping the overflow carry beyond the sign bit we obtain the correct result, namely $0 * 0101_2 = 5_{10}$. A negative result would appear in 2s-complement form with the sign bit equal to 1.

Example 16.3–1
Parallel adder/subtractor

Figure 16.3–6 diagrams a logic circuit capable of adding *or* subtracting two 5-bit binary words, either of which, or both, might represent negative numbers. (Leading sign bits are identified by an asterisk subscript.) A control variable *SUB* selects between subtraction ($SUB = 1$) or addition ($SUB = 0$). If $SUB = 0$, the circuit functions as a parallel binary adder (Fig. 16.3–3) with the additional provision that S_* conveys the correct sign of the sum—barring an actual magnitude overflow. If $SUB = 1$, the

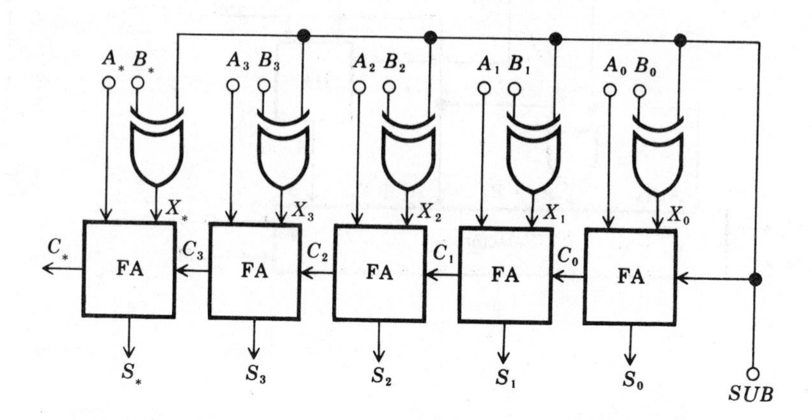

FIGURE 16.3–6 Parallel adder/subtractor.

circuit subtracts by adding the 2s-complemented B word to the A word. Another logic unit, not shown, would detect any magnitude overflow in addition or subtraction.

Exercise 16.3–3 Obtain the 2s-complement representation of -5_{10}. Then perform each of the following calculations as binary additions, and check the result: $5 - 9$, $-5 - 4$, $5 - 5$.

Exercise 16.3–4 Let $SUB = 1$, $A_* A_3 A_2 A_1 A_0 = 00000$, and $B_* B_3 B_2 B_1 B_0 = 11100 = -4_{10}$ in Fig. 16.3–3. Show that $S_* S_3 S_2 S_1 S_0 = 00100 = +4_{10}$.

ARITHMETIC/LOGIC UNITS

We have seen that a modified parallel binary adder adds or subtracts, depending on the state of a control variable. An *arithmetic/logic unit* (ALU) has even more versatility. It includes auxiliary gates and several control variables that actuate *logic* operations, as well as addition and subtraction. Provision may also be made for performing numerical comparisons.

Illustrating this idea, Fig. 16.3–7a shows one section of an adder/subtractor with four outputs delivered to a selector (like that in Fig. 15.2–8) controlled by $M_2 M_1$. The control bit M_0 plays the same role

FIGURE 16.3–7 (a) Section of arithmetic/logic unit. (b) Function table.

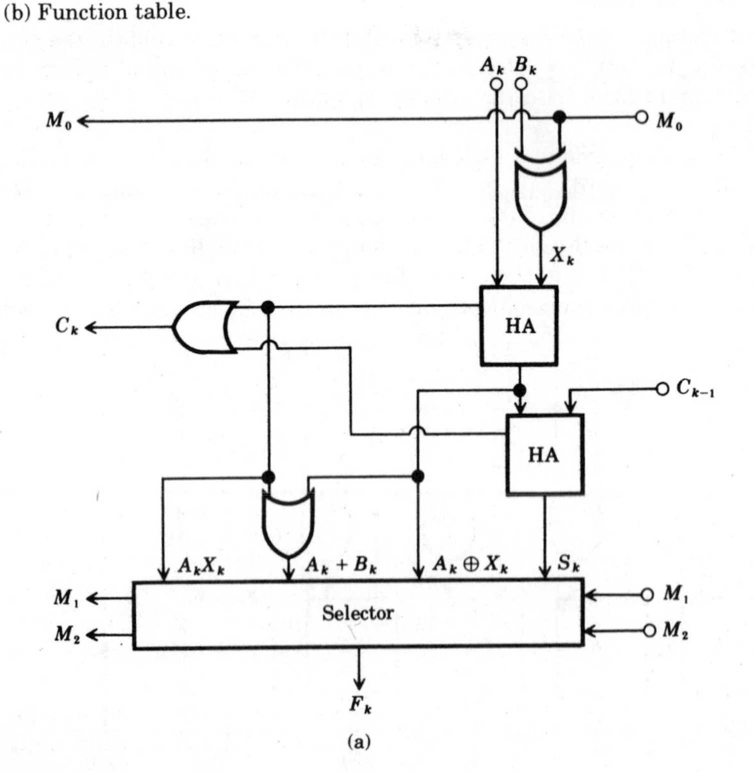

(a)

M_2	M_1	M_0	Function
0	0	0	$A \cdot B$
0	0	1	$A \cdot \overline{B}$
0	1	0	$A + B$
0	1	1	$A + \overline{B}$
1	0	0	$A \oplus B$
1	0	1	$A \oplus \overline{B}$
1	1	0	A plus B
1	1	1	A minus B

(b)

as SUB in Fig. 16.3–6, so that $X_k = B_k$ when $M_0 = 0$ and $X_k = \overline{B}_k$ when $M_0 = 1$. Besides S_k, the other outputs from the full adder consist of the partial carry $A_k \cdot X_k$ and the partial sum $A_k \oplus X_k$, which are applied to the auxiliary OR gate to obtain $A_k + X_k$. Thus, $M_2M_1M_0 = 000$ produces $F_k = A_kB_k$ (the logic AND), while $M_2M_1M_0 = 011$ produces $F_k = A_k + \overline{B}_k$ (the logic OR with B_k negated), and so forth. Figure 16.3–7b lists the output functions corresponding to the eight possible control words. The words "plus" and "minus" stand for the arithmetic operations, while an expression such as $F = A + B$ means that each bit of the F word equals the logic OR of the corresponding bits of the A and B words.

Now suppose we form another output by applying all the F_k to a NOR gate, producing $Z = \overline{F_0 + F_1 + \cdots}$. If we select the subtract mode ($M_2M_1M_0 = 111$) and if the A word numerically equals the B word, then the F word will be all zeros and $Z = 1$ indicates numerical equality, $A = B$. If $A \neq B$, then $Z = 0$ and we can distinguish between $A > B$ and $A < B$ simply by looking at the sign bit of the F word.

Although Fig. 16.3–7 represents part of a hypothetical ALU, commercially available IC units have all the features we've discussed, plus a few more, and are capable of as many as 32 arithmetic and logic functions. The implementation for 4-bit words involves about 80 gates, equivalent to some 800 discrete components, all fabricated on one MSI chip.

MULTIPLICATION AND DIVISION

Multiplication of positive binary numbers turns out to be relatively easy, involving only addition and position shifting. This is illustrated by the following example, where sign bits have been omitted for simplicity:

```
  13                1101     Multiplicand
×  6    ⇒     ×     0110     Multiplier
----          ----------
  78                0000 ←┐
              +    1101  ←┤
              ----------  │
                 011010   ├ Partial products
              +  1101    ←┤
              ----------  │
                1001110   │
              + 0000     ←┘
              ----------
               01001110      Product
```

Here the partial products are formed by multiplying the entire multiplicand by single bits of the multiplier, so they either repeat the multiplicand when the multiplier bit is 1, or they are all 0s when the multiplier bit is 0. Each partial product is shifted left to line up with its multiplier bit. Note that we have added the partial products two at a time rather than all at once. Also note that, in general, the product of two 4-bit words will be an 8-bit word.

These operations can be accomplished by a logic circuit comprising a shift register, parallel adder, and control unit as in Fig. 16.3–8, The shift

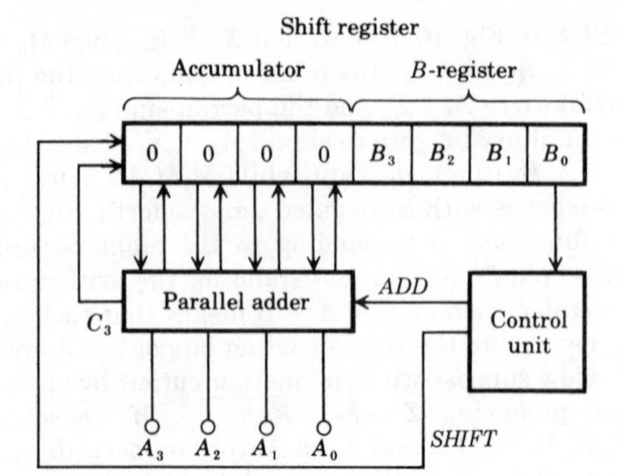

FIGURE 16.3–8
Multiplier.

register has two sections of equal length, labeled the accumulator and B register. Initially, the accumulator contains all 0s, and the B register holds the multiplier word $B_3B_2B_1B_0$. The multiplicand $A_3A_2A_1A_0$ is available as one input to the adder—the other input and the output being available to and from the accumulator. The *control unit* looks at the right-most bit in the B register to determine what actions should be taken.

If that bit equals 0, the control unit issues the *SHIFT* command, causing all bits in the shift register to shift one position to the right, with a 0 fed into the leftmost cell. If the rightmost bit in the B register equals 1, the control unit issues the *ADD* command, causing the multiplicand to be added to the word in the accumulator. The sum is then copied into the accumulator, and the shift command is given, with the carry bit C_3 fed into the leftmost cell of the shift register. These steps are repeated four times in a 4-bit multiplier, after which the entire shift register holds the final 8-bit product.

Modifications to Fig. 16.3–8 to account for signed numbers are relatively straightforward. Any negative numbers represented in 2s-complement form must be converted to their true magnitude values, and an additional logic circuit is incorporated to examine the sign bits of multiplicand and multiplier and to generate the corresponding sign bit for the product.

Obviously, binary multiplication takes much more time than addition or subtraction, for it requires several *sequential* steps rather than simultaneous parallel operations. Binary *division* is even more time consuming and complicated. We'll not go into the details here, but will point out that division is done by repeated subtraction and shifting, whereas multiplication involves repeated addition and shifting. After each step in the division process, logic circuitry must test a trial quotient bit by making a trial subtraction. If the trial quotient is wrong, the step must be repeated.

Should the need arise, one might build a "hard-wired" circuit designed specifically for binary multiplication or division. More commonly, however, multiplication and division are *software operations* performed by controlling a general-purpose circuit similar to that of Fig. 16.3–8. As a matter of fact, with an ALU replacing the adder, Fig. 16.3–8 contains most of the functional elements found in the central processor unit of a digital computer. Such a unit uses various numerical algorithms based on addition, subtraction, and comparison to carry out more complicated calculations (powers, roots, trigonometric functions, etc.), as well as multiplication and division.

Exercise 16.3–5

Taking $A_3A_2A_1A_0 = 1101 = 13_{10}$ and $B_3B_2B_1B_0 = 0110 = 6_{10}$, find the contents of the shift register in Fig. 16.3–8 after each step of the multiplication process and show that the final result is $01001110 = 78_{10}$.

16.4 COMPUTERS AND MICROPROCESSORS †

This closing section provides an overview of digital computers, primarily from the hardware viewpoint. Some attention is also given to the relationships between hardware and software. Following a survey of general concepts, we take a closer look at the integrated-circuit microprocessor and the semiconductor memory units that go hand in hand with the microprocessor in a microcomputer system.

COMPUTER CONCEPTS

Every programmable digital computer has five *functional sections* represented by the blocks in Fig. 16.4–1. The *input section* delivers the *data* to be processed along with the *instructions* that constitute the program for the task in question. The data and instructions, encoded as binary words, are held in the *storage section* until needed. The *arithmetic/logic section* (or ALU) receives stored data, performs processing operations, and produces *results* to be stored. When called for by the program, results are

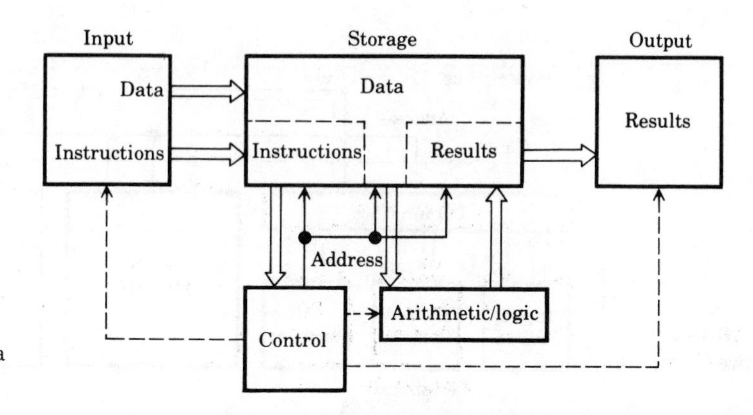

FIGURE 16.4–1 Functional sections of a digital computer.

transferred from storage to the *output section* where they are converted to the desired forms—printed copy, graphics display, tape recording, digital signal, and so forth.

The *control section* is the decision-making center of the computer. It coordinates the activities of all other sections by decoding the instruction words and generating appropriate *control signals*. In addition, it issues *address words* that tell the storage section the location of the next instruction or data word to be processed and where to place the results.

Although all computers must have these five *functions*, the actual *hardware units* do not necessarily correspond to the functional blocks. Underscoring the distinction between functional roles and physical layout, Fig. 16.4–2 shows the organization of a typical computing system. The heart of this system is a hardware assembly called the *central processing unit* (CPU), which encompasses the ALU, the control function, several storage registers, and some of the input/output (I/O) functions. Besides the registers in the CPU, the storage section is divided between an internal *main memory* (an array of magnetic cores or semiconductor cells) and the external *file memory* devices (magnetic disks, drums, and tape units). These devices provide auxiliary mass storage for data and programs that can be transferred to or from the main memory when required. An I/O *interface* unit serves as the buffer between the rest of the system and the *peripherals*, which include the file memory as well as input and output devices (teletypewriters, CRT terminals, line printers, etc.).

This system is said to be *bus-organized* in that the CPU, main memory, and I/O interface all connect to a set of wires known as the *system bus*. The bus, in turn, consists of three different wire groups: a data bus that carries data and instructions in both directions, a one-way address bus for the address words issued by the CPU, and a control bus. Most of the lines in the control bus convey CPU commands to the other units, but a few carry "interrupt" and "ready" signals from the peripherals back to the CPU.

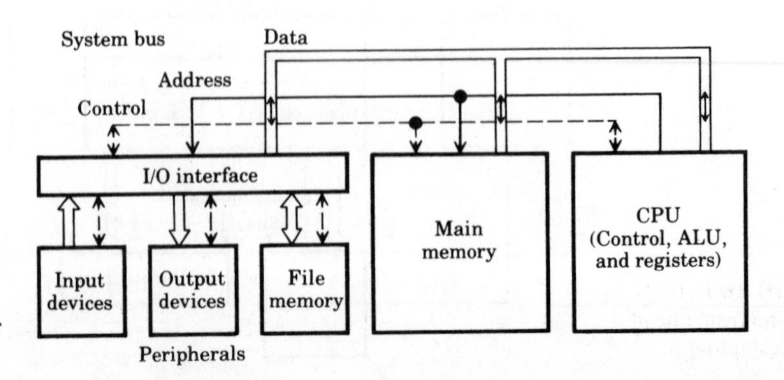

FIGURE 16.4–2
Bus-organized computer system.

With minor modifications, Fig. 16.4–2 could equally well represent a tiny *microcomputer,* a giant *mainframe computer,* or an intermediate-size *minicomputer.* In the case of a microcomputer, the CPU would be a *microprocessor*—a single IC chip comprising all the circuitry needed for the ALU, control unit, storage registers, and even perhaps part or all of the main memory. Except for peripherals, the rest of the microcomputer involves a few more IC chips and a power supply. All this hardware could fit inside a shoebox and sell for a few hundred dollars. On the other hand, the CPU and memory of a multimillion-dollar mainframe computer system typically fill several cabinets, and the I/O interface may itself consist of a number of minicomputers. The greater size and cost of a large computer reflect its wider range of capabilities, compared to a smaller model.

Fundamentally, the capabilities of a particular computer depend on two major factors: the CPU *word length* and the *capacity* of the main memory. A third factor is the *speed,* measured in terms of the time required to execute a basic operation such as adding two numbers. It turns out, however, that execution times depend as much on word length and memory capacity as they do on the actual speed of the electronic circuits.

To see the significance of word length, consider the fact that a given n-bit binary word represents one of 2^n possible combinations. If a microprocessor has a word length of 8 bits (or one byte) and uses 1-word instructions, then it could have an *instruction set* of up to $2^8 = 256$ basic operations—arithmetic/logic functions, data transfers, and register manipulations. More complicated operations, such as multiplication, can be carried out as a sequence of basic operations, so an 8-bit instruction word poses no serious limitations. But an 8-bit *data* word corresponds to only about 0.4% resolution, or two decimal digits in the case of BCD data, and would not be sufficiently accurate for many applications. Greater accuracy then demands either longer data words or *serial processing* of two or more words for each quantity.

The latter strategy must be employed in a small computer and involves considerable "overhead" time for the extra word transfers and the carry-bit handling. Consequently, the time required for an 8-bit microcomputer to execute the addition of two numbers encoded as 32 bits each will be much more than four times that of a mainframe computer with 32-bit words. Large computers often have other hardware provisions that further increase their effective speeds. For instance, the CPU might contain separate units dedicated to such specific tasks as addition/subtraction, multiplication, and floating-point conversion. This configuration permits *parallel processing,* in which several operations are performed simultaneously rather than one after the other.

The length of the *address words* also affects execution time, as well as usable *memory capacity.* If, for example, a microprocessor were limited to 8-bit address words then, regardless of actual memory capacity, there could be only 256 addressable storage locations because each location must have a unique address. And unlike data processing, serial addressing using two or more sequential words is not a practical solution because it unduly increases execution time and necessitates additional hardware in the storage section. For these reasons, an 8-bit microprocessor usually has 16-bit address words, permitting a memory capacity of $2^{16} = 65{,}536$ data and instruction words. But computer programs often store address words as part of their instructions, so a 16-bit address takes up two 8-bit locations and involves extra steps when addresses are transferred from memory to the microprocessor via the data bus. Such problems are not as serious in a large computer that has, say, 24-bit address words that fit within the 32-bit word format.

In view of these considerations, microcomputers are limited to relatively low-speed peripheral devices and to simple programs that involve small amounts of data. On the other hand, the low cost and programmability of microcomputers make them well suited to such *special-purpose applications* as instrumentation systems, industrial controllers, and electronic cash registers, where the size and price of a larger computer would be prohibitive. For *general-purpose applications,* a mainframe computer accommodates all types of peripherals and the gamut of computing languages for scientific and business purposes. Such a large system obviously should be kept operating at near capacity around the clock to justify the capital outlay for it. Intermediate price and performance characterize the extensive family of minicomputers. A small minicomputer has sufficient capability to handle banking terminal transactions or real-time experimental data analysis, for instance, whereas a larger model would better serve the computing needs of a laboratory or small business.

Although we've cited differences between micro-, mini-, and mainframe computers, their dividing lines are becoming less and less distinct as IC technology continues to increase the performance and decrease the price of digital hardware. The 16-bit microprocessor already bridges the gap between micros and minis, and 32-bit units show potential for larger computers. Going in the other direction, a mass-produced 4-bit microprocessor plus memory and I/O interface, all on one IC chip, now costs so little that it can be built into a variety of consumer products such as electronic ovens, television games, and automobile ignition systems. The versatile microprocessor therefore deserves further attention.

MICROPROCESSORS

Microprocessors come in several word lengths and have different capabilities, depending upon the number of registers, the amount of on-chip memory, and the intended applications. Here we introduce general microprocessor characteristics by describing the internal structure and functions of a simplified hypothetical 8-bit unit. This treatment should provide sufficient background for you to be able to pursue the details of a particular model. It will also shed some light on the workings of a CPU in a larger computer and the relationship between a computer program and machine operations.

The structure or *architecture* of our hypothetical microprocessor is given by the block diagram in Fig. 16.4–3. Three buffers interface with the system bus, and a two-way internal bus carries data and instructions between the arithmetic/logic section on the left, the control section in the middle, and the register stack on the right. But it must be emphasized at the start that very few of the units have direct or "hard-wired" connections to other units. Instead, interconnections take the form of *gate circuits* that the control section *enables* or *disables,* according to the program instructions. To put this another way—programming at the machine-language level boils down to *wiring with software*.

As to the specific units in Fig. 16.4–3, the arithmetic/logic section consists of an ALU similar to Fig. 16.3–7, an *accumulator* (A), and a *status register*. One input word for the ALU always comes from the accumulator, which also holds the output upon completion of the operation. The status register contains condition-code bits or *flags* that indicate when the result of an ALU operation generated a carry ($C = 1$), was

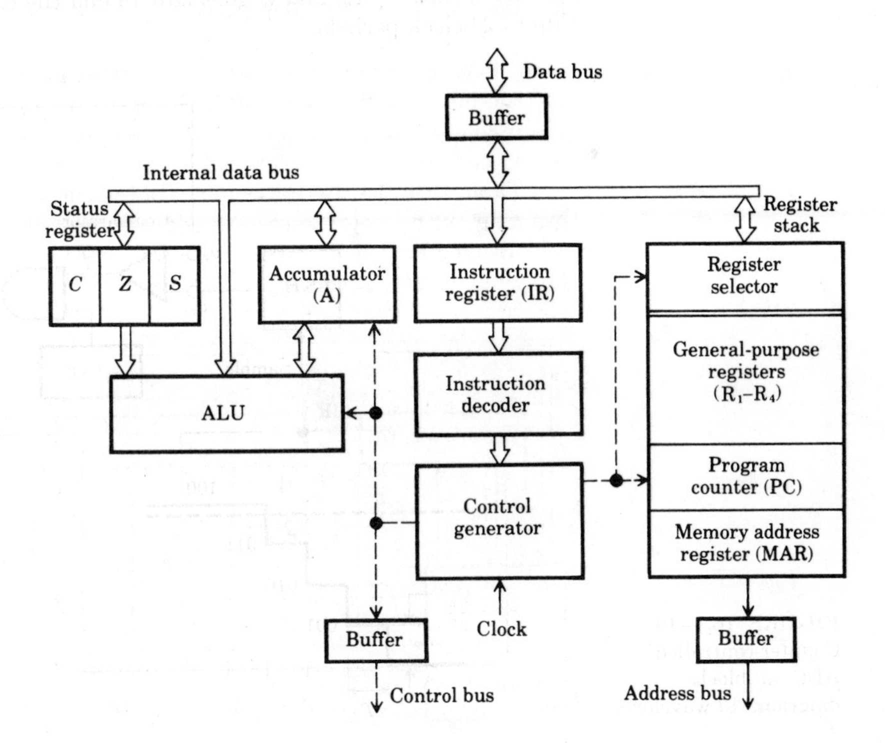

FIGURE 16.4–3 Hypothetical microprocessor.

equal to zero ($Z = 1$), or was negative ($S = 1$). The control section includes an *instruction register* (IR) for holding instruction words, a *decoder* that interprets those words, and a *control generator* that produces the corresponding control signals in synchronism with the clock. The register stack has four *general-purpose registers* (R_1–R_4) and a *selector* that permits data transfers to or from any register. There are also two special registers: a *program counter* (PC) to keep track of the program sequence, and a *memory address register* (MAR) that feeds the address bus.

For simplicity's sake, we assume a memory capacity of just 256 locations so the MAR and PC will be 8-bit registers. All other registers also hold 8 bits, except the status register. An actual 8-bit microprocessor has additional status bits, other special registers, and a PC and MAR that each holds 16 bits.

Now consider a hypothetical instruction word such as 10110001, whose first two bits might signify an ALU operation, the next three bits identify the operation to be addition, and the last three bits mean that the word to be added to the accumulator word resides in register R_1. When this instruction is transferred from memory to the instruction register, the decoder and control generator enable gates that yield the connections of Fig. 16.4–4. The ALU then adds the words in A and R_1, and the sum appears in A at the next clock pulse. Similarly, another instruction causes the A word to be copied into R_2, a "store" operation, or vice versa for a "load" operation. But loading or storing a memory word requires *two* instructions, the first to identify the operation and the second to provide the memory address of the word in question. (A third instruction word would be needed to complete a 16-bit address.)

Table 16.4–1 lists the sequence of instructions in an illustrative program for our microprocessor. All addresses are given as decimal numbers, and instructions have been expressed in English. The actual machine language would, of course, consist entirely of binary digits. The purpose

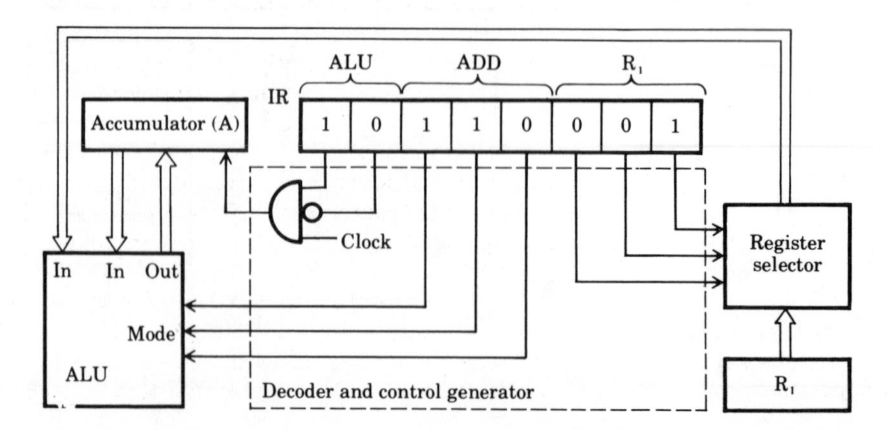

FIGURE 16.4–4

TABLE 16.4–1 Microprocessor Program

Address	*Instruction*	*Clock cycles*	*Comment*
000	Clear A	4	Puts *SUM* = 0 in R_1.
001	Store A in R_1	5	
002	Load R_2 from address	4	Puts *COUNT* = *N* in R_2.
003	050	3	
004	Load R_3 with	4	Puts *NADR* = 051 in R_3.
005	051	3	
006	Load R_4 from address in	4	Puts *NUMB* in R_4.
007	R_3	6	
008	Load A from R_1	5	Adds *NUMB* to *SUM*.
009	Add A and R_4	5	
010	Store A in R_1	5	
011	Increment R_3	5	Adds 1 to *NADR*.
012	Decrement R_2	5	Subtracts 1 from *COUNT*.
013	Load A from R_2	5	Loops back to instruction 006 if *COUNT* $\neq$ 0.
014	Jump if A $\neq$ 0 to address	4	
015	006	4	
016	Store R_1 at address in	4	Puts result in memory.
017	R_3	6	

of this program is to sum N numbers stored in memory locations 051 to $051 + N - 1$ and place the result in location $051 + N$, the value of N being found in location 050. The program uses the general-purpose registers for the following variables: R_1 holds *SUM*, the sum in progress; R_2 holds *COUNT,* which counts down from N to zero; R_3 holds *NADR*, the address of the next number to be added to *SUM*; and R_4 holds *NUMB*, the number addressed in R_3. Instructions 000–005 initialize the program variables; instructions 006–015 constitute the summing loop; and instructions 016–017 place the result in memory.

With the help of the program description and the comments in the table, you should be able to follow the sequence of instructions. But bear in mind the distinction between the *address* of a memory location and the *contents* of that location! For example, instruction 002 calls for R_2 to be loaded with the contents of the location whose address is given in instruction 003, namely at location 050. We call this a *direct address* instruction, as contrasted with the *indirect address* sequence in instructions 006–007 where the address of the data to be loaded is found in R_3. (Some microprocessors have a special *index register* for the role played by R_3 in this program.) Instruction 004 illustrates *immediate addressing,* since the word to be loaded in R_3 is identical to the next instruction word.

Now we turn to the actual *execution* of this program. The instructions first must be stored in memory at the locations shown in Table 16.4–1, and the data must be stored in locations 050 to 051 + N. Pressing the START button sets the program counter to 000 and initiates the sequence of operations listed in Table 16.4–2. At the first clock pulse, the control unit transfers the contents of the program counter to the memory address register and increments the program counter, as symbolized by PC → MAR and PC + 1 in the table. At the next clock pulse, the word addressed in the MAR is transferred to the instruction register, symbolized by (MAR) → IR, so the IR now holds the binary word meaning "Clear A." This instruction is decoded and carried out during the next two clock pulses. Complete execution of instruction word 000 requires a total of 4 *clock cycles*. Proceeding similarly, instruction 001 takes 5 clock cycles, while instructions 002–003 require 7 cycles.

Note that the second cycle of instruction 003 is not a decode operation because the previously decoded word told the control unit to look for an address in the next word. By means such as this, the microprocessor distinguishes between instructions, data, and addresses, even though they are all just strings of binary digits stored at various memory locations. Also note how the "jump" in instructions 014–015 resets the PC to 006 if the status register has $Z = 0$, meaning that $A \neq 0$.

Table 16.4–2 does not give the complete breakdown of all the operations in our simple program, but the clock cycles required for each instruction have been listed in Table 16.4–1. Since the loop is repeated N times, once for each number added, program execution takes a total of $33 + N \times 48$ clock cycles. Therefore, if the clock frequency is 1 MHz and $N = 200$, the program execution time will be $(33 + 200 \times 48)/1\text{ MHz} = 9633\ \mu s \approx 10$ ms.

Because we have been dealing with a *hypothetical* microprocessor, the program instructions, execution sequence, and clock-cycle requirements differ in various respects from those of an *actual* microprocessor. Nonetheless, Tables 16.4–1 and 16.4–2 correctly imply that even a very simple program becomes quite detailed at the machine-language level. When you write a program in a higher-level language such as BASIC or FORTRAN, it is first processed by another program called a *compiler* that translates your program statements into a sequence of basic operations represented by binary code words. The compiler thereby spares you most of the tedious bookkeeping involved in wiring with software.

But like any other program, the compiler must be stored in memory, and might use up so much space that there would be insufficient memory left for your encoded program. In that event, you must resort to *assembly language programming* along the lines of Table 16.4–1, which can be translated into machine language by a much shorter program called the *assembler*. We therefore see that the software capability of a microcom-

TABLE 16.4–2 Microprocessor Program Execution

Address	Clock cycle	Operation	Register contents PC	MAR	IR	A	R_1	R_2
000	1	PC → MAR, PC + 1	001	000	—	—	—	—
	2	(MAR) → IR			"Clear A"			
	3	Decode						
	4	Execute CLEAR				0		
001	1	PC → MAR, PC + 1	002	001				
	2	(MAR) → IR			"Store A in R_2"			
	3	Decode						
	4	Select R_1						
	5	Execute STORE					0	
002	1	PC → MAR, PC + 1	003	002				
	2	(MAR) → IR			"Load R_2 from"			
	3	Decode						
	4	Select R_2						
003	1	PC → MAR, PC + 1	004	003				
	2	(MAR) → MAR		050				
	3	Execute LOAD						N
⋮	⋮	⋮	⋮	⋮	⋮	⋮	⋮	⋮
014	1	PC → MAR, PC + 1	015	014	—	COUNT	SUM	COUNT
	2	(MAR) → IR			"Jump if A = 0 to"			
	3	Decode						
	4	Select Z						
015	1	PC → MAR, PC + 1	016	015				
	2	(MAR) → MAR		006				
	3	Execute Z TEST						
	4	MAR → PC if Z = 0	(006)					
⋮	⋮	⋮	⋮	⋮	⋮	⋮	⋮	⋮

puter, or any other computer, depends on the memory capacity. It's appropriate, then, that this section concludes with a brief consideration of memory units.

MEMORY UNITS

The main memory of a computer usually contains two kinds of units: a *read/write random-access memory* (RAM) and a *read-only memory* (ROM). Both the RAM and ROM permit access to any desired location without passing through other locations. (This contrasts with the sequentially accessed locations in a shift-register memory or magnetic tape recording.) A ROM differs from a RAM in that the CPU or microprocessor can read the contents of a location but cannot write a new word into any location. A ROM is, thus, designed for *permanent* storage of frequently used programs and data such as assemblers, code converters, and look-up tables. The more expensive RAM *temporarily* holds the instructions and data to be read for a user's program and provides a place to write the results.

Whether a memory unit is a ROM or a RAM, its three important parameters are its size, organization, and access time. The *size* simply equals the maximum number of bits that can be stored, while *organization* reflects the number of words and bits per word. To clarify, let's suppose a certain unit stores up to 2^{13} bits organized into 8-bit or one-byte words, allowing for $2^{10} = 1024$ words and requiring a 10-bit address. Since $1024 \approx 1000$, we let 1 K stand for 1024 and say that this unit holds $1\text{ K} \times 8$ bits or 1 K-byte. The time required to complete a read or write operation is called the *access time.* Access times range from about 400 ns for a MOS unit down to 40 ns for a BJT unit.

As an introduction to memory circuitry, consider the diode matrix diagrammed in Fig. 16.4–5a, a primitive 4 word × 3-bit ROM. An address word A_1A_0 applied to the decoder gates causes one of the *word lines* (W_0, W_1, . . .) to become HIGH while the others remain LOW. Any diode connected from the HIGH word line to an output *bit line* then conducts and makes that bit HIGH. Figure 16.4–5b is the truth table for the diode connections shown. Like all ROMs, the output bit pattern $B_2B_1B_0$ is permanent and *nonvolatile,* meaning that the contents are not lost if the power is turned off.

Clearly, the word length is easily increased by adding more bit lines and diodes, but increasing the number of words requires a longer address word, as well as more word lines and decoder gates. Thus, a 256-word ROM needs an 8-bit address and $2^8 = 256$ gates. Alternatively, a *two-dimensional addressing* scheme like that shown in Fig. 16.4–6 could be used to save gates. Here, the first 6 address bits are applied to a decoder with $2^6 = 64$ gates and word lines. Each word line crosses 32 bit lines that are divided into 4 groups. The remaining two address bits are applied to a bank of 4-to-1 selectors to choose a particular 8-bit output word.

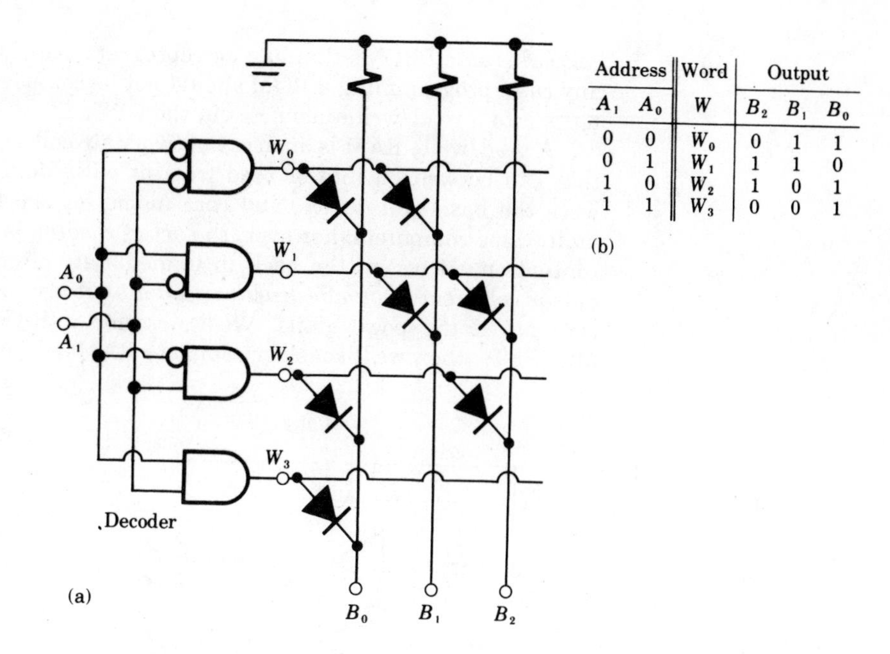

Address		Word	Output		
A_1	A_0	W	B_2	B_1	B_0
0	0	W_0	0	1	1
0	1	W_1	1	1	0
1	0	W_2	1	0	1
1	1	W_3	0	0	1

FIGURE 16.4–5
Diode matrix ROM with address decoder.

Going back to Fig. 15.2–8 shows us that each selector needs 5 gates, for a total of $64 + 8 \times 5 = 104$ gates rather than the 256 gates for one-dimensional addressing.

Mass-produced ROMs with contents preprogrammed by the manufacturer have low unit costs, but a single custom-made unit would be quite expensive. For a *user programmable* ROM (or PROM), the manufacturer locates a diode link or the equivalent at each matrix intersection so users can burn out the ones they don't want. Having once been programmed, a PROM becomes a ROM. EPROMs and EAROMs are *erasable* or *electri-*

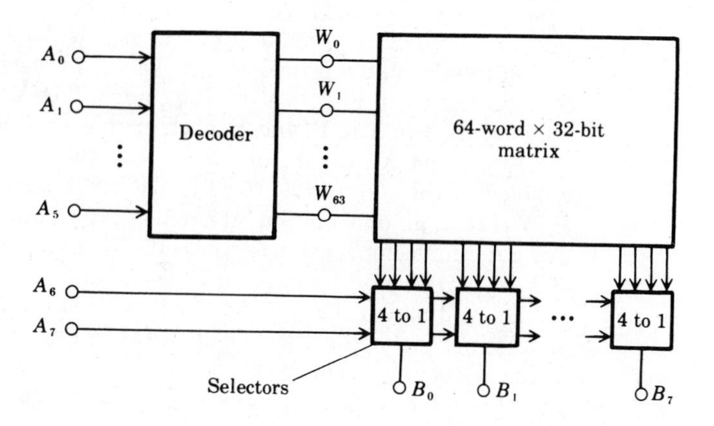

FIGURE 16.4–6
256 word × 8 bit ROM with two-dimensional addressing.

cally alterable PROMs that can be reprogrammed if the user desires. In any case, programming a ROM should not be confused with writing contents into a read/write memory via the CPU.

A read/write RAM is an array of two-state cells, similar to a flip-flop, that can be written into or read from at will. *Magnetic cores* formerly were the basic cell device, and core memories are still found in some mainframe computers; however, the act of reading is *destructive* and the contents must be written back in immediately after readout. *Semiconductor* cells feature *nondestructive readout* but are *volatile* and lose their contents if the power fails. We'll describe a RAM cell composed of MOSFETs; then we'll consider complete RAMs.

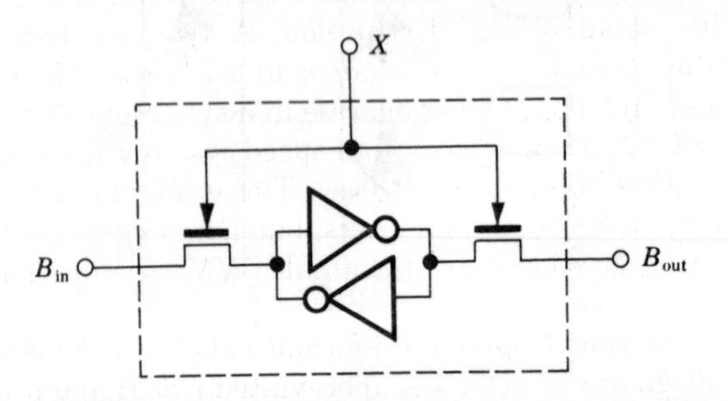

FIGURE 16.4–7
MOS memory cell.

Figure 16.4–7 is the circuit diagram of a *static* MOS cell whose cross-coupled inverters form a bistable flip-flop. The cell is enabled by making X go HIGH so the FET switches conduct; this allows B_{in} to set or reset the flip-flop, producing $B_{out} = \overline{B}_{in}$. The switches become open circuits when X goes LOW, but the flip-flop retains its state for subsequent readout. Complete implementation of this cell requires six transistors, including two for each inverter. Using capacitance for information storage reduces the requirement to three transistors per cell and permits more memory per chip. Such cells are said to be *dynamic,* and must be *refreshed* periodically to make up for charge leakage from the capacitor.

Now let's build a primitive 2 × 2 array or *memory plane* using static cells, as shown in Fig. 16.4–8. The control bits X_0 and X_1 each enable a *row* of cells, whereas Y_0 and Y_1 each enable a *column*. Thus, if X_0 and Y_1 are HIGH, and X_1 and Y_0 are LOW, only the cell marked X_0Y_1 will be enabled. If the *chip enable* (CE) is also HIGH while the *read/write select* $R/\overline{W}$ is LOW, an information bit can be written into cell X_0Y_1 through the NOT gate that compensates for the cell's inversion. Similarly, the contents of X_0Y_1 can be read by making $R/\overline{W}$ go HIGH.

Expanding our array to 64 × 64 and adding decoder gates for the X and Y lines yields the 4 K × 1-bit RAM in Fig. 16.4–9. A 12-bit address word selects any one of the $2^{12} = 4 \times 1024 = 4$ K locations, each holding 1 bit. For a 4 K × 8-bit RAM, we could put 8 chips in a row and apply the

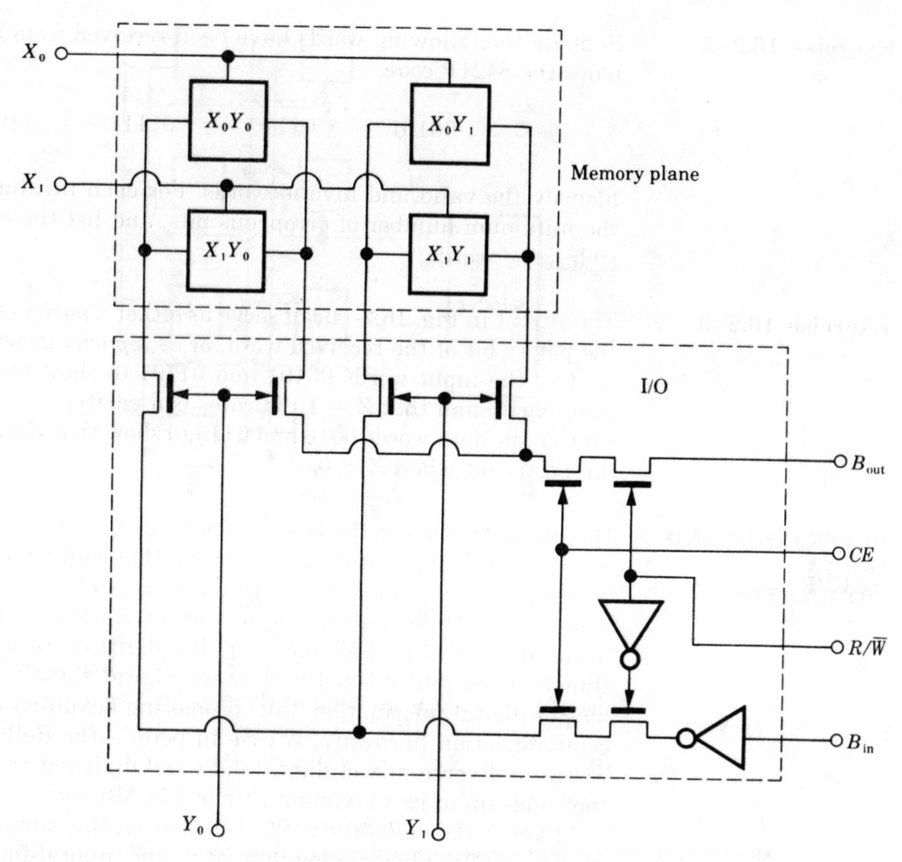

FIGURE 16.4–8
Four-bit RAM.

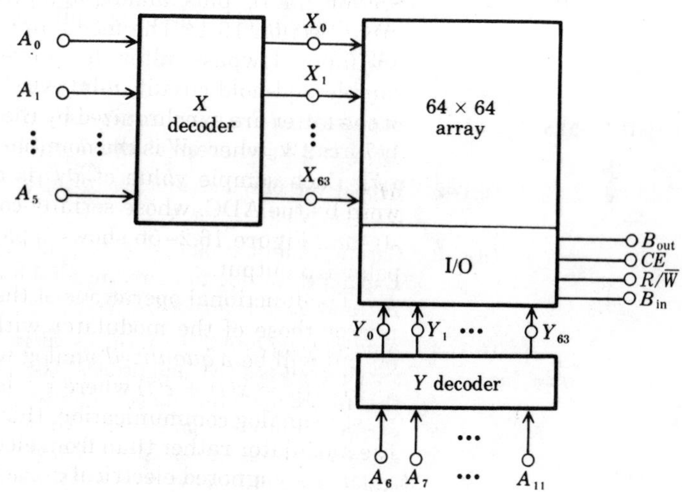

FIGURE 16.4–9
4 K × 1 bit RAM.

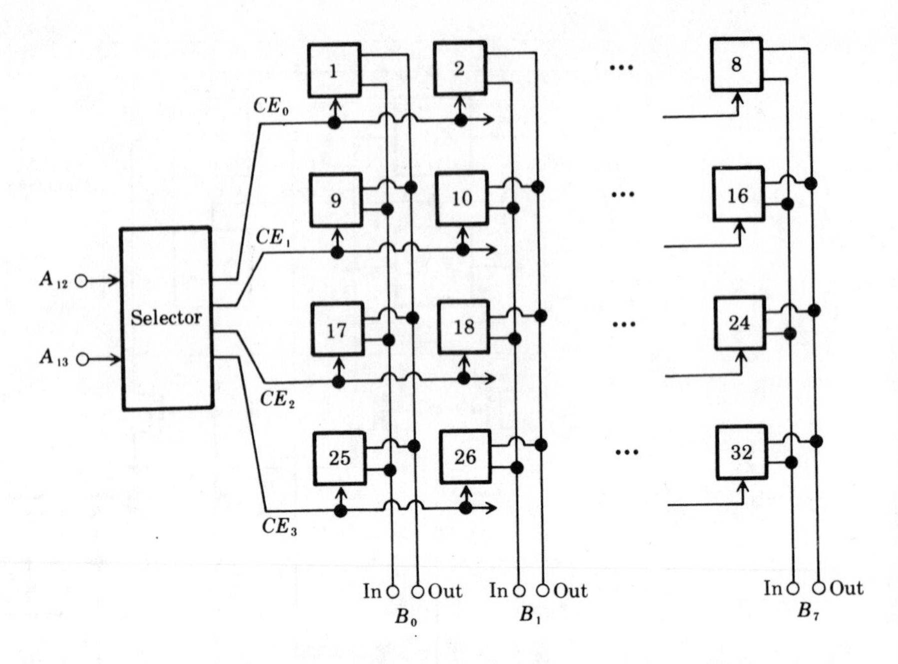

FIGURE 16.4–10
16 K × 8 bit RAM.

same address word to all of them. Each chip then provides the location for one of the 8 bits from each of the 4 K words.

Finally, Fig. 16.4–10 represents an array of 32 chips organized into a 16 K × 8-bit RAM. All chips receive the first 12 address bits (not shown) while A_{12} and A_{13} are decoded to select a *row* of chips via the *CE* terminals. This arrangement stores a total of 2^{14} words × 8 bits per word = 131,072 bits (or 128 K-bits). The astonishing advances of IC technology have made it possible to put all the circuitry of Fig. 16.4–10 on just one chip!

PROBLEMS

16.1–1 Suppose v_a in Fig. 16.1–2a represents the output of a temperature sensor. Modify the circuit to sound an alarm ($X = 1$) when $v_a > 10$ V or $v_a < 3$ V.

16.1–2 Repeat the previous problem with the additional provision to turn on a heater ($H = 1$) when $v_a < 7$ V.

16.1–3 Careful examination of Fig. 16.1–2b reveals that $B_1 = Q_2$ and $B_0 = Q_3 + \overline{Q}_2 Q_1$. Expand this table to represent a 3-bit converter, and obtain expressions for B_2, B_1, and B_0 in terms of Q_1–Q_7.

‡ **16.1–4** A *pulse-height analyzer* receives, at its input, a nonperiodic train of pulses with varying but positive amplitudes. The analyzer includes a bank of ripple counters that record the total number of pulses and the number of pulses whose amplitudes exceed various threshold levels. Draw the diagram and pertinent waveforms for a system that has thresholds at 2, 4, and 6 V. Use another comparator and a monostable (Fig. 11.3–7) to generate the sampling and timing pulses.

16.1–5 The *Gray code* in Fig. 16.1–4 partly compensates for the fact that the brushes on the shaft encoder have nonzero width. Suppose, for instance, that the brushes have an angular width of 5°; the voltage at the B_1 terminal will then be HIGH when $85° < \theta < 275°$ rather than $90° < \theta < 270°$. Make a table listing $B_2B_1B_0$ as they actually appear at the brush terminals, the corresponding quantized angle θ_q, and the error $\theta - \theta_q$ when $\theta = 86°$, 180°, and 274°. Repeat this listing assuming the segments are arranged in a standard binary pattern. For both cases, take θ_q at the midpoint of each quantization interval.

16.1–6 Consider an n-bit DAC built with an R-$2R$ ladder in which $R = 10$ kΩ. Find n, V_R, and R_F so that $\Delta v_q = 0.02$ V, $v_{q_{max}} \approx 5$ V, and $R_F \approx 25$ kΩ.

16.1–7 Repeat the previous problem with $v_{q_{max}} \approx 20$ V.

16.1–8 If V_R in Fig. 16.1–6 is replaced by a time-varying analog voltage $v_{in}(t)$, we obtain a *multiplying* DAC or *programmable attenuator* with output $v_q(t) = -Kv_{in}(t)$, where the multiplying factor K depends on $B_3B_2B_1B_0$ and could be controlled by computer. Find R_F/R such that $K_{max} = 1.2$. What are the other possible values of K?

16.1–9 Let the control signal for the switch in Fig. 16.1–5a be S_1S_0 such that S connects to ground when $S_1S_0 = 00$, to v_a when $S_1S_0 = 01$, and to $-V_R$ when $S_1S_0 = 11$. Devise a complete switch circuit using CMOS gates (like Fig. 16.1–7) for the connections.

‡ **16.1–10** Construct a timing diagram that shows all the signals pertinent to the control unit in Fig. 16.1–5a, including the switch control S_1S_0 as described in the previous problem. Then devise an implementation of the control unit using two *SR* flip-flops, a monostable, and other gates. You can assume that the duration of the *START* pulse is slightly longer than T and that the overflow signal is a short pulse like the *CARRY* in Fig. 15.3–10b.

16.1–11 A *tracking* ADC differs from Fig. 16.1–10a in that the counter starts from its previous value and counts up or down, depending on the state of Q. Modify Fig. 16.1–10a to implement this strategy using an up/down counter with a control input *UP/DOWN*. Then modify Fig. 16.1–10b to explain why v_r never reaches a constant value but "hunts" above and below v_{as}.

16.1–12 A digital processor edits questionable data as follows: It keeps any data point whose value falls within ±50% of the average of the previous two points; otherwise, it replaces the value with the average of the previous two. Write a set of equations relating the processor's output sequence y_k to the input x_k.

16.1–13 Repeat Problem 16.1–12, assuming the processor compares a given data point with the values just before and after it. Do not use x_{k+1} to compute y_k. (Why?)

16.1–14 The tapped-delay-line equalizer in Fig. P14.2–9 is to be replaced by a digital processor with input $x_k = v(kT)$ and output y_k. The time delay t_Δ equals five times the sampling period T. Obtain the expression for y_k.

16.1–15 Obtain an expression similar to Eq. (6a) that approximates a first-order highpass filter whose differential equation is $\tau\, dy/dt + y = \tau\, dx/dt$.

16.1–16 The second derivative d^2y/dt^2 may be approximated by $(y_k - 2y_{k-1} + y_{k-2})/T^2$. Use this expression to approximate Eq. (4b), Sect. 5.3, with $y = i$ and $x = v$.

16.2–1 How many 300-b/sec data signals can be multiplexed on a 3-kHz line if the composite binary waveform is to be self-synchronizing?

16.2–2 A data source produces 600 b/sec. An error-control bit is inserted after every 3 data bits and the composite signal is represented by an RB waveform. What is the minimum required bandwidth?

16.2–3 A digitized TV signal has $r = 90$ Mb/sec. If it is converted to a multilevel NRZ waveform, how many levels would be required to reduce the transmission bandwidth to 12 MHz?

16.2–4 Give a simple argument, based on the Fourier-series theorem, that supports the assertion that binary signaling cannot have a rate $r > 2B$.

16.2–5 If one of the data bits in the ASCII code is reserved for parity checking, how many special characters can be represented in addition to the decimal digits and the lower- and upper-case letters of the alphabet?

16.2–6 A certain code consists of separate words for each decimal digit and each letter of the alphabet (upper-case only). Each word has one parity bit for every two data bits. How many bits are there per word, and how many unassigned words?

16.2–7 A crude *error-correcting code* consists only of the words 000 and 111, representing the data bits 0 and 1. Explain how the receiver should generate a "corrected" bit from each incoming word. Under what conditions will the "corrected" bit be incorrect?

16.2–8 Suppose 16 data bits are arranged in a square array, with a parity bit added at the end of each row and at the bottom of each column. The transmitted code word then consists of 16 + 4 + 4 = 24 bits. Explain how parity checking at the receiver can correct a single error per word, and detect most double errors.

16.2–9 A signal with $W = 4$ kHz is transmitted by PCM on a system having $B = 50$ kHz. What is the maximum possible value of S/N_q?

16.2–10 Repeat the previous problem with $W = 12$ kHz.

16.2–11 Suppose a binary PCM signal is converted, before transmission, into a multilevel NRZ waveform with eight levels. Find the minimum value of B/W so that $S/N_q \geq 4000$.

‡ **16.2–12** Suppose $x(t)$ in Fig. 16.2–6 consists of sinusoids with amplitudes not greater than A_x and frequencies not greater than W. Obtain the condition on f_s to avoid slope overload if $\Delta x_r = A_x/10$.

16.3–1 Use Eq. (1), Section 15.2, to verify the following X-OR relations:

(a) $A \oplus B = B \oplus A$ (b) $B \oplus 1 = \overline{B}$

(c) $\overline{A \oplus B} = A \oplus \overline{B} = \overline{A} \oplus B$

(d) $A \oplus (B \oplus C) = (A \oplus B) \oplus C$ Hint: Use (c).

16.3–2 Use the results of Exercise 16.3–1 to implement a full adder entirely with NAND gates.

16.3–3 A *look-ahead-carry adder* minimizes carry-ripple delay by forming the carries directly from the input bits. Devise a gate circuit that adds $A_2A_1A_0$ and $B_2B_1B_0$ with direct carry generation. Use the fact that $C_k = A_kB_k + (A_k + B_k)C_{k-1}$ to ar-range your circuit so the longest path from any input terminal to any output goes through not more than four gates.

16.3–4 Suppose a sequence of no more than nine BCD integers is to be summed. The BCD words are available at the output terminals of a register like Fig. 15.3–14, and a new one appears immediately after each load pulse. Draw the block diagram of a system that accomplishes this task and includes two BCD adders.

16.3–5 Construct the truth table and gate circuit for a *half subtractor* with input digits X and Y. The outputs are the difference $D = |X - Y|$ and the borrow B, which equals zero unless $Y > X$.

16.3–6 Perform the calculations $8 - 16$ and $-8 - 8$ in 5-bit 2s-complement notation to verify that -16 is correctly represented by $1 * 0000$.

16.3–7 The adder/subtractor in Fig. 16.3–6 produces incorrect results, because of overflow, if the sum falls outside the range -16 through $+15$. Devise a gate circuit with inputs C_* and C_3 that generate $X = 1$ to indicate overflow. Hint: You can infer the general rule by performing the calculations $9 + 7, 9 + 6, 9 - 7, 7 - 9, -9 - 7$, and $-9 - 8$.

16.3–8 The 1s *complement* of a signed BCD integer is obtained simply by complementing all bits, so that -9 is represented as $\overline{0*1001} = 1*0110$. If a carry overflow results when you add two such numbers, it must be added to the least significant bit. This is called an *end-around carry;* it never changes the sign bit of the sum.

(a) Repeat Exercise 16.3–3 using 1s-complement arithmetic.

(b) Modify Fig. 16.3–6 for operation with 1s-complement numbers. Hint: You will need additional half adders.

16.3–9 Suppose a fourth mode bit M_3 is incorporated into the ALU in Fig. 16.3–7 such that A_k becomes $\overline{A}_k$ when $M_3 = 1$. Augment the function table, and identify any duplicate functions. (See Problem 16.3–1.)

16.3–10 Suppose the signed 2s-complement words $D_* D_3 D_2 D_1 D_0$ and $E_* E_3 E_2 E_1 E_0$ are to be multiplied by the system in Fig. 16.3–8. Devise a gate circuit that produces the correct magnitude word $A_3 A_2 A_1 A_0$ from the D word, and another circuit that produces the sign bit P_* of the product. Hint: Use half adders and X-OR gates.

16.3–11 Set up the multiplication of $A_1 A_0$ times $B_1 B_0$ in symbolic form, and devise a gate circuit for the product $P_3 P_2 P_1 P_0$. Hint: Use two half adders.

PART IV

The stator of a 3600-rpm steam turbine generator rated for 640 MVA at 22 kV. (Courtesy of General Electric)

ENERGY SYSTEMS

[illegible] from the first full adder onward, moving from right to left. [illegible] the sum of the two 4-bit words is [illegible] as illustrated in the previous example. This produces [illegible] of the particular system is limited to 4-bit words [illegible] could be expanded to [illegible]

[illegible] is called a *parallel binary adder* because all [illegible] simultaneously. However, it does take a finite amount of time for carries to propagate from right to left, a phenomenon known as the *ripple-carry problem*. The more complicated look-ahead-carry [illegible] overcome this problem and attain greater speed, typically taking [illegible] Alternatively, [illegible]

[illegible] computation time one could use the serial adder of Fig. [illegible] which involves just one FA and a D flip-flop for one-bit time delay. [illegible] the input bits A and B are entered serially from shift registers [illegible] sum bits S being delivered to an output shift register [illegible] synchronous timing provided by the clock.

[illegible] circuits that operate on BCD words, as in a calculator, must [illegible] structure. To demonstrate why, consider the following additions in BCD:

$$\begin{array}{r} 0110 \\ +0011 \\ \hline 1001 \end{array} \qquad \begin{array}{r} 6 \\ +3 \\ \hline 9 \end{array} \qquad \begin{array}{r} 0110 \\ +0111 \\ \hline 1101 \end{array} \qquad \begin{array}{r} 6 \\ +7 \\ \hline 13 \end{array}$$

The first result is correct but the second is wrong because 13_{10} is represented in BCD by two words, namely 0001 0011, not 1101. [illegible] 1101 is not even a proper code word (see Table [illegible]).

These difficulties stem from the fact that BCD uses only 10 of the 16 possible 4-bit words, and the remaining 6 words are forbidden or meaningless. To compensate, the adding circuit must detect whenever a sum exceeds [illegible]. When that happens, the circuit must add $6_{10} = 0110$ to the [illegible] the 6 forbidden words and to produce a carry. Figure [illegible] diagrams a parallel adder with these provisions.

Exercise [illegible] Construct a truth table for the full adder. Then show that $S = [illegible]$ and that $C_{out} = AB + AC_{in} + BC_{in}$.

17

MAGNETICS AND ELECTRO-MECHANICS

Electric *energy systems* are designed to generate, transmit, convert, or control large amounts of energy. Such systems often involve devices whose operation depends on a magnetic field, a topic alluded to in our earlier discussion of inductance, but not given explicit consideration. Now we must give direct attention to the magnetic field, its relationship to circuit elements, and its role in energy conversion.

We start with an overview of magnetics and electromechanics, including magnetic quantities, induction phenomena, and stored energy. We then examine magnetic circuits and the influence of magnetic materials. These concepts are combined to develop the principles of various electromechanical transducers.

OBJECTIVES

After studying this chapter and working the exercises, you should be able to do each of the following:

- Name the units and state the relationships for magnetic flux, flux density, magnetomotive force, and reluctance (Sections 17.1 and 17.2).
- Calculate the induced voltage and force for a current-carrying conductor in a uniform magnetic field (Section 17.1).

- Given a simple magnetic circuit with or without an air gap, find the flux, stored energy, and inductance (Section 17.2).
- Describe the nonlinear effects and losses in a magnetic circuit (Section 17.2).
- Analyze a simple electromechanical transducer that has a rotating or translating coil or a moving-iron member (Section 17.3).
- Use the impedance method to determine the dynamic behavior of a transducer (Section 17.4). †

17.1 MAGNETICS AND INDUCTION

A magnet has the ability to exert force on another magnet without any mechanical connection between them. We visualize this force-at-a-distance effect in a terms of a *magnetic field* in the space surrounding the magnet. Such fields are created not only by permanent magnets but also by moving charges or, more significantly, by currents. Furthermore, a current-carrying wire in a magnetic field experiences a *magnetic force* that can be used to do mechanical work. Many important devices accomplish electrical to mechanical energy conversion, thanks to magnetic forces.

Besides force, a magnetic field produces *induced voltage* in a moving wire; in fact, force and voltage always go hand in hand when a conducting body travels through a magnetic field. This phenomenon forms the basis for electrical generators and related devices that carry out mechanical to electrical energy conversion. Even a fixed conductor experiences induced voltage if the magnetic field around it changes with time, an effect that gives rise to inductance in electrical circuits and is exploited in electrical transformers and certain electromechanical devices.

This section introduces the concept of the magnetic field and its relationship to force and induced voltage. However, since magnetic field is a *vector* quantity, a complete development of the topic requires mathematical techniques beyond the scope of this textbook. We will therefore limit our discussion to a few specific configurations that occur frequently in engineering applications. (You may wish to refer to a physics text for the omitted details.) Subsequent sections will then apply our results to practical electromagnetic and electromechanical devices.

MAGNETIC FIELD AND FLUX

Figure 17.1–1a qualitatively depicts the magnetic field around a simple bar magnet. Lines emanating from the magnet's north pole and returning to the south pole represent the direction of the field at various points in space. Thus, for instance, a compass needle or other small magnet placed somewhere in the field tends to become tangent to the field line at that point. The alignment force is strongest close to the magnet

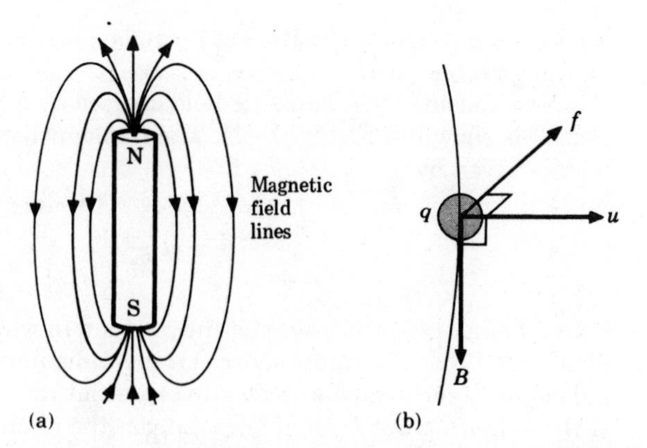

FIGURE 17.1–1
(a) Magnetic field.
(b) Force on a moving charge.

and decreases with distance, as represented by the spacing of the field lines.

Quantitatively, we describe the strength and direction of a magnetic field by a vector called the *flux density*. The flux-density magnitude B is proportional to the force exerted by the field on a charge at the point in question and corresponds to the line spacing (hence the name "density"). If a charge q moves with velocity component u perpendicular to the field, as in Fig. 17.1–1b, the charge experiences a force of magnitude

$$f = Buq \tag{1}$$

Note carefully that this force is not in the same direction as the field but, instead, is mutually perpendicular to both u and B so the directions of u, B, and f are related to each other like the directions of a right-hand coordinate system.

The unit of flux density is the *tesla* (T) which, from Eq. (1), equals force in newtons divided by charge times velocity; that is, $1\text{ T} = 1\text{ N}/(\text{C}\cdot\text{m/sec})$. Equivalently, and more conveniently for later work, we express this in terms of current in amperes by writing

$$1\text{ T} = 1\text{ N/A}\cdot\text{m}$$

having used $1\ C\cdot\text{m/sec} = 1\ (\text{C/sec})\cdot\text{m} = 1\text{ A}\cdot\text{m}$. Whenever flux density is uniform and perpendicular to some cross-sectional area A—as approximated at the ends of the bar magnet in Fig. 17.1–1a—we say that the total *flux* ϕ passing through that area is

$$\phi = BA \tag{2}$$

measured in *webers* (Wb), where

$$1\text{ Wb} = 1\text{ T}\cdot\text{m}^2 = 1\text{ N}\cdot\text{m/A}$$

Flux is not a vector quantity, but it can be positive or negative depending on the direction of B.

Now consider the magnetic field created by a long wire carrying current i, as shown in Fig. 17.1–2a. The flux density at distance d from the wire is given by

$$B = \mu \frac{i}{2\pi d} \tag{3}$$

with μ being the *permeability* of the medium in which the field exists. The direction of B at the points shown is perpendicular out of or into the page, indicated by the symbols • and × to represent the "tip" and "tail feathers" of the vector arrow. It then follows from the circular symmetry that the field lines form closed circles around the wire, as seen in the endwise view of Fig. 17.1–2b where ⊗ symbolizes the *current arrow* going into the page. The clockwise rotational direction of B relative to i may be visualized with the help of the *right-hand rule:*

> Point your right thumb in the direction of the current and your fingers curl in the direction of the magnetic field.

The radially increasing spacing of the circles in Fig. 17.1–2b reflects that flux density decreases as $1/d$.

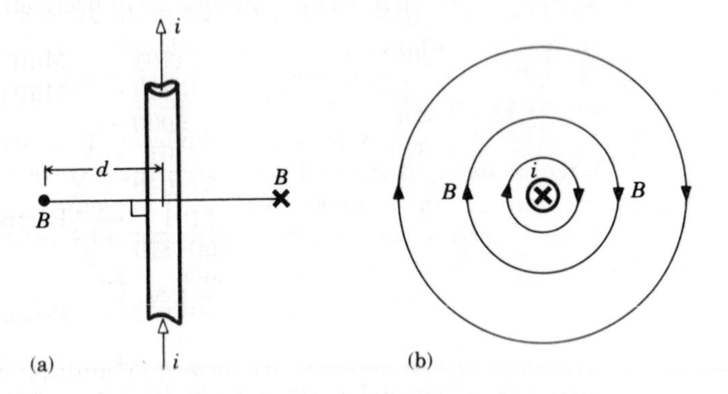

FIGURE 17.1–2 Magnetic field around a long current-carrying wire: (a) side view; (b) end view.

To obtain a more uniform field desired for many applications, the wire can be wound into a helical coil or *solenoid* of radius r and height h, as in Fig. 17.1–3a. If $h \gg r$, the interior flux density is approximately

$$B = \frac{\mu N i}{h} \tag{4a}$$

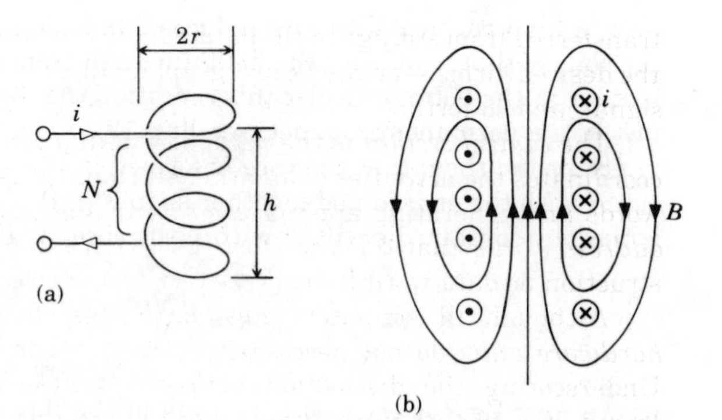

FIGURE 17.1–3
(a) Solenoidal winding.
(b) Magnetic field.

where N is the number of turns in the coil. The total flux will then be

$$\phi = \frac{\mu N i \pi r^2}{h} \tag{4b}$$

Equation (4a) is derived by integrating the field contributions from each differential current element, and Eq. (4b) comes from $\phi = BA$ with cross-sectional area $A = \pi r^2$.

Both of these expressions are valid anywhere inside the solenoid except close to the ends where the field gets somewhat weaker due to flux *leakage* between the turns. This leakage is shown in the axial field map, Fig. 17.1–3b.

The permeability μ in Eqs. (3) and (4) represents the magnetic property of the medium in question. The units are webers per ampere-meter (Wb/A · m) or, equivalently, henrys per meter (H/m). The permeability of free space is

$$\mu_0 = 4\pi \times 10^{-7} \tag{5}$$

and many ordinary materials have $\mu \approx \mu_0$. *Ferromagnetic* materials such as iron and steel have much larger values of μ, usually expressed in terms of the *relative permeability*

$$\mu_r \triangleq \mu/\mu_0$$

Typically, $\mu_r \approx 1000$ for a good magnetic material, but such materials are quite *nonlinear* in the sense that the value of μ depends on the flux density.

Returning to Eq. (4), we see that inserting a ferromagnetic *core* into a solenoid will substantially increase the internal magnetic field. Moreover, a high permeability core reduces flux leakage between the turns of the coil, so that nearly all the field lines exit from the top of the core and reenter at the bottom. Under this condition the solenoid looks externally just like a permanent magnet (see Fig. 17.1–1a).

Suppose, then, that the core is formed in the shape of a doughnut or *toroid* with the ends joined together as in Fig. 17.1–4. The magnetic field is thereby contained entirely within the core, and the flux becomes

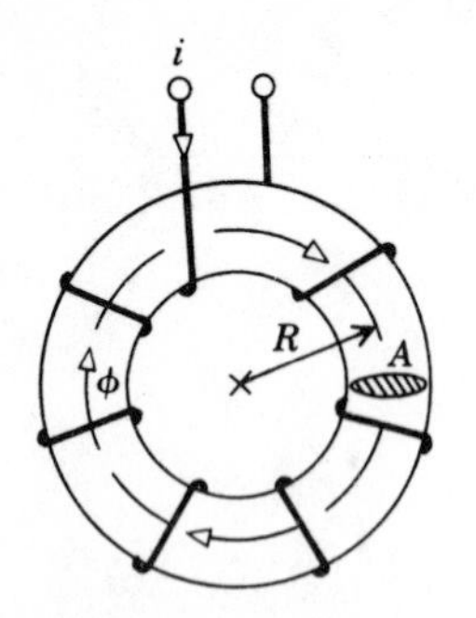

FIGURE 17.1–4
Toroidal winding.

$$\phi = \frac{\mu N i A}{\ell_\phi} \tag{6}$$

where $\ell_\phi = 2\pi R$ is the *average length* of the flux path through the core and A is the cross-sectional area (but not necessarily circular). High-μ toroids and similar configurations play important roles in devices that require a large confined flux.

Example 17.1–1

Consider a 400-turn coil wound on a toroidal iron core with average radius $R = 5$ cm $= 0.05$ m, cross-section radius $r = 2.5$ cm $= 0.025$ m, and relative permeability $\mu_r = 800$ so $\mu = 800\mu_0 \approx 10^{-3}$. The interior flux produced by $i = 0.4$ A will be

$$\phi = 10^{-3}\,\frac{\text{Wb}}{\text{A}\cdot\text{m}}\,\frac{400 \times 0.4\ \text{A} \times \pi(0.025\ \text{m})^2}{2\pi \times 0.05\ \text{m}}$$
$$= 10^{-3}\ \text{Wb} = 1.0\ \text{mWb}$$

with flux density

$$B = \frac{10^{-3}\ \text{Wb}}{\pi(0.025\ \text{m})^2} \approx 0.5\ \text{T}$$

These values are typical for coils of moderate dimension with ferromagnetic cores. Had the core been of nonmagnetic materials, the current required for the same flux value would have been $800 \times 0.4 = 320$ A!

Exercise 17.1–1

Suppose a solenoid has 10 turns per centimeter wrapped around a nonmagnetic core of radius 5 cm. What current is required to produce $\phi =$ 1 mWb inside the solenoid? And what is the corresponding flux density?

MAGNETIC FORCE AND INDUCED VOLTAGE

Having examined the magnetic field created by a magnet or current coil, let's now investigate its ability to produce force and voltage. To that end, we start with a conductor of length ℓ carrying current i perpendicular to a uniform magnetic field B as in Fig. 17.1–5. Recalling that current consists of moving charges, let $u_q = d\ell/dt$ be the average charge velocity in the conductor, and let dq be the amount of charge at any instant in length

$d\ell$. The force df on length $d\ell$ will be

$$df = Bu_q dq = Bi\,d\ell$$

in which $u_q dq = i\,d\ell$ follows from Eq. (3a), Section 2.1. Integrating over the entire length then yields

$$f = B\ell i \tag{7}$$

for the total magnetic force on a conductor carrying current i.

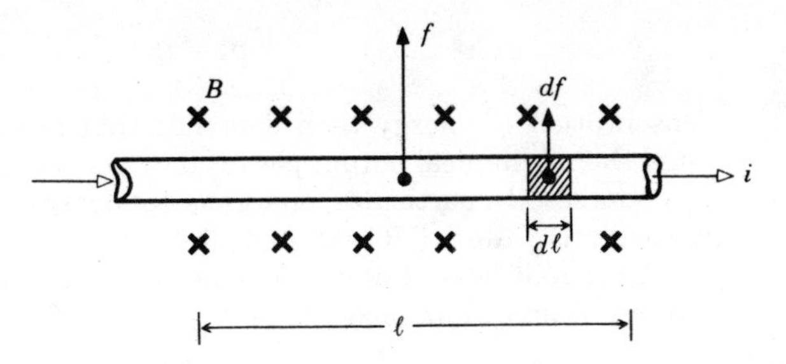

FIGURE 17.1–5
Magnetic force on a current-carrying wire.

We can put this force to use in the arrangement of Fig. 17.1–6a, where the conductor is free to slide along rails supplying current i from an electrical source. The force caused by the magnetic field then pushes the conductor to the right, thereby enabling such mechanical work as

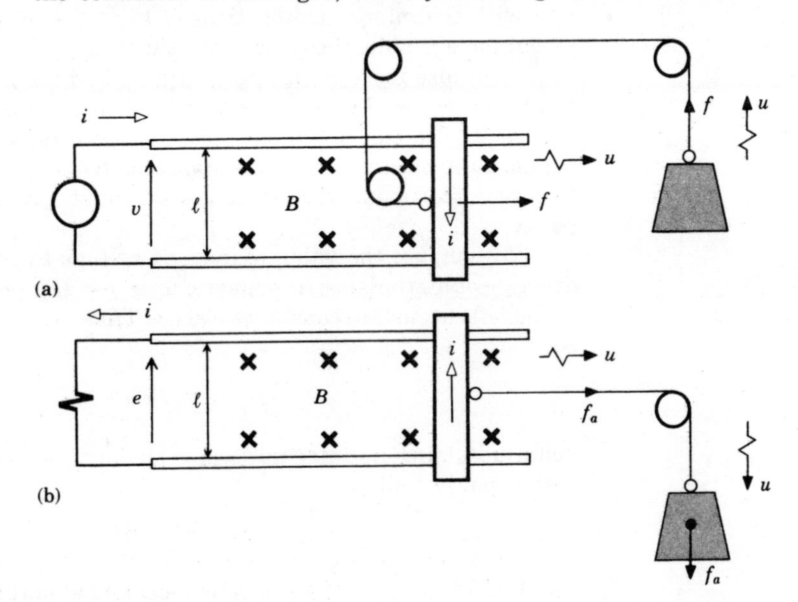

FIGURE 17.1–6
(a) Primitive motor.
(b) Primitive generator.

lifting a weight. (The rails also experience a force, but they are presumed to be fixed in place.) The conductor moves with velocity u, depending on the mechanical load, and the mechanical output power is

$$p_m = fu = B\ell iu \tag{8}$$

The system therefore acts as a primitive *motor* and converts electrical energy to mechanical form.

The electrical energy comes from the source, which provides input power

$$p_e = vi$$

Conservation of energy then demands that the electrical input power equal the mechanical output power, $p_e = p_m$, since there are no other energy sources. (In particular, energy is not drawn from the magnetic field, although the value of B is altered slightly due to the small field produced by the current loop.) Combining our expressions for p_e and p_m gives the required source voltage as

$$v = B\ell u$$

where the equality holds in the absence of Ohmic heating, friction, and other losses within the system.

This same system becomes a mechanical-to-electrical conversion device with the minor modifications of Fig. 17.1–6b. Here, an applied external force f_a pulls the conductor at velocity u, so the charges inside the conductor also move horizontally with it and the magnetic field exerts a force on them in the upward direction. This, in turn, creates an *induced voltage* e, also called the *electromotive force* (emf). Current then flows through the electrical load resistance, and the system acts like a primitive *generator,* converting mechanical input power to electrical output power.

We compute the value of induced voltage by observing that f_a must exactly counterbalance a magnetic force $f = B\ell i$ (whose direction is now to the left) to sustain constant velocity. Thus, $p_m = f_a u = B\ell iu = p_e = ei$, so

$$e = B\ell u \tag{9}$$

again assuming no internal losses. Under this condition, the electrical output power will be

$$p_e = ei = B\ell iu$$

which is identical to the previous mechanical output power.

Any device capable of electromechanical energy conversion in either direction is called a *bilateral transducer*. In the case of moving-conductor devices, such as in Fig. 17.1–6, the conversion medium is the interaction between a magnetic field and the moving conductor. Regardless of the conversion direction, there will always be *both* a magnetic force $f = B\ell i$ and an induced voltage $e = B\ell u$. We obtain motor action by applying a voltage $v \geq B\ell u$ which overcomes the induced voltage; this permits current flow from the electrical source, and interaction with the magnetic field produces mechanical output power (Fig. 17.1–6a). Conversely, we obtain generator action by applying a force $f_a \geq B\ell i$ which overcomes the magnetic force; this permits conductor motion whose interaction with the magnetic field produces electrical output power (Fig. 17.1–6b). Most motors, generators, and various other electromechanical devices are based on this two-way interaction. We also infer from Fig. 17.1–6 that motors and generators differ in particulars but not in general structure. In fact, some motors can be hooked up to operate as generators, and vice versa; and the only clear distinction between modes of operation is the direction of power flow.

Example 17.1–2

Suppose the motor system in Fig. 17.1–6a has $B = 1$ T and $\ell = 0.5$ m and we want to lift a 100-kg mass at a rate of 1.5 m/sec (about 5 km/hr). The minimum required source voltage is $v = B\ell u = 0.75$ V, a surprisingly small value. But the magnetic force must raise the mass against the gravitational force Mg, so $f = 100 \text{ kg} \times 9.8 \text{ m/sec} = 980$ N, and the source must provide a very large current, $i = f/B\ell = 1960$ A. Clearly, a gear-reduction system allowing faster conductor motion would yield more practical values in this case. For instance, with 200-to-1 reduction we could use $v = 150$ V and $i = 9.8$ A.

Exercise 17.1–2

Determine the mass M and its velocity u_M, such that the generator system of Fig. 17.1–6b produces 120 V across a 20-Ω resistance when $B = 1$ T and $\ell = 0.5$ m. Assume a gear arrangement such that $u = 10\, u_M$.

INDUCTION, INDUCTANCE, AND STORED ENERGY

Voltage induction takes place when a moving conductor "cuts across" magnetic field lines and, in so doing, changes the amount of flux passing through the circuit. It also takes place when *time-varying* flux passes through a *stationary* circuit. Specifically, if $\phi(t)$ passes through an N-turn coil or solenoid, as in Fig. 17.1–7a, a time-varying voltage $e(t)$, proportional to the rate of change of flux, is induced across the terminals; namely

$$e(t) = N\frac{d\phi}{dt} \tag{10}$$

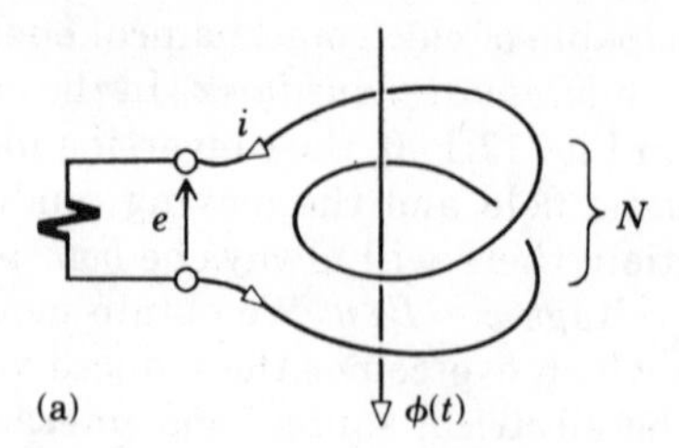

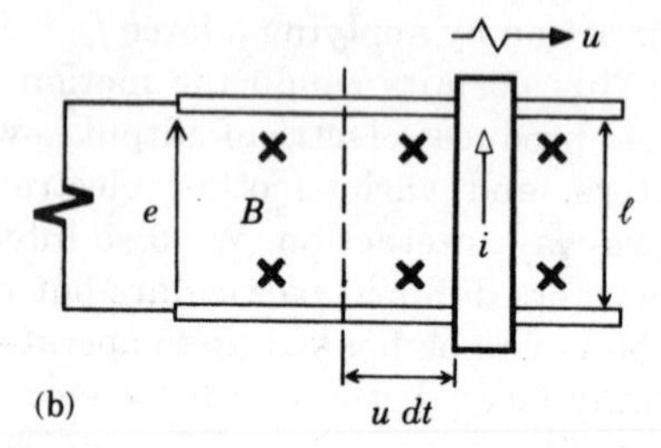

FIGURE 17.1–7

This is the famous *Faraday's law;* it applies in general, whether $d\phi/dt$ comes from a time-varying field or is caused by moving conductors.

To relate Faraday's law to our primitive motor, consider the moving conductor's position after it has traveled a distance $u\ dt$ in time dt, as in Fig. 17.1–7b. Clearly, the flux encircled by the current i has increased by an amount

$$d\phi = B\ dA = B\ell u\ dt$$

But $B\ell u$ equals the induced voltage e, so we can rewrite Eq. (9) as $e = d\phi/dt$, which says that the induced voltage simply equals the rate of change of flux through the circuit. If the total current had been Ni rather than i, we would have obtained $e = N\ d\phi/dt$–identical to Faraday's law.

Lenz's law, an equally general companion to Faraday's law, states

> Any induced quantity has a polarity or direction that tends to oppose the cause that induces it.

This law confirms that the polarity of e and direction of i are correctly shown in Fig. 17.1–7a when $d\phi/dt > 0$, for the induced current flow would produce flux through the coil in the upward direction (remember the right-hand rule) to oppose the increasing external flux in the downward direction. If you find this somewhat confusing, just keep in mind the principle that induction processes cannot be self-sustaining.

Faraday's law also holds if we *apply* voltage to a coil, causing a time-varying flux through it. By doing so, we will have created an *in-*

ductor. Take the toroid in Fig. 17.1–8 as a case in point: We know from Eq. (6) that the flux at any instant of time is $\phi = (\mu NA/\ell_\phi)i$. Therefore, the voltage required to produce a changing current and, as a result, a changing flux is

$$v = N\frac{d\phi}{dt} = N\frac{\mu NA}{\ell_\phi}\frac{di}{dt}$$

This expression has the familiar form $v = L\, di/dt$ with the inductance L given by

$$L = \frac{\mu N^2 A}{\ell_\phi}$$

Alternatively, since $\mu NA/\ell_\phi = \phi/i$, we can write

$$L = \frac{N\phi}{i} \tag{11}$$

with the interpretation that inductance equals the number of *flux linkages* $N\phi$ per unit current.

When we first introduced inductance as a circuit element, we stated that there was *stored energy* in a magnetic field equal to

$$w = \tfrac{1}{2}Li^2$$

Now we can relate energy storage directly to the flux density $B = \phi/A$ inside the toroid, by noting that $Ni = \phi\ell_\phi/\mu A$ and

$$Li^2 = \frac{N\phi}{i}i^2 = \phi Ni = \frac{\phi^2\ell_\phi}{\mu A} = \frac{1}{\mu}B^2 A\ell_\phi$$

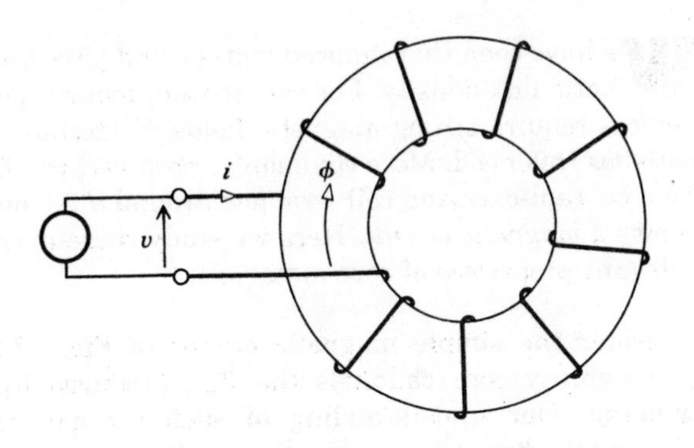

FIGURE 17.1–8

Thus,

$$w = \frac{1}{2\mu} B^2 A \ell_\phi \tag{12}$$

where the product $A\ell_\phi$ equals the *volume* of the toroid. This magnetic stored energy comes, of course, from the electrical power $p = vi$ delivered to the coil. We can recover the energy by replacing the source with a load resistance, so that $\phi \to 0$ as $t \to \infty$ and the voltage induced by the decreasing flux delivers power to the load.

Despite their specialized derivations, the preceding results are universally valid. Indeed, Eq. (11) serves as the *definition* of inductance for any N-turn coil enclosing flux ϕ produced by current i. Similarly, Eq. (12) gives the magnetic energy stored by a uniform magnetic field B in any region with volume $A\ell_\phi$. However, care must be exercised in the case of a ferromagnetic material whose permeability μ is not constant.

Example 17.1–3

The 400-turn toroidal in Example 17.1–3 was found to have $\phi =$ 1.0 mWb when $i = 0.4$ A. Thus the inductance is

$$L = 400 \times \frac{1.0 \text{ mWb}}{0.4 \text{ A}} = 1000 \text{ mH} = 1 \text{ H}$$

and the stored energy will be $w = 0.08$ J, calculated either from Eq. (12) or from $w = \frac{1}{2}Li^2$.

Exercise 17.1–3

Since the number of turns N is a dimensionless quantity, Eq. (10) implies that the units of flux may be written as 1 Wb = 1 V · sec. Confirm this expression by using the facts that force equals work per unit distance, and voltage equals work per unit charge.

Exercise 17.1–4

Find the value of h for the solenoid in Exercise 17.1–1 such that $L =$ 10 mH. Then calculate the stored energy when $\phi = 1$ mWb.

17.2 MAGNETIC CIRCUITS AND MATERIALS

We have seen that induced voltage and force are proportional to magnetic flux density. For this reason, motors, generators, and similar devices require strong magnetic fields. Sometimes a permanent magnet satisfies this need. More commonly, however, the field must be produced by a current-carrying coil wrapped around a ferromagnetic core that becomes a *magnetic circuit.* Here we study these magnetic circuits and the relevant properties of core materials.

MAGNETIC CIRCUITS

Consider the simple magnetic circuit of Fig. 17.2–1a, where a high-permeability core channels the flux produced by an N-turn current-winding. Our understanding of such configurations is enhanced by viewing the flux ϕ in a magnetic circuit as analogous to the current in an electric circuit. This viewpoint reflects the property that flux, like current, always "travels" around a closed path. Then, since flux production depends on both the number of turns and the current through the winding, we define a *magnetomotive force* (mmf)

$$\mathcal{F} \triangleq Ni \tag{1}$$

measured in ampere-turns (A-t). Mmf thereby plays the role of the source in a magnetic circuit, similar to the source voltage (electromotive force) in an electric circuit. Although the number of turns is a dimensionless quantity, we write the unit as A-t to underscore the presence of N in $\mathcal{F} = Ni$.

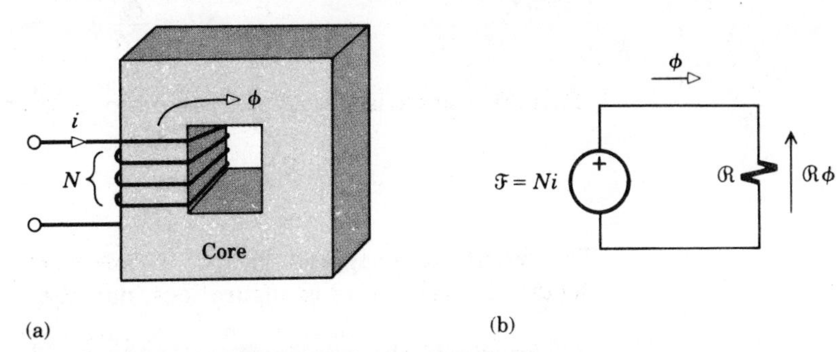

FIGURE 17.2–1
(a) Magnetic circuit.
(b) Analogous electrical circuit.

Completing the analogy, we define the magnetic *reluctance* $\mathcal{R}$ of the flux path through the core such that

$$\mathcal{F} = \mathcal{R}\phi \tag{2}$$

Reluctance, in ampere-turns per weber (A-t/Wb), is the magnetic equivalent of electrical resistance; and $\mathcal{F} = \mathcal{R}\phi$ becomes the magnetic version of Ohm's law, $v = Ri$. Table 17.2–1 summarizes our analogy and includes some additional quantities.

TABLE 17.2–1

Magnetic quantity	*Symbol*	*Unit*	*Electric analog*
Flux	ϕ	Wb	Current
mmf	$\mathcal{F} = Ni$	A-t	Voltage (emf)
Reluctance	$\mathcal{R}$	$\frac{\text{A-t}}{\text{Wb}}$	Resistance
Permeability	μ	$\frac{\text{Wb}}{\text{A} \cdot \text{m}}$	Conductivity
Flux density	B	T	Current density
Magnetizing force	H	$\frac{\text{A-t}}{\text{m}}$	Field strength

Figure 17.2–1b illustrates the concept of an equivalent electrical circuit for the magnetic circuit of Fig. 17.2–1a. The source $\mathcal{F}$ causes flux ϕ to "flow" around the closed loop while the core reluctance tends to resist it. Thus, ϕ equals $\mathcal{F}/R$ and an "mmf drop" $\mathcal{R}\phi = \mathcal{F}$ exists across $\mathcal{R}$. In principle, magnetic-circuit analysis using $\mathcal{F} = \mathcal{R}\phi$ exactly mimics the analysis of resistive electrical circuits, *if* we assume linear (constant μ) core material. In practice, however, we encounter a major difficulty when it comes to calculating reluctances for specific cores.

If the core happens to be a leakage-free toroid, Eq. (6), Section 17.1, shows that

$$Ni = \frac{\ell_\phi}{\mu A}\,\phi$$

Thus, by comparison with the definitions of $\mathcal{F}$ and $\mathcal{R}$, the reluctance is

$$\mathcal{R} = \frac{\ell_\phi}{\mu A} \tag{3}$$

This result then suggests an *approximate method* for calculating the reluctance of other core configurations, namely:

> Subdivide the core into sections having constant cross-sectional area A_k and average flux path length ℓ_k, and take $\mathcal{R}_k = \ell_k/\mu_k A_k$ as the reluctance of each section.

The approximation method yields fair results providing $\mu_r \gg 1$ (minimizing leakage error), and the inherent inaccuracy is usually tolerable in view of the fact that we seldom know the value of μ with better than 5% accuracy. Therefore, magnetic-circuit calculations provide us with useful approximations, while precise values generally require experimental measurements.

Putting these ideas to work, consider Fig. 17.2–2 where a high-μ core is broken by a short *air gap*. Here the same flux ϕ passes through both core and gap, and we have a *series* magnetic circuit with total reluctance $\mathcal{R}_c + \mathcal{R}_g$ and

$$\mathcal{F} = (\mathcal{R}_c + \mathcal{R}_g)\phi$$

where the core and gap reluctances are proportional to $1/\mu_c$ and $1/\mu_0$, respectively. Since $\mu_c \gg \mu_0$, it frequently turns out that $\mathcal{R}_c \ll \mathcal{R}_g$, with most of the mmf appearing across the gap. In any case, the gap mmf is

$$\mathcal{F}_g = \mathcal{R}_g\phi = \frac{\ell_g}{\mu_0 A_g}\,\phi \tag{4}$$

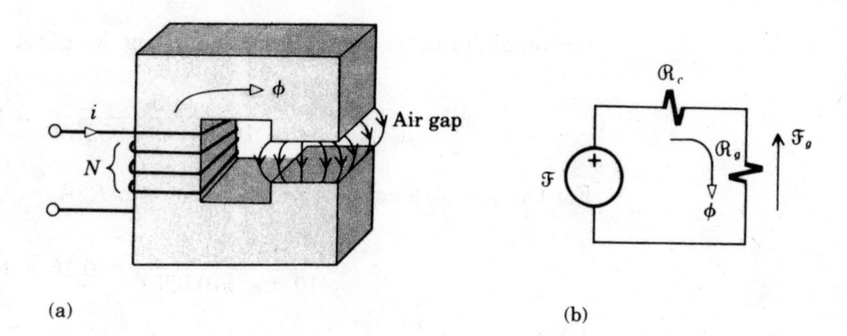

FIGURE 17.2–2 Magnetic circuit with an air gap.

To improve accuracy, we account for the "bulging out" or *fringing* of the field lines in the gap by adding ℓ_g to each of the two dimensions of the gap's cross-sectional area.

Figure 17.2–3 shows a more complicated core configuration with two windings. The equivalent circuit has two mmf sources and both series and parallel reluctances to represent the various flux paths. Note that the "polarity" of $\mathcal{F}_2$ opposes ϕ_2, as determined from the right-hand rule.

Finally, combining the definitions of reluctance $\mathcal{R}$ and inductance L, we obtain a simple and useful relationship between them,

$$L = \frac{N\phi}{i} = \frac{N}{i}\frac{\mathcal{F}}{\mathcal{R}} = \frac{N}{i}\frac{Ni}{\mathcal{R}} = \frac{N^2}{\mathcal{R}} \tag{5}$$

providing the core has just one winding. With two or more windings, an interactive effect known as *mutual inductance* occurs. This interaction is implied in Faraday's law and forms the basis of transformer operation, as we will show later.

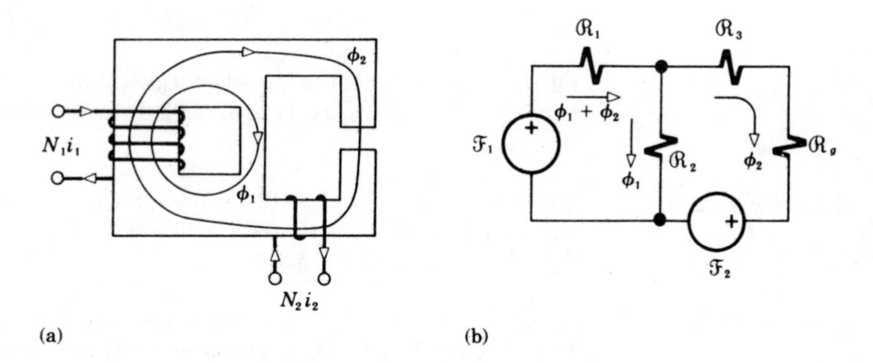

FIGURE 17.2–3

Example 17.2–1 Suppose the toroid of Example 17.1–1 is cut and pried apart slightly, creating an air gap with $\ell_g = 0.5$ cm. Adding ℓ_g to the gap's cross-sectional diameter to account for fringing, we get $A_g = \pi(2r + \ell_g)^2/4$ and

$$\mathcal{R}_g = \frac{\ell_g}{\mu_0 A_g} = \frac{0.005}{4\pi \times 10^{-7} \times \pi(0.055^2/4)} = 1.67 \times 10^6$$

For the core's reluctance, we have $h_c = 2\pi R$, $A_c = \pi r^2$, and $\mu_c = 10^{-3}$, so

$$\mathcal{R}_c = \frac{2\pi \times 0.05}{10^{-3} \times \pi(0.025)^2} = 0.16 \times 10^6$$

Thus

$$\mathcal{R} = \mathcal{R}_c + \mathcal{R}_g = 1.83 \times 10^6 \text{ A-t/Wb}$$

and the gap contributes about 90% of the total reluctance.

If we desire the same flux value as before (1.0 mWb), we must have

$$\mathcal{F} = \mathcal{R}\phi = (1.83 \times 10^6 \text{ A-t/Wb})(10^{-3} \text{ Wb}) = 1830 \text{ A-t} = Ni$$

Since $N = 400$, the required current becomes

$$i = \frac{1830}{400} \approx 4.6 \text{ A}$$

the increase due entirely to the air gap.

Presumably, larger current means more stored energy, as easily confirmed from

$$L = \frac{N^2}{\mathcal{R}} = \frac{400^2}{1.83 \times 10^6} = 87 \text{ mH}$$

and

$$w = \tfrac{1}{2}Li^2 = \tfrac{1}{2}(87 \times 10^{-3}) \times 4.6^2 = 0.92 \text{ J}$$

Furthermore, most of the stored energy is now in the air gap since the energy stored in the core is still 0.08 J, as we determined in Example 17.1–3.

Exercise 17.2–1 The core in Fig. 17.2–1 is 12-cm high, 8-cm wide, and 3.5-cm deep. Each leg of the core is 2-cm across. Calculate $\mathcal{R}_c$, L, and ϕ, taking $\mu_c = 1.8 \times 10^{-3}$, $N = 100$, and $i = 2.5$ A.

Exercise 17.2–2 Suppose a 0.5-cm air gap is cut across one leg of the core in Exercise 17.2–1. Calculate $\mathcal{R}_g$, $\mathcal{R}$, L, and ϕ, making allowance for fringing.

MAGNETIZATION AND HYSTERESIS

Up to now we have ignored the complicated behavior of ferromagnetic core materials hidden behind the permeability μ. Such materials not only are nonlinear but also have a magnetic "memory," as exemplified by permanent magnets and magnetic recording tape. We will not go into the underlying *domain theory* of magnetism here, but we will describe the external properties of magnetic materials. For that purpose we introduce the *magnetizing force* or *magnetic field intensity,*

$$H \triangleq \frac{\mathcal{F}}{\ell_\phi} = \frac{Ni}{\ell_\phi} \tag{6}$$

where ℓ_ϕ, again, stands for flux path length. Magnetizing force H represents the *mmf per unit length* (A-t/m) and is analogous to electric field strength $\mathcal{E}$—the emf per unit length.

Let's proceed by analogy, recalling that when an electric field exists in a medium with conductivity σ and cross-sectional area A, the resulting current will be $i = \sigma\mathcal{E}A$. When a magnetizing force H exists in a toroidal core with permeability μ and cross-sectional area A, the resulting magnetic flux will be $\phi = \mu(Ni/\ell_\phi)A = \mu HA$. Permeability μ, therefore, is the "magnetic conductivity" of the material. Furthermore, dividing by the area converts flux ϕ to flux density B and yields the important relation

$$B = \mu H \tag{7}$$

Thus, a plot of B versus H—called the *normal magnetizing curve*—reveals whether the material in question has constant permeability $\mu = B/H$.

Figure 17.2–4 shows typical magnetizing curves for cast iron and sheet steel. Also shown, but barely visible, is the curve for nonmagnetic material, the latter being a straight line with very small slope $\mu \approx \mu_0$. The magnetizing curve for cast iron has much greater initial slope (implying $\mu_r = \mu/\mu_0 \gg 1$) that gradually decreases with increasing H. The curve for sheet steel rises even more sharply and then flattens off rather abruptly. (The slope actually approaches μ_0 as $H \to \infty$.) We therefore say that such materials become quickly *saturated,* since increasing H beyond the "knee" of the curve yields little further increase in B.

Clearly, these ferromagnetic materials are *nonlinear* in the sense that the value of μ depends on the magnetizing force and, in particular, decreases with increasing H. This nonlinearity means that flux ϕ is not directly proportional to current i and that, therefore, inductance $L = N\phi/i$ and reluctance $\mathcal{R} = \mathcal{F}/\phi$ are not truly constants. Fortunately, approximate calculations based on an average value of μ often prove satisfactory, providing the final result is below saturation.

When magnetizing force varies with time, we must also take account of the *hysteresis* effect as shown in Fig. 17.2–5a. Here, the material is initially saturated at some point P on the B-H curve and the magnetizing force is removed ($H = 0$); but, instead of going to zero flux density along

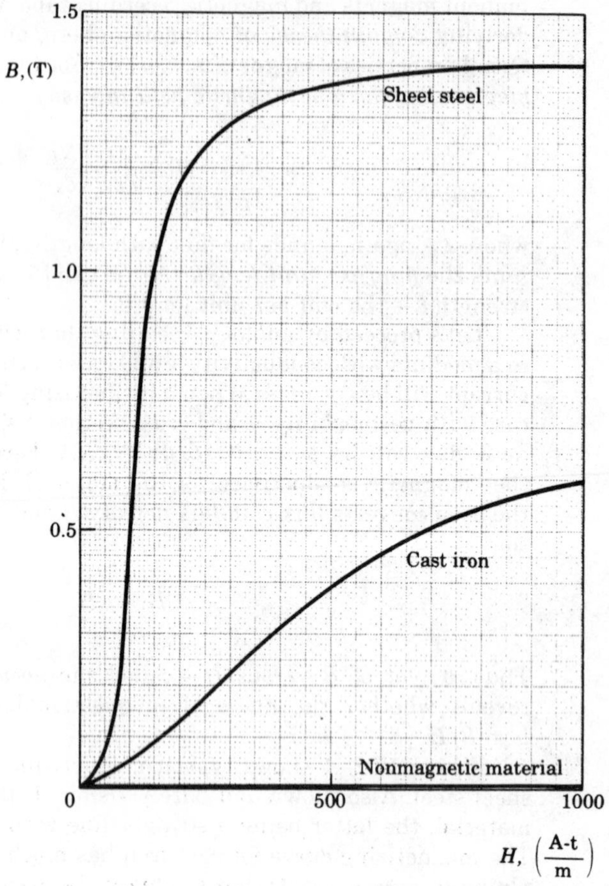

B, (T) 1.5 1.0 0.5 0 500 1000 $H, \left(\frac{\text{A-t}}{\text{m}}\right)$
Sheet steel
Cast iron
Nonmagnetic material

FIGURE 17.2–4
Magnetizing curves.

the curve, B drops only slightly to a nonzero value B_r known as the *retentivity*. We have now created a *permanent magnet* that retains flux density $B = B_r$ without any magnetizing force. This material can be demagnetized only when we apply a reverse magnetizing force of strength H_c, known as the *coercivity*. The special alloys used for such permanent magnets are called "hard" magnetic materials and typically have $B_r \approx$ 1 T and $H_c \approx 10^4$ A-t/m. "Soft" magnetic materials, on the other hand, have relatively little hysteresis.

If a hard magnetic material is subjected to a magnetizing force that oscillates over a range exceeding $\pm H_c$, the value of B lags behind H and the B-H relationship takes the form of a *hysteresis loop,* Fig. 17.2–5b. Materials having nearly square hysteresis loops will then act as *binary memory devices* that can be set into the $+B_r$ state by applying and then

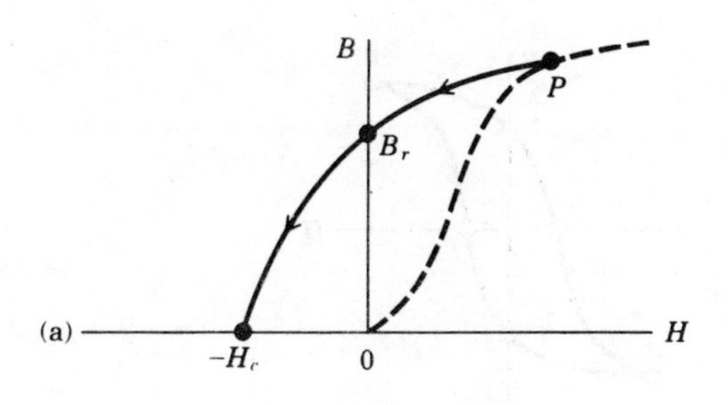

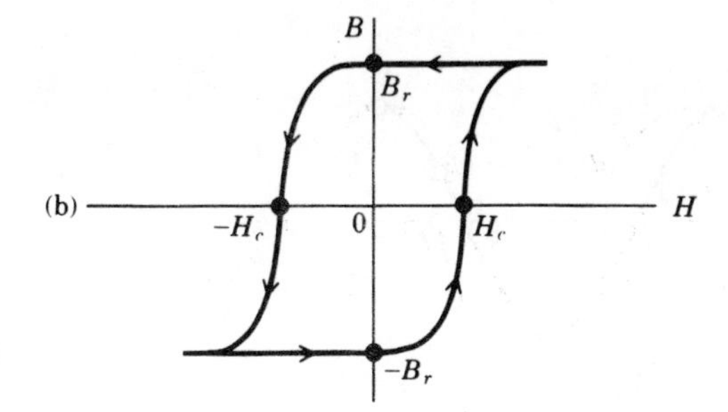

FIGURE 17.2–5
(a) Hysteresis effect in "hard" magnetic material. (b) Hysteresis loop.

removing $H > H_c$, or that can be set into the $-B_r$ state by applying and then removing $H < -H_c$. Many large digital computers have core memories comprising thousands of tiny toroids, each storing one binary digit.

But magnetic nonlinearity and hysteresis create problems in transformers and similar applications when we want undistorted waveforms. To illustrate, we consider the moderate *B*-*H* loop in Fig. 17.2–6a. If the magnetizing force $H(t)$ is sinusoidal, produced by an AC current, then the flux density and any induced voltage would have the flattened waveshape sketched in Fig. 17.2–6b. Conversely, to get a sinusoidal flux waveform $B(t)$, the magnetizing force $H(t)$ would have to assume the strange waveshape shown in Fig. 17.2–6c.

Finally, note that hysteresis prevents us from directly measuring the normal magnetizing curve. Instead, we have to work from loops like the one in Fig. 17.2–6a whose extreme tips define the curves given in Fig. 17.2–4. The resulting plots are called *average* magnetizing curves because the actual value of *B* for a given value of *H* may be either above or below the curve, depending upon the material's past history.

Example 17.2–2

To illustrate the use of *B*-*H* curves in magnetic circuit design, we'll estimate the mmf needed to produce $\phi = 1.0$ mWb in a toroidal core having $\ell_c = 0.1\pi m$ and $A_c = \pi(0.025\text{ m})^2$. The required flux density is $B = \phi/A_c \approx 0.5$ T. Figure 17.2–4 shows that the corresponding magnetizing

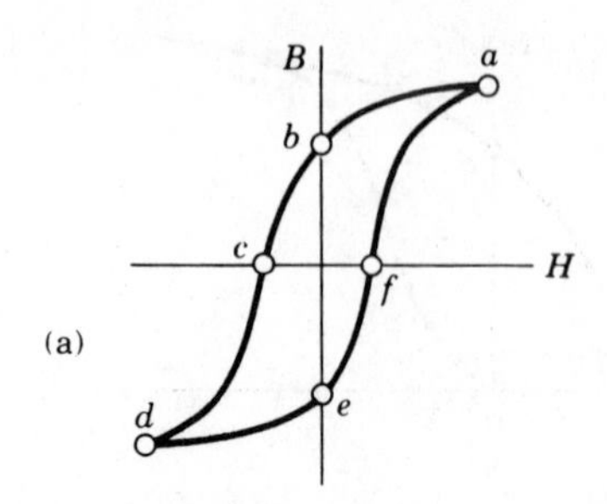

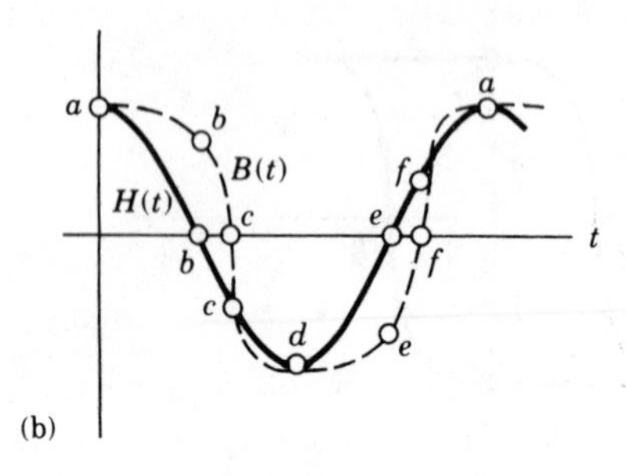

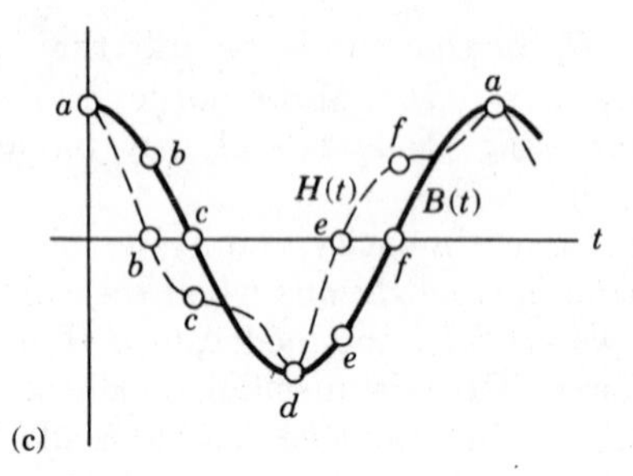

FIGURE 17.2–6
(a) Moderate hysteresis loop. (b) Distortion of $B(t)$ when $H(t)$ is sinusoidal.
(c) Distortion of $H(t)$ when $B(t)$ is sinusoidal.

force must be $H \approx 100$ A-t/m for a sheet-steel core, or $H \approx 700$ A-t/m for a cast-iron core. The mmf would then be $\mathfrak{F} = H\ell_c \approx 31$ A-t or 220 A-t, respectively. These values may be compared with the results of Example 17.1–1, where we had $\mathfrak{F} = Ni = 400 \times 0.4 = 160$ A-t, assuming $\mu \approx 10^{-3}$ or $400 \times 320 = 128{,}000$ A-t if $\mu = \mu_0$.

Now let there be a 0.5-cm air gap, as in Example 17.2–1. We then separate the total required mmf into two terms

$$\mathfrak{F} = \mathfrak{F}_c + \mathfrak{F}_g$$

where $\mathfrak{F}_c = H_c\ell_c$ and $\mathfrak{F}_g = \mathcal{R}_g\phi = 1.67 \times 10^6 \times 10^{-3} = 1670$ A-t. Thus, $\mathfrak{F} = 31 + 1670 = 1701$ A-t for the steel core, or $220 + 1670 = 1890$ A-t for the iron core. In either case, the air gap dominates and minimizes the inaccuracies of the core calculations.

Exercise 17.2–3 Let the preceding core be of sheet steel without an air gap. Assuming $N = 60$, calculate ϕ, $\mathcal{R}$, L, and w when (a) $i = 1$ A and (b) $i = 2$ A.

LOSSES

A magnetic device always has *losses* in the sense that some fraction of the input energy will be converted to unwanted heat. The most obvious loss is ohmic heating in the windings resulting from the small but inevitable winding resistance R_w. Often referred to as *copper loss,* this $R_w i^2$ power dissipation occurs with either constant or time-varying current. Time-varying current also produces two other forms of heating in the *core* itself, due to hysteresis and eddy currents.

Hysteresis loss represents the energy required to go around the hysteresis loop (Fig. 17.2–6a), when $H(t)$ has a cyclical time variation that alternately magnetizes and demagnetizes the core. For a sinusoidal variation at frequency $f = \omega/2\pi$, the average hysteresis power loss is given by the empirical formula

$$P_h = k_h f B_m^n A_c \ell_c \tag{8}$$

where B_m is the maximum flux density and k_h and n are characteristics of the core. The value of n, called the *Steinmetz exponent,* ranges from about 1.5 to 2.0, with 1.6 being a typical value.

Eddy-current loss comes from localized currents induced in the core by a time-varying flux. Illustrating this phenomenon, Fig. 17.2–7a shows $\phi(t) = B(t)A_c$ passing through the cross-section of a magnetic core. Since the iron or steel core material is a conductor, we have a conducting path encircling the flux (similar to that of Fig. 17.1–7a). The corresponding induced voltage produces an eddy current i_e around the path, and the current, in turn, causes ohmic heating. From Faraday's law, it can be shown that a sinusoidal flux variation produces an average eddy-current power loss

$$P_e = k_e \frac{(A_c f B_m)^2}{R_e} \tag{9}$$

where R_e is the equivalent core resistance and k_e another core constant. Unless special steps are taken, P_e may account for well over half of the total input power of a magnetic circuit.

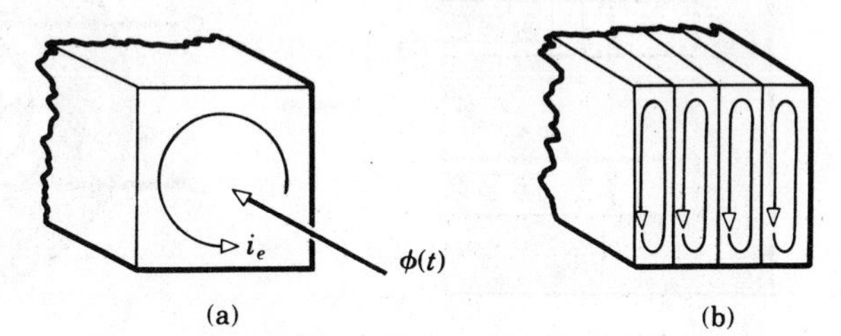

FIGURE 17.2–7
(a) Eddy current.
(b) Laminated core.

Fortunately, we can minimize eddy-current losses simply by constructing the core from thin sheets or *laminations* parallel to the flux, with insulating varnish between them, as in Fig. 17.2–7b. If there are m laminations, each one has a smaller area $A' = A_c/m$ and a larger resistance $R_e' \approx mR_e$; therefore, the total loss will be less, by a factor of about $1/m^2$, than the loss of a solid core.

Values of the core constants k_h and k_e are not usually tabulated. Rather, they are inferred from power and efficiency measurements. The efficiency of a magnetic device intended for power transfer (a transformer, for example) typically exceeds 95%, including core and copper losses.

17.3 ELECTROMECHANICAL TRANSDUCERS

The concepts developed in the previous two sections are now combined to derive the operating relations of common electromechanical transducers—devices that perform electromechanical conversion in one direction or the other. Actually, in view of the key role played by the magnetic field, we should refer to such devices as electro-*magneto*-mechanical transducers, a more descriptive but tongue-twisting name.

We first generalize from our introductory analysis of a simple bilateral transducer to encompass the broader category of *moving-coil* devices, distinguishing between those with lateral or *translational* motion and those with *rotational* motion. Then we examine *moving-iron* devices, in which force or torque is produced by allowing a ferromagnetic member to move under the influence of magnetic attraction.

Our treatment will be limited to static or steady-state conditions, the dynamic behavior of transducers being taken up in the next section. We will, however, include a discussion of mechanical and electrical losses.

TRANSLATING-COIL DEVICES

Most translating-coil devices contain a cylindrical coil wound around a sleeve, permitting back-and-forth motion perpendicular to a radial magnetic field in the air gap of a permanent magnet. Figure 17.3–1 gives cross-sectional and end views of this configuration. (Such devices serve,

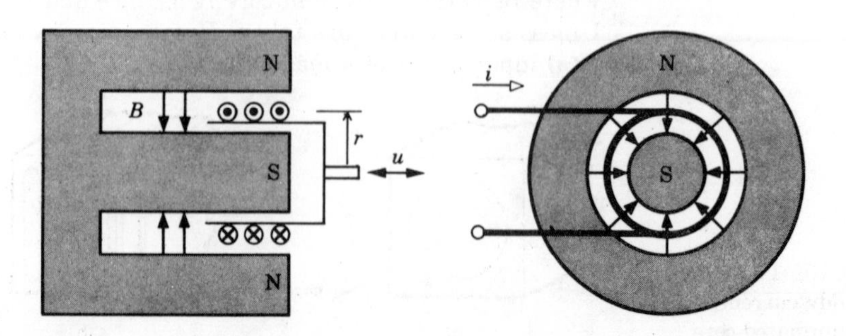

FIGURE 17.3–1
Translating-coil device.

for instance, as acoustic transducers when the sleeve is connected to the diaphragm of a microphone or loudspeaker.) The coil has N turns, each with circumference $c = 2\pi r$, for a total moving-conductor length $\ell = Nc$. Since every element of the conductor is mutually perpendicular to the radial B field and to the direction of the velocity u, the magnetic force and induced voltage will be

$$f = B\ell i = BNci = \alpha i \qquad e = B\ell u = BNcu = \alpha u \tag{1a}$$

where

$$\alpha \triangleq BNc \tag{1b}$$

which is the *electromechanical coupling coefficient* for a translating-coil device.

When the device performs electrical-to-mechanical energy conversion (as in a loudspeaker), it can be represented by Fig. 17.3–2a. This electrical equivalent circuit consists of the coil's winding resistance R, with applied source voltage v opposed by the induced voltage e. The accompanying mechanical diagram shows f acting on the moving unit, opposed by a *frictional drag force* $f_D = Du$ and the external load force f_a. Power flows from the electrical input to the mechanical output, with a portion lost as ohmic and frictional heat.

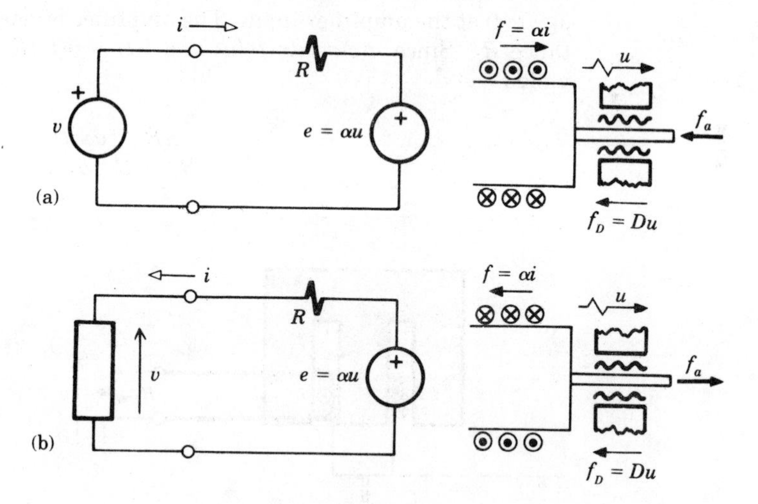

FIGURE 17.3–2
(a) Electrical-to-mechanical conversion.
(b) Mechanical-to-electrical conversion.

To analyze this case, we write a loop equation for the electrical circuit and sum the mechanical forces:

$$v = e + Ri = \alpha u + Ri \qquad f_a = f - Du = \alpha i - Du \tag{2}$$

Note that the coefficient α couples the electrical and mechanical equations. The transducer's power efficiency is, then,

$$\frac{p_m}{p_e} = \frac{f_a u}{vi} = \frac{\alpha iu - Du^2}{\alpha iu + Ri^2}$$

which underscores the dissipative nature of R and D, and reveals that good efficiency requires strong coupling.

If we reverse the direction of power flow in Fig. 17.3–2a, we have a mechanical-to-electrical conversion device represented by Fig. 17.3–2b. Now f_a is an applied force, opposed by f and f_D, and the electrical output is delivered to some load element. Proceeding as before, we obtain the coupled equations

$$v = \alpha u - Ri \qquad f_a = \alpha i + Du \tag{3}$$

which differ from Eq. (2) only in the signs of the dissipative terms. Completing the analysis from this point is a routine task, after you have identified the input and output variables for a particular application.

Example 17.3–1
Magnetic phonograph pickup

Figure 17.3–3 shows a translating-coil magnetic phonograph pickup. The coil is attached to a stylus, and the "bumps" in the groove of the rotating record impart a vertical displacement $x(t)$ that, in turn, produces a voltage $v(t)$ at the amplifier input. The amplifier is represented by load resistance R_L. Since $u(t) = dx(t)/dt$ and $i(t) = v(t)/R_L$, Eq. (3) immediately yields

$$v(t) = \frac{\alpha R_L}{R_L + R}\frac{dx(t)}{dt}$$

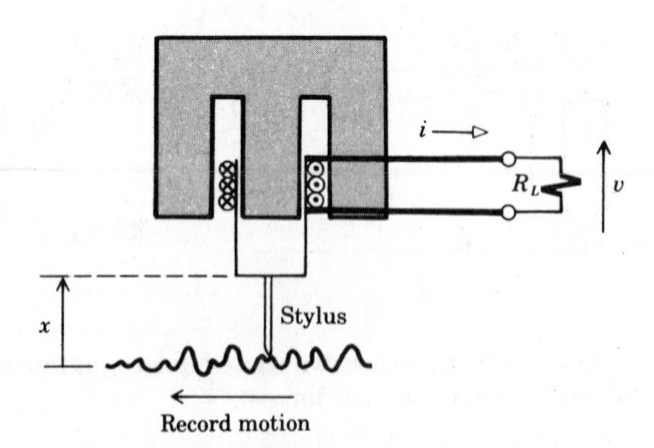

FIGURE 17.3–3
Magnetic phonograph pickup.

If a constant-amplitude tone at frequency ω is recorded as $x(t) = A_x \cos \omega t$, then $dx(t)/dt = -A_x\omega \sin \omega t = A_x \cos(\omega t + 90°)$, so the output voltage has a 90° phase shift, and its amplitude increases linearly with ω. The amplifier must therefore include a frequency-compensating network to *equalize* the overall frequency response.

Example 17.3–2

Let Fig. 17.3–2a represent an 8-Ω loudspeaker whose input is the voltage v and whose acoustical output is proportional to the diaphragm's velocity u. The acoustical load force is approximately proportional to u, given by $f_a = 30u$, and the internal drag coefficient is $D = 2$. We want a relationship between the output u and the input v in terms of the coupling $\alpha = BNc$.

Inserting $f_a = 30u$ and the other numerical values into Eq. (2) gives

$$v = \alpha u + 8i \qquad 30u = \alpha i - 2u$$

from which we eliminate i by noting that $\alpha i = 32u$. Hence, $v = \alpha u + 8(32u/\alpha)$ and

$$u = \frac{v}{\alpha + (256/\alpha)} = \frac{\alpha}{\alpha^2 + 256}\, v$$

This expression implies that maximum output requires an intermediate coupling value, for $u \approx \alpha v/256$ if $\alpha^2 \ll 256$, whereas $u \approx v/\alpha$ if $\alpha^2 \gg 256$.

Exercise 17.3–1

Let f_a be the acoustic input on the diaphragm of a microphone. Obtain an expression for the resulting output voltage v across a load resistance R_L by eliminating u from Eq. (3).

ROTATING-COIL DEVICES

Perhaps the simplest rotating-coil device is the common *d'Arsonval meter movement* illustrated in Fig. 17.3–4a. This transducer consists of a flat coil suspended in a magnetic field in such a way that magnetic force creates *torque* and rotates the coil about its vertical axis. Opposing torque from a pair of spiral springs results in a net angular deflection θ indicated by a pointer attached to the coil assembly. The operating principles of this meter illustrate the basic concepts of many rotating-coil devices.

As shown in Fig. 17.3–4b, the coil is more or less rectangular with length ℓ, width $2r$, and area $A = 2r\ell$. Its movement is restricted to the air-gap region of Fig. 17.3–4c, where a uniform radial B field has been established between a permanent magnet and a stationary iron cylinder. With current i through the coil, each axial length of conductor experiences force $f = B\ell i$ at radius r from the center. Since there are $2N$ axial

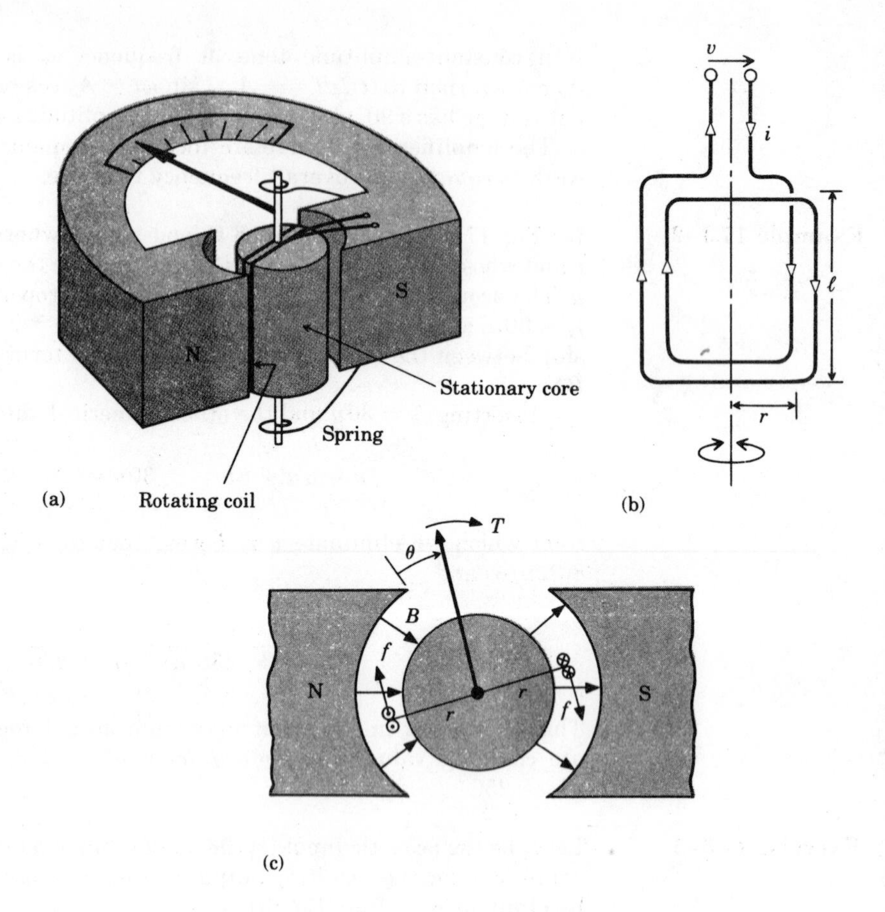

FIGURE 17.3–4 (a) d'Arsonval meter movement. (b) Coil detail. (c) Air-gap detail.

conductors in an N-turn coil, the total *magnetic torque* will be

$$T = 2N \times f \times r = 2N(B\ell i)r = \beta i \tag{4a}$$

where we have introduced the *rotational coupling coefficient*

$$\beta \triangleq 2NB\ell r = BN(2r\ell) = BNA \tag{4b}$$

This important torque expression is the rotational version of $f = \alpha u$; it also applies to nonrectangular coils with area A.

Temporarily ignoring the opposing spring torque, let's examine what happens when T rotates the coil at *angular velocity* $\omega = d\theta/dt$. Just as in the case of a single conductor cutting across flux lines, this rotation results in an *induced voltage* or emf given by

$$e = \beta\omega \tag{5}$$

which is the rotational version of $e = \alpha u$. To derive Eq. (5), recall that mechanical power in a rotating system equals torque times angular velocity, so

$$p_m = T\omega = \beta\omega i$$

if we assume no other torques than T. Likewise if we neglect any electrical losses, the applied voltage v must equal e, and the electrical power $p_e = vi$ equals p_m. Therefore, $e = v = p_e/i = p_m/i = (\beta\omega i)/i = \beta\omega$, as stated. (Note that this derivation follows the argument used in conjunction with Fig. 17.1–6.)

Now let a constant voltage $v = V_s$ be applied to the meter; the resulting current produces torque T that tends to accelerate the coil in the clockwise direction, as in Fig. 17.3–4c. But the springs impose a counterclockwise torque $T_K = K\theta$, with K being the "stiffness" constant of the springs. The meter movement eventually reaches mechanical and electrical equilibrium at a *steady-state deflection angle* θ_{ss} where $T = T_K$ and $e = 0$ since $\omega = d\theta/dt = 0$. Under these conditions the current is determined entirely by the winding resistance R, so $i = V_s/R = I_s$, $T = \beta I_s = T_K = K\theta_{ss}$, and

$$\theta_{ss} = \frac{\beta}{K} I_s = \frac{\beta}{KR} V_s \tag{6}$$

We thus have a DC *ammeter* or *voltmeter*. Maximum current or voltage is limited by the full-scale deflection angle, typically 90–120°, but external circuitry discussed in Section 3.4 permits larger measurement ranges.

Angular deflections of 180° or more are prevented by mechanical stops; otherwise, the torque would actually reverse its direction because the conductors would be in a B field pointed the opposite way. (You can confirm this effect by reversing the current directions in Fig. 17.3–4c, an equivalent of a deflection exceeding 180°.) Such *field and torque reversals* are inherent in all rotating-coil devices, and they present a major complication for the design of DC machines. On the other hand, field reversals lend themselves quite naturally to the production of *alternating* voltages in an AC generator. Similarly, the periodic reversals of current direction in an AC motor are synchronized with the field reversals to create unidirectional torque and continuous rotation. (Chapter 19 pursues these matters in greater detail.)

Another type of meter movement is the *electrodynamometer*, Fig. 17.3–5a, whose magnetic field comes from current through a stationary solenoidal coil divided into two parts, b and b'. This coil is wound on a nonmagnetic core in order to avoid magnetic nonlinearities, so B and β are directly proportional to i_b. One terminal of each coil bears the label $\pm$, indicating that proper "up-scale" deflection will be obtained when i_a and i_b *simultaneously* enter or leave their respective terminals.

For the general case of time-varying currents through both coils, mechanical inertia provides an *averaging* effect that results in the steady-state deflection

$$\theta_{ss} = K_M(i_a i_b)_{av} \tag{7}$$

where K_M is the meter's proportionality constant. If the coils are connected in series, as in Fig. 17.3–5b, and if we apply a DC source such that $i_a = i_b = I_s$, then $\theta_{ss} = K_M I_s^2$; we therefore have a DC ammeter or voltmeter that always deflects up-scale. However, because of the weaker magnetic field, the sensitivity will be rather poor compared to that of a d'Arsonval meter. On the other hand, the electrodynamometer also measures AC voltage or current, since an alternating current $i = I_m \cos \omega t$ will produce $\theta_{ss} = K_M(I_m^2 \cos^2 \omega t)_{av} = K_M I_m^2/2$. Thus, the meter can be

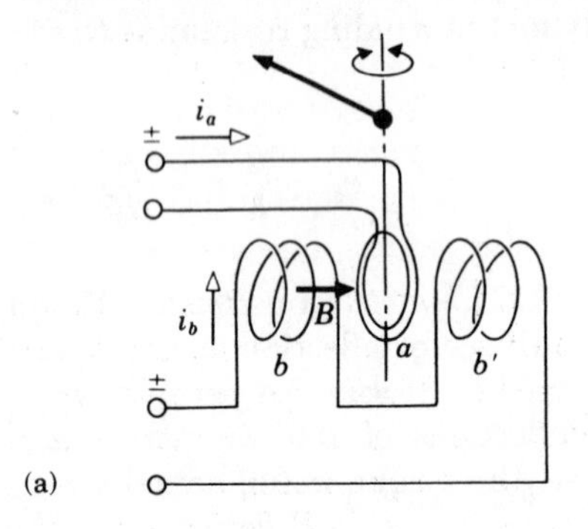

(b)

FIGURE 17.3–5
(a) Electrodynamometer movement. (b) Winding connection for AC/DC meter.

calibrated with a nonlinear scale to read the *rms value* $I_{\text{rms}} = I_m/\sqrt{2}$. Furthermore, an electrodynamometer has the distinctive capability of measuring *average power,* a topic discussed in Section 18.1.

Exercise 17.3–2

Suppose you want a d'Arsonval meter to have a full-scale deflection of 90° when $I_s = 10\ \mu\text{A}$. If $N = 25$, $r = 5$ mm, $\ell = 15$ mm, and the springs have $K = 5 \times 10^{-9}\ \text{N} \cdot \text{m/rad}$, what is the required value of B? Then calculate the full-scale voltage, assuming the winding resistance to be 4000 Ω/m.

Exercise 17.3–3

Obtain Eq. (5) by summing the induced voltages in each of the axial conductors. Hint: Consider the tangential velocity u.

MOVING-IRON DEVICES

In addition to force on current-carrying conductors, a magnetic field also acts on all nearby magnetic materials and tends to draw them into the densest part of the field—as demonstrated when a magnet picks up loose pieces of iron. Such forces exist at every air gap in a magnetic circuit, and are utilized by a variety of practical devices with movable iron members that convert the magnetic attraction into mechanical work. Because this gap force differs in nature from the induced force previously discussed, we must first develop a general expression for it that can be applied to specific configurations.

Figure 17.3–6 shows a portion of a magnetic circuit containing two plane-parallel iron faces with area A_g separated by a gap of length ℓ_g. The B field in the gap creates a force f_g that pulls the movable member toward the fixed piece and, consequently, alters the circuit's properties. For the purpose of analysis, we limit the movement to a very small *virtual displacement* dx and consider the resulting *virtual work* $dw_m = f_g\,dx$. If dw represents all other changes in energy caused by the displacement dx, conservation of energy then requires that $f_g\,dx + dw = 0$ or

$$f_g = -\frac{dw}{dx}$$

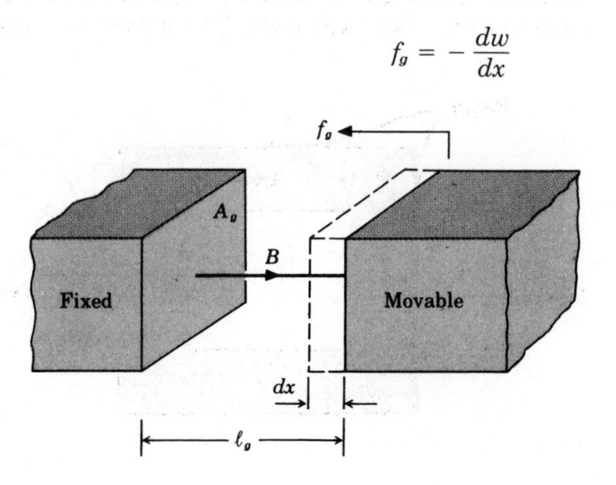

FIGURE 17.3–6
Magnetic force at an air gap.

Assuming B remains constant, dw consists entirely of the reduced stored energy in the air gap, namely

$$dw = -\frac{B^2 A_g}{2\mu_0}\,dx$$

which follows from Eq. (12), Section 17.1, since the volume of the gap has decreased by an amount $A_g\,dx$ (ignoring fringing). Therefore,

$$f_g = \frac{B^2 A_g}{2\mu_0} = \frac{\phi^2}{2\mu_0 A_g} \tag{8}$$

and the gap force is proportional to the square of the flux $\phi = BA$. This means that f_g is always inward, regardless of the direction of the flux.

An alternative expression for f_g is obtained by writing the stored energy in terms of the *gap reluctance* $\mathcal{R}_g = \ell_g/\mu_0 A_g$. Specifically, $w = B^2 A_g \ell_g/2\mu_0 = \phi^2 \ell_g/2\mu_0 A_g = \frac{1}{2}\phi^2 \mathcal{R}_g$ and

$$f_g = -\tfrac{1}{2}\phi^2 \frac{d\mathcal{R}_g}{dx} \tag{9}$$

where the minus sign agrees with the fact that $\mathcal{R}_g$ decreases as the gap gets smaller; that is, $d\mathcal{R}_g/dx$ is a negative quantity. This equation has particular value for studying moving-iron devices when the core reluctance and magnetic nonlinearities are negligible.

As a case in point, consider the magnetically operated switch or *relay* illustrated in Fig. 17.3–7. Switching contacts for an external circuit are attached to a fixed core and a hinged armature held open by the spring a distance ℓ_o in absence of magnetic excitation. Current i through the operating coil then creates a flux and downward force, drawing the contacts together. We'll calculate f_g for a fixed coil current and arbitrary gap

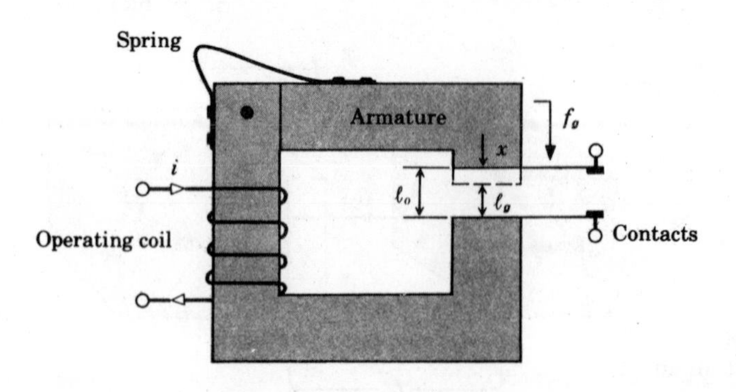

FIGURE 17.3–7
Magnetic relay.

spacing $\ell_g = \ell_o - x$, subject to the reasonable condition that the core has small reluctance compared to the gap.

The gap reluctance is $\mathcal{R}_g = \ell_g/\mu_0 A_g = (\ell_0 - x)/\mu_0 A_g$ so $d\mathcal{R}_g/dx = -1/\mu_0 A_g$ independent of the spacing. However, the flux depends on $\mathcal{R}_g$ and, therefore, on ℓ_g, since $\phi \approx \mathcal{F}/\mathcal{R}_g = \mu_0 NiA_g/\ell_g$. Substituting in Eq. (9) yields

$$f_g = \tfrac{1}{2}\left(\frac{\mu_0 NiA_g}{\ell_g}\right)^2\left(-\frac{1}{\mu_0 A_g}\right) = \frac{\mu_0 (Ni)^2 A_g}{2\ell_g^2} \tag{10}$$

and clearly brings out that f_g varies inversely with ℓ_g^2, whereas the opposing spring force would be relatively constant. Based on this result, we would expect the *pull-in* current required to close the relay, starting with $\ell_g = \ell_o$, will be greater than the *hold-in* current when the relay is closed and $\ell_g < \ell_o$. Also note that an AC coil current can be used since f_g is proportional to i^2.

Figure 17.3–8 illustrates another type of moving-iron device, called a magnetic actuator. The magnetic field of the actuator draws an iron plunger into the air gap and operates a valve or similar mechanical device. The plunger slides against brass (nonmagnetic) spacers of thickness ℓ_g, so that the combined series reluctance of the upper and lower spacers is $\mathcal{R}_g = 2(\ell_g/\mu_0 bx)$, assuming all the flux passes through the area $A = bx$ corresponding to that portion of the plunger within the gap. Thus, $d\mathcal{R}_g/dx = -2\ell_g/\mu_0 bx^2$, $\phi = \mu_0 Nibx/2\ell_g$, and we have

$$f_g = \frac{\mu_0 (Ni)^2 b}{4\ell_g} \tag{11}$$

independent of the plunger position x, a sometimes desirable feature. In practice, this device usually has a cylindrical configuration with a solenoidal coil around the plunger, thus the more common name *solenoid valve* or simply *solenoid.*

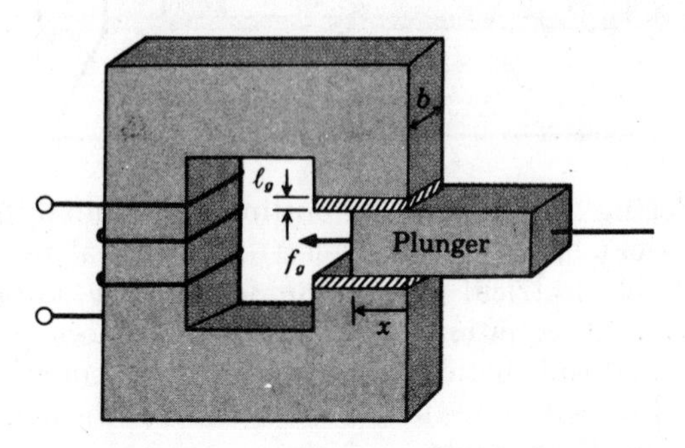

FIGURE 17.3–8 Magnetic actuator (or solenoid).

Example 17.3–3
Electromagnet

Figure 17.3–9 provides the cross-sectional dimensions of a large disk-shaped *electromagnet* of the type used to lift heavy pieces of iron or steel—a junked automobile, for instance. The inner and outer pole faces have areas A_1 and A_2, each equaling 0.283 m², and are protected by a brass sheet with $\ell_g = 2$ mm. Maximum force is attained when the load completely covers both faces, so that $\phi = Ni/\mathcal{R}_g$ where

$$\mathcal{R}_g = \frac{\ell_g}{\mu_0 A_1} + \frac{\ell_g}{\mu_0 A_2}$$

Equation (10) then gives the lifting force as

$$f_g = \frac{\phi^2}{2\mu_0 A_1} + \frac{\phi^2}{2\mu_0 A_2} = 0.0222(Ni)^2$$

Lifting a 1000-kg mass requires $f_g \geq 1000 \text{ kg} \times 9.8 \text{ m/sec} = 9800$ N, or an mmf of

$$Ni \geq \sqrt{\frac{9800}{0.0222}} = 664 \text{ A-t}$$

Of course a larger mmf would be needed in practice to compensate for imperfect contact with the load.

Exercise 17.3–4

Calculate the pull-in and hold-in mmf values for the relay in Fig. 17.3–7 when $\ell_g = 8$ mm and 3 mm, respectively; and $A_g = 1$ cm². The opposing spring force is essentially constant at $f_K = 10^{-6}$ N.

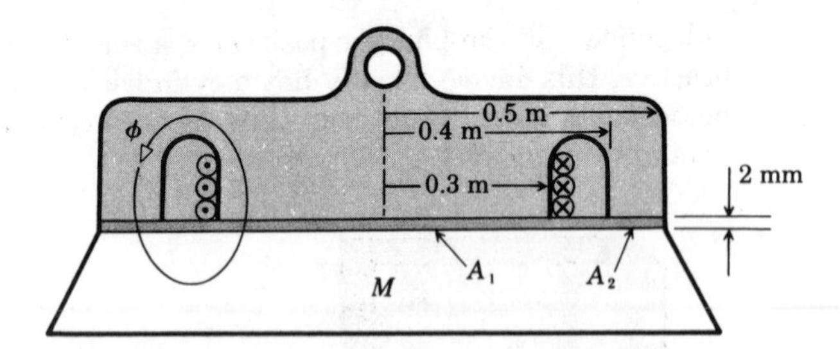

FIGURE 17.3–9
Electromagnet.

17.4 TRANSDUCER DYNAMICS †

Closing this chapter, we outline how to analyze the dynamic or transient behavior of an electromechanical transducer, including the effects of electrical and mechanical energy storage. We formulate the differential equations that describe devices with translational and rotational coil motion, and convert those equations into transfer functions. Linearity is assumed throughout for the mechanical, magnetic, and electrical components.

COUPLED DIFFERENTIAL EQUATIONS

The dynamic behavior of a translating-coil device may be investigated with the help of the symbolic representations in Fig. 17.4–1. These are based on Fig. 17.3–2, with the addition of electrical energy storage in the form of the coil's inductance L, mechanical energy storage in the form of the mass M (representing the coil and everything that moves with it), and a spring with stiffness constant K. The latter exerts a force $f_K = Kx$ when the coil has displacement x relative to its rest position. The coil's velocity is, of course, related to the displacement via $u = dx/dt$.

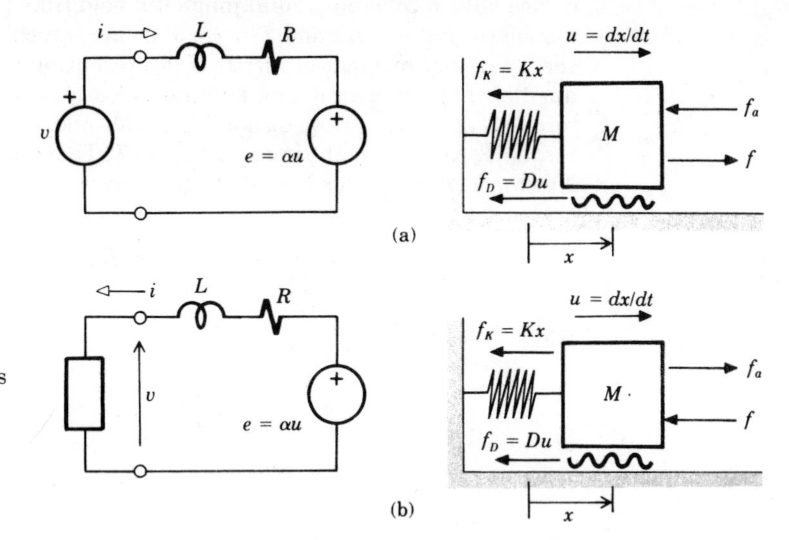

FIGURE 17.4–1 Translating-coil devices with energy storage. (a) electrical-to-mechanical; (b) mechanical-to-electrical.

Tackling the case of an electrical-to-mechanical transducer in Fig. 17.4–1a, we first write equations for the electrical and mechanical variables. Specifically, summing voltages around the loop according to Kirchhoff's voltage law gives

$$v = L\frac{di}{dt} + Ri + e$$

and summing forces according to Newton's third law gives

$$M\frac{du}{dt} = f - (f_a + Du + Kx)$$

where $du/dt = d^2x/dt^2$ is the acceleration of the mass. Substituting $e = \alpha u$ and $f = \alpha i$ and rearranging them yields

$$L\frac{di}{dt} + Ri = v - \alpha u \qquad M\frac{du}{dt} + Du + Kx = \alpha i - f_a \tag{1}$$

so we have a pair of coupled differential equations, with electrical and mechanical variables in both. We must also keep in mind the third equation, $u = dx/dt$.

Identical equations hold for a mechanical-to-electrical transducer, as in Fig. 17.4–1b, except that the signs are reversed on the right-hand side. Moreover, similar equations apply to rotating transducers, with rotational terms in place of translation terms.

Figure 17.4–2 shows the symbolic representations of a rotating-coil device with a constant radial magnetic field (like that of the d'Arsonval meter) so that $e = \beta\omega$ and $T = \beta i$. The mechanical terms here consist of the rotational moment of inertia J (representing the rotating parts), the angular displacement θ, and angular velocity $\omega = d\theta/dt$, frictional drag torque $T_D = D\omega$, spring torque $T_K = K\theta$, and external applied or load torque T_a, as well as the induced torque T. Thus, the rotational version of Newton's third law for Fig. 17.4–2a gives

$$J\frac{d\omega}{dt} = T - (T_a + D\omega + K\theta)$$

while, from KVL,

$$v = L\frac{di}{dt} + Ri + e$$

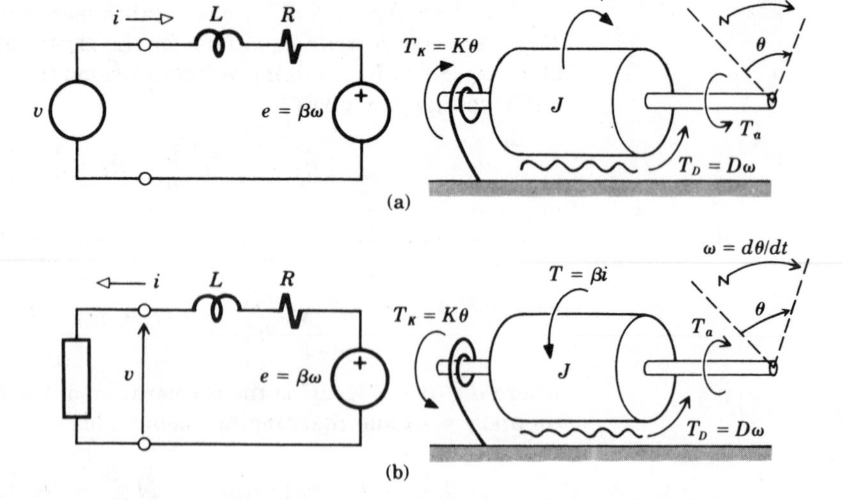

FIGURE 17.4–2 Rotating-coil devices with energy storage: (a) electrical-to-mechanical; (b) mechanical-to-electrical.

Rearranging then yields

$$L\frac{di}{dt} + Ri = v - \beta\omega \qquad J\frac{d\omega}{dt} + D\omega + K\theta = \beta i - T_a \tag{2}$$

which are directly comparable to Eq. (1). Again, reversing the signs on the right-hand side yields the equations for Fig. 17.4–2b.

ELECTRO-MECHANICAL TRANSFER FUNCTIONS

Our next step involves extending the *impedance* concept to include *mechanical* as well as electrical elements. We do this by assuming all variables to be exponential time functions of the general form Ae^{st}. Then, since $d(Ae^{st})/dt = sAe^{st}$, the differential equations reduce to algebraic equations in terms of s along with the element values and the coupling coefficient. Simple manipulations lead to a *transfer function* $H(s)$ that relates any two variables of interest. The dynamic behavior is determined from $H(s)$ as was done for electrical circuits in Chapters 7 and 8.

The method is illustrated below in an example. We use this method rather than attempting a generalized solution because the input and output variables depend on the specific applications; moreover, some of the energy-storage elements may justifiably be ignored in many cases of interest, thereby simplifying the analyses—an important consideration in view of the mathematical complications that arise with systems higher than second-order.

Example 17.4–1

Consider a d'Arsonval meter with negligible inductance and rotational inertia, so energy storage is limited to the pair of springs and we have a first-order system. We therefore set $L = J = 0$ in Eq. (2), and we also take $T_a = 0$ because there is no external torque to be included. We then let $v = Ve^{st}$, $i = Ie^{st}$, $\theta = \Theta e^{st}$, and $\omega = d\theta/dt = s\Theta e^{st}$ to get the algebraic equations

$$RIe^{st} = Ve^{st} - \beta s\Theta e^{st} \qquad (Ds + K)\Theta e^{st} = \beta Ie^{st}$$

in which e^{st} cancels out completely—as must be true for a linear system.

Assuming voltage v to be the input and angle θ the output, the transfer function $H(s) = \Theta e^{st}/Ve^{st}$ is obtained by eliminating I, which results in

$$H(s) = \frac{\beta/R}{K + (D + \beta^2/R)s} = \frac{\beta/KR}{1 + s\tau}$$

where

$$\tau = \frac{DR + \beta^2}{KR}$$

This function has the same form as that of a simple lowpass filter with DC gain $H(0) = \beta/KR$ and time constant $2\pi/f_{co} = \tau$, which follows by letting $s = j2\pi f$ and comparing the result with $\underline{H}(f)$ in Eq. (7), Section 7.1.

Drawing upon the lowpass-filter analogy, we conclude that a constant voltage $v = V_s$ applied at $t = 0$ produces the output step response

$$\theta(t) = \frac{\beta V_s}{KR}(1 - e^{-t/\tau})$$

as plotted in Fig. 17.4–3 for two values of the coupling coefficient. The steady-state deflection $\theta_{ss} = \beta V_s/KR$ agrees with our prior result, but we now see a trade-off between rapid response (small τ) and large deflection, because both τ and θ_{ss} increase with β. Meter design thus involves a potential compromise to achieve acceptable response time and deflection sensitivity.

Example 17.4–2

As a second example, let's find the steady-state AC frequency response of a loudspeaker represented by Eq. (1) with $R = 8$, $D = 2$, and load force $f_a = 30u$, as in Example 17.3–2. The other parameter values are $L \approx 0$, $\alpha = 16$, $M = 0.01$, and $K = 12{,}000$, all in SI units. The spring provides the restoring force that returns the diaphragm to its rest position.

The input is the voltage v, of course, and the diaphragm's velocity u will be taken as the output. To eliminate the displacement x, we first note that if $x = Xe^{st}$ then $u = dx/dt = sXe^{st}$; turning this around, if $u = Ue^{st}$ then $x = (U/s)e^{st}$, which better suits our purposes. Thus, writing $v = Ve^{st}$, etc., and substituting numerical values for the constants, we have

$$8I = V - 16U \qquad \left(0.01s + 2 + \frac{12{,}000}{s}\right)U = 16I - 30U$$

and routine algebra yields the transfer function

$$H(s) = \frac{Ue^{st}}{Ve^{st}} = \frac{200s}{s^2 + 6400s + 1.2 \times 10^6} = \frac{200s}{(s + 193)(s + 6207)}$$

The presence of s^2 signifies a second-order system, with the spring and mass storing energy.

The AC transfer function $\underline{H}(f)$ is then obtained by letting $s = j2\pi f$. After some more rearrangement, our final result becomes

$$\underline{H}(f) = 0.0322\,\frac{j(f/30.7)}{1 + j(f/30.7)}\,\frac{1}{1 + j(f/988)}$$

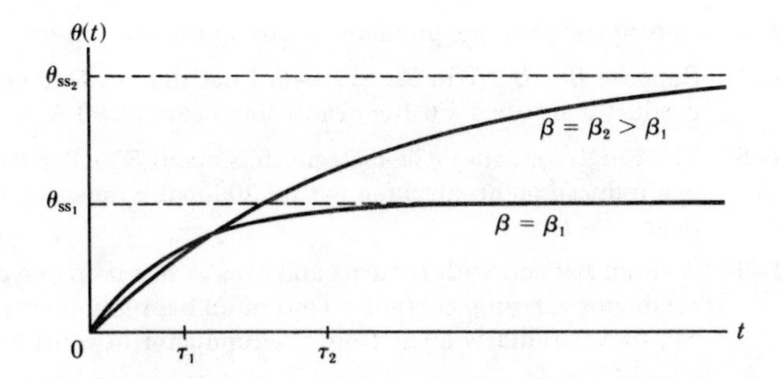

FIGURE 17.4–3
Step-response of a d'Arsonval movement.

in which the first term represents a midfrequency "gain" of 0.0322, the second term is a *highpass* function with cutoff frequency 30.7 Hz, and the third term is a *lowpass* function with cutoff frequency 988 Hz. Consequently, the loudspeaker has a *bandpass* characteristic over a frequency range of about 30 Hz to 1 kHz. This would make an excellent low-frequency woofer for a high-fidelity system that also has a tweeter to handle frequencies above 1 kHz.

Exercise 17.4–1 A moving-coil or dynamic microphone might be modeled by Fig. 17.4–1b, with $L = M = D \approx 0$. The input is acoustical force f_a, and the output voltage is applied to a load resistance R_L. Taking $f_a = Fe^{st}$, $x = Xe^{st}$, etc., obtain an expression for the AC transfer function $\underline{H}(f)$ from $H(s) = Ve^{st}/Fe^{st}$. Don't forget to reverse the signs on the right of Eq. (1).

PROBLEMS

17.1–1 The *mixed English* system of units expresses length in inches and flux in *kilolines,* where 1 kiloline = 10^{-5} Wb.

(a) Convert $B = 1$ T to English units.

(b) Calculate the value of μ_0 in English units.

17.1–2 Two parallel copper bars in a power plant are 1-m long and 0.5-m apart. Use Eqs. (3) and (7), Section 17.1, to find the force on each bar if, because of a circuit fault, they each carry 10,000 A.

17.1–3 Repeat the previous problem for the blades of a knife switch. The blades are 10-cm long, 4-cm apart, and each carries 100 A.

17.1–4 Consider two adjacent turns of the coil in Fig. 17.1–3b. Find the approximate direction of the force on one turn caused by the field of the other. Then describe what happens when a loosely wound coil carries a large current.

17.1–5 The pulley is disconnected from the moving bar in Fig. 17.1–6a, and the bar starts at rest and accelerates through a distance of 12 m along the rails. Calculate the final velocity if $B = 2$ T, $\ell = 0.1$ m, $i = 15$ A, the bar has mass $M = 0.5$ kg, and there are no losses. Also find the average power $P = \Delta w/\Delta t$ drawn from the source. Hint: First find the duration of the acceleration interval.

17.1–6 Repeat the previous problem assuming the bar has an initial velocity of 9 m/sec.

17.1–7 Suppose $B = 0.2$ T in Fig. 17.1–6. Find the net field on each side of the moving conductor if it has a 0.1-cm radius and carries 50 A.

17.1–8 The Earth's magnetic field strength is about 50 μT. Estimate the maximum voltage induced on an airplane with a 200-foot wingspan traveling at 600 miles per hour.

17.1–9 A small flat coil with N turns and area A has its plane aligned with the axis of a conductor carrying current i. Obtain an expression for the induced voltage if the coil moves radially away from the conductor at constant velocity u.

‡ **17.1–10** Suppose the wire used for the coil in Fig. 17.1–3a has resistance R_0 per unit length. If $N \gg h/r$, show that the total winding resistance is related to the inductance by $R_w \approx R_0\sqrt{4\pi hL/\mu}$.

17.2–1 Let $\mathcal{R}_{cm}$ denote reluctance calculated with core dimensions in *centimeters,* and let ϕ_{mWb} be the flux in *milliwebers.* Show that $\phi_{mWb} = 10\,\mathcal{F}/\mathcal{R}_{cm}$ and $B = 10\phi_{mWb}/A_{cm^2}$, where A_{cm^2} is the area in square centimeters.

17.2–2 The core dimensions given in Fig. P17.2–2 are in centimeters, and the core is 5-cm thick. The material has $\mu = 5 \times 10^{-3}$ and negligible nonlinearity. Estimate the values of B in the other two vertical legs if a coil with $Ni = 50$ A-t is around the left leg. (You may use the technique in Problem 17.2–1.)

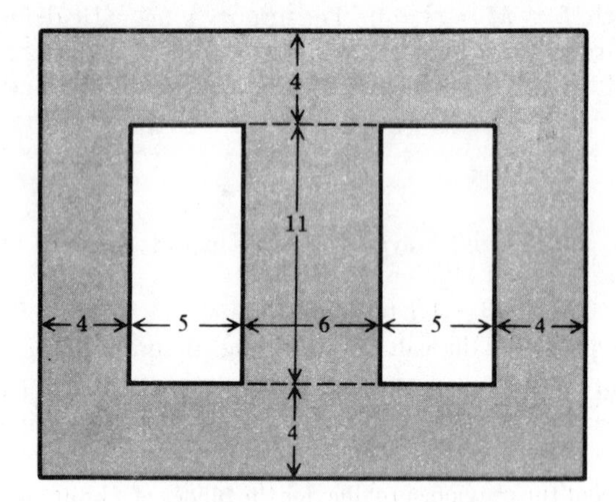

FIGURE P17.2–2

17.2–3 Repeat the previous problem assuming the middle leg, between the dashed lines, is a different material with $\mu = 10^{-3}$.

17.2–4 Suppose a 50-turn coil is around the middle leg of the core described in Problem 17.2–2. Estimate the current required to obtain $B = 0.1$ T in each of the outer legs.

17.2–5 Suppose the core described in Problem 17.2–2 has a 0.2-cm air gap cut through the middle leg, and identical coils on the outer legs. Estimate the total mmf needed to produce $B = 0.1$ T in the gap.

17.2–6 Repeat the previous problem with a 0.5-cm gap.

17.2–7 Calculate the ratio of energy stored in the core to total stored energy for the magnetic circuit in Exercise 17.2–2.

17.2–8 Use Eq. (7) to find and plot $\mu_r = \mu/\mu_0$ versus H for the sheet-steel material in Fig. 17.2–4.

17.2–9 When a coil has a nonlinear core and $i = I_0 + \Delta I \cos \omega t$, where $\Delta I \ll I_0$, the flux will be $\phi \approx \phi_0 + \Delta\phi \cos \omega t$ and the equivalent inductance is $L \approx N\, \Delta\phi/\Delta I$. Calculate L for a 25-turn coil with $I_0 = 1$ A and $\Delta I = 0.2$ A if the core is sheet steel with area $A = 10^{-4}$ m^2 and length $\ell = 0.05$ m. Compare your result with the value of $N\phi_0/I_0$.

17.2–10 Measurements indicate that a certain magnetic circuit has a total AC core loss of 25 W at 50 Hz and 32.4 W at 60 Hz, the voltage amplitude being held constant. Find the total core loss at 120 Hz.

17.2–11 A magnetic circuit with $N = 100$ and $\ell_c = 0.5$ m is found to have an AC hysteresis loss of 10 W when the peak current is $I_m = 1$ A. Estimate P_h when $I_m =$ 0.5 A and 2.0 A if the material is cast iron. Take $n = 1.5$ in Eq. (8).

17.2–12 Repeat the previous problem for a sheet-steel core.

17.3–1 A certain translating coil device has $B = 0.4$ T, $N = 100$ turns, $c = 0.2$ m, $R =$ 4 Ω, and $D = 9$ N · s/m.

(a) What value of applied voltage v holds the coil stationary when $f_a = 500$ N?
(b) If the terminals of the coil are shorted, what is the resulting velocity u when $f_a = 500$ N?

17.3–2 The device in the previous problem is used as a velocity sensor with a load resistance $R_L = 10$ Ω connected to the coil's terminals. Find v in terms of u, and calculate the efficiency p_e/p_m.

17.3–3 Verify that $\alpha = 16$ is the optimum coupling value in Example 17.3–2, and calculate the resulting efficiency p_m/p_e.

17.3–4 Suppose the iron cylinder is removed from Fig. 17.3–4c so the field lines are parallel and horizontal between the pole faces. Let θ now be measured from the horizontal center line to the pointer. Obtain an expression for the torque T at angle θ.

17.3–5 Let the coil in the device described in the previous problem be rotated at constant angular velocity so that $\theta = \omega t$. Find the conductors' velocity perpendicular to the field and obtain an expression for the induced emf.

17.3–6 Calculate the force across the gap in the magnetic circuit of Exercise 17.2–2.

17.3–7 Rewrite Eq. (11) in terms of the field B in the core. If $B \leq 1.4$ T (because of saturation) and $\ell_g \leq b/10$ (to minimize fringing), what is the condition on b to obtain $f_g = 100$ N?

17.3–8 Suppose the field in Fig. 17.3–6 is produced by a coil with inductance $L = N^2/\mathcal{R}$ and constant mmf $\mathcal{F} = Ni$. Show that Eq. (9) becomes $f_g = \frac{1}{2} i^2\, dL/dx$ if $\mathcal{R} \approx \mathcal{R}_g$.

17.3–9 An *electrostatic transducer* is created by letting $B = 0$ in Fig. 17.3–6 and applying a voltage v across the gap. We then have a *capacitor* C with an electric field $\mathcal{E} =$ v/ℓ_g (neglecting fringing). Show that the stored energy is $w = \frac{1}{2}\epsilon_0\mathcal{E}^2 A_g \ell_g$. Then apply the virtual-work method to obtain the gap force $f_g = \frac{1}{2}\epsilon_0\mathcal{E}^2 A_g$.

17.3–10 Estimate the maximum practical value of f_g/A_g for the electrostatic transducer in the previous problem. Compare with that of a moving-iron transducer. Hint: See Problem 5.1–6 and Fig. 17.2–4.

17.4–1 Let Fig. 17.4–2a represent a *motor* with input v, output ω, load torque $T_a = D_a \omega$, and $L = K = 0$.

(a) Obtain expressions for $H(s)$ and the time constant τ.

(b) Let $s = 0$ to find ω_{ss} when $v = V_s$.

(c) What is the condition on β for constant-speed operation despite load variations?

17.4–2 A *vibration table* is represented by Fig. 17.4–1a with $D = K = f_a = 0$, $L = 0.2$ H, $M = 5$ kg, and $R = 1\ \Omega$.

(a) Find $H(s) = Ue^{st}/Ve^{st}$ in terms of α.

(b) What value of α makes this device resonant at 60 Hz?

17.4–3 Let $s = j\omega$ and find the input impedance $Z(j\omega) = V/I$ of the loudspeaker in Example 17.4–2.

17.4–4 Find $H(s)$ when the loudspeaker in Example 17.4–2 has $M = 0$ and $L = 0.05$ H. Is the system overdamped or underdamped?

17.4–5 Repeat Example 17.4–1, including the rotational inertia J. Then obtain the condition on D for critical damping.

18

AC POWER SYSTEMS

Electric-power engineering deals with the generation, delivery, and use of electrical energy. The present chapter serves as an introduction to the AC-power aspects of this important field. We give considerable attention to power calculations and measurements, transformers, three-phase circuits, and power transmission and distribution. The discussion of generators and rotating loads will be deferred until Chapter 19.

OBJECTIVES

After studying this chapter and working the exercises, you should be able to do each of the following:

- Distinguish between real, reactive, and apparent power in an AC circuit; and carry out power calculations in terms of these quantities and the power factor (Section 18.1).
- State the need for and technique of power-factor correction (Section 18.1).
- Determine the voltages, currents, and efficiency of a transformer, given its equivalent circuit parameters (Section 18.2).

- Find the voltages, currents, and power in a balanced three-phase circuit with a wye or delta load, and state the advantages of a three-phase system (Section 18.3).
- Explain how power can be measured in a single-phase or three-phase system, and how power measurements are used to obtain a transformer's parameter values (Sections 18.1, 18.2, and 18.3).
- Analyze power transmission and dissipation on an AC transmission line, and explain why three-phase transformers play a critical role in power distribution (Section 18.4). †

18.1 AC POWER

Back in Section 6.3 we saw that instantaneous power in an AC circuit has an oscillatory behavior caused by the joint effects of power dissipation and of energy flowing back and forth between the source and reactive elements. This section pursues the details of that behavior and introduces the concepts of reactive and apparent power and the power factor—all of which help an engineer focus on the critical task of efficient power delivery from source to load.

REAL AND REACTIVE POWER

Let's investigate the power delivered to the series AC load impedance in Fig. 18.1–1a. In such investigations, it proves convenient to work exclusively with the effective or rms values of AC voltage and current, rather than the peak values V_m and I_m. Accordingly, we henceforth express phasor magnitudes in rms terms as

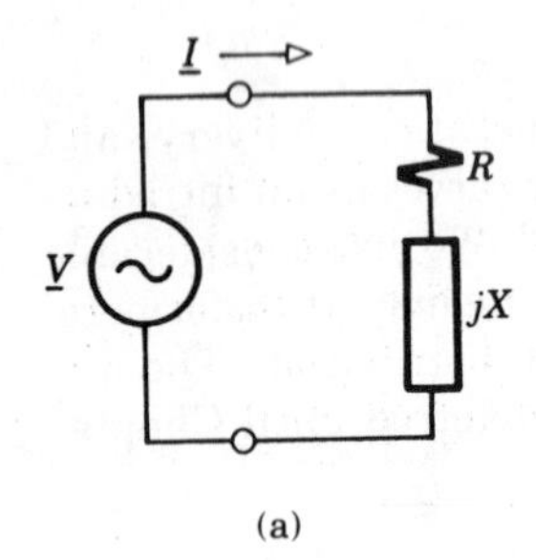

(a)

$$|V| \triangleq \frac{V_m}{\sqrt{2}} \qquad |I| \triangleq \frac{I_m}{\sqrt{2}} \tag{1a}$$

such that the phasors

$$\underline{V} = |V| \underline{/\theta_v} \qquad \underline{I} = |I| \underline{/\theta_i} \tag{1b}$$

represent the sinusoids

$$v(t) = \sqrt{2}\,|V| \cos(\omega t + \theta_v) \qquad i(t) = \sqrt{2}\,|I| \cos(\omega t + \theta_i) \tag{1c}$$

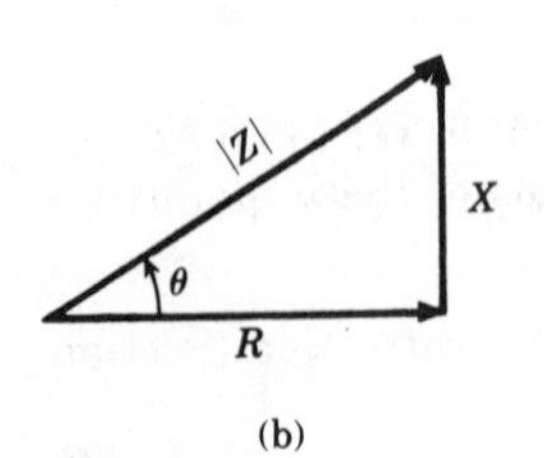

(b)

FIGURE 18.1–1

Ohm's law for AC circuits still holds in the form

$$\underline{V} = \underline{Z}\underline{I}$$

since $\sqrt{2}$ cancels out on both sides when we convert to rms values. Thus, $|V| = |Z|\,|I|$ and $\theta_v = \theta + \theta_i$ where, for simplicity, we have dropped the subscript z from the impedance angle $\theta = \measuredangle \underline{Z}$.

Using our rms notation, the instantaneous power $p(t)$ will be

$$p(t) = v(t)i(t) = 2|V|\,|I| \cos(\omega t + \theta_v) \cos(\omega t + \theta_i)$$
$$= |V|\,|I| \cos\theta\,[1 + \cos(2\omega t + 2\theta_v)] + |V|\,|I| \sin\theta \sin(2\omega t + 2\theta_v)$$

which has been expanded with the help of standard trigonometric identities and the fact that $\theta_i = \theta_v - \theta$. The terms $|V|\,|I| \cos\theta$ and $|V|\,|I| \sin\theta$ have important interpretations that are brought out by substituting $|V| = |Z|\,|I|$, noting from the impedance diagram of Fig. 18.1–1b that

$$|Z| \cos\theta = R \qquad |Z| \sin\theta = X$$

We then define

$$P \triangleq |V|\,|I| \cos\theta = |I|^2 R \tag{2a}$$

$$Q \triangleq |V|\,|I| \sin\theta = |I|^2 X \tag{2b}$$

so that

$$p(t) = p_R(t) + p_X(t) \tag{3a}$$

where

$$p_R(t) = P[1 + \cos(2\omega t + 2\theta_v)] \qquad p_X(t) = Q \sin(2\omega t + 2\theta_v) \tag{3b}$$

We have now decomposed the instantaneous power into the two components sketched in Fig. 18.1–2.

The first component $p_R(t)$ oscillates between 0 and $2P$ and has the average value P; this component represents all the *power dissipation* in the load *resistance,* and we call P the *real power.* The second component $p_X(t)$ oscillates between $-Q$ and $+Q$ and has zero average value; this component represents *energy storage* in the load's *reactance,* and we call Q the *reactive power.* Both P and Q have the units of volts × amps, or watts, but the special designation *volt-amperes reactive* (VAr) is used for Q to emphasize that it corresponds to reactive energy storage rather than actual power dissipation.

If the load happens to be purely reactive, it dissipates no power at all ($P = 0$) and $p(t) = p_X(t)$. For a single inductance or capacitance, Eq. (2b) becomes

$$Q_L = \omega L\,|I|^2 = \frac{|V|^2}{\omega L} \tag{4a}$$

$$Q_C = \frac{-|I|^2}{\omega C} = -\omega C\,|V|^2 \tag{4b}$$

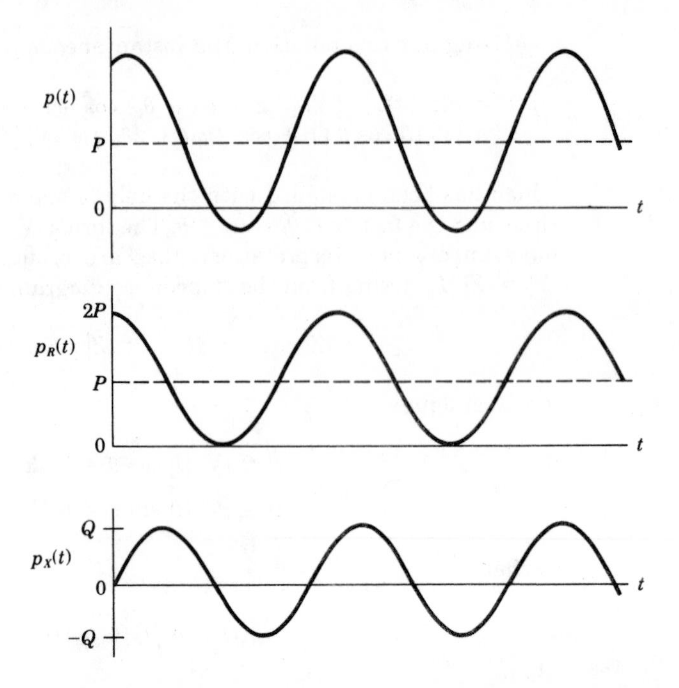

FIGURE 18.1–2 Instantaneous power and its real and reactive components.

where the negative value of Q_C means that the waveform $p_X(t)$ is inverted. More generally, an inductive load with $X > 0$ has $Q > 0$, and a capacitive load with $X < 0$ has $Q < 0$. We therefore say that inductive loads "consume" reactive power, whereas capacitive loads "produce" reactive power.

Although Q has no bearing on real power dissipation, it does have significant adverse impact on power transfer from source to load. Specifically, for fixed values of P and $|V|$, the required rms current increases with Q as

$$|I| = \frac{\sqrt{P^2 + Q^2}}{|V|} \tag{5}$$

Thus, for instance, if $Q = 0.5\,P$ then $|I| = \sqrt{1.25}P/|V|$ and the current must be about 12% more than it would be if $Q = 0$. This increased value reflects the flow of "reactive current" back and forth between source and load as the reactance alternately stores and returns energy. Increased current also means increased power wasted as ohmic heat in connecting wires. Equation (5) is easily derived by squaring and adding Eqs. (2a) and (2b).

The preceding results and conclusions apply in the case of a parallel load impedance or a combination of series and parallel elements. In these cases, however, R and X must be interpreted as the equivalent AC resistance and reactance.

Example 18.1–1

Figure 18.1–3 represents a parallel RC load connected to a generator with $|V| = 240$ V (rms) by a cable having $R_w = 2\ \Omega$ wire resistance. The total impedance seen by the generator is

$$\underline{Z} = 2 + (15 \| -j20) = 11.6\ \Omega - j7.2\ \Omega = 13.65\ \Omega\ \underline{/-31.8^\circ}$$

Hence, $|I| = (240\ \text{V}/13.65\ \Omega) = 17.6$ A, $|I|^2 = 309\ \text{A}^2$ and applying Eq. (2) gives

$$P = 309 \times 11.6 = 3584\ \text{W} \qquad Q = 309 \times (-7.2) = -2224\ \text{VAr}$$

The power dissipated in the cable is $P_w = 309 \times 2 = 618$ W leaving $P_L = 3584 - 618 = 2966$ W delivered to the 15-Ω load resistance. The transmission efficiency is $2966/3584 \approx 83\%$.

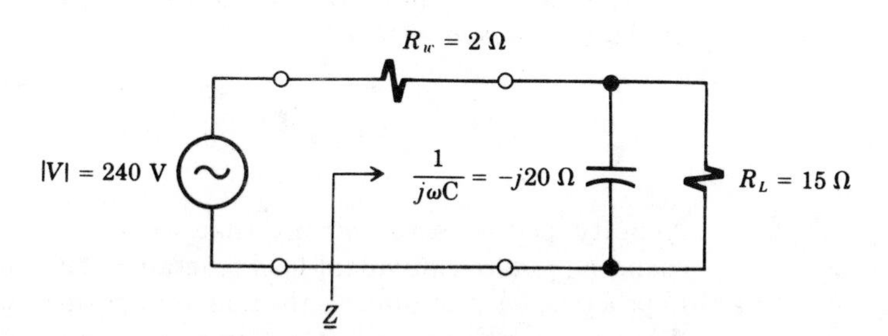

FIGURE 18.1–3 Circuit for Example 18.1–1.

But suppose the capacitive reactance were eliminated. Then the current goes down to $|I| = 240\ \text{V}/(2 + 15)\ \Omega = 14.1$ A, while the load power actually increases somewhat to $P_L = 14.1^2 \times 15 = 2989$ W. The efficiency is now about 88% since the wire loss has decreased to $P_w = 14.1^2 \times 2 = 399$ W.

Exercise 18.1–1

Repeat the preceding calculations with $j\omega L = j10\ \Omega$ in series with R_w.

APPARENT POWER AND POWER FACTOR

The product of the rms voltage and current at the terminals of a load is called the *apparent power,* measured in *volt-amperes* (VA). Apparent power is related to the real and reactive powers by Eq. (5), when written as

$$|V|\,|I| = \sqrt{P^2 + Q^2}$$

The form of this expression suggests that we could diagram apparent power as the hypotenuse of a right triangle whose sides are P and Q. For that purpose, we take the complex conjugate of the current phasor

$$\underline{I}^* \triangleq |I| \underline{/-\theta_i} = |I| \underline{/\theta - \theta_v}$$

and introduce the *complex power*

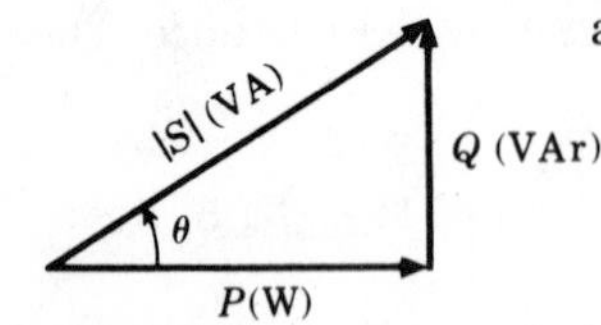

FIGURE 18.1–4
The power triangle.

$$\underline{S} \triangleq \underline{V}\underline{I}^* = |V|\,|I| \underline{/\theta} \tag{6a}$$
$$= |V|\,|I| \cos\theta + j\,|V|\,|I| \sin\theta$$
$$= P + jQ \tag{6b}$$

The resulting complex-plane diagram is known as the *power triangle,* Fig. 18.1–4, with hypotenuse $|S| = |V|\,|I|$. This triangle has the same shape as the impedance diagram (Fig. 18.1–1b), since

$$\underline{S} = |I|^2R + j\,|I|^2X = |I|^2\,\underline{Z}$$

which follows from Eq. (2) and $\underline{Z} = R + jX$.

The horizontal projection $|S| \cos\theta$ clearly equals the real power P, so we define the *power factor*

$$\text{pf} \triangleq \frac{P}{|S|} = \cos\theta \tag{7}$$

A unity power factor means that $|S| = P$, $\theta = 0$, and $Q = 0$, corresponding to zero equivalent load reactance. This condition is sought after in practice, for it minimizes the apparent power and the required current. When $\theta \neq 0$, the power factor is less than unity and $Q \neq 0$. The designations *lagging* and *leading* power factor are used to distinguish between $\theta > 0$ (inductive load: current lags voltage) and $\theta < 0$ (capacitive load: current leads voltage) when pf < 1. Most industrial loads have a lagging power factor resulting from the inductance of motor windings. A capacitance may then be added in parallel to "correct" the power factor—that is, to make pf ≈ 1.

A handy property of complex power is its *conservation,* in the sense that

> If two or more loads are fed from the same supply, the total complex power equals the sum of the individual components.

Thus, if two loads draw $\underline{S}_1 = P_1 + jQ_1$ and $\underline{S}_2 = P_2 + jQ_2$ individually, the total load draws $\underline{S} = \underline{S}_1 + \underline{S}_2 = (P_1 + P_2) + j(Q_1 + Q_2)$. Use of this conservation property, together with the power triangle, is illustrated by the following example.

Example 18.1–2 Consider the composite load circuit in Fig. 18.1–5a, with the individual components described by their real power consumption and power factor. (From the power-factor values we infer that $\underline{Z}_1$ is highly inductive, whereas $\underline{Z}_2$ is slightly capacitive.) We compute the apparent and reactive power drawn by $\underline{Z}_1$ using Fig. 18.1–5b, as follows:

$$|S_1| = 30\text{ kW}/0.6 = 50\text{ kVA} \qquad Q_1 = \sqrt{|S_1|^2 - P_1^2} = 40\text{ kVAr}$$

Thus

$$|I_1| = |S_1|/|V| = 208\text{ A}$$

Similarly, one finds that $|S_2| = 25$ kVA, $Q_2 = -7$ kVAr, and $|I_2| = 104$ A. The total load then has

$$P = 30 + 24 = 54\text{ kW} \qquad Q = 40 - 7 = 33\text{ kVAr}$$

$$|S| = \sqrt{54^2 + 33^2} = 63.3\text{ kVA} \qquad \text{pf} = 54/63.3 = 0.85$$

$$|I| = 63.3\text{ kVA}/240\text{ V} = 264\text{ A}$$

The power factor is lagging because $Q > 0$.

You should carefully observe those terms that add algebraically (P and Q), and those that add vectorially ($\underline{S}$ and $\underline{I}$). Also note that the total current is about 17% higher than the minimum of 225 A needed to get $P = 54$ kW with $|V| = 240$ V. We could reduce the current to 225 A by

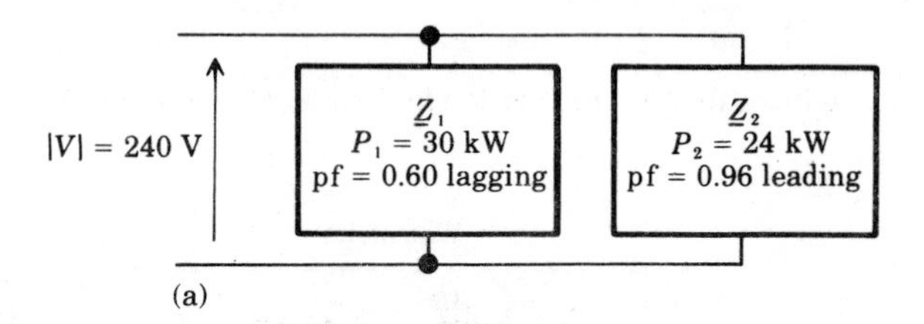

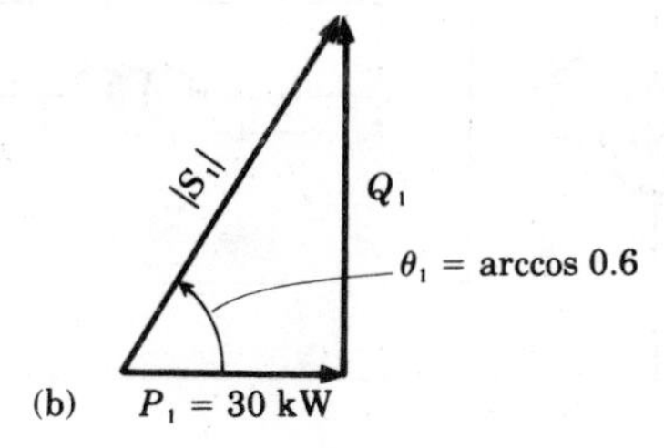

FIGURE 18.1–5 Circuit and power triangle for Example 18.1–2.

adding a parallel capacitance with $\omega C = 33\ \text{kVAr}/(240\ \text{V})^2$ so that $Q + Q_C = 0$ and pf = 1.0.

Exercise 18.1–2

Measurements on a certain 240-V AC motor reveal that $P = 12$ kW and $|I| = 60$ A. Find pf, Q, and $\underline{Z}$, making the reasonable assumption that the motor is inductive.

Exercise 18.1–3

Suppose that the motor in the previous exercise is put in parallel with a 12-Ω resistive load, and that a capacitance is also added in parallel to correct the power factor.
(a) Find P, Q, $|S|$, and $|I|$ for the total load without the capacitance.
(b) Determine the value of C needed to get pf = 1 if $\omega = 377$, and then recalculate $|I|$.

POWER MEASUREMENTS

Figure 18.1–6 shows an electrodynamometer connected as a *wattmeter* measuring the real power consumption of a load. The stationary and moving coils have negligible impedance, but a large multiplier resistance R_M has been added in series with the moving coil so that $i_a(t) = v(t)/R_M$, while $i_b(t) \approx i(t)$. From Eq. (7), Section 17.3, the meter's steady-state deflection will be

$$\theta_{ss} = K_M \frac{\sqrt{2}\,|V|}{R_M} [\cos(\omega t + \theta_v) \times \sqrt{2}\,|I| \cos(\omega t + \theta_v - \theta)]_{av}$$

$$= \frac{K_M}{R_M} |V|\,|I|\,[\cos\theta + \cos(2\omega t + 2\theta_v - \theta)]_{av}$$

whose first term inside the brackets is a constant and whose second term

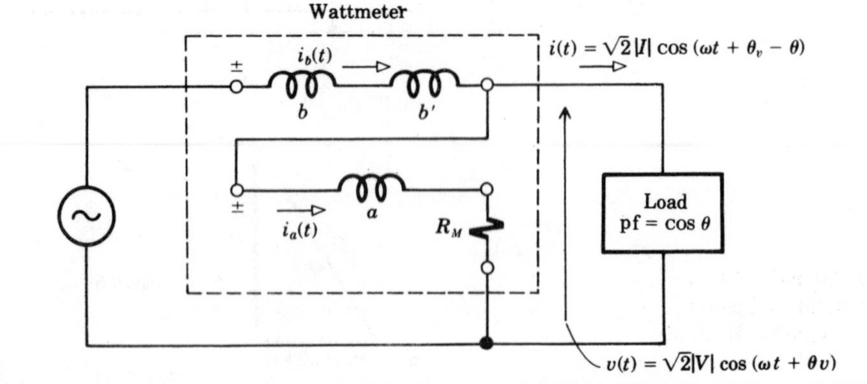

FIGURE 18.1–6
Power measurement with an electrodynamometer connected as a wattmeter.

has zero average value. Thus,

$$\theta_{ss} = \frac{K_M}{R_M} |V|\,|I| \cos \theta = \frac{K_M}{R_M} P \tag{8}$$

and the meter can be calibrated to read directly in watts.

Actually, the meter reading will be slightly greater than the load power P because $i_b(t) = i(t) + i_a(t)$. If we require better accuracy, we simply disconnect the load and subtract the resulting no-load reading from the original value. More sophisticated wattmeters have a *compensating winding* that cancels the no-load term.

Having found P, we can use separate ammeter and voltmeter measurements of $|V|$ and $|I|$ to calculate the apparent power $|S| = |V|\,|I|$, the reactive power $|Q| = \sqrt{|S|^2 - P^2}$, and pf $= P/|S|$. Determining the sign of Q (or of θ) entails additional measurement information from, say, an oscilloscope display to see whether $i(t)$ leads or lags $v(t)$. Alternatively, simple modifications convert an electrodynamometer into a *varmeter* that will measure Q directly.

18.2 TRANSFORMERS

One of the most important components in an AC power system is the transformer, which makes it possible to step-up or step-down voltage wherever necessary. The use of stepped-up high voltages for efficient power transmission will be clarified in Section 18.4. Here we'll review the concept of the *ideal* transformer, first introduced in Section 4.1, and show how its characteristics arise from magnetic coupling in an idealized magnetic circuit. We then discuss the differences between real and ideal transformers, and develop appropriate AC equivalent circuits.

MAGNETIC COUPLING IN THE IDEAL TRANSFORMER

The behavior of an ideal transformer is approximated by the magnetically coupled circuit in Fig. 18.2–1a. There are two lossless windings (the *primary* with N_1 turns and the *secondary* with N_2 turns), and an ideal core whose hypothetical material has a linear *B-H* curve and very large permeability μ. Under these conditions no leakage flux escapes from the core, and the total flux ϕ is proportional to the net magnetomotive force.

When current i_1 flows into the dotted terminal of the primary, it establishes an mmf $\mathfrak{F}_1 = N_1 i_1$ and, by the righthand rule, produces flux in the clockwise direction. Conversely, current i_2, flowing out of the dotted terminal of the secondary, creates an opposing mmf $\mathfrak{F}_2 = N_2 i_2$. So if the core has reluctance $\mathcal{R} = \ell_\phi/\mu A$ then $\phi = (\mathfrak{F}_1 - \mathfrak{F}_2)/\mathcal{R}$ or

$$N_1 i_1 - N_2 i_2 = \mathcal{R}\phi$$

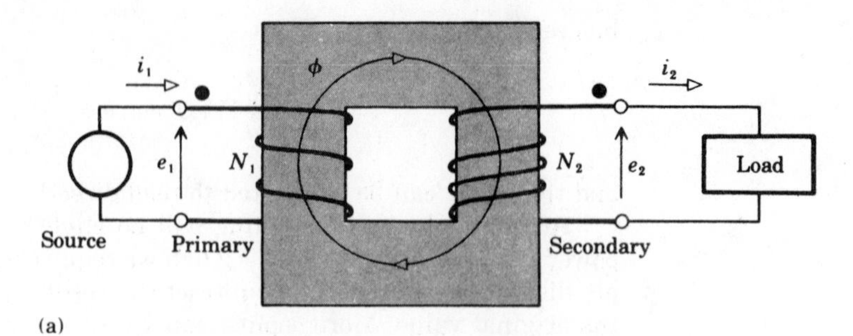

(a)

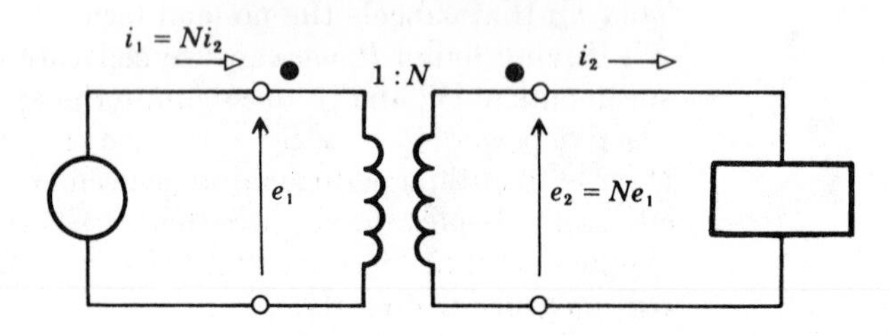

(b)

FIGURE 18.2–1
(a) Ideal transformer.
(b) Circuit representation.

However, we are assuming μ to be so large that, in essence, $\mu \to \infty$ and $\mathcal{R} \to 0$. Since $\mathcal{R}\phi = 0$, the applied mmf's must balance out in the sense that $N_1 i_1 - N_2 i_2 = 0$. Therefore,

$$i_1 = \frac{N_2}{N_1} i_2 = Ni_2 \tag{1}$$

where we have introduced the secondary-to-primary *turns ratio*

$$N \triangleq \frac{N_2}{N_1}$$

Equation (1) gives the relationship between the currents in an ideal transformer.

For the voltage relationship, observe that ϕ completely links both windings and, if time-varying, induces the emf's

$$e_1 = N_1 \frac{d\phi}{dt} \qquad e_2 = N_2 \frac{d\phi}{dt}$$

in accordance with Faraday's law. Therefore,

$$e_2 = \frac{N_2}{N_1} e_1 = Ne_1 \tag{2}$$

which, together with $i_1 = Ni_2$, leads to the circuit representation for an ideal transformer as diagrammed in Fig. 18.2–1b. This diagram simply repeats Fig. 4.1–4a except that we have used the induced-voltage notation e_1 and e_2 rather than v_1 and v_2.

To relate Fig. 18.2–1b to the magnetic circuit in Fig. 18.2–1a, suppose that a voltage source v_s is applied across the primary, forcing $e_1 = v_s$. Then $d\phi/dt = e_1/N_1$, which induces $e_2 = N_2 d\phi/dt = Ne_1$. A load resistance R_L across the secondary will draw $i_2 = e_2/R_L = Ne_1/R_L$, and i_2 in turn produces the mmf $\mathfrak{F}_2 = N_2 i_2$. The applied source must therefore provide current i_1 for the balancing mmf $\mathfrak{F}_1 = N_1 i_1 = \mathfrak{F}_2$, so $i_1 = Ni_2 = N^2 e_1/R_L$, and the equivalent input resistance seen by the source is

$$R_i = e_1/i_1 = R_L/N^2 \tag{3}$$

The input power $p_1 = e_1 i_1$ precisely equals the output power $p_2 = e_2 i_2$, the ideal transformer being a lossless device. (You might find it interesting to carry out similar arguments for the case of an applied current source such that $i_1 = i_s$.)

With a constant or DC excitation, the primary and secondary windings act like *short circuits* because $d\phi/dt = 0$ and $e_1 = e_2 = 0$. With an AC excitation, all voltages and currents will be sinusoidal and can be represented by their phasors. This important case deserves closer scrutiny—particularly when an arbitrary load impedance $\underline{Z}_L$ has been connected across the secondary, as in Fig. 18.2–2a.

We begin the analysis of Fig. 18.2–2a by assuming a sinusoidal flux variation with *peak* value Φ, angular frequency $\omega = 2\pi f$, and phase

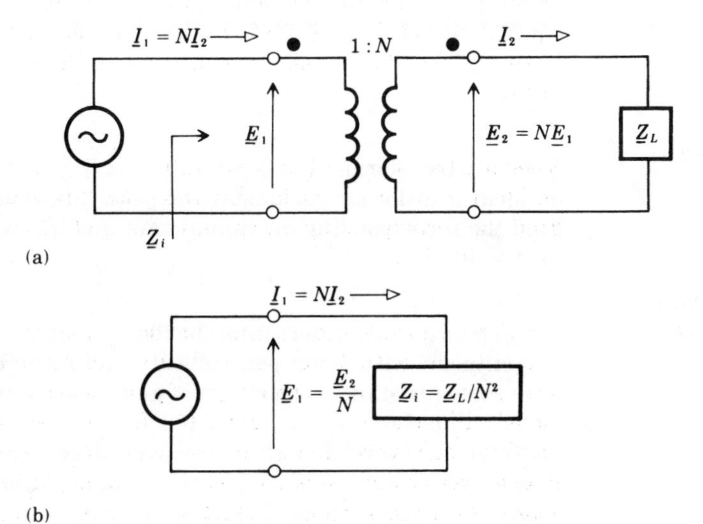

FIGURE 18.2–2
(a) Ideal transformer with load impedance.
(b) Equivalent circuit viewed from the source.

angle θ_ϕ, so that

$$\phi = \Phi \cos(\omega t + \theta_\phi)$$

Taking the time derivative and applying standard trigonometric identities yields

$$e_1 = N_1 \frac{d\phi}{dt} = N_1 \Phi[-\omega \sin(\omega t + \theta_\phi)]$$
$$= \omega N_1 \Phi \cos(\omega t + \theta_\phi + 90°)$$

revealing that the induced emf leads the flux by 90° and has the rms value

$$|E_1| = \frac{\omega N_1 \Phi}{\sqrt{2}} = \sqrt{2}\,\pi f N_1 \Phi = 4.44\, f N_1 \Phi \tag{4}$$

It likewise follows that if $e_1 = \sqrt{2}\,|E_1| \cos \omega t$, corresponding to the phasor $\underline{E}_1 = |E_1| \underline{/0°}$, then $\theta_\phi = -90°$ and $\Phi = \sqrt{2}\,|E_1|/\omega N_1$.

Clearly, the secondary emf will also be sinusoidal with the phasor representation $\underline{E}_2 = (N_2/N_1)\underline{E}_1 = N\underline{E}_1$. Thus $\underline{I}_2 = \underline{E}_2/\underline{Z}_L = N\underline{E}_1/\underline{Z}_L$, $\underline{I}_1 = N\underline{I}_2 = N^2\underline{E}_1/\underline{Z}_L$, and the equivalent input impedance is

$$\underline{Z}_i = \frac{\underline{E}_1}{\underline{I}_1} = \frac{\underline{Z}_L}{N^2} \tag{5}$$

which gives the AC version of Eq. (3). Figure 18.2–2a then becomes equivalent to Fig. 18.2–2b from the viewpoint of the source; note that we have *referred* the secondary quantities to the primary side of the transformer.

Exercise 18.2–1

A certain transformer has $N_1 = 100$ and $N_2 = 500$. It acts essentially like an ideal transformer as long as the peak flux Φ does not exceed 3 mWb. Find the corresponding maximum values of $|E_2|$ when (a) $f = 60$ Hz, and (b) $f = 400$ Hz.

TRANSFORMER EQUIVALENT CIRCUITS

A real transformer differs from the ideal in several respects. The iron core is nonlinear, with finite permeability and hysteresis, and the windings have some resistance. The effects of nonlinearity and hysteresis were previously illustrated by the distorted *B-H* waveforms in Fig. 17.2–6. An analysis of severe distortion involves describing the waveform by a Fourier series and examining each harmonic term separately—a tedious chore! With less extreme distortion, we can neglect all higher harmonics and use an approximate *linear equivalent circuit* to investigate the voltage and current components at the fundamental frequency $f = \omega/2\pi$. This approach gives reasonable results providing that the peak flux density stays below the nominal saturation level B_{max} of the core material. Correspondingly, from Eq. (4), we require that

$$|E_1| < 4.44\, fN_1 B_{max} A \tag{6}$$

for a uniform core of area A.

Under the condition of Eq. (6), a real transformer with AC excitation behaves approximately like the linear circuit diagrammed in Fig. 18.2–3a. All departures from ideal behavior have been lumped into four elements on the primary side. The resistance R_w represents the *ohmic* or *copper losses* of both windings, while R_c accounts for *core losses* caused by eddy currents and hysteresis. The *leakage reactance* X_ℓ and *magnetizing reactance* X_m reflect the core's finite permeability. Of course, the internal variables such as $\underline{I}_c$ and $\underline{V}_w$ associated with these elements do not exist as distinct physical quantities. However, they can be used to calculate external variables. For instance, Fig. 18.2–3b shows the phasor construction of $\underline{V}_1$ and $\underline{I}_1$ when we take $\underline{E}_1$ as the reference and assume that the load has a lagging power factor (so $\underline{I}_2$ lags $\underline{V}_2$).

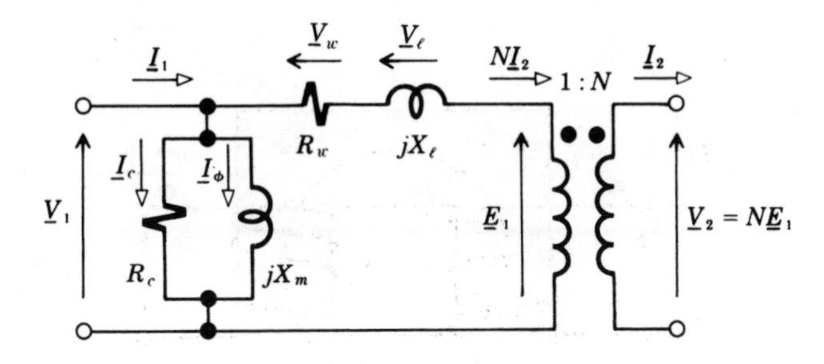

(a)

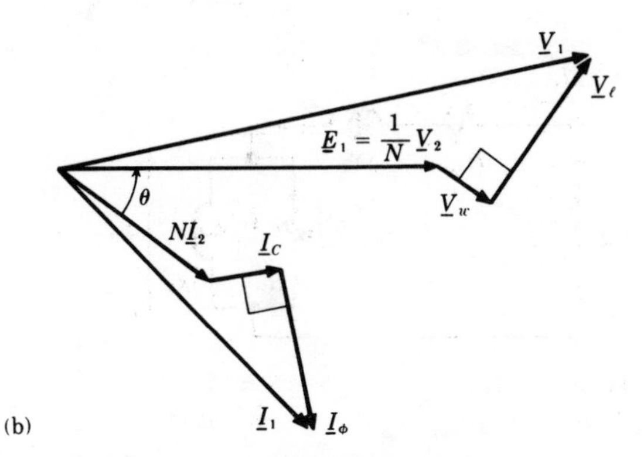

(b)

FIGURE 18.2–3
(a) Transformer equivalent circuit.
(b) Phasor diagram for lagging power factor.

This linear model expedites transformer-circuit analysis. It has the further advantage that the parameter values can be determined from simple measurements on an actual transformer, a matter to be pursued after we justify the model itself.

Suppose we deal first with an *unloaded* transformer ($i_2 = 0$), as in Fig. 18.2–4a. Because of the finite permeability and nonzero core reluctance $\mathcal{R}_c$, the core flux ϕ requires a *magnetizing current* i_ϕ such that $\phi = N_1 i_\phi/\mathcal{R}_c$. The stored magnetic energy is equivalent to an *inductance* $L_m = N_1\phi/i_\phi$ or an inductive reactance $X_m = \omega L_m$ in the circuit model of Fig. 18.2–4b. The voltage across this reactance equals the induced emf $e_1 = N_1\, d\phi/dt$, and $e_2 = N_2\, d\phi/dt = Ne_1$. Since i_ϕ represents energy storage, any power loss P_c in the core must be represented by a second current component i_c with rms value $|I_c| = P_c/|E_1|$. We therefore put the equivalent resistance $R_c = |E_1|/|I_c|$ in parallel with X_m so that $\underline{I}_1 = \underline{I}_\phi + \underline{I}_c$.

Besides core flux ϕ, *leakage* flux $\phi_{\ell p}$ links the primary winding and induces additional voltage that can be modeled by another reactance $X_{\ell p} = \omega L_{\ell p} = \omega N_1 \phi_{\ell p}/i_1$. We expect the leakage inductance $L_{\ell p}$ to be smaller and more linear than the magnetizing inductance L_m, whose flux is confined to the iron core. Adding the primary winding resistance R_{wp} in series with $X_{\ell p}$ completes the equivalent circuit, where $|V_1|$ must exceed $|E_1|$ due to the drop across the series elements.

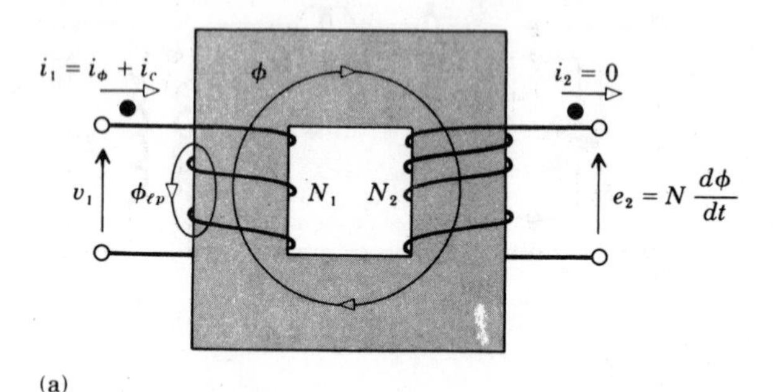

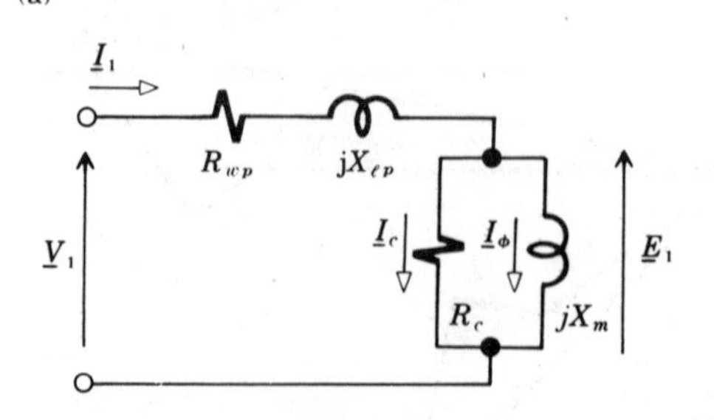

FIGURE 18.2–4
(a) Unloaded transformer. (b) Circuit model.

Now let the transformer be *loaded,* as in Fig. 18.2–5a. The secondary current i_2 produces a *demagnetizing* flux $\phi_d = N_2 i_2/\mathcal{R}_c$ that opposes ϕ and tends to decrease e_1. But if v_1 is fixed, the primary current increases by a *balancing* component i_b producing $\phi_b = N_1 i_b/\mathcal{R}_c$ that cancels ϕ_d. Such balancing is precisely the action of an *ideal* transformer with input current $i_b = N_2 i_2/N_1 = N i_2$, as shown in Fig. 18.2–5b. The equivalent circuit also includes the winding resistance R_{ws} and leakage reactance $X_{\ell s} = \omega L_{\ell s}$ of the secondary, meaning that $|V_2|$ will be less than $|E_2|$.

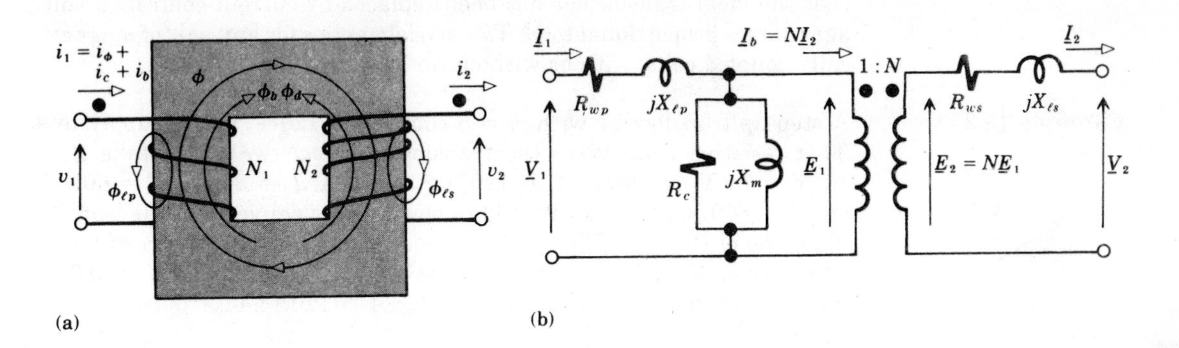

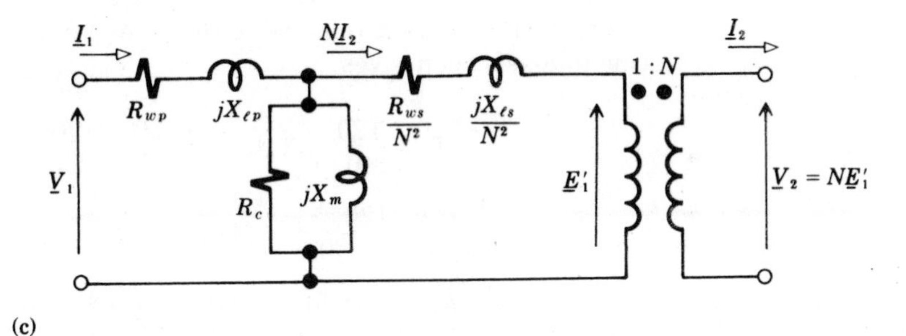

FIGURE 18.2–5
(a) Loaded transformer.
(b) Circuit model.
(c) Equivalent circuit with secondary elements referred to the primary.

But the two secondary elements can just as well be referred to the *primary* side of the ideal transformer with their values divided by N^2, as in Fig. 18.2–5c. Furthermore, any well-designed real transformer has $N|I_2| \approx |I_1|$ under loaded conditions, and little error results if we combine R_{wp} and $X_{\ell p}$ with the referred secondary elements. We thereby arrive at the model back in Fig. 18.2–3, where $R_w = R_{wp} + R_{ws}/N^2$ and $X_\ell = X_{\ell p} + X_{\ell s}/N^2$. Often we find the shunt elements to be much larger than the series elements, so that R_c and X_m can be omitted entirely.

Finally, Fig. 18.2–6 gives another equivalent circuit that may be used when core losses are negligible. The parameters of this circuit are the *self-inductances* L_1 and L_2 of the primary and secondary, the *mutual inductance* M, and the winding resistances $R_1 = R_{wp}$ and $R_2 = R_{ws}$. Note

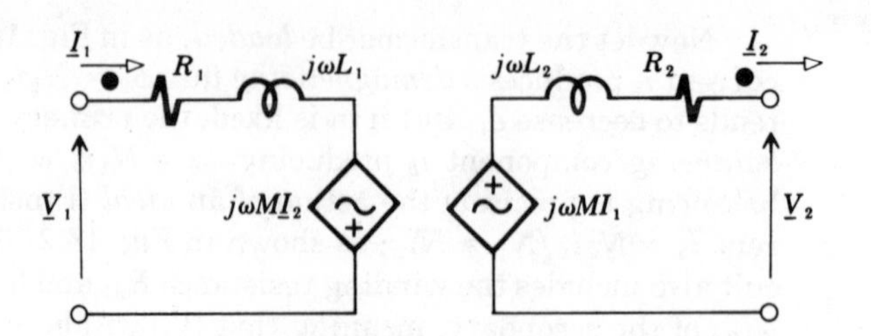

FIGURE 18.2–6 Transformer equivalent circuit with self-inductances and mutual inductance.

that the ideal transformer has been replaced by current-controlled voltage sources proportional to M. This model applies for any pair of magnetically coupled coils, with or without an iron core.

Example 18.2–1

A step-up transformer with $N = 5$ connects a 120-V, 60-Hz supply to a 30-Ω resistive load. Were the transformer *ideal,* we would take $\underline{Z}_i = 30/5^2 = 1.2\ \Omega$ to obtain $\underline{I}_1 = 120/1.2 = 100$ A, $\underline{I}_2 = 20$ A, $\underline{E}_2 = 600$ V, and $P = 600 \times 20 = 12$ kW, all the power being delivered to the load. To illustrate of the use of Fig. 18.2–3, let's assess the performance of a *real* transformer with $R_c = 40\ \Omega$, $X_m = 24\ \Omega$, $R_w = 0.05\ \Omega$, and $X_\ell = 0.3\ \Omega$. We assume the transformer has $N_1 B_{\max} A > 120/(4.44 \times 60)$ to satisfy Eq. (6).

Figure 18.2–7 shows the equivalent circuit with the parameter values and with the load referred to the primary as a 1.2-Ω resistance. Routine analysis then gives

$$\underline{I}_c = \frac{120}{40} = 3\text{ A} \qquad \underline{I}_\phi = \frac{120}{j24} = -j5\text{ A}$$

$$5\underline{I}_2 = \frac{120}{1.25 + j0.3} = 90.8 - j21.8 = 93.4\text{ A}\ \underline{/-13.5^\circ}$$

$$\underline{I}_1 = \underline{I}_c + \underline{I}_\phi + 5\underline{I}_2 = 93.8 - j26.8 = 97.6\text{ A}\ \underline{/-15.9^\circ}$$

Thus, the load has $\underline{I}_2 = 18.68$ A $\underline{/-13.5^\circ}$, $\underline{V}_2 = 30\ \underline{I}_2 = 560.4$ V $\underline{/-13.5^\circ}$, and $P_L = 10.47$ kW—all values that are somewhat lower than those of the ideal transformer because of the losses and reactances in the real transformer.

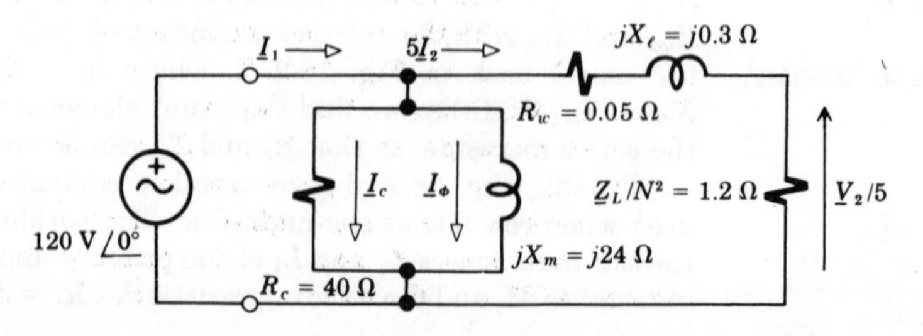

FIGURE 18.2–7 Circuit for Example 18.2–1.

Going to the primary side, we note that $\underline{I}_1$ lags $\underline{V}_1$ by $\theta = 15.9°$. The apparent, real, and reactive powers from the source are then

$$|S| = 120 \text{ V} \times 97.6 \text{ A} = 11.71 \text{ kVA}$$

$$P = |S| \cos\theta = 11.26 \text{ kW}$$

$$Q = |S| \sin\theta = 3.21 \text{ kVAr}$$

This means that the transformer dissipates $P - P_L = 800$ W and yields a reasonable power-transfer efficiency of $P_L/P \approx 93\%$.

Exercise 18.2–2 Repeat the foregoing calculations for a step-down transformer with $N = \frac{1}{2}$, when $\underline{V}_1 = 240 \text{ V} \angle 0°$ and $\underline{Z}_L = 1.0\ \Omega - j0.075\ \Omega$. The transformer parameter values are the same.

Exercise 18.2–3 Write equations for $\underline{V}_1$ and $\underline{V}_2$ in terms of $\underline{I}_1$ and $\underline{I}_2$ for Fig. 18.2–5b with R_c omitted. Then do the same for Fig. 18.2–6 to show, by comparison, that $\omega L_1 = X_{lp} + X_m$, $\omega L_2 = X_{ls} + N^2 X_m$, and $\omega M = N X_m$.

TRANSFORMER MEASUREMENTS †

The data listed on the nameplate of a transformer usually consist of the voltage ratings and the rated (or maximum) apparent output power at a specified frequency. From these we can calculate the turns ratio and the current ratings. For instance, the nameplate legend "60 Hz, 30 kVA, 1200/240 V" means that $|V_2|_{max} = 240$ V, $|I_2|_{max} = 30 \text{ kVA}/240 \text{ V} = 125$ A, $N = 240/1200 = 1/5$, $|V_1|_{max} \approx 1200$ V, and $|I_1|_{max} \approx 25$ A. The actual primary values are somewhat higher in view of the internal losses. (We could also turn this transformer around and use it as a step-up device, with $N = 5$, $|V_2|_{max} = 1200$ V, $|I_2|_{max} = 25$ A, etc.) However, our equivalent circuit involves additional parameters whose values must be determined experimentally. A convenient measurement procedure employing a watt-meter (WM), ammeter (AM), and voltmeter (VM), is as follows.

First, we *open-circuit* the secondary and adjust the primary voltage for its nominal rated value (Fig. 18.2–8a). The meter readings give us the real and reactive powers, P_{oc} and Q_{oc}, from which we find the shunt element values

$$R_c = \frac{|V_1|^2_{max}}{P_{oc}} \qquad X_m = \frac{|V_1|^2_{max}}{Q_{oc}} \tag{7a}$$

Next we *short-circuit* the secondary and adjust the primary current for its rated value (Fig. 18.2–8b). In terms of the measured powers P_{sc} and Q_{sc},

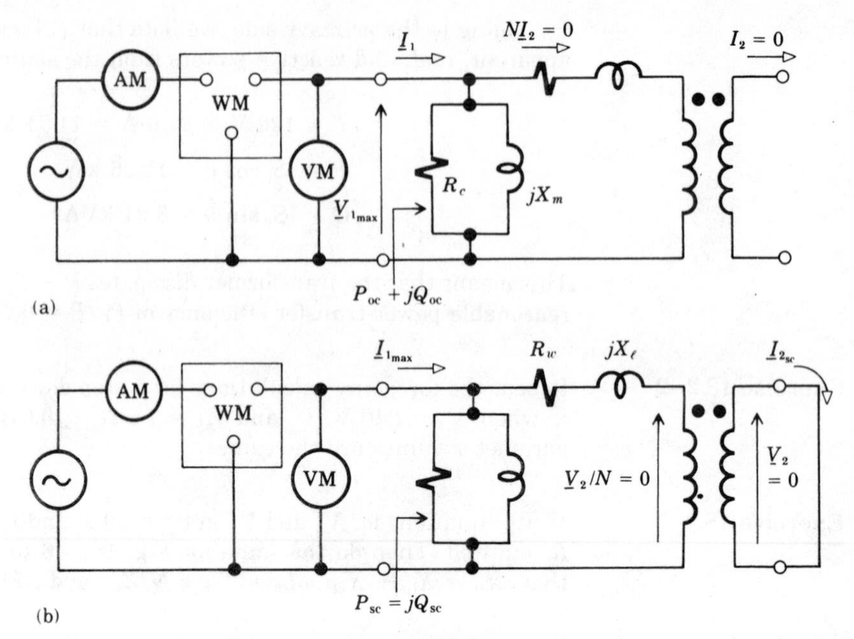

FIGURE 18.2–8 Transformer measurements: (a) open circuit; (b) short circuit.

the series element values are given approximately by

$$R_w \approx \frac{P_{sc}}{|I_1|^2_{\max}} \qquad X_l \approx \frac{Q_{sc}}{|I_1|^2_{\max}} \tag{7b}$$

since almost all of $\underline{I}_1$ should flow through the much smaller series elements. If the values of R_c and X_m do not support this approximation, the additional measurement of $|I_2|_{sc}$ is needed.

18.3 THREE-PHASE CIRCUITS

AC power has decided practical advantages over DC power in generation, transmission, and distribution. Its one major drawback is the oscillatory nature of the instantaneous power flow $p(t)$ in the *single-phase* circuits we have been discussing. However, a *three-phase* circuit can have *constant* instantaneous power, which eliminates the pulsating strain on generating and load equipment. As a further bonus for power transmission, a balanced three-phase system delivers more watts per kilogram of conductor than an equivalent single-phase system. For these reasons, almost all bulk electric power generation and consumption takes place in three-phase circuits. The properties of such circuits certainly deserve our attention.

THREE-PHASE SOURCES AND SYMMETRICAL SETS

The left-hand portion of Fig. 18.3–1a represents a three-phase generator. It consists of three AC voltage sources with equal amplitudes $\sqrt{2}\ V_p$, frequency ω, and relative phase shifts of 120°, so that

$$\begin{aligned} v_a &= \sqrt{2}\ V_p \cos \omega t \\ v_b &= \sqrt{2}\ V_p \cos (\omega t - 120°) \\ v_c &= \sqrt{2}\ V_p \cos (\omega t + 120°) \end{aligned} \tag{1}$$

all with respect to the *neutral point n*. The waveforms are illustrated in Fig. 18.3–1b. We call Eq. (1) a *symmetrical three-phase set* with *phase sequence a-b-c*, the latter referring to the time sequence of the waveform peaks. Other phase sequences, such as *a-c-b*, would have the same properties with the terminals relabeled.

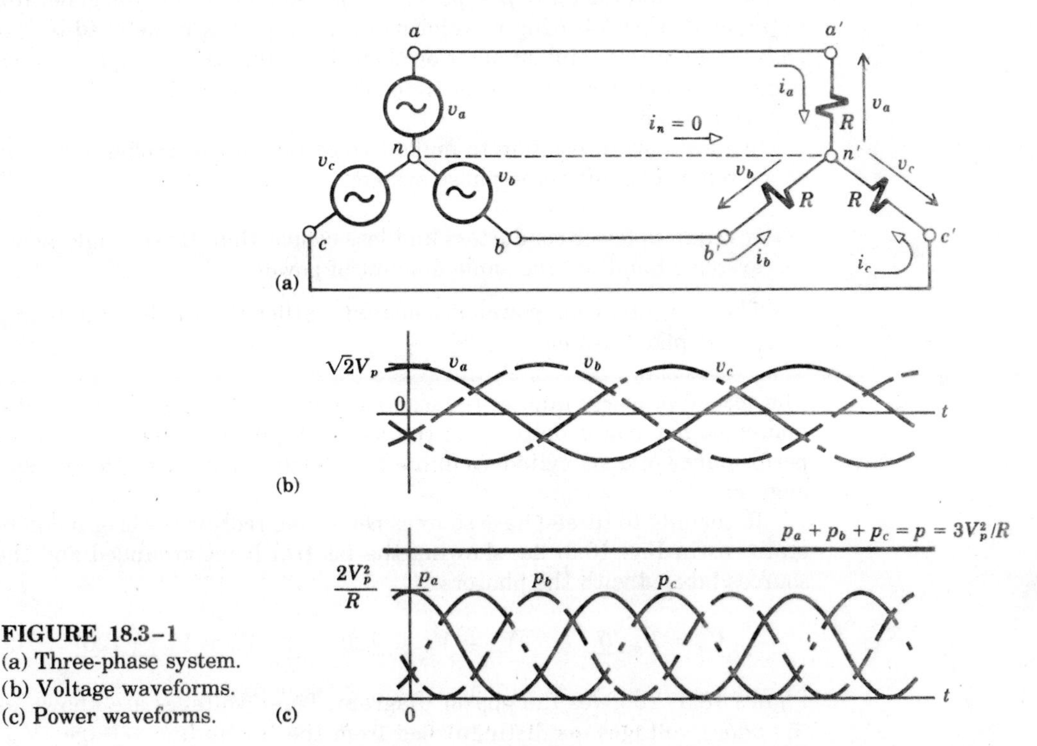

FIGURE 18.3–1
(a) Three-phase system.
(b) Voltage waveforms.
(c) Power waveforms.

The generator has been connected to a load consisting of three equal resistances tied together at n', as shown to the right of Fig. 18.3–1a. Much can be gleaned from studying this elementary three-phase system.

In particular, the load currents i_a, i_b, and i_c are directly proportional to their respective voltages, and their waveforms will have the same shapes. Hence, these currents also constitute a symmetrical three-phase set. Looking more carefully at Fig. 18.3–1b, we discover the important property that $v_a + v_b + v_c = 0$ at each and every instant of time! This property holds for any symmetrical three-phase set; it corresponds to the trigonometric identity

$$\cos \alpha + \cos (\alpha - 120°) + \cos (\alpha + 120°) = 0 \tag{2}$$

for any angle α. Its impact here emerges when we note that the neutral current is $i_n = -(i_a + i_b + i_c) = -(v_a + v_b + v_c)/R = 0$. Accordingly, the wire connecting n to n' may just as well be omitted.

Next we examine the instantaneous powers $p_a = v_a^2/R = (2\,V_p^2/R)\cos^2 \omega t$, etc., and the total, $p = p_a + p_b + p_c$, delivered from the generator to the load. The pulsating waveforms p_a, p_b, and p_c are sketched in Fig. 18.3–1c. Point-by-point summation then shows that $p = 3\,V_p^2/R$—a constant value for every instant of time! The mathematical proof is left for you as an exercise.

We are now in position to make two statements of profound significance regarding this three-phase system:

- It requires fewer conductors and less copper than three single-phase systems handling the same amount of power.
- The instantaneous power is constant, rather than pulsating as in a single-phase system.

The second property implies reduced mechanical strain at the generator and smoother power delivery to the load equipment—analogous to the performance of a six-cylinder engine as compared with a single-cylinder engine.

Returning to three-phase sources per se, we redraw the original generator as in Fig. 18.3–2a, showing the neutral point grounded and the sources labeled with the phasors

$$\underline{V}_a = V_p \underline{/0°} \qquad \underline{V}_b = V_p \underline{/-120°} \qquad \underline{V}_c = V_p \underline{/+120°} \tag{3}$$

Figure 18.3–2b gives the phasor diagram. These voltages are known as the *phase* voltages, as distinguished from the line-to-line voltages $\underline{V}_{ab}$, $\underline{V}_{bc}$, and $\underline{V}_{ca}$. Three-phase generators are usually rated in terms of their line voltages, and the relationship between line and phase voltage is easily developed by phasor analysis.

Clearly, $\underline{V}_{ab} = \underline{V}_a - \underline{V}_b$ in Fig. 18.3–2a, and the phasor construction of Fig. 18.3–2c immediately reveals that $\sphericalangle \underline{V}_{ab} = 30°$. Letting V_ℓ denote

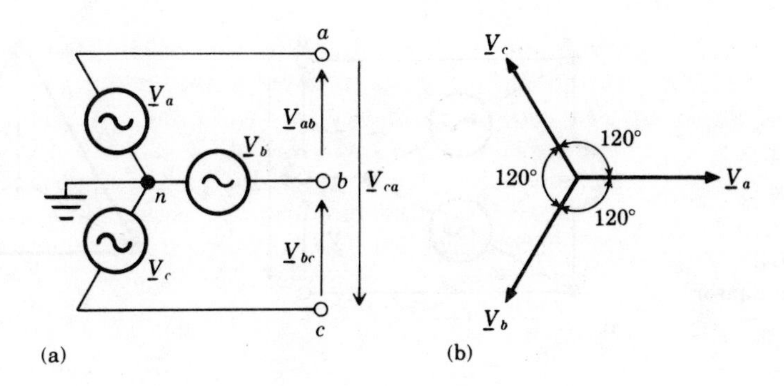

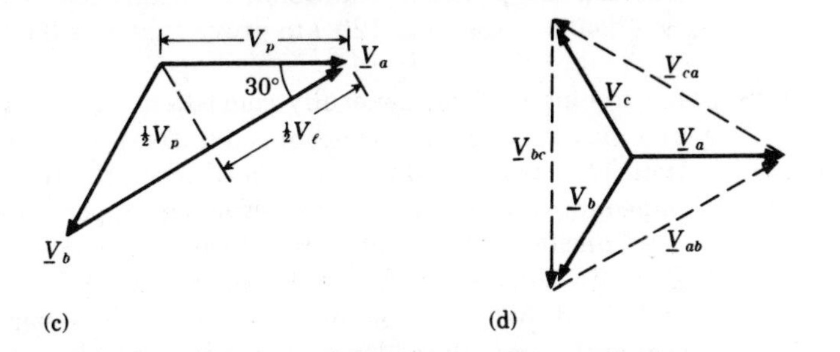

FIGURE 18.3–2 Wye-connected generator and phasor diagrams.

the rms value of the line voltage (so $V_\ell = |\underline{V}_{ab}|$), we see that $V_p^2 = (V_\ell/2)^2 + (V_p/2)^2$ and solving for V_ℓ yields

$$V_\ell = \sqrt{3}\, V_p \tag{4}$$

Therefore, the rms line voltage equals $\sqrt{3} \approx 1.7$ times the rms phase voltage. By symmetry, the other line voltages have the same magnitude, but are shifted $\pm 120°$ relative to $\underline{V}_{ab}$. Thus

$$\underline{V}_{ab} = V_\ell \,\underline{/30^\circ} \qquad \underline{V}_{bc} = V_\ell \,\underline{/-90^\circ} \qquad \underline{V}_{ca} = V_\ell \,\underline{/150^\circ} \tag{5}$$

which is another symmetrical three-phase set.

Figure 18.3–2d depicts the relationship between these two sets and brings out the fact that $\underline{V}_{ab} + \underline{V}_{bc} + \underline{V}_{ca} = 0$—as it should according to KVL. Moreover, by redrawing Fig. 18.3–2b with the vectors "tip-to-tail," we see that $\underline{V}_a + \underline{V}_b + \underline{V}_c = 0$, thus confirming the identity in Eq. (2).

The source configuration in Fig. 18.3–2a is called a *wye* connection. Occasionally, one encounters a *delta*-connected generator whose line voltages equal the source voltages as in Fig. 18.3–3a. The equivalent neutral point and the phase voltages are defined by the phasor diagram, Fig. 18.3–3b, where $\underline{V}_a = (V_\ell/\sqrt{3})\,\underline{/-30^\circ}$, and so forth.

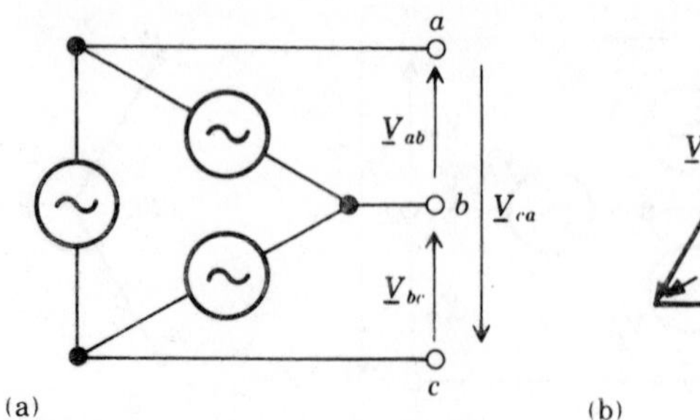

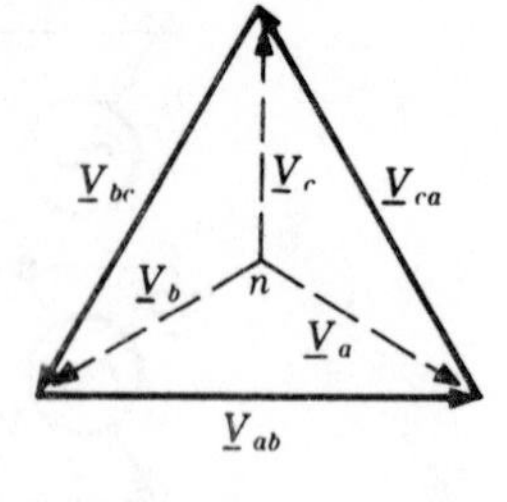

FIGURE 18.3–3 Delta-connected generator and phasor diagram.

Exercise 18.3–1 Use Eq. (2), plus the additional relations $\cos^2 \beta = \frac{1}{2} + \frac{1}{2}\cos 2\beta$ and $\cos(\gamma \mp 240°) = \cos(\gamma \pm 120°)$ to prove that $p = 3V_p^2/R$ in Fig. 18.3–1c.

BALANCED LOADS

A three-phase load generally consists of both resistance and reactance, and may be wye- or delta-connected irrespective of the source configuration. The load is said to be *balanced* when the three branches have equal impedances, a matter of great practical importance.

Consider the balanced wye load in Fig. 18.3–4a, where $\underline{Z}_a = \underline{Z}_b = \underline{Z}_c = \underline{Z}_Y$ with $\underline{Z}_Y = |Z_Y| \underline{/\theta}$. From symmetry, the potential at point n' will equal that at the generator's neutral, whether grounded or not. Consequently, each impedance has a phase voltage across it, and the resulting line currents $\underline{I}_a$, $\underline{I}_b$, and $\underline{I}_c$, have equal rms values,

$$I_\ell = \frac{V_p}{|Z_Y|} = \frac{V_\ell}{\sqrt{3}\,|Z_Y|} \tag{6}$$

The phase angle of $\underline{I}_a$ is then $\sphericalangle \underline{V}_a - \sphericalangle \underline{Z}_Y = -\theta$, since $\underline{I}_a = \underline{V}_a / \underline{Z}_Y$. The other current angles differ by $\pm 120°$ as diagrammed in Fig. 18.3–4b.

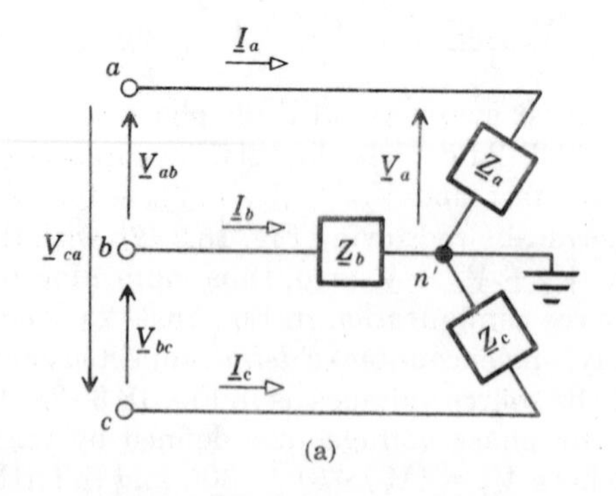

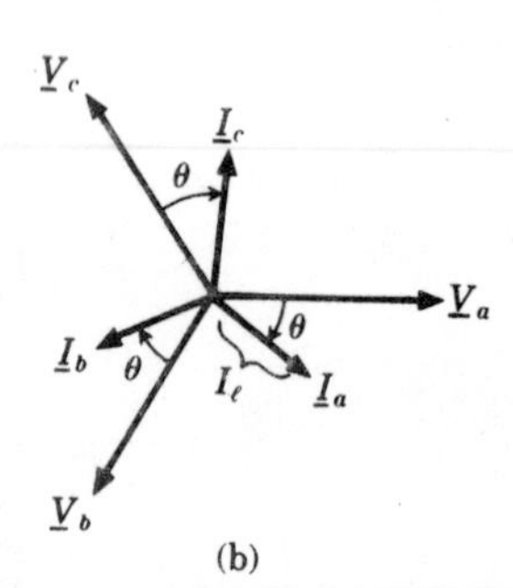

FIGURE 18.3–4 (a) Balanced wye load, $\underline{Z}_a = \underline{Z}_b = \underline{Z}_c = |Z_Y| \underline{/\theta}$. (b) Phasor diagram.

Having determined the voltages and currents, we turn to the topic of power. Specifically, the real and reactive power *per phase* will be

$$P = V_p I_\ell \cos\theta = \frac{1}{\sqrt{3}} V_\ell I_\ell \cos\theta$$

$$Q = V_p I_\ell \sin\theta = \frac{1}{\sqrt{3}} V_\ell I_\ell \sin\theta$$

The total delivered from generator to load is then exactly three times the phase values, namely

$$P_{\text{total}} = 3P = \sqrt{3}\, V_\ell I_\ell \cos\theta \qquad Q_{\text{total}} = 3Q = \sqrt{3}\, V_\ell I_\ell \sin\theta \tag{7a}$$

and therefore

$$|S_{\text{total}}| = \sqrt{(P_{\text{total}})^2 + (Q_{\text{total}})^2} = \sqrt{3}\, V_\ell I_\ell \tag{7b}$$

since $\cos^2\theta + \sin^2\theta = 1$. In other words, the apparent power to a balanced three-phase load equals $\sqrt{3}$ times the product of the rms line voltage and current. Just like the balanced resistive load (Fig. 18.3–1), the total instantaneous power is constant at $p = P_{\text{total}}$ because the reactive powers p_{aX}, p_{bX}, p_{cX} constitute a symmetrical set.

We now turn to the balanced delta load configuration of Fig. 18.3–5, where the line voltage appears across each impedance $\underline{Z}_{ab} = \underline{Z}_{bc} = \underline{Z}_{ca} = \underline{Z}_\Delta$. The *equivalent wye impedance* can be found by writing $\underline{I}_a = \underline{I}_{ab} - \underline{I}_{ca} = (\underline{V}_{ab} - \underline{V}_{ca})/\underline{Z}_\Delta$, and noting from Fig. 18.3–3b that $\underline{V}_{ab} - \underline{V}_{ca} = (\underline{V}_a - \underline{V}_b) - (\underline{V}_c - \underline{V}_a) = 2\underline{V}_a - \underline{V}_b - \underline{V}_c = 3\underline{V}_a - (\underline{V}_a + \underline{V}_b + \underline{V}_c) = 3\underline{V}_a$. Thus, $\underline{I}_a = 3\underline{V}_a/\underline{Z}_\Delta = \underline{V}_a/(\underline{Z}_\Delta/3)$, so a balanced delta load is equivalent to a balanced wye load with

$$\underline{Z}_Y = \tfrac{1}{3}\underline{Z}_\Delta \tag{8}$$

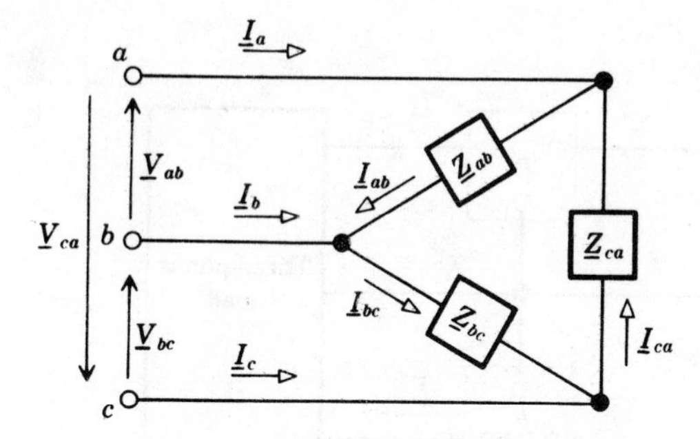

FIGURE 18.3–5 Balanced delta load, $\underline{Z}_{ab} = \underline{Z}_{bc} = \underline{Z}_{ca} = |Z_\Delta| \underline{/\theta}$.

Then, combining Eqs. (6) and (8), the rms line current becomes

$$I_\ell = 3\,\frac{V_p}{|Z_\Delta|} = \sqrt{3}\,\frac{V_\ell}{|Z_\Delta|} \tag{9}$$

as contrasted with the rms delta current $|\underline{I}_{ab}| = |\underline{I}_{bc}| = |\underline{I}_{ca}| = V_\ell/|Z_\Delta|$. Equations (7a) and (7b) still hold as they stand, with $\theta = \measuredangle \underline{Z}_\Delta$. Observe that a delta load draws three times the current and power of a wye load that has the same branch impedances.

To summarize, we have seen that the symmetry of a balanced three-phase system leads to very simple expressions for voltage, current, and power, whether the load has a wye or delta configuration. This symmetry allows us to carry out analyses by focusing on just *one* of the three phases.

Three-phase power measurements are only slightly more complicated than single-phase measurements, requiring two wattmeters arranged as in Fig. 18.3–6. Although the individual meter readings do not correspond to physical quantities, their *sum* equals the total real power delivered to the load. We prove this be letting θ stand for either $\measuredangle \underline{Z}_Y$ or $\measuredangle \underline{Z}_\Delta$, as the case may be, and taking $\underline{V}_{ab}$ for the reference phasor. Thus, from Figs. 18.3–4b and 18.3–3b,

$$\underline{V}_{ab} = V_\ell \,\underline{/0^\circ} \qquad \underline{I}_a = I_\ell \,\underline{/-30^\circ} - \theta$$
$$-\underline{V}_{bc} = V_\ell \,\underline{/60^\circ} \qquad \underline{I}_c = I_\ell \,\underline{/90^\circ} - \theta$$

(The results, of course, are independent of the phase reference.) The upper meter then reads $P_1 = |\underline{V}_{ab}|\,|\underline{I}_a| \cos (\measuredangle \underline{V}_{ab} - \measuredangle \underline{I}_a) = V_\ell I_\ell \cos (0^\circ + 30^\circ + \theta) = V_\ell I_\ell \cos (\theta + 30^\circ)$, while the lower meter reads $P_2 = |-\underline{V}_{bc}|\,|\underline{I}_c| \cos (\measuredangle - \underline{V}_{bc} - \measuredangle \underline{I}_c) = V_\ell I_\ell \cos (60^\circ - 90^\circ + \theta) = V_\ell I_\ell \cos (\theta - 30^\circ)$. Invoking the identity for the sum of cosines yields

$$\cos (\theta + 30^\circ) + \cos (\theta - 30^\circ) = 2 \cos 30^\circ \cos \theta = \sqrt{3} \cos \theta$$

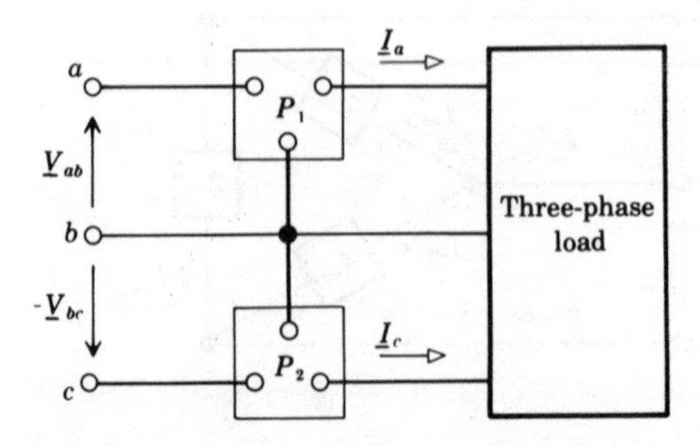

FIGURE 18.3–6 Three-phase power measurement with two wattmeters.

and, therefore, $P_1 + P_2 = V_\ell I_\ell \sqrt{3} \cos \theta = P_{\text{total}}$. Replacing the wattmeters with varmeters would give Q_{total}.

Example 18.3–1 Figure 18.3–7a depicts a three-phase source with $V_\ell = 45$ kV (*kilo*volts) connected to two balanced loads—a wye load with $\underline{Z}_w = (10 + j20)\ \Omega$ and a delta load with $\underline{Z}_d = 50\ \Omega$. Each of the three connecting wires has $2\ \Omega$ resistance. Our task is to find the line current and the various powers. For this purpose, we'll take advantage of the symmetry and concentrate on one phase, say the phase that includes terminals a and a'

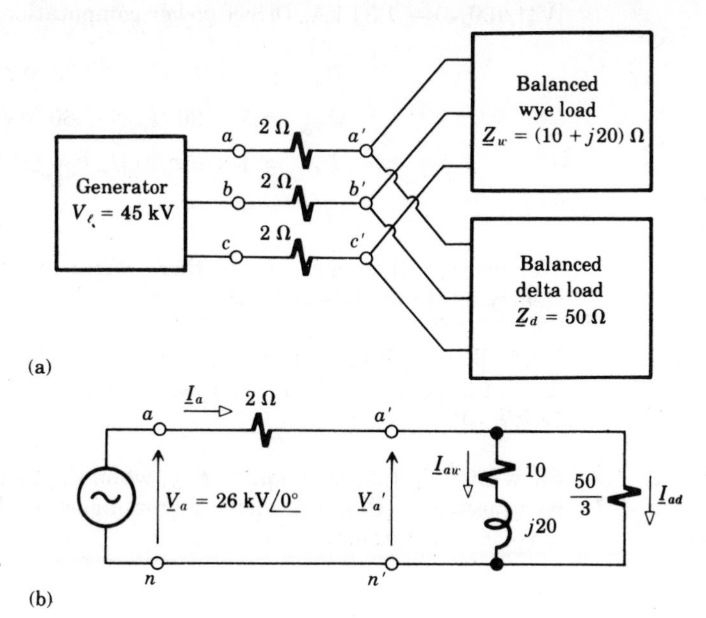

FIGURE 18.3–7
(a) Three-phase system for Example 18.3–1.
(b) Equivalent circuit of one phase.

First, we recognize that the two loads are in *parallel,* in the sense of having the same voltage across $a' - n'$, etc. Second, we apply Eq. (8) to obtain the equivalent wye impedance of the delta load as (50/3) Ω. Third, we calculate the phase voltage at the source, $V_p = 45\ \text{kV}/\sqrt{3} = 26$ kV, from Eq. (4). Fourth, we draw the equivalent circuit diagrammed in Fig. 18.3–7b, taking $\underline{V}_a$ for the reference phasor. This circuit includes one of the three equivalent wye sources, although the generator might actually be delta connected. The circuit also includes a resistanceless wire from n to n', reflecting the fact that in a balanced system those points must be at the same potential. (If you question the parallel load across $a' - n'$, you should go back to Fig. 18.3–7a and draw the complete wye load and then the complete wye equivalent for the delta load.)

The work having progressed this far, the rest moves along rapidly: We combine the parallel elements to obtain $(50/3)\|(10 + j20) = 10 +$

$j5$, and the total impedance seen by the generator is $12 + j5 = 13\ \Omega\ \angle 22.6°$ so that $I_\ell = |\underline{I}_a| = 26\text{ kV}/13\ \Omega = 2\text{ kA}$ (kiloamps) and $\theta = 22.6°$. Since $V_\ell = 45$ kV, Eq. (6) tells us that the generator must supply $|S_{\text{total}}| = 3 \times 45\text{ kV} \times 2\text{ kA} = 156 \times 10^6 = 156$ MVA, $P_{\text{total}} = 156 \cos 22.6° = 144$ MW, and $Q_{\text{total}} = 60$ MVAr. Note that powers come out in *mega* values when we deal with kilovolts, ohms, and kiloamps. These units are typical for a very large three-phase system.

As for the individual loads, we find that $\underline{V}'_a = \underline{V}_a - 2\underline{I}_a = 22.4\text{ kV}\ \angle -3.9°$, $|I_{aw}| = |V'_a|/|10 + j20| = 1.00$ kA, and $|I_{ad}| = |V'_a|/(50/3) = 1.34$ kA. Direct power computations then yield

$$P_{w_{\text{total}}} = 3 \times 10\ |I_{aw}|^2 = 30\text{ MW}$$

$$Q_{w_{\text{total}}} = 3 \times 20\ |I_{aw}|^2 = 60\text{ MVAr}$$

$$P_{d_{\text{total}}} = 3 \times (50/3)\ |I_{ad}|^2 = 90\text{ MW}$$

$$Q_{d_{\text{total}}} = 0$$

We also see that each of the three wires dissipates $2|I_a|^2 = 8$ MW— clearly an intolerable situation.

Exercise 18.3–2

Find all the voltages, currents, and powers when a wye load with $\underline{Z}_Y = (2 + j1.5)\ \Omega$ is supplied by a delta-connected generator with $\underline{V}_{ab} = 12\text{ kV}\ \angle 0°$.

Exercise 18.3–3

Show that $P_1 + P_2$ still equals P_{total} when the meters in Fig. 18.3–6 are reconnected so that the upper meter measures $\underline{I}_a$ and $-\underline{V}_{ca}$, and the lower meter measures $\underline{I}_b$ and $\underline{V}_{bc}$.

UNBALANCED LOADS †

Most of the lovely symmetry goes out the window in the case of a three-phase load with *unequal* impedances. We must then resort to brute-force analysis involving all three branches.

Consider, for instance, an *unbalanced wye load* with a "floating" (ungrounded) neutral as in Fig. 18.3–8. It is still true that $\underline{V}_{ca} = -(\underline{V}_{ab} + \underline{V}_{bc})$ and $-\underline{I}_b = \underline{I}_a + \underline{I}_c$, so we can write two KVL equations for $\underline{I}_a$ and $\underline{I}_c$ in terms of $\underline{V}_{ab}$ and $\underline{V}_{bc}$. Simple rearrangement yields

$$(\underline{Z}_a + \underline{Z}_b)\underline{I}_a + \underline{Z}_b\underline{I}_c = \underline{V}_{ab}$$
$$\underline{Z}_b\underline{I}_a + (\underline{Z}_b + \underline{Z}_c)\underline{I}_c = -\underline{V}_{bc} \tag{10a}$$

which can be solved for $\underline{I}_a$ and $\underline{I}_c$ and thence $\underline{I}_b$. The voltage at the load's neutral relative to the generator's neutral is found via

$$\underline{V}_{n'n} = \underline{V}_a - \underline{Z}_a\underline{I}_a \tag{10b}$$

and, unlike a balanced system, $\underline{V}_{n'n} \neq 0$. If we have a *four-wire wye-wye* system with a connection between n and n' forcing $\underline{V}_{n'n} = 0$, the previous analysis must be modified to account for the fourth current $\underline{I}_n$.

An *unbalanced delta load* can be analyzed directly from Fig. 18.3–5 if we note that $\underline{I}_a = \underline{I}_{ab} - \underline{I}_{ca}$ with $\underline{I}_{ab} = \underline{V}_{ab}/\underline{Z}_{ab}$ and $\underline{I}_{ca} = \underline{V}_{ca}/\underline{Z}_{ca}$—and so forth for $\underline{I}_b$ and $\underline{I}_c$. *Or,* we can convert the delta load into an equivalent wye load using the *delta-wye transformation* given in Eq. (4), Section 5.4. (That transformation, incidentally, yields the equivalent impedance $\underline{Z}_Y = \underline{Z}_\Delta/3$ for a balanced load.) Although the transformation approach appears to involve more work, it is necessary if there are series impedances in the lines connecting the generator and load.

Finally, we state without proof that the wattmeter connections in Fig. 18.3–6 will correctly read $P_1 + P_2 = P_{\text{total}}$ for an unbalanced wye or delta load. However, a third meter is required to measure the power in an unbalanced four-wire system.

Example 18.3–2

Suppose the load in Fig. 18.3–8 has $\underline{Z}_a = 12\ \Omega$, $\underline{Z}_b = 6\ \Omega$, and $\underline{Z}_c = (9 + j13)\ \Omega$; and the line voltage is 208 V. It is convenient to take $\underline{V}_{ab}$ as the phase reference, in which case $-\underline{V}_{bc} = 208\ \underline{/60^\circ} = 104 + j180$. Inserting values into Eq. (10a) gives

$$18\underline{I}_a + 6\underline{I}_c = 208 \qquad 6\underline{I}_a + (15 + j13)\underline{I}_c = 104 + j180$$

and solving yields

$$\underline{I}_a = 9.6\ \text{A}\ \underline{/-3^\circ} \qquad \underline{I}_b = 15.6\ \text{A}\ \underline{/-176^\circ} \qquad \underline{I}_c = 6.2\ \text{A}\ \underline{/15^\circ}$$

Then, using Eq. (10b),

$$\underline{V}_{n'n} = (104 - j60) - 12(9.6 - j0.5) = 54.7\ \text{V}\ \underline{/-101^\circ}$$

since $\underline{V}_a = (208/\sqrt{3})\ \underline{/-30^\circ}$.

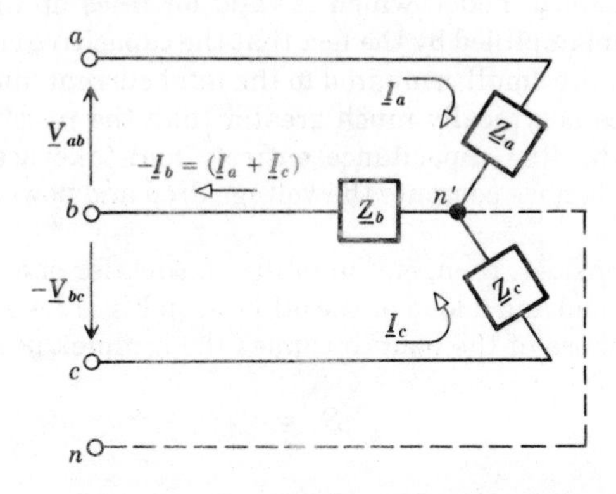

FIGURE 18.3–8
Unbalanced wye load.

The real power consumed by the load is $P_{\text{total}} = 12\,|I_a|^2 + 6\,|I_b|^2 + 9\,|I_c|^2 = 1.1 + 1.5 + 0.3 = 2.9$ kW, and $Q_{\text{total}} = 13\,|I_c|^2 = 0.5$ kVAr. The wattmeters in Fig. 18.3–6 would read $P_1 = 208 \times 9.6 \cos 3° = 2.0$ kW and $P_2 = 208 \times 6.2 \cos(60° - 15°) = 0.9$ kW, confirming that $P_1 + P_2 = P_{\text{total}}$ even though P_1 and P_2 have no meaning by themselves.

Exercise 18.3–4 Repeat the preceding calculations for a delta load with $\underline{Z}_{ab} = 12\ \Omega$, $\underline{Z}_{bc} = 6\ \Omega$, and $\underline{Z}_{ca} = (7 + j24)\ \Omega$, taking $\underline{V}_{ab} = 300\ \text{V}\ \underline{/0°}$. *Don't* use the delta-wye transformation.

18.4 POWER TRANSMISSION AND DISTRIBUTION †

This section traces the route of AC power from the generating plants to the consumers. It will be seen that the limitations of power transmission lines are overcome through the use of very high voltages. Three-phase power transformers are thus needed to step-up the voltage at the generating station and, at the other end, to step it down for distribution to individual users.

POWER TRANSMISSION LINES

Virtually all electric power is conveyed from the generating plants to consumers by AC three-phase overhead transmission lines. Such lines constitute *distributed circuits* in the sense that resistance, capacitance, and inductance are distributed continuously along their entire length. Energy flows in the form of a *traveling* wave moving with finite but very high velocity, nearly 3×10^8 m/sec.

Fortunately, most transmission lines are electrically "short" compared to the wavelength of the traveling wave. Each phase may then be approximated by the *lumped* circuit of Fig. 18.4–1, with parameter values proportional to line length. A representative line rated for 100 kV and 1 kA might have

$$R \approx 50\ \text{m}\Omega/\text{km} \qquad L \approx 1\ \text{mH/km} \qquad C \approx 10\ \text{nF/km}$$

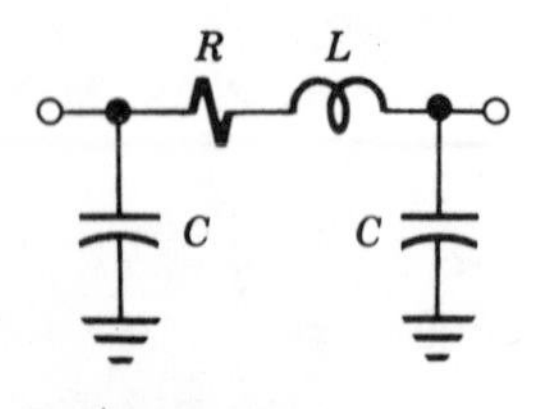

FIGURE 18.4–1 Lumped circuit model for one phase of a power transmission line.

The lumped model, which is valid for lines up to about 200-km long, is further simplified by the fact that the capacitive current to ground is typically quite small compared to the total current, and that the inductive reactance is typically much greater than the resistance. Thus, we'll disregard the line capacitance entirely and take account of the resistance only when we compute the voltage drop and power dissipation on the line itself.

Consider, then, our simplified model (for one phase) with a generator at one end and a load at the other as in Fig. 18.4–2. In a balanced system, each phase of the load consumes the complex power

$$\underline{S}_2 = \underline{V}_2\underline{I}^* = P_2 + jQ_2 \tag{1a}$$

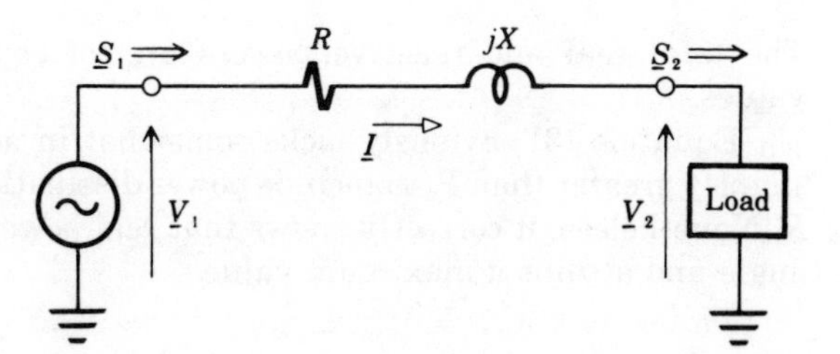

FIGURE 18.4–2
Simplified per-phase model.

and each phase of the generator produces

$$\underline{S}_1 = \underline{V}_1 \underline{I}^* = P_1 + jQ_1 \tag{1b}$$

where $\underline{V}_1$ and $\underline{V}_2$ are phase voltages. These voltages differ by the voltage drop along the line, so

$$\underline{V}_1 = \underline{V}_2 + (R + jX)\underline{I}$$

Normally, the line impedance $R + jX$ is small enough that

$$|V_1| \approx |V_2| \approx V_p$$

where we interpret V_p as the system's average or *nominal phase voltage*. However, we cannot ignore the phase shift

$$\delta \triangleq \sphericalangle \underline{V}_1 - \sphericalangle \underline{V}_2 \tag{2}$$

which is called the *power angle*. The significance of the power angle is brought out in the following analysis.

First, let $\underline{V}_2 = V_p \underline{/0°}$ be the reference phasor so that $\underline{V}_1 \approx V_p \underline{/\delta} = V_p e^{j\delta}$. Next, drawing upon the property $R \ll X$, we can write $\underline{I} \approx (\underline{V}_1 - \underline{V}_2)/jX$ and

$$\underline{I}^* \approx \frac{\underline{V}_1^* - \underline{V}_2^*}{-jX} = j\frac{V_p e^{-j\delta} - V_p}{X} = j\frac{V_p}{X}(e^{-j\delta} - 1)$$

Inserting $\underline{I}^*$ into Eq. (1) we obtain, after a few manipulations

$$P_1 \approx P_2 \approx \frac{V_p^2}{X}\sin\delta \tag{3}$$

$$Q_1 \approx Q_2 + 2\frac{V_p^2}{X}(1 - \cos\delta) \tag{4}$$

The total real and reactive powers are, of course, three times these values.

Equation (3) obviously lacks somewhat in accuracy, for P_1 must be slightly greater than P_2 to include power dissipation in the line resistance R. Nonetheless, it correctly shows that real power varies with the power angle and attains a maximum value

$$P_{max} \approx \frac{V_p^2}{X} \tag{5}$$

when $\delta = 90°$—that is, $\underline{V}_1$ is perpendicular to $\underline{V}_2$. We thus see that the maximum real power capacity of a transmission line is inversely proportional to its series reactance and proportional to the square of the nominal voltage. Equation (3) also predicts the possibility of *negative* real power when $\delta < 0$. Physically, this corresponds to power flow in the reverse direction, as might be the case if the line connects two nodes of a power grid with generators and loads at both ends. The flow of real power is then controlled by adjusting the relative phase angles of the generators, as will be discussed in Section 19.2.

Operating a line at close to its maximum real power capacity does not turn out to be an attractive proposition because of the accompanying large values of *reactive* power. Specifically, if $\delta = 90°$ then $\cos\delta = 0$ in Eq. (4) and $Q_1 \approx Q_2 + 2P_{max}$. The motivation for keeping reactive power at relatively low values is implicit in our earlier discussions, but we will make it explicit here by calculating the *power dissipation* along the line. Since $\underline{I}$ flows through the line resistance R in Fig. 18.4–2, the power dissipated is $P_R = R|\underline{I}|^2$. But $|\underline{I}| = |S_1|/|\underline{V}_1| \approx |S_1|/V_p$, so

$$P_R \approx R\frac{|S_1|^2}{V_p^2} = \frac{R}{V_p^2}(P_1^2 + Q_1^2) \tag{6}$$

which means that power lost as line heating increases as the reactive power increases.

Another important observation from this result relates to the power transmission *efficiency*

$$\text{Eff} = \frac{P_2}{P_1} = 1 - \frac{P_R}{P_1} \tag{7}$$

Clearly, efficiency increases with V_p because P_R decreases inversely with V_p^2. The combined implications of Eqs. (3)–(7) explain why large amounts of power must be transmitted at *high voltages*. Invariably, doing so requires a step-up transformer between the generator and line, and a step-down transformer at the other end.

Example 18.4–1

A generating station having phase voltage $V_s = 20$ kV is to be connected to a nominal 600-MW resistive load by a three-phase transmission line with $R = 0.5\ \Omega$ and $X = 10\ \Omega$ at $\omega = 377$. We immediately see the need for a step-up transformer since, if $V_p \approx V_s$, $P_{\max} \approx (20\text{ kV})^2/10\ \Omega = 40$ MW, whereas the real power per phase must be $600/3 = 200$ MW. Even if the line reactance were sufficiently reduced to get $P_{\max} > 200$ MW, the ohmic power dissipation would still be $P_R \approx 0.5\ \Omega \times (600\text{ MW})^2/(20\text{ kV})^2 = 50$ MW—which corresponds to a very poor transmission efficiency of $1 - (50\text{ MW}/200\text{ MW}) = 75\%$. With a 1:10 step-up transformer, however, $V_p \approx 200$ kV, so $P_{\max} \approx 4{,}000\text{ MW} \gg 200$ MW, and $P_R \approx 0.5$ MW, giving Eff $\approx 99.7\%$!

For a more detailed examination of the system with transformer, we take $\underline{V}_2 = 200\text{ kV}\ \underline{/0^\circ}$ and $\underline{I} = 1\text{ kA}\ \underline{/0^\circ}$ because the load description implies $P_2 = 200$ MW and $Q_2 = 0$. The voltage drop along the line is then $(R + jX)\underline{I} = (0.5 + j10)$ kV and $\underline{V}_1 = 200.5 + j10 = 200.7\text{ kV}\ \underline{/2.86^\circ}$ which requires a turns ratio $N = 10.035$. The nominal phase voltage and power angle are

$$V_p = \tfrac{1}{2}(200.7 + 200) \approx 200.4\text{ kV}$$

$$\delta = 2.86^\circ - 0^\circ = 2.86^\circ$$

Substituting these values into Eqs. (4) and (6) finally yields

$$Q_1 \approx 10\text{ MVAr} \qquad P_R \approx 0.499\text{ MW}$$

A capacitance across the line's input terminals could cancel the reactive power Q_1, but in this system would not appreciably reduce P_R.

Exercise 18.4–1

Repeat the calculations in Example 18.4–1 with $\underline{V}_2 = 500\text{ kV}\ \underline{/0^\circ}$.

DISTRIBUTION SYSTEMS AND THREE-PHASE TRANSFORMERS

The key role of transformers in AC power distribution is emphasized in Fig. 18.4–3, a schematic diagram of a portion of a power network with the three-phase transmission lines represented by single lines and transformers represented by a double-sawtooth symbol. Massive *busbars* serve as the interconnecting nodes at the generating stations and distribution substations. The diagram also shows a shunt capacitance at the substation for power-factor correction; similar corrections are made at generating stations by adjusting the generator's phase. Not shown at the ends of each transmission line are the circuit breakers that protect the system from overload fault conditions.

Although a very large load—a steel mill, for instance—might be fed directly from the grid, the power to your home undergoes several transformations from the generating sources to the eventual load. Here we will

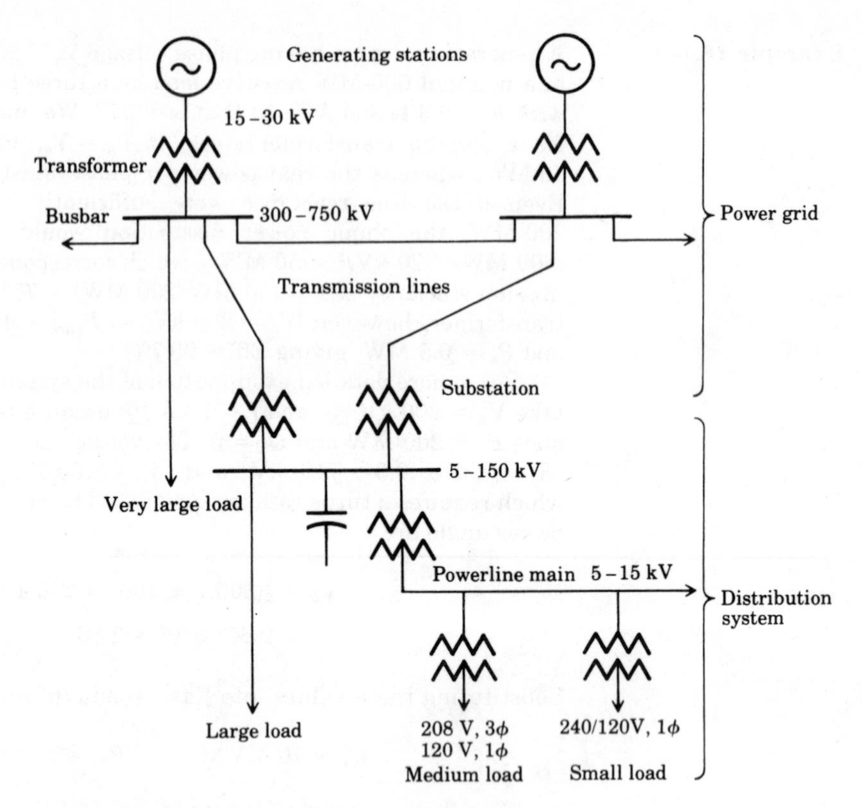

FIGURE 18.4–3 Schematic diagram of a portion of a power network.

look more closely at these power transformers, starting with the step-up transformer at a generating station.

Usually, a three-phase step-up transformer has a delta-wye configuration with equal turns ratios N on each of its three "legs," as in Fig. 18.4–4a. This configuration combats the nonlinear *distortion* as mentioned in Section 18.2, for it permits the dominant third-harmonic component of the distorted primary current to circulate harmlessly around the delta loop without inducing third-harmonic components in the secondary. We will also see that this configuration provides an additional step-up factor of $\sqrt{3}$. The core itself is built something like Fig. 18.4–4b, except that the high-voltage secondary is wound on top of the primary to maximize coupling. Individual flux return paths are not needed for each of the three legs, because the flux phasors constitute a symmetrical three-phase set and, therefore, $\phi_a + \phi_b + \phi_c = 0$ at every instant of time.

To analyze the output voltages, let V_ℓ be the rms generated line voltage and take $\underline{V}_{ab}$ as the reference phasor so that

$$\underline{V}_{ab} = V_\ell \underline{/0^\circ} \qquad \underline{V}_{bc} = V_\ell \underline{/-120^\circ} \qquad \underline{V}_{ca} = V_\ell \underline{/+120^\circ}$$

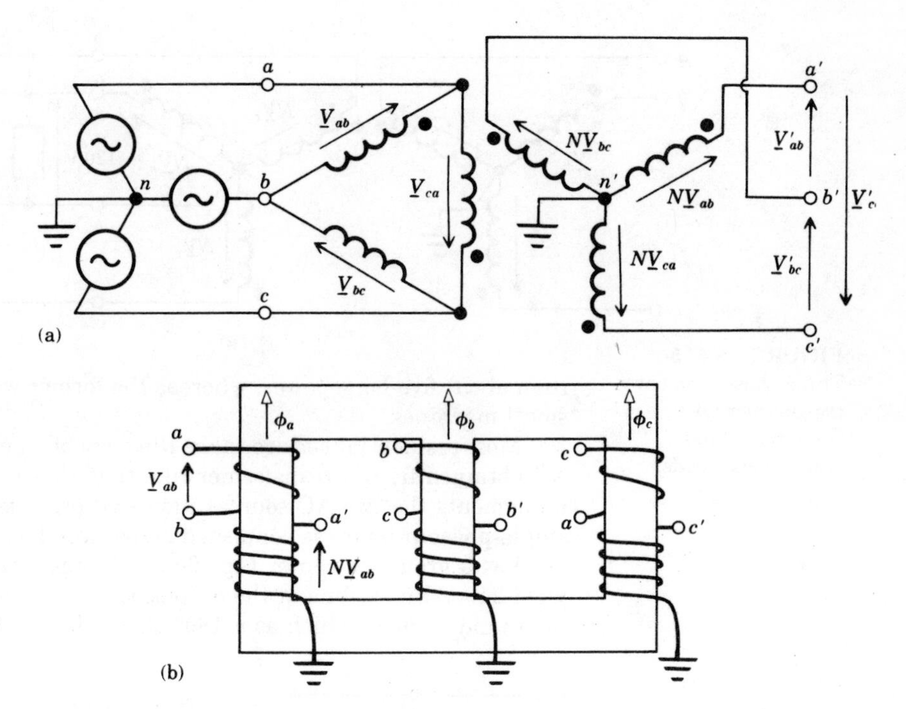

FIGURE 18.4–4
(a) Three-phase step-up transformer. (b) Core configuration.

assuming an *a-b-c* phase sequence. Then,

$$\underline{V}'_{ab} = N\underline{V}_{ab} - N\underline{V}_{bc} = N\sqrt{3}\,V_\ell \;\underline{/30^\circ} = V'_\ell \;\underline{/30^\circ}$$

which is obtained via a phasor construction similar to Fig. 18.3–2c. Therefore, in view of the symmetry, all three output line voltages have the rms value

$$V'_\ell = \sqrt{3}NV_\ell \tag{8}$$

and are shifted in phase by 30° relative to the corresponding inputs.

Three-phase step-down transformers may be delta-wye or wye-wye connected. (In the latter case, an "idle" delta winding can be added to absorb harmonic current components.) Figure 18.4–5 shows a wye-wye configuration with *four* output terminals for the purpose of supplying both *three-phase* (3-ϕ) and *single-phase* (1-ϕ) power to a medium load, such as an office building, store, or small factory. If V_p is the input phase voltage and $N = 120/V_p$, we then have 120-V, 1-ϕ AC between any of the "hot" output terminals (a', b', or c') and the grounded neutral n'. But we also have three-phase power with line-to-line voltage $\sqrt{3} \times 120 \approx 208$ V. The latter would be used for large heating elements or motors bigger

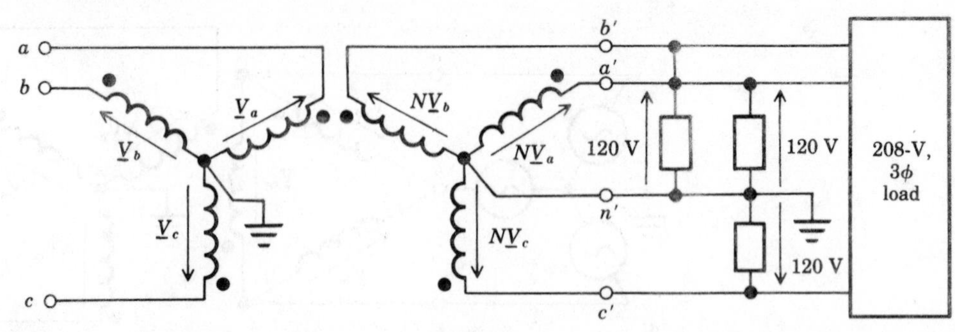

FIGURE 18.4–5
Three-phase step-down transformer with three-phase and single-phase loads.

than about five horsepower, whereas the former would supply lights and small machines.

Most residential service takes the form of single-phase, dual-voltage AC obtained from a transformer like that shown in Fig. 18.4–6, which implements the two AC sources indicated previously in Fig. 6.4–1. The single-phase primary is connected across any two lines, and the secondary has a grounded center tap. Thus, an appropriate turns ratio N will yield 240-V line-to-line at the output, with two 120-V terminals relative to ground—one of which as a 180° phase shift.

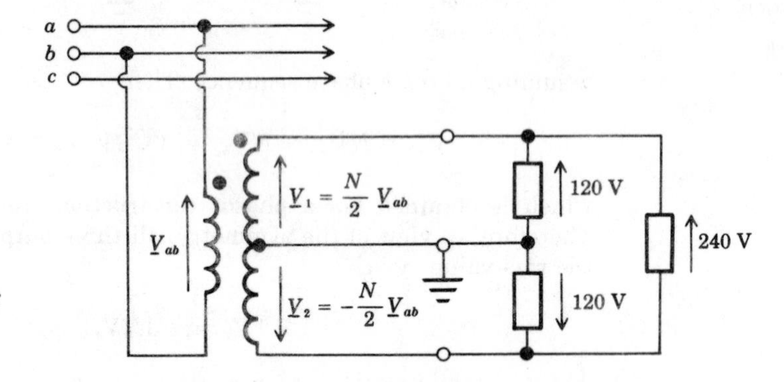

FIGURE 18.4–6
Step-down transformer for single-phase dual-voltage load.

You may have noticed that this arrangement presents an *unbalanced* load to the three-phase network, and so does that of Fig. 18.4–5b unless the 120-V load elements happen to be equal. But with literally thousands of residential loads connected across each of the three line pairs, and with many medium loads also connected, the law of averages takes effect and the total composite load on the distribution system appears to be virtually balanced.

Example 18.4–2 A step-up transformer is needed to deliver $P_L = 200$ MW per phase at $V'_\ell = 550$ kV in the delta-wye configuration of Fig. 18.4–4, starting from a generator with $V_\ell \approx 20$ kV. Equation (8) suggests that an *ideal* three-phase transformer having $N = 550/(\sqrt{3}\ 20) = 15.9$ would do the job. The output phase voltage and line current would be

$$V'_p = NV_\ell = 318 \text{ kV} \qquad I'_\ell = P_L/V'_p = 0.629 \text{ kA}$$

assuming unity power factor for the balanced load. Each primary winding would carry an rms current of $NI'_\ell = 10$ kA. (What would be the value of the line current from the generator?)

The *real* transformer proposed for this application bears the nameplate ratings 400/25 kV, 300 kVA per phase—implying that $N = 16$, $|V_1|_{max} = 25$ kV, and $|I_1|_{max} = 12$ kA in the step-up mode. Open-circuit and short-circuit measurements yield

$$P_{oc} = 1.6 \text{ MW} \qquad Q_{oc} = 3.1 \text{ MVAr} \qquad P_{sc} = 1.4 \text{ MW} \qquad Q_{sc} = 70 \text{ MVAr}$$

per phase, and Eq. (7), Section 18.2, gives the equivalent element values as

$$R_c = 391\ \Omega \qquad X_m = 202\ \Omega \qquad R_w = 0.00972\ \Omega \qquad X_l = 0.486\ \Omega$$

Our per-phase equivalent circuit thus has the form of Fig. 18.4–7 where, for convenience, we have let $\underline{V}_1 = \underline{V}_{ab}$, $\underline{V}_2 = N\underline{V}_{ab} = \underline{V}'_a$, and so forth.

To assess this transformer's suitability, we assume $\underline{V}_2 = 318 \text{ kV } \angle 0°$ and $\underline{I}_2 = 0.629 \text{ kA } \angle 0°$, and work backward to the generator. Specifically,

$$\underline{V}_1 = (R_w + jX_l)\, N\underline{I}_2 + \frac{\underline{V}_2}{N} = 20.56 \text{ kV } \angle 13.8°$$

so the generator must have $V_\ell = |\underline{V}_1| = 20.56$ kV to compensate for the voltage drop in the transformer. We then compute the internal power loss

$$P_{loss} = R_w\,|N\underline{I}_2|^2 + \frac{|\underline{V}_1|^2}{R_c} = 0.98 + 1.08 = 2.06 \text{ MW}$$

so the input power per phase is $P = P_{loss} + P_L = 202.06$ MW, and the

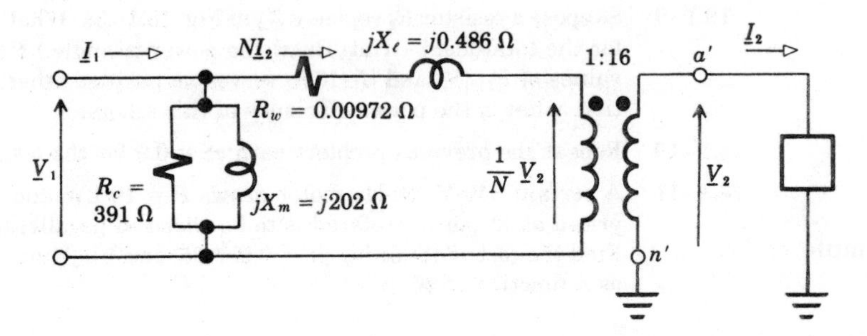

FIGURE 18.4–7 Circuit for Example 18.4–2.

transformer has an efficiency of $200/202.06 \approx 99\%$. Despite this impressive-looking value, the transformer will dissipate a total of $3 \times 2.06 = 6.18$ MW, and adequate cooling certainly must be provided to prevent an excessive temperature rise.

Exercise 18.4–2 Continue Example 18.4–2 by calculating the reactive power and apparent power per phase. Use your results to find the rms current $|I_1|$.

Exercise 18.4–3 Let the transformers in Figs. 18.4–5 and 18.4–6 be fed from the same three-phase main with $V_\ell = 7.6$ kV. Find the turns ratio required in each case to give the indicated output voltages. Neglect internal drops.

PROBLEMS

18.1–1 Use Eq. (2) to obtain expressions for P and Q in Fig. 18.1–1a in terms of R, X, and $|V|$.

18.1–2 Let R and jX be reconnected in parallel in Fig. 18.1–1a. Use Eq. (2) to obtain expressions for P and Q in terms of $|V|$. Then express P and Q in terms of $|I|$.

18.1–3 Measurements on a certain load yield $P = 500$ W and $|I| = 6$ A at 120 V and 60 Hz. The current is observed to decrease when the frequency increases.
(a) Model the load as two parallel elements.
(b) Model the load as two series elements.

18.1–4 Repeat the previous problem for a load whose current increases with frequency.

18.1–5 A third load with $P_3 = 26$ kW and pf = 1.0 is added in parallel in Fig. 18.1–5a.
(a) Calculate P, Q, $|S|$, $|I|$, and pf for the total load.
(b) If $f = 60$ Hz, what value of C in parallel yields $Q = 0$, and what is the resulting value of $|I|$?

18.1–6 Repeat the previous problem with $P_3 = 16$ kW and pf = 0.8 lagging.

18.1–7 Consider a load impedance $\underline{Z} = (120 + j50)\ \Omega$ connected to a 2.6-kV power line.
(a) Calculate pf, $|I|$, $|S|$, P, and Q.
(b) A capacitor is connected in parallel with $\underline{Z}$ to make pf = 1. Calculate P, $|I|$, and the rms current through $\underline{Z}$.
(c) A capacitor is connected in series with $\underline{Z}$ to make pf = 1. Calculate P, $|I|$, and the rms voltage across $\underline{Z}$. Why might this arrangement be undesirable?

18.1–8 Repeat the previous problem with $\underline{Z} = (100 + j240)\ \Omega$.

18.1–9 Suppose a resistor R_2 replaces $\underline{Z}_2$ in Fig. 18.1–5a. What value of R_2 yields pf = 0.8 for the total load? (Hint: Draw the power triangle.) Now calculate the resulting values of P_2, $|S|$, and $|I|$. If R_2 serves no purpose other than power-factor correction, what is the power efficiency of this scheme?

18.1–10 Repeat the previous problem with pf = 0.9 for the total load.

18.1–11 A certain 240-V, 60-Hz motor draws $P = 10$ kW and $Q = 6$ kVAr. A capacitor priced at \$1 per microfarad is to be added in parallel to adjust the power factor. Find the cost of obtaining pf = 0.9, 0.95, and 1.0. Hint: Derive an expression for C as a function of pf.

18.1–12 A certain load draws $|I| = 10$ A at 5 kV and 60 Hz. Connecting a 4-μF capacitor in parallel reduces $|I|$ to 8 A. Find P and Q for the load alone.

18.1–13 Generalize the measurement procedure suggested by the previous problem. Specifically, write equations for P and Q that involve only voltage and current measurements and a known capacitor value.

‡ **18.1–14** The wattmeter in Fig. 18.1–6 becomes a *varmeter* if R_M is replaced by an inductor L, and a resistor R is added in parallel with coil a. The elements are proportioned so that $\underline{I}_a \approx \underline{V}/jX_M$ where X_M is a large inductive reactance. Use Eq. (7), Section 17.3, to show that $\theta_{ss} = K_M Q/X_M$.

‡ **18.1–15** Consider the varmeter described in the previous problem. If coil a has inductance L_a and negligible resistance, what is the required condition on R and L, and what is the corresponding expression for X_M?

18.2–1 Add a third winding to Fig. 18.2–1a. Let this *tertiary* winding have N_3 turns wound in the same sense as the secondary. Find i_1 in terms of i_2 and i_3 when $\mathcal{R} \to 0$, and find e_2 and e_3 in terms of e_1. Compare your results with the model of a tapped-secondary transformer in Fig. 4.1–7b.

18.2–2 Repeat the previous problem with the tertiary winding in the opposite sense of the secondary.

18.2–3 Figure P18.2–3 shows an ideal transformer connected as an *autotransformer*. Both windings have the same power rating, so $|E_1 I_1| = |E_2 I_2| \leq |S|_{max}$, but the *throughput power* can be greater. Let $|S|_{max} = 2400$ VA, $N_2 = N_1/4$, and $\underline{V}_{in} = 800$ V; and let the load be resistive, so all voltages and currents are in phase with $\underline{V}_{in}$. Find $\underline{V}_{out}$, $|V_{out} I_{out}|$, and $|V_{in} I_{in}|$ when the load draws the largest permitted output current.

18.2–4 Repeat the previous problem with the input and output terminals interchanged and $\underline{V}_{in} = 1000$ V.

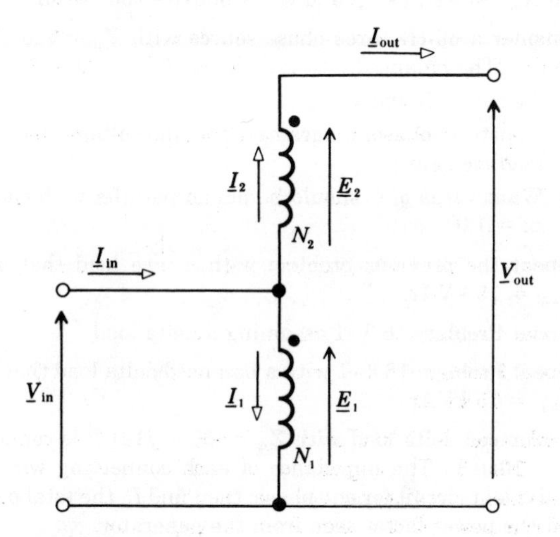

FIGURE P18.2–3

18.2–5 Consider the transformer and source in Example 18.2–1.

(a) Calculate $\underline{I}_1$, P, and Q when the secondary is open-circuited. Also construct a phasor diagram like Fig. 18.2–3.

(b) Repeat (a) for a short-circuited secondary.

18.2–6 Consider a transformer with $N = \frac{1}{10}$, $R_c = 3$ kΩ, $X_m = 1$ kΩ, $R_w = 5$ Ω, and $X_\ell = 20$ Ω. Use reasonable *approximations* to find $\underline{V}_2$, $\underline{I}_2$, P, Q, and the power-transfer efficiency when $\underline{V}_1 = 1200$ V and $\underline{Z}_L = 2.4\ \Omega\ \underline{/0°}$.

18.2–7 Repeat the previous problem with $\underline{Z}_L = (2.4 + j0.5)$ Ω.

‡ **18.2–8** The transformer model in Fig. 18.2–5c provides better accuracy for large load currents. The element values are usually taken to be $R_{wp} = R_{ws}/N^2 = \frac{1}{2}R_w$, and $X_{\ell p} = X_{\ell s}/N^2 = \frac{1}{2}X_\ell$. Use this model to estimate $\underline{I}_c$ and $\underline{I}_\phi$ when the transformer in Example 18.2–1 has $\underline{Z}_L = 10\ \Omega\ \underline{/0°}$. Compare your results with the values in the example, which are independent of $\underline{I}_2$.

18.2–9 Let a source $\underline{V}_s$ with resistance R_s be connected to the primary in Fig. 18.2–3a. Obtain expressions for the open-circuit secondary voltage and the short-circuit secondary current. You may assume that $R_w \ll R_c$, $X_\ell \ll X_m$, and $R_s \ll R_c$. Then draw the Thévenin equivalent circuit seen looking back into the secondary.

18.2–10 A certain transformer can be modeled by Fig. 18.2–6 with $R_1 = 1$ Ω, $\omega L_1 = 2$ Ω, $\omega M = 8$ Ω, $\omega L_2 = 50$ Ω, and $R_2 = 6$ Ω. Find $\underline{V}_2/\underline{V}_1$ and $\underline{V}_1/\underline{I}_1$ when a 30-Ω resistor is connected across the secondary.

18.2–11 Repeat the previous problem with $\omega M = 10$ Ω.

‡ **18.2–12** Show that the transformer model in Fig. 18.2–6 is equivalent to the tee network in Fig. 5.4–6a with $\underline{Z}_a = R_1 + j\omega(L_1 - M)$, $\underline{Z}_b = R_2 + j\omega(L_2 - M)$, and $\underline{Z}_c = j\omega M$.

‡ **18.2–13** Derive exact expressions for R_w and X_ℓ in Fig. 18.2–8b in terms of $|V_1|_{sc}$, $|I_2|_{sc}$, R_c, and X_m, as well as P_{sc} and Q_{sc}. Compare your results with Eq. (7b).

18.3–1 Consider a 60-Hz three-phase source with $\underline{V}_a = 120\ \text{V}\underline{/90°}$ and *a-b-c* phase sequence. The balanced wye load draws $P_{\text{total}} = 48$ kW with pf = 0.8 lagging.

(a) Calculate I_ℓ and $|S|_{\text{total}}$.

(b) Construct phasor diagrams of the line voltages and the currents through each load element.

(c) What value of C should be put in parallel with each load element to obtain pf = 1.0?

18.3–2 Repeat the previous problem with a wye load that draws $P_{\text{total}} = 36$ kW and $Q_{\text{total}} = 15$ kVAr.

18.3–3 Repeat Problem 18.3–1 assuming a delta load.

18.3–4 Repeat Problem 18.3–1 with a balanced delta load that draws $P_{\text{total}} = 36$ kW and $Q_{\text{total}} = 15$ kVAr.

18.3–5 A balanced delta load with $\underline{Z}_\Delta = (66 + j12)$ Ω is connected to a generator with $V_\ell = 1040$ V. The impedance of each connecting wire is $(2 + j3)$ Ω. Draw the equivalent circuit for one phase; then find I_ℓ, the total power delivered to the load, and the power factor seen from the generator.

18.3–6 Repeat the previous problem with $\underline{Z}_\Delta = (18 - j27)\ \Omega$.

18.3–7 When a load has a low power factor, one of the meters in Fig. 18.3–6 may have a *negative* reading. However, $P_1 + P_2$ still equals P_{total}. Verify these assertions by calculating P_1 and P_2 when $\underline{V}_{ab} = 600\ \text{V}\underline{/0^\circ}$, $-\underline{V}_{bc} = 600\ \text{V}\underline{/60^\circ}$, and the load draws $P_{total} = 58$ kW at pf $= 0.28$.

‡ **18.3–8** *Reactive power* is easily measured in a balanced three-phase system using the electrodynamometer in Fig. 18.1–6. In particular, suppose coil b carries the line current $i_b(t)$, and coil a with R_M is connected so that $v_{ca}(t)$ is applied across them. Use Eq. (7), Section 17.3, to show that θ_{ss} will be proportional to Q_{total}. Take $\sphericalangle \underline{V}_b = 0^\circ$ for the phase reference.

‡ **18.3–9** A balanced load obviously becomes unbalanced if there is a fault in one of the branches. Let $\underline{Z}_b \to \infty$ (open circuit) in Fig. 18.3–8, and let $\underline{Z}_a = \underline{Z}_c = (4 + j3)\ \Omega$ and $\underline{V}_a = 120\ \text{V}\ \underline{/0^\circ}$. Find $\underline{I}_a$, $\underline{I}_c$, P_{total}, and Q_{total} if n' is not connected to n. Compare your results with the values for a balanced load. Also calculate $\underline{V}_{n'n}$ from a phasor diagram.

‡ **18.3–10** Repeat the previous problem with n' connected to n. Also calculate $\underline{I}_{n'n}$ from a phasor diagram.

‡ **18.3–11** Utilize phasor diagrams to find $\underline{I}_a$, $\underline{I}_b$, and $\underline{I}_c$ in Fig. 18.3–5 when $\underline{Z}_{ab} = \underline{Z}_{ca} = (24 + j7)\ \Omega$, $\underline{Z}_{bc} \to \infty$ (open circuit), and $\underline{V}_a = 600\ \text{V}\ \underline{/0^\circ}$. Also calculate P_{total} and Q_{total}, and compare them with the values for a balanced load.

18.4–1 Consider a transmission line with $V_p = 500$ kV, $X = 250\ \Omega$, and $R = 40\ \Omega$. Find δ such that $P_2 = 200$ MW. Then estimate the corresponding values of Q_1, P_R, and Eff if the load has unity power factor.

18.4–2 Repeat the previous problem with $P_2 = 600$ MW.

18.4–3 Repeat Problem 18.4–1 with pf $= 0.95$ at the load.

18.4–4 Suppose the model of the transmission line in Example 18.4–1 includes capacitors with impedance $-j2000\ \Omega$ as in Fig. 18.4–1. Use lumped-circuit analysis to find $\underline{V}_1$, δ, $\underline{S}_1$, and P_R when $\underline{V}_2 = 200\ \text{kV}\underline{/0^\circ}$, $P_2 = 200$ MW, and $Q_2 = 0$.

‡ **18.4–5** Insert the given expression for $\underline{I}^*$ into Eq. (1) to obtain Eqs. (3) and (4).

18.4–6 Construct a phasor diagram to obtain $\underline{V}'_{ab}$ in Fig. 18.4–4a when the generator has an *a*-*b*-*c* phase sequence and $\sphericalangle \underline{V}_{ab} = 0^\circ$.

18.4–7 Repeat the previous problem for an *a*-*c*-*b* phase sequence.

‡ **18.4–8** Suppose the three-phase load in Fig. 18.4–5 draws $P_{total} = 28.8$ kW with pf $= 0.8$. The transformer is ideal and has $N = \frac{1}{20}$. Find $|I_a|$, $|I_b|$, and $|I_c|$ when a 10.4-kW resistive load is connected across a'-b', and there are no other loads. Take $\sphericalangle \underline{V}_a = 0^\circ$ and an *a*-*b*-*c* phase sequence.

‡ **18.4–9** Repeat the previous problem with the addition of a 6-kW resistive load across c'-n'.

19 ROTATING MACHINES

The vast bulk of "man-made" electrical energy comes from rotating generators driven by mechanical prime movers such as steam or hydro turbines. And, after transmission, a sizable portion of the generated output is converted back to mechanical energy by electrical motors. It seems fitting, therefore, to close our study of electric energy with a look at the rotating machines typically found at either end of a system.

Following an introductory overview of rotating-machine concepts, we examine the three-phase synchronous machine operated as a generator or a motor. Then we consider the workhorse induction motor—both three-phase and single-phase—and other AC motors. A concluding section deals with DC machines. Our emphasis throughout will be on broad operating principles and characteristics, with only brief mention of some of the second-order effects.

OBJECTIVES

After studying this chapter and working the exercises, you should be able to do each of the following:

- Explain how rotating fields and unidirectional torque are produced in single-phase and three-phase AC machines (Sections 19.1 and 19.2).

- Construct the phasor diagram and analyze a synchronous machine operating as a generator, motor, or capacitor (Section 19.2).
- State the differences between synchronous and induction motors in terms of equivalent circuits, structural features, and torque-speed curves (Sections 19.2 and 19.3).
- Explain the role of the commutator in a DC machine, and draw the equivalent circuit and typical performance curve for a DC motor or generator (Section 19.4).
- Compare methods for controlling the speed of an AC or DC motor (Sections 19.3 and 19.4).

19.1 ROTATING-MACHINE CONCEPTS

From a giant three-phase generator producing hundreds of megawatts to a tiny battery-powered motor driving a portable tape recorder, nearly every rotating machine operates on the same basic principles. We discuss those principles here, starting with the "Blu" and "Bli" laws as they apply to an elementary machine. We then introduce the concept of rotating fields to explain how an AC machine develops unidirectional torque. Attention is also given to a number of practical matters, including structural features, losses and efficiency, and machine capacity and performance curves.

AN ELEMENTARY MACHINE

Our prior study of rotating electromechanical transducers began with the assumption that the "working" coil or *armature* rotated in a fixed, uniform magnetic field. But a more common and more practical arrangement for continuous rotation is to rotate the field and keep the armature fixed, permitting direct (nonrotating) electrical connections to the coil. Figure 19.1–1a illustrates a simplified generator with this configuration. The armature coil has N turns fastened in slots along the inner face of a stationary frame, called the *stator,* with length ℓ and radius r. A *rotor*—assumed for the moment to be a permanent magnet—revolves within the stator at the mechanical angular frequency ω_m, driven by a prime mover coupled to its shaft. The north and south poles of the rotor alternately sweep past the armature conductors, inducing a time-varying emf just as though the conductors moved through a fixed but nonuniform field. (The spinning rotor also creates a time-varying flux through the stator, which should be laminated to minimize the resulting eddy currents.)

To analyze the induced emf, let $B_1(t)$ and $B_2(t)$ be the outward-directed radial fields at the upper and lower conductors, and note that the rotor has a tangential velocity $u = r\omega_m$. Application of the "Blu" law then gives the total emf as

$$e(t) = NB_1(t)\ell r\omega_m - NB_2(t)\ell r\omega_m = NA\omega_m \tfrac{1}{2}[B_1(t) - B_2(t)] \qquad (1)$$

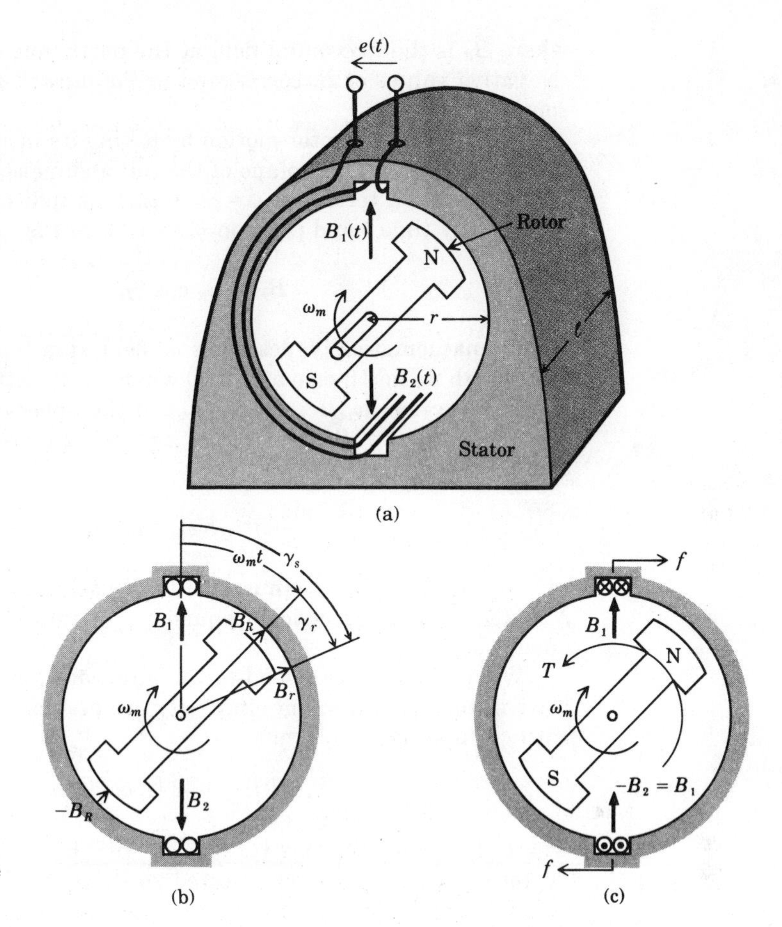

FIGURE 19.1–1
(a) An elementary generator. (b) Rotor field. (c) Magnetic force and reaction torque.

where $A = 2r\ell$ is the area of the armature coil. The minus sign here reflects Lenz's law for the two sets of N conductors, but we'll soon see that $B_1(t)$ and $B_2(t)$ have opposite signs and that, therefore, they reinforce rather than cancel each other. Moreover, since the fields reverse direction periodically from pole to pole, $e(t)$ alternates in polarity. Thus, we have an *AC generator* or *alternator*.

For proper operation, however, the fields must vary sinusoidally with time. This requirement will be satisfied if the rotor is shaped in such a way that the air-gap flux density has a sinusoidal pattern in space around the rotor. Specifically, let B_r in Fig. 19.1–1b be the outward-directed radial field at an arbitrary angle γ_r relative to the rotor's axis, and assume that

$$B_r = B_R \cos \gamma_r \tag{2}$$

where B_R is the maximum field at the north pole of the rotor, at $\gamma_r = 0$. (Negative values of B_r correspond to the *inward*-directed field near the south pole.)

We incorporate rotor motion by taking its instantaneous angle to be $\omega_m t$ with respect to the plane of the coil and measuring angular position on the stator in terms of $\gamma_s = \gamma_r + \omega_m t$, as indicated in the figure. The field at any time t and position $\gamma_r = \gamma_s - \omega_m t$ is

$$B_r = B_R \cos(\gamma_s - \omega_m t) \tag{3}$$

which mathematically describes a field rotating clockwise (with the rotor) with its north pole at $\gamma_s = 0$ when $t = 0$. Setting $\gamma_s = 0$ then yields $B_1(t) = B_R \cos(-\omega_m t) = B_R \cos \omega_m t$ at the upper conductors while, with $\gamma_s = 180°$, $B_2(t) = B_R \cos(180° - \omega_m t) = -B_R \cos \omega_m t = -B_1(t)$. Equation (1) therefore becomes

$$e(t) = NB_R A\omega_m \cos \omega_m t \tag{4}$$

so $e(t)$ has the desired sinusoidal time variation with rms value $|E| = NB_R A\omega_m/\sqrt{2}$ and electrical frequency ω, identical to the mechanical frequency ω_m.

When we connect a load to the alternator, a sinusoidal current $i(t)$ flows in the armature. For simplicity, we assume the current is in phase with $e(t)$ and has the form

$$i(t) = \sqrt{2}\,|I| \cos \omega_m t$$

where $|I|$ denotes the rms value. The "Bli" law then says that each conductor experiences a force directed as shown in Fig. 19.1–1c. Since the stator is immobile, the rotor experiences an equal and opposite *reaction torque*

$$\begin{aligned} T(t) &= NB_1(t)\ell i(t)r - NB_2(t)\ell i(t)r \\ &= \sqrt{2}\,NB_R A\,|I| \cos^2 \omega_m t \\ &= \frac{NB_R A\,|I|}{\sqrt{2}}(1 + \cos^2 \omega_m t) \end{aligned} \tag{5}$$

whose time variation is plotted versus $\omega_m t$ in Fig. 19.1–2, together with $B_1(t)$ and $i(t)$. This torque *pulsates but does not alternate* (change sign), and has an average value $T_{av} = NB_R A\,|I|/\sqrt{2}$ that opposes the torque applied to the rotor by the prime mover. The potential *torque reversals,* noted in our discussion of the d'Arsonval meter movement, do not appear here simply because the current reverses direction in perfect synchronism with the alternations of $B_1(t)$.

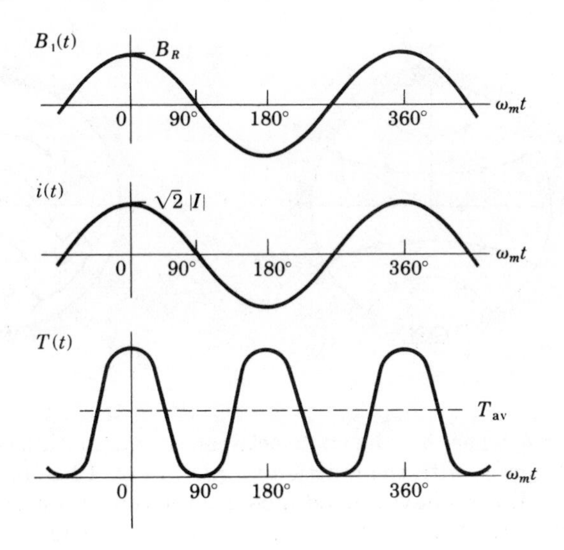

FIGURE 19.1–2

Recognizing a bilateral device when we see one, we immediately deduce that if an external source supplies an AC current in the opposite direction through the armature, the resulting magnetic torque will sustain clockwise rotor motion; we then have an AC *motor* rather than a generator. Such a motor is said to be *synchronous* because the mechanical rotation frequency and the electrical frequency of the source are in synchronism.

Exercise 19.1–1

Show that the average mechanical power $P_m = T_{av}\,\omega_m$ equals the average electrical power in a lossless synchronous motor when applied terminal voltage equals the induced emf.

ROTATING FIELDS AND MAGNETIC POLES

To gain further insight on the operation of a synchronous machine, and to prepare for subsequent consideration of other machines, we next examine the interaction between the rotor's field and the magnetic field established by the armature current. Obviously, the rotor field rotates, being bound to the mechanical movement of the rotor. Not so obvious is the fact that the immobile armature coil also has rotating fields.

Figure 19.1–3a gives the reference armature-current direction for clockwise generator operation and its corresponding field pattern based on the right-hand rule (see Fig. 17.1–3b). The stator thus has an apparent south pole at $\gamma_s = -90°$ and a north pole directly opposite. If we place the permanent-magnet rotor in this field as in Fig. 19.1–3b, its north pole will be drawn towards the south pole of the stator until the two magnetic axes become aligned. The magnetic alignment torque T is proportional to

γ_s

B_s

S N

(a)

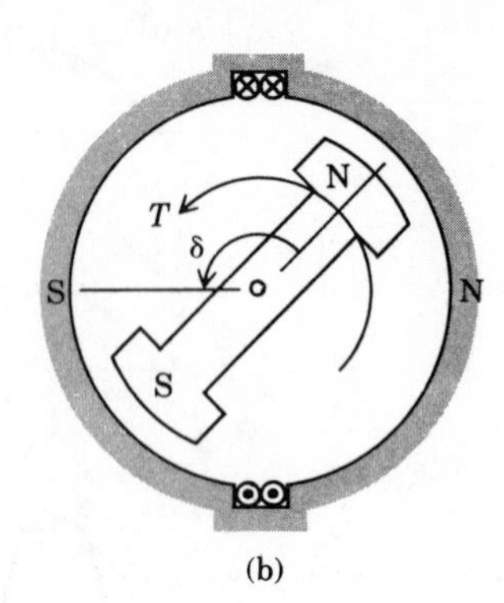

FIGURE 19.1–3
(a) Stator field.
(b) Angle δ between magnetic axes.

sin δ, where δ is the angle between the magnetic axes, and the direction of T agrees with the reaction torque in Fig. 19.1–1c.

But we have not yet included the time variation of the armature current $i(t)$. For that purpose, we ignore any magnetic nonlinearity and assume that the outward radial stator field in the air gap is

$$B_s = -k_s i(t) \sin \gamma_s = k_s i(t) \cos (\gamma_s + 90°)$$

where the constant k_s depends on the armature turns N and the reluctance of the flux path through stator, air gap, and rotor. Inserting $i(t) = \sqrt{2}\,|I| \cos \omega_m t$ and expanding yields

$$B_s = B_{S_1} \cos (\gamma_s + 90° - \omega_m t) + B_{S_2} \cos (\gamma_s + 90° + \omega_m t) \qquad (6)$$

with $B_{S_1} = B_{S_2} = k_s\,|I|/\sqrt{2}$. Comparison with the expression for B_r in Eq. (3) leads to the conclusion that B_s consists of *two rotating vectors* moving in opposite directions. The vector $\underline{B}_{S_1}$ rotates clockwise and has an apparent south pole at $\gamma_s = -90° + \omega_m t$, while $\underline{B}_{S_2}$ rotates counterclockwise and has an apparent south pole at $\gamma_s = -90° - \omega_m t$. Figure 19.1–4a shows how these two vectors sum to produce a horizontally alternating field with maximum value $B_S = B_{S_1} + B_{S_2} = \sqrt{2}\, k_s|I|$. Bear in mind that the rotating stator fields are caused by the oscillations of $i(t)$ at frequency ω_m, not by any mechanical movement.

Adding the rotor field vector $\underline{B}_R$ to our picture gives Fig. 19.1–4b, which explains the torque variation $T(t)$ in Fig. 19.1–2. Specifically, the clockwise stator field $\underline{B}_{S_1}$ continually lags $\underline{B}_R$ by $\delta_1 = 90°$ and tries to pull the rotor back into alignment with it. This accounts for the constant torque component T_{av} exerted on the prime mover. On the other hand, $\underline{B}_{S_2}$ revolves past the rotor with instantaneous angle $\delta_2 = 90° + 2\omega_m t$ and alternately tugs it backward and forward. This accounts for the torque variations $T_{\text{av}} \sin \delta_2 = T_{\text{av}} \cos 2\omega_m t$ superimposed on T_{av} in the total torque $T(t) = T_{\text{av}}(1 + \cos 2\omega_m t)$.

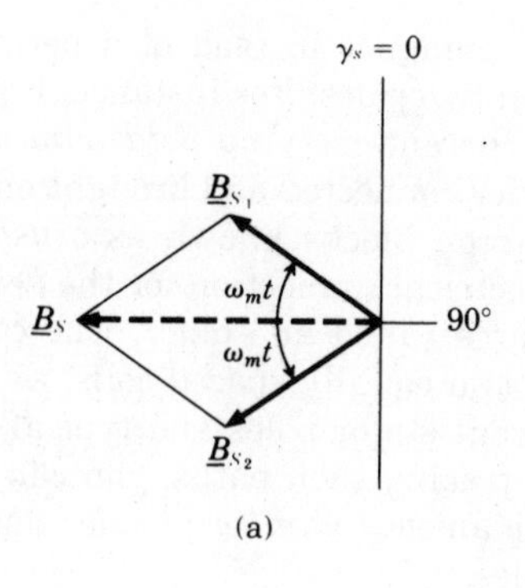

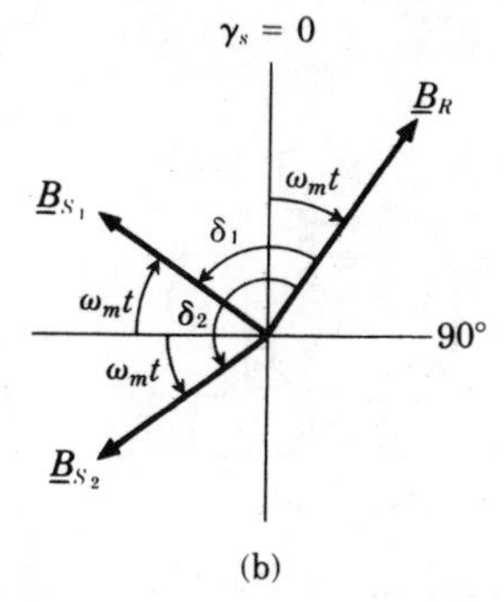

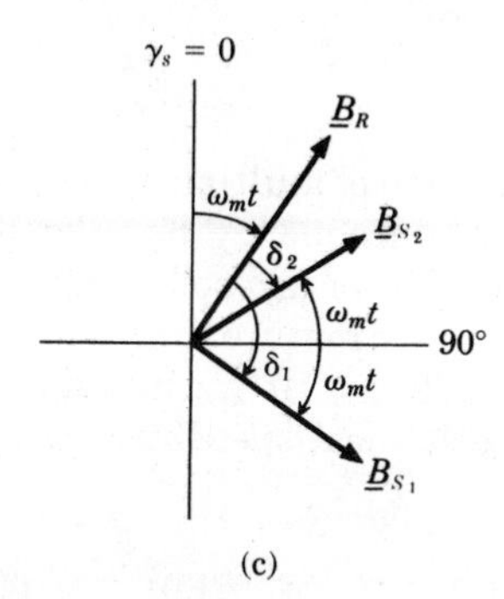

FIGURE 19.1–4
(a) Stator field decomposed into two rotating vectors.
(b) Rotor and stator field vectors for generator operation.
(c) Motor operation.

Reversing the direction of the armature current, corresponding to motor operation, reverses the stator field vectors and gives Fig. 19.1–4c. Now $\underline{B}_{S_1}$ leads the rotor by 90° and pulls it along in synchronous rotation, the developed motor torque again being T_{av}. Nonetheless, $\underline{B}_{S_2}$ still rotates in the opposite direction and produces alternating torque variations. Identical results would be obtained for motor action had we reversed the rotor field or the direction of mechanical rotation in Fig. 19.1–4b.

A practical implementation of our elementary machine could have a number of structural modifications. In particular, the rotor would almost certainly be an electromagnet instead of a permanent magnet, and it might have more than two poles. For instance, Fig. 19.1–5a illustrates a *four-pole* rotor with current-carrying *field windings* around each pole. The windings are series connected and brought out to a pair of *slip rings* upon which press carbon blocks known as *brushes*. This arrangement permits a rotating electrical connection for the *field current* I_f that comes from a small DC source called an *exciter*. The corresponding armature winding would have two equally spaced coils, as shown in Fig. 19.1–5b along with the apparent stator poles and typical flux paths through the rotor and stator. By tracing such paths, you can convince yourself that there must always be an *even number of poles* and one armature coil for each pair of rotor poles.

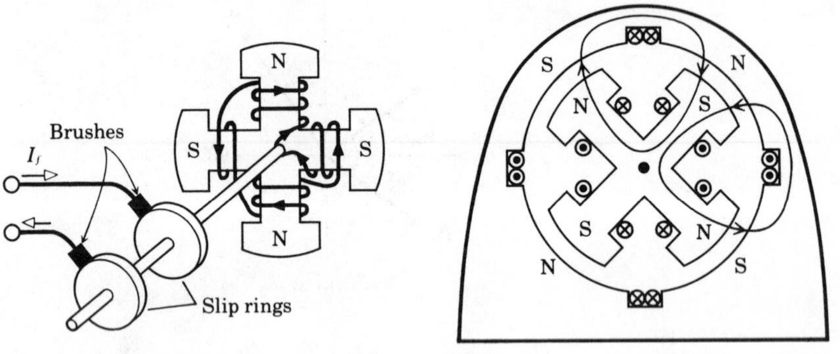

FIGURE 19.1–5 (a) Four-pole rotor with field windings and slip rings. (b) Four-pole stator.

The advantage of multiple pairs of poles is *lower mechanical speed* for a given electrical frequency. Specifically, if p is the number of poles, then one full revolution of a p-pole rotor produces $p/2$ complete cycles of induced emf in any armature conductor. Therefore, the electrical frequency will be $\omega = (p/2)\,\omega_m$. If we measure electrical frequency in *hertz*, $f = \omega/2\pi$, and mechanical speed in *revolutions per minute* (rpm) denoted by n_s, then

$$n_s = \frac{60}{2\pi}\,\omega_m = \frac{60}{\pi}\frac{\omega}{p} = 120\,\frac{f}{p} \tag{7}$$

where the subscript s emphasizes the fact that we are dealing with *synchronous* speed. For the usual case of $f = 60$ Hz, a hydro turbine or other low-speed prime mover might drive an 18-pole alternator at $n_s = 7200/18 = 400$ rpm, whereas a high-speed steam turbine unit would have $p = 2$ or 4 and $n_s = 3600$ or 1800 rpm.

In the high-speed case, the protruding or *salient* rotor poles of Fig. 19.1–5 could be subjected to extreme mechanical stress. That stress is minimized in a cylindrical rotor structure that has *nonsalient* poles formed by distributing the field winding around the rotor. This configuration is illustrated in Fig. 19.1–6a for $p = 2$. A distributed winding is also shown for the armature coil. Most machines, in fact, have *distributed* rather than *concentrated* armature windings because the former makes better use of materials. Figure 19.1–6b shows a simplified winding pattern for one coil that spans an angle of $2 \times 360°/p$ in a p-pole machine. Because of the spatial distribution of conductors, the emf is somewhat reduced by a *winding factor* $k_w \approx 0.82–0.96$ compared to the emf induced in a concentrated coil.

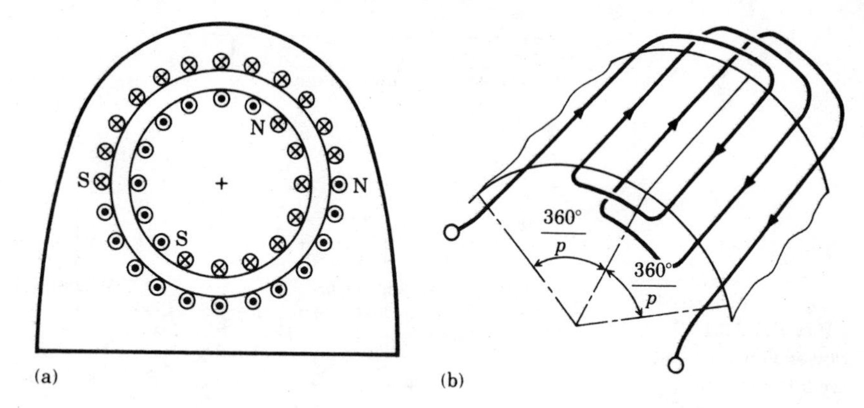

FIGURE 19.1–6 (a) Two-pole machine with nonsalient rotor poles and distributed windings. (b) Detail of winding pattern for p-pole machine.

Exercise 19.1–2 Redraw Fig. 19.1–4c with $\omega_m t = 0°$, 45°, and 90°. Then substitute the angle values in $T(t) = T_{av}(\sin \delta_1 + \sin \delta_2)$ to show that the shape of $T(t)$ in Fig. 19.1–2 also applies for motor action.

MACHINE PERFORMANCE CHARACTERISTICS

Although the "Blu" and "Bli" laws describe the interaction across the air gap, a number of additional factors enter the picture when we talk about the external characteristics of an actual machine. Here we briefly consider some of the major factors and their influence on machine performance.

Obviously, a real machine always has internal *losses* that result in less than 100% energy-conversion *efficiency*. Take the synchronous motor, for example. The total input power will consist of AC power to the armature plus the smaller DC power for the rotor field, from which we must deduct *copper losses* caused by ohmic heating in the winding resistances and brush-contact loss at the slip rings. There will also be *core losses* due to hysteresis and eddy currents (primarily confined to the iron through which passes time-varying flux), and *mechanical losses* due to brush and bearing friction and air drag or windage. Core and mechanical losses are usually considered together under the general heading of *no-load rotational loss* simply because it is easier to measure their joint effect. An additional term called the *stray load loss* arises from nonuniform current and field distributions, and adds to the total *rotational losses*.

Figure 19.1–7 diagrams the power flow from electrical input to mechanical "shaft" output. Similar diagrams apply for other types of motors. For generators, the direction of power flows from the mechanical input to the electrical output. Typical copper and rotational losses fall in the neighborhoods of 2–10% and 1–15%, respectively, for a net conversion efficiency of 75–97%.

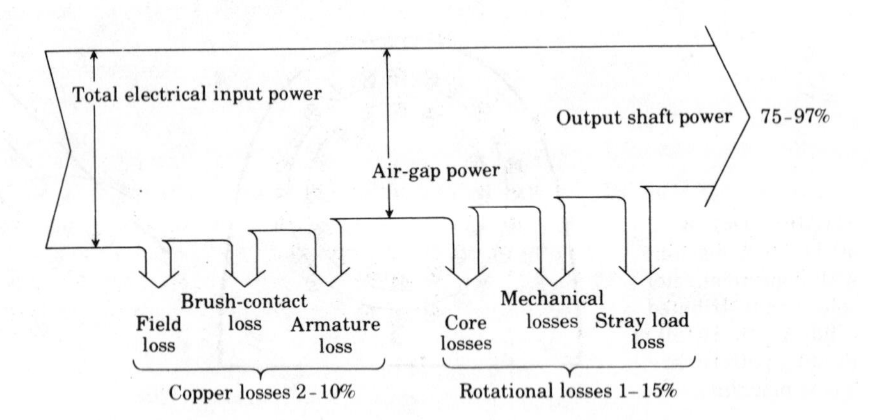

FIGURE 19.1–7 Power flow and losses for a typical motor.

Internal losses lead to *heating,* which must be considered in the design of any sizable machine. Excessive heating breaks down the winding insulation and eventually leads to failure. The National Electrical Manufacturers Association (NEMA) has therefore established standard *ratings* of allowable temperature rise for various classes of insulation. NEMA has also standardized many frame sizes. Special ventilation and cooling techniques may be needed to put a large-capacity machine in the smallest possible frame.

Our power-flow diagram (Fig. 19.1–7) identifies a quantity P_{ag} known as the *air-gap power,* which represents the rate of energy conversion in the air gap exclusive of any losses. This quantity is of fundamental importance in determining a machine's capacity. A general expression for P_{ag} that applies to most machines can be written as

$$P_{ag} = kN_aI_aB_fAn \tag{8}$$

where N_a is the number of armature conductors carrying the DC or rms

AC current I_a; B_f is the maximum air-gap flux density per field pole, $A = 2r\ell$ is the axial area of the rotor turning at n rpm, and k is a constant whose value depends on the individual machine. For a generator, P_{ag} equals the average electrical power in watts (or kilowatts or megawatts) before deducting internal copper losses. For a motor, P_{ag} equals the mechanical power before deducting rotational losses. Mechanical power is often expressed in *horsepower* (hp), defined by

$$1 \text{ hp} = 746 \text{ W} \tag{9}$$

Horsepower, of course, is not an SI unit.

Some important conclusions about machine capacity can be inferred from Eq. (8). In particular, we see that power increases with the armature mmf $N_a I_a$—which also implies increasing copper losses and ohmic heating. A high-capacity motor or generator may therefore require armature conductors of large cross-sectional area, special high-temperature insulation, and perhaps an elaborate cooling system. Power also increases with both flux density B_f and physical size corresponding to the area A, neither of which can be made arbitrarily large. Magnetic saturation limits the former (as seen in Fig. 17.2–4), while mechanical and economic factors and available space limit the latter. Last, P_{ag} increases directly with rotor speed n, suggesting the potential need in high-capacity machines for high-speed bearings, auxiliary gear boxes, and similar mechanical provisions.

But we seldom, if ever, expect a machine to operate at its maximum capacity. Instead, it should provide the power demanded of it by the particular load in a given application. As a matter of fact, the intersection of the load characteristics and machine characteristics determines the actual operating point. For example, the terminal voltage and output current of a generator might be related by the characteristic curve sketched in Fig. 19.1–8a. The dashed curve in the figure represents a load characteristic, not necessarily linear, that happens to intersect the generator's curve at the *rated* or *full-load* current I_{FL} and voltage V_{FL}. The corresponding full-load power $V_{FL} I_{FL}$ presumably would be less than the machine's maximum generating capacity to allow for occasional overload surges of brief duration.

Usually we want a voltage source to be "stiff," in the sense of maintaining constant voltage despite load-current variations. Loading effects, however, often result in a full-load voltage less than the *no-load* (open-circuit) voltage V_{NL}. As a measure of generator performance, we speak of the *voltage regulation* defined by

$$\text{VR} = \frac{V_{NL} - V_{FL}}{V_{FL}} \tag{10}$$

and expressed in percent. All practical generators include an automatic *regulator* that adjusts the field excitation to keep voltage constant, as illustrated by Fig. 19.1–8b.

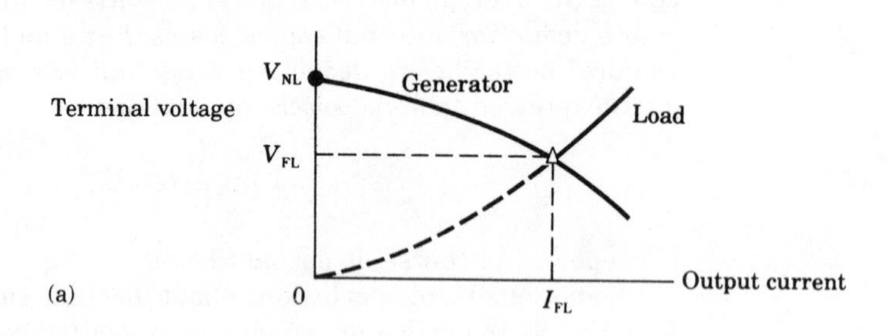

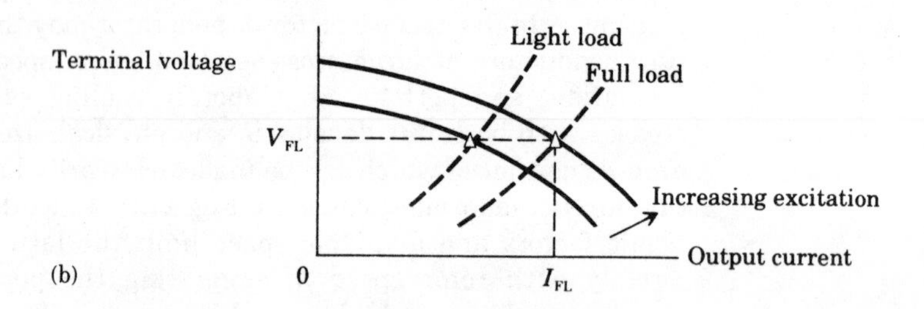

FIGURE 19.1–8
(a) Typical generator and load characteristics.
(b) Voltage regulation by adjustable field excitation.

The performance characteristics of a motor are described by its *speed-torque* (or torque-speed) *curve.* The synchronous motor, of course, is a constant-speed machine, but the more common *induction motor* has a speed-torque curve similar to the one shown in Fig. 19.1–9a. Since the full-load speed n_{FL} is less than the no-load speed n_{NL}, we define the *speed regulation*

$$\mathrm{SR} = \frac{n_{NL} - n_{FL}}{n_{FL}} \tag{11}$$

analogous to voltage regulation. The rated (full-load) horsepower may be only 30–50% of the maximum capacity, but there is a maximum or breakdown torque that cannot be exceeded without stalling the motor. The stalled torque or starting torque T_0 (at zero speed) will be greater than T_{FL} if the motor has been designed to come up to speed under loaded conditions.

Different motor applications have grossly different speed and torque requirements. For instance, think of the various requirements of an automobile starter, elevator drive, and magnetic-tape transport. Numerous types of motors have been developed to match various load categories—constant speed, adjustable speed, high starting torque, and so forth. Figure 19.1–9b illustrates the speed-torque characteristics of a hypothetical synchronous motor and a *universal motor* that have the same full-load and T_{max} values as the induction motor in Fig. 19.1–9a. The curve for the universal motor approximates a "constant-horsepower" characteristic with high starting torque and low torque at high speed. (The details behind these curves will be discussed in the remaining sections of this chapter.)

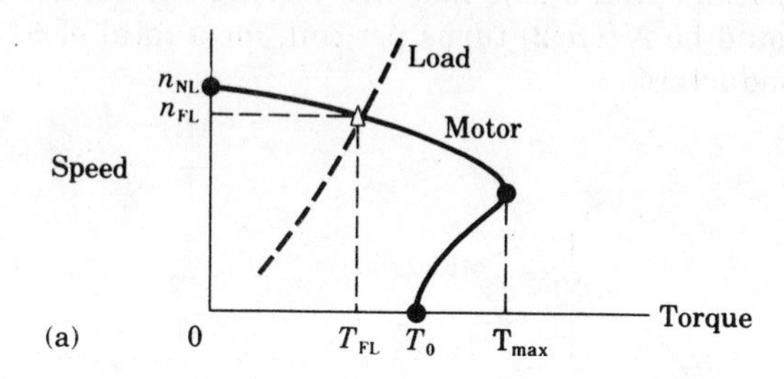

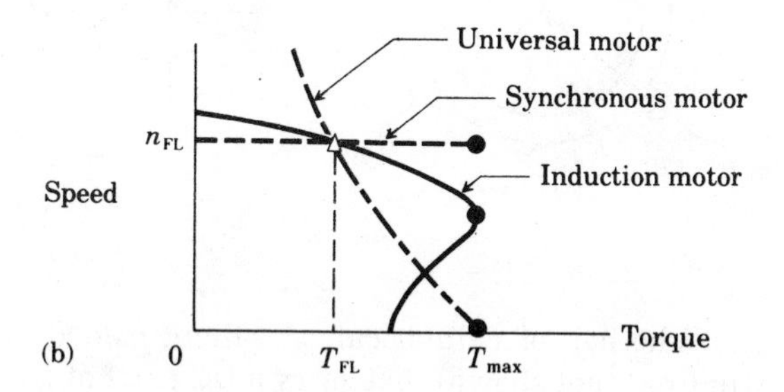

FIGURE 19.1–9 Typical speed-torque curves.

Exercise 19.1–3

A certain 6-pole synchronous motor operates at $f = 60$ Hz and develops 100 hp at the shaft. The copper and rotational losses are each 10%. Calculate the electrical input power, the air-gap power, and the output torque, all expressed in SI units.

Exercise 19.1–4

For a synchronous machine with a nonsalient rotor and distributed stator winding, the constant in Eq. (8) is $k \approx \sqrt{2}\pi 0.9/f$. Use this expression to calculate the armature mmf needed for the motor in the previous exercise if $B_f = 1$ T and $A = 2 \times 25$ cm $\times$ 1 m.

19.2 SYNCHRONOUS MACHINES

The elementary machine introduced in the previous section was found to be capable of operating as an AC generator or motor, but had the disadvantage of pulsating torque and instantaneous power. For improved performance, we turn to the balanced three-phase synchronous machine, in which three sets of rotating stator fields combine to produce constant torque and power. Similar results are attained by any *polyphase* machine with two or more symmetrical phase windings on the stator.

THREE-PHASE GENERATORS

All bulk AC power fed into a power grid comes from three-phase synchronous generators. These machines have the same general structure as a single-phase alternator except that there are three sets of armature coils per pair of rotor poles. The coils for successive phases have separate terminals and equal angular spacing of $120°/(p/2)$ around the stator, so the equivalent electrical spacing is 120°. Figure 19.2–1a shows the conductor positions in a 2-pole machine having one turn per coil. In general, there would be $N/(p/2)$ turns per coil, for a total of $6N$ uniformly distributed conductors.

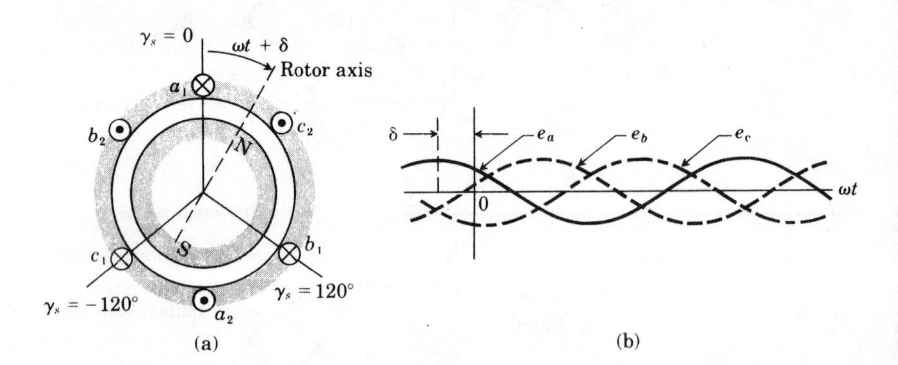

FIGURE 19.2–1
(a) Simplified two-pole three-phase machine.
(b) Waveforms of open-circuit emf.

Whether of cylindrical or salient-pole design, the rotor has field windings (not shown) that carry a DC current I_f that produces an approximately sinusoidal angular field pattern with maximum value B_R. The rotor turns at the synchronous speed $n_s = 60\ \omega_m/2\pi = 120f/p$, and has an arbitrary initial angle δ at $t = 0$. As the rotor field sweeps past successive coils, it induces the open-circuit emf's

$$
\begin{aligned}
e_a &= \sqrt{2}|E| \cos(\omega t + \delta) \\
e_b &= \sqrt{2}|E| \cos(\omega t + \delta + 120°) \\
e_c &= \sqrt{2}|E| \cos(\omega t + \delta - 120°)
\end{aligned}
\tag{1}
$$

with $\omega = 2\pi f$ being the electrical angular frequency. You should have no trouble understanding the phase shifts here by comparing the waveforms in Fig. 19.2–1b with the angular positions of the rotor and stator coils.

Drawing upon Eq. (4), Section 19.1, we find that the rms voltage per phase is

$$
|E| = \frac{\pi}{60\sqrt{2}} k_w NB_R An_s \tag{2a}
$$

where we have inserted the winding factor k_w and substituted $\omega_m = 2\pi n_s/60$. It can also be shown that the *flux per field pole* is $\phi = B_R A/p$ so, with $f = pn_s/120$,

$$|E| = \sqrt{2}\pi k_w f N\phi = 4.44\ k_w f N\phi \tag{2b}$$

similar to the transformer expression in Eq. (4), Sect. 18.2.

Now let the armature coils be connected in a wye configuration to a balanced three-phase load with lagging power factor pf = cos θ, as in Fig. 19.2–2a. The equivalent circuit for phase a (armature coil $a_1 - a_2$) is diagrammed in Fig. 19.2–2b, where δ has been chosen such that the terminal voltage phasor $\underline{V}_a = |V| \angle 0°$ is the phase reference and leads the line current phasor $\underline{I}_a = |I| \angle -\theta$. The terminal voltage differs from the open-circuit emf phasor $\underline{E}_a = |E| \angle \delta$ due to an internal voltage drop across the *synchronous reactance* X_s representing the inductance of the armature winding. (There would also be some winding resistance, but its effect is negligible compared to X_s in a large machine and the generator's Thévenin equivalent impedance essentially equals jX_s.) Identical diagrams hold for the other two phases with an added phase shift of ± 120°. We first justify this equivalent circuit by considering the stator fields, and then interpret the implications of Fig. 19.2–2.

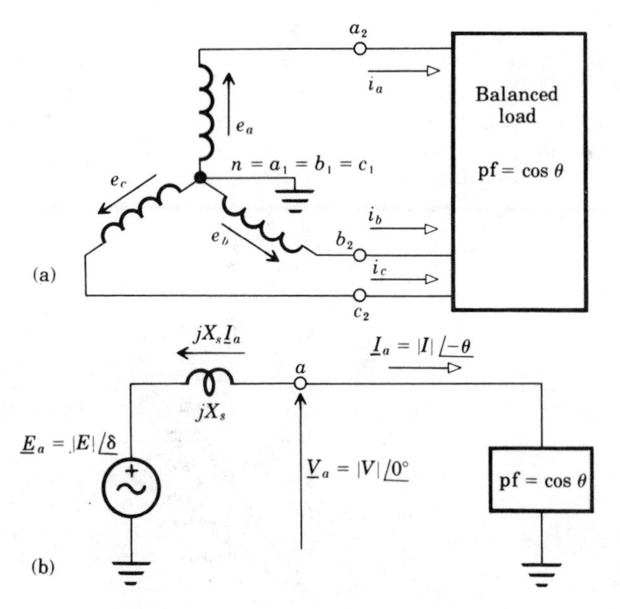

FIGURE 19.2–2 (a) Three-phase generator with balanced load. (b) Equivalent circuit of one phase.

From our previous study, we know that the current $i_a = \sqrt{2}\,|I| \cos(\omega t - \theta)$ creates two rotating stator fields, say $\underline{B}_{a_1}$ and $\underline{B}_{a_2}$, and likewise for i_b and i_c, so there are six stator field vectors all told. But when we account for the current phase angles and the spatial orientation of the coils, we arrive at the diagram of Fig. 19.2–3a, where the clockwise components fall in line and reinforce, while the counterclockwise components form a *symmetrical three-phase set* and completely cancel out. Hence, the balanced three-phase stator field acts like *one* clockwise rotating vector $\underline{B}_S = \underline{B}_{a_1} + \underline{B}_{b_1} + \underline{B}_{c_1}$! It immediately follows that the total magnetic reaction torque exerted on the rotor will be *constant,* rather than pulsating, agreeing with our conclusion in Sect. 18.3 that a balanced three-phase system has constant instantaneous power $p(t) = P_{\text{total}} = 3P$.

Adding $\underline{B}_S$ to the rotor field vector $\underline{B}_R$ yields the resultant vector $\underline{B}$ for the total air-gap field, Fig. 19.2–3b. We show $\underline{B}$ to be lagging $\underline{B}_R$ by the angle δ because, in fact, it is this total field that produces the terminal voltage $\underline{V}_a = |V| \angle 0°$ in phase a, whereas $\underline{B}_R$ produces the open-circuit emf $\underline{E}_a = |E| \angle \delta$. The difference between $\underline{E}_a$ and $\underline{V}_a$ comes from an additional emf $\underline{E}_{as}$ induced by the stator field $\underline{B}_S$. (We did not have this additional emf in the single-phase alternator because the net stator field perpendicular to the coil was always zero.) Since $\underline{B}_S$ is proportional to $|I|$ and lags $\underline{B}$ by $90° + \theta$, we conclude that $|E_{as}|$ must be proportional to $|I|$ and $\measuredangle \underline{E}_{as} = -90° - \theta$. We can thus write $\underline{E}_{as} = -jX_s\underline{I}_a$ and

$$\underline{V}_a = \underline{E}_a + \underline{E}_{as} = \underline{E}_a - jX_s\underline{I}_a$$

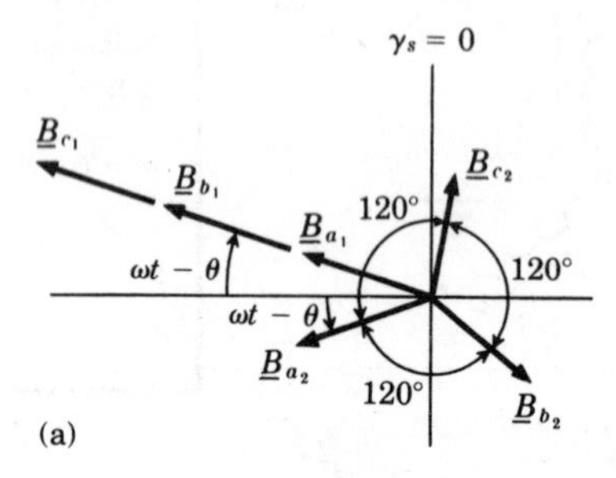

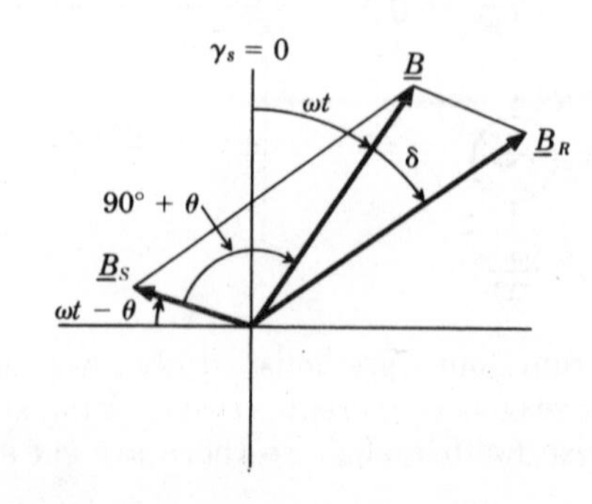

FIGURE 19.2–3
(a) Stator field vectors in a three-phase machine. (b) Resultant of stator and rotor field vectors.

thereby explaining the synchronous reactance X_s and the equivalent circuit in Fig. 19.2–2b.

To interpret the equivalent circuit we construct the phasor diagram of Fig. 19.2–4, where the subscript a has been dropped from $\underline{E}_a$, $\underline{I}_a$, etc., since the diagram equally well represents any one of the three phases. Note that the phasor $-\underline{E}_s = jX_s\underline{I}$ is perpendicular to $\underline{I}$ and that $X_s\,|I|\cos\theta = |E|\sin\delta$, a useful relation for examining the real (average) power P consumed by each phase of the load. Specifically,

$$P = |V|\,|I|\cos\theta = \frac{|V|\,|E|}{X_s}\sin\delta \tag{3}$$

which justifies naming δ the *power angle*. It is also called the *torque angle* in view of the fact that the torque from the prime mover must be $T = 3P/\omega_m$, neglecting any mechanical losses, and ω_m is supposed to be constant.

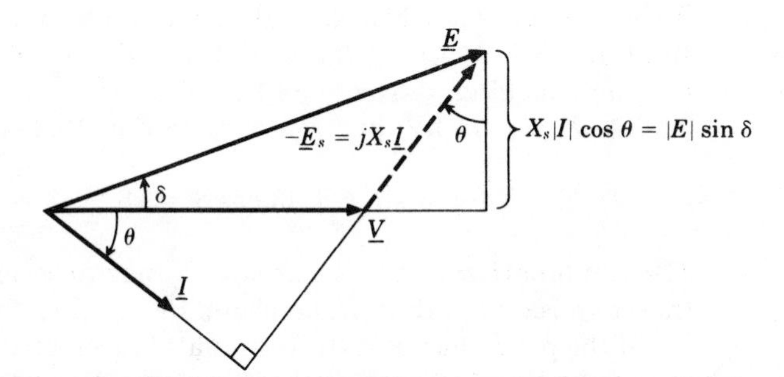

FIGURE 19.2–4
Phasor diagram for one phase.

If $|V|$ and $|E|$ have fixed values, an increasing power demand from the load requires increasing δ. Referring to Fig. 19.2–3b then reveals that the prime mover must produce greater torque to accelerate the rotor momentarily and advance $\underline{B}_R$ relative to $\underline{B}$. However, there exists an inherent upper limit or "pullout power"

$$P_{\max} = \frac{|V|\,|E|}{X_s}$$

when $\delta = 90°$—which is analogous to the maximum capacity of a transmission line in Eq. (5), Section 18.4. Figure 19.2–5 emphasizes this limitation in the plot of P as a function of δ based on Eq. (3). Actually, due to *reluctance-torque* effects, which are discussed in the next section, a salient-pole rotor has a somewhat higher pullout power at $\delta < 90°$, as represented by the dashed curve.

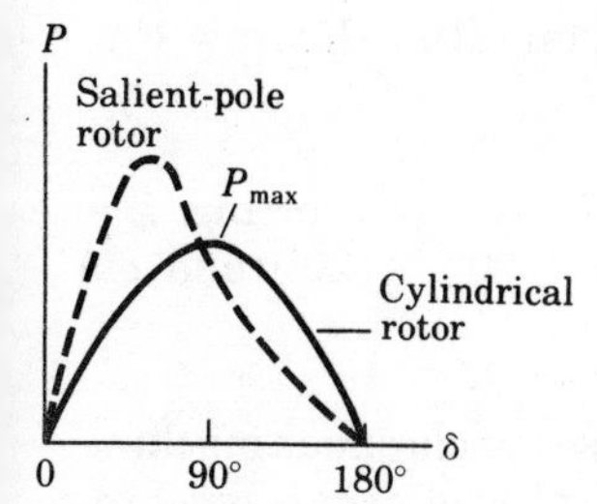

FIGURE 19.2–5
Average power as a function of power angle.

The rms value of the generator's terminal voltage is readily adjusted by changing the DC field excitation I_f, which increases or decreases B_R and $|E|$—see Eq. (2a). A plot of $|E|$ versus I_f therefore takes the shape of a *magnetization curve* like Fig. 17.2–4. If $|E|$ is held fixed, the curve for $|V|$ versus $|I|$ depends on the load's power factor. The plot previously shown in Fig. 19.1–8a would be typical for a load with $\theta = 0$, and underscores why generators need a built-in feedback regulator to maintain constant terminal voltage.

The necessary field current is usually produced in a large generator by a *brushless exciter* consisting of an auxiliary rotor and stator. The auxiliary rotor, mounted on the same shaft as the main rotor, serves as a small AC generator; its current is then rectified by a diode bridge and fed into the main rotor winding. The auxiliary stator provides the field for the exciter, and we control I_f by the AC or DC voltage applied to the auxiliary stator. This clever scheme eliminates the need for rotor slip rings and the accompanying brush-contact loss.

Example 19.2–1

A certain generator has $X_s = 10\ \Omega$ and its terminal voltage is held constant at $|V| = 24$ kV. If the load draws $P = 7.2$ MW per phase and has pf = 0.6 lagging, so $\theta = 53.1°$, then $|I| = 7.2\ \text{MW}/(0.6 \times 24\ \text{kV}) = 0.5$ kA, $X_s\,|I| = 5$ kV; and referring to Fig. 19.2–4 gives

$$\underline{E} = 24 + 5 \sin\theta + j5\cos\theta = 28 + j3 = 28.2\ \text{kV}\ \underline{/6.1°}$$

The combination of the synchronous reactance and the inductive load therefore requires that $|E|$ be about 18% higher than $|V|$.

If the power factor at the terminals is corrected to unity ($\theta = 0°$), then $|I| = 0.3$ kA and the required emf drops to $\underline{E} = 24 + j3 = 24.2\ \text{kV}\ \underline{/7.1°}$. Note that the torque angle δ increases slightly, from 6.1° to 7.1°, to maintain P with the smaller $|E|$.

Exercise 19.2–1

Repeat the preceding calculation for $\underline{E}$, taking $\theta = -53.1°$; show that $|E| < |V|$ when the load has a leading power factor. Hint: Redraw Fig. 19.2–4 with $\underline{I}$ leading $\underline{V}$.

Exercise 19.2–2

Justify the angles of $\underline{B}_{b_1}$ and $\underline{B}_{b_2}$ in Fig. 19.2–3a by inserting $i_b = \sqrt{2}|I| \cos(\omega t - \theta - 120°)$ into $B_b = k_s i_b \cos(\gamma_s + 90° - 120°)$ and expanding.

SYNCHRONOUS MOTORS AND CAPACITORS

If AC power is applied to the stator of a three-phase synchronous machine it becomes a *motor* capable of driving a mechanical load at the synchronous speed n_s. Such motors find application where the nature of the load demands *constant speed* and *nonpulsating torque,* justifying the added expense of the DC exciter. A synchronous motor without a mechanical load may also be used for *power-factor correction,* in which case it is called a *synchronous capacitor* or *condenser*—"condenser" being the original name for capacitor.

Figure 19.2–6a shows the equivalent circuit for one phase of a synchronous motor with applied terminal voltage $\underline{V} = |V| \angle 0°$. There is still an induced emf $\underline{E} = |E| \angle \delta$ with $|E|$ as given by our previous equations; but $\underline{E}$ now opposes the flow of stator current $\underline{I}$ and acts as a "back emf." Since $\underline{I}$ has a reversed direction compared to a generator (Fig. 19.2–2),

$$\underline{V} = \underline{E} + jX_s\underline{I}$$

and the phasor diagram of Fig. 19.2–6b reveals that $\underline{V}$ leads $\underline{E}$, meaning that the power angle δ must be *negative*. Negative power angle corresponds to the stator field revolving ahead of the rotor and pulling it along, as contrasted to a generator where the stator field trails behind and drags the rotor. Equivalently, $P = (|V|\,|E|/X_s) \sin \delta$ is a negative quantity when $\delta < 0$, reflecting the fact that the motor consumes rather than produces real power. By the same argument, replacing δ by $-\delta$ in Fig. 19.2–5 gives us the motor power (per phase) versus power angle, with the pullout power $P_{\max}$ being the point above which the rotor loses

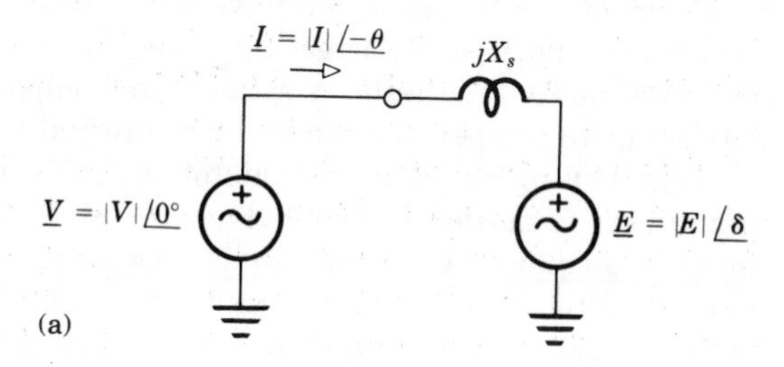

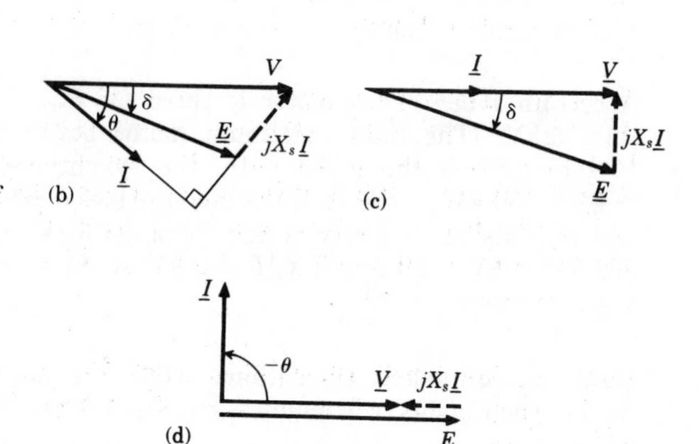

FIGURE 19.2–6
Synchronous motor: (a) equivalent circuit of one phase; (b) phasor diagram for underexcited motor; (c) phasor diagram for unity power factor; (d) phasor diagram for operation as a synchronous capacitor.

synchronism. The corresponding maximum torque will be

$$T_{\max} = \frac{3P_{\max}}{\omega_m} = \frac{3p\,|V|\,|E|}{2\omega X_s} \tag{4}$$

and is adjusted via the field excitation.

The motor also consumes reactive power if $\underline{V}$ and $\underline{I}$ have the relationship in Fig. 19.2–6b, for the equivalent input impedance will be $\underline{Z}_{eq} = \underline{V}/\underline{I} = (|V|/|I|)\ \underline{/\theta}$ with pf $= \cos\theta$. The motor is then said to be *underexcited*. But we can correct this lagging power factor simply by increasing the DC field current I_f until

$$|E|^2 - |V|^2 + (X_s\,|I|)^2$$

in which case $\underline{I}$ becomes colinear with $\underline{V}$ and $\theta = 0°$, as in Fig. 19.2–6c. Under this condition, a synchronous motor appears to be a purely resistive load with the desirable attribute of unity power factor.

Further increasing the excitation causes I to lead $\underline{V}$ $(\theta < 0)$, and the *overexcited* motor produces rather than consumes reactive power. In the extreme case where $\theta = -90°$ and $\delta = 0°$, as in Fig. 19.2–6d, the motor looks like a *capacitor* and cannot drive a mechanical load since $P = 0$. However, $Q = -|V|\,|I|$, so we have a synchronous capacitor that serves the valuable purpose of correcting a lagging power factor caused by other inductive loads. (Actually, a synchronous capacitor still requires some real power to compensate for its own internal losses.)

Whether operated in the motor or capacitor mode, an additional mechanism is needed to bring the rotor up to synchronous speed, since the average magnetic torque equals zero at any lower speed. For this purpose, the rotor usually has an auxiliary winding that acts like that of an *induction* motor. The induction winding produces no torque at synchronous speed but does help damp out unwanted mechanical "hunting" around n_s and is therefore called a *damper* or *amortisseur* winding.

Example 19.2–2

A certain synchronous motor is rated for $P_{\max} = 250$ kW when $|E| = |V| = 5$ kV. (The total maximum motor power is 750 kW, or about 1000 hp.) From the given data, the synchronous reactance is $X_s = (5\text{ kV})^2/250\text{ kW} = 100\ \Omega$. If the motor drives a 300-kW mechanical load, and is adjusted for unity power factor with $|V| = 5$ kV, then $|I| = \frac{1}{3} \times 300\text{ kW}/5\text{ kV} = 20$ A and $X_s|I| = 2$ kV, so $|E| = \sqrt{5^2 + 2^2} \approx 5.4$ kV and $\delta = -\arctan\frac{2}{5} \approx -22°$.

Exercise 19.2–3

Find the equivalent capacitance when the above motor is operated as a synchronous capacitor with $|E| = 6$ kV, $|V| = 5$ kV, and $\omega = 377$ rad/sec.

19.3 AC MOTORS

Routine applications—blowers, pumps, conveyors, and the like—seldom require the constant-speed property of a synchronous motor and do not warrant the expense. For these applications the preferred machine is almost always an *induction motor,* three-phase or single-phase depending on the power level. Indeed, the AC induction motor serves as a veritable rotating workhorse in home and industry. Its characteristics are detailed in this section, along with some special-purpose AC motors.

THREE-PHASE INDUCTION MOTORS

Structurally, the simplest, most rugged, most reliable rotating machine is the three-phase *squirrel-cage induction motor*. The stator is identical to that of a synchronous machine with distributed windings, but the rotor winding consists of a "squirrel-cage" arrangement of conducting bars shorted together by two end rings, as in Fig. 19.3–1a. There are no rotating electrical connections nor a DC exciter supplying field current. Instead, the rotor field needed to develop mechanical torque comes from the stator field via induction.

For a qualitative description of this process, consider the simplified rotor in Fig. 19.3–1b and assume it to be initially at rest. The stator has an applied voltage of frequency f and its field $\underline{B}_S$ rotates at the synchronous speed $n_s = 120f/p$ rpm. As $\underline{B}_S$ sweeps past the rotor bars, it induces an emf which causes current to circulate in the rotor and thereby produces a rotor field $\underline{B}_R$. Torque develops from the interaction between $\underline{B}_S$ and $\underline{B}_R$, so the rotor begins to turn and eventually reaches a steady-state speed n.

This rotor speed must be less than the synchronous speed, for there would be no induced emf if $n = n_s$ (why?), so no rotor current or field and no torque to sustain the rotation against a mechanical load. We therefore define the *slip*

$$s \triangleq 1 - \frac{n}{n_s} = \frac{n_s - n}{n_s} \tag{1}$$

corresponding to rotor speed $n = (1 - s)n_s$. The *slip speed*

$$sn_s = n_s - n$$

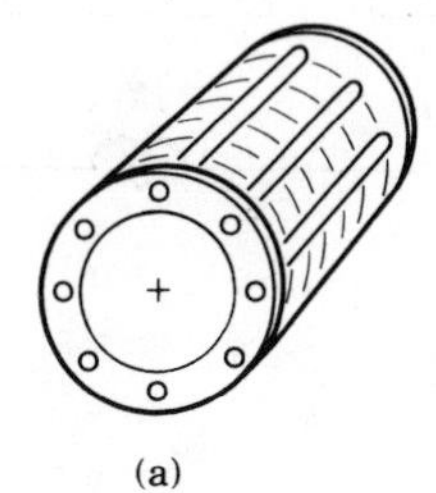

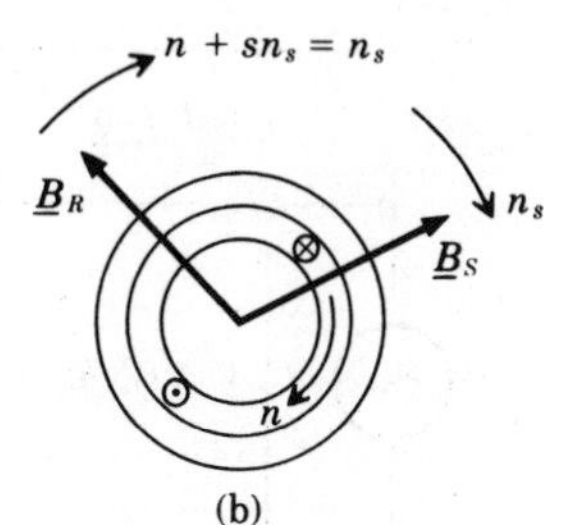

FIGURE 19.3–1
(a) Squirrel-cage rotor.
(b) Rotor and stator fields.

represents the relative velocity between $\underline{B}_S$ and the rotor bars, and the electrical frequency of the induced emf will be $(p/120)\, sn_s = sf$ Hz. Thus, upon including the rotor bars for the other two phases, we find that $\underline{B}_R$ rotates at speed sn_s relative to the rotor. Adding sn_s to the mechanical rotor speed n shows that $\underline{B}_R$ rotates at $n + sn_s = n_s$, in synchronism with $\underline{B}_S$. Therefore, the developed torque will be constant and free of pulsations, just as in a synchronous motor, even though the rotor turns at $n < n_s$.

For analytic purposes, we view the induction motor as a three-phase *transformer* with a fixed primary (the stator) and a rotating secondary (the rotor). We take account of the rotor's motion by observing that the secondary voltages and currents have electrical frequency $s\omega$ rad/sec and that the induced secondary voltage will be proportional to s as well as to the effective turns ratio N. Figure 19.3–2a diagrams the resulting equivalent circuit for one phase, including winding resistance and reactance in both the primary and secondary. (Magnetizing inductance and core-loss resistance have been neglected.) The phasors $\underline{V}$ and $\underline{E}_1$ are the applied voltage and back emf, respectively. This circuit will lead to a quantitative expression for the developed torque after two manipulations.

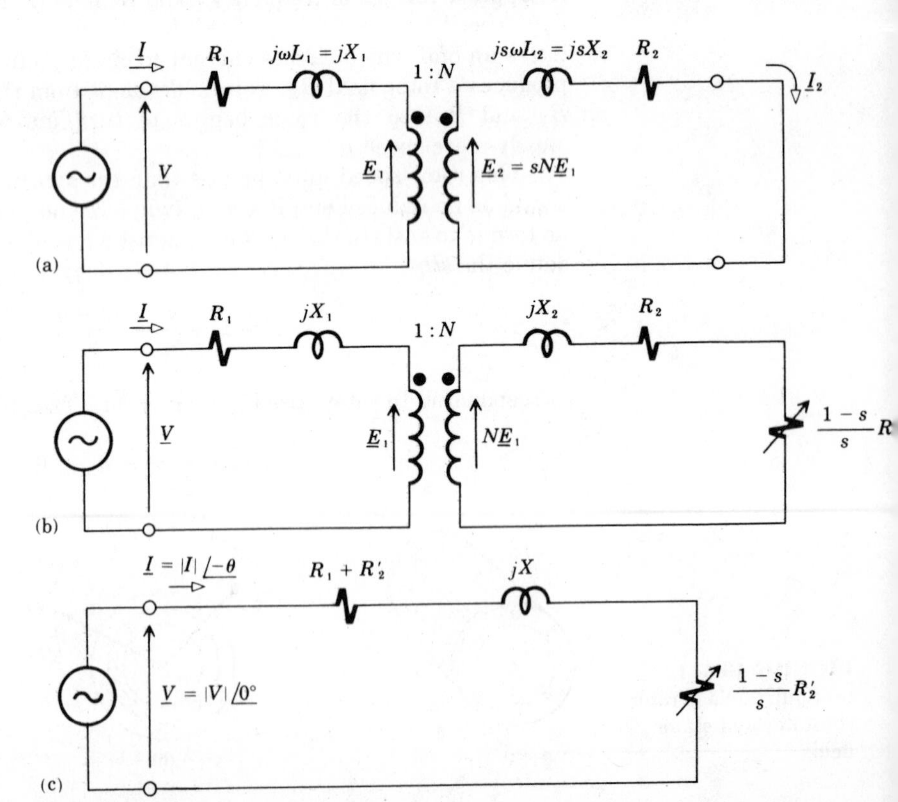

FIGURE 19.3–2 Equivalent circuits for one phase of an induction motor.

First, since $sN\underline{E}_1 = (R_2 + jsX_2)\underline{I}_2$ in the shortcircuited secondary, dividing through by s yields

$$N\underline{E}_1 = \left(\frac{R_2}{s} + jX_2\right)\underline{I}_2 = \left(R_2 + jX_2 + \frac{1-s}{s}R_2\right)\underline{I}_2$$

and we can redraw the secondary as shown in Fig. 19.3–2b. Note that R_2/s has been broken up into two series resistances: R_2 itself, which represents power actually lost as heat in the rotor; and the speed-dependent resistance $(1 - s)R_2/s$, which represents electrical-to-mechanical energy conversion. Next, the total secondary impedance $\underline{Z}_2$ is referred to the primary by writing

$$\begin{aligned}\underline{V}/\underline{I} &= (R_1 + jX_1) + \underline{Z}_2/N^2 \\ &= (R_1 + R_2/N^2) + j(X_1 + X_2/N^2) + (1 - s)R_2/sN^2 \\ &= (R_1 + R_2') + jX + \frac{1-s}{s}R_2' \end{aligned} \tag{2}$$

where

$$R_2' = \frac{R_2}{N^2} \qquad X = X_1 + \frac{X_2}{N^2}$$

giving the reduced circuit of Fig. 19.3–2c. This circuit model clearly brings out the fact that an induction motor acts as an *inductive* load whose AC resistance and power factor varies with the slip s. The model's parameter values R_1, R_2', and X can be determined by measurements similar to those used for a power transformer.

Now we calculate the torque by finding the electrical power delivered to the speed-dependent resistance:

$$\begin{aligned}P &= |I|^2\left(\frac{1-s}{s}R_2'\right) = \frac{|V|^2\,(1-s)\,R_2'/s}{[R_1 + R_2' + (1-s)\,R_2'/s]^2 + X^2} \\ &= (1-s)\,|V|^2\,\frac{sR_2'}{(sR_1 + R_2')^2 + s^2X^2}\end{aligned}$$

As this represents the electromechanical power conversion per phase, the total developed torque will be

$$T = \frac{3P}{\omega_m} = \frac{3p|V|^2}{2\omega}\,\frac{sR_2'}{(sR_1 + R_2')^2 + s^2X^2} \tag{3}$$

after having inserted $\omega_m = (2/p)(1 - s)\omega$ for the rotor's angular velocity.

Plotting T from this formula versus slip s or rotor speed $n = (1 - s)n_s$ gives the *torque-speed curve,* illustrated for a typical motor in Fig. 19.3–3, and reveals three important points:

- An induction motor develops torque at any rotor speed less than the synchronous speed, but is normally operated at a *rated speed* slightly below n_s with a slip of 1–5% corresponding to $s = 0.01$–0.05.
- The *starting torque* T_0 at $n = 0$ (or $s = 1$) generally exceeds the rated torque, permitting the motor to accelerate under load and come up to rated speed n_{FL}.
- There is a *maximum torque* T_{max} developed below rated speed, allowing the motor to drive an overload torque $T > T_{FL}$ at reduced speed for short intervals. But a load torque exceeding T_{max} will stall the motor or even drive it in the reverse direction.

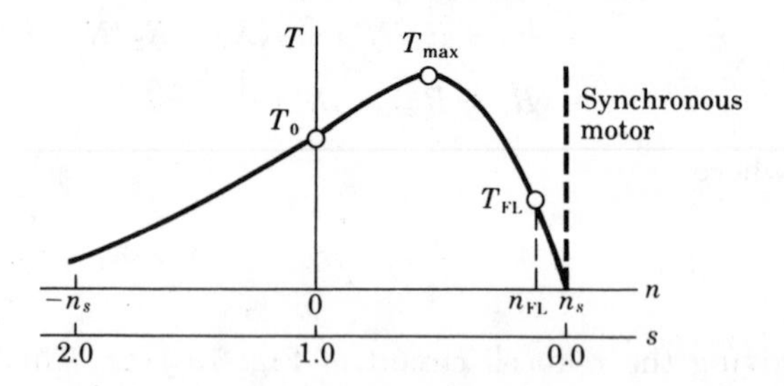

FIGURE 19.3–3
Torque-speed curve of an induction motor.

Contrast these induction-motor characteristics with those of synchronous motor, represented by the dashed curve, which also has a maximum torque but operates only at n_s and has no starting torque of its own. On the other hand, the induction motor lacks the adjustment flexibility made possible by the variable excitation of a synchronous motor.

The maximum torque of an induction motor has the value

$$T_{max} = \frac{3p|V|^2}{4\omega(R_1 + \sqrt{R_1^2 + X^2})} \tag{4a}$$

and occurs when

$$s = \frac{R_2'}{\sqrt{R_1^2 + X^2}} \tag{4b}$$

These expressions are derived by differentiating Eq. (3) with respect to s, solving $dT/ds = 0$ for s, and inserting the result back into Eq. (3).

Example 19.3–1

A 6-pole, 208-V (line-to-line), 60-Hz induction motor has a rated torque of 100 N · m. Electrical measurements yield the following parameter values:

$$R_1 = 0.26\ \Omega \qquad R_2' = 0.14\ \Omega \qquad X = 0.40\ \Omega$$

From these data we will predict the operating characteristics.

The synchronous speed is $n_s = 1200$ rpm, but the rated speed will be somewhat less. Assuming sR_1 and sX are small compared to R_2' at rated speed (as is usually true), Eq. (3) becomes

$$T_{FL} \approx \frac{3p|V|^2 s}{2\omega R_2'}$$

where $p = 6$, $|V| = 208/\sqrt{3} = 120$ V, and $\omega = 377$. Solving gives $s \approx 0.04$ at the rated torque, and $n_{FL} = 0.96\ n_s = 1152$ rpm. This corresponds to an angular velocity of $2\pi \times 1152/60 = 121$ rad/sec, so the rated mechanical output power is $100\ \text{N} \cdot \text{m} \times 121$ rad/sec $= 12.1$ kW or about 16 hp.

Taking $s = 1$ in Eq. (3) gives the starting torque $T_0 = 151\ \text{N} \cdot \text{m}$, which is 51% greater than T_{FL}. Equation (4) then shows that $T_{max} = 233\ \text{N} \cdot \text{m}$—more than twice the rated torque—occurring at slip $s = 0.29$ or $n = 818$ rpm. Under this condition the mechanical output power will be nearly 20 kW, as compared to the 12.1-kW rated power.

Exercise 19.3–1 Use Fig. 19.3–2c to find $|I|$ and the power factor of the above motor when the torque equals

(a) T_0 (b) T_{max} (c) T_{FL}

STARTING AND SPEED CONTROL

A large induction motor must have a *controller* designed to limit inrush current and excessive starting torque, to protect against overheating and, perhaps, to permit reversing direction. A schematic diagram of a simplified controller appears in Fig. 19.3–4, where the symbol —(F)— stands for the operating coil of a magnetic relay (as in Fig. 17.3–7) that closes all pairs of contacts represented by —| |$_F$—. This controller functions as follows. Pressing the forward button (FWD) energizes relay F and applies the line voltage to the motor through resistances R_S, whose voltage drops reduce $|V|$ at the motor and thereby limit the starting torque and inrush current. Relay F also closes auxiliary contactor F_a so one can release the push button. After the motor comes up to the reduced-voltage speed, a time-delay contactor T_F closes and energizes relay A whose contacts then close and bypass the starting resistances. (In practice, this would be done in several steps using additional resistances, relays, and time-delay contactors.)

The normally closed contactor labeled OLR is part of an *overload relay* whose heater elements in series with the motor cause the relay to

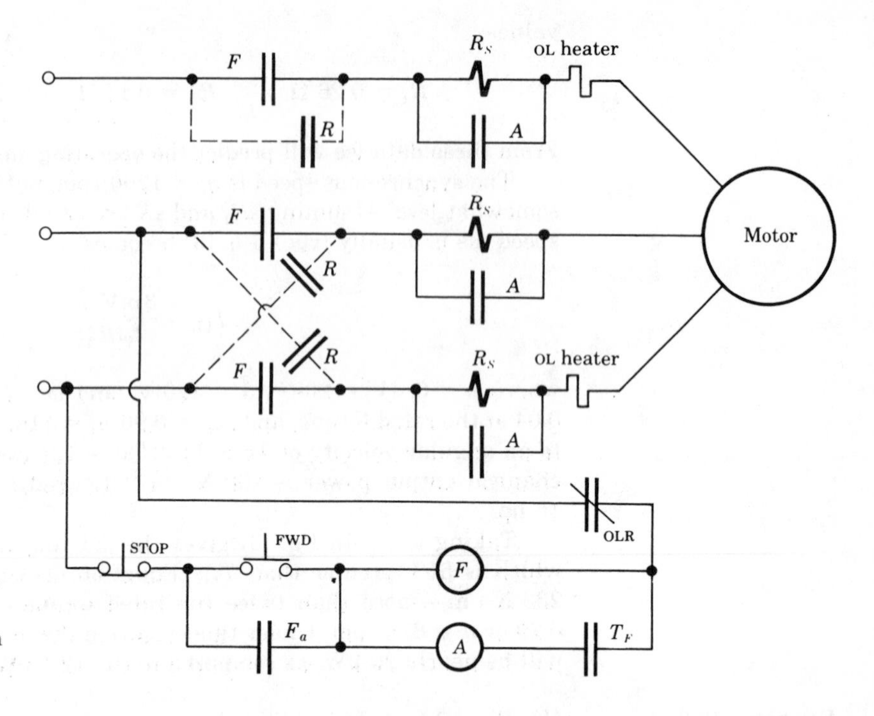

FIGURE 19.3–4 Controller circuit for an induction motor.

open in the event of prolonged or extreme overload conditions. Three other contacts labeled R will *reverse* the phase sequence applied to the motor and thus reverse the direction of rotation. For clarity, we have omitted the reverse button and the interlocking arrangement that opens the F contactors when the R contactors are closed.

Starting resistances do not provide effective nor efficient running-speed control. For applications requiring such speed control, the *wound-rotor motor* may be appropriate. Instead of short-circuited conducting bars, the rotor has a symmetrical three-phase winding with the same number of poles as the stator; the three phases are connected by slip rings to a set of external variable resistors (or three-phase *rheostat*) that increases the equivalent value of the secondary resistance.

Figure 19.3–5 gives the rotor diagram and the torque-speed curves for three values of resistance. The curves are obtained from Eq. (4) with the per-phase external resistance R_{ext} added to R_2'. The maximum torque remains unchanged, since it does not depend on R_2', but $T_{\max}$ now occurs at the increased slip

$$s = \frac{R_2' + R_{\text{ext}}}{\sqrt{R_1^2 + X^2}}$$

The rheostat, therefore, controls the starting torque and the running speed at rated torque. The price of this control, compared to the squirrel-cage motor, is not only its greater initial cost, but also its increased maintenance problems and its decreased efficiency, the latter attributable to power wasted as heat in the rheostat.

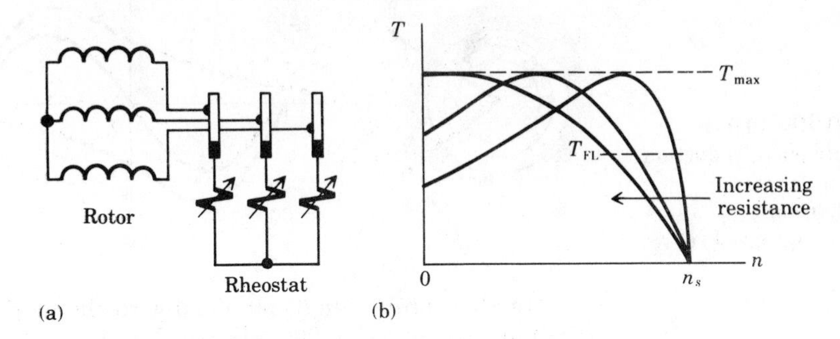

FIGURE 19.3–5 (a) Wound-rotor motor. (b) Torque-speed curves.

Recent advances in semiconductor power electronics have led to improved systems for adjusting and controlling the speed of an induction (or synchronous) motor. These systems employ controlled rectifiers and other circuitry to produce variable-frequency, variable-voltage three-phase AC power. A feedback path from a tachometer may be included to facilitate automatic speed control.

Exercise 19.3–2

Suppose the motor in Example 19.3–1 has a wound rotor.
(a) What external resistance (per phase) gives maximum starting torque?
(b) Compute T at $s = 0.04$ with this resistance in the rotor circuit.

SINGLE-PHASE INDUCTION MOTORS

Single-phase induction motors are used in residential and other applications where three-phase AC is not available and the mechanical power requirement does not exceed about 5 hp. These motors have a squirrel-cage rotor but, of course, only one phase in the stator winding. They also must have special provision for starting purposes.

To explain the operating principles and starting problem of a single-phase induction motor, recall that the magnetic field of a single-phase stator consists of two components rotating at synchronous speed in opposite directions. These are represented in Fig. 19.3–6a by $\underline{B}_{S_1}$ and $\underline{B}_{S_2}$, and we assume the rotor to be turning at rate n in the forward direction. The interaction of $\underline{B}_{S_1}$ with the rotor produces a torque-speed characteristic just like that of a three-phase motor (Fig. 19.3–3), and $\underline{B}_{S_2}$ has the same effect but with the torque and speed in the opposite direction. The resulting average torque is then the sum as plotted in Fig. 19.3–6b. Thus, the motor can run in either direction, but has zero starting torque. When the motor is running in a given direction, the instantaneous torque consists of the average torque plus a pulsating component caused by the oppositely rotating stator field, similar to the conditions in a single-phase alternator. Of course, the torque *must pulsate* if the speed is constant, since a single-phase AC circuit delivers pulsating instantaneous power.

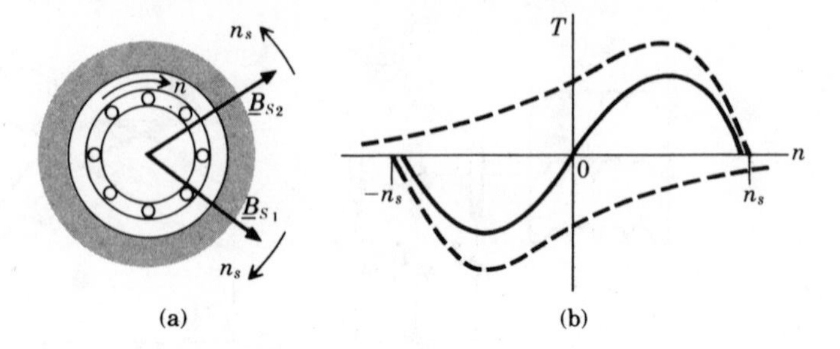

FIGURE 19.3–6
Single-phase induction motor: (a) stator field vectors; (b) torque-speed curve.

Starting torque can be obtained with the help of an auxiliary stator winding displaced by 90° (electrically) from the main winding, Fig. 19.3–7a. This smaller winding is designed so that its current is phase-shifted relative to the current through the main winding. Consequently, the forward rotating fields from both windings partially reinforce and produce net torque at zero rotor speed. A centrifugal switch mounted in the shaft disconnects the auxiliary winding after starting, and the machine normally runs on the main winding alone. Figure 19.3–7b shows a typical torque-speed curve of this machine, which is called a *split-phase motor*.

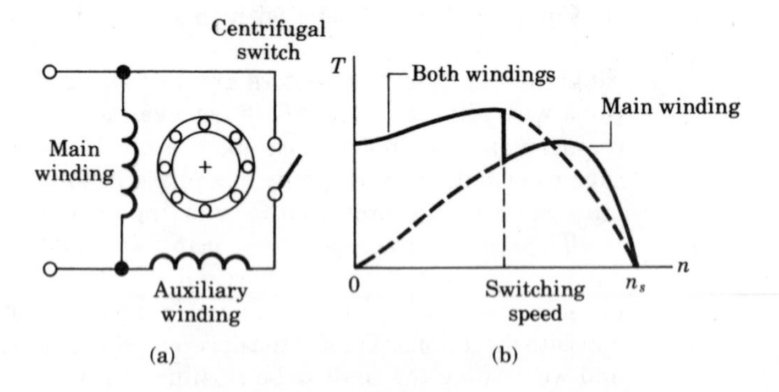

FIGURE 19.3–7
Split-phase motor: (a) windings; (b) torque-speed curve.

Other variations of the split-phase strategy employ one or two capacitors to enhance the current phase shift and to improve running performance. Alternatively, very small motors (under $\frac{1}{20}$ hp) may have copper rings called *shading coils* around a portion of each of the two salient stator poles, as in Fig. 19.3–8. Induced currents in the shading coils distort the field and produce starting and running torque in the direction

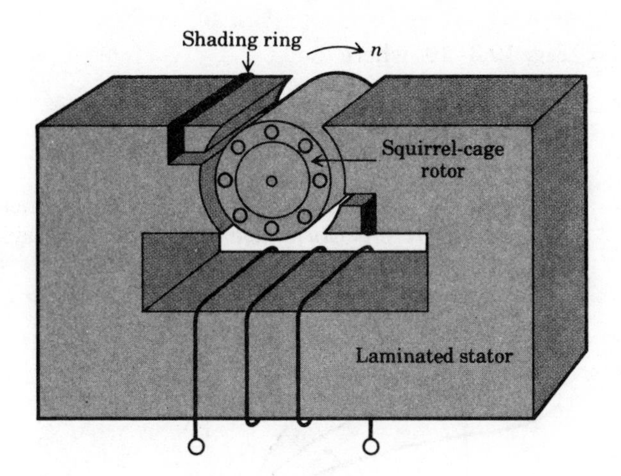

FIGURE 19.3–8
Shaded-pole motor.

from the unshaded to the shaded portion. The economical structure of this *shaded-pole motor* makes it a popular choice for hair dryers and similar small appliances.

Regardless of starting method, the running torque of a single-phase induction motor is proportional to the square of the line voltage $|V|$—see Eq. (3). Crude but effective speed control is therefore possible by changing the line voltage, as illustrated in Fig. 19.3–9 with a typical load characteristic. This technique is implemented in low-power two-speed or three-speed devices, notably fans, through the simple expedient of inserting a voltage-dropping resistance. Such motors must have high winding resistance.

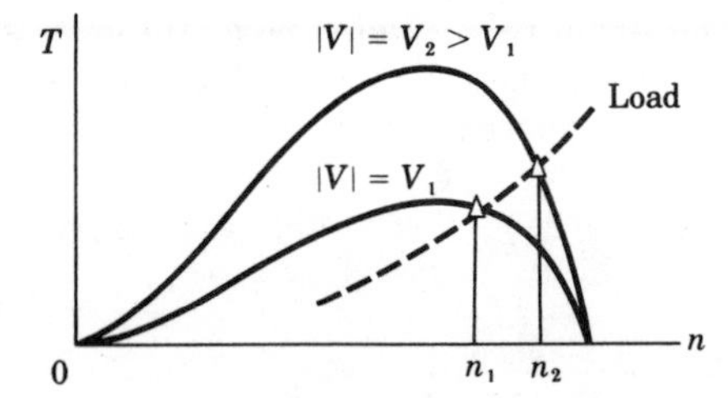

FIGURE 19.3–9
Line-voltage speed control.

SPECIAL-PURPOSE AC MOTORS

Numerous special-purpose AC motors have been developed to fit particular needs. Three interesting types are described below. Another type, the important *universal motor,* will be covered in the next section.

Servomotors are *two-phase* induction motors designed for use in servomechanisms and automatic control systems. Fixed AC voltage of amplitude V_F is applied to one stator winding and an AC *control voltage* with variable amplitude V_C and 90° phase shift is applied to the other winding. Adjusting the control voltage yields a family of torque-speed curves like

Fig. 19.3–10, where

$$T \approx k\left(\frac{V_C}{V_F} - \frac{n}{2n_s}\right) \tag{5}$$

at speeds up to about 25% of n_s. This nearly linear relationship is valuable for effective performance of many control systems.

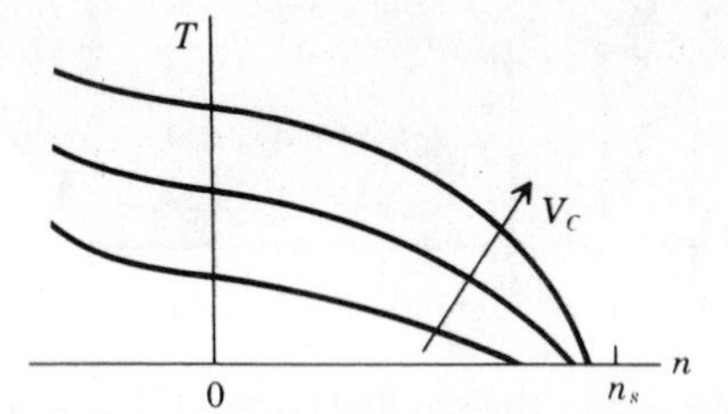

FIGURE 19.3–10 Torque-speed curves of two-phase servomotor.

Hysteresis motors have a split-phase stator winding and a smooth cylindrical rotor of hard magnetic material. The rotating stator field magnetizes the rotor, but hysteresis causes the rotor magnetization to lag behind by a constant angle δ. The motor therefore develops constant torque at any speed up to n_s and will accelerate from standstill to steady-state synchronous rotation providing that the load does not exceed the power rating—normally less than $\frac{1}{10}$ hp. Electric clocks, timing devices, and phonograph turntables are common applications of this simple synchronous motor.

Reluctance motors have a cogwheel-shaped rotor of soft magnetic material and, usually, a shaded-pole stator, as shown in Fig. 19.3–11a. The rotor's shape causes the total reluctance of the flux path to vary with the rotor angle γ approximately as

$$\mathcal{R} = \mathcal{R}_0 - \Delta\mathcal{R} \cos p\gamma$$

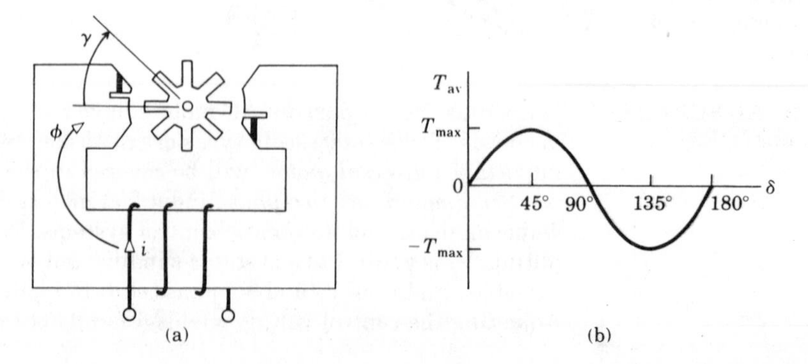

FIGURE 19.3–11 (a) Reluctance motor. (b) Average torque as a function of power angle.

where p is the number of rotor "teeth." What we have here is a *rotating-iron* device, and the rotational version of Eq. (9), Section 17.3, gives the *reluctance torque*

$$T = -\tfrac{1}{2}\phi^2 \frac{d\mathcal{R}}{d\gamma} = \tfrac{1}{2}\phi^2 \,\Delta\mathcal{R} p \sin p\gamma \tag{6a}$$

If flux ϕ is proportional to the current i, and the instantaneous rotor angle is $\gamma = (2/p)(\omega t + \delta)$, then the torque consists of alternating components, plus a constant term

$$T_{av} = T_{max} \sin 2\delta \tag{6b}$$

plotted versus δ in Fig. 19.3–11b. Thus, if $0 < \delta < 90°$, the rotor will turn at the synchronous speed $n_s = 120\, f/p$. The stator shading coil and a squirrel-cage winding on the rotor provide the initial torque needed to get things going. The advantage of the reluctance motor over the hysteresis motor is its lower mechanical speed since p can be a large integer.

Reluctance-torque effects will occur in *any* AC machine with a *salient-pole* rotor. This observation, together with Fig. 19.3–11b, accounts for the difference between the two curves of P versus δ for synchronous generators in Fig. 19.2–5.

Exercise 19.3–3

Carry out all the steps between Eqs. (6a) and (6b), taking $i = \sqrt{2}\,|I| \cos \omega t$ and assuming $\mathcal{R}_0 \gg \Delta\mathcal{R}$ so $\phi \approx Ni/\mathcal{R}_0$.

19.4 DC MACHINES

Because of their mechanical reliability, plus the widespread availability of AC power, AC motors are preferred for rotational drive units whenever possible. Some applications, however, require operating characteristics beyond the scope of AC motors with fixed source voltage and frequency. A DC motor may then be the machine of choice, especially for its ease of speed control. This section describes the general features and characteristics of DC machines.

COMMUTATION AND STRUCTURAL FEATURES

DC machine operation depends upon a *commutator,* whose principles are illustrated by Fig. 19.4–1 for a simplified two-pole machine. Here, a concentrated armature coil is on the rotor and the field windings are on the stator, just the opposite of a synchronous machine. The field windings consist of many turns of relatively thin wire, all connected in series and usually carrying a small current compared to the low-resistance, high-current armature. The salient stator poles are shaped to establish a fixed and nearly uniform radial field B_f in the air gap, a field that reverses direction from pole to pole. The potential torque reversals are canceled

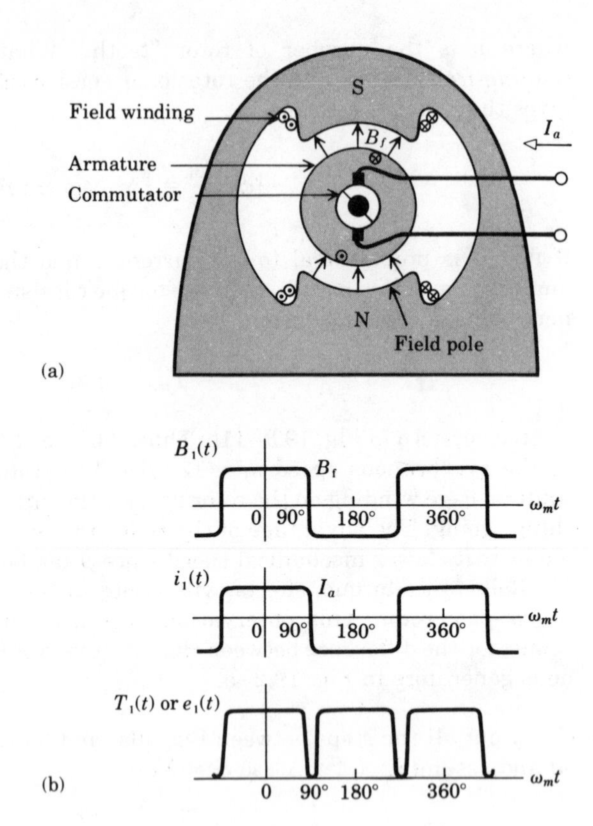

FIGURE 19.4–1
(a) DC machine with split-ring commutator.
(b) Waveforms.

out by a *split-ring commutator* on the rotor shaft that reverses the direction of the DC armature current I_a when the plane of the coil crosses the field reversals.

Now imagine traveling with one of the armature conductors while it rotates at frequency ω_m. The field we see varies as shown in Fig. 19.4–1b, but—thanks to the commutator—the current reverses direction simultaneously with the field reversals. Consequently, the torque $T_1(t)$ experienced by that conductor and the emf $e_1(t)$ induced across it stay almost constant, save for the dips at the instants of commutation. Similar curves hold for a p-pole machine if one substitutes the equivalent electrical angle $(p/2)\omega_m t$ for the mechanical angle $\omega_m t$ in Fig. 19.4–1b.

A *distributed* armature winding with a *multisegmented commutator* essentially eliminates the commutation dips and gives rise to the vector picture in Fig. 19.4–2a, where the armature field $\underline{B}_a$ remains fixed and perpendicular to the stator field $\underline{B}_f$. (The brushes are omitted from this drawing for clarity, as are the rather complicated connections to the commutator segments.) Since this picture applies at any mechanical speed, controlling the speed of a DC motor only requires adjusting the value of the armature current. Whether the machine functions as a motor or a generator, the total induced emf equals the sum of the contributions from the many series-connected armature wires and looks like Fig. 19.4–2b—a large constant voltage E with a small *ripple* due to commutation. A plot of the total magnetic torque would have the same shape.

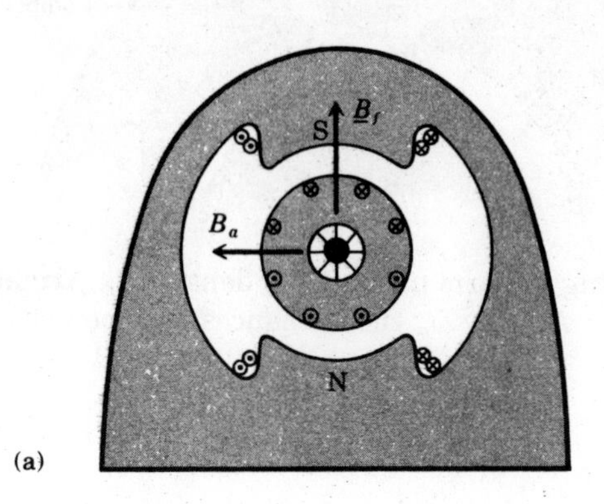

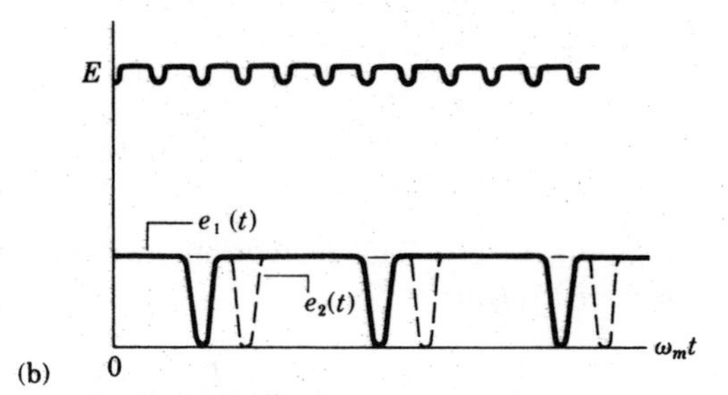

FIGURE 19.4–2
(a) DC machine with multi-segmented commutator.
(b) Induced emf.

Large machines often have two other sets of windings as illustrated in Fig. 19.4–3. *Commutating windings* on small *interpoles* midway between the field poles provide flux that assists commutation action and reduces sparking at the brushes caused by current reversals. (Recall that $v_L = L\,di/dt$, so sudden current reversal in an inductive armature winding generates a voltage pulse.) *Compensating* or *pole-face windings* carry current in the opposite direction of the adjacent armature conductors and thereby cancel the unwanted magnetic field caused by armature current that would otherwise distort the air-gap field—an effect known as *armature reaction*. Both of these auxiliary windings are connected in series with the armature.

INDUCED EMF AND DC GENERATORS

To calculate the value of E, let the armature have radius r and length ℓ, so the area of each pole is $A_p \approx 2\pi r\ell/p$ and the *flux per pole* will be

$$\phi = B_f A_p = B_f \frac{2\pi r\ell}{p}$$

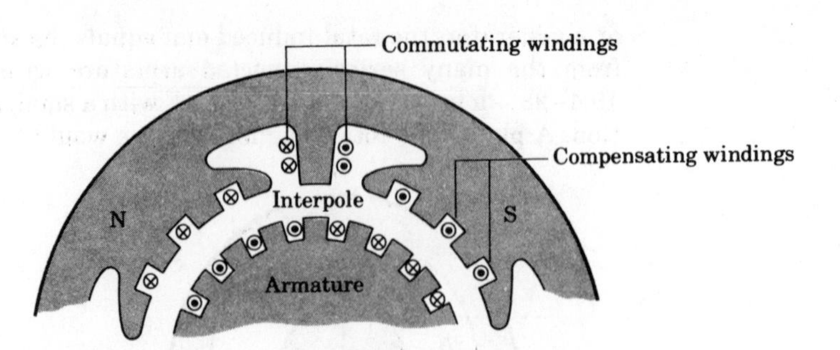

FIGURE 19.4–3
Commutating and compensating windings.

assuming uniform air-gap flux density B_f. Armature rotation at constant angular velocity ω_m then produces the open-circuit emf

$$E = N_a B_f \ell r \omega_m = N_a \frac{p\phi}{2\pi} \omega_m = k_a \phi \omega_m \tag{1a}$$

where N_a is the number of series-connected armature conductors between brushes and $k_a = N_a p/2\pi$. Equivalently, expressing E in terms of the armature speed $n = 60\ \omega_m/2\pi$,

$$E = \frac{\pi}{30} k_a \phi n \tag{1b}$$

By analogy with previous derivations, we immediately surmise that the total induced torque is

$$T = \frac{EI_a}{\omega_m} = k_a \phi I_a \tag{2}$$

where I_a is the DC armature current.

The value of ϕ in the foregoing expressions depends on the field-winding current I_f in accordance with the magnetization properties of the complete magnetic circuit, Fig. 19.4–4 being a representative example. This has the same shape as a *B-H* curve (Fig. 17.2–4) since B is proportional to ϕ, and H is proportional to the field mmf $\mathcal{F} = N_f I_f$. For moderate values of field current we can use the linear approximation $\phi \approx N_f I_f/\mathcal{R}$ and write

$$E = k_f I_f \omega_m \qquad T = k_f I_f I_a \tag{3}$$

with $k_f = k_a N_f/\mathcal{R} = N_a N_f p/2\pi\mathcal{R}$. Saturation voids linearity at high field currents, while residual magnetism (or retentivity) causes a residual flux ϕ_r when $I_f = 0$. Both effects turn out to have importance for certain types of DC machines.

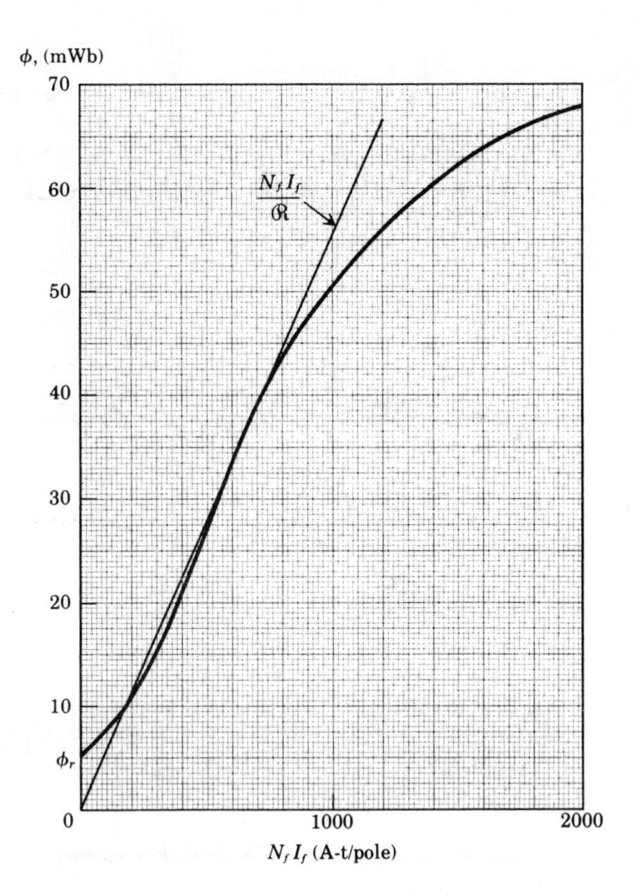

FIGURE 19.4–4 Flux per pole versus field mmf for a typical DC machine.

Figure 19.4–5 gives schematic diagrams for the four common ways of connecting the armature and field. In the *separately-excited* mode (Fig. 19.4–5a), the field current comes from a separate source, whereas I_a and I_f are interdependent in the remaining three *self-excited* modes. With armature and field connected in *series* (Fig. 19.4–5b), I_f equals I_a, and the field winding must have lower resistance and fewer turns than the other configurations, unless most of I_a is bypassed through a small *diverter resistance* R_D. The parallel or *shunt* connection in Fig. 19.4–5c puts the same voltage across both windings so $I_f \ll I_a$ in view of the high field resistance. A variable resistance R_C called a *control rheostat* may be included to adjust I_f. The *compound* connection (Fig. 19.4–5d), has part of the field winding in series but mostly in parallel with the armature, so there are actually two different field currents. The relative merits of these various connection schemes depend on the particular application.

In the case of generator operation, a mechanical prime mover (an AC induction motor, for instance) drives the armature at angular frequency

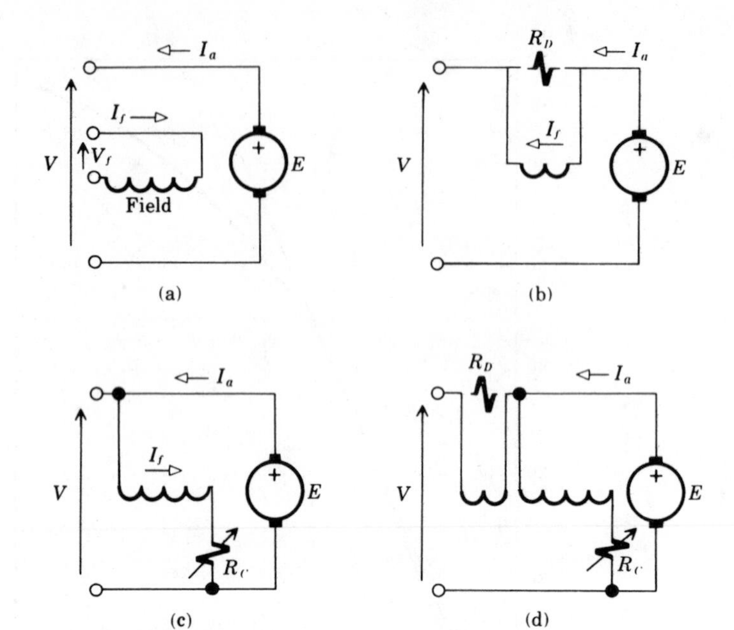

FIGURE 19.4–5 (a) Separately-excited machine. (b) Series connection. (c) Shunt connection. (d) Compound connection.

ω_m and a voltage $V \leq E$ appears at the armature terminals. We usually want this terminal voltage to be sensibly independent of the output load current I. For a *separately-excited generator* having the equivalent circuit of Fig. 19.4–6, we see that $I_a = I$ and

$$V = E - R_a I = k_a \phi \omega_m - R_a I \tag{4}$$

Thus the V-I characteristic is a straight line whose slope equals the negative of the small armature resistance R_a, as plotted in Fig. 19.4–7.

The rated full-load voltage $V_{FL} = E - R_a I_{FL}$ will be somewhat less than the no-load voltage $V_{NL} = E$ when $I = 0$. The entire curve may be shifted up or down by changing the field voltage V_f, since ϕ depends on $I_f = V_f/R_f$. The total input power consists of $P_m = T\omega_m$, plus the small field power $P_f = V_f I_f$.

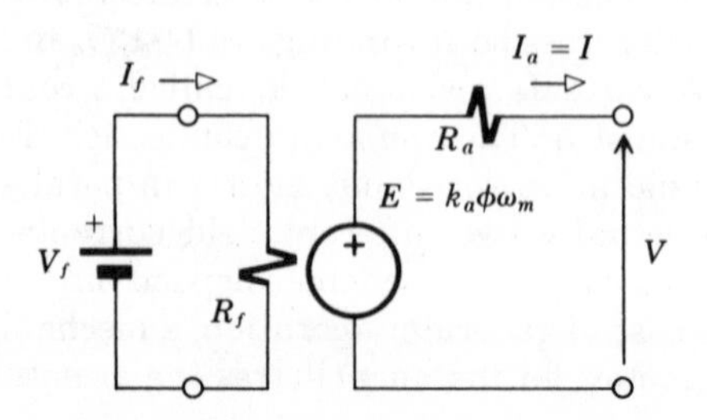

FIGURE 19.4–6 Separately-excited generator.

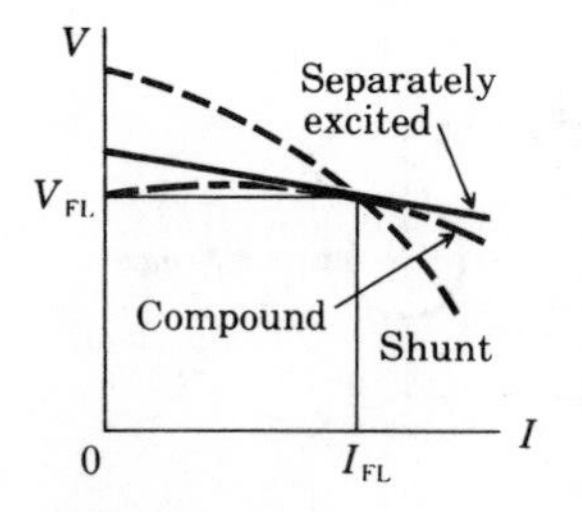

FIGURE 19.4–7
DC generator characteristics.

The self-excited *shunt generator* needs no separate field supply but its V-I curve in Fig. 19.4–7 falls off rapidly with increasing load, so a higher no-load voltage would be required for the same full-load voltage as that of a separately-excited generator. You can see why this happens by referring to Fig. 19.4–5c and noting that any decrease of the terminal voltage also decreases the field current I_f which, in turn, further reduces the terminal voltage. The opposite effect takes place in a *series generator* whose rising V-I curve (not shown) would have little practical value by itself. On the other hand, appropriate combination of the shunt and series characteristics in a *compound generator* yields a nearly constant V-I curve, with the series field providing increased flux when the shunt-field flux decreases. The windings can be proportioned such that $V_{FL} = V_{NL}$, in which case the generator is said to be *flat-compounded* and has the typical characteristics sketched in Fig. 19.4–7.

Example 19.4–1

A 4-pole separately-excited generator is rated for 20 kW at $V_{FL} = 250$ V when driven at 1200 rpm. There are $N_a = 60$ armature conductors with total resistance $R_a = 0.35\ \Omega$, so $k_a\omega_m = (60 \times 4/2\pi)(2\pi \times 1200/60) = 4800$, $I_{FL} = 20\text{ kW}/250\text{ V} = 80$ A, and $E = V_{FL} + R_a I_{FL} = 250 + 0.35 \times 80 = 278$ V. The flux per pole must then be $\phi = E/k_a\omega_m = 0.058$ Wb = 52 mWb, which requires $N_f I_f \approx 1300$ A-t/pole if the magnetic characteristics are as plotted in Fig. 19.4–4.

Taking $I_f = 2$ A and $V_f = 250$ V, we obtain the field-circuit parameters $N_f = 650$ turns per pole and $R_f = 125\ \Omega$. A control rheostat would be used to get the latter value, assuming the actual field winding resistance is less than 125 Ω.

The copper loss at full load is now found to be $R_a I_a^2 + R_f I_f^2 = 2240 + 500 \approx 2.7$ kW, dominated by the armature resistance. Exclusive of any mechanical losses, the generator's efficiency is 20 kW/22.7 kW ≈ 88%.

Exercise 19.4–1

Draw a circuit diagram like Fig. 19.4–6 for the case of a series generator without a diverter resistance so $I = I_a = I_f$. Assuming the magnetic characteristics of Fig. 19.4–4, and that $k_a\omega_m = 4800$, $R_a = 0.35\ \Omega$, and $R_f = 0.45\ \Omega$, find N_f such that $V = 220$ V when $I = 25$ A. Then calculate E and V if the load draws 50 A.

DC MOTORS AND THE UNIVERSAL MOTOR

Like a DC generator, a DC motor may be self-excited or separately-excited. The latter offers a combination of good speed regulation and flexible control unmatched by any other type of motor. The series motor, however, has unique advantages as a universal motor. These properties, along with those of the shunt motor, will be explored below.

Consider the equivalent armature circuit of a separately-excited motor with applied voltage V, Fig. 19.4–8a. This differs from a generator in that I_a flows in the opposite direction and is opposed by the back emf

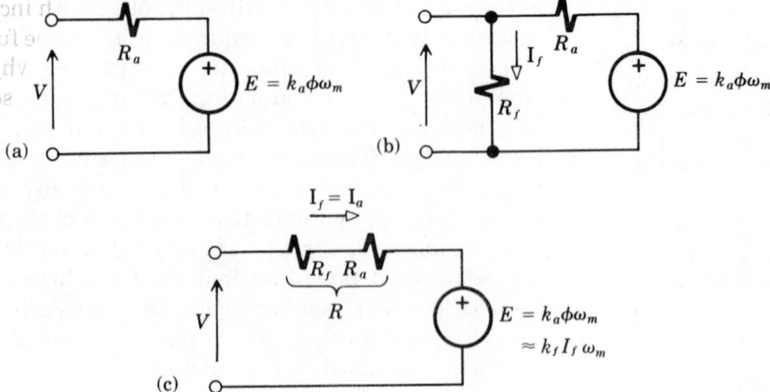

FIGURE 19.4–8 Equivalent armature circuits: (a) separately-excited motor; (b) shunt motor; (c) series (or universal) motor.

$E = k_a\phi\omega_m$. Since $V = R_aI_a + k_a\phi\omega_m$, the angular velocity will be $\omega_m = (V - R_aI_a)/k_a\phi$. But $T = k_a\phi I_a$ and converting to speed in rpm yields

$$n = \frac{30}{\pi}\left[\frac{V}{k_a\phi} - \frac{R_a}{(k_a\phi)^2}\,T\right] = n_{\rm NL} - bT \tag{5}$$

where, as before, ϕ depends upon the field current I_f. The speed-torque characteristic is then a straight line, as in Fig. 19.4–9, with no-load speed $n_{\rm NL} = 30\ V/\pi k_a\phi$ and negative slope of magnitude $b = 30R_a/\pi(k_a\phi)^2$.

Usually, the flux is such that $bT_{\rm FL} \ll n_{\rm NL}$ so $n_{\rm FL} \approx n_{\rm NL}$ and we have a nearly *constant-speed* motor. The key to this behavior is the induced voltage $E = \pi k_a\phi n/30$, which is just slightly less than V at full load. Consequently, a small change in n corresponds to a large change in current $I_a = (V - E)/R_a$ and a large change in torque $T = k_a\phi I_a$. Excessive load, of course, stalls the motor ($n = 0$) when $T_0 = n_{\rm NL}/b$.

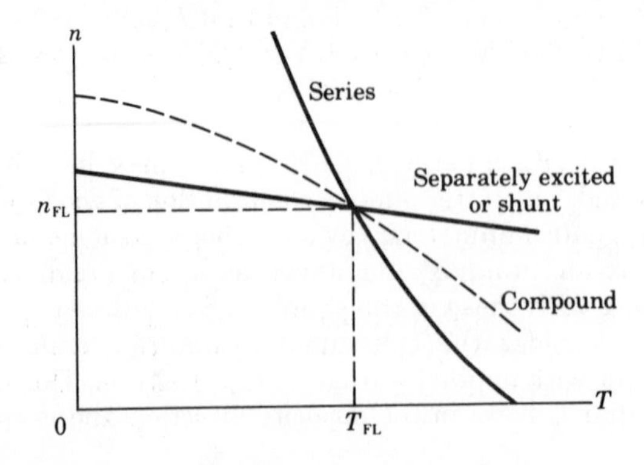

FIGURE 19.4–9 DC motor characteristics.

Closer examination of the expression for n_{NL} reveals that we can control the motor speed simply by adjusting the armature voltage V. Furthermore, since n_{NL} is directly proportional to V, smooth *reversal* of rotation is possible by reducing V to zero and then reversing its polarity. An electronic DC power supply performs these tasks quite well, although a three-phase rectifier is needed for power levels above about 5 kW. Prior to the development of high-power semiconductor devices, DC motors were supplied from a motor-generator (MG) set consisting of a DC generator driven by an induction motor. This arrangement, known as the *Ward-Leonard system,* also provides *regenerative braking* that slows down or stops the DC motor by reducing the generator's field current to the point that power flows back to the generator. Regenerative braking has special value for electric locomotives and similar applications where the coasting or braking motor actually returns power to the system.

Equation (5) and the corresponding speed-torque curve also hold for the *shunt motor* diagrammed in Fig. 19.4–8b. But bear in mind that $I_f = V/R_f$, so ϕ depends on the applied voltage and rotation reversal via voltage polarity is not possible. By the same token, a control rheostat in series with the field winding provides speed control with a ratio of about 2:1—as long as care is taken not to reduce the field too much (Why?).

The *series motor* in Fig. 19.4–8c turns out to have dramatically different characteristics because its flux varies with the armature current. For an analytic investigation we use the linear approximations of Eq. (3) and let $I = I_f = I_a$ and $R = R_f + R_a$. Thus,

$$V = RI + E \qquad E = k_f I \omega_m$$

and the torque is proportional to current *squared,* namely,

$$T = k_f I_f I_a = k_f I^2 \tag{6}$$

Solving for n in terms of T yields

$$n = \frac{30}{\pi}\left(\frac{V}{\sqrt{k_f}}\frac{1}{\sqrt{T}} - \frac{R}{k_f}\right) \tag{7}$$

as plotted in Fig. 19.4–9.

We clearly see that the series motor is a *varying-speed* device that has large starting torque at $n = 0$ but tends to "run away" under no-load conditions, since $n \to \infty$ if $T \to 0$. This behavior would be suitable for a constant-power application that needs low torque at high speed and high torque at low speed. Other applications might call for the intermediate speed-torque characteristics of a *compound motor,* also plotted in Fig. 19.4–9.

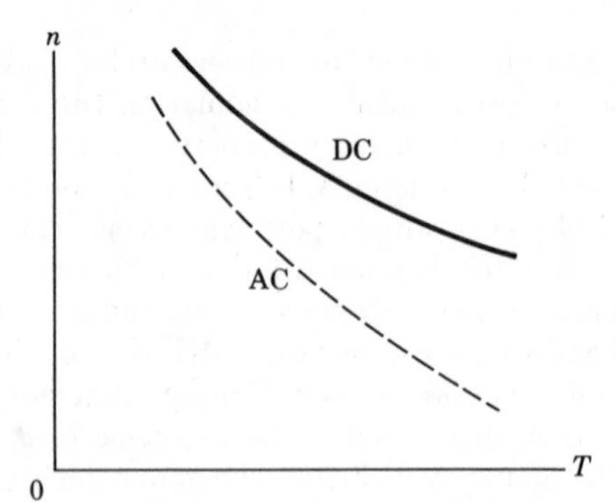

FIGURE 19.4–10
Speed-torque curves for a universal motor with DC and AC excitation.

A significant implication of Eq. (6) is that a series motor with *alternating* current $i = \sqrt{2}\,|I| \cos \omega t$ will develop pulsating but *unidirectional* torque $T(t) = 2k_f\,|I|^2 \cos^2 \omega t$. It is therefore a *universal motor,* capable of operating with AC or DC excitation (providing that both stator and rotor are laminated). The typical speed-torque curves in Fig. 19.4–10 reveal lower mechanical output power in the AC mode due to reactive voltage drops and magnetic saturation at the AC current peaks. Nonetheless, most universal motors are intended for use with an AC supply for the following reason.

The universal motor is a *high-speed* machine normally running around 5,000–15,000 rpm. As such, in view of Eq. (8), Section 19.1, it has a high ratio of output *power to weight* and best suits the needs of many hand tools and portable appliances—electric drills, vacuum cleaners, and so forth. Moreover, a simple thyristor circuit suffices for torque/speed control if required.

Example 19.4–2

Suppose the machine described in Example 19.4–1 is operated as a shunt motor at 250 V. The field mmf will be constant at $N_f I_f = 1300$ A-t, so $\phi = 0.058$ Wb and inserting values into Eq. (5) gives

$$n = 1078 - 0.681\ \text{T}$$

Thus, $n_{\text{NL}} = 1078$ rpm and $T_0 = 1078/0.681 = 1583$ N · m. The rated full-load conditions are found by writing $\omega_{\text{FL}} T_{\text{FL}} = 20$ kW and solving the resulting quadratic, which yields $T_{\text{FL}} = 203$ N · m and $n_{\text{FL}} = 941$ rpm. The speed regulation of this motor is then SR $= (1078 - 941)/941 \approx 15\%$. Under stalled conditions with $n = 0$ and $E = 0$, the motor draws $I_a = 250\ \text{V}/0.35\ \Omega = 714$ A and could be damaged by such a large current.

Exercise 19.4–2 Consider a series motor with $V = 120$ V, $R = 20\ \Omega$, and $k_f = 0.09$. Calculate the current, torque, and horsepower developed at 10,000 rpm, and the corresponding electrical-to-mechanical conversion efficiency.

Exercise 19.4–3 Use the linear approximations from Eq. (3) to obtain the speed-torque characteristics of a shunt motor in the form

$$n = \frac{30}{\pi}\left[\frac{R_f}{k_f} - R_a\left(\frac{R_f}{k_f V}\right)^2 T\right]$$

What does this expression imply about controlling speed by adjusting voltage?

PROBLEMS

19.1–1 Let the field of the four-pole rotor in Fig. 19.1–5b be $B_r = B_R \cos 2\gamma_r$, where γ_r is with respect to the centerline of one of the north poles. Use a drawing similar to Fig. 19.1–1b with N conductors at $\gamma_s = 0°, 90°, 180°$, and $270°$ to show that $e(t) = NB_R A\omega \cos \omega t$ where $\omega = 2\omega_m$.

19.1–2 Repeat the previous problem with conductors only at $\gamma_s = 0°$ and $180°$, in which case $e(t) = 0$.

19.1–3 Suppose the stator in Fig. 19.1–1 has a primitive *distributed* winding with $N/2$ conductors at $\gamma_s = -45°, 45°, 135°$, and $-135°$. Use Eq. (3) to show that $e(t) = \frac{1}{2}NB_R A\omega_m[\cos(\omega_m t - 45°) + \cos(\omega_m t + 45°)]$. Then let $t = 0$ to find e_{max} and the winding factor $k_w = e_{max}/NB_R A\omega_m$.

19.1–4 The voltage-current relationship of a certain generator is $V = \sqrt{25 \times 10^4 - 100I^2}$, and $I_{FL} = 30$ A. Sketch the curve; find V_{NL} and V_{FL}; and compute VR.

19.1–5 The speed-torque relationship of a certain motor is $n = 5000 - 8T^2$, and $T_{FL} = 10$. Sketch the curve; find n_{NL}, n_{FL}, and T_{max}; and compute SR.

19.1–6 The speed-torque relationship of a certain motor is $n = 1200(1 - 5T/V^2)$, where V is the source voltage. The motor drives a load that requires $n = 3T$. Find the operating speed and torque when $V = 120$ V and $V = 240$ V.

19.1–7 Repeat the previous problem for a load that requires $n = 0.01T^2$.

19.1–8 A certain 60-Hz synchronous motor with $k = 0.07$, $N_a I_a = 1000$ A-t, $B_f = 1$ T, and $A = 6$ cm $\times$ 10 cm develops 0.5 hp. How many poles do you think it has? Explain your answer.

19.2–1 Consider a three-phase generator with $|E|_{max} = 2000$ V, $X_s = 10\ \Omega$, and $|I|_{FL} = 100$ A. The generator supplies a load with pf $= 1.0$. Use a phasor diagram to find the values of $|E|$ and δ, so that $|V| = 1600$ V when $|I| = 0$, 50, and 100 A. What is the largest possible value of $|V|$ when $|I| = 150$ A?

19.2–2 Consider the generator and load in Problem 19.2–1. Use a phasor diagram to calculate $|V|$, δ, and P for $|I| = 0$, 50, 100, and 150 A, assuming $|E|$ is held constant at $|E|_{max}$. Then plot $|V|$ versus $|I|$, similar to Fig. 19.1–8.

19.2–3 Repeat the previous problem with $|E| = 1600$ V.

19.2–4 Let the generator in Problem 19.2–1 have $|E| = 2000$ V and $|I| = 100$ A. Find $|V|$, δ, and P for the following load power factors: 1.0, 0.8 lagging, 0.8 leading.

19.2–5 Let the generator in Problem 19.2–1 have $|E| = 2000$ V and $|V| = 1600$ V. Find $|I|$ and P when the load's power factor is 0.8 lagging. Compare with the values when pf = 1.0.

19.2–6 Repeat the previous problem with pf = 0.8 leading.

19.2–7 A certain load has pf = 0.96 lagging and draws $P = 24$ kW (per phase) at 400 V. A synchronous machine with $\delta = 0$ and $X_s = 2\ \Omega$ is connected in parallel with the load and adjusted to obtain unity power factor for the combination. Find the values of $|I|$ and $|E|$ for the synchronous machine.

19.2–8 A certain synchronous motor is rated for $|V| = 400$ V, $|I| \leq 50$ A, and pf = 0.8 leading. If this machine is used for the application in the previous problem, what is the available mechanical power?

19.2–9 Consider a synchronous motor with $|V| = 2$ kV, $|E| \leq 3$ kV, and $X_s = 60\ \Omega$. Find P_{max}. Then use a phasor diagram to calculate the corresponding values of $|I|$, θ, and $Q = |V|\,|I| \sin \theta$.

19.2–10 Suppose the motor in the previous problem is operated with θ held fixed at 0°. Find the maximum possible output power, and sketch $|I|$ and $|E|$ versus P.

‡ **19.2–11** A synchronous motor with $X_s = 12\ \Omega$ and $|V| = 400$ V supplies constant load power $P = 10$ kW. Use a phasor diagram to find δ, $|I|$, and θ for $|E| = 300$, 500, and 700 V. Then sketch $|I|$ versus $|E|$, which is called a *vee curve* because of the shape. Hint: Since P is constant, $|E| \sin \delta$ must be constant.

19.3–1 Consider a 12-pole induction motor with $f = 60$ Hz and 4% slip. Find the rotor speed, the slip speed, and the frequency of the rotor currents.

19.3–2 An induction motor has a rated speed of 860 rpm at 60 Hz. Determine the number of poles and the full-load slip.

19.3–3 A four-pole, 60-Hz induction motor develops 50 hp at $n_{FL} = 1710$ rpm. Calculate T_{FL} and the full-load slip. Then estimate n when $T = 1.2T_{FL}$ and $T = 0.8T_{FL}$.

19.3–4 Suppose the motor in the previous problem is connected to a 50-Hz source, the voltage being unchanged. Use the small-slip approximation in Example 19.3–1 to estimate the value of n at which $T = 200$ N · m.

19.3–5 Suppose the stator windings of the motor in Example 19.3–1 are reconnected in a delta configuration. Sketch and label the resulting torque-speed curve, and compare with the original one.

19.3–6 Suppose the motor in Example 19.3–1 has a wound rotor, and the rheostat is adjusted so that T_{max} occurs at $s = 0.5$. Sketch and label the resulting torque-speed curve, and compare it with the original one.

19.3–7 Consider an induction motor with $R_2' = R_1$. If T_0 equals the torque at $s = 0.1$, show that $X = 3R_1$. Then find T_{max}/T_0.

19.3–8 Use Fig. 19.3–2a to show that the efficiency of an induction motor is $P_m/P_e = (1 - s)/(1 + R_1 s/R_2')$, which neglects core losses and mechanical losses. Then evaluate the efficiency of the motor in Example 19.3–1 when $s = 0.04$, 0.1, and 0.29.

‡ **19.3–9** Derive Eq. (4). Hint: Divide the numerator and denominator in Eq. (3) by s^2; let $y = R_2'/s$, and find dT/dy.

19.3–10 A *two-phase motor* with two poles has two identical field coils displaced by 90°. The voltages applied to the coils are $v_A(t) = \sqrt{2}\,|V| \cos \omega t$ and $v_B(t) = \sqrt{2}|V| \cos(\omega t - 90°)$. Use sketches similar to Figs. 19.1–4 and 19.2–3 to show that a constant rotating field can be produced.

19.3–11 Consider the voltages required for the two-phase motor in the previous problem. Use a phasor diagram to show how these voltages could be obtained by connecting appropriate transformer windings to a three-phase source with $\underline{V}_{ab} = |V| \underline{/0°}$ and an *a-b-c* phase sequence. (The *Scott transformer* is designed especially for three-phase–to–two-phase conversion.)

19.4–1 The stator of a DC tachometer generator is a two-pole permanent magnet with $\phi = 0.15$ mWb. The commutator has four pairs of bars, as in Fig. 19.4–2a. If $E = 5$ V at $n = 1000$ rpm, what is the value of N_a?

19.4–2 Find the ripple frequency in hertz and the peak-to-peak ripple voltage when the tachometer generator in the previous problem runs at 2000 rpm.

19.4–3 Consider the generator in Example 19.4–1 with $V_f = 250$ V and $N_f = 650$. Find the value of R_f needed so that $V_{FL} = 250$ V when the speed of the prime mover is $n = 960$ rpm. Then calculate the efficiency at full load.

19.4–4 Repeat the previous problem with $n = 1800$ rpm.

19.4–5 Draw a diagram like Fig. 19.4–6 for a self-excited shunt generator. Assume $R_a \ll R_f$ and $\phi \approx k_f' \sqrt{I_f}$, and show that the current-voltage relationship is $I \approx (K\sqrt{V/R_f} - V)/R_a$, where $K = k_a k_f' \omega_m$.

19.4–6 Suppose the generator in the previous problem has $R_a = 1\ \Omega$, $R_f = 100\ \Omega$, and $K = 200$. Calculate I when $V = 400$, 300, 200, and 100 V; then sketch the *V-I* curve.

19.4–7 A separately-excited motor is rated for $T_{FL} = 100$ N · m. It produces $n_{NL} = 1200$ rpm and $n_{FL} = 1000$ rpm when operated with $V = 200$ V and $\phi = 40$ mWb. Find n_{NL} and n_{FL} when

(a) $V = 250$ V and $\phi = 40$ mWb;

(b) $V = 200$ V and $\phi = 50$ mWb.

19.4–8 What values of V and ϕ would be needed to obtain $n_{NL} = 1000$ rpm and $n_{FL} = 950$ rpm from the motor in the previous problem?

19.4–9 Suppose the speed and torque data in Problem 19.4–7 pertain to a shunt motor with $V = 200$ V and $R_f = 400\ \Omega$. Use the expression in Exercise 19.4–3 to find n_{NL} and n_{FL} when

(a) $V = 250$ V and $R_f = 400\ \Omega$.

(b) $V = 200$ V and $R_f = 500\ \Omega$.

19.4–10 What values of V and R_f would be needed to obtain $n_{NL} = 1000$ rpm and $n_{FL} = 950$ rpm from the motor in the previous problem?

19.4–11 Show that a separately-excited motor produces maximum output power when $n = \frac{1}{2} n_{NL}$.

SUPPLEMENTARY READING

The textbooks listed below, by subject area, are recommended as supplements to this book. The symbol ‡ identifies more advanced treatments of particular subjects.

GENERAL ELECTRICAL ENGINEERING

C. Belove and M. Drossman, *Systems and Circuits for Electrical Engineering Technology.* McGraw-Hill, New York, 1976.

V. Del Toro, *Electrical Engineering Fundamentals.* Prentice-Hall, Englewood Cliffs, 1972.

A. Fitzgerald, D. Higginbotham, and A. Grabel, *Basic Electrical Engineering,* 4th edition. McGraw-Hill, New York, 1975.

S. Senturia and B. Wedlock, *Electronic Circuits and Applications.* Wiley, New York, 1975.

R. Smith, *Circuits, Devices, and Systems,* 3rd edition. Wiley, New York, 1976.

CIRCUIT ANALYSIS

C. Close, *The Analysis of Linear Circuits*‡. Harcourt, Brace & World, New York, 1966.

D. Johnson, J. Hilburn, and J. Johnson, *Basic Electric Circuit Analysis.* Prentice-Hall, Englewood Cliffs, 1978.

J. O'Malley, *Circuit Analysis.* Prentice-Hall, Englewood Cliffs, 1980.

G. Kirwin and S. Grodzinsky, *Basic Circuit Analysis*‡. Houghton Mifflin, Boston, 1980.

ELECTRONIC CIRCUITS

W. Hayt and G. Neudeck, *Electronic Circuit Analysis and Design.* Houghton Mifflin, Boston, 1976.

A. Malvino, *Electronic Principles,* 2nd edition. McGraw-Hill, New York, 1979.

J. Millman, *Microelectronics: Digital and Analog Circuits and Systems*‡. McGraw-Hill, New York, 1979.

D. Schilling and C. Belove, *Electronic Circuits: Discrete and Integrated*‡. McGraw-Hill, New York, 1979.

A. Seidman and J. Waintraub, *Electronics: Devices, Discrete and Integrated Circuits.* Merrill, Columbus, 1977.

ELECTRONIC DEVICES

B. Streetman, *Solid State Electronic Devices*‡, 2nd edition. Prentice-Hall, Englewood Cliffs, 1980.

E. Yang, *Fundamentals of Semiconductor Devices*‡. McGraw-Hill, New York, 1978.

ELECTRICAL MEASUREMENTS

P. Kantrowitz, G. Kousourou, and L. Zucker, *Electronic Measurement.* Prentice-Hall, Englewood Cliffs, 1979.

S. Wolf, *Guide to Electronic Measurements and Laboratory Practices.* Prentice-Hall, Englewood Cliffs, 1973.

FEEDBACK AND SYSTEM DYNAMICS

C. Close and D. Frederick, *Modeling and Analysis of Dynamic Systems.* Houghton Mifflin, Boston, 1978.

R. Dorf, *Modern Control Systems*‡, 3rd edition. Addison-Wesley, Reading, 1980.

D. Frederick and A. Carlson, *Linear Systems in Communication and Control*‡. Wiley, New York, 1971.

COMMUNICATION SYSTEMS

A. Carlson, *Communication Systems*‡, 2nd edition. McGraw-Hill, New York, 1975.

H. Stark and F. Tuteur, *Modern Electrical Communications: Theory and Systems*‡. Prentice-Hall, Englewood Cliffs, 1978.

F. Stremler, *Introduction to Communication Systems*‡. Addison-Wesley, Reading, 1977.

DIGITAL LOGIC AND CIRCUITS

A. Malvino and D. Leach, *Digital Principles and Applications,* 2nd edition. McGraw-Hill, New York, 1975.

F. Mowle, *A Systematic Approach to Digital Logic Design*‡. Addison-Wesley, Reading, 1976.

R. Sandige, *Digital Concepts Using Standard Integrated Circuits*‡. McGraw-Hill, New York, 1978.

C. Strangio, *Digital Electronics: Fundamental Concepts and Applications*. Prentice-Hall, Englewood Cliffs, 1980.

MICROPROCESSORS AND COMPUTERS

T. Bartee, *Digital Computer Fundamentals,* 4th edition. McGraw-Hill, New York, 1977.

D. Givone and R. Roesser, *Microprocessors/Microcomputers: An Introduction*. McGraw-Hill, New York, 1980.

L. Leventhal, *Introduction to Microprocessors: Software, Hardware, Programming*. Prentice-Hall, Englewood Cliffs, 1978.

ELECTRIC POWER SYSTEMS

O. Elgerd, *Basic Electric Power Engineering*. Addison-Wesley, Reading, 1977.

W. Stevenson, *Elements of Power System Analysis,* 3rd edition. McGraw-Hill, New York, 1975.

ELECTROMECHANICS

V. Del Toro, *Electromechanical Devices for Energy Conversion & Control Systems*. Prentice-Hall, Englewood Cliffs, 1968.

L. Matsch, *Electromagnetic and Electromechanical Machines,* 2nd edition. Harper & Row, New York, 1977.

H. Skilling, *Electromechanics*. R. E. Krieger, Huntington, 1979.

TABLES OF MATHEMATICAL RELATIONS

The following tables summarize most of the mathematical relationships used in the text and in the problems.

DERIVATIVES AND INTEGRALS

$$\frac{d}{dt} e^{st} = s e^{st} \qquad \int e^{st}\, dt = \frac{1}{s} e^{st}$$

$$\frac{d}{dt} \cos \omega t = -\omega \sin \omega t \qquad \int \cos \omega t\, dt = \frac{1}{\omega} \sin \omega t$$

$$\frac{d}{dt} \sin \omega t = \omega \cos \omega t \qquad \int \sin \omega t\, dt = -\frac{1}{\omega} \cos \omega t$$

$$\int_0^{\theta/\omega} \sin \omega t\, dt = \frac{1}{\omega}(1 - \cos \theta)$$

COMPLEX NUMBERS

$$\underline{A} = A_r + jA_i = |A| \underline{/\theta_A}$$

$$\mathrm{Re}[\underline{A}] = A_r = |A| \cos \theta_A$$

$$\mathrm{Im}[\underline{A}] = A_i = |A| \sin \theta_A$$

$$|A| = \sqrt{A_r^2 + A_i^2}$$

$$\sphericalangle \underline{A} = \theta_A = \begin{cases} \arctan (A_i/A_r) & A_r \geq 0 \\ \pm 180^\circ - \arctan (A_i/-A_r) & A_r < 0 \end{cases}$$

$$\underline{A}^* = A_r - jA_i = |A|\angle{-\theta_A}$$

$$e^{\pm j\phi} = \cos\phi \pm j\sin\phi = 1\angle{\pm\phi}$$

$$e^{j\phi} + e^{-j\phi} = 2\cos\phi \qquad e^{j\phi} - e^{-j\phi} = j2\sin\phi$$

EXPONENTIAL AND LOGARITHMIC FUNCTIONS

$$e^{\alpha}e^{\beta} = e^{(\alpha+\beta)} \qquad \frac{e^{\alpha}}{e^{\beta}} = e^{(\alpha-\beta)}$$

$$\log xy = \log x + \log y \qquad \log\frac{x}{y} = \log x - \log y$$

$$\log x^n = n\log x \qquad \log_a x = \frac{\log_b x}{\log_b a}$$

TRIGONOMETRIC FUNCTIONS

$$\cos\theta = \sin(\theta + 90^\circ) \qquad \sin\theta = \cos(\theta - 90^\circ)$$

$$\cos^2\theta = \tfrac{1}{2}(1 + \cos 2\theta) \qquad \sin^2\theta = \tfrac{1}{2}(1 - \cos 2\theta)$$

$$\sin(\alpha \pm \beta) = \sin\alpha\cos\beta \pm \cos\alpha\sin\beta$$

$$\cos(\alpha \pm \beta) = \cos\alpha\cos\beta \mp \sin\alpha\sin\beta$$

$$\sin\alpha\sin\beta = \tfrac{1}{2}\cos(\alpha - \beta) - \tfrac{1}{2}\cos(\alpha + \beta)$$

$$\cos\alpha\cos\beta = \tfrac{1}{2}\cos(\alpha - \beta) + \tfrac{1}{2}\cos(\alpha + \beta)$$

$$\sin\alpha\cos\beta = \tfrac{1}{2}\sin(\alpha - \beta) + \tfrac{1}{2}\sin(\alpha + \beta)$$

SERIES EXPANSIONS AND APPROXIMATIONS

$$(1 + x)^n = 1 + nx + \frac{n(n-1)}{2!}x^2 + \cdots$$

$$(1 + x)^{-n} = 1 - nx + \frac{n(n+1)}{2!}x^2 - \cdots$$

$$e^x = 1 + x + \frac{1}{2!}x^2 + \cdots$$

$$\sin x = x - \frac{1}{3!}x^3 + \frac{1}{5!}x^5 - \cdots$$

$$\cos x = 1 - \frac{1}{2!}x^2 + \frac{1}{4!}x^4 - \cdots$$

ANSWERS TO EXERCISES

2.1–1 $i = 4.17$ A, $T = 10^5$ sec = 27.8 hours.
2.1–2 15 mW
2.2–1 $i = 8$ A, $p = 1.6$ kW.
2.3–1 $v = 15$ V, $p = 45$ mW.
2.3–2 1.59 Ω
2.4–1 $i_E = 5$ mA, $v_{CE} = 20$ V.
2.4–2 (a) $i = 4$ A, (b) $v_R = 0$, (c) $i = 2$ A, (d) $v_R = -15$ V.
2.4–5 $v_1 = 14$ V, $i_2 = -3$ A.

3.1–1 120 Ω, 120 W, 240 W, 360 W.
3.1–3 $R_{eq} = 4.8$ kΩ, $i = 2.5$ mA, $p = 30$ mW, 1.0 mA through the 7-kΩ resistance.
3.1–4 (a) 9 V, 8.57 V; (b) 2.45 V, 2.34 V.
3.2–1 $R_s = 20$ Ω, $v_s = 14$ V, $v = 12.73$ V.
3.2–2 $R \geq 9$ kΩ
3.2–3 When $R_L = 15$ Ω, $i = 6$ A, $v = 90$ V, $p_L = 540$ W, $p_s = 180$ W, Eff = 75%.
3.2–4 $R = 193$ Ω, $I_m = 0.881$ A.
3.3–1 4.04 Ω
3.3–2 10 kΩ
3.3–3 $v_{oc} = 60$ V, $i_{sc} = 5$ A, $R_o = 12$ Ω.
3.3–4 2.5 mA
3.4–2 $R_v = 597$ Ω, $R_a \approx 0.075$ Ω.

4.1–1 $p_1 = 83.3$ mW, $p_2 = 3.82$ mW, $p = 40.8$ mW.
4.1–2 (a) $N = 2$, $v_2 = 10 \cos \omega t$ V, $P = 0.25$ W; (b) $N = \sqrt{0.4}$, $v_2 = 5.75 \cos \omega t$ V, $P = 0.0826$ W.
4.1–3 $N_2 = 300$, $P = 25.8$ W.
4.1–4 $h_i = R_a$, $h_f = -1$, $h_o = 1/R_b$, $h_r = 1$.
4.2–4 $A_i \approx 30$ dB, $G \approx 24$ dB.
4.3–1 $v_{out}/v_{in} = A/(A + 1) = 0.99999$.
4.3–2 $K < 22.5$, $R_1 \geq 2$ kΩ, $R_2 = 3R_1$, $R_F = KR_2$.

5.1–1 $i(t) = 37.4$ mA for $0 < t < 53.5$ μs, $i(t) = -200$ mA for $53.5 < t < 63.5$ μs.
5.1–2 $v_2 = -RC\, dv_1/dt$
5.1–4 $C_{eq} = 8$ μF obtained by four parallel branches with two series capacitors; $T = 20$ sec.
5.2–1 $v_L = -40 \sin 100t$ V, $w_{total} = 0.8 + 0.8 \sin^2 100t$ J.
5.2–3 $w = 0.76$ J
5.3–1 $dv_C/dt = 10{,}000$ V/sec
5.3–3 $v_R = \dfrac{RV}{s^2LCR + sL + R} e^{st}$
5.3–4 $H(s) = \dfrac{sL}{s^2LC + sCR + 1}$
5.4–1 $Z(s) = 1000 \dfrac{s + 200}{11s + 200}$
5.4–2 $H(s) = \dfrac{R}{s^2LC + sCR + 1}$
5.4–3 $i = -\dfrac{5}{21} e^{-5000t}$

6.1–1 $\underline{V}_R = 70\ \underline{/0^\circ}$, $\underline{V}_L = 126\ \underline{/90^\circ}$, $v_L(t)$ peaks at $t_0 = -2.5$ ms.
6.1–2 $\underline{V} = 24.5 + j8.1 = 25.8\ \underline{/18.3^\circ}$
6.1–3 $j\underline{A} = 5\ \underline{/36.9^\circ}$, $\underline{AA} = -7 - j24$, $1/\underline{A} = 0.12 + j0.16$.
6.2–2 $X(\omega) = \dfrac{R^2\omega L}{R^2 + (\omega L)^2}$
6.2–3 $\underline{Z} = 167 + j49$, $i(t) = 0.35 \cos(2000t - 16.4^\circ)$.
6.3–1 $P = 2.25$ W
6.3–3 $\omega_0 = 5 \times 10^4$, $|\underline{V}_L| = 600$ V.
6.3–4 3.18 nF, 7.96 μH.

7.1–1 $\underline{V} \approx 20\ \underline{/0^\circ}$ at $\omega = 10$, $\underline{V} = 14.1\ \underline{/-45^\circ}$ at $\omega = 10^3$, $\underline{V} \approx 0.2\ \underline{/-90^\circ}$ at $\omega = 10^5$.
7.1–2 $R_a \approx 1.6$ kΩ, $R_b \approx 320$ Ω.
7.1–3 $R_1 = 1$ kΩ, $C_1 = 3.2\ \mu$F.
7.2–3 $Q = 42.8$, $R = 28.8$ kΩ, $C = 22.1$ pF.
7.2–4 42.2 kΩ
7.3–1 14 dB
7.3–2 $f_\ell = 150$, $f_u = 1200$
7.4–3 $y(t) \approx 1 + 0.45 \cos(\omega_0 t + 45^\circ) + 0.14 \cos(2\omega_0 t + 27^\circ)$

8.1–1 $\tau \approx 43$ sec, $R = 4.3$ MΩ, $i(0^+) = -23\ \mu$A.
8.2–1 $I_0 = -3$ mA, $I_{ss} = 5$ mA, $\tau = 0.1$ ms.
8.2–2 $t_r \approx 0.22$ ms, $v = 99.99546$ V.
8.2–4 $i(t) = 10.6 \cos(377t - 42^\circ) + 6.2e^{-t/0.003}$
8.3–2 $i_N(t) = 0.5e^{-80t} \cos(60t + 90^\circ)$

9.1–2 $v_Q = 57$ V, $i_Q = 0.43$ A.
9.2–1 $R_f \approx 87$ Ω, $V_T \approx 18$ V, $i \approx 0.48$ A.
9.2–3 9.12 V, 9.35 V.
9.3–1 $n_i = 6.9 \times 10^{-15}\ m^{-3}$, $R = 50.4$ MΩ.
9.3–2 20 parts per billion.
9.3–3 (a) $I_0 = 1.57\ \mu$A, $I_D = 8.43$ A; (b) $I_0 = 6.28\ \mu$A, $I_D = 40.84\ \mu$A, $i = 34.56\ \mu$A.

10.1–3 $v_{GS} = 7$ V, $v_{DS} > 3$ V.
10.1–4 $V_{GS} = 5$ V, $I_D = 1$ mA, $V_{DS} = 9$ V.
10.2–1 $A_v = -10$, $A_i = 100$.
10.2–2 (a) $i_C = 0$, (b) $i_C = 4$ mA, (c) $i_C = 6$ mA.
10.2–3 $R_B = 18.8$ kΩ, $V_B \approx 1.1$ V.
10.3–2 $V_{GS} = -2$ V, $I_D = 2$ mA, $V_{DS} = 6$ V.

11.1–1 $a_3 = \frac{1}{7}$, $V_3 = 3$ V, $R_3 = 5$ kΩ.
11.1–2 3.9 V, 4.1 V.
11.2–1 (b) $\beta \geq 7.47$
11.2–2 (a) $v_2 = 4.9$ V, (b) 3.08 V.
11.2–4 9 ms, 32 ms.
11.3–4 $f = \dfrac{1}{2RC \ln 4} \approx \dfrac{1}{2.77RC}$
11.4–1 Center-tapped rectifier: 1260 PIV, 37 + 37 turns, $P \approx 2$ kW.
11.4–3 $C \approx 4200\ \mu$F, $I_{max} \approx 3.2$ A.
11.4–4 $V_{DC} = V_m/\pi$

12.1–1 $R_D < 17.5$ kΩ, $A_v \approx -14$.
12.1–2 $I_D = 1.27$ mA, $V_{GS} = 5.125$ V; Take $R_K = 1$ kΩ, $V_{DD} = 12$ V, $R_1 = 4.6$ MΩ, $R_2 = 0.2$ MΩ
12.1–4 $I_B = 8.1\ \mu$A
12.2–1 $R_i \approx 9.1$ kΩ
12.2–2 0.904
12.2–4 $V_{DS} = V_{GS} = 4$ V, $V_{CE} = 3.76$ V, $\mu = 32$.
12.3–1 $C_o \approx 5\ \mu$F, $C_i \approx 0.8\ \mu$F
12.3–3 $f_u \approx 2.6$ MHz
12.4–1 $P_{out} = 2.22$ W, $P_{DC} = 26.7$ W, $P_{CO} = 8.89$ W.
12.4–2 $N \approx 2.8$, $V_{CC} = 25$ V, $I_C = 0.4$ A.
12.4–3 $I_C = \dfrac{V_{CC}}{N^2R_L + R_E}$
12.4–4 (b) $V_{CC} = 25$ V, $N = 1.25$.

13.1–1 4.94, 4.88, 3.33, −5.06, −130, +120.
13.2–1 $G_1 = 5$, $H = 0.9$.
13.2–2 $B_F = 3.18$ kHz
13.3–3 $f_0 = 60/2\pi$, $K < 468{,}000$.

14.1–2 (a) $W \approx 67$ kHz, (b) $W \approx 167$ kHz.
14.2–1 $P_{out} \approx 20$ mW
14.2–2 (b) $D \geq 1/B = 50\ \mu$s
14.3–1 Frequencies: 0, 50, 200, 250, 950, 1000, 1200, 1250 Hz.
14.3–3 (a) $B \geq 12$ kHz, (b) $B \geq 36$ kHz.
14.4–1 (a) 0.23 mV, (b) 0.19 mV.
14.4–2 40 mV
14.4–3 870 K
14.4–5 80 sec
14.5–1 Radio: 66 dB; transmission line: 120 dB.
14.5–3 540 kHz and 540 kHz + $2f_{IF}$ = 1450 kHz.
14.5–4 1.5%

15.1–1 $H = \overline{[U + (\overline{C} \cdot L)]}$
15.1–3 (a) 4,294,967,295 ≈ 4×10^9, (b) 99,999,999 ≈ 10^8.
15.1–4 $B_1 = D_2 + D_3 + D_6 + D_7$, $D_4 = \overline{B}_3 \cdot B_2 \cdot \overline{B}_1 \cdot B_0$.
15.2–1 (c) $\overline{L} = S_1 + \overline{S}_2\overline{S}_3$
15.2–3 $X = \overline{A}\overline{B} + AB$
15.2–4 $A_0 = \overline{S}_1\overline{S}_0X$, $A_1 = \overline{S}_1S_0X$, etc.
15.2–6 $X = \overline{B} + \overline{A}C = (\overline{B} + C)(\overline{A} + \overline{B})$
15.3–3 (a) 6 + 6 + 5 + 3 = 20, (b) 2 + 2 + 2 + 1 = 7.

16.1–1 $2^3 - 1 = 7$ comparators; $\Delta v_q = 1.5$ V; $v_q = \pm 0.75, \pm 2.25, \pm 3.75, \pm 5.25$ V.
16.1–3 $R_0 = 327.68$ MΩ, $R_F = 5$ kΩ.
16.1–4 $R_1 = R_F$, $R_{0.1} = 10R_F$, $R_{0.01} = 100R_F$.
16.2–1 (a) 2 sec, (b) $\frac{1}{3}$ sec, (c) 178 μs, (d) 100 sec.
16.2–2 01110 could result from one error in 00110, 01010, 01100, or 0111; 11000 could result from two errors in 00000, 01001, 01010, 01100, 10001, or 10010.
16.2–4 $B \geq 108$ kHz
16.3–2 $C_{out} = 1$ and $S_3S_2S_1S_0 = 0101$, implying $1 \times 10^1 + 5 \times 10^0 = 15_{10}$
16.3–3 $-5_{10} = 1{*}1011$

17.1–1 101 A, 0.127 T.
17.1–2 $M = 3.06$ kg, $u_M = 24$ m/sec.
17.1–4 $h = 1.01$ m, $w = 51$ J
17.2–1 $\mathcal{R} = 0.25 \times 10^6$, $\phi = 1$ mWb.
17.2–2 $\mathcal{R}_g = 3.98 \times 10^6$, $\phi = 59\ \mu$Wb
17.2–3 (a) $\phi \approx 2.3$ mWb, $L \approx 138$ mH; (b) $\phi \approx 2.6$ mWb, $L \approx 78$ mH.

17.3–1 $v = \dfrac{\alpha R_L f_a}{\alpha^2 + (R_L + R)D}$

17.3–2 $B = 0.21$ T
17.3–4 1.01 A-t, 0.38 A-t.

17.4–1 $H(f) = \dfrac{R_L}{\alpha} \dfrac{j(f/f_{co})}{1 + j(f/f_{co})}$, $f_{co} = \dfrac{K(R + R_L)}{2\pi\alpha^2}$

18.1–1 $|I| = 20.1$ A, $P = 4695$ W, $Q = 1133$ VAr, Eff $\approx$ 83%.
18.1–2 pf = 0.833 lagging, $Q = 7.96$ kVAr, $\underline{Z} = 4\ \Omega\ \underline{/33.6^\circ}$.
18.1–3 (a) 16.8 kW, 7.96 kVAr, 18.6 kVA, 77.5 A; (b) 367 μF, 70 A.
18.2–1 (a) 399.6 V, (b) 2664 V.
18.2–2 $\underline{V}_2 = 119$ V $\underline{/-4.3^\circ}$, $P_L = 14.05$ kW, $\underline{I}_1 = 66$ A $\underline{/-8.7^\circ}$
18.3–2 $I_\ell = 2.77$ kA, $\underline{V}_a = 6.93$ kV$\underline{/-30^\circ}$, $|S_{total}| = 57.6$ MVA.
18.3–4 $I_a = 18.8$ A$\underline{/-27.5^\circ}$, $P_{total} = 23.5$ kW.
18.4–1 $P_R \approx 0.08$ MW, $\delta \approx 0.46^\circ$, $Q_1 \approx 1.6$ MVAr.
18.4–2 $Q = 51.3$ MVAr, $|I_1| = 10.14$ kA.
18.4–3 0.0273, 0.0316.

19.1–3 93.3 kW, 83.9 kW, 594 N·m.
19.1–4 1,866 A-t
19.2–1 $\underline{E} = 20.2$ kV $\underline{/8.5^\circ}$
19.2–3 5.31 μF
19.3–1 (a) 212 A, 0.707; (b) 142 A, 0.881; (c) 31.8 A, 0.994.
19.3–2 (a) 0.337 Ω, (b) 27.6 N·m.
19.4–1 $N_f = 40$ turns per pole, $E \approx 326$ V, $V \approx 286$ V.
19.4–2 $I = 1.05$ A, $P_m = 0.139$ hp.

INDEX